The Chemistry of Cement and Concrete

The Chemistry of Cement and Concrete

Third edition

F. M. Lea, KT, C.B., C.B.E., D.SC., F.R.I.C.
Formerly Director of Building Research

Edward Arnold (Publishers) Ltd

First Published in 1935
Reprinted 1937, 1940
Second Edition 1956
Third Edition 1970

SBN: 7131 2277 3

Printed in Great Britain by Bell and Bain Ltd., Glasgow.

Preface

Since the publication of the previous edition of this book in 1956 there have been many advances in knowledge of the chemistry, and physics, of cements and of the circumstances that affect the performance of concrete in practice. The present edition has therefore been substantially revised and extended. Increased attention has, *inter alia*, been given to physical properties, such as shrinkage and creep, and the nature of the bonding action; to the mechanism of setting and hardening; and to the durability of concrete.

The purpose of the book is to deal with the chemical and physical properties of cements and concretes and their relation to the practical problems that arise in manufacture and use. As such it is addressed not only to the chemist and those concerned with the science and technology of silicate materials, but also to those interested in the use of concrete in building and civil engineering construction. While some of the earlier chapters are devoted to matters which the engineer and architect will probably consider the private concern of the chemist, the latter chapters are intended to provide information, which is often otherwise difficult of access, on many problems of importance to them. Much attention is therefore given to problems arising in the use of concrete, to the suitability of materials, to the conditions under which concrete may deteriorate, and to the precautionary or remedial measures that can be adopted. The literature on such subjects is widely scattered and often not readily traced, and there is much in records of experience that is not available in print. The book is not intended to be a manual of instruction in the operation of a cement plant or of the production of concrete. These subjects have been covered in other well-known books.

The plan of the book remains unchanged. The constitution of cements, the nature and properties of the compounds they contain, and the chemistry of the changes which the raw materials undergo in the course of cement manufacture, are discussed, and an account is given of the influence of changes in composition and other factors on the properties of the product. The hydration of cements and the hydrated compounds formed come under review, followed by a consideration of the atomic structures of the cement compounds and their relation to cementing properties. The setting and hardening of cements, and the bearing of the physical structure of the set material on typical properties of concretes, such as strength and dimensional change, are outlined.

The testing of cements is discussed only in general outline, attention being directed primarily to an examination of the value and significance of the various tests and the principles underlying them. Detailed descriptions of the techniques of cement testing are to be found in the appropriate standard specifications.

Pozzolanic cements, the various types of cement made from blast-furnace slag, high-alumina cement, and the properties of concretes made from these materials, are reviewed in detail and experience in their use at home and abroad is outlined.

In another chapter an account is given of masonry cements, oil-well cements, non-shrinking cements, non-calcarious cements, and other materials for specialised uses. Various types of concrete and some special properties of cements are also discussed. Concrete for radiation shields, the crazing of cements, carbonation shrinkage, the effect of alkalis in cements in causing efflorescence, the choice of paints for application to concrete, and the corrosion of metals in concrete are examples.

The mechanical properties of mortars and concretes are freely used to illustrate the properties of cements, but a detailed description of concrete-making, or of the grading of aggregates, falls outside the limits of this book and such subjects are only discussed in a general manner. The nature of concrete aggregates, and any special properties inherent in them which favour or militate against their successful use in concrete, form, however, the subject of a chapter. Here heavy and lightweight aggregates and other important matters, such as the reaction of aggregates with alkalis from cement, alkali-carbonate rock reaction and shrinkable aggregates are discussed.

The behaviour of concrete in use forms the subject of the latter part of the book. The factors that determine the quality of concrete and the influence of air-entraining agents, water-reducing agents, and other admixtures, are first examined. Then such topics as the resistance of concrete to frost and fire come under review and a subject of much importance, the resistance of concrete to chemical attack. The use of concrete in seawater, in soils containing soluble sulphate salts or in acid waters, in pipes for the conveyance of water and sewage, and for many other purposes where attack may occur, raises many problems for the engineer. Decisions have to be made as to the risk of deterioration, the type of cement most suited to particular conditions, and the extent to which other protective measures are essential or desirable. In its industrial use in factories for floors, tanks, etc., similar problems of the permanence of concrete, and of the means which may be adopted for its protection, often arise. Information is given of the action on concrete of a wide variety of potentially destructive agents and of the protective measures that can be used. This information is derived from both practical experience and from the results of many large-scale trials and exposure tests. From the nature of the subject it is inevitable that the definition of the conditions that may be expected to affect concrete deleteriously must lack precision, and that no recommendations can be applied independently of the standards of workmanship attained in construction. It is hoped nevertheless that the information given will guide and assist engineers and users in the assessment of risk, the design of new constructions, and the repair of old ones. It covers the major hazards and many of the less common ones encountered in practice.

The coming change to the International System of Metric Units (SI Units) in Great Britain while many countries retain the older form of metric units and others the yard-pound-second system has presented a problem. Data has to a large extent been expressed in the units in which it was originally published but metric conversions from the existing British units have been freely given. For strength data the metric unit of kg per sq. cm. has been retained rather than the SI unit of Newtons per sq. mm. except for those British Standards where SI units have already been introduced.

The author has to thank Dr. R. W. Nurse for undertaking much of the work of revision of Chapters VI, XV and XVI; Dr. W. Gutt for that of Chapter IV; Dr. M. H. Roberts that of Chapter IX; and Dr. H. G. Midgley that of Chapter XI and Appendix III. The responsibility for the views expressed rests with the author.

For the new Plates the author has to thank Dr. W. Gutt for the photographs reproduced in Plate I, Dr. H. G. Midgley for that in Plate X, and Dr. A. J. Majumdar for those in Plates XI and XII, and Dr. H. F.W. Taylor for permission to reproduce Figs. 91 and 92 from his book "The Chemistry of Portland Cement" published by the Academic Press.

F. M. L.

Contents

List of Plates

Abbreviated Formulae

The following abbreviated formulae are used in the text:

C = CaO, A = Al_2O_3, S = SiO_2, F = Fe_2O_3, T = TiO_2,
M = MgO, K = K_2O, N = Na_2O, H = H_2O

Thus, for example:

C_3S = $3CaO.SiO_2$
C_2S = $2CaO.SiO_2$
C_3A = $3CaO.Al_2O_3$
CA = $CaO.Al_2O_3$
C_2AS = $2CaO.Al_2O_3.SiO_2$
C_2F = $2CaO.Fe_2O_3$
C_4AF = $4CaO.Al_2O_3.Fe_2O_3$
C_3MS_2 = $3CaO.MgO.2SiO_2$
$KC_{23}S_{12}$ = $K_2O.23CaO.12SiO_2$
NC_8A_3 = $Na_2O.8CaO.3Al_2O_3$

In the case of the hydrated calcium silicates there is a special nomenclature (see p. 195).

Notes

All temperatures are quoted in degrees Centigrade unless otherwise stated.
All refractive indices (R.I.) refer to sodium light unless otherwise stated.

Reference

Sym., Stockholm 1938 = *Symposium on the Chemistry of Cements, Stockholm,* 1938. Published by Ingeniörsvetenskapsakademien.

Sym., London 1952 = *Symposium on the Chemistry of Cements, London, 1952.* Published by Cement and Concrete Association.

Sym., Washington 1960 = *Symposium on the Chemistry of Cements, Washington,* 1960. Published by U.S. National Bureau of Standards, Monograph 43.

Sym., Tokyo 1968 = *Symposium on the Chemistry of Cements, Tokyo, 1968.* To be published by the Cement Association of Japan.

Sym., Moscow 1956 = *U.S.S.R. Symposium on the Chemistry of Cements.* Published by State Publication of Literature on Structural Materials, Moscow 1956.

BS = British Standard.

ASTM = American Society for Testing and Materials.

I The History of Calcareous Cements

Cements may be defined as adhesive substances capable of uniting fragments or masses of solid matter to a compact whole. Such a definition embraces a large number of very different substances having little in common with one another but their adhesiveness, and the very unequal technical and scientific importance of different members of the class has tended to bring about a restriction of the designation to one group of adhesive substances, namely, to the plastic materials employed to produce adhesion between stones, bricks, etc. in the construction of buildings and engineering works. Cements of this kind also bear a chemical relationship to each other, consisting as they do of mixtures which contain compounds of lime as their principal constituents. The term 'cements' in this restricted sense then becomes equivalent to 'calcareous cements,' but it may be allowed to include certain allied compounds of magnesium.

The use of cements in building is not met with below a relatively advanced stage of civilisation. The earlier structures are composed of earth, sometimes raised in the form of walls or domes by ramming successive layers, or of stone blocks, set one above another without the aid of any cementing material, as in prehistoric megalithic structures, and in the Cyclopean masonry of Greece. The stability of walls of the latter kind is derived entirely from the regular placing of heavy masses of stone without any assistance from adhesion. Although remarkable works have been accomplished by such a method of construction, notably in the domed chambers of Mycenæ, where small stone wedges are driven between the large blocks in order to tighten the joints, yet Cyclopean work has always given place in later times to masonry or brickwork, erected with the aid of some plastic material.[1]

The simplest plan is that found in the brick walls of ancient Egyptian buildings. The bricks are dried in the sun without baking, and each course is covered with a moist layer of the loam (Nile mud) used for making the bricks, with or without the addition of chopped straw. The drying of this layer makes the wall a solid mass of dry clay. Such a mode of construction is only possible in a rainless climate, as the unburnt material possesses little power of resistance to water. It

[1] See N. Davey, Building materials in antiquity, *Chemy Ind.*, 1950, 43; *A History of Building Materials*, Phoenix House, London, 1960.

has nevertheless persisted throughout the ages, and towns such as Kuwait on the Persian Gulf have been entirely built of mud up to recent years. Burnt bricks and alabaster slabs were employed by the Babylonians and Assyrians, and were cemented together with bitumen. This method is very efficacious, but being necessarily confined to those regions in which natural deposits of the material occur, it was not copied elsewhere.

In the massive masonry constructions of the Egyptians we meet with our present-day system of uniting blocks and slabs of stone with a mortar, consisting of a mixture of sand with a cementitious material. Whilst the typical Egyptian mortar has been generally described by writers on Egypt as burnt lime, even where found in buildings as old as the Great Pyramid, chemical examination shows[1] that the Egyptians never used lime until the Roman period and that the cementing material was always obtained by burning gypsum. As the gypsum was quarried in a very impure state, it usually contained calcium carbonate, which might be partly decomposed in the process of burning, or even if found in an undecomposed state in the mortar would produce the impression that lime had been used. The gypsum was very roughly burnt, so that a mixture of the unchanged mineral with 'dead-burnt' plaster was generally obtained. Such a mortar must have been irregular in setting, and troublesome to make use of successfully.

According to the valuable work of A. Lucas,[2] the reason for using gypsum instead of lime, although limestone was more abundant and more accessible than gypsum, was the scarcity of fuel, lime requiring a much higher temperature, and consequently more fuel for its calcination.

Though the early Egyptians were not acquainted with the use of lime it was used at a very early period by the Greeks, and earlier still in Crete, and the Romans must have borrowed it from Greece. The mortar was prepared in the modern fashion, by slaking the lime and mixing with sand, and the examples of Roman brickwork which still exist are sufficient evidence of the perfection which the art attained in ancient times. The remarkable hardness of the mortar in walls of Roman workmanship puzzled many engineers and has sometimes led to the assumption that some secret was possessed by the workmen which is now lost, but a comparison of the analyses of the mortar with the descriptions of the method by ancient authors gives no ground for such a supposition.

The subject attracted special attention at the time of the construction of the great water works at Versailles. The ancient writers insisted on storing lime in the slaked condition for a long time, preferably several years, before use, but in 1765 Loriot, a French engineer engaged on this work, maintained that the passages in question had been misunderstood, and proposed the addition of quicklime to the mortar at the time of mixing, in order to obtain increased

[1] A. Choisy, *L'Art de Bâtir chez les Égyptiens*, Paris, 1904. An analysis of mortar from the Great Pyramid by W. Wallace, *Chem. News* **11,** 185, 1865, shows it to contain 81·5 per cent calcium sulphate, and only 9·5 per cent of the carbonate. Analyses of early Greek and Phoenician mortars are also given by W. Wallace.

[2] *Ancient Egyptian Materials and Industries*, London, 1934. Analyses of Ancient Egyptian lime mortars (Roman period) and gypsum mortars are given in this work.

strength and impermeability.[1] De la Faye[2] contested this, and recommended that egg-sized lumps of lime should be immersed in water, transferred after a time to a cask, and there allowed to slake, evolving steam. This practice is borrowed from St Augustine. Faujas de Saint-Fond,[3] in the course of a very careful study of pozzolanic materials, favoured this latter plan of slaking, and also called attention to an Indian process, employed on the Malabar Coast, of mixing molasses with lime concrete in order to increase the hardness. Finally, Rondelet,[4] after a careful examination of Roman buildings and after making many experiments with the methods proposed by Loriot and others, came to the conclusion that the excellence of Roman mortars depended, not on any secret in the slaking or composition of the lime, but on the thoroughness of mixing and ramming.

The subsequent history of building abounds in examples of differences in the permanence of brickwork and masonry due to differences in the attention paid to this all-important point.

Rondelet's explanation is without doubt correct. Analyses show nothing abnormal in Roman mortars, but the texture is very close, and the interior is often found to contain lime which has not become carbonated, showing the impermeability of the mortar to gases. The practice of long-continued ramming is confirmed by Indian experience. In Bengal, where finely ground brick replaced the whole or part of the sand, fat lime and ground brick (*surki*) were mixed wet in an edge-runner until a sticky mass was formed, and this was then added to the aggregate, the whole being mixed thoroughly and rammed into place. After this, tamping was kept up for several hours, until on scooping a hole and filling it with water this was not absorbed.

Both the Greeks and the Romans were aware that certain volcanic deposits, if finely ground and mixed with lime and sand, yielded a mortar which not only possesses superior strength, but was also capable of resisting the action of water, whether fresh or salt. The Greeks employed for this purpose the volcanic tuff from the island of Thera (now called Santorin) and this material, known as Santorin earth, still enjoys a high reputation on the Mediterranean. The mortar used by the peasants of Santorin—an island destitute of wood for building—long remained identical in its composition and preparation with that of ancient times.[5]

The corresponding material of the Roman builders was the red or purple volcanic tuff found at different points on and near the Bay of Naples. As the best variety of this earth was obtained from the neighbourhood of Pozzoli or Pozzuoli (in Latin *Puteoli*), the material acquired the name of Pozzolana[6] and

[1] *Mémoire sur une découverte dans l'art de bâtir*, faite par le Sr. Loriot, Paris, 1774.
[2] de la Faye, *Recherches sur la préparation que les Romains donnoient à la chaux dont ils se servoient pour leurs constructions, et sur la composition et l'emploi de leurs mortiers*, Paris 1777.
[3] Faujas de Saint-Fond, *Recherches sur la pouzzolane, sur la théorie de la chaux, et sur la cause de la dureté du mortier*, Grenoble and Paris, 1778.
[4] Rondelet, *L'Art de Bâtir*, Paris, 1805.
[5] M. Gary, *Mitt. K. MaterPrüfAmt.* **25,** 11 (1907).
[6] This spelling is to be preferred, as being the modern Italian form. It is adopted by, and will be employed throughout the present work. Other forms, due mainly to French influence, are *pozzualana* and *puzzuolana*.

this designation has been extended to the whole class of mineral matters of which it is a type. Vitruvius says of it: 'There is a species of sand which, naturally, possesses extraordinary qualities. It is found under Baiæ and the territory in the neighbourhood of Mount Vesuvius; if mixed with lime and rubble, it hardens as well under water as in ordinary buildings.'[1]

If volcanic earth did not happen to be available, the Romans made use of powdered tiles or pottery, which produced a similar effect. To quote Vitruvius again, 'If to river or sea sand, potsherds ground and passed through a sieve, in the proportion of one-third part, be added, the mortar will be the better for use.'[2] It is remarkable that the word 'cement' in its Late Latin and Old French forms was first employed to designate such materials, now classed as artificial pozzolanas; its meaning then changed to denote the mortar prepared by mixing the three ingredients, and it is only in recent times that it assumed its modern meaning. There is evidence that crushed potsherds were added to lime mortar to give it hydraulicity in the Minoan civilisation of Crete and it seems that the Romans may have used crushed tile additions before they discovered the natural pozzolanas that occur near Rome.

The Romans carried their knowledge of the preparation of mortar with them to the remoter parts of their empire, and the Roman brick-work found in England, for example, is equal to the best of that in Italy. Ground tiles were the most commonly used ingredient, but in a few districts deposits bearing some resemblance to the natural pozzolanas of the Bay of Naples were found. The use of Rhenish volcanic tuffs known as Trass[3] was probably introduced at this time, and this material, like pozzolana, is still employed at the present day.

A gradual decline in the quality of the mortar used in buildings set in after Roman times, and continued throughout the Middle Ages. Saxon and Norman buildings, for instance, show evidence of badly mixed mortars, often prepared from imperfectly burnt lime. The conclusion appears certain, from the examination of French buildings,[4] that during the ninth, tenth, and eleventh centuries the art of burning lime was almost completely lost, the lime being used in badly burnt lumps, without the addition of ground tiles. From the twelfth century onwards the quality improved, the lime being well burnt and well sifted. After the fourteenth century excellent mortar is found, and the precaution was evidently taken of washing the sand free from adhering dirt or clay. References to 'tarrice' or 'tarras' in documents of the seventeenth-century[5] indicate that the use of pozzolanas in mortar must by then have been established again in England.

The term 'cement' was commonly applied in the Middle Ages to the mortar, as for instance in the work, so much used as an authority in medieval and later times, the *De Proprietatibus Rerum* of Bartholomew Anglicus, in which we read:

[1] Bk. ii, chap. vi, English translation by Joseph Gwilt, London, 1826. There is a later translation by M. H. Morgan, Cambridge, Mass., 1914, and another by F. Granger, London, 1931. Pliny's account is mainly a reproduction of that of Vitruvius.

[2] Ibid., chap. v.

[3] Formerly *Tarras*, a word of Dutch origin.

[4] E. E. Viollet-le-Duc, *Dict. raisonné de l'Architecture Française*, vol. vi, p. 402, 1863.

[5] J. C. Rogers, Trans. St. Albans and Hertfordshire Architectural and Archaeological Society **76,** 99 (1933).

'Lyme . . . is a stone brent; by medlynge thereof with sonde and water sement is made.'[1] The word 'mortar' was, however, also employed as early as 1290.[2]

The Roman mixture of lime and natural or artificial pozzolana long retained its position as the only suitable material for work under or exposed to water. Thus Belidor, for a long time the principal authority on hydraulic construction, recommends an intimate mixture of tiles, stone chips, and scales from a blacksmith's forge, carefully ground, washed free from coal and dirt, dried and sifted, and then mixed with freshly slaked lime.[3] The same author mentions the use of pozzolana or trass where such materials are available. A very thorough and illuminating survey of the subject which will be referred to later is that in Rondelet's great work on building.[4] It is remarkable that even at this late date most of the authorities quoted are ancient, including Pliny, Vitruvius, and St Augustine.[5]

When we come to more recent times, the most important advance in the knowledge of cements, the forerunner of all modern inventions and discoveries in this connection, is undoubtedly the investigation carried out by John Smeaton. On being called upon in 1756 to erect a new lighthouse on the Eddystone Rock after the destruction of the previously existing building by fire, he proceeded to make inquiries as to the best building materials for work under such severe conditions.[6] He found that the usual mortar for work under water was composed of 'two measures of quenched or slaked lime, in the dry powder, mixed with one measure of *Dutch Tarras*,[7] and both very well beat together to the consistence of a paste, using as little water as possible.' The results with this mixture not being always satisfactory, he attempted to discover the effect of using limes of different origin, comparing the mixtures by a cold-water test (immersing a stiffly-worked ball of mortar in water immediately after setting). Finding that lime from the Aberthaw limestone, in Glamorgan, gave better results than ordinary lime, he compared the chemical behaviour of different limestones, and discovered that those which gave the best results as mortars agreed in containing a considerable proportion of clayey matter. This was the first occasion on which the properties of hydraulic lime were recognised. He also compared several varieties of natural and artificial pozzolana as substitutes for trass, including burnt ironstone and forge scales. Ultimately, mortar prepared with blue Lias hydraulic lime, and pozzolana from Civita Vecchia, in equal quantities very thoroughly mixed, was used for the work.

In spite of the success of Smeaton's experiments, the use of hydraulic lime made little progress, and the old mixture of lime and pozzolana retained its supremacy for a long period.[8] The discovery that a hydraulic cement could be

[1] English translation by John Trevisa, 1397.
[2] Date of the first quotation of the word in *The Oxford English Dictionary*.
[3] Belidor, *Architecture Hydraulique*, vol. ii, bk. i, chap. ix, Paris, 1788.
[4] J. Rondelet, *L'Art de Bâtir*, Paris, 1805.
[5] *City of God*, bk. xxi.
[6] J. Smeaton, *A Narrative of the Building . . . of the Eddystone Lighthouse*, 2nd edn., London, 1793. The experiments in question form chap. iv of bk. iii.
[7] So called from being shipped from Holland. Smeaton found in 1775 that the material really came from Andernach, on the Rhine.
[8] Bryan Higgins, *Experiments and Observations made with the view of improving the Art of composing and applying Calcareous Cements* (London, 1780) studied the effect of the addition of a large number of different substances to lime mortar, and patented that of bone ash.

made by calcining nodules of argillaceous limestone, known as septaria, found in certain Tertiary strata, was made in 1796.[1] About 1800 the product thus obtained was given the inappropriate and misleading name of Roman cement, from its hydraulic properties, although it in no way resembles the Roman mortar. Being a quick-setting cement, it was found useful in work in contact with water and steadily grew in favour.[2] Its heyday for civil engineering work lasted until about 1850 after which it was gradually driven out by Portland cement. About the same time as Roman cement was introduced a similar natural cement was prepared in France from similar concretions found near Boulogne, and deposits of 'cement rock,' capable of yielding a hydraulic cement on calcination, were found at Rosendale, and at Louisville, in the United States. It was not long before the American natural cement industry attained great importance.

The investigations of L. J. Vicat[3] on hydraulic lime led him to prepare an artificial hydraulic lime by calcining an intimate mixture of limestone (chalk) and clay, ground together in a wet mill. This process may be regarded as the principal forerunner of the manufacture of Portland cement.[4] James Frost also patented a cement of this kind in 1811, and established works at Swanscombe, the first in the London district. He only calcined his mixture lightly, and the product was evidently inferior to Roman cement, as it was sold at a lower price.

The story of the invention of Portland cement is not easy to disentangle.[5] The usual attribution to Joseph Aspdin, a Leeds builder or bricklayer, is only partly true. Aspdin's first patent is dated 21st October, 1824. He used a hard limestone as used for the repairing of roads, crushed and calcined it,[6] and mixed the lime with clay, grinding to a fine slurry with water. 'Then I break the said mixture into suitable lumps and calcine them in a furnace similar to a lime kiln till the carbonic acid is expelled. The mixture so calcined is to be ground, beat, or rolled to a fine powder, and is then in a fit state for making cement or artificial stone.' Like Frost, he probably used a low temperature, and the product must have been of poor quality. He appears to have been guided by the idea that artificial heat applied to clay must produce a material similar in properties to the natural volcanic earths. His first works were at Wakefield. His son, William Aspdin, continued the manufacture, both on the Thames and at Gateshead-on-Tyne. In the meantime, Isaac Charles Johnson, who died in 1911 at the age of 100, had observed that over-burned lumps found in the kilns, although slow-setting when ground, made a better cement than the usual product.

[1] Patent by James Parker, of Northfleet. No. 2170 (1796).

[2] See A. P. Thurston, *Engineering* **147,** 757 (1939) for history of Roman cement; A. J. Francis, *Cem. Lime Mf.* **37** (6), 113 (1964).

[3] *Recherches expérimentales*, Paris, 1818; *Mortiers et Ciment Calcaires*, Paris, 1828, English translation by Capt. J. T. Smith, London, 1837.

[4] A contemporary of Vicat, who arrived independently at the same conclusions, was J. F. John, whose dissertion on *Lime and Mortar*, first published in Berlin in 1819, and overlooked by later writers, was republished in English in 1925 by the Verein Deutscher Portland-Cement Fabrikanten.

[5] P. Gooding and P. E. Halstead, *Sym., London, 1952*, p. 1, have reviewed the early history of Portland cement; also H. H. Steinour, *J. Res. Dev. Labs Portld Cem. Ass.* **2** (2), 4 (1960); A. W. Skempton, *Trans. Newcomen Soc.* **35,** 117 (1962–3).

[6] The limestone being hard, calcining was probably merely the most convenient means of reducing it to powder.

He had some difficulty at first in finding the correct proportions of clay and chalk, but in 1851 he set up works at Rochester, and later took over Aspdin's abandoned works at Gateshead. A higher temperature of firing must have been introduced in Aspdin's works before this, as on the resumption of the construction of the Thames Tunnel in 1838, Brunel employed Portland cement in spite of the fact that its price was double that of Roman cement, and in the face of strong opposition.[1] Aspdin long kept his process secret, and according to Johnson, who claimed to have suggested the firing at a temperature high enough to produce vitrification, he used to carry trays of copper sulphate into the kilns when charging, in order to convey an impression that the process depended on the addition of salts. William Aspdin spent his last years in Germany, where he had set up works in 1856.

The name Portland cement was given to the product from a fancied resemblance of the colour of the cement after setting to Portland stone. Most earlier works were on the Thames and Medway, chalk and Thames mud being found to be convenient raw materials, but the manufacture later became world-wide.

The use of concrete, an artificial conglomerate of gravel or broken stone with sand and lime or cement, is also of great antiquity. Vitruvius describes it, and Pliny[2] thus refers to the construction of cisterns: 'Cisterns should be made of five parts of pure gravelly sand, two of the very strongest quicklime, and fragments of silex[3] not exceeding a pound each in weight; when thus incorporated, the bottom and sides should be well beaten with iron rammers.' A form of concrete made with broken tiles was much employed for pavements, and cement mixed with oil or other organic matter was often applied as a surface coat for water-proofing purposes. Much of the best concrete was made from broken brick, lime, and pozzolana, whilst in large works volcanic tuff generally took the place of brick. The great vaults of the Thermæ and of the Basilica of Constantine are cast in concrete.[4,5]

The most famous of Roman buildings erected in concrete is the Pantheon, the walls of which, 20 feet thick, are of tuff and pozzolana concrete thinly faced with brick, whilst the dome, 142 feet 6 inches in span, is cast solid in concrete containing pumice and pozzolana. Wooden boards were used as moulds, and the concrete was filled in in a semi-fluid condition. The present condition of many Roman buildings of this class is a sufficient testimony to the excellence of the material. That the process was well understood is further shown by the existence of great masses of Roman concrete on the coast between Naples and Gaeta, polished by the sea but uninjured.

Concrete was also employed in building walls throughout the Middle Ages, but less systematically, and with less knowledge of the material, than under the Romans. The early Christian churches of Rome have concrete walls, whilst

[1] See a pamphlet, issued in 1854 by Aspdin, Ord and Co., of Gateshead, *A Concise Account of Patent Portland Cement.*

[2] *Hist. Nat.*, bk. xxxvi, chap. 52.

[3] Not true silex, or flint, but a hard lava from the Alban Hills.

[4] J. H. Middleton, *Encyc. Brit.*, 9th edn., vol. xx, p. 809.

[5] Details of such constructions are given by A. Leger, *Les Travaux publics, les mines et la Métallurgie aux temps des Romains*, 2 vols., Paris, 1875.

examples of the same mode of construction in England are Kendal Castle[1] and Corfe Castle, a Saxon structure, from which Smeaton derived the idea of using concrete in engineering works. Much of the concrete in medieval buildings is of very inferior quality. It is not until quite modern times that it again assumes importance, the first instance being the construction of the West India Docks in 1800. A great impetus was given to its use by the introduction of Portland cement, and since that time the use of concrete has grown until it has become the most versatile constructional material. The invention of reinforced concrete, a material in which the resistance to compression of well-mixed and hardened concrete is combined with the tensile strength of steel, increased the use of concrete and had a far-reaching effect in bringing about a steady improvement in the quality of cement.

As the demand for Portland cement, especially as an ingredient of concrete, increased, and the requirements of engineers called for a more perfect material, it became increasingly desirable that standards of quality should be set up, by which any consignment could be judged after the performance of certain agreed tests. In this way standard specifications have arisen in most countries, either under official auspices, or as the work of voluntary organisations of engineers and consumers, or of associations of cement manufacturers. Such an association was founded in Germany so far back as 1877, and shortly afterwards established rules for controlling the quality of the product. The first German Standard Specification was drawn up by this body. The British Standard Specification was drawn up in 1904 by the Engineering Standards Committee (now the British Standards Institution) and its ninth revision appeared in 1958. A specification for Portland blastfurnace cement was issued in 1923 and its fifth revision appeared in 1958. The first ASTM specification was issued in 1904. All such specifications are liable to revision from time to time, the alterations made being almost invariably in the direction of increasing the stringency of the requirements. At the same time, most commercial cements commonly more than fulfil the conditions of the official specifications.

The scientific study of cements is of relatively recent date. Even the earliest authors dealt with the theory of setting, but their explanations were naturally of an extremely hypothetical character. Thus Vitruvius, who probably only recorded the current opinion of his time, and did not make any original contribution to the subject, is only able to suggest, in explanation of the properties of mortar, that

> stones . . . having passed through the kiln, and having lost the property of their former tenacity by the action of intense heat, their adhesiveness being exhausted, the pores are left open and inactive. The moisture and air which were in the body of the stone having, therefore, been extracted and exhausted, the heat being partially retained, when the substance is immersed in water before the heat can be dissipated, it acquires strength by the water rushing into all its pores, effervesces, and at last all the heat is excluded. . . . The pores of limestone, being thus opened, it more easily takes up the sand mixed with it,

[1] Schaffhäutl, *Dinglers polytech. J.* **122,** 186 (1851).

and adheres thereto, and thence, in drying, binds the stones together, by which sound work is obtained.[1]

An ancient belief that the quality of a lime depended on the texture of the limestone from which it was made, a harder limestone giving a more durable mortar, persisted until long after the introduction of Portland cement.

Smeaton's remarkable experiments, in which he showed that the hydraulic limes owed their special properties to the clayey constituents of the limestone, were of fundamental importance for the understanding of the nature of cements, but they received little attention from chemists. The hypothesis of Bergmann,[2] assigning the hydraulic properties of cements to the presence of manganese salts, although based on the accidental finding of some manganese in a hydraulic lime, was generally accepted, and was only overthrown by the work of Collet-Descotils,[3] who proved that the burning converted the silica into a soluble form, and especially by the very thorough theoretical and practical investigations of Vicat.[4] The great range of materials studied by this author rendered his work particularly valuable. His principal theoretical conclusion was that the silica of the clay was the essential agent in the hardening process. On the other hand, Frémy,[5] who failed to prepare a calcium silicate with hydraulic properties, but succeeded in obtaining artificially a hydraulic calcium aluminate, assigned the principal share to the alumina. Frémy's work also contains an interesting anticipation of the thesis maintained later by Michaëlis, that the hardening of Portland cement and the reaction between lime and pozzolana are processes of the same chemical nature. The proof by J. N. Fuchs[6] that quartz and other forms of crystalline silica are inactive, whilst the amorphous and hydrated forms behave as pozzolanas, marked a further step in advance. The view that basic silicates are formed in burning, and are then hydrolysed by water, yielding lime and hydrated lower silicates, was propounded by A. Winkler[7] and has since fully established itself.

It is unnecessary, in such a brief survey of the history of cement investigations, to mention the hypotheses and modifications of hypotheses that have seen the light since the publication of Vicat's work. Two early authors who materially contributed to the solution of the problems involved are W. Michaëlis and H. Le Chatelier, of whom Michaëlis is the earlier in point of date, his first paper on the subject of cements having appeared in 1867,[8] whilst the French chemist's first publication dates from 1883.[9] Another major contribution came from Törnebohm in Sweden in 1897.

Systematic work on the constitution of Portland cement was begun in the United States in the Geophysical Laboratory of the Carnegie Institution at Washington, as a development of the investigations of igneous rocks which have formed a principal part of the work of that institution. A brilliant application of thermal and petrological methods to the problem from 1906 onwards set the

[1] Bk. ii, chap. v.

[2] *Opusc. chim. phys.*, ii, 229.

[3] *Mines* **34,** 308 (1813).

[4] Op. cit.

[5] *C. r. hebd. Séanc. Acad. Sci., Paris* **60,** 993 (1865).

[6] *Dinglers polytech. J.* **49,** 271 (1883).

[7] *Prakt. Chem.* **67,** 444 (1856).

[8] Ibid. **10,** 257 (1867).

[9] *Compt. rend.* **96,** 1056 (1883).

knowledge of cements on an entirely new scientific basis. The study of setting was undertaken shortly afterwards in the laboratories of the Bureau of Standards, and since 1926 much work on Portland cement has also been conducted by the Portland Cement Association. In Germany scientific work on cements continued to be conducted in an increasing number of laboratories while fundamental contributions to the chemistry of cements began to grow from France, Italy, Sweden and other European countries and later from Canada, the U.S.S.R., and elsewhere. Contributions from Great Britain in the early years of the twentieth century were not proportionate to the magnitude of the industry, nor to its pioneer work in the development of Portland cement in the preceding century, but with the establishment by the Department of Scientific and Industrial Research of the Building Research Station in 1921 systematic research on cements commenced as part of its programme and many major contributions have since come from it. Following the war of 1939–1945 a research laboratory was established by the Cement and Concrete Association, while schools of research, interested in the nature of the cement compounds and products, grew in some of the Universities. The literature of cement has now become so large that, without the aid afforded by a specialised journal of abstracts, it is difficult to keep in touch with it. *Building Science Abstracts* published monthly by the Building Research Station since 1928 is one key to the literature. Another is to be found in the *Documentation Bibliographique* issued quarterly since 1948 by the Centre d'Études et de Recherches de L'Industrie des Liants Hydrauliques in France. Wecke's *Handbuch der Zement Literatur*, which is a compilation of abstracts of papers up to 1925, facilitates reference to the older literature.

The following works and memoirs, in addition to those already cited, may be referred to for the history of the manufacture and investigation of calcareous cements:

RAUCOURT DE CHARLEVILLE, *Traité sur l'Art de faire de bons Mortiers*, 2nd edn., Paris, 1828.

L. J. VICAT, *Treatise on Calcareous Mortars and Cements*, translated with additions by J. T. Smith, London, 1837.

SIR C. W. PASLEY, *Observations on Lime, Calcareous Cements, etc.*, London, 1838.

Q. A. GILLMORE, *Practical Treatise on Limes, Hydraulic Cements and Mortars*, New York, 1874 (for American cements).

W. MICHAËLIS, *Die Hydraulischen Mörtel, Leipzig*, 1869.

A. C. DAVIS, *A Hundred Years of Portland Cement*, London, 1924.

C. SPACKMAN, *Some Writers on Lime and Cement from Cato to the Present Time*, Cambridge, 1929.

R. W. LESLEY, *History of the Portland Cement Industry in the U.S.A.*, International Trade Press, Chicago, 1924.

H. F. GONNERMAN, Development of cement performance tests and requirements, *Bull. Res. Dept. U.S. Portld Cem. Ass.*, No. 93, 1958.

A. W. SKEMPTON, Portland cements 1843–1887, *Trans. Newcomen Soc.* **35**, 117, 1962–3.

2 Classification of Cements

The development of hydraulic cements has proceeded in a series of stages some of which have been outlined in the foregoing chapter. In this gradual evolution there have arisen some cements which attained at one time to a considerable degree of importance, but have now fallen partly or entirely into disuse. The natural cements are perhaps the best example of this group. The growth of new varieties of cements has, however, far outweighed any decrease due to the disappearance of former products until to-day their number is apt to become bewildering. Many of these cements are used only for certain limited purposes, and their output is a very small fraction of that of ordinary Portland cement, which still represents the great bulk of the world's production. In the following pages the types of hydraulic cements are summarised, and opportunity is taken to describe briefly limes, plasters and magnesium oxychloride cement, which otherwise fall outside the general scope of this book and are not discussed in detail in later pages.

Limes

The burning of lime consists essentially in the calcination of one of the naturally occurring forms of calcium carbonate at a sufficiently high temperature to decompose the calcium carbonate and drive off the carbon dioxide as a gas. Calcium carbonate occurs naturally in forms of varying purity as marble, chalk, and limestones.

The reaction $CaCO_3 \rightarrow CaO+CO_2$ attains a dissociation pressure equal to atmospheric pressure at 894°. The decomposition is carried out in shaft and rotary kilns, or by more primitive forms of burning, to produce the various classes of limes. The composition and properties of the resulting product depend both on the composition of the chalk or limestone used and on the efficiency with which the burning process is carried out.

Fat (*high-calcium*) *limes* are obtained from the calcination of limestones of a high degree of purity and contains 95 per cent and upwards of calcium oxide. On the addition of water they slake rapidly with the evolution of much heat, the lumps breaking down to form a lime putty.

Hydraulic limes are obtained from the burning of limestones that contain a proportion of clay. At the burning temperature of 1000–1200° the reaction products[1] include β $2CaO.SiO_2$, and less basic silicates, $2CaO.Al_2O_3.SiO_2$, $4CaO.Al_2O_3.Fe_2O_3$ and calcium aluminates. The hydraulic properties are to be attributed to the dicalcium silicate and in some degree to the aluminates. There exists an almost continuous series of limes varying from fat limes with a content of alumina and silica below 1–2 per cent to eminently hydraulic limes with up to 50 per cent of these constituents. The intermediate materials, sometimes known as lean limes, are classified according to BS 890:1966 as semi-hydraulic limes. Eminently hydraulic limes which are not widely available in the United Kingdom are not included in BS 890. There are various ASTM specifications for limes such as C5–59 and C207–49 (1961) for high-calcium quick limes and hydrated limes and C141–67 for hydrated hydraulic lime. As the content of alumina and silica increases, the rapidity of slaking and the evolution of heat, so characteristic of fat limes, decrease until with the eminently hydraulic limes no appreciable action occurs unless they are finely ground. The nature of the product obtained after long contact with water varies from a putty to a set cement.

Fat limes when used as mortars harden only by absorption of carbon dioxide from the air with formation of calcium carbonate, a process which is very slow to penetrate beyond the surface. They will not set under water. Hydraulic limes, by virtue of the alumina and silica compounds they contain, harden slowly without such adventitious aid and can be used under water.

The temperature reached in the calcining of limes is normally below that at which any sintering of the materials occurs, but in the burning of hydraulic limes, if the temperature is too high, a portion of the product may sinter. Such sintered lumps are sometimes separated and form, after grinding, the French 'grappier' cement, a product closely akin to the natural cements. The French 'ciment prompts' and 'ciments romain' are quick-setting materials produced from certain argillaceous limestones, but again, in properties, they should probably be classed as natural cements.

Natural cements

Natural cements are materials formed by calcining a naturally occurring mixture of calcareous and argillaceous substances at a temperature below that at which sintering takes place. The British 'Roman' cement which preceded Portland cement, but which is still manufactured on a small scale, and the former American rock cements, belong to this class. The Belgian 'natural Portland cements' were also similar, but burnt at a somewhat higher temperature. These materials represent a group intermediate between the hydraulic limes and Portland cement. Since the composition is governed by that of the naturally occurring raw material, it is more variable than that of artificial mixtures. The importance of this class of cements has now much diminished and in many countries their manufacture has ceased or continues only on a very small scale.

[1] M. H. Roberts, *Cem. Lime Mf.* **29,** 27 (1956); U. Ludwig, H. Muller-Hesse and H. E. Schwiete, *Zement-Kalk-Gips.* **13,** 449 (1960).

Portland cements

Portland cement may be defined as in the British Standard as a product obtained by intimately mixing together calcareous and argillaceous, or other silica, alumina, and iron oxide-bearing materials, burning them at a clinkering temperature, and grinding the resulting clinker. A typical analysis is shown in Table 2.

Rapid-hardening Portland cement is similar to ordinary Portland cement, but is normally ground finer and slightly altered in composition. Its setting time is similar, but it develops its strength more rapidly. These cements overlap with the ordinary Portland cements. In some countries there are more than two strength grades covering the range of the British ordinary and rapid-hardening Portland cements. Thus the Belgian standard specification includes 'ciment Portland normal,' 'ciment Portland à haute résistance' and 'ciment Portland à durcissement rapide,' while in the French specifications there are four grades. In some of the European specifications the cements are classified by numbers representing in kg/cm^2 the minimum compressive strength required in a standard mortar at 7 and 28 days. The four French classes, for example, are

160–250 (ciment ordinaire)
210–325 (ciment à haute résistance)
315–400 (ciment à haute résistance initiale)
355–500 (super ciment)

The French specification also allows for a class of Portland cements containing up to 20 per cent granulated slag or pozzolana (ciments Portland au latier, ciments Portland à la pouzzolana) in the first two of the above strength classes, and up to 10 per cent in the third. Other countries, such as Germany, classify the cements by numbers representing in kg/cm^2 the compressive strength required in a standard mortar at 28 days. Thus the German classes are 275, 375, 450 and 550. The numbers used in this way in the specifications of different countries cannot be compared because of differences in the test methods. The ASTM specifications for Types I and III correspond to ordinary and rapid-hardening Portland cement, while Type II is slightly slower in rate of strength development than Type I but has additional limitations on the chemical composition.

Quick-setting Portland cement differs only from a normal Portland cement in that its setting time is less. Its rate of hardening may be similar to that of ordinary or rapid-hardening Portland cement.

White Portland cement is an ordinary Portland cement containing only a low proportion of iron oxide, so that its colour is white instead of grey.

Waterproofed Portland cements are ordinary Portland cements to which has been added in grinding a small proportion of calcium stearate or of a non-saponifiable oil. A proprietary brand of waterproof cement sold in many countries under the name of 'super' cement is a normal Portland cement which has been ground with the product obtained by treating gypsum with tannic acid.

Hydrophobic cement is a material obtained by grinding Portland cement clinker with a water-repellent film-forming substance such as a fatty acid in order to reduce the rate of deterioration under unfavourable storage or transport conditions.

Low-heat Portland cement is a material in which the chemical composition has been so adjusted as to reduce the heat of hydration. Its rate of strength development, though not its ultimate strength, is lower than that of ordinary Portland cement. In the U.S.A. it is termed Type IV cement.

Sulphate-resisting Portland cement is a material with a composition so adjusted as to give it an increased resistance to sulphate-bearing waters. In the U.S.A. it is termed Type V cement.

Kühl cement is a Portland cement of low silica and high alumina and iron oxide content that has been made in a number of European countries and Japan. In strength it corresponds to rapid-hardening Portland cement.

Iron-ore cement, or erz cement, was a type of Portland cement at one time manufactured near Hamburg in Germany with iron ore replacing the normal clay. It originally had a high iron oxide (about 8 per cent) and low alumina (about 2 per cent) content and was light to chocolate brown in colour and with a higher specific gravity, about 3·3, than Portland cement. Later the $Al_2O_3:Fe_2O_3$ ratio was somewhat increased and now its place has been taken by Ferrari cement.

Ferrari cement is a Portland cement with, originally, a ratio of alumina to iron oxide of 0·64, but now often approaching unity, and having improved resistance to chemical attack. It falls in the class of sulphate-resisting Portland cements.

Expanding, or non-shrinking, cements are cements which expand slightly on hardening, or have no net shrinkage on subsequent air-drying. They are manufactured in the U.S.A. and the U.S.S.R. and were formerly made also in France.

Air-entraining cements are Portland, or Portland blastfurnace, cement to which a small amount of an air-entraining agent has been added during grinding.[1] Air-entraining agents may also be used as additions to a concrete mix instead of being interground with the cement.

The agents used in the U.S.A., either as powder, flakes, or liquid, have been classified[2] as:

A. Alkali salts of wood resins, soluble in coal tar hydrocarbons (i.e. benzene, toluene, etc.) and insoluble in petroleum solvents.

B. Synthetic detergents of the alkyl-aryl sulphonate type, the alkyl group being derived from a petroleum distillate such as kerosene, while the aryl group is usually a sulphonated benzene or naphthalene ring. The usual product is the sodium salt.

C. Calcium lignosulphonate derived from the sulphite process in paper-making. Calcium chloride may also be added.

D. Sodium salts of *cyclo*-paraffin carboxylic acids—e.g. naphthenic—obtained in petroleum refining.

[1] Relevant ASTM specifications are C175–68, and C226–68; C233–66T and C260–66T are also of interest.

[2] W. J. Halstead and B. Chaiken, *Public Roads* **27,** 268 (1954); F. H. Jackson and A. G. Timms, ibid. **27,** 259 (1954).

E. Calcium salts of glues and other proteins obtained in the treatment of animal hides.
F. Alkali, or triethanolamine, salts of fatty acids derived from fats and vegetable oils, and of resinous acids from the alkaline process of paper-making.
G. Triethanolamine salts of sulphonated aromatic hydrocarbons derived from petroleum refining.

These agents are added to the extent of 0·025–0·1 per cent of the cement with the exception of types C and E, where 0·25–1 per cent and 0·25–0·5 per cent respectively are used. Some commercial products are mixtures of more than one type. There are other additives, including animal and vegetable fats and oils and their fatty acids, other wetting agents, aluminium powder, hydrogen peroxide, foaming agents, etc., which entrain air in some degree and act as workability aids, or in high degree to produce aerated lightweight concrete.

High-alumina cement

Aluminous, or high-alumina, cement is manufactured by heating until molten, or more rarely by sintering, a mixture of limestone and bauxite. The product is cooled and finely ground. High-alumina cement is characterised by a very rapid rate of development of strength and approaches closely to its final strength in twenty-four hours after gauging. Its setting time is similar to that of Portland cement. In colour it is black.

Cements containing granulated blastfurnace slag

Granulated blastfurnace slag is the product obtained by the rapid chilling of a basic (high-lime) blastfurnace slag as it emerges from the blastfurnace. It is a light, friable and porous product.

Slag cement, or cold-process slag cement as it is sometimes called, is a mixture of hydrated lime and granulated blastfurnace slag. Certain salts may be added to accelerate the set. This type of cement is known in Belgium as 'ciment de laitier' and in France as 'ciment de laitier à la chaux.'

Portland blastfurnace cement is made by grinding together a Portland cement clinker and granulated blastfurnace slag in proportions, as defined in the British Standard, such that the content of granulated slag does not exceed 65 per cent of the whole. Similar cements, but with various permitted proportions of granulated slag, are manufactured in other countries as shown in Table 1. Typical analyses are given in Table 2.

Supersulphated cement is composed essentially of granulated blast-furnace slag, calcium sulphate and a small percentage of Portland cement or lime. In Belgium it is known as 'ciment métallurgique sursulfaté', in France by this term, or, more shortly, as 'ciment sursulfaté,' and in Germany as 'sulfathüttenzement.'

TABLE 1 Cements based on Portland cement and granulated blastfurnace slag

Country	Name of cement	Per cent granulated slag
Great Britain	Portland blastfurnace cement	Not above 65
U.S.A.	Portland blastfurnace slag cement	25–65
Belgium	Ciment de haut-fourneau	30–70
	Ciments permétallurgiques	Above 70
France	Ciment Portland de fer	25–35
	Ciment métallurgique mixte	45–55
	Ciment de haut fourneau	65–75
	Ciment de laitier au clinker	Not less than 80
Germany	Eisenportlandzement	Not above 40
	Hochofenzement	41–85

TABLE 2 Composition of cements

Cement	CaO	Al_2O_3	Fe_2O_3	FeO	SiO_2	MgO	SO_3	S
Portland	64·1	5·5	3·0	—	22·0	1·4	2·1	—
Kühl	64·6	8·2	6·0	—	17·0	1·6	1·5	—
Erz	64·0	2·8	6·2	—	22·0	0·9	1·8	—
Ferrari	64·5	3·6	5·6	—	22·5	1·6	1·5	—
High-alumina	37·7	38·5	12·7	3·9	5·3	0·1	0·1	—
Portland blastfurnace	59·0	8·1	1·0	0·5	22·8	3·5	1·7	0·5
Eisen-Portland	59·4	8·5	0·8	1·7	23·5	3·3	2·0	0·5
Hochofen	53·2	11·0	0·5	1·3	26·5	3·0	1·6	1·0
Slag (cold process)	49·0	12·8	—	1·3	27·4	2·8	1·0	1·0
Supersulphated	45·0	13·1	1·0	0·9	25·2	3·5	7·0	1·0

Pozzolanas and pozzolanic cements

A pozzolana may be defined as a material which is capable of reacting with lime in the presence of water at ordinary temperature to produce cementitious compounds. Italian pozzolana, trass and Santorin earth are examples of naturally occurring pozzolanas of volcanic origin. Artificial pozzolanas are prepared by burning at suitable temperatures certain clays, shales, and diatomaceous earths containing a proportion of clay. Diatomaceous silica and some natural amorphous silica deposits may also form pozzolanas, either with or without a heat treatment. Pulverised fuel ash (fly-ash) is also used as a pozzolana. Pozzolanic cements are produced by grinding together Portland cement clinker and a pozzolana, or by mixing together a hydrated lime and a pozzolana.

Oil-well cements

These are cements specially produced for cementing the steel casing of gas and oil wells to the walls of the bore-hole and to seal porous formations. The cement slurry has to be pumped into position, before it sets, under conditions of high temperature and pressure. Portland cements, more coarsely ground than normal, with the addition of special retarders such as starches, sugars, or organic hydroxy acids, are used, also slow-setting Portland cements of a composition between the Ferrari and Erz types, and pozzolanic cements.

Masonry cements

This group of cements consists of materials intended for use in mortar. Their purpose is to provide a cement which gives a more plastic mortar than ordinary Portland cement. They are often produced by grinding more finely than usual a mixture of Portland cement and limestone together with a plasticiser that entrains air. They are also made by intergrinding mixtures of Portland cement with hydrated lime, granulated slag, or inert fillers with the addition of calcium stearate or some other waterproofing agents. Natural cements and cold-process slag cements are also found in this group, and also materials such as 'grappier' cement in France. Masonry cements are discussed in Chapter 17.

Magnesium oxychloride or Sorel cement

Magnesium oxychloride cement is the product obtained when magnesia and a solution of magnesium chloride react together. Magnesite is calcined so as to give a lightly burnt reactive product and the ground material mixed as required with a strong solution (about 20 per cent anhydrous salt) of magnesium chloride. Combination of the magnesia and magnesium chloride takes place with the development of heat and results in the formation of magnesium oxychloride,[1] $3MgO.MgCl_2.11H_2O$. In all probability the magnesium oxide and chloride first react in solution and a supersaturated solution of the oxychloride is formed from which the solid separates out. The aged cement appears to be composed of particles of magnesium hydroxide of varying size from which radiate large numbers of fine needle crystals of the oxychloride binding the material together.

The product is hard and strong, but is attacked by water. It is used mainly as a flooring material with an inert filler and a pigment to colour it, and is often known as magnesite flooring. The surface is protected against water by polishing with wax in turpentine. It has a strongly corrosive action on iron pipes laid in or passing through it. It is also used for cementing glass and metal. There is a British Standard (BS 776:1963) defining the properties of the materials.

A similar magnesium oxy-sulphate cement is also used as a flooring material and as a binder for wood-wool slabs. Other analogous mixtures, such as zinc oxychloride, are known and find a limited application in dental cements, etc. Thus zinc oxide yields a hard mass with a solution of zinc chloride or syrupy phosphoric acid. The product in these cases is probably crystalline, although the crystals may be so minute and so closely interlocked as to resemble a glassy mass.

[1] Cf. C. R. Bury and E. R. H. Davies, *J. chem. Soc.*, 1932, 2008; H. S. Lukens, *J. Am. chem. Soc.* **54,** 2372 (1932); E. S. Newman, *J. Res. natn. Bur. Stand.* **54,** 347 (1955); T. Demediuk, W. F. Cole and H. V. Hueber, *Aust. J. Chem.* **8** (2), 215 (1955).

Gypsum plasters

There are a number of cementing materials consisting essentially of calcium sulphate that are produced by the partial or complete dehydration of gypsum, $CaSO_4.2H_2O$. Natural deposits of gypsum provide the raw material for manufacture and it is from one of these, the hill of Montmartre in Paris, that the name Plaster of Paris is derived. Gypsum produced as a by-product in the manufacture of phosphate fertilisers from phosphate rock and sulphuric acid is also used to make hemihydrate gypsum plasters, while anhydrous calcium sulphate formed as a by-product in the manufacture of hydrofluoric acid is used with an accelerator to form a flooring plaster.

The various types of gypsum plaster are classified in BS 1191:1955 as follows: Class A, plaster of Paris; Class B, retarded hemihydrate gypsum plaster; Class C, anhydrous gypsum plaster; Class D, Keene's plaster.

When finely ground gypsum is heated to about 150° in shallow open iron pans or in deeper kettles with a vent at the top the product obtained is the hemihydrate $CaSO_4.\frac{1}{2}H_2O$, usually mixed with some unchanged gypsum and some hard-burnt material. This is the ordinary plaster of Paris. Its speed of setting is normally too fast for ordinary use as a plastering material, owing to the presence of unchanged gypsum which accelerates the set. The addition of a protein retarder, normally keratin, to the extent of about 0·1 per cent retards the onset of crystallisation and holds up the set for an hour or two. This forms the retarded hemihydrate gypsum plaster. Another method of dehydrating gypsum to the hemihydrate is to heat it in an autoclave under pressure at about 130° and this gives a plaster requiring less water for mixing and having a higher strength.

If calcination is carried out at a higher temperature the remaining water of crystallisation is lost and first soluble and then insoluble anhydrite is formed. Soluble anhydrite, which forms the main basis of the anhydrous gypsum plasters (Class C) is predominant when the temperature is allowed to rise to 190–200°. It is very hydroscopic and absorbs water vapour very rapidly to form the hemihydrate. When soluble anhydrite is heated to higher temperatures its reactivity steadily diminishes until at about 600° the product is relatively inert and is known as insoluble anhydrite. It requires the addition of a suitable catalyst, an accelerator, to render it reactive. Keene's plaster is usually prepared by calcining gypsum in lump form in a kiln at a dull red heat. About 0·5–1 per cent of potash alum or potassium sulphate is added, but mixed accelerators such as ferrous or zinc sulphates with potassium sulphate are also occasionally used. Lime also acts as a promoter, though a less active one. When gypsum is calcined at 1100–1200° some dissociation into sulphur trioxide and lime occurs, leaving free lime dispersed in the product to act as an accelerator of set. This forms the traditional very slow-setting German flooring plaster known as Estrich Gips. However, a considerable proportion of the flooring plasters now used in Germany are derived from by-product anhydrite.

The chemistry of the gypsum transformation products is still, despite extensive work, not entirely agreed. The transformation point of gypsum to hemihydrate in water is 97°, but in an atmosphere of lower relative humidity the dehydration proceeds at much lower temperatures, and correspondingly also in solutions of salts in which the water has a reduced vapour pressure. Thus in saturated magnesium chloride solution the temperature is as low as 11°. There has been considerable controversy as to whether hemihydrate, $CaSO_4.\frac{1}{2}H_2O$, exists as a definite compound or whether it is a member of a zeolitic system extending over the composition range $3CaSO_4.2H_2O$–$CaSO_4$. When the end member of this series, soluble anhydrite, is heated to higher temperatures, or for longer periods at 200°, there is a steady increase in density from about 2·50 until a value of 2·94 corresponding to that of natural anhydrite is reached. There have been suggestions that various soluble anhydrites exist, but the evidence suggests that they are only stages in the condensation of the open lattice, left when water is removed from gypsum, to that of anhydrite. Under commercial conditions of production which do not represent equilibrium states, the temperatures of formation of the hemihydrate and soluble anhydrite are raised.

The hemihydrate $CaSO_4.\frac{1}{2}H_2O$ can be prepared in a crystalline form as needles, probably monoclinic with pseudo-trigonal symmetry, by warming gypsum at 40–50° for 18 hours with nitric acid (specific gravity 1·4) or by heating finely divided gypsum in superheated water, under pressure, for some hours followed by rapid filtration and washing with absolute alcohol. The needles formed have straight extinction and positive elongation, refractive indices $\omega = 1{\cdot}558$, $\varepsilon = 1{\cdot}586$ and specific gravity 2·735. The commercial Plaster of Paris and the hemihydrate prepared by dry ignition of gypsum have refractive indices around 1·55–1·57 and specific gravity about 2·55–2·65. Gypsum has refractive indices $\alpha = 1{\cdot}520$, $\gamma = 1{\cdot}530$, specific gravity 2·32 and shows oblique extinction. On heating gypsum the outward form of the crystals does not change on dehydration to the hemihydrate, but they become striated showing the presence of groups of fine needles having straight extinction. Anhydrite forms needles with straight extinction and refractive indices $\alpha = 1{\cdot}570$, $\gamma = 1{\cdot}612$.

Though during the setting of plaster the total solid volume decreases, there is an overall expansion of the mass, caused by the manner of crystal growth and this is a valuable property in obtaining sharp casts in moulds. For some uses, however, a material with a low setting expansion is desirable and 'low expansion' hemihydrate plasters are made by suitable additions.

3 Portland Cements: Raw Materials and Processes of Manufacture

Portland cement is prepared by igniting a mixture of raw materials, one of which is mainly composed of calcium carbonate and the other of aluminium silicates. The most typical materials answering to this description are limestone and clay, both of which occur in nature in a great number of varieties. Marls, composed of a mixture of chalk and clay, and shales are also common raw materials. The industry has tended to establish itself in neighbourhoods where two raw materials occur in close proximity, as, in Great Britain, on the banks of the lower Thames, Medway and Humber where chalk and alluvial clay are used, in the Midland counties where the blue lias and oolitic limestones occur together with shales, and in the more northern counties where carboniferous limestones and clay or shale are found. Suitable raw materials are also worked in other districts. Analyses of some raw materials and a typical cement raw mix are given in Table 3.

Chalk is a soft limestone of organic origin, being almost entirely composed of the calcareous remains of marine organisms, and is an exceedingly convenient source of lime. As found in the south-east of England it contains bands of flints, which are easily removed in the wet process, the flints simply accumulating at the bottom of the wash-mills, and being removed periodically.

The Cambridgeshire marl occupies a special position, as it approximates to the composition of an artificial Portland cement. It was therefore at one time used for the production of a natural cement, but has for long been mixed with materials from neighbouring strata to give a correct composition.

A great variety of limestones, marls and clays is used on the Continent. Cement works on the Picardy coast use chalk, identical in composition with that of Kent. In the U.S.A.[1] the greater part of the cement made was originally produced from an argillaceous limestone known as 'cement rock,' found in the Lehigh district of Pennsylvania. This was at first converted into a natural cement, but when deficient in lime a small proportion of limestone was added. Though the Lehigh district is still an important centre of cement making the

[1] See C. F. Clausen, *Portld Cem. Ass. Rep.*, No. MP–95 (1960).

industry is now widely developed across the North American continent. Whilst chalks and marls are used in some parts the majority of the raw materials used in the U.S.A. are limestones and shales, and occasionally marine shells.

The variety of raw materials used in the different countries throughout the world is very great, and for further details reference must be made to textbooks on cement manufacture, and to the accounts of new plants in the technical press. In addition to natural materials use is also made in some plants of artificial

TABLE 3 Analyses of typical raw materials

	Chalk	Clay	Lime-stone	Shale	Marl	Typical raw mix
SiO_2	1·14	60·48	2·16	55·67	16·86	14·30
Al_2O_3†	0·28	17·79	1·09	21·50	3·38	3·03
Fe_2O_3	0·14	6·77	0·54	9·00	1·11	1·11
CaO	54·68	1·61	52·72	0·89	42·58	44·38
MgO	0·48	3·10	0·68	2·81	0·62	0·59
S	0·01	n.d.	0·03	0·30	nil	nil
SO_3	0·07	0·21	0·02	nil	0·08	0·07
Loss on ignition	43·04	6·65	42·39	4·65	34·66	35·86
K_2O	0·04	2·61	0·26	4·56	0·66	0·52
Na_2O	0·09	0·74	0·11	0·82	0·12	0·13
	99·97	99·96	100·00	100·20	100·07	99·99
$CaCO_3$	97·6		94·1		76·0	79·3

† Includes also P_2O_5, TiO_2 and Mn_2O_3

products. Blastfurnace slag, which is the most important of these, is discussed in Chapter 15. Alkali waste, a form of precipitated calcium carbonate obtained as a by-product in the alkali and synthetic ammonium sulphate industries, is also used as a substitute for limestone. Sand, waste bauxite and iron oxide are also sometimes used in small amounts to adjust the composition of the mix. Raw materials that are too low in lime content can also be upgraded, or the alumina-silica ratio adjusted, by a flotation process.

The process of cement manufacture consists in the incorporation of the raw materials to form a homogeneous mixture, the burning of the mix in a kiln to form a clinker, and the grinding of the clinker with the addition of a small proportion of gypsum to a fine powder. Two processes, known as the wet and dry processes according as to whether the raw materials are ground and mixed ia n wet or dry condition, are used. In a variant of these processes, the semi-dry process, the raw materials are ground dry and then mixed with 10–14 per cent water and formed into nodules. The wet process was originally used for very friable materials such as chalk and clay and later extended to the harder limestones and shales. For many years the wet process was preferred in Great Britain because of the more accurate control of the raw mix which was possible with it. With improved control, there has been a swing back to the dry process, for less fuel is required for burning than in the wet process. Modern developments in the dry mixing of powdered materials have, moreover, now removed the disadvantage of the dry process. Originally the dry process was preferred in the U.S.A. where most of the raw materials are hard rocks, but later the wet process was widely adopted because of its advantages in control of cement composition.

In the dry process the raw materials are crushed, dried in rotary driers, proportioned, and then ground in tube mills consisting of rotating steel cylinders

TABLE 4

	Average composition of Lehigh cement rock	Limestone used for addition to cement rock
SiO_2	16–21	1–4
Al_2O_3	4–8	0·7–2
Fe_2O_3	1–2	
CaO	35–41	51–55
MgO	2	0·5–1·5

containing a charge of steel balls or other grinding media. They are often divided into two, or three, or even four compartments containing balls of different size. The mills are continuous in operation, being fed at one end and discharging the ground material at the other. Two mixes, one high and one low in lime, are often prepared and blended in the required proportions in a silo with vigorous air circulation. The dry powder is fed to the kiln.

In manufacture by the wet process the method differs somewhat with the nature of the raw materials. When chalks and marls are used the raw materials are broken up and incorporated in wash mills. These usually consist of circular pits lined with bricks or concrete and containing gratings in the walls through which the raw materials can pass when reduced to a sufficiently fine condition. The chalk and clay are fed in the required proportions to the wash mill together with sufficient water to form a liquid of creamy consistence. A number of radial arms attached to a vertical spindle carry iron rakes which break up the lumps of solid matter as the arms revolve. The solid material is reduced to a fine state of division and passes as a slurry through the screens in the walls of the pit. An flints present in the chalk remain at the bottom of the wash mill and are removed periodically. In modern practice the raw materials are usually further reduced in size by treatment in another wash mill with finer screens, in centrifugal screening mills, or by passage through a tube mill. When the raw materials are harder limestones and shales the wash mill is inadequate to effect the reduction. The raw materials are in such cases crushed and fed to large tube mills and water added to the mill in the amount required to form the slurry. When hard limestone and clay are the raw materials the clay is usually fed to the tube mill already dispersed in the water. The finished slurry does not usually contain more than a few per cent of material remaining on a 170 mesh, and its water content varies from 35 to 45 per cent with different raw materials. The slurry is pumped to slurry tanks or basins in which both rotating arms and agitation by compressed air are used to keep the mixture homogeneous. The proportion of lime in the mix is controlled by analysis and the charge fed to the wash mills or tube mills is adjusted periodically as required. Final adjustments of composition is often obtained by blending the slurry from two basins, one of which is kept slightly high and one slightly low in lime. In some plants more elaborate blending systems are practised, using a number of slurry basins, blending tanks, and final storage tanks.

The rotary kiln in which the cement is burnt at 1300–1500° is a long cylinder rotating on its axis and inclined so that the materials fed in at the upper end travel slowly to the lower end. Here the fuel, pulverised coal, oil or natural gas, is blown in by an air blast and ignited. In the upper part of the kilns chains are fixed to assist in the transfer of heat from the kiln gases to the raw materials. The slurry is dried in the upper part of the kiln and the water driven off as steam, and then, as it descends the kiln, the dried slurry undergoes a series of reactions, forming in the most strongly heated zone hard granular masses, mostly from $\frac{1}{8}$ inch to $\frac{3}{4}$ inch diameter, known as clinker. At the lower end of the kiln the clinker passes into coolers. In the older plants these consist of rotating steel cylinders arranged usually underneath the rotary kilns, but in most modern plants they form a composite part of the kiln. In these cases they consist of a series of cylinders parallel with the body of the kiln and arranged around its periphery. The clinker after passing through the hottest zone in the kiln falls into them through openings in the walls of the kiln. In these cylinders, which revolve with the kiln, are hung loose chains which cause the particles of clinker to fall in a cascade, thus being brought into contact with the current of cold air which is drawn through. Grates serving both as a transporter and cooler are often favoured. These are enclosed in an air-tight casing and a controlled amount of air is drawn through the bed of clinker. The heated air drawn from the coolers is used for the combustion of the coal, so providing an economical exchange of heat. The comparatively cool clinker then falls on to conveyers and is transferred to storage hoppers or passed direct to the grinding mills. The grinding is usually done in modern plants in large three- or four-compartment tube mills which are usually water-cooled in order to help to dissipate the heat generated. A small quantity of gypsum is added during grinding to control the setting, as described later, and the finely ground cement passes to silos from which it is drawn for packing. The packing is performed by automatic machines. The general course of the wet process is indicated by the following flow sheet from which minor mechanical operations have been omitted:

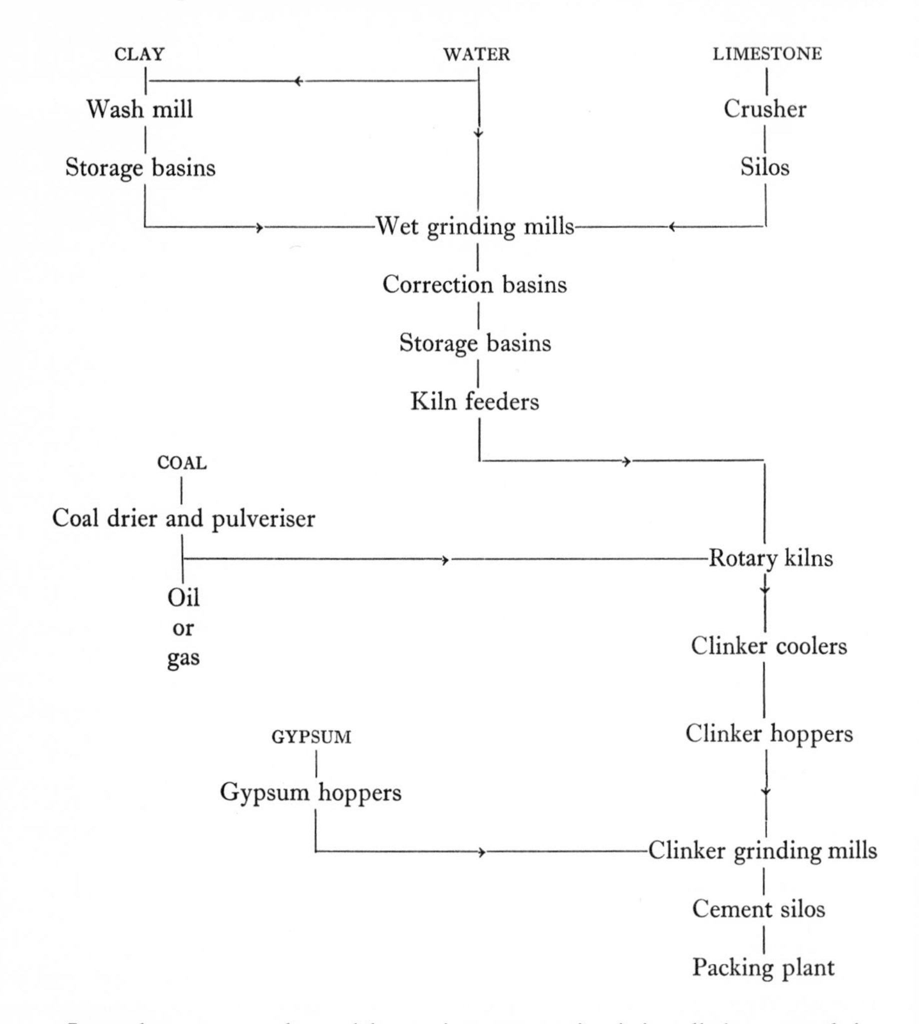

In modern cement plants elaborate instrumentation is installed to control the burning process, combined sometimes with automatic control of the kiln. The raw materials in some plants are sampled automatically and the raw-mix composition required is calculated by a computer.

This short description will apply to many plants with only slight modification, but for details of the numerous variations that may be introduced and of the types of plant employed reference should be made to a textbook on cement manufacture. The tendency over the years has been to increase the length of the rotary kiln, particularly when working on the wet process, in order to increase its thermal efficiency and often to add some form of pre-heater. Thus in dry-process plants it was formerly common practice to pass the exit gases through waste heat boilers but a cyclone preheater, in which the feed material is heated, is now preferred as a means of recovering heat from the hot gases, the temperature of which can be reduced to about 300°. Wet-process kilns go up to over 700 ft in length and to 25 ft in diameter in order to obtain the most efficient heat interchange in the kiln and reduce the flue gas temperature to 200° or even lower.

Numerous methods are used to improve fuel economy in wet-process plants. Thus the water content of the slurry may be reduced by slurry filters consisting of large rotating discs or drums covered with filter cloth. The water content of the slurry, normally about 35–45 per cent by weight, is reduced to 20–30 per cent in the filter cake produced. The amount of water required in the slurry for it to be pumpable can in some cases be reduced a few per cent by adding very small amounts of agents such as sodium carbonate or silicate.

Various devices are also employed to improve the interchange of heat between raw feed and hot gases so as to use as much as possible of heat that would otherwise be lost up the chimney stack. These may take the form of spirals or guide vanes in the back-end of the kiln or of additional rotating cylinders through which the slurry passes before being fed to the kiln. Another well-known development is the Lepol kiln in which a travelling chain grate enclosed by a casing is installed in advance of the kiln. The raw materials are prepared in the form of small nodules by the addition of a small percentage of water to the dry mix. These nodules are piled in a layer some 6 to 8 inches thick on the travelling grate which conveys them to the rotary kiln. The hot waste gases from the kiln are caused to pass into this chamber and are sucked through the raw mix, transferring their heat to it. The travelling grate can be used in combination with a short kiln having a high exit-gas temperature, and it will reduce the gas temperature to as low as 120°. A considerable economy in fuel consumption can thus be effected, combined with the advantages of a short kiln. In another method, nodules from the filter pressing of slurry are preheated on a horizontal rotating hearth through which the flue gases pass and which forms part of a chamber which is fed from above and discharges through a central orifice in the hearth. Various trials have also been made on fluidised beds formed by feeding the dry raw materials, fuel and preheated combustion air into the bottom of a reactor vessel.[1]

The gases from rotary kilns carry with them a certain amount of dust which if allowed to escape can cause a nuisance. In wet-process plants the dust at the kiln exit varies up to about 5 per cent of the clinker output and in dry-process plants it can rise as high as 15 per cent. Electrostatic precipitators, cyclone separators, scrubbing towers and bag filters are used to trap the dust, which is usually returned to the kiln. The dust is relatively rich in the more volatile constituents of the charge, specially potash, and has sometimes been used as a source of this valuable salt.[2]

The heating of kilns in Great Britain is mainly by means of pulverised coal, and this very efficient fuel was used for this purpose long before it was introduced

[1] R. Pyzel, U.S. Pat. 1961 **2** (977), 105; Y. Susukawa et al., *Sym., Tokyo 1968*; S. Hata and T. Sanari, *Zement-Kalk-Gips* **57,** 509 (1968).
[2] A. W. G. Wilson, *J. Soc. chem. Ind., Lond.* **38,** 314T (1919).

into other industries. Waste blastfurnace gas has also been used in Scotland. In America both oil and natural gas are in common use in addition to pulverised coal. Electric firing of rotary kilns was also tried in Switzerland during the war[1] of 1939–1945 using graphite electrodes and a low-tension arc. About 1·1 kWh was required per kilo of clinker.

The choice of a refractory lining for rotary cement kilns presents certain difficulties. The lining has not only to withstand heavy wear due to abrasion by the raw materials and the clinker, but also to be resistant to chemical attack by the cement mix at the high temperatures prevailing in the clinkering zone. Ordinary firebricks, containing a large proportion of silica, react with the more basic materials of cement, causing partial destruction of the lining in the hotter parts and building up of rings of sintered material which obstruct the passage down the kiln. They are used in the back part of the kiln where temperatures are relatively low, but as the burning zone is approached more refractory and resistant materials are required. High-alumina bricks containing 60 per cent and upwards of alumina are widely used; chrome-magnesite and sometimes dolomite bricks are also favoured. Portland cement clinker itself has also been used, fragments of clinker being bonded with cement or a little refractory clay.

Another development[2] in cement burning has been the suction grate in which the raw materials and fuel, mixed together, are fed on to a travelling chain grate on which they are burnt. A typical grate may be 40 feet long, 6 feet wide and travel at about 2 feet per minute. A layer of already burnt clinker is fed on to the grate, to protect it, followed by a layer of nodulised raw material and coke fuel. The underside of the grate is enclosed by a suction box to draw air through it. The grate travels under an ignition hood to start combustion which then proceeds as the grate travels along. The burning process is rapid and the maximum clinkering temperature of 1300–1400° is only maintained for one or two minutes.

Stationary kilns of many different types were at one time used in cement manufacture, but with the exception of the automatic shaft kiln they have now fallen into disuse. The automatic shaft kiln is a continuous kiln of the vertical type some 30–50 feet high and 8–10 feet in diameter. It is fed at the top with the raw materials and fuel briquetted together. The raw mix is burnt to a clinker as it descends the kiln and an automatic discharge arrangement effects the removal of the clinker from the bottom of the kiln. The automatic shaft kiln has found a certain amount of favour in Europe on account of its low fuel consumption.

A number of processes are also worked for the combined production of Portland cement and other products. In one[3] of these, operated in England, France and Germany and some eastern European countries, sulphuric acid and

[1] H. Gygi, *Revue Matér. Constr. Trav. publ.* **381,** 247 (1947); V. D. Lysenko, *Tsement* 1962, 7.
[2] K. Meyer and H. Wendeborn, *Sym., London 1952*, 782; B. M. Pearson, *Rock Prod.* **55** (10), 102 (1952); P. Rolfsen, *Tids. Kjemi* **19,** 129 (1939).
[3] W. L. Bedwell, *Lect. R. Inst. Chem.*, No. 3 (1952); J. Manning, *Proc. Fertil. Soc.*, No. 15 (1951); B.I.O.S. Final Report, No. 678, H.M.S.O., London (1946); W. Q. Hull, F. Schon and H. Zirngibl, *Ind. Engng Chem.* **49,** 1204 (1957).

Portland cement are made by burning in a rotary kiln a raw mix of calcium sulphate (anhydrite or gypsum), clay and coke together with sand and iron oxide as required to give correct mix proportions. Sulphur dioxide is produced from the calcium sulphate, the main reactions being:

$$\text{(i)}\quad CaSO_4 + 2C = CaS + 2CO_2$$

$$\text{(ii)}\quad 3CaSO_4 + CaS = 4CaO + 4SO_2$$

The lime liberated reacts in the usual way with the alumina, silica and iron oxide constituents to form cement clinker. The presence of coke in the mix is necessary to obtain rapid decomposition of the calcium sulphate. The burning temperature is about 1400°, pulverised coal being used for firing. The gases leaving the kiln contain about 9 per cent SO_2 which is converted to sulphuric acid by normal processes.

Pig iron and Portland cement are produced together by the Bassett process[1] used in a number of countries. The raw materials, limestone, coke and iron ore, are burnt in a rotary kiln having a section following the burning zone shaped to collect the molten iron in a pool from which it flows out through a tap hole at each revolution. The clinker passes on to the end of the kiln and the coolers. A reducing atmosphere is maintained in the kiln.

Alumina and Portland cement can be made together by the Séailles process.[2] The raw materials are limestone and an alumina-containing material such as siliceous bauxite, and clays, slags, and coal ashes of high alumina content. The mix is proportioned so as to yield $5CaO.3Al_2O_3$ and $2CaO.SiO_2$ and burnt in the normal way in the rotary kiln. The product, which is self-pulverising from the dicalcium silicate inversion, is extracted with water or sodium carbonate solution to dissolve the alumina. The solid residue, mainly dicalcium silicate and any iron compound, is mixed with additional limestone and reburnt to give cement clinker, while the alumina is recovered from the extract solution.

[1] D. Choffat, *Rock Prod.* **39** (7), 48 (1936).
[2] J. C. Séailles. Conference Maison de la Chimie, Paris, 1943; *Cem. Lime Mf.* **20,** 1 (1947); H. Anders, *Ber. dt. keram. Ges.* **29,** 204 (1952); M. von Euw, *Silic. ind.* (Belgium) **15,** 181, 202, 240 (1950).

4 Cement Components and their Phase Relations

Although most of the substances of which hydraulic cements are composed contain three or more elements in a state of combination, we may introduce a considerable simplification into their study by regarding them as produced by the association of simpler binary compounds. As most of the elements present, both metallic and non-metallic, are combined with oxygen we may represent the complex compounds as being built up of oxides, some of which are basic and others acidic in character. In dealing with silicates and other minerals it has been the custom of mineralogists to represent them either in this way as the sum of a number of oxides, or as salts of silicic, or aluminosilicic, acids. Thus calcium orthosilicate can be represented as $2CaO.SiO_2$ made up by the combination of the separate oxides or as Ca_2SiO_4 in which it is regarded as the calcium salt of an orthosilicic acid H_4SiO_4. The study of the structure and mode of combination of silicate minerals has made great strides since methods of X-ray analysis were applied to them, and it is now known that neither of these views succeeds in representing their structure correctly. The molecules of the silicate minerals have been found to be built up of silicon-oxygen complexes which may be extended indefinitely, an oxygen atom always forming the link between two silicon atoms. In some cases silicon may be replaced by aluminium. The metallic atoms appear to be linked to oxygen atoms which have only one link to silicon. The silicon-oxygen groups only exist in the solid state and do not form free ions. The silicate compounds cannot therefore be regarded as similar to salts which dissociate in water, forming positive and negative ions; they are either unattacked or decomposed by water. While the structure of silicate molecules is not correctly described as the sum of a number of oxides, their composition is correctly described thus, and the oxide form remains the most convenient method of representing their composition.

The compounds present in cements are formed by the interaction during burning of the lime, silica, alumina and ferric oxide compounds which constitute the bulk of the raw mix, together with the various minor components, such as magnesia and the alkalis, which are also present. Portland cement is composed of over 90 per cent lime, alumina and silica, and a knowledge of the compounds which may be formed amongst these oxides is thus of major importance. The effect of the minor components on this basic cement system has also to be considered. In high-alumina cement ferric, or ferrous, oxide is a more important component than silica.

In applying the conclusions of physical chemistry, based on the study of systems in equilibrium, to cements it will be necessary to consider in a later chapter to what extent they may be treated as equilibrium products. Portland cement is produced at a temperature below that of complete fusion and is only clinkered, that is, heated to a temperature at which only a minor proportion of the reacting mixture is in the molten condition. High-alumina cement, like blastfurnace slag, is usually produced in a liquid state, but if complete crystallisation does not occur on cooling some departure from the equilibrium state arises.

The remainder of this chapter contains an account of the oxides which may be considered as the components of cements and of the compounds which can be formed between them.

Calcium oxide

Calcium oxide (CaO) or lime is an essential component of all the cements with which we have to deal, and is mostly derived from the decomposition of calcium carbonate ($CaCO_3$). It is a white, amorphous and friable material with a specific gravity varying from 3·08 to 3·30, the value being greater the higher the temperature at which the lime is burnt. The chemical reactivity of lime, like its specific gravity, depends on the temperature at which it is prepared. Lime which has not been heated above 1000° 'slakes' instantly on the addition of water, whilst a more strongly ignited product requires a considerable time for hydration. A similar inactivity is observed towards steam or carbon dioxide.

The melting-point of pure lime is 2614° and it crystallises to give colourless, or pale yellow, transparent cubes belonging to the isometric system and showing perfect cubic cleavage. Its specific gravity is 3·32, hardness 3–4 on Mohs's scale, and refractive index 1·83.

A lump of crystalline lime, placed in water, is dissolved very slowly at the surface, the hydration being so gradual that no apparent rise of temperature is observed. The finely powdered crystals on the other hand appear to be inert when mixed with water, but after a few minutes the reaction sets in with explosive violence. It appears, therefore, that the decrease in the reactivity of the lime with increasing temperature of ignition is due less to a molecular change than to shrinkage and consequent reduction of surface. Loose powdered lime, if kept for any length of time at 1400°, agglomerates to form crystals which continually increase in size. Lime prepared at a low temperature is loose and porous, so that a large surface is presented to the action of the water, and hydration takes place very rapidly. The more strongly burnt lime is denser, offering a smaller surface in proportion to its mass, and the action is therefore confined to a smaller area and proceeds more slowly. Crystalline lime, being dense and non-porous, is naturally the least reactive.

Lime only occurs to a small extent in the free state in Portland cement.

Calcium hydroxide

The product of the hydration of lime by water is calcium hydroxide, $Ca(OH)_2$. It decomposes into lime and water vapour at about 400° in a free atmosphere; the vapour pressure[1] reaches 1 atmosphere at 512°:

$$Ca(OH)_2 \rightleftharpoons CaO+H_2O$$

It is therefore not found in cement clinker immediately after burning. It is usually produced to a small extent during the grinding of cements by the action of water released from the added gypsum.

It is obtained in the slaking of lime as a white, amorphous powder. If crystallised slowly, as in the hydration of Portland cement, it forms large, well-developed crystals. Hexagonal plate crystals of calcium hydroxide can be observed in set Portland cements, and from old cement briquettes very good crystals have been obtained.

Calcium hydroxide reacts with carbon dioxide to form calcium carbonate:

$$Ca(OH)_2+CO_2 = CaCO_3+H_2O$$

and this reaction is the cause of the hardening of high-calcium lime mortars. It only takes place in the presence of moisture.

Magnesium oxide

Magnesium oxide or magnesia, MgO, is only present in small quantities in Portland cement, being derived from magnesium carbonate present in the original limestone in the form of dolomite, $CaCO_3.MgCO_3$, or to a small extent from the clay or shale.

Pure magnesium oxide, if not heated to too high a temperature, possesses distinct hydraulic properties. Light, porous magnesia combines with water to form the hydroxide $Mg(OH)_2$, but without setting, whilst dense magnesia prepared at full red heat sets, yielding a coherent mass, although of low tensile strength. Magnesite, $MgCO_3$, when dead burnt at 1400–1500° yields a product which reacts only very slowly with water; it is used as a refractory.

Magnesia is soluble to some extent in lime[2] at high temperatures and lime is rather less soluble in magnesia. These solid solutions decompose at lower temperatures and for all practical purposes it may be assumed that the compounds crystallise independently from their fused mixtures. If the double carbonate, dolomite,[3] is ignited at 650–750° a mixture of magnesium oxide and calcium carbonate is obtained.

The melting-point of MgO is 2852°. It is isometric, shows perfect cubic cleavage, and has a refractive index of 1·737. It occurs in nature as the mineral periclase.

[1] P. E. Halstead and A. E. Moore, *J. chem. Soc.*, 1957, 1383.
[2] R. C. Doman, J. B. Barr, R. N. McNally and A. M. Alper, *J. Am. Ceram. Soc.* **46,** 313 (1963).
[3] Y. Schwob, *Revue Matér. Constr. Trav. publ.* **413,** 33; **414,** 85 (1950).

Magnesium hydroxide decomposes into magnesia and water at a comparatively low temperature, the water vapour pressure reaching 1 atmosphere[1] at about 350–380°. It is only very slightly soluble in water, about 0·01 gm per litre at 20°, and in the solid state is very little acted on by carbon dioxide.

Silicon dioxide

Silicon dioxide, or silica, is the most abundant of all minerals and is a component of the vast group of silicates, among which are the clays. It is present in cements in a state of combination, and is derived from the clay or shale used as a raw material. It is the main component of pozzolanas and is used in the form of sand as a constituent of mortar.

Pure crystalline silica occurs in nature as quartz, forming hexagonal crystals belonging to the trigonal system $a:c = 1:1{\cdot}10$. It is uniaxial positive with refractive indices $\omega = 1{\cdot}544$, $\varepsilon = 1{\cdot}553$. The specific gravity is 2·651 and hardness 7. Quartz, otherwise known as low-quartz or α quartz, is transformed at 573° into high-quartz or β quartz with a 2 per cent increase in volume. At 870° tridymite becomes the stable form and at 1470° this is further transformed to cristobalite which melts at 1710°. These transformations to tridymite and cristobalite are very sluggish and, unless heated very slowly, quartz simply passes into a viscous glass at about 1500–1600°; the true melting-point of quartz is not known accurately, but is probably below 1470°. There are also a number of other minor modifications of silica which need not be discussed here and numerous variations in the order in which the transformations can occur.[2] The question whether tridymite is a stable phase or whether alkali or water are needed to stabilise the structure remains controversial. Hydrothermal studies have suggested that tridymite is stable[3] but other studies[4] indicate that impurities are needed. The tridymite and cristobalite structures are closely related.

Tridymite has a specific gravity of 2·26, cristobalite 2·32, and vitreous (glassy) silica 2·203.

When cooling from the liquid state silica has a great tendency to assume a glassy condition, without crystallising. As this form is much less dense than the crystalline modification its crystallisation is accompanied by a marked contraction. If cooled under favourable conditions crystals are formed; these prove to be cristobalite, but a further change to tridymite may occur. Quartz has never been observed to crystallise from pure vitreous silica, but may be obtained in the presence of fluxes as an inversion product of cristobalite or tridymite.

Quartz is chemically a very inert substance at ordinary temperatures, but when strongly heated it reacts vigorously with bases. Sand behaves in mortars as an indifferent material, being bound together mechanically by the particles of lime or cement, but not entering into chemical reaction with them. At high temperatures silica behaves as an acid oxide and is capable of combining with bases to form silicates, and of expelling other acids from their compounds.

[1] S. J. Gregg and R. L. Razouk, *J. chem. Soc.*, 1949, Suppl. 1, 536.
[2] Cf. R. B. Sosman, *The Properties of Silica.*
[3] V. J. Hill and R. Roy, *Acta crystallogr.* **10,** 833 (1957).
[4] S. B. Holmquist, *Z. Kristallogr.* **111,** 71 (1958); O. W. Flörke, *Ber. dt. keram. Ges.* **32,** 369 (1955); F. M. Wahl, R. E. Grim and R. B. Graf, *Am. Miner.* **46,** 196 (1961).

Quartz will react with lime in the presence of water, if heated under pressure; this is the reaction which occurs in sand-lime brick manufacture. The readiness with which this reaction takes place is dependent on the surface presented by the quartz.

A distinction is commonly drawn between soluble and insoluble silica, the former being readily attacked by caustic alkalis or lime water, but there is no rigid line of demarcation between the two. Such substances as kieselguhr (infusorial earth) owe a large part of their reactivity to their highly porous character, presenting a large surface to the action of the reagent. Some forms of soluble silica contain combined water, but this fact apart, the differences of reactivity are often to be referred rather to physical differences of texture and porosity than to actual differences of chemical character.

The combination of silica with water gives rise to the silicic acid gels. There is no definite evidence that a crystalline silicic acid has ever been prepared. The work of van Bemmelen, and of many later investigators, has indicated that silicic acid gel is composed of silica and water in proportions that vary with the age and previous history of the gel, and with the vapour pressure of the atmosphere with which it is in contact. Some authors have suggested that definite hydrates such as $SiO_2.2H_2O$ and $SiO_2.H_2O$ may exist. Whether silicic acids of definite composition exist or not, it is nevertheless convenient, when considering the silicates which result from the combination of silica with bases, to represent them as salts of silicic acids. It is thus customary to speak of metasilicic acid, H_2SiO_3, and orthosilicic acid, H_4SiO_4, and, for example, to call $CaSiO_3$ and Ca_2SiO_4 calcium meta- and ortho-silicate respectively.

Aluminium oxide

Aluminium oxide or alumina, Al_2O_3, occurs in nature as corundum ($\alpha\ Al_2O_3$). Its melting point is 2045°. Another form, $\beta\ Al_2O_3$, has been found to crystallise from mixes of alumina with various bases, but it does not seem to exist as a pure oxide but only as compounds such as $CaO.6Al_2O_3$ [1] and alkali compounds. The latter have generally been considered to have the composition $Na_2O.11Al_2O_3$ and $K_2O.11Al_2O_3$ but evidence has been advanced[2] to indicate that the true composition of these alkali compounds is $R_2O.6Al_2O_3$. There are a number of other forms of anhydrous, or nearly anhydrous, alumina, of which $\gamma\ Al_2O_3$ is one, that are obtained on heating hydrated alumina. $\alpha\ Al_2O_3$ crystallises as hexagonal plates with refractive indices ω 1·768, ε 1·760; $\beta\ Al_2O_3$ has a mean refractive index of about 1·68.

Alumina in a combined state is an important constituent of cements, in which it behaves as an acid. It occurs combined with silica in all clays and in these it may be regarded as a base. It also occurs free in a hydrated form, mixed with a proportion of ferric oxide and smaller amounts of titania and silica, in bauxite. This is used in the manufacture of high-alumina cement. The natural hydrated forms of alumina are gibbsite (or hydrargillite) $Al_2O_3.3H_2O$, diaspore

[1] A. L. Gentile and W. R. Foster, *J. Am. Ceram. Soc.* **46,** 74 (1963) and many earlier papers referred to by these authors.

[2] R. Scholder and M. Mansmann, *Z. anorg. Chem.* **321,** 246 (1963).

($Al_2O_3.H_2O$) and boehmite (or bauxite) $Al_2O_3.H_2O$. Gibbsite is the main constituent of American bauxites and boehmite of European bauxites. There is some confusion in nomenclature. Some American authors[1] call diaspore and boehmite β and α forms respectively of $Al_2O_3.H_2O$, and gibbsite (or hydrargillite) an α form of $Al_2O_3.3H_2O$. A synthetic trihydrate known as bayerite is correspondingly styled β $Al_2O_3.3H_2O$. Another nomenclature, due to Weiser,[2] and one which is adopted in Great Britain, is α $Al_2O_3.H_2O$ (diaspore), γ $Al_2O_3.H_2O$ (boehmite), α $Al_2O_3.3H_2O$ (bayerite) and γ $Al_2O_3.3H_2O$ (gibbsite). Alumina gel[3] precipitated from an aluminium salt solution in the hot is hydrous γ $Al_2O_3.H_2O$ (boehmite). In the cold the hydrous γ $Al_2O_3.H_2O$ first formed changes spontaneously, first into α $Al_2O_3.3H_2O$ (bayerite) and then into γ $Al_2O_3.3H_2O$ (gibbsite). The changes that occur on dehydration are complex and a variety of intermediate forms of alumina are involved.[4] γ Al_2O_3 is one of these, the final product being α Al_2O_3; γ Al_2O_3 is also formed on heating clays to about 900°.

Hydrated alumina is not formed during the setting of Portland cement but is a product, in the form of gibbsite, of the setting of high-alumina cement.

Ferric oxide

The oxides of iron only occur to a relatively small extent in Portland cements, being derived from the clay or shale, but are an important constituent of high-alumina cements where they are derived from the bauxite. Ferrous oxide (FeO) does not usually occur in more than small amounts, ranging from a trace to about 0·4 per cent, in Portland cement, but it is present in high-alumina cement in amounts from 2 to 3 per cent upwards.

Ferric oxide (Fe_2O_3) is a constituent of both Portland and high-alumina cements. It resembles alumina in character and acts as an acid radical in cements, being combined with the bases (mainly CaO) present. The melting-point of pure Fe_2O_3 is unknown, as it dissociates under atmospheric pressure before its melting-point is reached,[5] giving oxygen and solid solutions of Fe_3O_4 in Fe_2O_3. Measurable dissociation of Fe_2O_3 occurs in air from 1100° upwards, the amount of dissociation increasing with the temperature.[6]

According to a summary by Rooksby,[7] there are three forms of hydrated ferric oxide, $Fe_2O_3.H_2O$. The α form occurs in nature as goethite and it is also produced on ageing of gels and sols of hydrous basic ferric oxide. The β form has not been found in nature, but can be produced from ferric chloride solutions.

[1] F. C. Frary, *Chemy Ind.*, 1946, 14; H. C. Stumpf, A. S. Russell, J. W. Newsome and C. M. Tucker, *Ind. Engng Chem.* **42,** 1398 (1950).
[2] H. B. Weiser and W. O. Milligan, *J. phys. Chem.* **38,** 1175 (1934).
[3] H. B. Weiser, *Colloid Chemistry*, 2nd Ed. 1949, 312 (Wiley, New York); H. B. Weiser and W. O. Milligan, *J. phys. Chem.* **44,** 1081 (1940).
[4] D. J. Stirland, A. G. Thomas and N. C. Moore, *J. Br. Ceram. Soc.* **57,** 69 (1958); H. Saalfield, *Clay Miner. Bull.* **3,** 249 (1958); G. W. Brindley and M. Nakahira, *Z. Kristallogr.* **112,** 136 (1959); T. Sato, *J. appl. Chem.* **12,** 9, 553 (1962); **14,** 303 (1964).
[5] R. B. Sosman and H. E. Merwin, *J. Wash. Acad. Sci.* **6,** 532 (1916).
[6] J. White, *Carnegie Scholarship Memoirs of the Iron and Steel Institute* **27** (1938); G. L. Malquori and V. Cirilli, *Sym., London 1952,* 120.
[7] *X-Ray Identification and Crystal Structure of Clay Minerals*, p. 244. Mineralogical Society, London, 1951. Editor G. W. Brindley.

The γ form occurs naturally as lepidocrocite and it can also be produced artificially by the oxidation of ferrous compounds. The crystal structures of the α and γ forms are similar to that of diaspore and boehmite respectively. No trihydrate of Fe_2O_3, analogous to $Al_2O_3.3H_2O$, appears to exist. Limonite, which has sometimes been supposed to have a higher content of combined water than the monohydrate, appears to be composed of finely divided goethite with adsorbed water.

Binary and ternary compounds of cements

The compounds which may be formed between the various oxides which are present in cements have been the subject of numerous investigations. As the result of such work the major compounds are now known, though our knowledge is still not complete as regards the minor constituents. Some of the work which has contributed to this knowledge has been carried out by mineralogists interested in the constitution of rocks and it was indeed from this source that the original, and now classical, contributions, which did so much to stimulate the study of the constitution of cement, came.

When two oxides such as lime and silica, or lime and alumina, are heated together at a high temperature, and if feasible one at which fusion occurs, the product obtained may appear to be either a mixture or homogeneous. It is not, however, sufficient to conclude from the homogeneous appearance of a mixture after fusing or sintering that a definite compound has been obtained. A homogeneous solid solution may have been formed, or the structure may be so minutely heterogeneous and the constituents so intimately mixed as to escape observation. The study of compound formation, and the determination of their composition, are based on the measurement of melting-points for various mixes. From these a curve may be constructed giving the melting-point of all mixes of the two constituents and indicating by its form the nature of the solids which crystallise out on cooling. Such work depends on microscopic or X-ray examination of the heated materials, as well as on the determination of melting-points. The microscopic examination is made on thin sections of the products ground until transparent and examined by transmitted light as in the petrological study of rocks, by reflected light on polished surfaces, or often more usefully on powdered preparations.

One method of thermal analysis consists in allowing heated mixtures of the two components to cool and recording temperature-time curves. The breaks on these curves indicate when solid first starts to separate out from the liquid, the eutectic points, and any inversions which may occur. This method cannot usually be applied to mixtures containing silica, for the crystallisation of compounds is so slow that supercooling occurs to a very great extent and no true breaks are obtained. The reverse method of heating curves can, however, be used and often provides very useful information, though its results may be difficult to interpret without additional help from other methods. It is usual to use the differential method of thermal analysis. In this the test sample is heated side by side with a reference substance and the temperature difference between the two substances is recorded against time or temperature. Very small samples can be investigated and in the extreme the specimen may consist of a tiny droplet fused on to the thermocouple used to measure the temperature. Methods for the study of systems with volatile components have also been developed. The samples are held in sealed platinum or other suitable capsules.[1,2]

The method which forms a most important line of attack in almost all investigations on the crystallisation of silicate mixes, is known as the quench method. A mix of any desired composition is prepared and heated until combination is complete. A small portion, about 0·1 g, is placed in a small platinum bucket and heated in a suitable furnace at a constant temperature for a sufficient period to allow equilibrium to be attained. It is then quenched by dropping rapidly into mercury or water and the product powdered and examined under a petrographic microscope. The rapid cooling 'freezes' the equilibrium which existed at the temperature of the furnace; any liquid present forms a glass in which are imbedded crystals of compounds which had not melted at the temperature of the furnace. By quenching from progressively higher temperatures a point is reached at which only glass is obtained, indicating that the charge has been entirely molten. The final melting-point is thus fixed. Quenches from lower temperatures show points at which successive solid phases crystallise out. These solid compounds occurring in the quench can be identified under a polarising microscope by means of their refractive indices and other optical properties. Examination by reflected light or by X-ray diffraction is also helpful. These methods are described in more detail in Chapter 6.

In determining solubility at room temperature it is a commonplace to filter off a solution from the solid in equilibrium with it for separate chemical analysis. Many attempts have been made to extend this technique to high temperatures, the most successful being a method developed by Newkirk in which separation is effected by centrifuging the charge at high temperature.[3]

Another technique of importance which can supplement, or in many cases replace, the quench method is high-temperature microscopy[4] in which the melt is observed directly on a microscope hot-stage. The most convenient form of apparatus is that in which a droplet of the melt is formed in the V of a thermocouple; by means of suitable electrical devices this thermocouple serves both to heat the specimen and measure its temperature. This method has the advantages of speed and directness over the quench method and is especially useful for the study of systems in which phases or compounds occur which can not be preserved to room temperature by quenching.[5] Optical changes accompanying polymorphic inversions can be observed and single crystals grown for X-ray structure analysis.[6] Suitable choice of thermocouples permits the determination of melting points and other phenomena by this method up to 2150°[7] and this can be extended to 2450° by replacing the thermocouple with a pure iridium heater and measuring the temperature pyrometrically. A high-temperature X-ray

[1] W. Gutt and G. J. Osborne, *Trans. Br. Ceram. Soc.* **65,** 521 (1966).
[2] A. D. Russell, *J. scient. Instrum.* **44,** 399 (1967).
[3] T. F. Newkirk and F. Ordway, *Bull. Am. Ceram. Soc.* **31,** 116 (1952).
[4] J. H. Welch, *J. scient. Instrum.* **31,** 458 (1954); **38,** 402 (1961); Br. Pat. 961, 019.
[5] W. Gutt, *Silic. ind.* **27,** 285 (1962).
[6] F. Ordway, *J. Res. natn. Bur. Stand.* **48,** 152 (1952).
[7] W. Gutt, *J. scient. Instrum.* **41,** 393 (1964).

camera based on the same thermocouple heater has also been developed.[1] High-temperature photomicrographs of tricalcium silicate and dicalcium silicate crystals growing as primary phases in $CaO-SiO_2$ mixes at a temperature above 2000° are shown in Plate 1. Extensive phase equilibrium data has been accumulated with the help of these direct methods. The variety of methods now available has much increased the speed and certainty with which complex phase equilibria may be unravelled.

It is necessary to distinguish clearly between the true melting-point and the softening-point. In mixtures of two substances which do not form solid solutions with one another, the variation of the freezing-point with the composition of the mixtures is shown in Fig. 1. The freezing-point of the substance

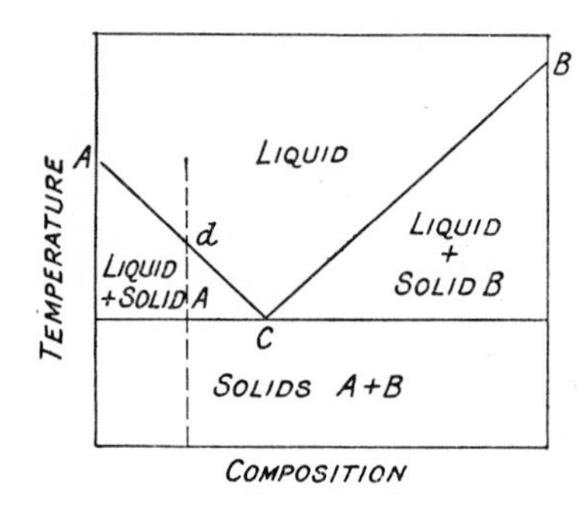

FIG. 1 Solidification of a mixture forming a eutectic.

A is lowered by the addition of the substance *B*, and that of *B* by the addition of *A*. The mixture *C*, which has the lowest freezing-point of the series, is known as the eutectic. A mixture, the composition of which lies between *A* and *C*, begins to freeze at a temperature indicated by the point *d*, depositing crystals of the substance *A*. The separation of these crystals as the temperature falls alters the composition of the portion remaining liquid until this reaches the eutectic composition, when it solidifies at the temperature represented by the point *C* where solid *B* also separates. Conversely if a solid mixture of composition *d* is heated, a part of it begins to melt as soon as the temperature represented by *C* is reached. At this temperature all the solid *B* present in the mix passes into liquid together with some of solid *A* to give the liquid composition *C*. As the temperature rises, more of solid *A* passes into solution and the composition of the liquid moves along the line *CA*. When the temperature *d* is reached the last remnants of solid *A* dissolve and a liquid of composition *d* remains. The liquid composition has thus traversed the path *Cd*. Now, if we have a solid mixture consisting of crystals *A* with only a small amount of the eutectic mixture *C*, the melting of the latter alone will not be sufficient to render the mass liquid, but will cause some degree of softening. The larger the proportion of eutectic present, the softer the mass will become when heated to a temperature slightly above *C*. The softening-point of a mixture is therefore the

[1] F. Aruja, J. H. Welch and W. Gutt, *J. scient. Instrum.* **36,** 16 (1959).

temperature at which the eutectic present melts, and the degree of softening is an indication of the proportion of eutectic in the mixture.

Before passing on to consider the phase equilibrium diagrams for the various oxide combinations of interest in cement chemistry, a brief résumé of the main features of such diagrams may be useful. The simplest form of the melting-point composition diagram is that already shown in Fig. 1, which represents a simple eutectic mix. Other forms of binary diagrams are shown in Figs. 2 to 5. Fig. 2

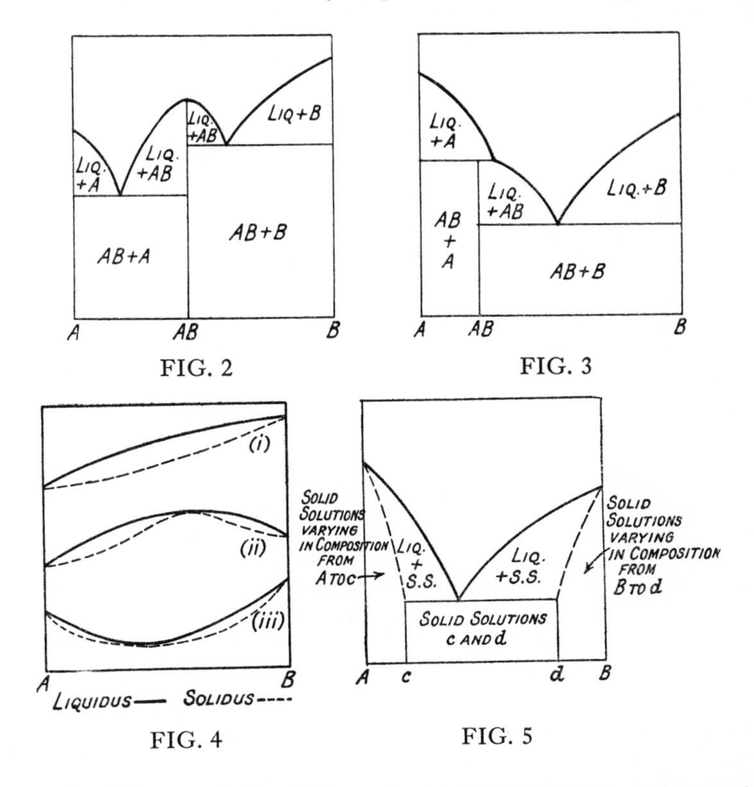

FIG. 2

FIG. 3

FIG. 4

FIG. 5

represents a binary system in which a compound stable at its melting-point is formed, while in Fig. 3 a compound is formed, but becomes unstable before its melting-point is reached and dissociates into liquid and one of the components. Such a compound is called an incongruent compound and said to melt incongruently. Fig. 4 shows the three different possible forms of a complete solid solution series, (i) showing no maximum or minimum in melting temperatures, (ii) with a maximum, and (iii) with a minimum. Fig. 5 represents a simple binary eutectic mix of two components, each of which has a limited solubility in the other. When we come to consider systems of three components it is no longer possible on a plane diagram to plot temperature as one of the variables, and a new method of representation has to be employed. The percentage composition of any ternary mix can be represented by a point in an equilateral triangle, the apices of which each represent 100 per cent of one of the pure components. Temperatures on such a plane diagram can only be represented by drawing isothermal lines, or indicating by arrows or other means the directions of falling temperatures. Only in a solid model obtained by erecting a prism on the triangular base can the temperature be actually represented as a co-ordinate. When plotting phase equilibrium diagrams in ternary oxide systems the method commonly used with aqueous systems of plotting solubility curves at a definite temperature, and showing by tie lines the solid phases in equilibrium therewith, is not as a rule employed. In its place we plot what are known as primary phase fields, that is, the composition regions in which any one solid is the first to separate when a completely liquid mix is allowed to cool. Fig. 6 shows the simplest diagram of this

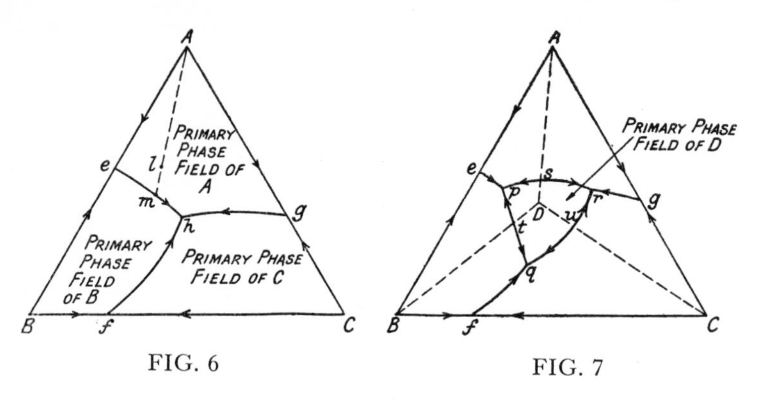

FIG. 6

FIG. 7

type consisting of three components, any pair of which form a simple eutectic system and the three together a simple ternary eutectic.

The binary eutectics are represented by the points *e*, *f*, *g* on the sides of the triangle. The addition of *C* to a mix of composition *e* lowers the binary eutectic temperature and the composition of the eutectic liquid moves into the ternary regions within the triangle. We thus obtain a line running from *e* into the triangle. Similar lines are obtained running from *f* and *g* and all three meet in the point *h* which is the ternary eutectic for the system. Any mix whose composition is represented by a point in the area *Aehg* will, on cooling from the liquid state, first deposit crystals of *A*. Thus the mix *l* on cooling deposits solid *A* and the liquid composition moves away from *A* along *Al* produced until it meets the line *eh* at *m*. The line *eh* is the boundary line between the primary phase fields of *A* and *B*, described above as their eutectic line. Hence at *m* the solid *B* also commences to separate out, and on further cooling *A* and *B* separate together while the liquid composition moves along the boundary line in the direction of falling temperature until the ternary eutectic *h* is reached. Here the third solid *C* commences to separate out and crystallisation of *A*, *B* and *C* from the liquid continues at constant temperature until solidification is complete. In general, any mix first deposits on cooling the solid in whose primary phase field it falls; the crystallisation path sooner or later intersects a boundary curve which is the line separating two primary phase fields and a second solid phase commences to separate. The crystallisation path follows the boundary curve until the ternary eutectic, which is the point of contact of three primary phase fields, is reached. The third solid separates in addition and the entire liquid goes solid at this point. The temperatures plotted on such a diagram are the temperatures at which any mix finally becomes completely liquid, or, regarded in the reverse manner, first begins to deposit a solid phase. The direction of falling temperatures along the sides of the triangle and the boundary lines is indicated by the arrows. The diagram, it will be noticed, is not an isothermal one, but from it the ordinary aqueous system type of isotherm can easily be derived for any temperature if the temperatures of final melting are obtained over the whole triangle and temperature contours drawn in. For any given temperature it can at once be seen if any particular mix is completely liquid, partially crystallised so as to consist of a liquid and either one or two solids, or completely solid.

In Fig. 7 is shown a ternary system in which a ternary compound *D*, of composition $A_xB_yC_z$, stable at its melting-point, is formed. We now have four primary phase fields, one for each of the components *A*, *B*, *C*, and an extra one, *pqr*, for the ternary compound *D*; *p*, *q* and *r* are respectively the ternary eutectics between *A*, *B* and *D*; *B*, *C* and *D*; *A*, *C* and *D*. The composition of the compound *D* falls within its own primary phase field and is a temperature maximum within it; the points where the lines connecting *D* to *A*, *B* and *C* intersect the boundary curves (at *s*, *t*, *u* respectively) of this field are temperature maxima on these boundaries. This is indicated by the arrows showing the temperatures of final melting falling away on either side of the points *s*, *t* and *u* along the respective boundary curves on which they occur. Since the point *s* lies on the line *AD* at the point where both solids *A* and *D* are present, it must represent the binary eutectic between *A* and *D* which form a simple binary system within the ternary system. The point *s* is thus a minimum temperature along the line *AD*. Similar considerations apply to *t* and *u*. The temperatures of all points along the boundary lines are lower than those of points on either side. Any mix whose composition falls within the triangle *ABD* must on final solidification consist of the three solids, *A*, *B* and *D*; its crystallisation path must end at the ternary eutectic *p*. For the regions *BDC* and *ADC* the corresponding final crystallisation end-points are *q* and *r*.

A third type of diagram is that shown in Fig. 8, which represents the case in which a ternary compound *D* is formed which is unstable at its melting-point and melts incongruently to form a liquid and another solid. The composition of *D* in this case falls outside its primary phase field. The line *AD* does not cut the boundary *pr* between the primary fields of *A* and *D* and no temperature maximum occurs along it, but temperatures fall continuously from *p* to *r*. The line *CD* cuts the boundary between the field of *C* and *D* and a temperature maximum occurs on *qr* at *u*. The line *BD* produced cuts *pq*, the boundary between the *B* and *D* fields, at *t* and a temperature maximum on *pq* occurs at this point. Points *q* and *r* are ternary eutectics as before, but *p* is only an invariant point as the temperature falls away from it towards *r*. No temperature maximum occurs in the field of *D*, but temperatures rise towards the boundary *pq*, and continue to do so to the left of this boundary where *B* is the stable phase. The boundary line *pq* is not therefore a minimum temperature line with respect to points on either side of it as it was in Fig. 7. Both *pr* and *qr* are, however, minimum temperature lines with respect to points on either side of them as was the case in Fig. 7.

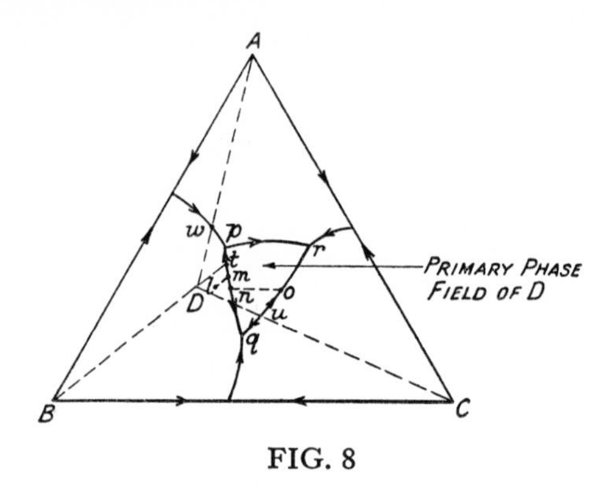

FIG. 8

Some of the crystallisation paths in a system of this type are more complex than those previously considered. Thus while in Fig. 7 a liquid of composition *D* would solidify entirely at one temperature to form the solid *D* this does not occur in Fig. 8. In this case *D* falls in the primary field of *B* and a liquid of composition *D* first deposits solid *B* on cooling. The liquid composition moves along *BD* produced until it meets the boundary *pq* of the primary phase field of *D* at *t*. Here *D* starts to crystallise and the previously formed solid *B* commences to redissolve in the liquid *t* so as to maintain its composition constant. This process continues until the last crystal of *B* disappears and the last drop of liquid simultaneously solidifies. The final solid product is thus *D* alone. An example of a longer crystallisation path is afforded by the point *l*. A liquid of this composition crystallises out *B* and moves along *Bl* produced until it cuts the boundary *pq* at *m*. *D* now separates out alongside of the previously formed *B* and the liquid moves along *tq* to *n*. The mean composition of the total solid separated up to any point along *mn* is given by the intersection of the line drawn from the particular point on *mn* through the original composition *l* with *BD*. When the point *n* is reached such that *Dln* forms a straight line the composition of the total solid separated must be *D*. Hence when moving along *mn* the pre-existing solid *B* must have been dissolving while *D* separated out. At *n* the last of solid *B* redissolves in the liquid, leaving *D* as the single solid phase, and the crystallisation path now leaves the boundary and crosses the primary phase field of *D* on further cooling. The path across this field is given by the prolongation of *Dn*. This intersects the boundary *qr* at *o*, where *C* commences to separate out together with *D*. The boundary is followed to the ternary eutectic *r*, where *A* separates out in addition, and the mix goes solid at this temperature, forming a mixture of *A*, *C* and *D*. Any point lying in the region *Dtpw* will on crystallisation first deposit *B* until a boundary curve is reached. Either *A* or *D* now separates, depending on which boundary curve is encountered, and the liquid composition moves along the boundary curve until the invariant point *p* is reached. A little consideration will show that the mix cannot go solid here since its original composition fell within the triangle *ADC*, and on final solidification the three solids, *A*, *D* and *C*, must be present. This is not the case at *p*, where *A*, *D* and *B* co-exist together. The reaction at *p* consists, then, of separation of solids *A* and *D* while *B* redissolves. The liquid composition remains at *p*, and the temperature remains constant, until all *B* has redissolved. The crystallisation path then follows along the boundary *pr* on further cooling, with *A* and *D* separating, and the mix finally goes solid at *r*, where *C* also separates. Some mixes, in contrast to this, will go completely solid when *p* is reached. This is the case for any mix whose composition falls in the triangle *ABD*.

Many more complex ternary diagrams occur, but an application of the same general principles will indicate their significance and will allow the crystallisation path of any mix to be deduced. The general rules governing crystallisation paths which have been utilised in the foregoing examples can be briefly summarised as follows:

1. When one solid phase only is present the crystallisation path lies within a primary phase field and is given by the prolongation of a line drawn from the composition of the solid separating through the original mix composition. This is necessarily a direction of falling temperature.

2. When two solid phases are present, the crystallisation path follows the boundary line between the primary phase fields of the two solids in the direction of falling temperature. The mean composition of the total solid separated down to any point is given by the intersection of the line drawn through the point on the boundary curve considered and the original mix composition with the line joining together the composition of the two solids present. The composition of the solid actually separating at any point is given by the intersection of the tangent to the boundary curve at that point with the same composition line or its prolongation. If the intersection is with this line produced, then one solid is redissolving and the other separating out. If one solid entirely redissolves, the crystallisation path leaves the boundary line and crosses an adjoining primary phase field.

3. When three solids are present together, the liquid composition must be at an invariant or eutectic point. In the latter case the mix will go solid at this point. In the former case the mix will go solid at this point if the original mix composition is such as to fall within the triangle formed by joining together the compositions of the three solids co-existing at this point. If this is not the case, one solid will entirely disappear by redissolving and the crystallisation path then follows abo undary curve to another invariant or eutectic point.

The assumption is made in tracing such crystallisation paths that equilibrium between solid and liquid is always maintained. The path followed by a solid mix on heating is, with the same assumption of equilibrium, the exact reverse of the crystallisation path on cooling.

Abbreviation of compound formulae

In the remainder of this chapter and in following chapters frequent reference is made to certain compounds whose full molecular formulæ occupy much space. The convention has been adopted of using a shortened notation and it will be convenient to adopt the same practice here. When first appearing the compounds are shown in the ordinary chemical notation, but subsequently the shortened notation is frequently used. The shortened formulæ used are:

$$
\begin{aligned}
C_3S &= 3CaO.SiO_2 \\
C_2S &= 2CaO.SiO_2 \\
C_3S_2 &= 3CaO.2SiO_2 \\
C_3A &= 3CaO.Al_2O_3 \\
\left.\begin{aligned} C_5A_3 &= 5CaO.3Al_2O_3 \\ C_{12}A_7 &= 12CaO.7Al_2O_3 \end{aligned}\right\} \\
MA &= MgO.Al_2O_3 \\
M_2S &= 2MgO.SiO_2 \\
C_3MS_2 &= 3CaO.MgO.2SiO_2 \\
CMS &= CaO.MgO.SiO_2 \\
C_3T_2 &= 3CaO.2TiO_2
\end{aligned}
$$

$$
\begin{aligned}
\left.\begin{aligned} C_3A_5 &= 3CaO.5Al_2O_3 \\ CA_2 &= CaO.2Al_2O_3 \end{aligned}\right\} \\
C_2AS &= 2CaO.Al_2O_3.SiO_2 \\
C_2F &= 2CaO.Fe_2O_3 \\
CF &= CaO.Fe_2O_3 \\
C_4AF &= 4CaO.Al_2O_3.Fe_2O_3 \\
C_6A_2F &= 6CaO.2Al_2O_3.Fe_2O_3 \\
C_6AF_2 &= 6CaO.Al_2O_3.2Fe_2O_3 \\
KC_{23}S_{12} &= K_2O.23CaO.12SiO_2 \\
NC_8A_3 &= Na_2O.8CaO.3Al_2O_3 \\
C_3P &= 3CaO.P_2O_5
\end{aligned}
$$

CALCIUM SILICATES

The system $CaO-SiO_2$

Lime and silica when heated together can unite to form four distinct compounds:[1] (1) the metasilicate $CaO.SiO_2$, (2) the compound $3CaO.2SiO_2$, (3) the orthosilicate $2CaO.SiO_2$, (4) the compound $3CaO.SiO_2$. All these compounds appear in more than one crystal form. The phase diagram is shown in Fig. 10.

Calcium Metasilicate, $CaO.SiO_2$, appears in three forms. The high-temperature form, α $CaO.SiO_2$ pseudowollastonite, is very rare as a natural mineral. Two low-temperature forms occur naturally; they are known as wollastonite and parawollastonite. The term β $CaO.SiO_2$ is used indiscriminately for either of the low-temperature forms; the product obtained from the inversion of α $CaO.SiO_2$ may be wollastonite, parawollastonite, or an intimate inter-

[1] A. L. Day, E. S. Shepherd and F. E. Wright, *Am. J. Sci.* (iv) **22,** 265 (1906); N. L. Bowen, J. F. Schairer and E. Posnjak, ibid. (v) **26,** 206 (1933); J. H. Welch and W. Gutt, *J. Am. Ceram. Soc.* **42,** 11 (1959); see also J. W. Greig, *Am. J. Sci.* (v) **13,** 1 (1927), for formation of the two immiscible liquids in the high SiO_2 region.

growth of the two. Pseudowollastonite is obtained when a liquid mixture of lime and silica in this proportion solidifies and also when either of the low-temperature forms is heated to above 1125°,[1] the change taking place with absorption of heat, but without any appreciable change in volume, all forms having specific gravity about 2·90[2].

α *CaO.SiO*$_2$, pseudowollastonite, is triclinic strongly pseudo-hexagonal, and has optical properties:[3] $\alpha = 1{\cdot}610$; $\gamma = 1{\cdot}654$; optical axial angle small, nearly uniaxial, optical character positive.

β *CaO.SiO*$_2$, wollastonite, is triclinic and usually occurs as fibrous aggregates. Optical properties: $\alpha = 1{\cdot}620$; $\beta = 1{\cdot}632$; $\gamma = 1{\cdot}634$. Nearly parallel extinction, biaxial negative, 2 V = 39°.

β *CaO.SiO*$_2$, parawollastonite, is monoclinic with parallel extinction and has optical properties: $\alpha = 1{\cdot}620$; $\beta = 1{\cdot}631$; $\gamma = 1{\cdot}633$; biaxial negative, 2V = 44°.

Water is practically without chemical action on the metasilicate. It is not a constituent of Portland cement clinker.

The compound $3CaO.2SiO_2$, rankinite, melts incongruently, dissociating[4] into $2CaO.SiO_2$ and liquid at 1464°. It is monoclinic and has optical properties: $\alpha = 1{\cdot}641$; $\beta = 1{\cdot}645$; $\gamma = 1{\cdot}650$; biaxial positive, optic axial angle large. Density 2·97. It is not a constituent of Portland cement clinker. Another polymorphic form, known as kilchoanite or phase Z, can be prepared hydrothermally and occurs as a natural mineral.

Calcium Orthosilicate $2CaO.SiO_2$ is present in cements and slags. It melts at 2130° and exists in four well-established forms.[5]

α *2CaO.SiO*$_2$ (hexagonal) is stable above 1420–1447°;[5,6] on cooling it changes reversibly to the α' form. Specimens stabilised at room temperature appear as clear colourless grains, basal sections sometimes showing a hexagonal outline. Optical properties are variable according to stabiliser present; with Na_2O and Al_2O_3 $\alpha = 1{\cdot}702$, $\gamma = 1{\cdot}712$; positive 2V small or uniaxial. Density (on crystals from basic slag) 3·04. This form of dicalcium silicate is completely miscible[6] with the highest temperature form[7] of $3CaO.P_2O_5$ and can be stabilised to room temperature by addition of 34 per cent by weight $3CaO.P_2O_5$. When stabilised by tricalcium phosphate this polymorph has no cementing properties.[8] The solubility of Al_2O_3 is between 2–3 per cent by weight in the α form (1400°–1500°) and that of Fe_2O_3 is 1·5 per cent at 1400° and 2·5 per cent at 1500°.[9] The solubility of MgO in α C_2S is 1·5 per cent MgO at 1600°.[10]

1 A. Muan and E. F. Osborn, *Ind. Heat.*, 1952, p. 1293.
2 J. H. Howison and H. F. W. Taylor, *Mag. Concr. Res.* **9** (25), 13 (1957); L. Heller and H. F. W. Taylor, *Crystallographic Data for the Calcium Silicates*, H.M.S.O. London (1956).
3 Unless otherwise stated, all refractive indices quoted in this book refer to sodium light.
4 E. F. Osborn, *J. Am. Ceram. Soc.* **26,** 321 (1943).
5 R. W. Nurse, *Sym., London 1952*, 56; A. Guinier and M. Regourd, *Sym., Tokyo 1968.*
6 R. W. Nurse, J. H. Welch and W. Gutt, *J. chem. Soc.*, 1959, 1077.
7 Ibid., *Nature, Lond.* **182,** 1230 (1958).
8 J. H. Welch and W. Gutt, *Sym., Washington 1960*, 59.
9 G. Yamaguchi, Y. Ono and S. Kawamura, *Review of 16th General Meeting Japan Cement Engineering Association*, 1962, 35.
10 C. M. Schlaudt and D. M. Roy, *J. Am. Ceram. Soc.* **49,** 430 (1966); W. Gutt, Ph.D. Thesis, London 1966.

The hexagonal structure suggested for α2CaO.SiO$_2$ by Bredig[1] received support from recent diffractometric studies of dicalcium silicate when held at 1500°C.[2]

α' 2CaO.SiO$_2$, orthorhombic, bredigite, is stable from about 800° to 1447° on heating starting with the γ form; on cooling it persists down to about 650 to 670° where it inverts reversibly to the β form. Stabilised material forms irregular grains with cyclic or multiple twinning. α' 2CaO.SiO$_2$ can be stabilised to room temperature by the addition of 15 wt per cent of 3CaO.P$_2$O$_5$, the two compounds forming solid solutions. When stabilised in this way this polymorph is only weakly hydraulic. Up to 1 per cent Al$_2$O$_3$ or Fe$_2$O$_3$ dissolve in the α' form at 1300–1350°, at 1400° up to 2 per cent MgO. Optical properties, variable according to stabiliser. Crystals from slag $\alpha = 1{\cdot}713$; $\beta = 1{\cdot}717$; $\gamma = 1{\cdot}732$, biaxial positive 2V = 20–30°; synthetic solid solution K$_2$O.23CaO.12SiO$_2$ $\alpha = 1{\cdot}695$; $\gamma = 1{\cdot}703$, nearly uniaxial. Density (slag crystals) 3·40. The orthorhombic β K$_2$SO$_4$ structure suggested for α' 2CaO.SiO$_2$ by Bredig has not been so far experimentally confirmed.

β 2CaO.SiO$_2$ monoclinic, larnite, is generally regarded as metastable since it cannot be obtained by heating of the γ form. It consists of rounded grains or crystals with polysynthetic twinning formed by inversion from the α' form at 650° to 670° metastably. The pure mineral on cooling to about 520° inverts to the γ form. This inversion temperature can vary over a range. In the presence of impurities it may be delayed or restrained indefinitely. Optical properties: monoclinic: $\alpha = 1{\cdot}717$, $\gamma = 1{\cdot}735$; biaxial positive, 2V moderate to large. Density 3·28. It has been claimed[3] that the β form is stable between 760° and 900° on heating and between 650° and 450° on cooling but this has not been confirmed by other work. Suggestions[4] have also been made that slightly modified forms of the α' and β polymorphs designated α'_H and β_H, also exist but the evidence for them is not strong. The diagram shown in Fig. 9 represents the best agreement amongst the data available.

γ 2CaO.SiO$_2$ can take up small amounts of many other oxides in solid solution[5] and the belite in cement is an impure form of β C$_2$S (see p. 107) though the α and α' forms may occur to a small extent. γ2CaO.SiO$_2$, orthorhombic, can be readily prepared by ignition of finely ground CaCO$_3$ and quartz at 1600°. It is obtainable only as a fine powder. It is stable below 725–830° above which it inverts to the α' form. Colourless prisms with parallel cleavage; optical properties; extinction parallel to the prism axis; $\alpha = 1{\cdot}640$; $\gamma = 1{\cdot}654$; biaxial negative, 2V = 60°. Density 2·97. The olivine structure of γ 2CaO.SiO$_2$ has been directly established.[6]

A solidified mass containing calcium orthosilicate is at first very hard. It has, however, the remarkable property of cracking and rapidly disintegrating at the

[1] M. A. Bredig, *J. Am. Ceram. Soc.* 1950 **33** (6), 188.
[2] A. Guinier and M. Regourd, *Sym., Tokyo 1968.*
[3] N. A. Toropov, B. V. Volkonsky and V. I. Sadkov, *Dokl. Akad. Nauk SSSR* **112,** 467 (1957).
[4] D. K. Smith, A. J. Majumdar and F. Ordway, *J. Am. Ceram. Soc.* **44,** 405 (1961).
[5] R. W. Nurse, *Sym., London 1952,* 56; *Sym., Tokyo 1968*; E. L. Newman and L. S. Wells, *J. Res. natn. Bur. Stand.* **36,** 137 (1946).
[6] D. K. Smith, A. J. Majumdar and F. Ordway, *Acta crystallogr.*, 1965, **18,** 787.

ordinary temperature, falling into a fine white powder. This behaviour is observed even in mixes containing considerable proportions of other substances. It is due to an inversion to the γ form which occurs after cooling. While the specific gravity of the β form is 3·28, that of the γ form is only 2·97 so that its formation is accompanied by an increase in volume of nearly 10 per cent. Under ordinary conditions the cooling of the mass is too rapid to allow the change to γ to take place. At the ordinary temperature, therefore, the orthosilicate is in an unstable condition and tends to pass spontaneously into the stable γ form. If this change begins it spreads rapidly through the mass, heat being developed, and the great increase of volume which accompanies the change results in the shattering of the crystals into dust. The process is usually referred to as the dusting of the compound. The stability of the orthosilicate is much affected by the rate of cooling and by the presence of other substances in a state of solid solution; many mixtures, if rapidly cooled, may be preserved indefinitely without undergoing disintegration, and then possess hydraulic properties. The γ form has no hydraulic properties.

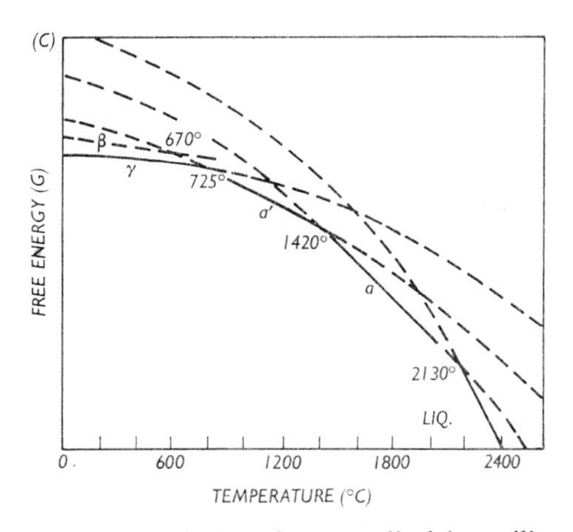

FIG. 9 Phase diagram for pure dicalcium silicate.

It is somewhat difficult to prepare β C_2S in the absolutely pure form as the mix of this composition inverts so readily on cooling to the γ form which is thus easily prepared. The presence of small amounts of other oxides, such as 0·1 per cent B_2O_3 or 1 per cent Cr_2O_3 or $3CaO.P_2O_5$, in the melt is, however, very effective in preventing the β → γ inversion. The preparation is carried out by heating a mix of $2CaCO_3+1SiO_2$ to 1450° for an hour, regrinding, and reheating for a further hour, and repeating this until the preparation is homogeneous. An addition of 0·1 per cent boric oxide is made if it is desired to obtain the β form, though it is possible with rapid cooling to manage without it. The β form may also be obtained by heating γ C_2S to 1000° for several hours and cooling rapidly.

The percentage composition of the eutectic between C_2S and C_3S at 2050° is 69·5CaO, 30·5SiO_2.

Tricalcium silicate, $3CaO.SiO_2$, occupied for many years a unique position in silicate chemistry. The earlier investigations indicated that the pure compound was formed only by a solid-state reaction between CaO and $2CaO.SiO_2$ at temperatures above 1250° and below 1900°. Above this temperature it again dissociated in the solid state into CaO and $2CaO.SiO_2$. Direct high-temperature microscopy has now shown that $3CaO.SiO_2$ does not decompose at 1900° into $2CaO.SiO_2$ and CaO but that it melts incongruently to CaO and liquid.[1] The phase diagram for the system $CaO–SiO_2$ incorporating the revised melting behaviour of C_3S is given in Fig. 10. The diagram[2] for the lime-rich portion

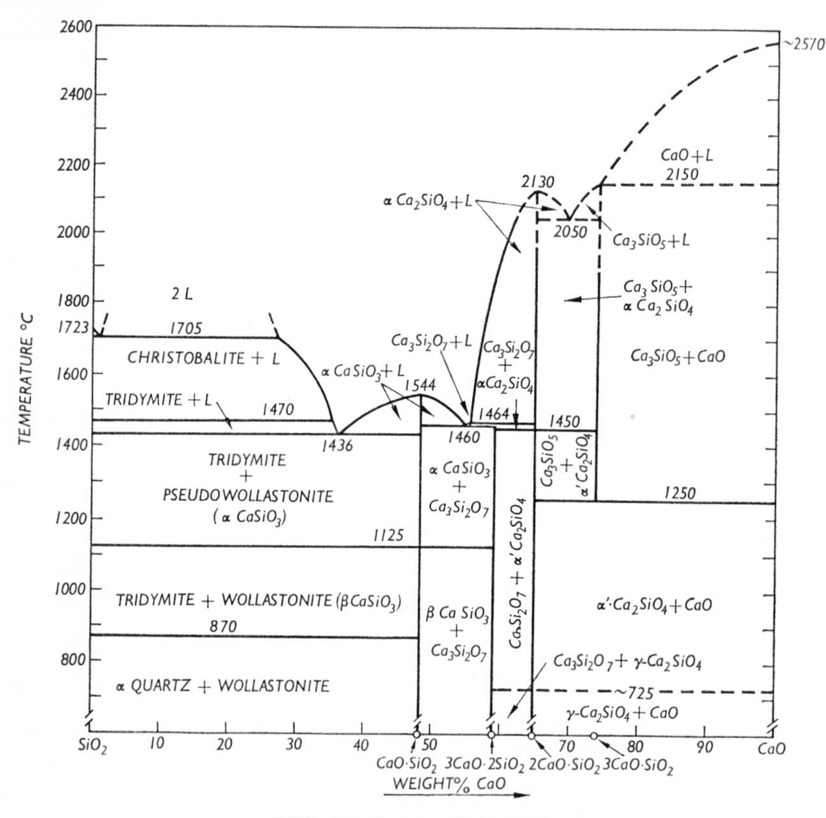

FIG. 10 System $CaO–SiO_2$

of the system $CaO–SiO_2$ in Fig. 11 gives detail of the melting behaviour of C_3S. Tricalcium silicate is unstable below 1250°.[3] If annealed below this tem-

[1] J. H. Welch and W. Gutt, *J. Am. Ceram. Soc.* **42,** 11 (1959). G. Trömel, W. Fix and R. Heinke, *Tonind Ztg.* **93,** 1 (1969) have maintained that C_3S decomposes above 1800° but the evidence against this is strong.

[2] W. Gutt, Ph.D. Thesis, London 1966.

[3] F. M. Lea and T. W. Parker, *Phil. Trans. R. Soc.* **234,** No. 731, 1 (1934) (A); S. L. Meyers, *Rock Prod.* **33** (8), 78 (1930).

perature it slowly decomposes into C_2S and CaO. This change is accelerated by the presence in the compound of either of the decomposition products C_2S and CaO, or of calcium sulphate and certain other salts, but compounds such as $2C_2S.CaSO_4$ may then be formed as well as C_2S and CaO. If a sample of Portland cement is held at 1150° the C_3S in it is decomposed and free lime is produced. This change is very slow, many hours being required even under the most favourable conditions to produce a few per cent of free CaO. This slow rate of decomposition, so slow that below about 700° C_3S can exist indefinitely, may have misled those authors who have questioned the validity of the breakdown at 1250°.[1]

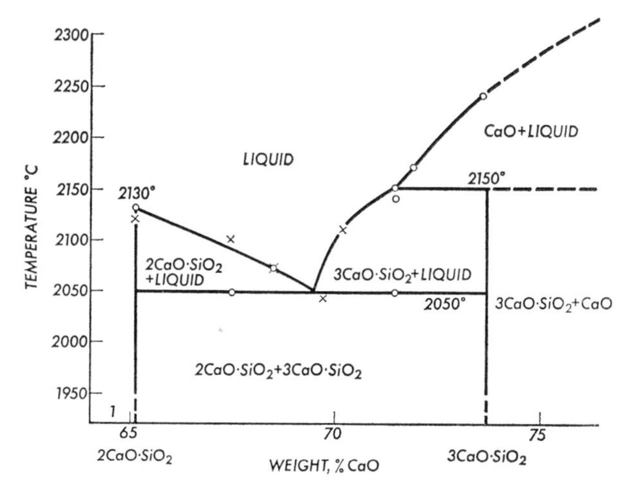

FIG. 11 Phase diagram for lime-rich portion of system $CaO–SiO_2$

Tricalcium silicate occurs in small equant grains without apparent cleavage. The refractive indices are $\alpha = 1{\cdot}718$, $\gamma = 1{\cdot}723$, the birefringence being very low, about 0·005. It is optically negative, the axial angle varying widely; on pure material it is very large but in crystals forming from melts it may be as low as 5°. Density 3·15. It is the principal cementing constituent in Portland cement. Very detailed X-ray work[2] showed that the phase formed in Portland cement contains MgO and Al_2O_3 and differs slightly from pure $3CaO.SiO_2$.

Tricalcium silicate exists in a number of polymorphs, three triclinic (T_1, T_{11}, T_{111}), two monoclinic (M_1, M_{11}) and the high-temperature trigonal (rhombohedral) form.[3] The differences between the separate triclinic and monoclinic forms are slight and not all are equally firmly established. The series of Transformation is $T_1 \xrightarrow{600°} T_{11} \xrightarrow{920°} T_{111} \xrightarrow{980°} M_1 \xrightarrow{990°} M_{11} \xrightarrow{1050°} R$. It is doubtful whether these various polymorphs differ very significantly in their cementing

[1] N. Yannaquis, *Revue Matér. Constr. Trav. publ.* **480,** 213 (1955); G. Yamaguchi and H. Miyabe, *J. Am. Ceram. Soc.* **43,** 219 (1960).
[2] J. W. Jeffery, *Sym., London 1952*, 30.
[3] A. Guinier and M. Regourd, *Sym., Tokyo 1968*, summarised the evidence and gives references to the many reports on the subject.

properties (see p. 85). The alite in Portland cement is usually in the monoclinic form but sometimes the trigonal and triclinic forms may occur. It is not pure C_3S but contains various other oxides in solid solution[1] (see p. 107) and these have some influence on its strength properties. The more significant of these are MgO, Al_2O_3 and Fe_2O_3. The important factor is their solubility in the trigonal form of C_3S stable at the clinkering temperature. The inversions of C_3S all take place below 1050° at a temperature where solution and exsolution will be very slow and where in any event they are metastable. The Mg^{2+} ion substitutes directly for Ca^{2+} with a limit of about 2 per cent MgO weight replacement at 1500°. The monoclinic form of C_3S is stabilised to room temperature by 1·5 per cent MgO. The limit of solubility of Al_2O_3 is about 0·9 per cent weight, $3Al^{3+}$ ions replacing $3Si^{4+}$ ions, the balance of charge being maintained by the introduction of Al^{3+} in interstitial positions. The substitution series may be represented by a solid solution between C_3S and a hypothetical $C_{4\cdot5}A$. The presence of alumina stabilised the triclinic forms of C_3S at room temperatures. When both MgO and Al_2O_3 are taken into solid solution only 1 per cent MgO is required to stabilise the monoclinic form of C_3S to room temperature but when the content of each is below about three-quarters of their individual saturation values the triclinic form is found. Ferric oxide dissolves in C_3S to the extent of 1 per cent Fe_2O_3 by weight at 1400–1500° as a solid solution of a hypothetical C_3F in C_3S. The limit of solubility is unaffected by the simultaneous presence of Al and Mg ions. The solid solution of some other minor components is discussed later.

The preparation of pure tricalcium silicate is a somewhat tedious process and necessitates a series of burns. The most rapid method for preparing it on a large scale (e.g. several hundred grammes) is as follows: A mix of $3CaCO_3+1SiO_2$ is moulded with water, pressed into a compact mass and burnt at 1200°. The product is ground, remoulded with water, and pressed and burnt at 1450–1500°. This is repeated until microscopic examination shows that over half the product is tricalcium silicate. The ground mass is then moulded and pressed with carbon tetrachloride and burnt at 1450–1500°. This process is continued until combination is complete. Some four burns after mixing with water and two after mixing with carbon tetrachloride usually suffice. The reaction is markedly accelerated if steam is blown into the furnace.

CALCIUM ALUMINATES

The system $CaO–Al_2O_3$

Four stable compounds occur in this system[2]: $3CaO.Al_2O_3$, $CaO.Al_2O_3$, $CaO.2Al_2O_3$ and $CaO.6Al_2O_3$. All these compounds melt incongruently and have primary crystallisation fields in the binary system. If the system is studied

[1] H. G. Midgley and K. E. Fletcher, *Trans. Br. Ceram. Soc.* **62,** 917 (1963); R. W. Nurse, *Sym., Tokyo 1968*; K. E. Fletcher, *Trans. Br. Ceram. Soc.* **64,** 372 (1965); E. Woermann, W. Eysel and Th. Hahn, *Zement-Kalk-Gips* **21,** 241 (1968); **22,** 235, 414 (1969).

[2] See T. D. Robson, *Sym., Tokyo 1968* for a review of investigations on this system.

in air of normal humidity and not in a strictly moisture-free atmosphere a further metastable compound $12CaO.7Al_2O_3$ which is not binary and contains some hydroxyl groups can also occur in addition to the four stable compounds. The phase diagrams shown in Figs. 12 and 13 are based on the most recent work.[1]

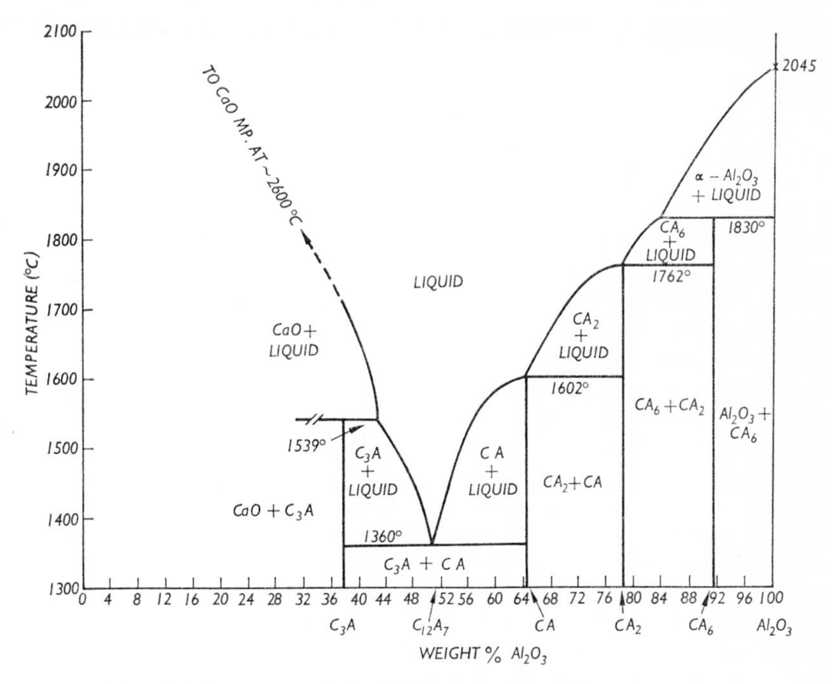

FIG. 12 The system $CaO-Al_2O_3$ in moisture-free atmosphere.

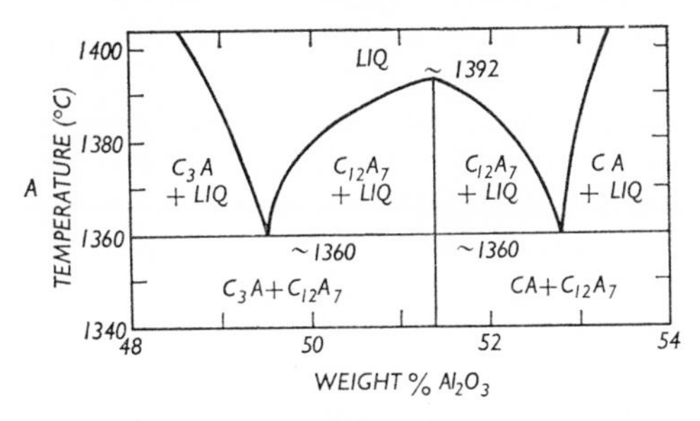

FIG. 13 Congruent melting of $12CaO.7Al_2O_3$ in atmosphere of ordinary humidity.

Tricalcium aluminate $3CaO.Al_2O_3$ is unstable at its melting point and melts incongruently at 1539° dissociating into CaO and liquid of composition (weight

[1] R. W. Nurse, J. H. Welch and A. J. Majumdar, *Trans. Br. Ceram. Soc.* **64,** 323, 409 (1965).

per cent) CaO 57·2, Al_2O_3 42·8. It is for this reason best obtained by annealing a mix of C_3A composition at about 1365° for many hours with intermediate grinding. It is present in Portland cement. $3CaO.Al_2O_3$ occurs in equant grains sometimes showing crystalline outline. It belongs to the cubic, or isometric system and has refractive index 1·710. Its hardness is 6, density 3·04.

The compound $12CaO.7Al_2O_3$ was formerly known as pentacalcium tri-aluminate $5CaO.3Al_2O_3$. The difference in composition is small:

C_5A_3 47·78CaO, 52·22Al_2O_3
$C_{12}A_7$ 48·53CaO, 51·47Al_2O_3

In the range 930°–1350° $12CaO.7Al_2O_3$ enters into reversible equilibrium with water molecules normally present in the ordinary furnace atmosphere. This is not merely a surface effect, because the unit cell size, density and refractive index are all affected. Hydroxyl groups are present in the structure and at about 950°C when sorption is maximum the constitutional formula is $Ca_{12}Al_{14}O_{32}(OH)_2$. It is not certain whether OH groups are completely removed on melting and it has not been possible to prepare completely anhydrous $C_{12}A_7$. The melting point of $C_{12}A_7$ in air is 1392°. In a dry atmosphere it melts incongruently to CA and liquid at 1374° and C_3A and CA form a eutectic at 1360°. Thermodynamic calculations also suggest that at all temperatures up to the melting point of $C_{12}A_7$ an assemblage of $CA+C_3A$ is more stable than $C_{12}A_7$. A strong case has been made for omitting the primary phase field of $C_{12}A_7$ from the $CaO-Al_2O_3$ phase diagram as shown in Fig. 12. In a furnace atmosphere of ordinary humidity $C_{12}A_7$ is found as a compound melting congruently as shown in Fig. 13. $C_{12}A_7$ occurs in rounded grains or octahedra and is cubic with refractive index 1·608. Its hardness is 5 and density 2·69. The compound is notable in that its refractive index is lower than that of the quenched glass (1·670) of the same composition, an unusual occurrence. It is occasionally present in Portland, and more frequently in high-alumina, cement.

Pentacalcium aluminate $5CaO.3Al_2O_3$ was formerly believed to be the composition of the phase known to be $12CaO.7Al_2O_3$, which is very probably metastable as discussed above. The early authors also described an unstable form of this compound obtained by annealing the glass and related this 'unstable C_5A_3' to the prismatic pleochroic compound found in high-alumina cement. These are now known to be quite separate compounds. The existence of un-stable C_5A_3 as described by Rankin and Wright has been verified.[1] The pleo-chroic phase in high-alumina cement was identified by Parker[2] as a solid solution $6CaO.4Al_2O_3(FeO, MgO)SiO_2$ but later work[3] has shown it to have a more complex formula which has yet to be firmly established.

Calcium monoaluminate $CaO.Al_2O_3$ melts incongruently at 1600° to CA_2 and a liquid of approximate composition (weight per cent) CaO 36, Al_2O_3 64 and not congruently as originally reported. It crystallises in irregular grains sometimes prismatic. Twinning is characteristic. Crystal system is probably

[1] E. Aruja, *Acta crystallogr.* **10,** 337 (1957).
[2] T. W. Parker, *Sym., London 1952,* 485.
[3] H. G. Midgley, *Trans. Br. Ceram. Soc.* **67,** 1 (1968).

monoclinic but sometimes the crystals are markedly pseudohexagonal. Optical properties $\alpha = 1{\cdot}653$, $\beta = 1{\cdot}655$, $\gamma = 1{\cdot}663$, biaxial negative $2V = 36°$. Density 2·98. Hardness 6·5. *CA* is one of the main constituents of high-alumina cement.

Tricalcium penta aluminate $3CaO.5Al_2O_3$ was thought to be the composition of a phase now identified as $CaO.2Al_2O_3$. An unstable monotropic form was also described but has not been reported in any modern work. C_3A_5 was said to be one of the cementing compounds in high-alumina cement; the true compound, CA_2, however, is inert and the cementing value of a melt of composition C_3A_5 can be accounted for by the *CA* which it contains. Impure CA_2 as occurs in high-alumina cement may be more reactive.

The compound $CaO.2Al_2O_3$ melts incongruently at 1762° to CA_6 and a liquid of approximate composition (weight per cent) CaO 22, Al_2O_3 78, and not congruently at 1750° as reported earlier. It crystallises in laths or sometimes rounded grains and belongs to the monoclinic system with optical properties α 1·6178, β 1·6184, γ 1·6516; biaxial positive $2V = 12°$, density 2·86.

The compound $CaO.6Al_2O_3$ melts incongruently at $1833° + 15°$ to corundum and a liquid of approximate composition (weight per cent) CaO16, $Al_2O_3$84. It crystallises in the hexagonal system as hexagonal plates $\omega = 1{\cdot}759$, $\varepsilon = 1{\cdot}752$; uniaxial negative.

The compounds in the system $CaO–Al_2O_3$ can all be prepared by heating the appropriate mix of calcium carbonate and alumina at 1350°–1550°. It is essential for the alumina to be very finely divided and free from alkali if good preparation of CA_2 and CA_6 are required. More rapid combination can be achieved by starting with the mixed nitrates. The weight percentage compositions of the eutectics and invariant points in the binary system lime-alumina illustrated in Fig. 12[1] are:

$CaO–C_3A$ at 1539°	$57{\cdot}2CaO, 42{\cdot}8Al_2O_3$
$C_3A–CA$ at 1360°	$49{\cdot}35CaO, 50{\cdot}65Al_2O_3$
$CA–CA_2$ at 1602°	$36CaO, 64Al_2O_3$
$CA_2–CA_6$ at 1762°	$22CaO, 78Al_2O_3$
$CA_6–\alpha Al_2O_3$ at 1830°	$16CaO, 84Al_2O_3$

The phase diagram in Fig. 12 differs from earlier diagrams by the absence of a primary phase field for $C_{12}A_7$ and in showing *CA* and CA_2 as melting incongruently.

Tricalcium aluminate forms solid solution with Fe_2O_3, SiO_2, alkalies and MgO. The pure compound is cubic with unit cell side[2] $a = 15{\cdot}263 \pm 0{\cdot}003$A and the extent of solid solution has been determined by measurement of the change in a. The solid solution of Fe_2O_3 can be regarded as that of a hypothetical 'C_3F' in C_3A with Al ions being replaced by Fe ions. Various values[3] have been reported for the limit of solid solution but the most probable values seems to be

[1] R. W. Nurse, J. H. Welch, A. J. Majumdar, *Trans. Br. Ceram. Soc.* **64,** 409 (1965).
[2] A. E. Moore, *Silic. ind.* **32,** 87 (1967).
[3] A. J. Majumdar, *Trans. Br. Ceram. Soc.* **64,** 105 (1965); P. Tarte, *Nature, Lond.* **207,** 973 (1965); *Silic. ind.* **31,** 343 (1966); C. M. Schlaudt and D. M. Roy, *Nature, Lond.* **206,** 819 (1965); D. H. Lister and F. P. Glasser, *Trans. Br. Ceram. Soc.* **66,** 293 (1967).

about 7 per cent (mol) 'C_3F' equivalent to 4·1 per cent weight Fe_2O_3. The value of the unit cell size, *a*, increases by 0·0032Å per mole per cent of Al^3 replaced by Fe^3.

The limit of solid solution of Si is about 5–6 atom per cent Si/Al+Si. For each Al^{3+} ion replaced by Si^{4+} ion the unit cell decreases by about 0·0032Å and one extra oxygen ion is accommodated for every 2Al ions replaced to balance the charge. Only traces of K_2O are taken into solid solution in C_3A in the absence of silica, but there is extensive solution[1] of Na_2O up to a limiting composition of 91 mole per cent C_3A 9 mole per cent N_3A. This composition is lower in Na_2O than that of the compound NC_8A_3 which does not enter into solid solution in C_3A despite their rather similar structures. The limiting solid solution of MgO in C_3A is 2·5 per cent by weight,[2] equivalent to 5·4 mole per cent replacement. The C_3A phase in Portland cement has a lower value of *a* than in pure C_3A, suggesting that solid solution of SiO_2, MgO and Na_2O, all of which decrease the cell size, is the dominant factor.

CALCIUM FERRITES

The system $CaO–Fe_2O_3$

The study of systems containing Fe_2O_3 as a component is complicated by loss of oxygen on heating in air so that compounds such as Fe_3O_4 in which iron is partly reduced to Fe^{2+} may be formed. This effect is pronounced in iron-rich compositions. The phase diagram of the system $CaO–Fe_2O_3$ has been redrawn several times since the first diagram was published in 1916 by Sosman and Merwin. Fig. 14 shows the latest results obtained by Phillips and Muan.[3] The iron-rich portion of the system is only pseudobinary owing to loss of oxygen but the lime-rich portion important in cement chemistry is essentially binary. Three calcium ferrites are now well established C_2F, CF and CF_2.

C_2F melts congruently at 1450°. The eutectic between C_2F and CaO, at 1438°, lies very close to the C_2F composition. It crystallises well, giving black crystals which in transmitted light are a yellow-brown colour. Optical properties: $\alpha_{Li} = 2{\cdot}200$, $\beta_{Li} = 2{\cdot}220$, $\gamma_{Li} = 2{\cdot}290$; $\alpha_{Na} = 2{\cdot}261$, $\gamma_{Na} = 2{\cdot}274$[4]; biaxial positive with moderate optic axial angle. Neither mono- nor di-calcium ferrite exists in Portland cement, but a solid solution lying on the composition line C_6AF_2–C_2F.

CF decomposes at 1216° to form C_2F and liquid. It can be formed by heating a finely powdered mix of $CaCO_3+Fe_2O_3$ at a temperature well below the incongruent melting-point. It crystallises well, giving black crystals which in transmitted light under the microscope show a deep red colour. Optical properties: $\omega_{Li} = 2{\cdot}465$, $\varepsilon_{Li} = 2{\cdot}345$; $\omega_{Na} = 2{\cdot}58$, $\varepsilon_{Na} = 2{\cdot}43$; nearly or quite uniaxial, negative.

[1] J. A. Conwicke and D. E. Day, *J. Am. Ceram. Soc.* **47,** 654 (1963); K. E. Fletcher, H. G. Midgley and A. E. Moore, *Mag. Concr. Res.* **17** (53), 171 (1965).
[2] M. Muller-Hesse and H. H. Schwiete, *Zement-Kalk-Gips* **9,** 386 (1956).
[3] B. Phillips and A. Muan, *J. Am. Ceram. Soc.* **42,** 413 (1959).
[4] R. J. Colony and D. L. Snader, *Columbia Univ. New York, U.S.A. Dept. of Civil Eng.*, Bull. No. 4 (1931).

CF_2 is only stable between 1155° and 1226° and is the primary phase in contact with liquid in compositions containing between 19 and 20 per cent CaO. At 1226° *CF_2* decomposes to give hematite (α Fe_2O_3) and liquid, and at 1155° to give hematite and *CF*. A compound $3CaO.Fe_2O_3$ was formerly thought to exist, but the balance of evidence now indicates that, if it exists at all, it is only when stabilised by other impurities such as OH, H_2O or SiO_2.

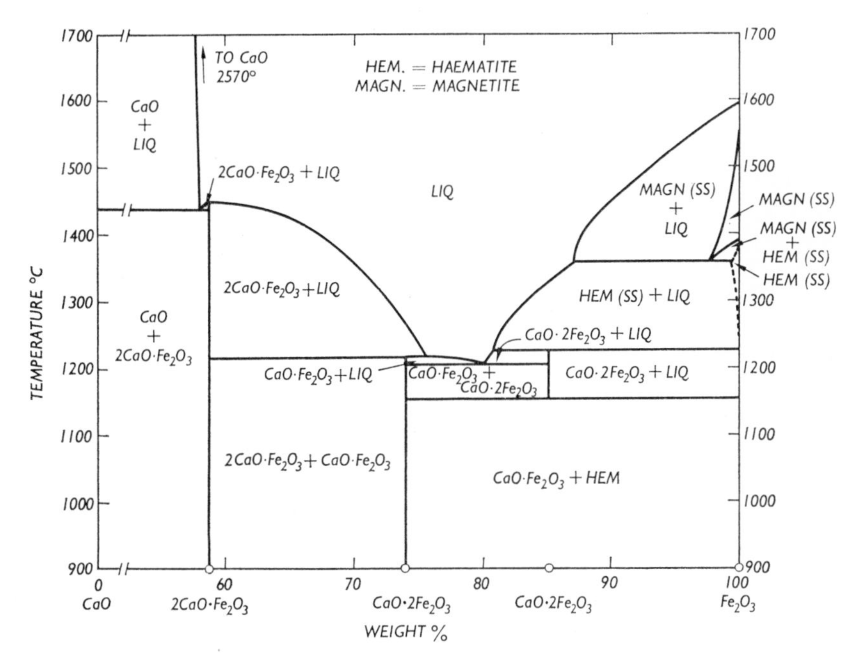

FIG. 14 System CaO–Fe_2O_3 in air.

ALUMINIUM SILICATES

The system $Al_2O_3SiO_2$

Three forms of an anhydrous aluminium silicate of composition $Al_2O_3.SiO_2$ occur in nature as sillimanite, andalusite and cyanite. It was formerly considered that sillimanite was the only stable alumina-silica compound and that andalusite and cyanite were monotropic with respect to it. Later work showed that none of these compounds are stable and that another silicate of formula $3Al_2O_3.2SiO_2$ is the only stable silicate.[1] Very soon afterwards it was found to occur in natural rocks from the Isle of Mull and was named mullite. Sillimanite and mullite are very similar in optical and crystallographic properties and the differences in their X-ray diffraction patterns are so slight that for a time they were considered identical. Synthetic mullites vary in their Al/Si ratio.

[1] N. L. Bowen and J. W. Greig, *J. Am. Ceram. Soc.* **7**, 238 (1924).

The composition and melting behaviour of mullite have been the subject of much controversy but it now seems probable that it approximates to $2Al_2O_3 . SiO_2$ and melts incongruently to form Al_2O_3 and liquid. The phase diagram according to Welch[1] is shown in Fig. 15. Nevertheless it has been suggested that mullite has

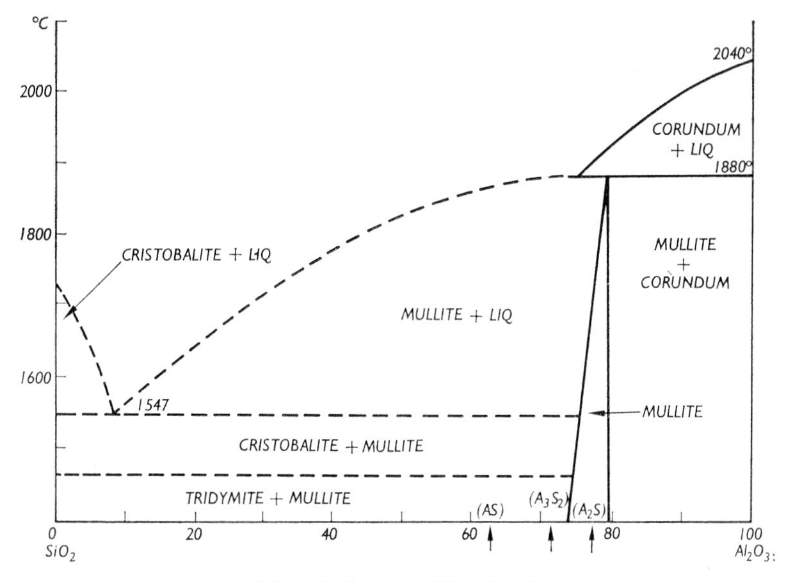

FIG. 15 System SiO_2–Al_2O_3.

the A_3S_2 composition and melts congruently at 1850°. Although natural sillimanite is unstable at high temperatures it is very inert and suffers no change on heating for several days below 1547°, the temperature of the eutectic between cristobalite and mullite.

Compounds of alumina and silica alone do not occur in Portland or high-alumina cement, but they form one of the best refractory materials. Of the alumina-silica refractories it was formerly considered that sillimanite (64 per cent Al_2O_3, 36 per cent SiO_2) formed the most refractory composition, but it is now evident that this is not so and that a mullite composition (over 70 per cent Al_2O_3) is superior. Certain hydrated aluminium silicates are the essential constituents of clay which forms so important a raw material in the cement industry.

The system $CaO–Al_2O_3–SiO_2$

A knowledge of the compounds existing in the system lime-alumina-silica is fundamental to all cement chemistry. It is the basic ternary system which we have to consider since the three oxides composing it form some 90 per cent

[1] J. H. Welch, *Nature, Lond.* **186,** 545 (1961); *Trans. 7th Int. Ceram. Congr., London 1960,* 197; A. J. Majumdar and J. H. Welch, *Trans. Br. Ceram. Soc.* **62,** 602 (1963).

Portland cement and over 80 per cent high-alumina cement. This system was the first ternary oxide system to be worked out and the methods developed in its investigation have since been successfully applied to many other oxide systems. This work we owe to the Geophysical Laboratory of the Carnegie Institute of Washington.

The equilibrium diagram of this system is shown in Fig. 16, where the zone of Portland cement compositions is also indicated. Reference to Fig. 36 (p. 88)

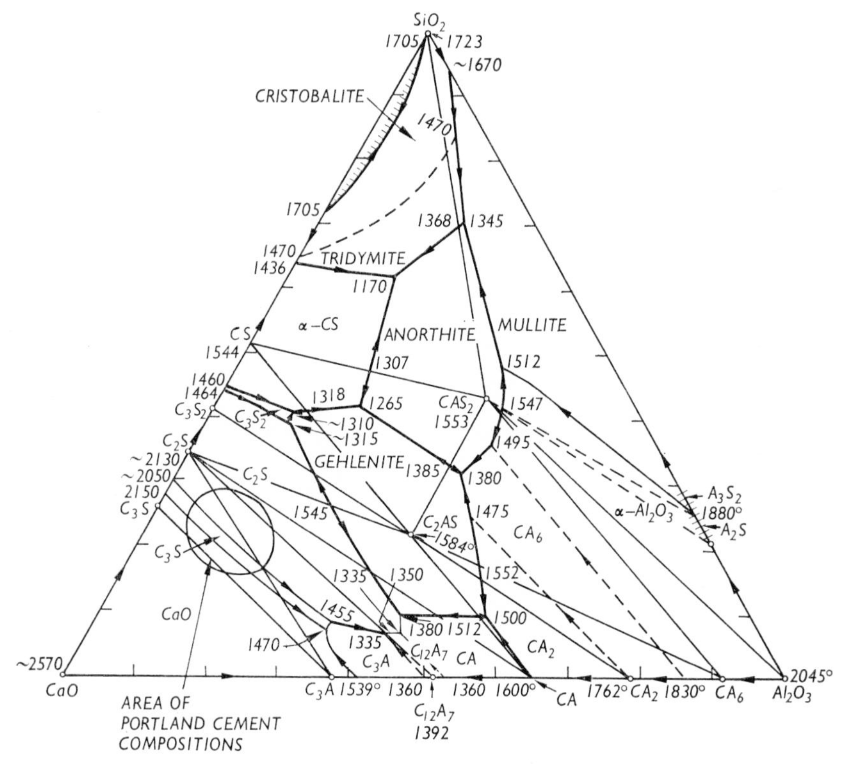

FIG. 16 System $CaO–Al_2O_3–SiO_2$. Shaded area denotes two liquids.

will also show the Portland and high-alumina cement zones more clearly in relation to the various triangular areas which indicate the compounds which will be present in any completely crystallised mix. The diagram in Fig. 16 differs from the original diagram[1] mainly in the phase field of C_3S which reaches the $CaO–SiO_2$ edge of the triangle. Other differences concern the path of the mullite corundum boundary, and the field of CA_6.

Tricalcium silicate forms invariant points, not eutectics with CaO, C_3A and with C_2S, C_3A at 1470° and 1455° respectively. The field of dicalcium silicate

[1] G. A. Rankin and F. E. Wright, *Am. J. Sci.* (iv) **39,** 1, 1915.

covers a large area and this compound forms invariant points with tricalcium silicate as mentioned above, and ternary eutectics with C_3A and $C_{12}A_7$ at 1335°; $C_{12}A_7$, CA at 1335°; also an invariant point with CA and a ternary compound C_2AS at 1380.° The phase relations concerning $C_{12}A_7$ apply only in the presence of moisture since this compound does not occur under strictly anhydrous conditions, as has been discussed earlier.

Temperatures fall from the incongruent melting-point of C_3S at 2150° to the invariant point between CaO, C_3S and C_3A at 1470° and then to the invariant point at 1455° at which solid C_3S, C_2S and C_3A coexist with liquid. Temperatures continue to fall between this invariant point and the first ternary eutectic to be encountered is that at which C_3A, C_2S and $C_{12}A_7$ coexist at 1335°. Temperatures also fall along all the boundary curves proceeding inwards from the sides of the triangle. Temperature maxima occur on the $C_2S/C_{12}A_7$ and C_2S/C_2AS field boundaries and the invariant points between C_3A, C_2S, $C_{12}A_7$, as noted above, and $C_{12}A_7$, C_2S, CA are ternary eutectics. The maxima on the boundary curves occur at the points where they are intersected by a line joining together the compositions of the two compounds whose primary phase fields they separate. The ternary compound $2CaO.Al_2O_3.SiO_2$ (gehlenite) melts congruently at 1584°. It crystallises as clear grains and belongs to the tetragonal system. Optical properties: ω 1·669, ε 1·658, uniaxial negative. Hardness about 6. Specific gravity 3·038. This compound does not occur in Portland cements, but it is a constituent of some, but probably not all, high-alumina cements. It has no cementing properties in the pure state.

The percentage compositions of some of the more important invariant points and eutectics are:

Invariant point CaO–C_3S–C_3A at 1470°
59·7CaO, 32·8Al_2O_3, 7·5SiO_2
Invariant point C_3S–C_2S–C_3A at 1455°
58·5CaO, 32·9Al_2O_3, 8·6SiO_2
Eutectic C_2S–C_3A–$C_{12}A_7$ at 1335°
52CaO, 41·2Al_2O_3, 6·8SiO_2
Eutectic C_2S–$C_{12}A_7$–CA at 1335°
49·5CaO, 43·7Al_2O_3, 6·8SiO_2
Invariant point C_2S–C_2AS–CA at 1380°
48·3CaO, 42Al_2O_3, 9·7SiO_2
Eutectic C_2AS–CA–CA_2 at 1500°
37·7CaO, 53Al_2O_3, 9·3SiO_2

The system $CaO–Al_2O_3–Fe_2O_3$

This system is of particular interest, since it contains a ternary phase, brownmillerite, which occurs in Portland and high-alumina cements.[1] Brownmillerite was considered for many years to be a compound of the definite composition $4CaO.Al_2O_3.Fe_2O_3$, melting at 1415°. Numerous investigations, using X-ray,

[1] W. C. Hansen, L. T. Brownmiller and R. H. Bogue, *J. Am. chem. Soc.* **50**, 396

magnetic, thermal and chemical methods, have now established the existence of a series of solid solutions between C_2F and a hypothetical C_2A, terminating in a limiting solid solution usually given as $6CaO.2Al_2O_3.Fe_2O_3$.[1] Optical properties for C_4AF and C_6A_2F are:

	C_4AF	C_6A_2F
α_{Li}	1·96	1·94
β_{Li}	2·01	—
γ_{Li}	2·04	2·02
Density	3·77	3·74

Both are pleochroic γ = brown, α = yellow; biaxial negative, 2V small; prismatic reddish-brown grains.

Solid solution to a limited extent, as shown by the hatched lines in Fig. 17, also occurs in C_3A, $C_{12}A_7$ and CA.

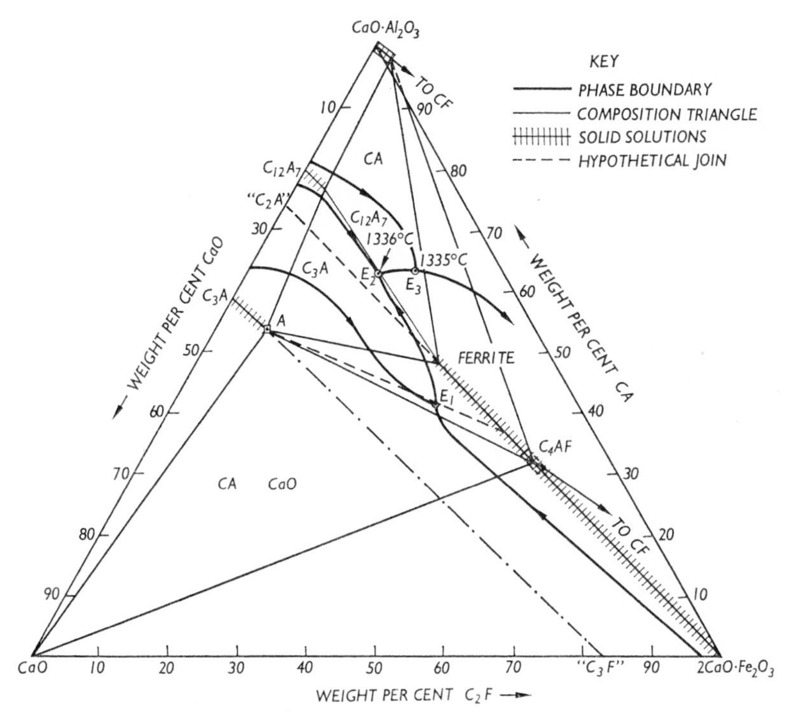

FIG. 17 System $CaO–CaO.Al_2O_3–2CaO.Fe_2O_3$.

Fig. 17 shows the phase relations in that part of the system of interest in cement chemistry.[2] The invariant point E_1 between CaO, C_3A and ferrite

[1] T. Yamauchi, *J. Jap. Ceram. Assoc.* **45,** 433, 614, 880 (1937); M. A. Swayze, *Am. J. Sci.* **244,** 1 (1946); G. L. Malquori and V. Cirilli, *Sym., London 1952*, 120.

[2] T. F. Newkirk and R. D. Thwaite, *J. Res. natn. Bur. Stand.* **61,** 233 (1958); A. J. Majumdar, *Trans. Br. Ceram. Soc.* **64,** 105 (1965).

occurs at 1389°. This is what is known as a peritectic point because the composition of the liquid phase cannot be represented in terms of the solids which are in equilibrium with it. The C_3A phase in equilibrium at E_1 is the solid solution represented by the point A and the composition of the ferrite phase is close to C_4AF. The invariant point E_1 lies outside the triangle $CaO–A–C_4AF$. On cooling of a mix of the composition E_1, at which CaO, C_3A, ferrite and liquid are in equilibrium, the CaO redissolves in the melt and the subsequent crystallisation path follows along the boundary curve between C_3A and ferrite to E_2 where the mix goes solid. This eutectic point E_2 between C_3A, $C_{12}A_7$ and ferrite occurs at 1336° and the invariant point E_3 between $C_{12}A_7$, CA and ferrite at 1335°. Although the solid solution series between C_2F and 'C_2A' covers a wide range and extends to compositions of higher $A:F$ ratio than C_4AF, the ferrite phase in equilibrium at E_1 has a composition 48 mol per cent 'C_2A' which is very close to C_4AF. As we note later (p. 103) the ferrite phase in Portland cements can vary in composition from C_6AF_2 to C_6A_2F though it can often approximate to C_4AF. C_3A takes up some 8–9 per cent weight of a hypothetical 'C_3F' in solid solution and its refractive index is raised to 1·723. CA takes up 13 per cent CF in solid solution, its refractive indices being raised to 1·70 for α and 1·72 for γ. CF also takes up 9 per cent CA in solid solution ω_{Li} decreasing to about 2·25 and ε_{Li} to 2·13. The compound CA_2 takes up Fe^3 ions in solid solution, the refractive indices and the optic axial angle increasing. In that part of the whole system $CaO–Al_2O_3–Fe_2O_3$ which lies on the low lime side of the $CA–C_2F$ composition line another ternary phase with a composition lying in the composition line 'CA_3'–'CF_3' has been discovered[1] but this could not be formed in Portland cement.

The system $CaO–SiO_2–Fe_2O_3$

Examination of this system, which is only pseudoternary because of loss of oxygen in the iron-rich compositions, shows that no ternary compounds are formed.[2] Mixes in the more basic portion crystallise (Fig. 18) to give only the binary compounds occurring in the bounding binary systems. The primary field of C_3S extends from the $CaO–SiO_2$ edge to two invariant points for CaO, C_2F and liquid at 1412° and for C_2S, C_2F and liquid at 1405°. It is evident that no ternary compounds composed of lime, silica and ferric oxide are to be expected in cements.

The system $CaO–SiO_2–FeO$

The more acidic region of this system has been explored.[3] In the binary system between C_2S and $2FeO.SiO_2$ a lime-iron-olivine $CaO.FeO.SiO_2$ is formed. This compound enters into solid solution with C_2S up to a maximum of 47 per cent of $2FeO.SiO_2$. $CaO.FeO.SiO_2$ is isomorphous with γ C_2S so the

[1] D. H. Lister and F. P. Glasser, *Trans. Br. Ceram. Soc.* **66,** 293 (1967); R. R. Dayal and F. P. Glasser, *Prog. ceram. Sci.* **4,** 19 (1967).
[2] B. Phillips and A. Muan, *J. Am. Ceram. Soc.* **42,** 413 (1959); A. Muan and E. F. Osborn, *Phase Equilibria among Oxides in Steelmaking.* Addison-Wesley, Reading, Mass., 1965.
[3] N. L. Bowen, J. F. Schairer and E. Posnjak, *Am. J. Sci.* (V) **26,** 193 (1933); W. C. Allen and R. B. Snow, *J. Am. Ceram. Soc.* **38,** 264 (1955).

solid solution series is presumably complete at low temperatures. The C_3S field extends to an FeO content of 65 per cent and down to a temperature of 1200°. From the phase diagram[1] for the complete system it can be deduced that the phase assemblages that coexist together which may have possible relevance to Portland cement constitution are:

$CaO–C_3S$—solid solution of FeO in CaO
$C_3S–C_2S$—solid solution of CaO in FeO

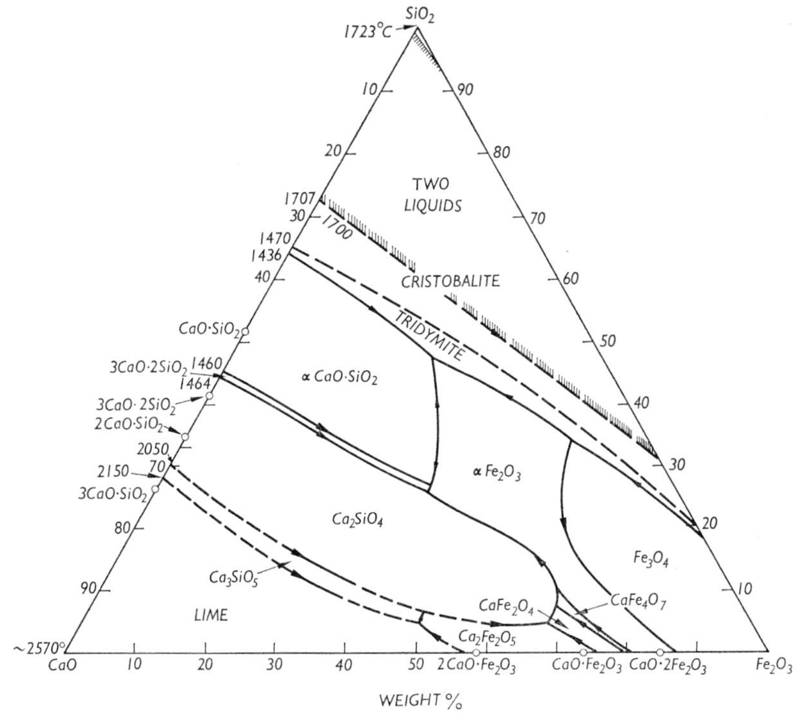

FIG. 18 System $CaO–Fe_2O_3–SiO_2$.

The quaternary system $CaO–Al_2O_3–SiO_2–Fe_2O_3$

Both Portland and high-alumina cement clinkers are composed, as to 95 per cent or more, of the four oxides lime, alumina, silica, ferric oxide. A knowledge of the phase relations in this system, including as it does the basic ternary system $CaO–Al_2O_3–SiO_2$ and the important ones $CaO–Al_2O_3–Fe_2O_3$ and $CaO–SiO_2–Fe_2O_3$, affords therefore decisive evidence regarding all the major compounds present at equilibrium in Portland and high-alumina cements. That

[1] E. F. Osborn and A. Muan, *Phase Diagrams of the Oxide Systems*, Am. Ceram. Soc., 1960.

portion of this system which covers Portland cement compositions has been worked out,[1] but the region of high-alumina cement compositions still remains to be fully investigated. Just as all possible compositions in a ternary system can be represented in an equilateral triangle the three apices of which each represents 100 per cent of one of the pure components, so can quaternary compositions be represented in a regular tetrahedron, each of the four apices representing one of the pure components. The four sides of this solid model are equilateral triangles and each represents one of the four 3-component systems which can be made up by taking the four components three at a time. The six edges represent the six binary systems which can be made up by taking the four components two at a time.

The quaternary system $CaO-Al_2O_3-SiO_2-Fe_2O_3$ is shown in outline in Fig. 19. The full lines lie in the plane of the paper and the dotted lines in the

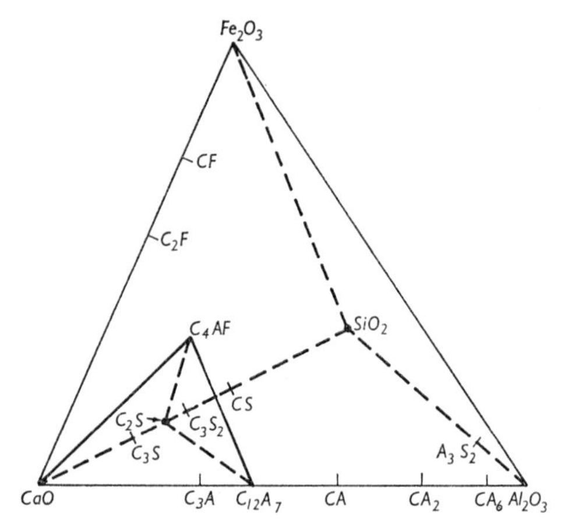

FIG. 19 Relationship of the system $CaO-C_2S-C_{12}A_7-C_4AF$ to the larger system $CaO-Al_2O_3-SiO_2-Fe_2O_3$.

solid below the plane of the paper. The system $CaO-2CaO.SiO_2-4CaO.Al_2O_3.Fe_2O_3-12CaO.7Al_2O_3$, which is a smaller quaternary system within the major system, includes the entire region of Portland cement compositions. As discussed previously, C_4AF seems to be only a point in a solid solution series terminating at the composition 70'C_2A'; 30C_2F; but if the quaternary system is worked out using this latter end point as a component then it is seen that in some respects C_4AF may conveniently be considered as a compound. Also, the iron phase crystallising in Portland cement is often close to the composition C_4AF.[2] It seems justified therefore to discuss cement constitution in terms of the familiar system $CaO-C_2S-C_{12}A_7-C_4AF$. This is shown enlarged in outline, as a separate system in Fig. 20. The phase equilibrium diagram of this system is shown in Figs. 21 and 22, which are drawings

[1] F. M. Lea and T. W. Parker, *Phil. Trans. R. Soc.* **234,** No. 731, 1, 1934 (A).
[2] H. G. Midgley, *Sym., London 1952,* 140; A. J. Majumdar, *Trans. Br. Ceram. Soc.* **64,** **105** (1965).

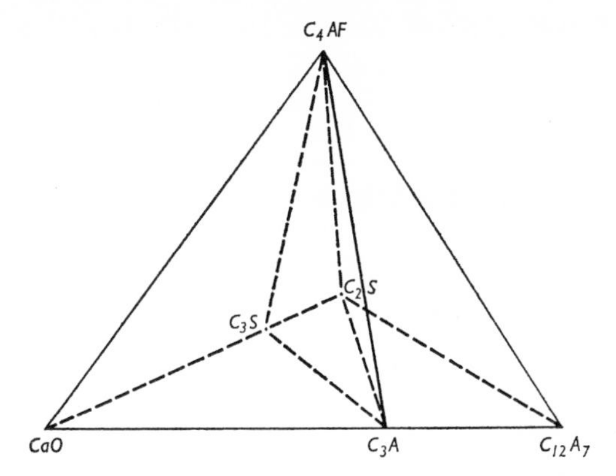

FIG. 20 System $CaO–C_{12}A_7–2CaO.SiO_2–4CaO.Al_2O_3.Fe_2O_3$.

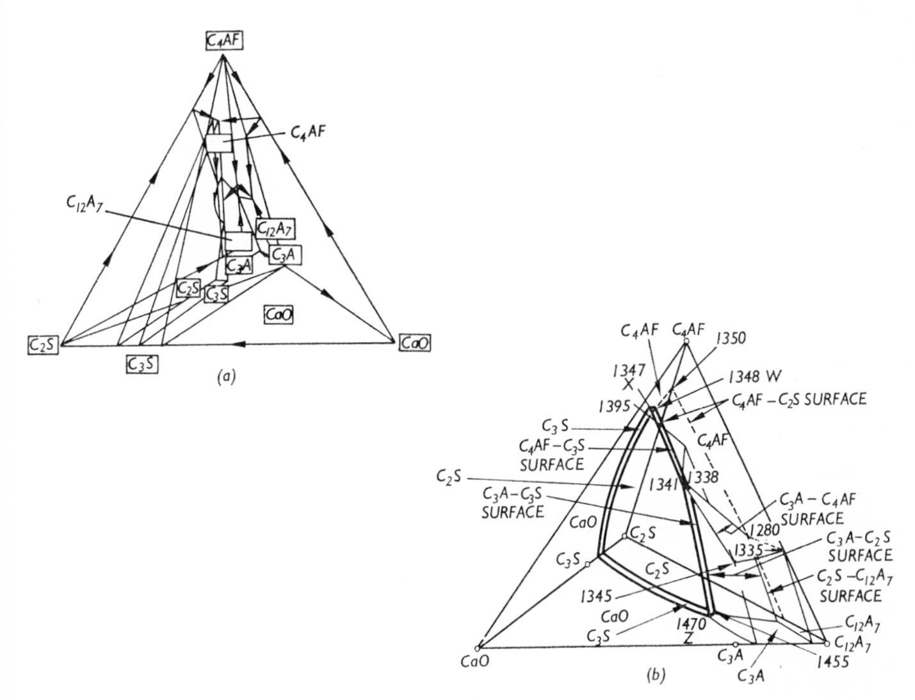

FIG. 21 Quaternary system $CaO–C_2S–C_{12}A_7–C_4AF$, showing two views of the solid model.

from the solid model. These diagrams are those found by Lea and Parker amended to take account of the incongruent melting of C_3S as subsequently found by Welch and Gutt. This system, and various other systems discussed in the following pages, were worked out by their authors in terms of 'C_5A_3' but it seems preferable to substitute the correct compound composition $C_{12}A_7$. The boundary lines between different primary phase fields which were encountered in the 3-component systems become bounding surfaces between different primary phase volumes in the quaternary system; the ternary eutectics and

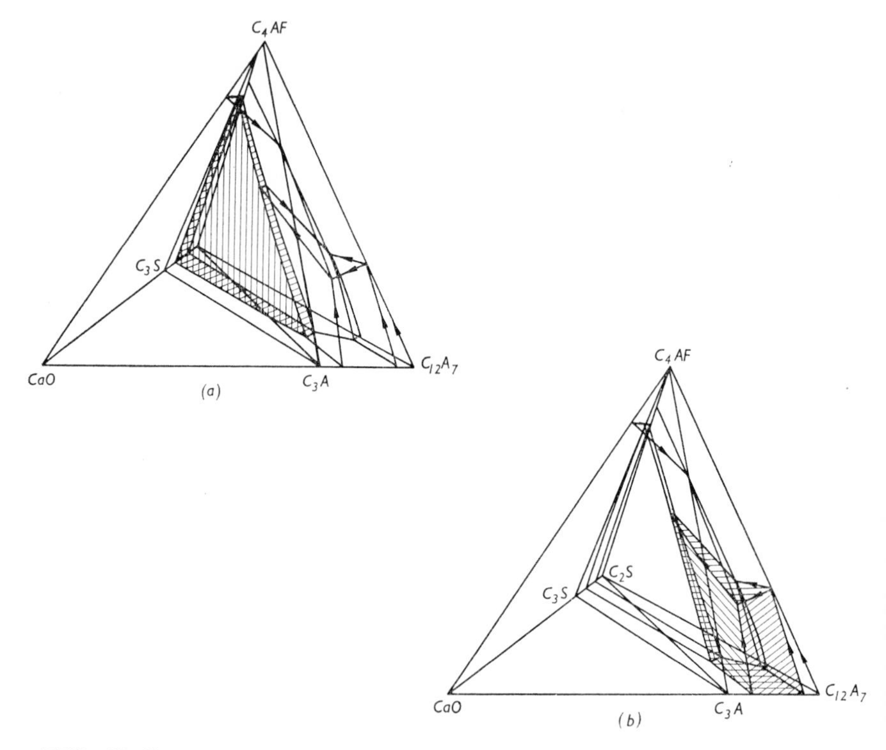

FIG. 22 Quaternary system $CaO–C_2S–C_{12}A_7–C_4AF$ showing (*a*) the C_3S primary phase volume shaded and (*b*) the C_3A primary phase volume shaded.

invariant points become lines in the solid and a new set of eutectic and invariant points at which four solids and one liquid co-exist in equilibrium are found. No new compound occurs in the system and at equilibrium any mix crystallises to yield the four solids which form the apices of the particular irregular tetrahedral volume in which it occurs. The solid model is shown divided up into three such volumes by the planes $C_4AF–C_3A–C_3S$ and $C_4AF–C_3A–C_2S$ in Figs. 20, 21 and 22. Almost all modern Portland cement compositions lie within the solid $C_4AF–C_3A–C_3S–C_2S$ and hence at equilibrium will crystallise to yield these four compounds. In practice, $C_{12}A_7$ may also be present, for reasons which are discussed in Chapter 7.

The more important invariant points and eutectics in the quaternary system are:

CaO–C_3S–C_3A–C_4AF at 1341°
55·0CaO, 22·7Al_2O_3, 5·8SiO_2, 16·5Fe_2O_3
C_3S–C_2S–C_3A–C_4AF at 1338°
54·8CaO, 22·7Al_2O_3, 6·0SiO_2, 16·5Fe_2O_3
C_2S–C_3A–$C_{12}A_7$–C_4AF at 1280°
50·0CaO, 34·5Al_2O_3, 5·6SiO_2, 10·0Fe_2O_3

If C_6A_2F is taken as the iron-containing component, the first two of these become:

CaO–C_3S–C_3A–C_6A_2F at 1342°
53·9CaO, 21·2Al_2O_3, 5·8SiO_2, 19·1Fe_2O_3
CaO–C_2S–C_3A–C_6A_2F at 1338°
53·5CaO, 22·3Al_2O_3, 6·0SiO_2, 18·2Fe_2O_3

The ternary systems CaO–C_4AF–C_2S and C_4AF–C_2S–$C_{12}A_7$, which form two of the sides of the quaternary system and which have not previously been described, are shown in Figs. 23 and 24.

The percentage composition of the invariant points and eutectics are as follows:

C_2S–C_4AF. Binary eutectic at 1350°
49·6CaO, 17·1Al_2O_3, 6·5SiO_2, 26·8Fe_2O_3
CaO–C_3S–C_4AF ternary eutectic at 1347°
52·8CaO, 16·2Al_2O_3, 5·6SiO_2, 25·4Fe_2O_3
C_2S–C_3S–C_4AF quintuple point at 1348°
52·4CaO, 16·3Al_2O_3, 5·8SiO_2, 25·5Fe_2O_3
C_2S–$C_{12}A_7$–C_4AF at 1280°
50·0CaO, 34·5Al_2O_3, 5·6SiO_2, 9·9Fe_2O_3

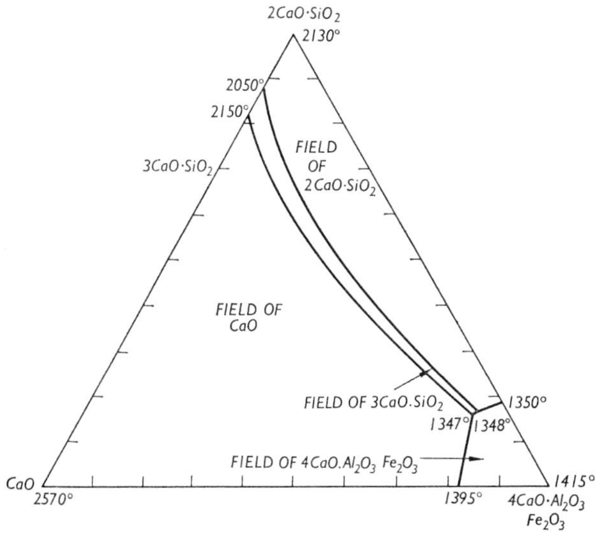

FIG. 23 System CaO–2CaO.SiO_2–4CaO.Al_2O_3.Fe_2O_3.

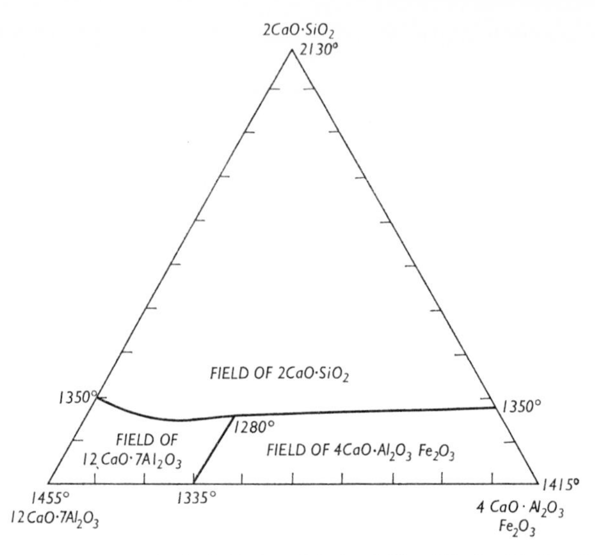

FIG. 24 System $12CaO.7Al_2O_3$–$2CaO.SiO_2$–$4CaO.Al_2O_3.Fe_2O_3$.

The system $CaO–Al_2O_3–MgO$

The ternary system $CaO–MgO–Al_2O_3$ was first studied by Rankin and Merwin[1] who did not detect any ternary compounds stable in contact with melt. Since then[2] two new ternary compounds have been described. One of these, originally[2] formulated as $C_{25}A_{17}M_8$ and later corrected[3] to $3CaO.2Al_2O_3.MgO$ exists at the liquidus. The second compound C_7A_5M is metastable. Fig. 25 shows the system $CaO–MgO–Al_2O_3$ revised in the light of these new findings and also of new data on the system $CaO–Al_2O_3$. The compound C_3A_2M melts incongruently at 1350·5° to a mixture of CA, MgO and liquid and its primary field lies between those of CA, MgO and $C_{12}A_7$. The invariant point MgO–C_3A–$C_{12}A_7$ is a ternary eutectic so that any mix definable in terms of these three compounds will finally crystallise to a solid at this point. The compound C_3A_2M would not be expected therefore to occur in Portland cement, but it would be a possible compound in high-alumina cement. The second compound of approximate composition C_7A_5M closely resembles the first in optical properties but is readily distinguished by its X-ray diffraction pattern. It is a metastable product in the system and melts at 1332°C.

MgO and Al_2O_3 form a binary compound spinel $MgO.Al_2O_3$ which melts congruently at 2105°C. Spinel forms invariant points with CA_2 and CA, and also with CA and MgO but does not in this system exist in equilibrium with either C_3A or $C_{12}A_7$.

[1] G. A. Rankin and H. E. Merwin, *J. Am. chem. Soc.* **38,** 568 (1916).
[2] J. H. Welch, *Nature, Lond.* **191,** 559 (1961).
[3] A. J. Majumdar, *Trans. Br. Ceram. Soc.* **63,** 347 (1964).

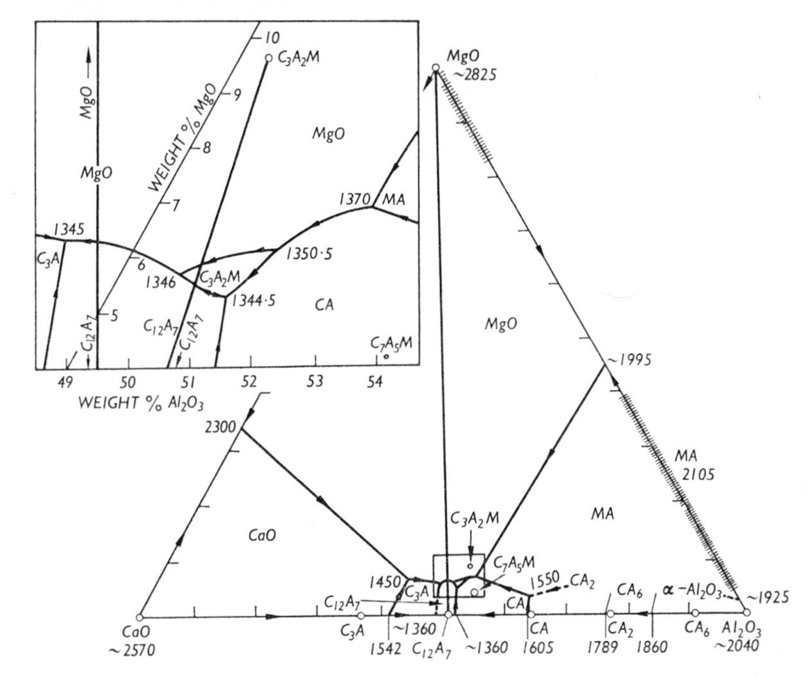

FIG. 25 System $CaO–MgO–Al_2O_3$. Inset: region surrounding the primary phase field of C_3A_2M on an enlarged scale.

Spinel crystallises in the cubic system and has a refractive index of 1·72 (values from 1·718 to 1·723 are reported). It takes up alumina in solid solution[1] and its index rises to 1·733. Hardness is about 8. Spinel and MgO show limited mutual solubility.[2] The identification of spinel optically in the presence of magnesia (refractive index 1·737) is difficult, but it can be easily differentiated by warming on a microscope slide with a few drops of mixed hydrochloric and nitric acids diluted with an equal amount of water. Magnesia dissolves easily, as do any magnesia-alumina glasses but the spinel is insoluble.

The system $CaO–MgO–SiO_2$

This system[3] is shown in Fig. 26. Three ternary compounds are formed: merwinite $3CaO.MgO.2SiO_2$, monticellite $CaO.MgO.SiO_2$ and the recently reported[4] calcium magnesio-silicate $(CaO)_{1\cdot7}(MgO)_{0\cdot3}SiO_2$. Merwinite first melts incongruently at 1580°C to give dicalcium silicate, periclase (MgO) and

[1] G. A. Rankin and H. E. Merwin, *Am. J. Sci.* **45,** 301 (1918).

[2] A. M. Alper, R. N. McNally, P. H. Ribbe and R. C. Doman, *J. Am. Ceram. Soc.* **45,** 263 (1962).

[3] J. B. Ferguson and H. E. Merwin, *Am. J. Sci.* (iv) **48,** 81 (1919); T. W. Parker and R. W. Nurse, *J. Iron Steel Inst.*, 1943, 475P; R. W. Ricker and E. F. Osborne, *J. Am. Ceram. Soc.* **37,** 133 (1954); W. Gutt, Ph.D. Thesis London 1966; C. M. Schlaudt and D. M. Roy, *J. Am. Ceram. Soc.* **49,** 430 (1966).

[4] W. Gutt, *Nature, Lond.* **190,** 339 (1961); **207,** 184 (1965).

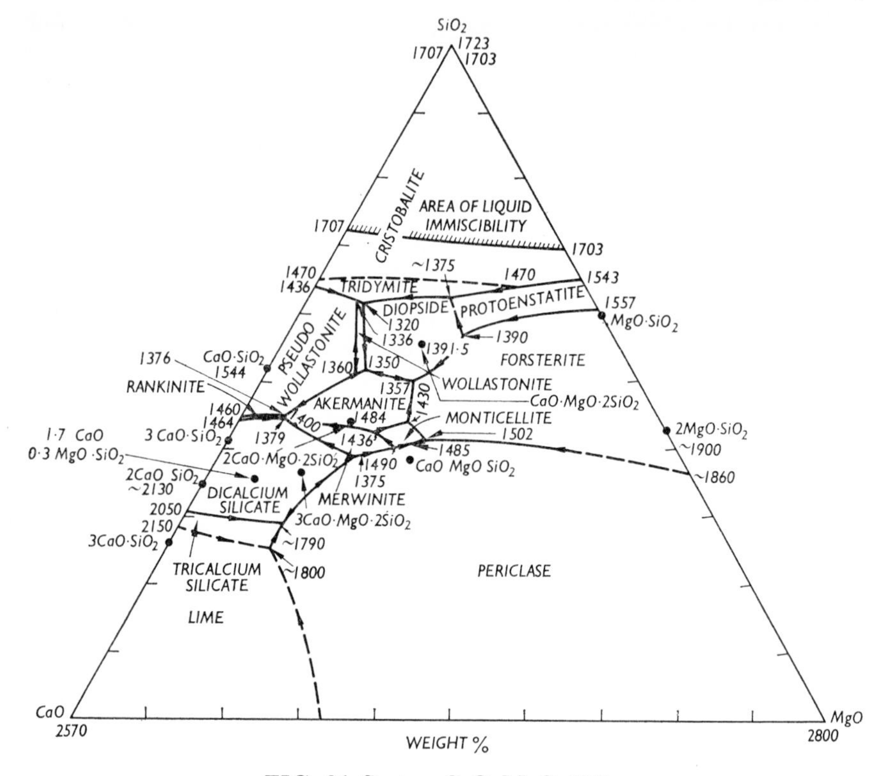

FIG. 26 System CaO–MgO–SiO$_2$.

liquid and then at 1590° gives dicalcium silicate and liquid. Monticellite melts incongruently just above 1490° to give merwinite, MgO and liquid. The compound $(CaO)_{1\cdot 7}(MgO)_{0\cdot 3}SiO_2$ is stable at room temperatures and decomposes in the solid state at 1460° forming α dicalcium silicate solid solution and merwinite. Since in cement burning a liquid phase is present down to 1280° this suggests that $C_{1\cdot 7}M_{0\cdot 3}S$ might be formed by crystallisation from the clinker liquid. Schlaudt and Roy considered that this compound was only stable between 979° and 1381° but this conclusion is open to question. Merwinite strongly resembles β and α' C_2S but has no hydraulic properties. Optical properties α 1·724, β 1·711, γ 1·708; positive 2V = 70°; characteristic polysynthetic twinning. Monticellite has α 1·639, β 1·646, γ 1·653; biaxial 2V = 85° to 90°. The compound $C_{1\cdot 7}M_{0\cdot 3}S$ has α 1·711, γ 1·725; biaxial positive 2V = 30°. Tricalcium silicate has a primary crystallisation field in the system CaO–SiO$_2$–MgO. In Fig. 26 this has been shown to reach the CaO–SiO$_2$ boundary to take account of the revised view of the melting behaviour of C_3S. The eutectic between C_3S, C_2S and MgO is at about 1790°.

The field of stability of merwinite lies entirely on the high-silica side of the line C_2S–M_2S. Mixtures of CaO, MgO and SiO$_2$ in the proportions in which they occur in Portland cement will lie within the triangle C_3S–C_2S–MgO and these will, therefore, be the phases formed on cooling under equilibrium conditions. Nevertheless, in more complex systems containing other components the possibility remains that the field of merwinite might extend further towards the CaO–MgO boundary and so render the occurrence of this compound, under suitable conditions, possible in Portland cement. The discovery of the calcium magnesiosilicate $(CaO)_{1\cdot7}(MgO)_{0\cdot3}.SiO_2$ has introduced a further factor since this compound may have a primary crystallisation field yet to be delineated in the ternary system $CaO-SiO_2-MgO$ and could also be a compatibility product present in cement if this field were to enter the zone of cement compositions.

In Fig. 27 is shown diagrammatically the four-component system CaO–MgO–$Al_2O_3-SiO_2$. The zone of Portland cement compositions lies on the CaO side

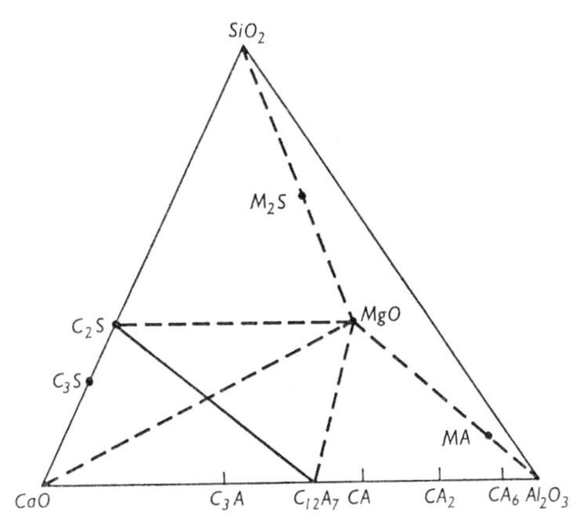

FIG. 27 System $CaO-Al_2O_3-SiO_2-MgO$.

of the $C_2S-C_{12}A_7$–MgO plane. It is evident that if any calcium magnesium silicate is to appear in a Portland cement, made up of the four oxides considered here, its primary phase volume must intersect and pass through the plane mentioned above. This plane was examined by Hansen and so the possibility can be tested.

The system $MgO-12CaO.7Al_2O_3-2CaO.SiO_2$

The equilibrium diagram of this system[1] is shown in Fig. 28. It is a simple eutectic system showing no compound formation and no solid solution of magnesia in either of the other two components. This system affords therefore evidence that neither merwinite nor $C_{1\cdot7}M_{0\cdot3}S$ nor CMS can occur as an equilibrium product under any conditions in a $CaO-Al_2O_3-SiO_2-MgO$ mix of Portland cement composition, and therefore probably not in Portland cement itself.

[1] W. C. Hansen, *J. Am. chem. Soc.* **50,** 3081 (1928).

The evidence is perhaps not entirely conclusive since C_3MS_2 and $C_{1\cdot7}M_{0\cdot3}S$ were unknown at the time of Hansen's investigation. Their optical properties are close to those of C_2S so that the presence of one or other of them could have

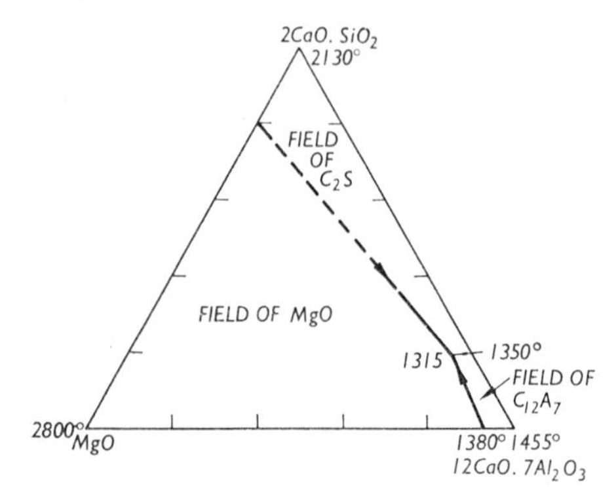

FIG. 28 System $MgO-12CaO.7Al_2O_3-2CaO.SiO_2$.

been missed. It will also be observed that spinel ($MgO.Al_2O_3$) does not occur along the $MgO-12CaO.7Al_2O_3$ boundary and hence is also improbable as a constituent of Portland cement for similar reasons.

The binary system $MgO-4CaO.Al_2O_3.Fe_2O_3$

No compound formation occurs in the system magnesia-tetracalcium alumino-ferrite, but there is evidence that C_4AF can take up a very small amount of magnesia in solid solution.[1] X-ray studies have indicated a limiting solubility of 1 to 2 per cent MgO in C_4AF. It has frequently been observed that the addition of MgO to a $CaO-Al_2O_3-Fe_2O_3-SiO_2$ mix darkens the colour after firing, and the pleochroism of C_4AF crystallised in the presence of MgO is more intense. Since in the preceding pages it has been shown that, except for solid solution in 'alite' or C_4AF, magnesia in Portland cement is not likely to be combined with either silica or alumina, it must occur as the free oxide in fully crystalline clinker.

The system $CaO-MgO-Al_2O_3-SiO_2$

This system has not yet been fully worked out, although sections through it containing between 5 and 35 per cent Al_2O_3 have been completed.[2] The relation of the subsidiary system $CaO-C_{12}A_7-C_2S-MgO$ to the larger quaternary is shown in Fig. 27; Fig. 29 shows the phase relations in the smaller system[3] including revisions to the C_3S field.

[1] R. H. Bogue, *Ind. Engng Chem. Anal. Ed.* **1,** 192 (1929); H. E. Schwiete and H. Z. Strassen, *Zement* **23,** 511 (1934); **25,** 843 (1936); H. Insley and H. F. McMurdie, *J. Res. natn. Bur. Stand.* **20,** 173 (1938).

[2] E. F. Osborn, R. C. de Vries, K. G. Gee and H. M. Kraner, *J. Metals* **6** (1), 3 (1954); A. T. Prince, *J. Am. Ceram. Soc.* **37,** 402 (1954).

[3] H. F. McMurdie and H. Insley, *Bur. Stand. J. Res.* **16,** 467 (1936).

The quaternary invariant points are:

CaO–C_3S–C_3A–MgO at 1395°
$CaO = 54{\cdot}0$, $MgO = 5{\cdot}5$, $Al_2O_3 = 33{\cdot}5$, $SiO_2 = 7{\cdot}0$
C_3S–C_2S–C_3A–MgO at 1380°
$CaO = 53{\cdot}0$, $MgO = 5{\cdot}5$, $Al_2O_3 = 34{\cdot}0$, $SiO_2 = 7{\cdot}5$
C_3A–$C_{12}A_7$–C_2S–MgO (eutectic) at 1295°
$CaO = 48{\cdot}5$, $MgO = 5{\cdot}0$, $Al_2O_3 = 41{\cdot}5$, $SiO_2 = 5{\cdot}0$

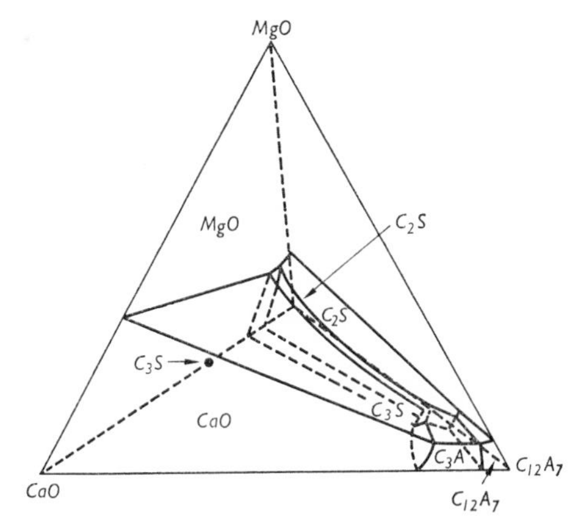

FIG. 29 System CaO–$C_{12}A_7$–C_2S–MgO.

Figure 30 shows the section through the CaO–Al_2O_3–SiO_2–MgO system at the 5 per cent MgO plane. This section is of importance in considering the constitution of high-alumina cement and contains a field of stability for the

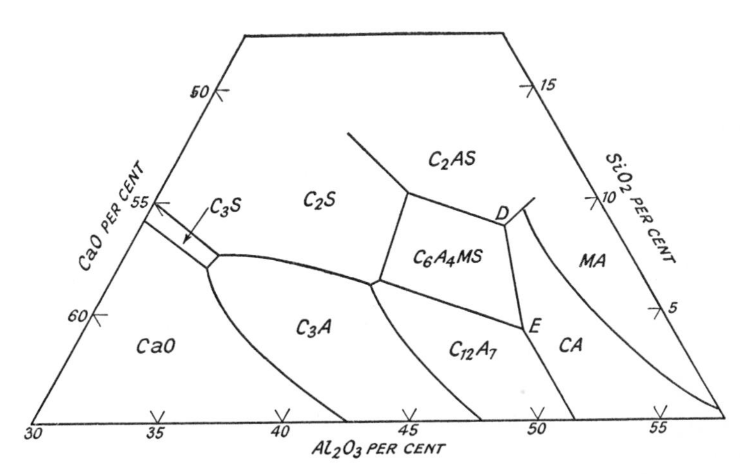

FIG. 30 Plane at 5 percent MgO through the CaO–Al_2O_3–SiO_2–MgO system.

quaternary compound first described by Parker,[1] who assigned it the composition $6CaO.4Al_2O_3.MgO.SiO_2$. Later work suggested[2] that this phase was a solid solution of C_7A_5M and C_2AS but neither this nor the composition C_6A_4MS could be confirmed.[3] X-ray electron microprobe analysis[4] has led to a more complex formula which has yet to be firmly established. The composition of the quaternary phase which occurs in high-alumina cement and which was originally identified by Parker as the analogue of the quaternary magnesian phase just described also seems to have a more complex composition. Optical properties of the quaternary magnesium phase are α 1·669, γ 1·673, biaxial positive 2V about 45°, strong dispersion of optic axes $v > p$. Density 3·01. Mineral from high-alumina cement α 1·676, γ 1·680, pleochroic pale blue green to blue. Density 3·14. Parker found that the quaternary magnesium phase melts incongruently forming $CaO.Al_2O_3$, $MgO.Al_2O_3$ and liquid at 1380° and $MgO.Al_2O_3$ and liquid at 1400°; final melting is at 1420°. These findings apply to the quaternary phase independently of the changed view of its formula. Possible phase assemblages will be discussed in the chapter on high-alumina cement. Other phase diagrams that have been worked out for the high-lime region of the quaternary system include:

C_2S–MgO–Al_2O_3,[5] C_2S–M_2S–$C_{12}A_7$,[6] C_2S–M_2S–Al_2O_3,[7] MA–C_3MS_2–C_2AS[8] and C_3MS_2–C_2MS_2–C_2AS.[9]

The system $CaO–2CaO.SiO_2–12CaO.7Al_2O_3–2CaO.Fe_2O_3$ modified by 5 per cent MgO

The liquids in the part of this quinary system which is of interest in cement technology are completely saturated with MgO if the latter exceeds 5 per cent of the total composition. It is, therefore, possible to represent the system CaO–C_2S–$C_{12}A_7$–C_2F in a tetrahedron of edge representing 95 per cent, and within this to draw phase spaces showing the second phase to crystallise, MgO always being present as the primary solid phase. In this way it is possible to depict the changes in the quaternary system brought about by the addition of 5 per cent MgO.[10]

The main change is in the position and nature of the invariant points. These become:

	Temp	CaO	Al_2O_3	MgO	Fe_2O_3	SiO_2
CaO, C_3S, C_3A, $C_6A_xF_y$, MgO .	1305°	50·9	22·7	5·0	15·8	5·6
C_3S, C_2S, C_3A, $C_6A_xF_y$, MgO .	1301°	50·5	23·9	5·0	14·7	5·9

[1] T. W. Parker, *Sym., London 1952*, 485.
[2] J. H. Welch, *Nature, Lond.* **191,** 559 (1961).
[3] A. J. Majumdar, *Trans. Br. Ceram. Soc.* **63,** 347 (1964).
[4] H. G. Midgley, *Trans. Br. Ceram. Soc.* **67,** 1 (1968).
[5] A. T. Prince, *J. Am. Ceram. Soc.* **34,** 44 (1951).
[6] E. R. Segnit, J. H. Weymouth, *Trans. Br. Ceram. Soc.* **56,** 253 (1957).
[7] W. Gutt, *J. Iron Steel Inst.* **201,** 532 (1963).
[8] W. Gutt, *J. Iron Steel Inst.* **202,** 770 (1964).
[9] W. Gutt, *J. Iron Steel Inst.* **206,** 840 (1968).
[10] M. A. Swayze, *Am. J. Sci.* **244,** 65 (1946).

Both these points move to the high-alumina side of the plane passing through C_3S, C_2S and C_6A_2F; it therefore becomes possible for C_6A_2F to crystallise together with C_3A, C_2S and liquid.

Alkalis

The alkalis (Na_2O and K_2O) are amongst the most widely distributed elements in nature and are found in small amounts in all the raw materials used for cement manufacture. Some volatilisation of the alkalis occurs during burning and the dust from cement kilns has a relatively higher alkali content than the raw mix fed to the kiln. Indeed, attempts have been made to utilise the dust as a source of potash (see p. 154) and to increase its amount by adding felspars to the raw mix. Portland cement contains from 0·5 to 1·3 per cent K_2O+Na_2O and high-alumina cement from 0·1 to 0·6 per cent.

Sodium oxide is taken up to a small extent,[1] about 0·3 per cent, in solid solution in the high-temperature trigonal form of C_3S. In the alite of Portland cement it substitutes for Ca and/or Mg in the Jeffery structure (see p. 319). Potassium oxide is probably also taken up similarly to judge from the electron probe analyses of alite (see p. 107). Both these alkalis are also found present from about 0·2 to 1 per cent in the C_2S in Portland cement.

Work[2] on the system $CaO–Na_2O–Al_2O_3$ has shown that a compound $8CaO.Na_2O.3Al_2O_3$ is formed in the lime rich mixes and hence that it is a possible constituent of Portland cement clinker. The equilibrium diagram shown in Fig. 31 was worked out before the existence of CA_6 was known or the formulae C_5A_3 and C_3A_5 corrected to $C_{12}A_7$ and CA_2 respectively. The compound $8CaO.Na_2O.3Al_2O_3$ can be regarded as made up of three molecules of $3CaO.Al_2O_3$ in which one CaO group has been substituted by Na_2O. Like $3CaO.Al_2O_3$, it decomposes below its melting-point and dissociates at 1508° into CaO and liquid. Its composition therefore falls outside the field in which it is a primary phase. It forms an invariant point with CaO, $3CaO.Al_2O_3$ and liquid at 1490°. The stability of this compound in the presence of calcium silicates has been demonstrated by the investigation of a series of planes in the system $CaO–SiO_2–Na_2O–Al_2O_3$ passing through the points CaO and C_2S and containing varying proportions of $Na_2O:Al_2O_3$.[3] It is therefore a possible phase in Portland cement and has been identified tentatively with the dark prismatic interstitial material of clinker. The optical properties of NC_8A_3 are: $\alpha = 1·702$, $\gamma = 1·710$, biaxial negative, 2V medium. Studies on the relation between C_3A and C_8NA_3, as noted earlier (p. 52), have shown that there is replacement of CaO by Na_2O to a limiting composition of 9 mole per cent C_3A 9 mole per cent N_3A (equivalent to about 6 per cent replacement of CaO by Na_2O). This is slightly lower in Na_2O than NC_8A_3 which lies at 89 mol per cent C_3A 11 mole per cent N_3A. There is no solid solution between the compounds NC_8A_3 and C_3A as such.

[1] G. Yamaguchi and H. Uchekawa, *Zement-Kalk-Gips* **14,** 497 (1961).
[2] L. T. Brownmiller and R. H. Bogue, *Am. J. Sci.* (v) **23,** 501 (1932).
[3] K. T. Greene and R. H. Bogue, *J. Res. natn. Bur. Stand.* **36,** 187 (1946).

In the quinary system containing ferric oxide as well, the invariant point C_3S, C_2S, C_3A, NC_8A_3, $C_6A_xF_y$ and liquid has been located[1] at 1 per cent Na_2O, 48 per cent CaO, 31 per cent Al_2O_3, 13·5 per cent Fe_2O_3 and 6·5 per cent SiO_2 and a temperature of 1310°.

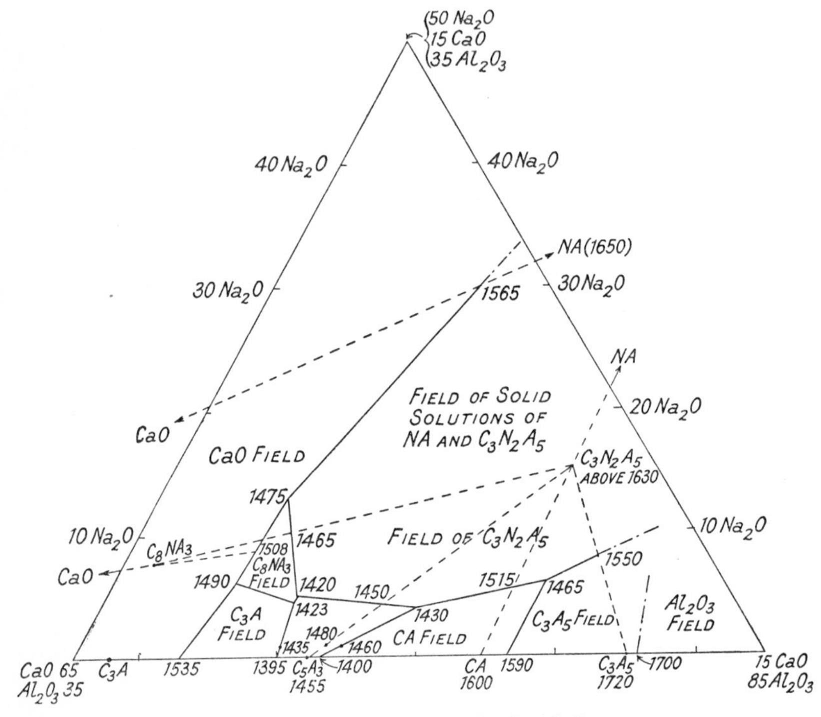

FIG. 31 System $CaO–Na_2O–Al_2O_3$.

Another soda-lime aluminate $3CaO.2Na_2O.5Al_2O_3$ is also formed in the ternary system, but, from the position of its field of stability, is hardly a possible compound in Portland cement. It is a possible compound in high-alumina cement since its field adjoins those of all the three less basic calcium aluminates, viz. $C_{12}A_7$, CA and CA_2. The compound $C_{12}A_7$ takes up about 1 per cent Na_2O into solid solution, its refractive index being lowered to 1·59 and its melting-point raised.

There are no ternary compounds in that part of the system $CaO–K_2O–Al_2O_3$ containing less than 30 per cent K_2O.[2] The field of $K_2O.Al_2O_3$ adjoins those of C_3A, $C_{12}A_7$ and CA and a eutectic is formed between C_3A, $C_{12}A_7$ and KA at 1425°. KA has been shown to be stable in contact with the C_6A_2F solid solution series.[3] It crystallises in the cubic system, refractive index, 1·603.

[1] W. R. Eubank, ibid. **44,** 175 (1950).
[2] L. T. Brownmiller, *Am. J. Sci.* **26,** 260 (1935).
[3] W.C. Taylor, *J. Res. natn. Bur. Stand.* **21,** 315 (1938).

In C–A–F–S–K mixes there is some evidence[1] that a metastable compound KC_8A_3, analogous to NC_8A_3, but containing a small amount of silica (<3 per cent) can be formed and that it may appear in the dark prismatic phase in Portland cement clinker. It appears not to be stable at ordinary temperatures and only to appear in a rapidly cooled clinker.

In the system CaO–K_2O–SiO_2 the ternary compound adjacent to the C_2S field is $K_2O.CaO.SiO_2$. The join C_2S–KCS has been investigated[2] and the formation of a compound $K_2O.23CaO.12SiO_2$ postulated. Later work has shown that $KC_{23}S_{12}$ is not a ternary compound but a point in a solid solution series.[3] In mixes of C_2S and KCS quenched from 1300°, β C_2S has been reported[4] to be stable with contents up to 5 per cent KCS, at 10 per cent α' C_2S is stabilised and at 25 per cent α C_2S. Tests of the stability of KA in the presence of C_3S, C_2S or mixtures of the two showed in every case that reaction took place forming $KC_{23}S_{12}$. The latter phase has also been shown to be stable in contact with C_4AF and $3CaO.Al_2O_3$. It appears likely therefore that K_2O in Portland cement mixtures can be combined as a solid solution of C_2S of approximate composition $KC_{23}S_{12}$.

Both NC_8A_3 and $KC_{23}S_{12}$ react with $CaSO_4$:

$$Na_2O.8CaO.3Al_2O_3+CaSO_4 = Na_2SO_4+3C_3A$$
$$K_2O.23CaO.12SiO_2+CaSO_4 = K_2SO_4+12C_2S$$

In cement clinker therefore alkali sulphates are formed preferentially by combination with sulphur trioxide in the kiln gases and it is only the alkalis in excess of this requirement which will form NC_8A_3 and $KC_{23}S_{12}$.

Sodium and potassium sulphates form a solid solution. The potash combines preferentially with the sulphur trioxide and in tests on laboratory clinkers burnt to equilibrium, Newkirk[5] found that the solid solution tended to have a composition approximating to $3K_2SO.Na_2SO_4$ to the extent that the mix composition allowed. Cements commonly have a higher content of potash than soda and in British cements the molar ratio $K_2O:Na_2O$ varies from two to five with occasional cases below one or up to eight. In commercial clinkers the molar ratio $K_2O:Na_2O$ in the solid solution has been found to average about twice that in the clinker as a whole.[6] Alkalis are usually present in excess of the SO_3 and the surplus alkali forms NC_8A_3 and $KC_{23}S_{12}$ with some taken up also in the calcium silicates. In the less usual cases where sulphur trioxide is present in excess of the alkalis, or at lesser SO_3 contents under non-equilibrium conditions in cements burning, the compounds $2CaSO_4.K_2SO_4$ and β $CaSO_4$ are also formed. Both have been observed in commercial clinkers.

It has often been considered that the alkalis play a considerable part in determining the rate at which the calcium silicates react with water. The rapid release of a portion of the alkalis into solution when a Portland cement is shaken

[1] Y. Suzukawa, *Zement-Kalk-Gips* **9,** 390 (1956).
[2] W. C. Taylor, *J. Res. natn. Bur. Stand.* **27,** 311 (1941); **29,** 437 (1942).
[3] Y. Suzukawa, *Zement-Kalk-Gips* **9,** 390 (1956).
[4] N. F. Federov and E. R. Brodinka, *Neorganicheskie Materialy* **2,** 7458 (1966); see R. W. Nurse, *Sym., Tokyo 1968.*
[5] T. F. Newkirk, *J. Res. natn. Bur. Stand.* **47,** 349 (1951); *Sym., London 1952,* 151.
[6] H. W. W. Pollitt and A. W. Brown, *Sym., Tokyo 1968.*

with water definitely indicates that they are present partly as sulphates but it is known that $KC_{23}S_{12}$ hydrates more rapidly than C_2S. The important part played by the water-soluble soda and potash of a cement in regard to certain properties of set cements is discussed in Chapter 17.

Sulphur compounds

The clay or shale used in the raw materials for Portland cement manufacture may contain sulphur compounds, while the powdered coal used to fire the kiln also contains sulphur in the form of pyrites (FeS) and organic sulphur compounds. Other fuels, oil, natural gas and charcoal also contain sulphur compounds. Part of the sulphur is eliminated in the burning process (see p. 153) and part appears in the cement clinker generally as alkali sulphate. The content varies usually from less than 0·1 per cent SO_3 to 0·5 per cent but in the unusual cases where charcoal is the fuel the SO_3 content may reach about 1–5 per cent. It may also reach this value in clinkers from the cement-sulphuric acid process.

In addition to alkali sulphate, which is formed preferentially from available sulphate, two other sulphate compounds relevant to cement production are known. The compound[1] $(2CaO.SiO_2)_2.CaSO_4$ has been detected in an impure form in clinker rings. This compound might also be an intermediate in C_2S and C_3S formation in the presence of sufficient sulphate in the raw materials.[2] The calcium alumino sulphate, $3(CaO.Al_2O_3).CaSO_4$, which is a component of expanding cement is also a possible intermediate phase in the burning of Portland cement mixes of high sulphate content. The compound $(2CaO.SiO_2)_2.CaSO_4$ has been found[3] to take part in the stable phase assemblages in the system $CaO–CaO.SiO_2–CaSO_4$ at 1000° and 1200°, but not to be present at 1310° nor to coexist with $3CaO.SiO_2$. This supports the suggestion that it may form as an intermediate phase in cement burning and be decomposed again before the clinkering temperature is reached. $3CaO.SiO_2$ is found in this system at 1350° up to high SO_3 concentrations (38 per cent). It takes up SO_3 in solid solution up to a maximum of 2·9 per cent at 1310°. One S^{6+} ion is assumed to substitute for $2Ca^{2+} + 0{\cdot}5Si^{4+}$ giving a formula for the limiting solid solution of 92 mole per cent C_3S, 8 mole per cent '$2CaSO_4.SiO_2$'. Small amounts (0·5 per cent) of $CaSO_4$ can stabilise β C_2S and larger amounts α' C_2S. The extent of solid solution in α' C_2S is 1·7 mole per cent $CaSO_4$ at 1000° and 1·1 mole per cent at 1200°.

Sulphate ions, unlike phosphate ions (p. 79), do not decompose C_3S, but the combined action of a aluminium and sulphate ions leads to preferential formation of a β C_2S solid solution, rich in aluminium and sulphate, in both the system $CaO–SiO_2–Al_2O_3–SO_3$ and in cement clinker without magnesia or alkalies. Magnesium ions counteract this adverse effect of aluminium and sulphate ions on C_3S formation. Ferric iron and sulphate ions together do not hinder C_3S formation even in the absence of magnesium ions.[4]

[1] W. Gutt and M. A. Smith, *Nature, Lond.* **210,** 408 (1966).
[2] W. Gutt, *Sym., Tokyo 1968.*
[3] W. Gutt and M. A. Smith, *Trans. Br. Ceram. Soc.* **66,** 557 (1967).
[4] W. Gutt and M. A. Smith, *Trans. Br. Ceram. Soc.* **67,** 487 (1968).

Sulphides are not normally present to any appreciable extent in Portland cement clinker owing to the oxidising atmosphere in the kiln, but amounts up to about 0·1 per cent are occasionally found. The estimation of sulphides in Portland cement cannot be carried out directly by the ordinary evolution method owing to the presence of ferric iron in the cement. A modified method involving a preliminary reduction of the ferric iron by addition of stannous chloride has been described.[1]

Titania

Titania (TiO_2) occurs to a small extent, about 0·2–0·3 per cent in all Portland cements, being derived from the clay or shale. It is a more important constituent of high-alumina cement, where about 1·5–2 per cent is usually present; it is derived from the bauxite. Kühl[2] found that in a Portland cement the substitution of silica by titania in small amounts slightly increased the strengths, but that in larger amounts a reduction in strength occurred. The optimum content was found to be about 4·5 per cent. Kühl suggests that a compound $3CaO.TiO_2$ is formed and early workers on the system $CaO–TiO_2$ claimed to have prepared both $3CaO.TiO_2$ and $2CaO.TiO_2$; there is, however, now good evidence[3] that the maximum amount of lime combined corresponds to a ratio $3CaO.2TiO_2$. Although behaving in many respects as a separate compound, it seems likely[4] that $3CaO.2TiO_2$ is a solid solution of CaO in one of the polymorphic forms of $CaO.TiO_2$. One other compound $4CaO.3TiO_2$ melting incongruently at 1755° has also been reported[5] but its existence has not been confirmed.[6] The system $CaO–CaO.Al_2O_3–CaO.TiO_2$ contains no ternary compounds, each of the aluminates being in equilibrium with CT solid solution; a few per cent of CA or $C_{12}A_7$ in solid solution in CT stabilises it in the optically isometric form in which it is usually found in nature, with refractive index 2·38, density 4·05. $3CaO.2TiO_2$ quenched from high temperature has optical properties: $\alpha = 2{\cdot}30$, $\varepsilon = 2{\cdot}32$, uniaxial negative; after annealing at 1200°, $\alpha = 2{\cdot}16$, $\beta = 2{\cdot}20$, $\gamma = 2{\cdot}22$, 2V about 70°, negative; density 3·75.

In the system $CaO–MgO–TiO_2$, CT solid solutions occur in equilibrium with CaO and MgO[7]; in the system $CaO–SiO_2–TiO_2$ there are invariant points at which CaO, C_3S and CT (s.s.), and C_3S, C_2S and CT (s.s.) coexist in equilibrium and at which the solid solution of CaO in $CaO.TiO_2$ reaches the composition $3CaO.2TiO_2$.[8]

[1] D. I. Watson, *Cement* **5,** 49 (1932); H. A. Bright, *J. Res. natn. Bur. Stand.* **18,** 137 (1937); *Analysis of Calcareous Materials,* Soc. Chem. Ind. Monogr. No. 18, 103 (1964).
[2] *Zement* **14,** 37 (1925).
[3] R. W. Nurse, M.Sc. Thesis, London, 1934.
[4] F. M. Lea and R. W. Nurse. Unpublished data which conflicts somewhat with the results for the binary system $CaO–TiO_2$ given by L. W. Coughanour, R. S. Roth and V. A. de Prosse, *J. Res. natn. Bur. Stand.* **52,** 37 (1954), but is more in agreement with R. C. de Vries, R. Roy and E. F. Osborn, *J. phys. Chem.* **58,** 1069 (1954).
[5] R. S. Roth, *J. Res. natn. Bur. Stand.* **61,** 437 (1958).
[6] J. A. Imlach and F. P. Glasser, *Trans. Br. Ceram. Soc.* **67,** 581 (1968).
[7] A. S. Berezhnoi, *Ogneupory* **15,** 350 (1950).
[8] A. S. Berezhnoi, ibid. **15,** 446 (1950); R. C. de Vries, R. Roy and E. F. Osborn, *J. Am. Ceram. Soc.* **38,** 153 (1955).

In a fully crystallised Portland cement TiO_2 could occur as the solid solution of composition C_3T_2 and in high-alumina cement as an aluminous CT solid solution. However electron probe analyses of clinkers has shown that TiO_2 is present in small amounts, presumably in solid solution, in all the four major compounds of Portland cement[1] and in the ferrite phase in high-alumina cement.[2] Only any residual amount of titania will therefore be available to form the C_3T_2 or the aluminous CT solid solution.

Manganese oxide

(Mn_2O_3) is a constituent of Portland cement clinkers produced by burning raw mixes of blastfurnace slag with limestone, but is not usually present, except in very small amounts, when the raw materials are a clay or shale and limestone. Contents of 1 per cent or more Mn_2O_3 are found in the cements derived from a slag raw material. The presence of manganese oxide gives a brown colour to a cement, but otherwise it appears to substitute for ferric oxide and to have little effect on the properties of a Portland cement. The introduction of Mn ions alone into the crystal lattice of C_3S does not affect the early strength but slows strength development at later ages. The effects of Mn and Fe^{3+} ions on the strength of C_3S are materially different since Fe^{3+} ions present alone in the C_3S crystal lattice lead to low early strength with recovery at later ages.[3] A cement[4] analogous to Kühl's cement can be produced by the substitution of manganese oxide for ferric oxide. Contents of Mn_2O_3 up to 4 per cent have been found to produce excellent cements, but with contents above 5 per cent the strength is somewhat reduced. For commercial reasons the content of manganese oxide in a cement is unlikely to exceed about 3 per cent. A compound $4CaO.Al_2O_3.Mn_2O_3$ analogous to $4CaO.Al_2O_3.Fe_2O_3$ exists, and the two form a continuous series of solid solutions.[5] It is likely that the manganic oxide in Portland cement is present in such a solid solution. The compound $4CaO.Al_2O_3.Mn_2O_3$ melts congruently at 1478°, but the liquid melt then commences to lose oxygen owing to dissociation of the manganic oxide. $4CaO.Al_2O_3.Mn_2O_3$ occurs as a dark opaque material and as transparent crystals showing marked pleochroism. The optical properties are $\alpha = 1{\cdot}86$, $\gamma = 2{\cdot}06$; pleochroism α colourless, γ almost opaque. The usual practice of calculating the manganic oxide along with the ferric oxide when representing the composition of a cement by various moduli seems therefore to be justified.

In one cement clinker examined, containing 0·86 per cent MnO, and which had apparently been burnt under reducing conditions (FeO 1·08, Fe_2O_3 1·80), the iron compound appeared to contain about 20 per cent $4CaO.Al_2O_3.Mn_2O_3$ but the β C_2S was also coloured a bright green. The high-temperature forms α and α', of C_2S can take up Mn_2SiO_4 in solid solution[6] to the extent of 20–25

[1] H. G. Midgley, *Sym., Tokyo 1968*; P. Terrier, H. Hornain and G. Socroun, *Revue Matér. Constr. Trav. publ.* **630,** 109 (1968).
[2] Mme Jeanne, *Revue Matér. Constr. Trav. pupl.* **629,** 53 (1968).
[3] W. Gutt and G. J. Osborne, *Trans. Br. Ceram. Soc.* **68**, 129 (1969).
[4] A. Guttmann and F. Gille, *Zement* **18,** 500, 537, 570 (1929); O. Goffin and G. Mussgnug, ibid. **22,** 218, 231 (1933); **26,** 809 (1937).
[5] T. W. Parker, *Sym., London 1952*, 143.
[6] F. P. Glasser, *Silikattechnik* **11,** 362 (1960); *Am. J. Sci.* **259,** 46 (1961).

per cent. The colour of the C_2S with limited amounts of manganese in solid solution can be blue, green or pink. A small amount of manganese oxide, up to 0·5 per cent, can be taken up in solid solution in alite.[1]

Manganic oxide is also present in Portland blastfurnace cement, usually both as a constituent of the cement clinker and of the granulated slag which is added before grinding.

Fluorides

Fluorspar, CaF_2, or fluoride wastes are sometimes added as a flux or a mineraliser to aid cement burning (see p. 156). CaF_2 may also occur sporadically as an impurity in the limestone component or it may be present in industrial by-products which may be considered as raw materials for cement manufacture. The presence of fluorides influences both the mechanism of formation of C_3S[2] and of C_2S in the clinker, and also the cementing strength of C_3S.[3] The phase diagram[4] of the system $2CaO.SiO_2$–CaF_2 (Fig. 32) shows that C_2S, unlike

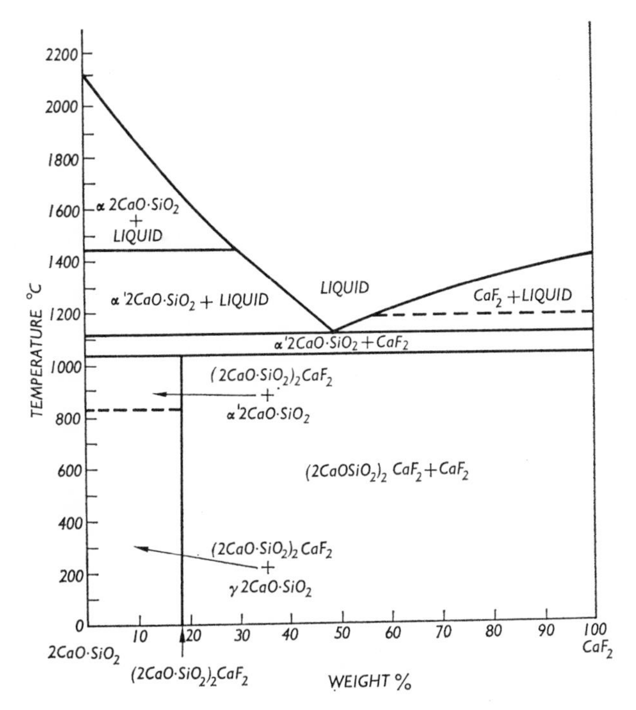

FIG. 32 System $2CaO.SiO_2$–CaF_2.

[1] T. Sakurai, *Japan Cement Engineering Association: Review of 14th General Meeting 1960*, p. 23.
[2] W. Gutt, *Sym., Tokyo 1968.*
[3] J. H. Welch and W. Gutt, *Sym., Washington 1960*, 59.
[4] W. Gutt and G. J. Osborne, *Trans. Br. Ceram. Soc.* **65,** 521 (1966).

C_3S, does not form solid solutions with fluoride ions but a calcium silicofluoride $(2CaO.SiO_2)_2CaF_2$ exists in the system, this being the fluoride analogue of calciochondrodite (substituting F^- for OH^-). It decomposes at 1040° into α' C_2S and CaF_2 and it has no cementing properties. Dicalcium silicate is the primary phase in compositions up to $51C_2S$, $49CaF_2$ (weight per cent) which marks the eutectic with CaF_2 at 1110°. In the ternary system $CaO–2CaO.SiO_2–CaF_2$ a compound tentatively assigned the formula $(3CaO.SiO_2)_3.CaF_2$ melting incongruently has been found to exist.[1] The phase diagram[2] for this system is shown in Fig. 33. It provides a good illustration of the disappearance

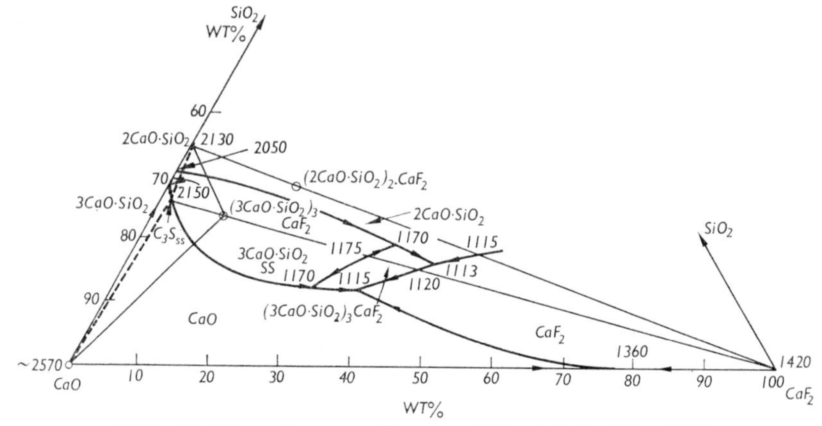

FIG. 33 Phase diagram of system $CaO–2CaO.SiO_2–CaF_2$.

of the C_3S field when melting temperatures fall below the decomposition point which lies at 1170° in the presence of fluorine.[3]

The compound of tentative composition $(3CaO.SiO_2)_3.CaF_2$ is unstable below 1100° giving C_2S, CaO and CaF_2 or $(C_2S)_2$, CaF_2 and CaO depending on the temperature. Its presence lowers the strength of cement. There is no C_3A field[4] in the system $CaO–CaF_2–12CaO.7Al_2O_3$. A simple eutectic is formed between CaO, $C_{12}A_7$ and CaF_2 at about 1340°. A compound having the composition $3CaO.3Al_2O_3.CaF_2$ exists.[5] It has been identified in aluminous slags in phosphorus manufacture. It has cementitious properties. This compound looses fluorine when heated in open vessels, but examination[6] in sealed platinum capsules showed congruent melting at 1506°. A compound of the composition $11CaO.7Al_2O_3.CaF_2$ ($Ca_{12}Al_{14}O_{32}F_2$ analogous with $Ca_{12}Al_{14}O_{32}(OH)_2$) has also been reported.[7]

[1] W. Gutt and G. J. Osborne, *Trans. Br. Ceram. Soc.* **67,** 125 (1968); M. Tanaka, G. Sudoh and S. Akaiwa, *Sym., Tokyo 1968.*
[2] W. Gutt and G. J. Osborne, *Trans. Br. Ceram. Soc.* **69**, 125 (1970).
[3] W. Gutt, *Sym., Tokyo 1968.*
[4] W. Eitel, *Zement* **30,** 17, 29 (1941).
[5] J. K. Leary, *Nature, Lond.* **194,** 79 (1962).
[6] W. Gutt, unpublished.
[7] J. Jeevaratnam, F. P. Glasser and L. Dent-Glasser, *J. Am. Ceram. Soc.* **47,** 105 (1964); C. Brisi and P. Rolando, *Industria ital. Cem.*, 1967 (1), 37; P. A. Williams, *J. Am. Ceram. Soc.* **51,** 531 (1968); F. Massazza and M. Pezzvoli, *Revev Matér. Constr.* publ., **642,** 81 (1969).

Addition of CaF_2 to the system $CaO–Al_2O_3–SiO_2$ leads to reduction in liquidus temperatures but does not modify the melting behaviour of individual compounds.[1] Liquidus temperatures are reduced by 50–70° by 5 per cent CaF_2 and by 100–120° in the presence of 10 per cent CaF_2. Boundaries of primary crystallisation fields, and therefore the positions of invariant points, are altered as well. The $3CaO.2SiO_2$ field disappears at 20 per cent CaF_2.

Phosphorus pentoxide

The content of P_2O_5 in most Portland cements is of the order of 0·2 per cent but phosphate occurs in significant amounts in some raw materials used for cement manufacture. In such cases it passes into the clinker. Satisfactory cements can be made[2] from cement clinker containing up to 2·5 per cent P_2O_5 by correct burning and proportioning but the rate of hardening is slower. This is because the P_2O_5 decomposes C_3S in favour of a C_2S solid solution containing the P_2O_5 and excess CaO. If larger amounts of P_2O_5 are present, free lime forms. In the system $CaO–P_2O_5$ the more basic compounds are $4CaO.P_2O_5$ and $3CaO.P_2O_5$ and the phase diagram in this region has been established.[3]

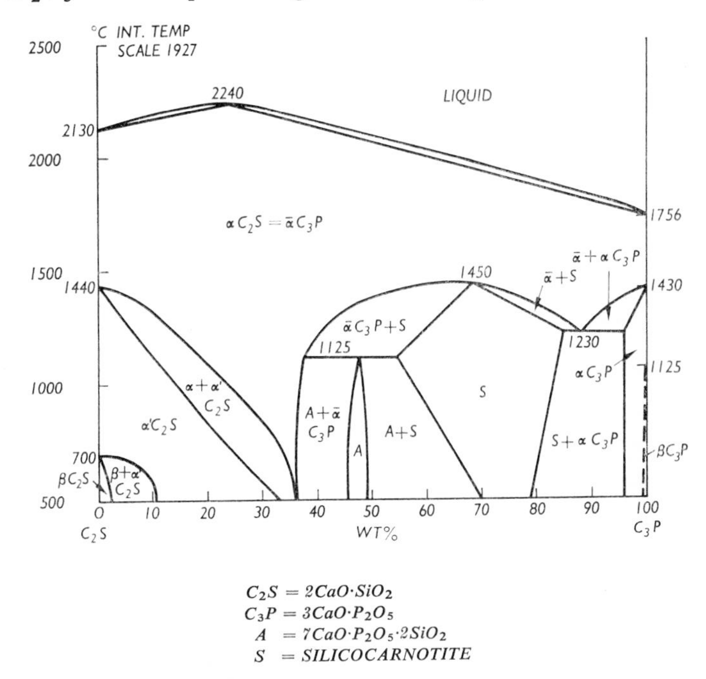

FIG. 34 System $2CaO.SiO_2–3CaO.P_2O_5$.

[1] N. A. Toropov and V. P. Barzakovskii, *High-temperature Chemistry of Silicates and Other Oxide Systems*, p. 9, Consultant Bureau New York (1966).
[2] R. W. Nurse, *J. appl. Chem.* **2,** 708 (1952).
[3] J. H. Welch and W. Gutt, *J. chem. Soc.* 1961, 4442; G. Tromel and W. Fix, Archiv. *Eisenhüttenwesen* **32,** 209 (1961); W. Fix, H. Heymann and R. Heinke, *J. Am. Ceram. Soc.*, **52,** 346 (1969).

Tricalcium phosphate forms a binary system with $2CaO.SiO_2$ in which at high temperature a continuous series of solid solutions between α' C_2S and $\bar{\alpha}$ C_3P is formed with a liquidus maximum[1] (see Fig. 34). Addition of $3CaO.P_2O_5$ to $2CaO.SiO_2$ stabilises in turn the β, α' and α polymorphic forms of dicalcium silicate. Two intermediate phases are formed by solid state reaction at lower temperatures. These have compositions near $5CaO.P_2O_5SiO_2$, previously accepted for silicocarnotite, and near $7CaO.P_2O_5.2SiO_2$ respectively. These phases exist as homogeneous solid solutions over a range of composition that varies with temperature as shown in Fig. 34. Phase A (C_7PS_2) is not to be confused with the mineral of the same composition,[2] nagelshmidtite, which has the structure associated with the continuous high temperature solid solution series α $2CaO.SiO_2$–$\bar{\alpha}$ $3CaO.P_2O_5$. Tricalcium phosphate exists in three polymorphic forms: β stable up to 1125°, α stable between 1125° and 1430°, and the recently discovered[3] $\bar{\alpha}$ form which melts at 1756°. It is the $\bar{\alpha}$ form which

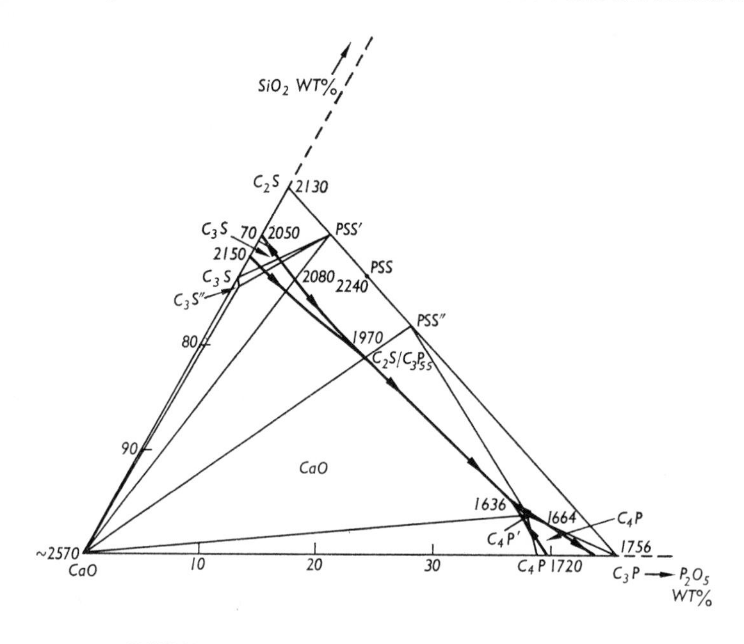

FIG. 35 Phase diagram of system $2CaO.SiO_2$–$3CaO.P_2O_5$–CaO.

[1] R. W. Nurse, J. H. Welch and W. Gutt, *J. chem. Soc.*, 1959, 1077.
[2] G. Nagelschmidt, *J. chem. Soc.*, 1937, 865.
[3] R. W. Nurse, J. H. Welch and W. Gutt, *Nature, Lond.* **182,** 1230 (1958).

is fully miscible with α $2CaO.SiO_2$. The cementing properties of phosphate stabilised polymorphs of $2CaO.SiO_2$ is discussed on p. 155.

In the ternary system CaO–SiO_2–P_2O_5 tricalcium silicate has a primary phase field extending to an invariant point at 13 per cent P_2O_5 at 1970° as shown[1] in Fig. 35. It takes up a small amount of Ca^{2+} and PO_4^{2-} ions to form a solid solution C_3S'' containing 0·5 per cent P_2O_5 which has cementing properties not much different from C_3S. In the phase assemblies stable at 1500°, however, C_3S or C_3S'' cease to appear at a lower P_2O_5 concentration in the mix. This may be seen from the set of three phase assemblies at 1500° into which Fig. 35 is divided. Only compositions on the low P_2O_5 side of the line joining CaO to the C_2S–C_3P solid solution marked PSS' will contain C_3S or C_3S'' as a solid phase on clinkering at 1500°. At higher P_2O_5 contents the stable phases become lime and C_2S–C_3P solid solutions. Since a Portland cement must contain a substantial proportion of tricalcium silicate the maximum proportion of P_2O_5 in a cement mix must not exceed 2·0–2·5 per cent.

Phosphatic limestones may contain calcium fluoride as well as P_2O_5 so that the formation of fluorapatite $3(3CaO.P_2O_5)CaF_2$ in the burning of a cement mix has also to be considered.

In the system CaO–SiO_2–P_2O_5–CaF_2 on the join $3CaO.SiO_2$–$3(3CaO.P_2O_5)CaF_2$ a tricalcium silicate solid solution with fluorapatite exists[2] containing up to 2 mole per cent at 1905°, the invariant temperature for co-existence of $3CaO.SiO_2$, CaO, $2CaO.SiO_2$ and liquid. In the system CaO–$2CaO.SiO_2$–fluorapatite the primary field of C_3S extends to 3 mole per cent of fluorapatite which is equivalent to 5·08 weight per cent P_2O_5. On the join C_3S–$(3CaO.P_2O_5)_3.CaF_2$ fluorapatite does not appear in mixes containing below 10 mole per cent of this compound. This is equivalent to 1·2 per cent fluoride and this compound need not be expected therefore in detectable amounts in phosphatic fluorinated cements which normally contain no more than 1·4 per cent fluoride.[3]

[1] W. Gutt, *Nature, Lond.* **197,** 142 (1963); *Sym., Tokyo 1968*; Ph.D. Thesis, London 1966.
[2] C. M. Schlaudt, Ph.D. Thesis, Pennsylvania State Univ. 1964.
[3] W. Gutt, *Sym., Tokyo 1968.*

5 Cementing Qualities of the Cement Compounds

After the work of Rankin and Wright on the system $CaO–Al_2O_3–SiO_2$ had shown definitely what compounds could occur in mixes of Portland cement composition, Bates undertook an investigation of the cementitious qualities of these compounds.[1] A number of similar studies have also since appeared.[2]

Tricalcium silicate has all the essential properties of Portland cement. It undergoes an initial and final set within a few hours after gauging and, when properly prepared, shows no unsoundness. Mixes of tricalcium silicate and water are less plastic than Portland cement and more water is required to obtain a workable paste. The addition of gypsum to the mass renders it more plastic and has some effect on the setting time. Both acceleration and retardation of the set have been reported.

Dicalcium silicate (β) exhibits no definite setting time and the gauged mass sets only slowly over a period of some days. The addition of gypsum produces little change.

Tricalcium aluminate gives a flash (i.e. almost instantaneous) set on gauging with water and this is accompanied by the evolution of so much heat as to lead to violent steaming. On further mixing, so as to break up the initial flash set, a plastic, easily workable mass is obtained. If it is kept in moist air the mass sets and hardens, giving fair strength, but on placing in water the set material disintegrates and crumbles away. A normal setting time can be obtained by the addition of 15 per cent gypsum in mixes with a high water content (w/c 0·50).

The strengths developed by the pure cement compounds, and of mixtures of them are shown in Tables 5 and 6.

The dicalcium silicate produces little strength at early ages, but gains steadily in strength at later ages until it approaches equality with tricalcium silicate. The

[1] P. H. Bates and A. A. Klein, *Tech. Pap. natn. Bur. Stand.*, No. 78 (1917); P. H. Bates, *Tech. Pap. natn. Bur. Stand.*, No. 197 (1921).

[2] R. J. Colony and D. L. Snader, *Bull. Columbia Univ. Dep. civ. Engng*, No. 4 (1931); S. Nagai and R. Naito, *J. Soc. chem. Ind., Japan Suppl., Binding* **33** (4), 133B, 164B, 316B (1930); **34** (5), 159B (1931); G. Haegermann, *Tagung Verein Deutscher Portland Zement Fabrikanten*, 1932, p. 5; E. Spohn, *Dissn tech. Hochsch.*, Berlin, 1932. R. H. Bogue and W. Lerch, *Ind. Engng Chem.* **26,** 837 (1934). A. Guttmann and F. Gille, *Zement* **22,** 204 (1933); Yu M. Butt and V. V. Timashev, *Tsement* **27** (2), 17 (1961).

tricalcium silicate attains the greater part of its strength in seven days. The tricalcium aluminate produces some strength at one day, but shows no subsequent development. The influence of this compound in the mixes seems erratic and in some cases it lowers the strength.

TABLE[1] 5

Cementing mix	1 : 3 Sand mortars. Water stored. Tensile strength (lb/in^2)		
	3 days	7 days	28 days
$3CaO.SiO_2$	114	199	270
$3CaO.SiO_2 + 5\%$ gypsum	156	199	313
$3CaO.Al_2O_3$	—	—	43
$4CaO.Al_2O_3.Fe_2O_3$	—	—	86
$75\%\ 3CaO.SiO_2 + 25\%\ 3CaO.Al_2O_3$	214	284	455
$75\%\ 3CaO.SiO_2 + 25\%\ 3CaO.Al_2O_3 + 2{\cdot}5\%$ gypsum	214	—	270

TABLE[2] 6

Cementing mix	1 : 3 Sand mortars. Water stored. Compressive strength (lb/in^2)			
	3 days	7 days	28 days	6 months
85% C_3S 15% $\beta\ C_2S$	2040	2620	4100	4960
85% C_3S 15% C_3A	1720	2650	3850	4600
85% C_3S 15% C_4AF	1600	2450	2900	4300

4·3 per cent gypsum in the cements

Bogue and Lerch found that the substitution of 15 per cent C_3A for C_3S slightly raised the strength of neat cement pastes up to three days, but decreased the later strengths. No favourable influence was found in mortars, as is seen from the data in Table 6. Some data on the four compounds reported by Butt[3] and given in Table 7 show the same general trends, with a strength for C_4AF which is surprisingly high compared with the results of other workers.

[1] Haegermann, loc. cit.
[2] Bogue and Lerch, loc. cit.
[3] Yu. M. Butt, V. M. Kolbasov and V. V. Timashev, *Sym., Tokyo 1968.*

TABLE 7 Neat cement specimens: compressive strengths (kg/cm^2)

	7 days	28 days	180 days	365 days
C_3S	322	466	512	584
$\beta\ C_2S$	24	42	193	325
C_3A	118	124	0	0
C_4AF	300	384	493	595

It is possible to obtain an indication of the relative contributions of the four cement compounds to strength from an analysis of the strengths of cements of varying composition (see p. 174). Such results confirm the general conclusions set out above.

It is doubtful if tricalcium aluminate is a true hydraulic cement, and its value in Portland cement seems to be mainly limited to its effect in rendering burning possible at the kiln temperatures attainable commercially. It probably makes no direct contribution to the cementing action other than accounting for the initial set and adding perhaps somewhat to the early strengths.[1] There is evidence[2] however that the presence of C_3A increases the rate of hydration and strength development of C_3S. The question may be asked, why it is that a Portland cement does not exhibit the expansion and disintegration shown by tricalcium aluminate on immersion in water, or the flash setting which is so characteristic of it. The answer to this is to be found, probably, not only in the small amount of it present, from 6 to 14 per cent, but also in the manner of its occurrence and the high proportion, relative to it, of the added gypsum. In a cement the tricalcium aluminate occurs intermixed with various other constituents which must partially surround it; its rate of reaction with water will thus be reduced owing to the decreased availability of its surface. An addition of 3–4 per cent gypsum to a cement, as is commonly made, corresponds to about 25–50 per cent on the tricalcium aluminate content, a very high proportion.

The compound C_4AF hydrates rapidly but its contribution to strength still seems uncertain. Though setting occurs in a few minutes, it does not show a flash set like C_3A. Marked heat evolution also accompanies the setting, but it is much less vigorous than with tricalcium aluminate.

Dicalcium ferrite[3] reacts with water and sets and hardens rapidly, but without the flash set or marked heat evolution shown by tricalcium aluminate. The hydration stops long before completion. A continuous expansion occurs after setting which eventually disrupts the set material and reduces the strength to a small value. When placed in water the set material rapidly disintegrates. The

[1] Colony and Snader obtained quite high compressive strengths from tricalcium aluminate, but since the test specimens were cracked by expansion, such values cannot be regarded as demonstrating that the material has much cementing value.

[2] A. Celani, P. A. Maggi and A. Rio, *Sym., Tokyo 1968*; K. A. Alexander, J. H. Taplin and J. Wardlaw, ibid.; L. E. Copeland and D. L. Kantro, ibid.

[3] S. Nagai and K. Asaoka, *J. Soc. chem. Ind., Jap. Suppl., Binding* **33** (6), 190B, 1930; Colony and Snader, loc. cit.

compound C_2F cannot therefore be considered to have any permanent cementing value. It is not present as such in Portland cement though, as we have seen, it is a constituent of the iron compound solid solution which can vary in $Al_2O_3:Fe_2O_3$ ratio on either side of C_4AF.

It is not to be expected that the setting and strength properties of the pure compounds should be identical with those of the minerals occurring in Portland cement, since the latter contain small amounts of other oxides in solid solution, which influence their reactivity and also their crystal form. Thus the addition of 0·3–0·5 per cent of BaO or P_2O_5 has been reported to increase the strength of a cement while similar amounts of SrO or TiO_2 had little effect.[1] Alumina, ferric oxide, magnesia, and the alkalis are also taken up in small amounts in solid solution in C_3S and C_2S. Hedin[2] found that alite containing 4 per cent C_3A reacted with water considerably faster than pure C_3S and Verbeck[3] has reported that alite containing 1 per cent Al_2O_3 and an equivalent amount of magnesia had a significantly higher early strength than pure C_3S. The crystal form of C_3S is changed by the presence of other elements in solid solution, and it has also been suggested that dislocations, or strain or distortion, set up in the crystal lattice by solid solution increases the reactivity.

Some data obtained by Nurse[4] on different crystal forms of pure C_3S and alites containing magnesia, or magnesia and alumina, are shown in Table 8.

TABLE 8 Compressive strength (lb/in²) of tricalcium silicate

Type of C_3S	Age (days)		
	1	3	28
Triclinic, T_1, pure	1720	2670	3400
Triclinic, T_{11} C_{154} M_2 S_{52}	1340	2790	3310
Monoclinic, C_{15}. M_5 S_{52}	2320	3590	4110
Monoclinic, C_{54} S_{16} MA	1800	4300	5100

Tested on ½ in. cubes of 1 : 3 mortar with 4 per cent gypsum by weight of cement

Though these data show higher strengths for the monoclinic than the triclinic C_3S, a further series of tests in which changes from the triclinic to the monoclinic form were produced by varying the MgO content did not confirm this. Other work[5] has also failed to show a significant difference between the triclinic and monoclinic forms of alite though the monoclinic form hydrated rather the more rapidly. It is difficult from these data to distinguish between the effect

[1] W. Kurdowski and A. Szummer, *Silic. ind.* **33,** 183 (1968).
[2] R. Hedin, *Proc. Swed. Cem. Concr. Inst.*, No. 3, 137 (1945).
[3] G. Verbeck, *J. Res. Dev. Labs Portld Cem. Ass.* **7** (3), 57 (1965).
[4] R. W. Nurse, *Spec. Rep. Highw. Res. Bd*, No. 90, 258 (1966).
[5] G. Yamaguchi, K. Shirasuka and T. Ota, *Spec. Rep. Highw. Res. Bd*, No. 90, 263 (1966); W. Gutt and G. J. Osborne, *Trans. Br. Ceram. Soc.* **68,** 129 (1969).

of the substitution of magnesia and alumina in C_3S on the reactivity of the compound and the effect of the crystal form. There is the added complication that the reactivity of alite crystals seems also to be influenced by crystal size.[1] In the case of C_2S it is necessary to stabilise the β form and marked difference in the rate of strength development has been found[2] with different stabilisers as shown in Table 9. The α and γ forms of C_2S have no cementing properties at normal temperatures.

TABLE 9 Compressive strength (lb/in^2) of β dicalcium silicate

Stabiliser	Age (days)		
	7	28	180
Fe_2O_3 Na_2O	212	350	1544
B_2O_3	401	726	2417
$Ca_3\ (PO_4)_2$	553	1402	3607

Tested on ½ in. cubes of 1 : 3 mortar

The less basic aluminates

Some early tests showing the cementing qualities of the calcium aluminates of lower lime content than C_3A were carried out by Schott, Killig, Spackman and others, but it is to Bates[3] in the U.S.A. and Endell[4] in Germany that we owe the first published systematic studies of the cementitious qualities of these compounds. Much work must also have been done in France by Bied, to whom the development of high-alumina cements is due, but the results have never been made public and were probably more concerned with mixes possible in commercial production than with the pure compounds.

The supposed compound C_5A_3 differs from the more basic tricalcium aluminate in showing no flash set. Initial set occurs in about 3–5 minutes and the final set in 15–30 minutes, the latter being accompanied by the evolution of much heat. The addition of plaster of Paris has little effect on the initial set, but retards the final set by about an hour.

Calcium monoaluminate rapidly attains high strengths and does not suffer from rapid setting. Bates found initial and final setting times of 25 minutes and 2 hours respectively. The addition of plaster, or gypsum, accelerates the set and, indeed, with some high-alumina cement compositions, the acceleration is so marked as to render the cement unusable. It also reduces the strength considerably, in contrast to its action with $C_{12}A_7$, the strength of which is increased by 50–100 per cent by the addition of gypsum.

[1] J. Grzymek, *Silikattechnik* **6,** 296 (1955); **10,** 81 (1959).
[2] R. W. Nurse, *Sym., London 1952,* 56.
[3] *Tech. Pap. natn. Bur. Stand.*, No. 197 (1921).
[4] *Zement* **8,** 319, 334, 347 (1919).

The supposed compound C_3A_5 also develops considerable strength much of which must be ascribed to its content of CA for the true compound CA_2 is relatively unreactive. Impure CA_2 as will occur in some high-alumina cements may be more reactive and contribute to strength.

The strengths of water-stored 1:3 mortars of these calcium aluminates, found by Bates, were:

Compound composition	Tension (lb/in²)				Compression (lb/in²)			
	Days				Days			
	1	7	28	90	1	7	28	90
C_5A_3	—	261	264	—	—	1750	1300	—
Approaching CA	370	403	435	440	3570	5945	3765	3945

Tensile briquettes of U.S.A. semi-plastic mortar type. For such briquettes compare the U.S.A. specification requirements for Type III Portland cement of 275 and 375 lb/in² at 1 and 3 days.

Tests[1] on pure CA in a 1:2 mortar with a w/c ratio of 0·6 have given compressive strengths of 8700 and 10 100 lb/in² at 1 and 7 days respectively. Some comparative data[2] on the strength of the neat aluminate compounds (w/c 0·3) at 21° are shown below.

Compound	Compressive strength (kg/cm²)			
	Days			
	1	3	7	14
$C_{12}A_7$	200	225	230	250
CA	700	850	1000	1200
CA_2	30	60	100	250

The pure ternary compound C_2AS reacts with water only very slowly and exhibits no cementitious properties. When mixed with 10 per cent lime it develops strength.

Cement zones in the system $CaO–Al_2O_3–SiO_2$

The different cement zones in the system $CaO–Al_2O_3–SiO_2$ are shown in Fig. 36. In representing the Portland cement area the ferric oxide has been calculated in with the alumina and the sum of the lime, silica, alumina and ferric oxide made up to 100 per cent. The calculation of the ferric oxide as alumina for representation in this diagram is not unobjectionable since their combining ratios with lime are different. The one acts, however, as a substitute for the other as a flux to reduce the burning temperature, and it therefore seems preferable to conform to the common practice and add the alumina and ferric oxide together when representing the Portland cement zone.

It will be observed that the area covered by Portland cement compositions is relatively small, and indeed almost all modern cements fall in the part of this zone lying on the high lime side of the 65 per cent CaO line. Cements lying

[1] F. M. Lea, *J. Soc. chem. Ind., Lond.* **59,** 18 (1940).
[2] H. Lehmann and K. J. Leers, *Tonind.-Ztg. keram. Rdsh.* **87,** 29 (1963).

within the zone, but on the low lime side of this line, are rather slow in hardening, owing to their small contents of tricalcium silicate; they yield good strengths at long ages. Some Portland cement specifications specify a lower limit to the weight ratio of lime to alumina and silica of about 1·7. This corresponds to a lime content in the system $CaO–Al_2O_3–SiO_2$ of 63 per cent. If allowance be made for the presence of about 2–3 per cent ferric oxide, the minimum lime content is about

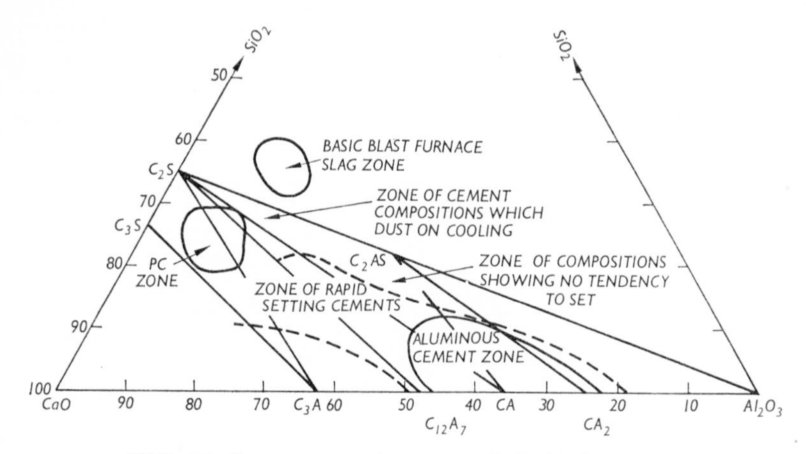

FIG. 36 Cement zones in system $CaO–Al_2O_3–SiO_2$.

60 per cent, which is seen in Fig. 36 to mark the lower lime limit of the Portland cement zone. The zone is limited on the high-lime side by the necessity of keeping the composition such that all the lime can be combined in the clinker. On the low-alumina side the zone is limited rather by difficulties of burning, due to the high clinkering temperatures involved, than by any composition requirements. The white Portland cements lie on this side of the zone. The limit on the high-alumina side is determined by the increasing rapidity of set of the products. Portland cements with $SiO_2:R_2O_3$ ratios much below 1·5:1 begin to show a rapid set which cannot be controlled by the addition of gypsum, owing to their large contents of tricalcium aluminate. If, however, the increased R_2O_3 content is made up mainly of ferric oxide, a somewhat lower $SiO_2:R_2O_3$ ratio is possible, for the compound C_4AF does not show such rapid setting properties as C_3A.

The zone is limited on the low-lime side, as mentioned earlier, by the decreasing cementing qualities, and also by increased trouble from dusting due to the transformation of β to γ C_2S on cooling. This dusting becomes more evident when the low lime content is combined with a low alumina content, and a correspondingly high silica content.

Modern Portland cements fall within the triangle $C_3S–C_2S–C_3A$ and only earlier cements fall outside into the $C_2S–C_3A–C_{12}A_7$ triangle or, at the limit, into the $C_2S–C_{12}A_7–CA$ triangle. The general trend in Portland cement manufacture has been towards higher lime contents, and cement compositions have thus moved into regions of higher tricalcium silicate content.

The high-alumina cement zone shown in Fig. 36 is based on various experimental studies[1] which have been made with pure lime-silica-alumina mixes. It is semi-elliptical in shape, the major semi-axis falling roughly on a line joining CA with C_2S. Within this field some of the compositions yield cements of higher strengths than even the pure aluminates CA and $C_{12}A_7$. In this ternary system of three oxides the high-alumina cement zone covers an area such that three of the compounds CA_2, C_2AS, CA, C_2S, $C_{12}A_7$ are present at equilibrium, the three being taken in succession. The shape of the zone seems to be conditioned by the fact that C_2S has good, though slow-developing, cementing qualities, whereas C_2AS has questionable, or much less, cementing value though in commercial cements its reactivity may be improved by other oxides in solid solution. Cements to the right of the $C_2AS–CA$ line will contain all their silica as C_2AS whereas cements to the left of this line will contain increasing proportions of C_2S.

From the high-alumina cement zone proper there spreads out a long tongue, indicated by dotted lines, towards the Portland cement zone. Cements can be obtained in this area, but their setting times are too rapid for practical use; as the Portland cement zone is approached a region is passed through in which the cementing value is almost zero. The compositions lying above the upper of these dotted lines have no cementing properties, as would be anticipated since they are in the region of a C_2AS composition. The cementing properties of compositions lying to the right of the high-alumina cement zone becomes more inferior as the alumina content rises. No adequate data are available for compositions lying below the lower of the two dotted lines, between it and the lime-alumina boundary side. It seems, however, that no useful cements are likely to be obtained from this area since they would contain a large proportion of C_3A and would show a very rapid or flash set.

Cementing properties in the system $CaO–SiO_2–Fe_2O_3$

The iron ore, or Erz cements which are composed essentially of lime, silica and ferric oxide, show all the physical properties of Portland cement with the exception of a somewhat lower rate of development of strength. While Erz cement normally contains a small amount of alumina, cements have been obtained composed only of lime, silica and ferric oxide, and showing similar strengths, setting times and soundness to Portland cement. The field from which such cements can be produced was found by Kühl[2] to be rather wider than that of the Portland cement field in the $CaO–Al_2O_3–SiO_2$ system. Thus with Portland cement the limiting $SiO_2:R_2O_3$ ratio is about 1·5:1, but when only ferric oxide is present it is possible to work down to a $SiO_2:Fe_2O_3$ ratio of 1·2:1 before setting becomes uncontrollable. This is to be attributed to the formation of $2CaO.Fe_2O_3$ in this system, which does not show the flash setting characteristic of $3CaO.Al_2O_3$. There is also an extension of the field in the direction of higher values of the $SiO_2:R_2O_3$ ratio than are possible with ordinary Portland cement. This may be due to the temperature of initial liquid formation being nearly

[1] Berl and Loblein, *Zement* **15,** 642 et seq. (1926); Bates, loc. cit.; Endell, loc. cit.
[2] *Zement* **13,** 25 et seq. (1924).

50° lower in $CaO–Fe_2O_3–SiO_2$ mixes than in those of $CaO–Al_2O_3–SiO_2$, though at higher temperatures the amount of liquid formed is less with ferric oxide mixes than with alumina. The limit of lower lime content is also somewhat decreased owing to the difference in the combining ratios of alumina and ferric oxide. In Portland cement the alumina is present as C_3A, one part of alumina requiring 1·65 parts of lime, but in the Erz cements where C_2F is probably formed, one part of ferric oxide requires only 0·70 part of lime. Hence more lime is left to combine with the silica and the minimum permissible lime content is lowered. The maximum lime content must also be similarly decreased.

6 The Constitution of Portland Cement

The problem we have to consider in this chapter is how the component oxides which make up Portland cement are combined together in the clinker after burning. In a previous chapter we have seen that in a mixture of lime, alumina, silica and ferric oxide of Portland cement composition which has attained equilibrium the compounds formed are C_3S, C_2S, C_3A, $C_{12}A_7$ and a ferrite often close in composition to C_4AF. We have further seen that the cementing qualities of these compounds are such as when mixed to reproduce in a general way the characteristic properties of Portland cement. Phase equilibrium studies have also indicated how the minor components present in commercial raw materials can enter, to a limited extent, into solid solution in the major compounds or form modified compounds such as NC_8A_3. It might seem that the evidence was almost sufficiently conclusive to render further discussion unnecessary, but we still need to show that these compounds can be identified in Portland cement in the relative proportions expected and to examine in what way commercial conditions of manufacture may lead to any departures from the equilibrium products. There was, indeed, for many years, a strong school of thought which denied the validity of the conclusions drawn from the phase equilibrium studies and asserted that Portland cement was not an equilibrium product, and that present in it there were compounds and solid solutions which are not met in such studies.

Prior to the publication of Rankin and Wright's pioneer work on the system lime-alumina-silica there was no definite evidence as to what compounds could be formed in mixes of Portland cement composition and hence all the earlier work on the constitution of cement was hindered by lack of this vital information. Speculative views as to the probable compounds formed were prevalent and these older views had much subsequent influence on the continental schools of thought. Chemists in the United States, more closely in touch with the work of Rankin and Wright, were, under the leadership of Bates, the first to study the detailed application of the phase equilibrium data to cement.

Portland cement is a product made by sintering the raw materials at a temperature such that only a portion, less than one-third, of the material becomes liquid. Now chemical reactions between solid substances, even when they are very intimately mixed, tend to proceed only slowly and it has often been argued that under such conditions equilibrium would not be approached. The possibility and desirability of making Portland cement by complete fusion of the mix has often been suggested, but, apart from the difficulty of attaining commercially the high temperatures that would be required, there are, as will be seen in the next chapter, certain objections to complete melting in manufacture.

There are two ways in which a cement might depart from the equilibrium state. The first may perhaps be termed a quantitative, and the second a qualitative, departure from equilibrium. The reactions during burning between the lime and the clay may form the normal compounds C_3S, C_2S, C_3A, $C_{12}A_7$ and C_4AF which have been found to be formed from mixes of Portland cement composition in the $CaO–Al_2O_3–SiO_2–FeO_3$ system, but, owing to their slowness, these reactions might not be quite complete. This would lead to a deficiency in the amounts of the most basic lime compounds formed, and particularly of tricalcium silicate, and to the presence of a corresponding amount of uncombined lime in the clinker. The amount of uncombined lime would be a measure of the inefficiency of the burning. An increase in the burning time or temperature, or a decrease in the proportion of lime present, would result in a closer approach quantitatively to equilibrium. Such quantitative departures from equilibrium would only change the properties of the cement in so far as the presence of free lime affected it, since all the other compounds present would be the normal equilibrium compounds, though their proportions would be somewhat changed. By burning so that the free lime content was quite low the cement would almost represent an equilibrium product. There could also be a departure from equilibrium during the cooling of the clinker from the burning temperature, but again this would not cause the formation of new compounds but only the substitution of a frozen liquid (glass) for some crystalline materials and an alteration in their relative proportions.

Alternatively it might happen, as was in the past contended by many investigators, that the compounds formed in cement burning differed radically from the equilibrium compounds, and that, owing to equilibrium not being attained, unstable compounds were present in the cement clinker which never occurred in studies on systems in equilibrium; that is, the cement clinker would differ qualitatively in the type of compounds present from the equilibrium state. In this connection it may be noted that studies on phase equilibria often show not only the compounds present at equilibrium, but also any unstable compounds which are formed during the approach to equilibrium. Further, if quite different compounds are formed in cement burning it might reasonably be supposed that they would tend to be low in lime, thus leaving much uncombined lime in the clinker. This is not true of a properly burnt cement. That equilibrium can be reached by merely sintering can also be shown by laboratory experiment.

There is another way in which cement clinker can depart from strict chemical equilibrium; because of inefficient grinding and mixing of the raw meal there may be local excess of the calcareous or silicious component, or absorption of ash or gaseous substances from the kiln atmosphere may produce surface skins of differing chemical composition to that of the bulk clinker. Such variations from homogeneity are easily detected by methods to be described.

Our purpose in this chapter then is to see whether Portland cement resembles the equilibrium product, differing perhaps slightly therefrom in the quantitative sense, or whether it is qualitatively different.

Portland cement clinker as ordinarily prepared is not a homogeneous substance, but a rather fine-grained mixture of several solid phases. It is therefore difficult to draw any conclusions from a study of its chemical reactions alone, since these may involve more than one constituent, and the only methods capable of yielding trustworthy results are those which enable us to deal with the individual constituents separately. The conditions have a close parallel in the case of the study of igneous rocks. It would be impossible to determine the structure of a granite by observing its gross behaviour towards reagents. The reactions with the quartz, felspar, and mica would be superimposed and confused and the resultant action would give only a meaningless average. A number of methods are available for overcoming these difficulties, such as those of mineral separation after crushing, optical examination of the crushed material or of thin section or of polished and etched surfaces, the use of X-rays, and electron microprobe analysis.

A rock may be crushed to such a degree of fineness as to release the constituent crystal grains and the powder thus obtained suspended in liquids of suitable densities, so bringing about a separation of the light and heavier minerals. In skilled hands the method is capable of giving very accurate results. It has been applied to cements, but with less success, as the close intermixture and friable character of the constituents render a separation impossible until the whole has been reduced to a fine powder, when the particles no longer settle satisfactorily after suspension in heavy liquids. An improvement may be obtained by effecting the separation of the crushed material by centrifugal or magnetic means, but even then a quite clean separation has not been effected. Microscopic examination of the crushed material shows that many of the grains are composed of one constituent, but that attached to their edges may remain fragments of a second constituent, thus rendering a perfect separation impracticable. Such partially successful separations, using chemical or physical methods, can be a useful preliminary to X-ray examination since the intensity of the reflexions obtained for a given constituent will be enhanced.

The microscopic examination of the crushed cement does, however, permit of certain optical properties, such as refractive index, being measured on the separate mineral constituents and their form or crystalline habit being observed. The methods for determination of refractive indices and other optical properties under the microscope afford a very useful means of attack. Examination under the microscope of a mineral grain immersed in a liquid of the same refractive index as itself shows that it almost, or entirely, disappears from view, but that if the liquid is of different refractive index the mineral grain stands out surrounded, under suitable conditions of illumination, by a thin band of light, known as the Becke line. If the mineral is of higher refractive index than the liquid, and the objective is racked downwards from the point of good focus, the band of light shifts outward towards the liquid. Conversely, if the mineral is of lower refractive index than the liquid, the band will move inward away from the liquid. By using a series of liquids of varying refractive index the refractive index of a mineral grain may be determined. The method requires practice.

The material to be examined should be crushed rather than ground to such a size as to pass a 100–180 mesh to the linear inch. A very small amount of the crushed material is placed on a microscope slide, a drop of an R.I. liquid placed on it, and then a cover glass which is pressed down lightly to spread out the powder and eliminate air bubbles. The slide is examined under a fairly high-power objective of 4 mm or 6 mm focal length in conjunction with a substage condenser, an iris diaphragm stopped well down, and a substage Nicol prism producing plane polarised light. In a preparation which is composed of a mixture of minerals, or as we shall see shortly of even a single anisotropic mineral, a number of liquids will be found in which the different minerals disappear, depending on their refractive indices. It will also often be possible to distinguish minerals one from another in a mixture by their form and behaviour under crossed Nicols.

Crystalline substances fall into two main groups. The first group consists of isotropic crystals which, however orientated, do not pass any light when placed between crossed Nicols. When isotropic materials, either crystals, glasses, or amorphous products, are examined by light which is plane polarised so that all its rays are vibrating to and fro along only one direction, the refractive index remains unchanged however the material is orientated with respect to the plane of polarisation of the incident light. Hence if an isotropic grain is immersed in a liquid in which it disappears, it will continue to do so if it is rotated by turning the microscope stage. The other main group of crystals are anisotropic, i.e. they allow light to pass when placed between crossed Nicols. Such crystals show different refractive indices as their orientation with respect to the plane of polarisation of the incident light is changed. An anisotropic grain which disappears in one position in some particular liquid will no longer do so if the stage on which it rests is rotated. It is found with such crystals that when orientated in one particular direction they show a maximum refractive index and in another direction at right angles to it a minimum index. This main group of anisotropic crystals is subdivided into two classes of uniaxial and biaxial crystals. For uniaxial crystals two refractive indices are obtained, one termed omega (ω) and one epsilon (ε). When $\varepsilon > \omega$ the crystal is said to be optically positive and, when the reverse holds, negative. For one particular direction of passage of light uniaxial crystals become isotropic and when so orientated show only their ω index. When in this position on the microscope stage the latter can be rotated in a plane at right angles to the incident light without altering the refractive index of the material, or it becoming anisotropic. Biaxial crystals are characterised by three refractive indices to which are assigned the letters α, β and γ. β always lies between α and γ and often therefore only these two extreme values are given. For biaxial crystals there are two directions of passage of light through them for which they become isotropic.

The difference between ω and ε, or α and γ is termed the birefringence of a crystal and is usually fairly small in inorganic compounds, though there are many notable exceptions, e.g. calcite. Anisotropic crystals mounted on a microscope slide will normally be found orientated in a variety of positions such that the difference between the two refractive indices exhibited varies from zero to a maximum value determined by the birefringence of the material. When the difference is zero the grain appears isotropic, while it is a maximum in grains showing, for their size, the highest interference colours under crossed Nicols. It is necessary to determine for a given mineral the maximum and minimum indices shown by any grain. The interference colour is determined by the product of the depth of a crystal and the difference between the two refractive indices characterising the plane of the crystal which is at right angles to the direction of the light transmitted through it. It thus varies with the size of the grain whereas the refractive indices do not.

The differentiation of minerals in a mixture by optical means[1] is only possible if their refractive indices differ sufficiently, or if their other properties, such as crystal habit or behaviour under crossed Nicols, are different.

Much the same methods are employed in examining thin sections of cement clinker, but the refractive indices of a grain can only be compared with that of adjacent grains or of the resin used in mounting the slice. If the thickness of the section is less than the grain size of a constituent, an accurate estimate of the retardation (thickness × birefringence) can be made provided that one of the constituents may be used as standard. A thin section examined at low magnification is valuable in revealing any homogeneity across the clinker pellet.

Preparation of thin sections of cement clinker

A small chip of material may be obtained by breaking a piece of clinker with a hammer or a thin slice by cutting with a diamond saw. The diamond saw consists of either a slow-running soft-iron disc which has diamond dust pressed into its rim, or a fast-running resin-bonded diamond saw. As the material to be cut is hydraulic, water cannot be used as a lubricant but liquid paraffin or ethylene glycol are suitable.

With porous materials, such as Portland cement clinker, it is necessary to impregnate with resin. Even with a dense material, setting in resin is a convenient method of obtaining a specimen which can be handled easily and marked for identification. Moreover it is often convenient to slice through the impregnated specimen, keeping one cut face for preparing a thin section and using the other for polishing and etching. The material, whether slice, lump or about 6 small chips, is placed in a glass specimen tube of $\frac{3}{4}$-inch diameter in a desiccator fitted with a two-holed rubber bung in the cap, through one hole of which a separating funnel fitted with a tap is inserted while the other is used for evacuating. The desiccator is held evacuated for 1 hour and then resin (Bakelite R0014) is run in to the specimen tube from the funnel until the sample is covered. The specimen tube and contents are then removed, and heated in an oven for 4 hours at 50°, followed by 24 hours at 100°. Cold setting epoxy resins may also be used according to the makers' instructions. The hardened resin, containing the

[1] It is not possible to discuss here in any detail the optical properties of crystals. For further information one of the numerous textbooks devoted to the subject should be consulted. Lists of liquids suitable for use in refractive index work can be found in the catalogues of chemical suppliers or in the textbooks.

specimen, is removed from the tube by breaking the glass. The next stage is to grind the specimen until one flat surface of sufficient area is obtained. A rotating metal lap can be used for this grinding with 60 minute carborundum, and then 120 minute carborundum, as abrasive. When the surface is judged to be flat, it is cleaned of all grinding powder, and is then ready for final polishing. This is done by hand on a glass plate, using 4F carborundum powder, or on a plastics plate using $\frac{1}{4}\,\mu$ diamond dust. Some workers prefer to use diamond impregnated laps for all stages. When a perfectly even surface is obtained the specimen is cleaned well and left to dry. It is then ready for mounting. To mount the specimen a hot-plate maintained at 100° is required. A 3 inch × 1 inch glass microscope slide is placed on the hot-plate, and a small quantity of filtered Canada Balsam is placed on the microscope slide and left to 'cook'. The state of the cooked balsam is critical and it should be tested at frequent intervals. A small drop of the cooking balsam is taken on the end of a needle and pressed against the thumb nail; when it shatters on slight pressure the balsam is cooked sufficiently. When the balsam is judged to be ready for mounting, the specimen is lowered into the balsam, polished face downwards, and then removed from the hot-plate and pressed to remove any entrapped air bubbles. It is then left to cool. Various synthetic substitutes for Canada balsam (e.g. Lakeside 70 C)[1] are available and are easier to use. The cooking temperature may be higher, perhaps 140°.

The mounted specimen is now ready for final grinding. This is done on the metal lap using coarse 60 minute and 120 minute carborundum powder. The section is ground until it is just transparent, great care being required to keep the thickness of the thin section even and not to break away the surrounding layer of balsam as this is the only protection for the edges of the now thin section.

When the section is just transparent, it is cleaned of all grinding powder and the final grinding carried on by hand as described above. As the section becomes thinner, repeated examinations on a petrological microscope are made, and when the minerals give the desired interference colours under crossed Nicols the grinding is ceased. It is usual to make sections about 20 μ thick; this is the thickness which will cause $3CaO.SiO_2$ to have white-grey polarisation colours.

A cover glass should be placed on top of the finished section for protection. This can be done by applying a small blob of Canada balsam or substitute to a thin cover glass whilst on a hot-plate; the finished slide is also placed on the hot-plate and the cover glass placed on top of the specimen, balsam side down. When all the balsam has covered the specimen the slide is removed from the hot-plate and finally pressed into position and then left to cool. The time the slide is on the hot-plate while mounting the cover glass is sufficient to fix the cover glass so that it is rigid when cold. It is therefore not necessary to cook the balsam in mounting the cover glass. When cool the surplus balsam may be removed by a solvent, such as methylated spirits or benzene. The section is then ready for examination on the petrological microscope.

[1] Obtainable from Hugh Courtright and Co., 7600 Greenwood Avenue, Chicago 19, Ill., U.S.A.

Preparation and examination of polished surfaces

An important method that has been used to examine cement clinker is that commonly used in metallography in which a polished and etched surface is examined by reflected light. The cement clinker is first mounted in a plastic resin, then ground flat by successively finer grades of carborundum powder on rotating steel laps, and polished on cloth or metal or paper-covered plastics laps with polishing powders such as rouge, diamond powder or freshly ignited magnesia. When polishing Portland cement clinker it is advisable to use alcohol or other inert liquid for the final stage of polishing. Alternatively a lap impregnated with $\frac{1}{4}$ μ diamond dust may be used for all stages following the first levelling of the surface. The surface of a thin section can also be polished, making it possible to identify the minerals by their optical properties under transmitted light and to compare the etch properties of the same grain under reflected light. A thin section, about 20 μ, is first prepared but without a cover glass. A cork or wood support is temporarily cemented to the back of the slide with shellac and the slide is cut so that it only extends to about $\frac{1}{4}$ in. beyond the actual section. The top surface is then polished, preferably on a diamond impregnated lap at not too high a speed.

The refractive indices of minerals in a polished surface may be estimated by measuring their reflectivity and light absorption, but normally, when the specimen has been polished, it is etched by various liquids before examining under a metallographic microscope.

Etching methods

The etch reagent is poured into a small watch glass, and the specimen rapidly immersed, polished surface down, care being taken to avoid trapping air bubbles on the surface. The watch glass is gently rocked during the etching period. The specimen is then removed and rinsed, in alcohol if the etch reagent is made up in alcohol, otherwise in water. Many of the etch reactions depend on the formation of a thin film on the surface of the crystals, which produces interference colours. Care has to be taken in drying the specimen not to scratch this film. The best method is to dab gently on lens cleaning tissue, or a soft cloth. Where the etch acts by removal of a surface film, the surface is not harmed by gentle wiping. A method sometimes used is to expose the surface to hydrofluoric acid vapour. The specimen is placed face down over a crucible of lead or platinum half filled with the acid.

When using high magnifications, trouble may be caused by the objective misting over. This can be avoided by a final drying of the specimen with a jet of warm air.

When an unetched specimen of Portland cement clinker is examined, two constituents, appearing white against the grey groundmass, are differentiated owing to their high reflectivity. The commoner is $4CaO.Al_2O_3.Fe_2O_3$ which occurs as small needle- or spear-shaped crystals. The other constituent, MgO, is rare. It tends to occur as squares or triangles, often rounded at the angles.

Free CaO may also be seen in unetched specimens, being darker in colour than the groundmass, and often scratched owing to its relative softness. If the specimen is buffed, the free CaO will be pitted. Free CaO is etched after about two seconds by distilled water, appearing pitted and displaying interference colours.

$3CaO.SiO_2$ and free CaO are etched by immersing the specimen for 3 hours in a mixture of equal volumes of ethylene glycol and absolute alcohol. Another method which is quicker, and therefore in some respects more convenient, is to immerse the specimen for 10 seconds in a 10 per cent solution of magnesium sulphate in water, followed by a rapid wash in water, and then alcohol, and drying. A disadvantage is that the reagent also etches the groundmass, but with experience this ceases to be a difficulty.

$2CaO.SiO_2$ is etched to a blue or red colour (depending on the time of etching) by holding the inverted specimen over a crucible containing hydrofluoric acid (40 per cent solution) and allowing the vapour to come in contact with the polished surface. About 10 or 20 seconds will bring about the blue stage, somewhat longer etching producing the red colour. At this later stage differentiation of the groundmass also occurs, whilst $3CaO.SiO_2$ is etched to a straw colour.

A useful general etching method is: the specimen is immersed in distilled water for 5 seconds, washed in alcohol, and then immersed in a 0·25 per cent solution of nitric acid in alcohol. The surface is finally washed in alcohol and dried by blotting. The acid colours the calcium silicates, while the water affects mainly the $3CaO.Al_2O_3$ and some glass compositions, the $4CaO.Al_2O_3.Fe_2O_3$ remaining unattacked. There is some doubt as to the effect on glass, but Nurse and Parker[1] concluded that glass compositions of high iron content are not etched, while those of low iron, but high alumina, content are etched by this reagent. Etching by KOH solution is claimed to etch successfully the high-iron glass, but not the crystalline ferrites.[2]

A typical polished surface of Portland cement clinker etched with water followed by 0·25 per cent nitric acid in alcohol is shown in Plate II (i). Large prismatic crystals of $3CaO.SiO_2$ etched light grey and rounded darker striated crystals of $2CaO.SiO_2$ may be seen. The groundmass is principally $4CaO.Al_2O_3.Fe_2O_3$ with a little grey interstitial matter that is probably glass. One crystal of $3CaO.SiO_2$ encloses a small triangular crystal of MgO. Plate II (ii) shows a polished surface etched with hydrofluoric acid vapour. The $3CaO.SiO_2$ occurs as irregular prismatic grains, light grey in colour. The $2CaO.SiO_2$ is darker and occurs in three types: large rounded and striated crystals, irregular masses 'fingered' at the edges, and groups of small dots. The latter have been formed by exsolution during crystallisation of the glass. The groundmass is crystalline, the light grey being $3CaO.Al_2O_3$ and the white $4CaO.Al_2O_3.Fe_2O_3$.

Mineral separation techniques

Many problems of the constitution of materials such as natural rocks have been solved by the physical or chemical separation of the constituents, and attempts have been made to apply such methods to cement clinker. The conventional

[1] T. W. Parker and R. W. Nurse, *J. Soc. chem. Ind., Lond.* **58,** 255 (1939).
[2] H. Insley, *J. Res. natn. Bur. Stand.* **25,** 295 (1940).

methods depend on making use of some property such as density, magnetic susceptibility, dielectric constant, solubility, which will enable a specific constituent to be purified and separately analysed. Only partial success has been achieved with cement because of the small crystals formed and their inhomogeneity and range of properties.

Tricalcium silicate grows to a larger size than any other constituent and Guttmann and Gille in their classical investigation of alite succeeded in separating enough for analysis by centrifuging a graded clinker powder in heavy liquids. Midgley[1] separated a sample of C_2S by partial hydration of a clinker, leaving the slowly hydrating C_2S to be obtained as a residue after acid extraction of the set cement. Many workers[2] have used an alkaline solution of a dimethylamine salt to dissolve the silicate phases, leaving a concentration of other constituents. A solution of salicylic acid in methanol has also been used.[3] These methods, or alkaline ammonium citrate or acetic acid,[4] have been used to intensify the X-ray reflections from the ferrite phase in order to obtain its composition from the d-spacings. These solubility methods depend on differences in the rate of solution of the various compounds and require therefore a nice choice of the extractions conditions. A bar magnet has also been used to separate the ferrite phase.[5]

The Portland cement minerals

The author whose publications had the greatest influence in determining the original ideas held as to the constitution of Portland cement is Le Chatelier,[6] whose first paper[7] appeared in 1883, yet whose views had been strikingly confirmed in the modern theory. It is to him that we owe the application to cement of the methods of petrography as employed in the study of rocks. The entirely independent work of Törnebohm,[8] who gave distinctive names to the component 'minerals' observed in thin sections of cement clinker, confirmed Le Chatelier's observations and provided the classification which was to form the basis of much subsequent work.

Both Le Chatelier and Törnebohm observed four different kinds of crystals in thin sections of cement clinkers and to these Törnebohm gave the names alite, belite, celite, and felite. Alite, the most abundant constituent, consisted of colourless biaxial crystals of weak birefringence, rectangular or hexagonal in outline. Belite consisted of small rounded grains of no definite crystal form having a darker colour and a higher birefringence, showing bright interference colours. The grains were biaxial and often striated. Celite was recognised by its

[1] H. G. Midgley, *J. appl. Phys.*, 1952, **3** (9), 277.
[2] F. Keil and F. Gille, *Zement* **28,** 429 (1939); F. Kiel, *Sym., London 1952*, 555.
[3] G. Yamaguchi and S. Takagi, *Sym., Tokyo 1968.*
[4] N. Fratini and R. Turriziani, *Ricerca scient.* **26,** 2747 (1956); N. Fratini, *Industria ital. Cem.*, 1956 (12) ,286; H. G. Midgley, D. Rosaman and K. E. Fletcher, *Sym., Washington 1960*, 69.
[5] H. G. Midgley, *Mag. Concr. Res.* **10** (28), 13 (1958).
[6] Le Chatelier's work was collected together in *Experimental Researches on the Constitution of Hydraulic Mortars.* Translated by J. L. Mack. New York, 1905.
[7] *C. r. hebd. Seanc. Acad. Sci., Paris* **96,** 1056 (1883).
[8] *Tonind. Zeit.*, 1897, 1148; *Uber die Petrographie des Portland Zements*, Stockholm, 1897.

dark orange-yellow colour and its high birefringence. In well-burnt clinker it formed the filling material and magma from which the alite had separated. Felite was described as a colourless biaxial compound with strong birefringence occurring in a rounded, or elongated, striated form. It was found in variable quantity in different clinkers and was often entirely absent. It seemed to replace belite. In addition to these four minerals Törnebohm observed an isotropic colourless glass of high refractive index which also occurred as a filling material between the alite grains.

Le Chatelier made the interesting observation that the 'grappiers,' or hard nodular masses which resist slaking, occurring in certain of the French hydraulic limes, and which yield a cement when ground, consist almost entirely of alite with a minimum quantity of inter-crystalline material. The analyses of these grappiers were:

	SiO_2 %	CaO %	Al_2O_3 %	Fe_2O_3 %	CaO : SiO_2 ratio
Grey grappiers. Paviers	26·0	66·0	3·0	1·2	2·75 : 1
Grey grappiers. Teil	26·0	66·0	3·5	0·8	2·75 : 1
Green grappiers. Senonches	25·5	68·0	3·6	0·7	2·85 : 1

the last column giving the molecular ratio of lime to silica. As this ratio was not far below the 3 the conclusion drawn was that alite was essentially tricalcium silicate, retaining only a small quantity of aluminate as an impurity.

The nature of the minerals classified by Törnebohm and particularly that of alite, attracted much attention. Le Chatelier's view that tricalcium silicate was the main cementing constituent of Portland cement did not meet with general acceptance, and alite was regarded by many investigators as a solid solution of calcium aluminates and silicates[1] because analyses of various samples, separated from Portland cements, showed alumina contents, varying from 3 to 8 per cent.[2] Prior to the final publication of the work of Rankin and Wright, Bates[3] had identified in Portland cements, by means of their optical properties, the compounds C_3S, C_2S, C_3A and, occasionally, $C_{12}A_7$. Despite this identification the applicability of the work of Rankin and Wright to Portland cement was not accepted by most European investigators and it was not until 1931 that anything like general agreement was reached.

Jänecke, following his claim in 1912 to have discovered a compound $8CaO.Al_2O_3.2SiO_2$, suggested that this was alite. In the succeeding twenty years many papers appeared in the European, and particularly the German, literature on the nature of alite, and in almost all it was regarded as containing alumina. Dyckerhoff,[4] however, suggested that it consisted of a solid solution of lime in β C_2S which slowly, but only partially, became converted to C_3S during burning.

[1] A detailed account of the earlier work on the constitution of Portland cement was given by R. H. Bogue in *Concrete* (U.S.A.), July, 1926, to February, 1927.
[2] A. Guttmann and F. Gille, *Zement* **16,** 921, 951 (1927); **18,** 912 (1929); **20,** 144 (1931).
[3] *Concrete Cement Age* (*Cement Mill Section*) **2** (3) (1913).
[4] *Zement* **11,** 245, 257 (1922); **14,** 174, 200 (1925); **16,** 731 (1927).

General agreement that alite was essentially C_3S was reached in the years around 1930. Guttmann and Gille obtained by centrifuge separation a purer sample of alite than had previously been obtained and found it to have a molecular composition, after correction for identified impurities, of 3·00 CaO, 0·04 MgO, 0·02 Al_2O_3, 1·0 SiO_2. While optical and X-ray examination confirmed the conclusion that alite was essentially C_3S, later work indicated that small amounts of other oxides are taken up in solid solution. This has been demonstrated by phase equilibrium studies by many authors and in recent years by electron microprobe analysis of the minerals present in cement clinkers (see p. 107).

A re-examination of Törnebohm's original slides which had been preserved by the Swedish Geological Survey led Assarsson and Sundius[1] to the conclusion that the alite corresponded optically to C_3S, belite and felite to C_2S. The isotropic material was for the most part C_3A and $C_{12}A_7$. There remained, however, doubt as to which form of C_2S was present, and Sundius[2] tended to the view that belite was α C_2S and felite β C_2S. In view of the confusion which then existed as to the polymorphic forms of C_2S it is hardly surprising that some uncertainty remained. The celite which occurred as one of the interstitial materials between the alite and belite crystals was clearly a compound containing iron and is now to be identified with the C_4AF–C_6A_2F solid solution. In the light of this work and the description given later of the constituents of cement clinker we can now identify Törnebohm's minerals as:

Alite	C_3S
Belite	C_2S, usually β
Felite	C_2S, usually β
Celite	C_6A_2F–C_6AF_2 solid solution
Isotropic matter	C_3A, $C_{12}A_7$ and glass

About the same time as these conclusions were being reached Brownmiller and Bogue[3] examined many cement clinkers by X-ray analysis. In 28 commercial clinkers examined these authors were able to identify C_3S and C_2S in all, and furthermore their patterns were always the most prominent, indicating that they were the most abundant constituents. One or more of the compounds C_3A, C_4AF and MgO were identified in most of the clinkers, but not all of these latter compounds were identified in all clinkers.

This short historical account has shown how the combination of evidence from phase equilibria studies, optical and X-ray examination, and chemical analysis had, by soon after 1930, established the main outlines of the constitution of Portland cement. Later developments have, as we shall now see, added much in detail to this general picture without disturbing its broad outline.

Microscopic examination of modern Portland cement clinker usually shows clearly several constituents. By transmitted light C_3S and β C_2S are seen embedded in interstitial material which, under crossed Nicols, can often be seen to contain the ferrite phase. In polished and etched surfaces the interstitial matter can be further resolved.

[1] Swedish Geol. Survey, Ser. C, No. 357, Årsbok 23 (1929), No. 2, Stockholm.
[2] *Z. anorg. allg. Chem.* **213,** 343 (1933).
[3] *Am. J. Sci.* (v) **20,** 241 (1930).

Photomicrographs by transmitted light are shown in Plates 3 and 4. Plate 3 (i) shows a clinker composed almost entirely of white prismatic grains of tricalcium silicate with scattered brown interstitial matter. Plate 3 (ii) shows a central mass of darker rounded dicalcium silicate grains and some white tricalcium silicate towards the edge of the picture. The dicalcium silicate grains are surrounded by, and set in, the dark brown interstitial matter; they appear much darker in the photomicrograph than the tricalcium silicate owing to their much higher light absorption. When examined visually under the microscope the contrast does not appear so definite, but there is a marked difference in transparency between the tri- and di-calcium silicates. The former are transparent and white, while the dicalcium silicate crystals have a yellow, or somewhat brownish, colour, depending on the thickness of the thin section, or occasionally a pale green colour. Plate 4 shows the interstitial matter particularly clearly differentiated from the primary crystals.

The size of the silicate crystals and the general appearance of the section vary considerably with different cements and processes. In some clinkers the C_3S and C_2S and sometimes free CaO are aggregated together—compare (i) and (ii) Plate 3—and if the section is examined under low power these aggregations can often be seen to be relicts of coarser particles of the raw meal; if the particle was originally calcite a clump of C_3S with possibly a core of free CaO is seen; if siliceous the relict is usually C_2S. Clinkers differ greatly in their porosity, and skins and cores are sometimes visible which may suggest the action of ash or changes in kiln conditions.

The largest crystals of C_3S or C_2S are about 40 μ, an average size would be 15–20 μ. Tricalcium silicate in thin sections shows grey to white interference colours under crossed Nicols. The optic axial angle is low, especially when Al_2O_3 is present in solid solution, but it is raised by increasing Fe_2O_3 in solution. The clinker mineral is always impure, but it can only be distinguished with certainty from pure C_3S by X-ray methods. In polished surfaces C_3S is specifically etched by ethylene glycol or magnesium sulphate; in some clinkers the edges of the crystals can be seen to be corroded and converted to C_2S by reaction with the liquid during cooling.

Dicalcium silicate rarely shows any crystal form, appearing as rounded grains which in transmitted light show yellow interference colours in a section of normal thickness (about 0·025 mm). The crystals frequently show polysynthetic or inversion twinning which is particularly clear in the polished and etched surface. The powder X-ray diagram, even after concentration of the C_2S fraction, usually only shows lines corresponding to the β form.[1] The α' form may be detected in clinkers usually high in K_2O, or containing P_2O_5. A small amount of the α, or α', form insufficient to show by X-rays could be present as suggested by Metzger,[2] but the optical properties of all three modifications are so strongly affected by solid solution that detection by optical means is very difficult. The optic axial angle found in the clinker mineral is often much smaller than that quoted for 'pure' β C_2S, but the refractive indices and bire-

[1] H. G. Midgley, *J. appl. Phys.* **3,** 277 (1952).

[2] A. A. T. Metzger, *Meddn finska KemSamf.* **62** (3–4), 104 (1953).

fringence support the X-ray findings. Insley[1] has distinguished three types of β C_2S according to their appearance in the polished etched specimen. It has been shown that these arise from the conditions of formation and that the braided appearance of C_2S clinker is paralleled in the natural mineral, larnite. A small amount of C_2S is sometimes found dispersed in the interstitial material as tiny rounded dots (see Plate 2 (ii)); this has crystallised from the liquid or glass during cooling.

Free CaO occurs in clinker as aggregates or single grains, rounded in shape and frequently as large as the C_2S grains. It is not easily distinguished in thin section but is readily seen in the polished surface, being soft and easily scratched. It is readily etched by water or moist air and then appears dark, and when over-etched, pitted. In powdered clinker or cement free CaO is readily detected under the microscope by White's test (see p. 108). It can be determined quantitatively by analysis (p. 108) or by X-ray methods.

MgO or periclase is not affected by the etches commonly employed and always appears bright in the polished surface. It is found in association with all the other clinker constituents and tends to exhibit rectangular or triangular forms. It is hard and if the surface is lightly polished on a cloth it stands out in relief. Free MgO can be determined analytically (see p. 111), or by X-ray methods.

The interstitial material is frequently difficult to resolve by microscopic examination. C_4AF or the corresponding ferrite solid solution is recognised in this section by its reddish colour, birefringence and pleochroism. It is usually prismatic in form. An estimate of the position of this compound in the ferrite solid solution series can be made either by measurement of the reflectance in polished specimens or more reliably by measurement of the 200 or 202 reflections in the powder X-ray pattern.[2] The composition of the ferrite can vary from C_6A_2F to C_6AF_2 but the median value is fairly close to C_4AF. The ferrite phase in reflected light appears very bright and is not etched by the reagents commonly used.

Dark interstitial material seen in the polished surface after water etching has been divided into three types; crystalline rectangular, crystalline prismatic and amorphous. The first is considered to be C_3A. Some observers have recognised C_3A in thin sections also. The prismatic dark material is associated with high alkali content and perhaps with abnormal cooling conditions.[3] It is usually included with C_3A in estimating compound compositions. Lines due to C_3A can almost always be seen in the powder X-ray diagram of clinker. Since the X-ray pattern is a fairly simple one of cubic symmetry a small amount of C_3A can be detected by this means; there may be some splitting of its lines when the prismatic phase is present.

Microscopic examination of clinker by the immersion method shows that glass fragments from the same piece of clinker vary in refractive index. In the polished specimen the reflectivity varies in the same way and is also affected by the Al_2O_3:Fe_2O_3 ratio, a high-iron glass tending to be confused with

[1] H. Insley, *J. Res. natn. Bur. Stand.* **17,** 353 (1936).
[2] H. G. Midgley, loc. cit.; D. L. Kantro, L. E. Copeland, C. H. Weise and S. Brunauer, *J. Res. Dev. Labs. Portld. Cem. Ass.* **6** (1), 20 (1964).
[3] S. Chromy and M. Gregor, *Zement-Kalk-Gips* **21,** 451 (1968).

C_4AF and low-iron glass with C_3A. Some workers have used a KOH etch to distinguish between glass and crystalline dark interstitial matter.

Potassium sulphate, or the solid solution with sodium sulphate, and the compound $2CaSO_4.K_2SO_4$ have been identified in polished section[1] of clinker and also in powdered clinker.[2]

In order to obtain a quantitative estimate of the amount of the various phases present, one of the many forms of integrating micrometer stage may be employed. The thin section or polished specimen is carried on a slide which can be traversed along a given line by any one of a number of micrometer screws. Each component is assigned to a definite screw and provision is made for spanning gaps in the specimen. The specimen is scanned in a regular manner, the stage being moved so that each screw measures the total of intercepts on the crosswires corresponding to its assigned constituent. The volume percentages of the constituents are thus obtained. In applying the method to Portland cement clinker, difficulties arise because of the high magnification which has to be employed and because of doubtful identification, particularly of interstitial matter. The 'point counter' system has many advantages. Here the stage is moved a definite distance by operating any one of a number of electrical contacts. Each contact is assigned to a definite compound and as soon as the area on which the crosswires intersect is identified the corresponding switch button is pressed, and the stage automatically moves on. There is much less weight on the stage than in the screw method and the specimen remains in focus. Also there is no occasion to retrace the traverse over a doubtful grain boundary and backlash is therefore eliminated. The difficulty of identification remains, of course.

Much attention has been given to the estimation of compound content by X-ray methods. The pattern obtained from an X-ray photograph or diffractogram of a crystalline substance consists of lines of varying intensity. In a mixture of two or more materials it may often happen that some, though not all, of the lines due to the various compounds present coincide. As the quantity of one of the crystalline materials present decreases the general intensity of the pattern due to it decreases until ultimately a point is reached when even its most intense lines can no longer be observed. The minimum quantity of a crystalline substance which can be detected by X-rays in a mixture depends therefore on the intensity of its strongest lines and the degree to which their positions differ from lines due to other compounds present. In general crystalline materials belonging to the crystallographic systems of higher symmetry, such as the isometric, tend to give strong patterns containing relatively few lines while materials of lower crystal symmetry give weaker patterns and a greater number of lines. A further complication arises from the fact that very high-resolution X-ray spectrometers will break down what is apparently a single strong reflection into a series of closely spaced weaker lines. This occurs with the impure forms of C_3S and C_2S which are found in cement clinker and furthermore the position of these complex reflections varies with the nature of the impurities present. X-ray analysis is therefore an operation requiring much skill and

[1] H. W. W. Pollitt and A. W. Brown, *Sym., Tokyo 1968*.

[2] W. C. Taylor, *J. Res. natn. But. Stand.* **29**, 437 (1942).

experience, especially when, as is the case with Portland cement, reflexions from different minerals coincide or overlap.

Estimates of the possible errors in the estimation of the contents of the compounds in a clinker by X-ray methods which have been given by various workers fall in the following ranges:

C_3S ± 2–5 per cent C_3S
C_2S ± 5–9 per cent C_2S
C_3A ± 0·5 to 1·5 per cent C_3A
Ferrite ± 0·5 to 2 per cent ferrite

A diffractomer is now almost invariably used. The principle is similar to the powder method employing a photographic film, but the specimen is flat. It is automatically rotated and the diffracted energy is measured by a Geiger or other counter. The result appears either as a trace of diffracted energy versus angle turned by the specimen, or as total counts at predetermined angular positions. In either case the diffracted energy for a selected reflection is compared with that of a standard preparation of the substance to be estimated. To check the intensity of the incident X-ray beam an internal standard such as Si or TiO_2 may be mixed with the sample, or a separate check run may be made with an external standard, usually α Al_2O_3. There are many difficulties to be overcome, such as errors arising from the mechanical performance of the equipment, the selection of monochromatic radiation, lack of uniformity in specimen preparation, unavoidable background radiation, and overlapping of lines in the spectra present. For cement clinker a particular error arises from the small variations in the composition of the phases which result in displacement of lines in the X-ray spectra. The output from the diffractometer may be processed directly by computer, and further refinement of such methods will probably lead to increased accuracy. At present many workers consider microscopic point counting more accurate for the silicate phases and prefer X-ray analysis for C_3A, ferrite, CaO and MgO.

Any glass present in the clinker will only appear in the diffractometer trace as a broad band which would normally be counted as part of the background radiation. However, in those methods in which each crystalline constituent is independently analysed, the presence of glass should be seen from the totals being less than 100 per cent. That this is so can be demonstrated by analysing a clinker which has been reheated in the laboratory and quenched in water. There is uncertainty in the case of industrial clinker because the accumulated error of the X-ray analysis is quite high, but most workers conclude that there is little or no evidence for glass among X-ray results. Nevertheless, cooling rate does affect the properties of the cement, particularly magnesia unsoundness and sulphate resistance. It has been suggested that the explanation lies in the 'texture' of the interstitial material; finely-crystalline intermixed interstitial material would seem isotropic in optical examination but crystalline to X-rays. Moreover, the increased rate of reaction of fine textured interstitial material when mixed with water would account for the physical effects of rapid cooling.

The heat of solution method which has been used to estimate glass would equally show up these textural differences because the increased surface energy of small grains would be determined by the calorimetry. Lerch and Brownmiller[1] determined the heat of solution in acid of glass compositions of varying $Al_2O_3:Fe_2O_3$ ratio. Their method for estimating the glass content of clinker was then as follows: the heat of solution of the clinker as received was determined and also the heat of solution of the same clinker after a heat treatment designed to produce complete crystallisation. The difference between these two values, divided by the known heat of solution of glass of the same $Al_2O_3:Fe_2O_3$ ratio as the clinker, gives the glass content of the clinker as received. Correction has to be made for the heat involved in the β–γ inversion of C_2S if this takes place during the annealing process. The results obtained by this method and by microscopic measurement are discussed in Chapter 7; agreement in detail is not very good, although the general conclusions drawn from both methods are the same.

Electron microprobe analysis is now giving much information on the content of minor elements in solid solution in the main clinker compounds. A polished surface or polished thin section of clinker is prepared and the surface is made conducting by evaporating on to it in a vacuum a carbon or metal film. It is then made the target for a focussed beam of electrons, which excites an area about 1 μ diameter and stimulates the emission of X-rays characteristic of the chemical elements present in the irradiated area. These X-rays are analysed at a fixed angle by suitable spectrometers which can be adjusted to give an electrical output proportional to the intensity of X-radiation characteristic of the elements present. The equipment is housed in a vacuum chamber and arranged so that the section may be viewed as a whole with a normal optical microscope; any area or points selected optically can be stimulated by the electronprobe. A general picture of the stimulated area can be displayed on a screen by causing the electron spot to traverse the specimen following a raster which is also locked to the display screen. This display can be a view of the section using scattered electrons, or a view in which intensity is proportional to element concentration at any point. Alternatively, the variation in chemical composition may be displayed along any line on the specimen surface, or for a selected spot. For numerical analyses, calibration with standard substances is necessary. Using this method, analyses of C_3S and C_2S in clinker have been made, but the structure of the interstitial matter is too fine for accurate resolution.

In one of the earliest reports of the electron probe method Miss Moore[2] investigated the prismatic dark phase and showed qualitatively that it contained little TiO_2, the latter being concentrated in the ferrite phase. Peterson[3] found K_2O as K_2SO_4 in the interstitial material and as solid solution in C_2S. He estimated the Al_2O_3 and MgO content of C_3S. Midgley[4] analysed the C_3S in one clinker and Fletcher[5] the C_2S in a normal and sulphate-resisting clinker. Fletcher[6] attempted also to analyse C_3A and C_4AF but to get accurate results

[1] W. Lerch and L. T. Brownmiller, *J. Res. natn. Bur. Stand.* **18,** 609 (1937); E. S. Newman, ibid. **38,** 661 (1947).
[2] A. E. Moore, *Silic. ind.* **30,** 455 (1965).
[3] O. Peterson, *Zement-Kalk-Gips* **20,** 61 (1967).
[4] H. G. Midgley, *Mag. Cancr. Res.* **20** (62), 41 (1968).
[5] K. Fletcher, *Mag. Concr. Res.* **20** (64), 167 (1968).
[6] K. Fletcher, *Mag. Concr. Res.* **21,** (66) 3 (1969).

had to prepare artificial clinkers and heat treat them to promote good crystallisation. The same technique was employed by Mme Jeanne[1] to investigate the ferrite in high-alumina cement. Terrier[2] and co-workers analysed four clinkers of varying types. Several reports state that greater differences were found between grains in a single clinker than between different samples of clinker. Midgley[3] found that the ratio of weights of a given element in C_3S and C_2S was fairly constant for adjacent grains, but the absolute values might vary in a single clinker sample by as much as four times.

These results are obviously of great interest, but a wider range is needed before elaborate statistical analysis is possible. The indications are that both the C_3S and C_2S phases contain an excess of CaO. The maximum contents of MgO, Al_2O_3 and Fe_2O_3 are in good agreement with the limits of solid solution in the temperature range for clinkering as summarised in Chapter 4. Alkalis and TiO_2 enter into the C_2S lattice to a greater extent than in C_3S. Assuming, following Fletcher, that all the foreign elements in C_3A substitute for Al, except for Na and K which substitute for Ca, an excess of cations is found. Better agreement would follow by substituting $2 \times$ Na for Ca. Both C_3A and the ferrite phase have important contents of SiO_2, which have been rather overlooked in phase studies; the microprobe work also draws attention to the accumulation of Ti in the ferrite phase. The analyses of the ferrite phase are as yet too inaccurate to draw firm conclusions about the position in the ferrite solid solution series. Where comparisons can be made there is fair agreement with X-ray analysis.

The range of solid solution in the four main clinker components indicated by microprobe analysis is given in Table 10 in weight per cent.

TABLE 10

Oxide	Al_2O_3	Fe_2O_3	MgO	SiO_2	Na_2O	K_2O	TiO_2
Component							
C_3S	0·7–1·7	0·4–1·6	0·3–1·0	—	0·1–0·3	0·1–0·3	0·1–0·4
C_2S	1·1–2·6	0·7–2·2	0·2–0·6	—	0·2–1·0	0·3–1·0	0·1–0·3
C_3A	—	4·4–6·0	0·4–1·0 13·0*	2·1–4·2	0·3–1·7	0·4–1·1	0·1–0·6
Ferrite	—	—	0·4–3·2	1·2–6·0	0·0–0·5	0·0–0·1	0·9–2·6

* Dark prismatic phase

These results, broadly agreeing with those obtained from phase-rule studies, open up the possibility of setting up improved formulae of the Bogue type possibly first sub-dividing the data according to $Al_2O_3:Fe_2O_3$ ratio.

[1] Mme. Jeanne, *Revue Matér. Constr. Trav. publ.* **629,** 53 (1968).
[2] P. Terrier, H. Hornain, G. Socroun, *Revue Matér. Constr. Trav. publ.* **630,** 109 (1968).
[3] H. G. Midgley, *Sym., Tokyo 1968.*

To summarise this discussion of the constitution of Portland cement, we can say that it contains the compounds to be expected from phase equilibrium studies especially when these take account of solid solution, but that their relative proportions do not represent exactly the equilibrium state for the cooled material. This may arise from incomplete burning leaving a small amount of free lime in the clinker, with corresponding effects on the proportions of C_3S and C_2S; from failure to maintain chemical equilibrium during cooling from the burning temperature; and from failure of all the clinker liquid to crystallise, some remaining as glass. Essentially we can regard Portland cement as a material which has closely approached equilibrium at the clinkering temperature, but which has cooled too rapidly to allow complete equilibrium to be established at the temperature of final solidification (see p. 135).

Free lime in Portland cement

Free, or uncombined, calcium oxide usually occurs to a small extent in Portland cement clinker, and in the ground cement. In a fresh clinker it is entirely present as calcium oxide, but in a ground cement a considerable proportion of the total free lime is usually found to be present in the hydrated form as calcium hydroxide. This is due to hydration of the free calcium oxide during grinding of the clinker by moisture contained in the added gypsum, or by combined water released from the gypsum owing to the relatively high temperature developed in the grinding mill. Clinker is also often stored for a long period before grinding and may absorb some moisture. Free lime may be detected qualitatively in the ground clinker or cement by the use of White's reagent, a solution of 5 g of phenol in 5 cc of nitrobenzene to which two drops of water are added. A few milligrams of the cement or of the finely powdered clinker are placed on a microscope slide, a drop of White's reagent added and then a cover-glass pressed lightly down. The slide is examined under crossed Nicols with a microscope fitted with polarising and analysing Nicols. A 16 mm, or higher power, objective should be used. If appreciable amounts of free lime are present, long needles of high birefringence are observed within a few minutes of preparing the slide, at first radiating from the small particles in which free lime is present and gradually spreading over the slide. Though the crystals, which consist of calcium phenoxide, develop rapidly if much free lime is present, the slide should be allowed to stand an hour to detect traces. The test is sensitive and will detect less than 0·1 per cent free lime in a cement. After long standing there is some indication of attack by the reagent on other constituents of the cement, but this is so slow that it does not interfere with the test. With a little practice an observer should find no difficulty in recognising the characteristic needles. Both free calcium oxide and hydroxide are detected by this test.

A quantitative estimation of the total free calcium oxide and hydroxide in a cement can be made by the extraction method of Lerch and Bogue which makes use of the solubility of lime in a solvent consisting of one part by volume pure glycerol to five volumes absolute alcohol. It was formerly customary, and is still a practice, to use as anhydrous a glycerol as possible, containing not less than 99·2 per cent glycerol, but the ASTM specification (C114—67) only requires that the water content of the glycerol shall not exceed about 5 per cent, corresponding to a minimum specific gravity at 25/25° of 1·249. The presence of some moisture accelerates the initial rate of solution of the lime; too much may cause attack on the cement compounds, but comparative trials carried out by the ASTM[1] showed little difference between the results with glycerols containing 0·5 and 5 per cent water. The alcohol is preferably anhydrous (99 per cent) ethyl alcohol, but anhydrous alcohol denatured with 5 per cent anhydrous methyl alcohol is permitted. The indicator used is 0·18 g phenolphthalein in 2160 ml of the mixed solvent. If the glycerol-alcohol solvent is acidic to the indicator an addition of a dilute solution of NaOH in absolute alcohol is made until a slight pink colour occurs, and, if alkaline, of the standard ammonium acetate solution. This solution, which is used for titrating the lime dissolved in the estimation, is prepared by dissolving 16 g dry crystalline ammonium acetate in 1 litre of absolute, or anhydrous denatured, alcohol. It is standardised as follows: 0·1 g of freshly ignited CaO is placed in a 200 ml Erlenmeyer flask and 60 ml of the glycerol–ethyl alcohol solvent added and the CaO dispersed by shaking. A reflux condenser is fitted and the mixture boiled for 5 to 20 minutes. It is titrated while still nearly boiling with the ammonium acetate solution. The container is replaced and the boiling continued for 5 to 20 minutes and the mix titrated again. This process is continued until the free CaO content does not increase by more than 0·05 per cent after 2 hours boiling.

The free-lime estimation is carried out by weighing 1 g of finely ground cement into a flask, adding 60 ml of the solvent, and proceeding as in the standardisation of the ammonium acetate solution. The free CaO content of the cement is calculated as follows:

$$\%\text{CaO} = \frac{100EV}{w}$$

where E = equivalence of standard ammonium acetate solution in g CaO per ml,
V = ml of ammonium acetate solution required, and
w = weight of sample taken.

The method estimates free CaO present both as oxide and hydroxide. Free MgO burnt at kiln temperatures does not react or affect the results. The end-point is not very good as a weak base, calcium glycerolate, is being titrated with a weak acid, ammonium acetate.

The method is somewhat tedious and it can be accelerated, by the addition of 1 g anhydrous barium chloride or 0·6 g anhydrous strontium nitrate to the 60 ml solvent. The solution can then be titrated at five-minute intervals. The results tend to be slightly high because of some attack on the cement compounds, particularly if glycerol containing five per cent water is used. The mechanism of the alcohol-glycerol method has been discussed by Swenson and Thorvaldson,[2] who have also shown that strontium chloride is effective as an accelerator.

[1] W. C. Hanna, T. A. Hicks and G. A. Saeger, *Bull. Am. Soc. Test. Mater.*, No. 94, 47 (1938); *Proc. Am. Soc. Test. Mater.* **39,** 314 (1939).
[2] *Can. J. Chem.* **30,** 257 (1952); **29,** 140 (1951).

Another method which is more rapid has been described by two Swiss investigators.[1] About 0·5–1·0 g of cement is placed in a small flask with 40–50 cc of ethylene glycol and a pinch of clean quartz grains. The flask is closed with a rubber bung and shaken in a water-bath at 60–70° for 30 minutes. Provided it is light in colour and free from sulphides, the mix may then be titrated directly with N/10 alcoholic benzoic acid solution or 0·1 N aqueous HCl. The indicator is made up by dissolving 0·1 g phenolphthalein and 0·15 g α-naphtholphthalein in 100 cc absolute alcohol. An alternative indicator is 0·05 g methyl red and 0·05 g bromocresol green in 100 ml alcohol.[2] Preferably, however, the sample is filtered through a sintered glass filter, or hardened filter paper on a porcelain plate, and the filter washed three times with about 10 cc absolute alcohol or preferably with glycol. The ethylene glycol must be pure, neutral, and free from water. Its purity can most easily be determined by the density which should be 1·109 at 20°. The method works very satisfactorily for fresh cements and the free calcium oxide or hydroxide present dissolves completely under the conditions stated. The results are in reasonable agreement with those obtained by the alcohol-glycerol method. Coarsely crystalline calcium hydroxide, such as occurs in set cements, does not, however, dissolve at all readily in ethylene glycol, and even when ground to pass a 240 mesh the solution is by no means complete in 30 minutes at 60–70°. For this, and other, reasons the method is therefore not satisfactory for estimation of calcium hydroxide in set cement.

Various other methods, or modifications of the alcohol-glycerol and ethylene-glycol methods, have been proposed, but the only one that need concern us is that of Franke.[3] This method makes use of an acetoacetic ester-*iso*butyl alcohol solvent. The pure ester attacks the cement compounds but the *iso*butyl alcohol reduces the pH and retards their reaction. About 0·05–1 g cement, depending on its free lime content, is placed in a 200 ml Erlenmeyer flask fitted with a ground glass joint and a reflux condenser with calcium chloride and soda-lime guard tubes; 3 ml acetoacetic ester and 20 ml isobutyl alcohol are added and the mixture boiled for 1 hour. After cooling the solution is filtered through a Hirsch funnel (porcelain filter disc) under vacuum, the filter and flask washed with 20 ml isobutyl alcohol and the calcium in the filtrate determined (see p. 247).

The Franke method gives results in good agreement with the other two main methods though Dubuisson[4] has claimed that more than 1 hour's boiling is required. Its chief advantage is that it can be applied to the estimation of calcium hydroxide in set cement.

It will be noticed that the above methods give the total free calcium oxide and hydroxide in a cement and do not differentiate between the two. An approximate determination of the calcium hydroxide present in a fresh cement, and hence, by subtracting it from the total free calcium oxide and hydroxide, of the free

[1] P. Schlapfer and R. Bukowski, *Rep. Swiss fed. Lab. Test. Mater., Zurich*, No. 63 (1933); P. Schlapfer, *Sym., Stockholm 1938*, 289; *Schweiz Zement Kalk und Gips Jahrbuch*, No. 33 (1943).

[2] *Sym. Analysis of Calcareous Materials: Monogr. Soc. chem. Ind., Lond.*, No. 18, 120 (1964); S. R. Bowden, ibid., 140.

[3] B. Franke, *Z. anorg. allg. Chem.* **247,** 180 (1941); E. E. Bresler, S. Brunauer and D. L. Kantro, *Analyt. Chem.*, **28,** 896 (1956).

[4] A. Dubuisson, *Revue Matér. Constr. Trav. publ.* **398,** 293 (1948).

calcium oxide may be obtained by an additional estimation[1] of the water combined as calcium hydroxide.

The small proportion of combined water in a fresh cement may be present either in calcium hydroxide or in hydrated cement compounds and gypsum. On heating the cement, the water combined in the hydrated cement compounds other than calcium hydroxide is mostly lost below 350° while that combined in calcium hydroxide is lost almost entirely between 350° and 550°. The loss of water from a fresh cement over the range 350–550° may, after a small correction, be used as a measure of the calcium hydroxide content. The correction is due to a small loss of water occurring between 350° and 550° from the hydrated cement compounds. It is found with set cements that the loss of water from the hydrated cement compounds at 550° is about 1·09 times that at 350°. This factor is rather variable (from about 1·07 to 1·12), but when the loss at 350° is low, as in fresh cements, the error introduced is fairly small. The method cannot for this reason be applied satisfactorily to hydrated cements in which the loss at 350° is large. The calcium hydroxide content is calculated as follows, about 5 g of the cement being heated for thirty minutes at the stated temperatures.

If loss on heating at 350° $= x_1$
and loss on heating at 550° $= x_2$

then loss due to decomposition of calcium hydroxide $= x_2 - 1{\cdot}09x_1$ This loss multiplied by the ratio of the molecular weight of CaO to H_2O, $\frac{56}{18}$, gives the calcium hydroxide content in terms of CaO. The method is probably only correct to about $\pm 0{\cdot}5$ per cent CaO.

Calcium hydroxide in a cement can also be determined by differential thermal analysis[2] or by thermogravimetric methods.[3]

Some typical results of estimations of the total free lime, and free calcium oxide and hydroxide contents, of fresh cements are shown in Table 11.

Rapid-hardening Portland cements tend on the whole to have rather higher total free-lime contents than normal Portland cements, but this is far from being general. Fresh Portland cement clinkers, taken direct from the kiln, contain only free CaO, but the content of this in different clinkers varies over the same range as the total free calcium oxide and hydroxide content of ground cements. Though it might have been anticipated that the water released from the added gypsum during grinding of the cement would produce additional free calcium hydroxide by attacking the cement compounds, it appears from the above and from other evidence, that this is not usually the case.[4] The released water at the temperatures prevailing in a grinding mill, which may rise to 150° or over, seems only to hydrate the free calcium oxide and not to attack the cement. The importance of this action in increasing the soundness of cements is shown in a later chapter.

Uncombined magnesia. Free MgO in Portland cement can be determined by a method developed by Taylor and Bogue, in which the free oxide is extracted

[1] G. E. Bessey, *J. Soc. chem. Ind., Lond.* **52,** 219 (1933).
[2] H. G. Midgley, *Sym., Washington 1960*, 479.
[3] P. Longuet, *Revue Matér. Constr. Trav. publ.* **537,** 139 (1960); F. M. Biffen, *Anal. Chem.* **28,** 1133 (1956).
[4] G. E. Bessey, loc. cit.

by treatment with a solution of ammonium nitrate in anhydrous alcohol and glycerol. The MgO forms a double salt according to the equation:

$$MgO+4NH_4NO_3 = Mg(NO_3)_4(NH_4)_2+2NH_3+H_2O.$$

TABLE 11

Cement		Total free lime content expressed as % CaO	Free calcium hydroxide expressed as % CaO	Free calcium oxide* expressed as % CaO
Rapid-hardening Portland cement	1	1·2	1·1	0
	2	3·5	2·9	0·5
	3	3·7	2·3	1·5
	4	3·9	1·6	2·5
	5	4·2	3·4	1·0
	6	4·6	4·2	0·5
Normal Portland cement	7	0·4	0·7	0
	8	0·7	0·9	0
	9	1·4	1·2	0
	10	1·8	1·2	0·5
	11	2·3	1·5	1·0
	12	3·6	2·6	1·0

* To nearest 0·5 per cent

Aqueous solutions of ammonium nitrate attack magnesia present in compounds, or in the glass, of the cement and are therefore unsuitable. The estimation is carried out in the same way as for free lime, using 1 g of the cement ground to pass a 300-mesh sieve and adding 4 g ammonium nitrate to the 50 ml absolute alcohol 10 ml glycerol mixture. The mixture is boiled gently on a steam bath or hot plate for 1 to 5 hours until evolution of ammonia ceases. The solution is filtered and the residue washed with hot absolute alcohol and the magnesia in the filtrate determined. X-ray analysis can also be used provided that more than about 0·5 per cent MgO is present.

Analysis of cements

Procedures for the analysis of cements are given in several standards[1] and in various collections of analytical procedures.[2] In recent years considerable effort has been devoted to the development of more rapid and improved separation schemes and analytical procedures. Titrimetric methods for estimating the main constituents of cements are becoming almost universally employed especially for routine work. Most of these methods are based on the use of the complexi-

[1] For example, ASTM C114–67; BS 4550 Part II, *Methods of Testing Cement.* Chemical Tests (1970).

[2] Compilation of basic methods of analysis, *Sym. Analysis of Calcareous Materials: Monogr. Soc. chem. Ind., Lond.,* No. 18, 3–143 (1964); *Analysengang fur Zemente.* Verein Deutscher Zementwerke E.V. Dusseldorf (1961).

metric reagent EDTA (diaminoethane tetra acetate—usually as the disodium salt). This reagent has the property of forming soluble octahedral complexes with many metal ions, selectivity being achieved through control of pH, by masking interfering elements by suitable reagents and by analytical separations. Complete schemes for the analysis of cements are available[1] and a comprehensive review of compleximetric methods applied to cement has been prepared.[2] Spectrophotometric methods are now widely employed for the estimation of the minor elements in cement[3] particularly TiO_2 Mn_2O_3 and P_2O_5 In many rapid schemes of analysis SiO_2 is determined as the yellow silicomolybdate, or sometimes after reduction to molybdenum-blue. Aluminium and iron can also be determined by spectrophotometry. Flame photometry[4] is virtually the only method now used to estimate the alkalis in cement; provided precautions are taken to eliminate interferences from other elements (particularly calcium) only simple equipment is required for these determinations. With a more sensitive spectrophotometer flame estimations can be extended to strontium[5] and magnesium.[6] Manganese can also be determined by this method.[7]

The more recent technique of atomic absorption spectrophotometry makes possible the accurate determination of elements present in major proportion in cement. In this technique a cement solution is atomised in a flame and the energy absorbed from a spectral lamp for each element to be determined is measured in turn. Very little chemical separation is required. Elements which have been determined in cement by this method include Si, Al, Fe, Ti, Mn, Ca, Mg, Sr, Na, K and Li.[8] Most of the newer rapid methods for the analysis of cement include provision for the estimation of SO_3 usually titrimetrically[9] or turbidimetrically[10]; the latter method is widely used in cement works practice. Carbon dioxide is still largely determined by conventional methods,[11] fluorine is usually separated by steam distillation or pyrohydrolysis and then estimated titrimetrically or colorimetrically.[12] Spectrographic analysis is particularly useful for the determination of elements present in cement in only small amounts,[13]

[1] R. G. Blezard, *Sym. Analysis of Calcareous Materials: Monogr. Soc. chem. Ind., Lond.* No. 18, 222 (1964); P. Stiglitz, J. Comet, M. T. Mounier and M. G. Gravland, *Revue Matér. Constr. Trav. publ.* **576,** 271 (1963).

[2] L. Burgland and P. Longuet, *Revue Matér. Constr. Trav. publ.* **604,** 1; **605,** 49; **606,** 17 (1966).

[3] J. A. Klinkman, *Zement-Kalk-Gips* **19,** 424 (1966); Z. T. Jugovic, *Spec. tech. Publ. Am. Soc. Test. Mater.,* No. 395, 65 (1965).

[4] *Sym. Analysis of Calcareous Materials: Monogr. Soc. chem. Ind., Lond.,* No. 18, 95 (1964); H. Otterbein, *Zement-Kalk-Gips* **19,** 497 (1966); ASTM C114–67.

[5] S. R. Bowden, *Sym. Analysis of Calcareous Materials,* loc. cit., 275.

[6] S. R. Bowden, loc. cit., 268; T. C. Wilson and N. J. Krotinger, *Bull. Am. Soc. Test. Mater.,* No. 189, 56 (1963).

[7] C. L. Ford, *Bull. Am. Soc. Test. Mater.,* No. 250, 25 (1960).

[8] L. Capacho-Delago and D. C. Manning, *Analyst, Lond.* **92,** 553 (1967); R. F. Crow, W. G. Hime and J. D. Connolly, *J. Res. Dev. Labs Portld Cem. Ass.* **9** (2), 60 (1967); H. L. Kahn, *Mater. Res. Stand.* **5,** 337 (1965); T. Takeuchi and M. Suzuki, *Talanta* **11,** 1391 (1964).

[9] C. Dadak, *Cement-Wagno-Gips 1965,* (5) 32.

[10] ASTM C114–67; F. Matouschek, *Zement-Kalk-Gips* **7,** 9 (1954); M. Wallraf, ibid. **12,** 55 (1959).

[11] G. E. Bessey, *Jour. Soc. Chem. Ind.* **58,** 178 (1939); F. E. Jones, ibid. **59,** 21 (1960).

[12] S. R. Bowden, *Soc. Chem. Ind.,* Monograph No. 18, 181 (1964); H. Lehmann, F. W. Locher and H. M. von Seebach, *Tonind. Ztg.* **89,** 49 (1965).

[13] Z. G. Hanna and J. M. Ibrahaim, *Zeit. analyt. Chem.* **208,** 276(1965).

but this technique has also been employed for the major elements with an accuracy better than 4 per cent. X-ray fluorescence spectroscopy is being increasingly employed, particularly where large numbers of cement analysis are required such as cement works control.[1] In this technique the sample is irradiated with a primary high-energy X-ray beam and the secondary (fluorescent) X-rays emitted are detected. Each element has its characteristic X-ray spectrum and thus it is possible to identify and determine any desired element. Radioisotopes may be used to provide the primary beam and when coupled with non-dispersive detectors the method can be used for the continuous monitoring of raw materials and products at a works.[2]

Calculation of the constituent compounds in cements

If the analytical composition of a cement is known, and also the compounds which have been formed in it during burning, it is clearly possible to calculate the amounts of these compounds which are present in the cement. The results of the work on the constitution of Portland cement have shown that it consists essentially of the compounds, C_3S, C_2S, C_3A, a ferrite close to C_4AF, MgO, and a little free CaO. To these must be added the gypsum introduced during grinding to control the setting time of the cement. The minor components such as TiO_2, Mn_2O_3, K_2O, Na_2O, do not normally amount to more than about 2 per cent, and it is usual to ignore them in calculations of the *potential* compound composition of a cement. The word *potential* is used because it is assumed in any such calculation that the cement is entirely crystalline and that no glass remained in the clinker after cooling. This is not entirely correct as cement may contain a perceptible amount of glass, though its amount is difficult to estimate accurately.[3] In any case the crystallisation or devitrification of the liquid present at the burning temperature is unlikely to be an equilibrium process.

The presence of glass will affect the percentage contents of the various compounds formed in two ways. In the first place it is obvious that it reduces the total amount of crystalline compounds present, but it has also another effect. As we shall see in the next chapter, the changes occurring during the period in which a cement clinker is cooling, while liquid is still present, are not, if equilibrium is to be maintained, limited to simple crystallisation of the liquid. They also include an interaction of the liquid with the crystalline compounds already present which may actually decrease rather than increase the amount of some one crystalline compound. For these reasons the direct calculation of compound composition of a cement from the analytical data by the method shown below can only be regarded as very approximate.

Methods by which improved values can be obtained are discussed in Chapter 7.

[1] L. Lahl, *Zement-Kalk-Gips* **18,** 78 (1965); R. Rabot and R. Alegre, *Silic. ind.* **27,** 181, 250, 305 (1962).

[2] R. Alegre, *Soc. Chem. Ind.*, Monograph No. 18, 289 (1964); *Rev. Mat. Constr.* **605,** 65 (1966); M. von Euw, ibid. **624,** 321 (1967).

[3] At a clinkering temperature a cement clinker contains about 20–30 per cent liquid. This, then, represents the maximum possible glass content, but owing to crystallisation on cooling, it is usually much lower.

PLATE I

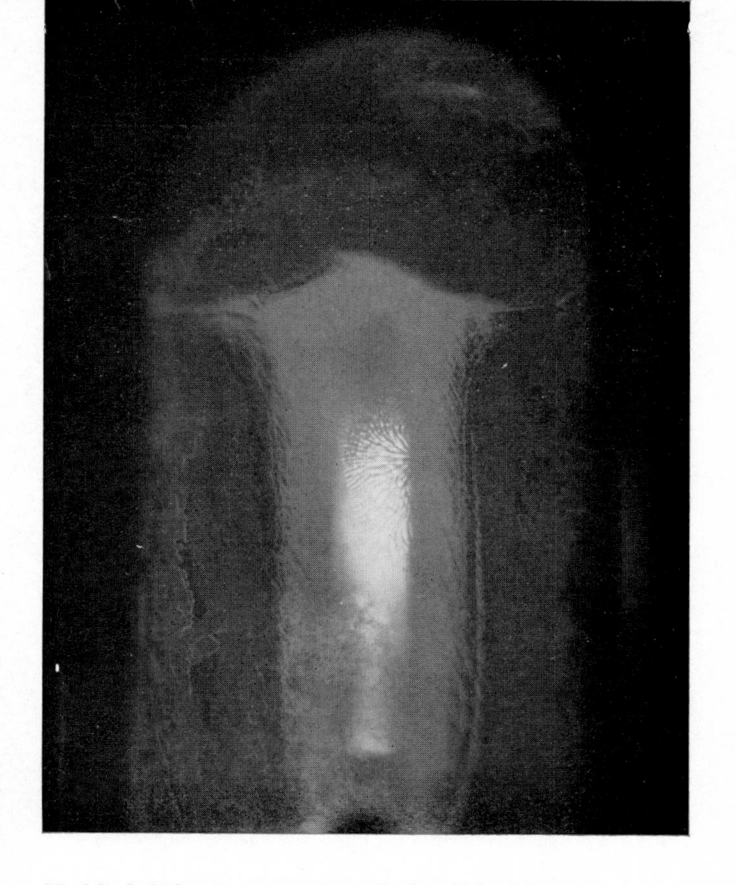

Crystals of (i) $3CaO.SiO_2$ and (ii) $2CaO.SiO_2$ Growing in a CaO—SiO_2 Melt at Above 2000° (×50)

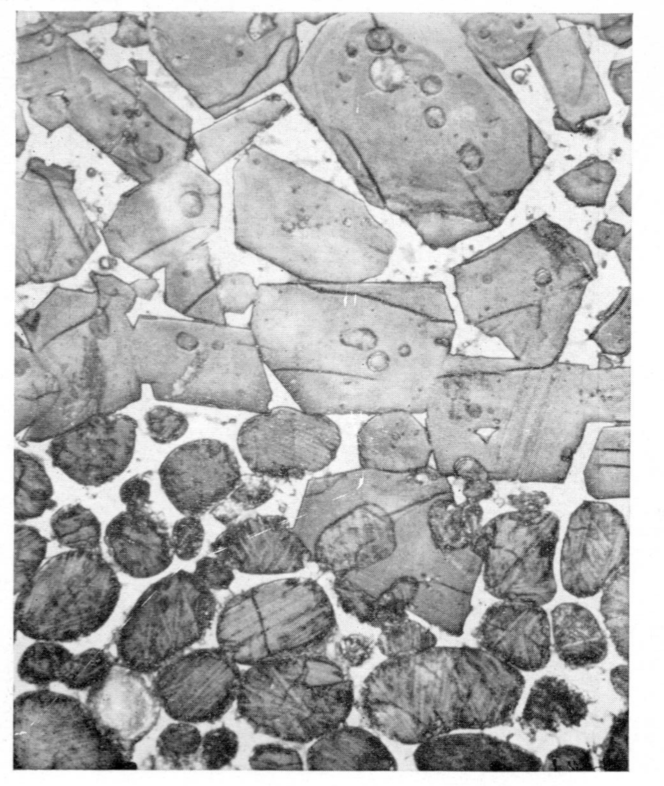

(i) Polished Surface of Portland Cement Clinker Etched with Water Followed by 0.25 per cent HNO_3 in Alcohol (×360)

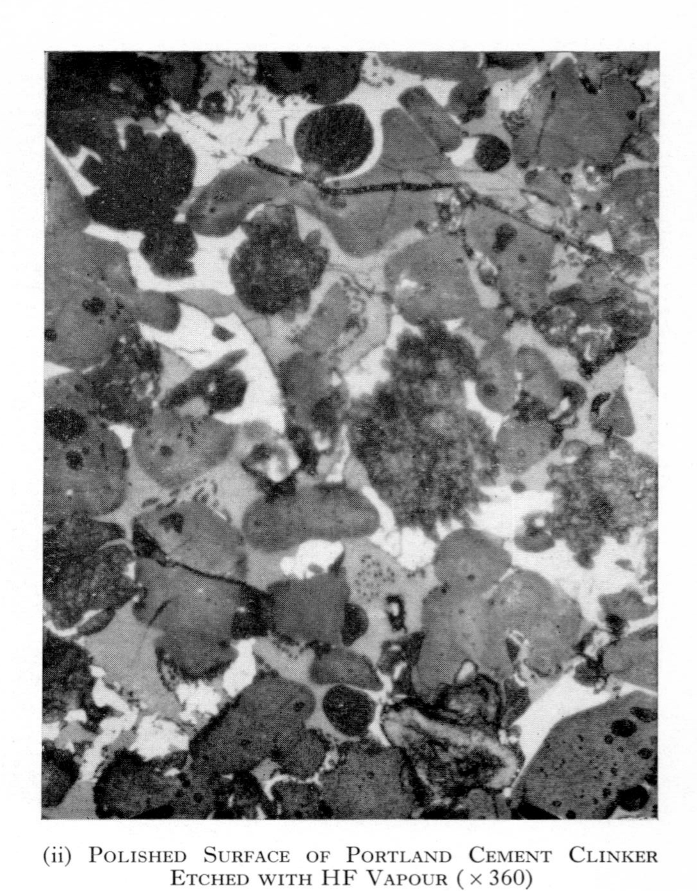

(ii) Polished Surface of Portland Cement Clinker Etched with HF Vapour (×360)

The compound composition of a Portland cement can be calculated as follows.[1] The SO_3 content is first considered and the equivalent amount of CaO required to form $CaSO_4$ calculated. This, and the free-lime content of the cement as obtained by estimation are subtracted from the total lime content. The remaining lime, together with the content of SiO_2, Al_2O_3 and Fe_2O_3 are now divided up to give the compounds C_4AF, C_3A, C_3S and C_2S. The amounts of CaO and Al_2O_3 required to combine with the Fe_2O_3 to form C_4AF are derived and the content of this compound obtained. The remaining alumina is now calculated to C_3A. The CaO remaining after subtraction of the amounts used in forming the above two compounds is finally allotted to the silica and the contents of C_3S and C_2S calculated. The method is best illustrated by an actual calculation, remembering that the molecular proportions of the various oxides in the different compounds are as follows:

$$\underset{136}{CaSO_4} = \underset{56}{CaO} + \underset{80}{SO_3}$$

$$\underset{485}{4CaO.Al_2O_3.Fe_2O_3} = \underset{224}{4CaO} + \underset{102}{Al_2O_3} + \underset{159}{Fe_2O_3}$$

$$\underset{270}{3CaO.Al_2O_3} = \underset{168}{3CaO} + \underset{102}{Al_2O_3}$$

$$\underset{172}{2CaO.SiO_2} = \underset{112}{2CaO} + \underset{60}{SiO_2}$$

$$\underset{228}{3CaO.SiO_2} = \underset{168}{3CaO} + \underset{60}{SiO_2}$$

$$\underset{172}{2CaO.SiO_2} + \underset{56}{CaO} = \underset{228}{3CaO.SiO_2}$$

Ratios

$$\frac{CaO}{SO_3} = 0{\cdot}70 \qquad \frac{3CaO}{Al_2O_3} = 1{\cdot}65$$

$$\frac{Al_2O_3}{Fe_2O_3} = 0{\cdot}64 \qquad \frac{4CaO}{Fe_2O_3} = 1{\cdot}40$$

$$\frac{3CaO.Al_2O_3}{Al_2O_3} = 2{\cdot}65 \qquad \frac{4CaO.Al_2O_3.Fe_2O_3}{Fe_2O_3} = 3{\cdot}04$$

$$\frac{2CaO}{SiO_2} = 1{\cdot}87 \qquad \frac{3CaO.SiO_2}{CaO} = 4{\cdot}07$$

[1] R. H. Bogue, *Ind. Engng Chem. analyt. Edn.* **1,** (4), 192 (1929).

A cement has the following analysis:

	per cent
CaO	64·10
SiO_2	22·90
Al_2O_3	4·50
Fe_2O_3	3·11
MgO	0·79
TiO_2	0·24
Na_2O	0·54
K_2O	0·64
SO_3	2·37
Loss by difference	0·81
	100·00

Free CaO 0·9 per cent

2·37 per cent SO_3 = 2·37 × 0·70 per cent CaO = 1·66 per cent CaO

Free CaO = 0·9 per cent

Total 2·56 per cent

Remaining CaO = 64·1 − 2·56 = 61·54 per cent

3·11 per cent Fe_2O_3 = 3·11 × 0·64 per cent Al_2O_3 = 1·99 per cent Al_2O_3

= 3·11 × 1·40 per cent CaO = 4·35 per cent CaO

= 3·11 × 3·04 per cent $4CaO.Al_2O_3.Fe_2O_3$

= 9·5 per cent $4CaO.Al_2O_3.Fe_2O_3$

Remaining Al_2O_3 = 4·50 − 1·99 = 2·51 per cent

2.51 per cent Al_2O_3 = 2·51 × 1·65 per cent CaO = 4·14 per cent CaO

= 2·51 × 2·65 per cent $3CaO.Al_2O_3$

= 6·7 per cent $3CaO.Al_2O_3$

Remaining CaO = 61·54 − 4·35 − 4·14 = 53·05 per cent

The calculation of the $3CaO.SiO_2$ and $2CaO.SiO_2$ contents may now be obtained directly from the formulae

$$\text{Per cent } 3CaO.SiO_2 = 4{\cdot}07\,P - 7{\cdot}60\,Q$$
$$\text{Per cent } 2CaO.SiO_2 = 8{\cdot}60\,Q - 3{\cdot}07\,P$$

where P is the percentage CaO content remaining for silicate formation and Q is the percentage silica content. Thus in the present example P = 53·05 and Q = 22·90.

The compound composition then is

	per cent
$4CaO.Al_2O_3.Fe_2O_3$	9·5
$3CaO.Al_2O_3$	6·7
$3CaO.SiO_2$	42
$2CaO.SiO_2$	34

There are present in addition 0·9 per cent free CaO, 4·0 per cent $CaSO_4$ and the magnesia, alkalis and titania, and moisture and carbon dioxide which are lost on ignition.

If the Al_2O_3:Fe_2O_3 ratio of the ferrite phase has been determined by X-rays, the above calculation can be made using the content of Fe_2O_3 corresponding to the ferrite composition. The result is sometimes called the 'modified Bogue calculation'. Various other corrections to the Bogue calculation have been proposed, some of which attempt to allow for the effects of solid solution. None of these have yet received any general acceptance, but with the information now becoming available from electron microprobe analysis it is possible that some corrected compound compositions might eventually be agreed for use in the Bogue calculation.

7 The Burning of Portland Cement

We have learnt in the previous chapter the nature of the compounds formed in Portland cement, but no consideration has as yet been given to the manner of their formation from the raw mix, or to any factors which in practice will restrict the raw materials that can be used. It is possible to produce in the laboratory from lime and silica alone a cement composed only of tricalcium and dicalcium silicate, but the rate of reaction between the two solid components is so slow that its commercial production is, and seems likely to remain, impracticable. The range of compositions from which a Portland cement can be produced in practice is thus only a part of the whole range that show cementing properties. When plotting the Portland cement area in Fig. 36 therefore the zone was not shown extending as far as the CaO–SiO_2 side of the diagram, but only down to a minimum content of about 5 per cent alumina and ferric oxide. The introduction of these other components does not apparently improve the properties of an adequately burnt cement, but their presence is a manufacturing necessity in order to permit of burning in commercial kilns.

Successful burning under commercial conditions depends not only on the composition of the raw mix, but also on the physical and chemical condition of the materials. Thus a mixture of calcium carbonate, ground sand (silica), alumina and ferric oxide, if sufficiently finely ground, can be burnt to form a cement clinker in a small laboratory furnace, but on a commercial kiln its burning would prove difficult. A mix of the same gross composition made up of finely divided limestone and clay can on the other hand be burnt commercially on account of the greater reactivity of its clay component. In studying the burning of cement we are concerned, then, not only with the various reactions which successively occur, but also in the rates at which they proceed.

The burning of cement consists of a series of reactions taking place between finely divided solids, and it is only in the final stages of burning that a liquid is formed and becomes the medium through which reaction occurs. When liquids or solutions react together the actions usually occur rapidly, and the products depend only on the composition of the reaction mixture and the temperature. With reactions between solids the conditions are very different. Action takes place at the surface of grains and diffusion of fresh solid to the surface proceeds but slowly. The reaction thus tends to proceed unevenly at different points and, if the action is arrested before completion, a series of reaction products may be found present. If a particle of clay is imagined surrounded by particles of lime the reaction will commence at the surface of the clay particle, forming first the less basic and then the more basic compounds. These in turn will react with further clay inside the particle, losing some of their combined lime to it. If complete combination is not attained, the stage to which the reaction has progressed depends not only on the temperature, but also on the time, or rate, of heating, and the general physical and chemical condition of the reacting materials.

The production of Portland cement by a clinkering process in which only a minor proportion of the mix becomes liquid is thus dependent on three factors:

1. The chemical composition of the mix.
2. The physicochemical state of the raw constituents.
3. The temperature and period of burning.

A fourth condition which influences the resulting clinker is the rate of cooling.

The closeness with which the chemical composition of cement can be made to approach the limits theoretically attainable must depend on the second and third factors and will vary somewhat with different raw materials.

The burning of Portland cement has been studied by various methods each of which has made a very definite contribution towards increasing our knowledge of the process. Small mixes of the component oxides, or of limestone and clay, can be burnt in small laboratory furnaces and the effect of burning under different conditions carefully studied. This method has both advantages and disadvantages. The mix can be made up to any desired composition and by the use of platinum boats any contamination with the furnace lining materials can be avoided. The use of suitable electric or gas-fired furnaces also avoids any contamination of the charge with coal ash as occurs in large coal-fired kilns. The temperature can be carefully measured and, within limits, the rate of heating and the period of any maximum temperature can be controlled. Only small amounts of cements can be prepared, but by the use of small-scale methods of testing some at least of the physical properties can be determined. On the other hand, the conditions existing in a commercial kiln in regard to the rate of heating and cooling cannot be entirely reproduced. Further, the mix is held in a container of some kind and does not undergo the continual turning-over and mixing which occur in the rotary kiln.

Tests can be, and often have been, made on full-size rotary kilns under commercial conditions. Samples can be taken at various points along the kiln after stopping it and allowing it to cool, or by suitable sampling devices while burning is actually in progress. Large samples can be obtained sufficient to permit of testing on any desired scale. The exact control of the mix which is possible in small-scale burns is, however, much more difficult to attain, and large quantities of raw materials are required. The measurement of the temperatures existing at different points along the kiln presents some difficulties. As an intermediate stage between small laboratory experiments and tests with commercial kilns, small rotary kilns of about 6–12 feet in length and about 5–10 inches internal diameter are sometimes employed. The study of the reactions occurring in raw cement mixes by methods of thermal analysis depending on heating curves has also proved very useful.

The remaining method of investigating the burning process depends on the application of the data obtained from the phase equilibrium studies. This throws little light on the early stages of the burning process, but affords very valuable evidence of the condition of the material at the clinkering temperatures after liquid has been formed, and of the nature of the changes occurring during cooling.

Solid reactions are much influenced by imperfections or strains in the lattice structure of materials which increase their reactivity in the solid state. Specially active states also arise at temperatures at which a transition from one crystalline form to another takes place.

Reactions occurring in cement burning

The raw materials fed to a cement kiln contain calcium carbonate, a little magnesium carbonate, clay or shale, and water. The reactions which occur during burning may be classified simply as follows:

1. Evaporation of free water.
2. Release of the combined water from the clay.
3. Dissociation of magnesium carbonate.
4. Dissociation of calcium carbonate.
5. Combination of lime and clay.

It will be seen later that these reactions do not all occur separately, but that (4) and (5) may often overlap and go on together. The essential reaction we shall have to consider in cement burning is (5), but the preliminary actions may first be considered.

Evaporation of free water takes place at, or below, 100°, but release of combined water from the clay only becomes appreciable above about 500°. Clays are composed of a number of different hydrated aluminosilicates (see p. 420) with ratios of $SiO_2:Al_2O_3$ varying from 2:1 to 4–5:1; some clays also contain ferric oxide as an essential constituent, but attention may be restricted here to one of the simplest of the clay compounds, kaolinite, $2SiO_2.Al_2O_3.2H_2O$. There has been much controversy about the changes that kaolinite undergoes when it loses water at about 500–600°. The main two alternative theories are that an anhydrous alumino-silicate $2SiO_2.Al_2O_3$ known as metakaolin, is produced or that the kaolin splits into an intimate mixture of amorphous silica and alumina. On further heating an exothermic reaction occurs at about 970° variously attributed to the formation of a silicon spinel phase approximating to $2Al_2O_3.3SiO_2$ which converts to mullite above 1000° by splitting off silica, or to the crystallisation of γ Al_2O_3 from amorphous alumina.[1] On either theory mullite is formed at higher temperatures, but in cement burning the reaction with lime takes place before this stage is reached and indeed commences before the exothermic reaction of kaolinite at 970°.

Magnesium carbonate decomposes at about 600–700° but the temperatures reported vary considerably with the source of the material used. The de-

[1] See W. Eitel, *Silicate Science*, Vol. V, 133, Academic Press, New York and London (1966).

composition pressure of calcium carbonate, when heated alone, reaches atmospheric pressure at 894°.

In mixes of calcium carbonate and finely divided quartz reaction occurs very faintly at as low a temperature as 600°, proceeds very slowly at 800°, at an appreciable speed at 1100°, and rapidly at 1400°. Dicalcium silicate, probably as the α' form, is first formed whatever the ratio of lime to silica in the mix for the formation of the monosilicate is normally slow and is generally a secondary reaction between the disilicate and silica.[1] In mixes of sufficient lime content all the combined silica is present as C_2S by 1200°. Though the formation of tricalcium silicate commences at about 1300° to 1400° it proceeds only very slowly even at 1500°, but the addition of alumina and particularly ferric oxide to the lime-silica mix increases markedly the rate of its production.

When mixes of lime, alumina and silica are heated the first compounds formed are monocalcium aluminate and α' dicalcium silicate. Products obtained from burning at 1150–1350° dust down (i.e. crumble) on cooling owing to the formation of γ C_2S. Initial melting commences at about 1400° and at higher temperatures dusting ceases to occur unless the cooling is slow. This may be due to the formation of the relatively more stable α C_2S or to the formation of solid solutions in the C_2S. Formation of tricalcium aluminate does not occur readily, nor that of tricalcium silicate commence, until a temperature of 1300° is reached.[2] In mixes[3] of CaO, Al_2O_3 and Fe_2O_3 the first iron compound to form at about 800° is C_2F, followed by a transient appearance of CF between 900° and 1100°. Formation of C_4AF spreads over a temperature range of 1100° to 1250°. Mixes of CaO, Al_2O_3 Fe_2O_3 and SiO_2 show some evidence of a transient appearance of C_2AS between 1000° and 1100° and formation of C_3A and C_4AF beginning at somewhat lower temperatures than in the ternary mixes.[4] The free CaO rapidly decreases above 1330° as C_3S is formed.

A somewhat similar progress of reactions is found when mixes of kaolin and calcium carbonate are heated,[5] but at corresponding temperatures the reactions proceed more vigorously than in the oxide mixes. Gehlenite (C_2AS) starts to form at about 900° and then decomposes again by 1100°. A double salt, spurrite, $2C_2S.CaCO_3$, may form[6] as a transient phase and it has also been observed in clinker rings in rotary kilns. The compound $C_{12}A_7$ is observed between 900° and 1100° and converts to C_3A at 1100° upwards. Tricalcium silicate starts to form between 1200° and 1300° and by 1400° the reaction products are C_3S, C_2S, C_3A, C_4AF and any remaining uncombined lime.

[1] W. Jander and E. Hoffman, *Z. anorg. allg. Chem.* **218,** 211 (1934); W. Jander and J. Petri, *Z. Elektrochem.* **44,** 748 (1938); I. Weyer, *Z. anorg. allg. Chem.* **209,** 409 (1932); W. L. de Keyser, *Bull. Soc. chim. Belg.* **62,** 235 (1953); W. Kurdowski, *Silic. ind.* **30,** 500 (1965).

[2] I. Weyer, *Zement* **20,** 560 (1931); W. L. de Keyser, *Bull. Soc. chim. Belg.* **60,** 516 (1951).

[3] W. L. de Keyser, *Bull. Soc. chim. Belg.* **64,** 395 (1955); V. Cirilli, *Ricerca scient.* **17,** 439 (1947).

[4] A. van Bemst, *Silic. ind.* **26,** 290 (1961).

[5] A. van Bemst, loc. cit.; U. Ludwig, H. Müller-Hesse and H. E. Schweite, *Zement-Kalk-Gips* **13,** 449 (1960), see also H. Lehmann, F. W. Locher and P. Thormann, *Tonind.-Ztg. keram. Rdsh* **88,** 489, 537 (1964); W. Quittkat, ibid. **89,** 351 (1965).

[6] F. Becker and W. Schrämli, *Cem. Lime Mf.*, **42** 91 (1969); M. Amafuji and A. Tsumagari, *Sym. Tokyo. 1968.*

First formation of liquid occurs at about 1250–1280° when the heated mass undergoes a pronounced shrinkage, sometimes known as the Nacken shrinkage.[1]

Montmorillonite and illite are less reactive than kaolinite and higher temperatures are needed for comparable extents of reaction.[2]

The progressive combination of calcium carbonate and clay on heating a cement mix is illustrated[3] in Fig. 37. These data were obtained in the laboratory

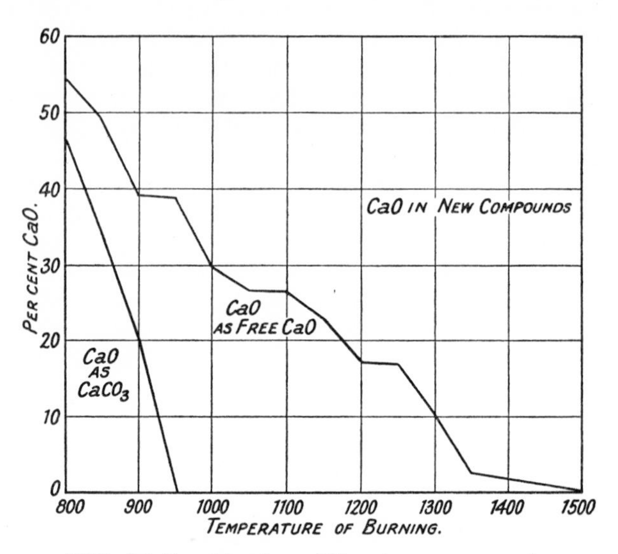

FIG. 37 Combination of lime in a cement mix.

and probably overestimate the maximum amount of free CaO which is produced in practice. On a commercial rotary wet process kiln Weber[4] found a maximum of under 20 per cent free CaO in the mix at a temperature in the region of 1000°.

The course of the reactions can probably best be summarised as follows:

Below 800°	Formation of $CaO.Al_2O_3$, $2CaO.Fe_2O_3$ and $2CaO.SiO_2$ begins.
800–900°	Formation of $12CaO.7Al_2O_3$ begins.
900–1100°	$2CaO.Al_2O_3.SiO_2$ forms and decomposes again. Formation of $3CaO.Al_2O_3$ and $4CaO.Al_2O_3.Fe_2O_3$ starts. All $CaCO_3$ decomposed and free CaO reaches a maximum.
1100–1200°	Formation of major part of $3CaO.Al_2O_3$ and $4CaO.Al_2O_3.Fe_2O_3$. Content of $2CaO.SiO_2$ reaches a maximum.
1260°	First liquid formation starts.
1200–1450°	Formation of $3CaO.SiO_2$ with progressive disappearance of free lime.

[1] R. Nacken and W. Dyckerhoff, *Zement* **11,** 245 (1922).
[2] W. Georg, *Tonind.-Ztg. keram. Rdsh* **80,** 219 (1956); U. Ludwig et al., loc. cit.
[3] H. Kuhl and H. Lorentz, *Zement,* **18,** 604 (1929).
[4] P. Weber, *Zement-Kalk-Gips,* Special Issue No. 9 (English Version), *Heat Transfer in Rotary Kilns* (1963).

This table can only be regarded as approximate, but it indicates broadly the mode of cement compound formation. Overlapping of some of the stages shown may well occur[1] and any inhomogeneity in the raw mix will tend to increase it. Formation of $3CaO.SiO_2$ does not occur to any appreciable extent until liquid is present in the mix at 1260° and above. No formation of this compound, even by solid reaction, is indeed to be anticipated much below this temperature in view of its instability at lower temperatures (cf. p. 46). The formation of tri-calcium silicate becomes complete about 1350–1450° when the free lime remaining is reduced to a small quantity. When the clinkering temperature is reached most, if not all, of the alumina, ferric oxide and minor components will have passed into the liquid and the only solids present are C_3S, C_2S and, often, a little free calcium oxide.

The crystallisation of this liquid on cooling produces again the compounds C_3A and C_4AF as well as the various compounds that arise from the minor components.

Thermo-chemistry of cement formation

The detailed discussion[2] of the heat balance of rotary kilns is outside the scope of this book, but we may consider the basic thermochemical data involved in these calculations. Heat is absorbed, or evolved, in the following reactions in cement formation.

Temperature	*Reaction*	*Heat change*
100°	Evaporation of free water	Endothermic
500° and above	Evolution of combined water from clay	Endothermic
900° and above	Crystallisation of amorphous dehydration products of clay	Exothermic
900° and above	Evolution of carbon dioxide from calcium carbonate	Endothermic
900–1200°	Reaction between lime and clay	Exothermic
1250–1280°	Commencement of liquid formation	Endothermic
Above 1280°	Further formation of liquid and completion of formation of cement compounds	Probably endothermic on balance

Evaporation of free water at 100° requires 538 cal/g, or 968 Btu per lb. The heat absorbed in the dehydration of kaolin[3] at 20° is 186 cal/g (335 Btu per lb) referred to liquid water and 262 cal/g referred to water vapour. In the case of montmorillonite the corresponding values are 58·5 and 94·6 cal/g and for illite 43 and 84·5. The product of the dehydration shows an exothermic reaction at about 900–950° arising from the crystallisation of the amorphous material and amounting in the case of kaolin to 72 ± 10 cal/g dehydrated kaolin. The heat absorbed in dissociation of calcium carbonate is 422 cal/g (760 Btu per lb) at 20° and 393 cal/g (708 Btu per lb) at 890°. For magnesium carbonate the values

[1] G. Mussnug, *Zement* **29,** 217, 233 (1940).
[2] See H. Gygi, *Sym., London 1952*, 750; W. Anselm, *Radex Rdsch.*, 1950, 50; *Revue Matér. Constr. Trav. publ.* **446,** 289 (1952) et seq.; W. Weber, loc. cit.; H. Zur. Strassen, *Zement-Kalk-Gips* **10,** 1 (1957); A. Folliot, *Centre d'Études et de Recherches de L'Industrie des Liants Hydrauliques Tech. Publ. No. 70* (1955), Paris.
[3] H. E. Schweite and G. Ziegler, *Zement-Kalk-Gips* **9,** 257 (1956).

found by different investigators vary from 284 to 324 cal/g at 20° and 250–290 at 590°.

The reaction of the dehydrated clay products with lime is associated with a considerable evolution of heat which varies with the cement composition and raw materials. It is about 100–120 cal/g for the normal raw materials, but only about 50–70 cal/g when slag and limestone are used.[1] A heat evolution of 100 cal/g, or 180 Btu per lb, is sufficient theoretically to raise the temperature of the reacting masses by over 300°.

The heats of formation of the principal cement compounds are shown in Table 12 and the specific heat data needed in thermochemical calculation in Table 13. The former are still subject to some degree of uncertainty, varying as they do with the physical form of the silica and alumina used in the synthesis of the compounds, and with the value assumed for the heat of formation of alumina.

TABLE 12 Heats of formation

Reaction		20°		1300°	
		kcal/mol	cal/g	kcal/mol	cal/g
$2CaO + SiO_2$ (gel)	$= \beta\ C_2S$[2]	33·24	193·0	—	—
$2CaO + SiO_2$ (gel)	$= \gamma\ C_2S$[2]	34·27	199·0	—	—
$3CaO + SiO_2$ (gel)	$= C_3S$[2]	32·77	143·5	—	—
$2CaO + SiO_2$ (aerosil)	$= \beta\ C_2S$[3]	27·5	159·6	—	—
$3CaO + SiO_2$ (aerosil)	$= C_3S$[3]	27·3	119·5	—	—
$2CaO + SiO_2$ (quartz)	$= \beta\ C_2S$[4]	29·8	173	25·2	146
$3CaO + SiO_2$ (quartz)	$= C_3S$[4]	29·4	129	25·2	111
$3CaO + \alpha\ Al_2O_3$	$= C_3A$[5]	4·3	16	5·8	21
$4CaO + \alpha\ Al_2O_3 + Fe_2O_3$	$= C_4\ AF$[5]	12	25	—	—
$6CaO + 2\alpha Al_2O_3 + Fe_2O_3$	$= C_6\ A_2F$[6]	18·3	37	—	—

It will be noted from the data in Table 12 that the formation of tricalcium silicate from β dicalcium silicate and lime is accompanied by an absorption of 0·47 kcal/mol according to Thorvaldson or 0·2 according to Gaulitz *et al.* The magnitude of this endothermic effect is still uncertain since values of 3·2 and 2·4 kcal/mol have been obtained by King[7] and Brunauer[8] respectively. The values obtained by King for the heat evolved on formation of $2CaO.SiO_2$ and $3CaO.SiO_2$ from quartz and lime were 30·19 and 26·98 kcal/mol at 25°. The

[1] H. E. v Gronow, *Zement* **25,** 437, 453 (1936).
[2] O. K. Johannson and T. Thorvaldson, *J. Am. chem. Soc.* **56,** 2327 (1934).
[3] R. Gaulitz, H. E. Schwiete and A. Tin Tiklen, *Zement-Kalk-Gips* **15,** 104 (1962). Values at 25°. The heat of transformation for SiO_2 (aerosil) → SiO_2 (cristobalite) is given as 1·57 kcal/mole.
[4] From value of 3·4 kcal/mole at 20° for SiO_2 (gel) → SiO_2 (quartz) ascribed by W. Eitel and H. Richter, *Zement* **31,** 505 (1942).
[5] T. Thorvaldson, private communication.
[6] E. S. Newman and R. Hoffman, *J. Res. natn. Bur. Stand.* **56,** 319 (1956). Data at 25°.
[7] E. G. King, *J. Am. chem. Soc.* **73,** 650 (1951).
[8] S. Brunauer, J. C. Hayes and W. E. Hass, *J. phys. Chem.* **58,** 279 (1954).

heat of formation of alite containing alumina and magnesia differs from that of C_3S. For a composition $C_{54}S_{16}AM$, Gaulitz obtained a value of 261·4 kcal/mol or 63·3 cal/g and other values for different alumina and magnesia contents. Different values have also been reported for the heat of formation of $3CaO.Al_2O_3$, e.g. 1·59 kcal/mol by Coughlin.[1] The heat evolved in the formation of $CaO.Al_2O_3$ according to Coughlin is 3·69 kcal/mol and for $12CaO.7Al_2O_3$ 2·71 kcal per 1/7 mol. Values for the heats of formation of $K_2O.23CaO.12SiO_2$ and $Na_2O.8CaO.3Al_2O_3$ from their elements have been given by Newman.[2]

In the calculation of the heat balance in cement production it is usual to use the heats of formation of the compounds from the crystalline oxides, because the heat of crystallisation of the oxides formed in the dehydrated kaolin is allowed for separately. It is assumed that this covers the conversion of amorphous alumina to α Al_2O_3 and of amorphous silica to quartz, but it is uncertain if it actually includes the conversion of γ Al_2O_3 to the α form, in which 7·8 kcal/mol are involved,[3] or the full crystallisation of quartz. The heat of crystallisation of amorphous silica to quartz is given by Eitel and Richter[4] as

TABLE 13 Values of mean specific heat (cal/g)*

Temp. range	CaO	Al_2O_3	$CaO.Al_2O_3$	$3CaO.Al_2O_3$	$2CaO.SiO_2$	$3CaO.SiO_2$
20– 300°	0·2010	0·2322	0·2223	0·2118	0·2069 (γ)	0·2070
20– 500	0·2057	0·2491	0·2342	0·2213	0·2185 (γ)	0·2176
20– 675	—	—	—	—	0·2272 (γ)	—
20– 675	—	—	—	—	0·2314 (β)	—
20– 700	0·2091	0·2586	0·2419	0·2261	0·2327 (β)	0·2273
20– 900	0.2129	0·2647	0·2501	0·2289	0·2415 (β)	0·2340
20–1100	0·2153	0·2706	0·2547	0·2319	0·2486 (β)	0·2406
20–1300	0·2170	0·2763	0·2579	0·2348	0·2542 (β)	0·2439
20–1400	—	—	0·2592	—	—	—
20–1500	0·2185	0·2810	—	—	—	0·2475

Temperature range	SiO_2	$CaCO_3$	Kaolin $2SiO_2.Al_2O_3.2H_2O$	Dehydrated kaolin $2SiO_2.Al_2O_3$	Portland cement clinker†
20– 450°	0·239	0·248	0·280	0·238	0·217
20– 500	0·246	0·251	—	0·241	0·220
20– 700	0·258	0·261	—	0·250	0·229
20– 900	0·263	0·266	—	0·258	0·236
20–1100	0·266	—	—	0·265	0·242
20–1400	0·271	—	—	0·270	0·262
20–1500	0·272	—	—	—	0·270

* E. v. Gronow, *Zement*, **25**, 437, 453 (1936).

† CaO 65.0, SiO_2 22.6, Al_2O_3 5.9, Fe_2O_3 3.0, MgO 2.2, SO_3 0.6, Alkalies 0.4, ignition loss 0.4 per cent

[1] J. P. Coughlin, *J. Am. chem. Soc.* **78**, 5479 (1956).
[2] E. S. Newman, *J. Res. natn. Bur. Stand.* **61**, 75 (1958); **62**, 207 (1959).
[3] According to W. Eitel and H. Richter, *Zement* **31**, 505 (1942). A later value by H. v. Wartenberg, *Z. anorg. allg. Chem.* **269**, 76 (1952) is 20·6 kcal/mole. A value of 5·3 kcal/mole at 705° is given by T. Yokokawa and O. J. Kleppa, *J. phys. Chem.* **68**, 3246 (1964).
[4] *Zement*, **31**, 505 (1942).

3·4 kcal/mol SiO_2 and this is used in Table 12 to derive the heats of formation of C_3S and C_2S from quartz.

The specific heat data given in Table 13 were obtained by v. Gronow and should be comparable amongst themselves, but later data for $3CaO.Al_2O_3$ and $2CaO.SiO_2$ differ slightly. For tricalcium aluminate Bonnickson[1] gives heat content data from 25° to 1500° from which mean specific heats may be derived as follows:

Temperature range from 25° to	327°	527°	727°	927°	1127°	1327°	1527°
Mean specific heat of C_3A	0·212	0·220	0·227	0·231	0·234	0·237	0·237

Similarly for dicalcium silicate the following mean specific heats may be derived from heat content data given by Coughlin and O'Brien.[2]

Temperature range from 25° to	327°		527°		727°	927°	1127°	1327°	1527°
Form of C_2S	γ	β	γ	β	α'	α'	α'	α'	α
Mean specific heat	0.202	0.208	0.213	0.219	0.229	0.236	0.242	0.247	0.267

α Al_2O_3 used as a standard for calometric measurements has specific heats[3] as follows

Temperature	64°	150°	250°	354°	459°	552°	550°
Specific heat	0.204	0.232	0.253	0.267	0.277	0.284	0.289

The relative order of the theoretical heat quantities involved in the various stages of cement burning may be illustrated by the following example in which 1·55 kg dry calcium carbonate and clay are used to form 1 kg clinker.

	kcal/kg clinker
Heat absorbed	
Heating raw materials from 20° to 450°	170
Dehydration of clay at 450°	40
Heating materials from 450° to 900°	195
Dissociation of calcium carbonate at 900°	475
Heating de-carbonated material from 900° to 1400°	125
Net heat of melting	25
	1030
Heat evolved	
Exothermic crystallisation of dehydrated clay	10
Exothermic heat of formation of cement compounds	100
Cooling clinker from 1400° to 20°	360
Cooling carbon dioxide from 900° to 20°	120
Cooling steam from 450° to 20° including condensation to water	20
	610

Net theoretical heat required to form 1 *kg clinker*, 420 *kcal*

For different raw mixes, and for slightly different data, values from 400 to 430 kcal/kg clinker have been calculated by different authors. The 'net heat of

[1] K. R. Bonnickson, *J. phys. Chem.* **59**, 220 (1955).
[2] J. P. Coughlin and C. J. O'Brien, *J. phys. Chem.* **60**, 767 (1957).
[3] B. E. Walker, J. A. Grand and R. R. Miller, *J. phys. Chem.* **60**, 231 (1956).

melting' arises from the failure of all the glass to crystallise on cooling, and is obviously a variable factor.[1] The value to be used for the exothermic heat of crystallisation of the dehydrated clay at about 900° is also, as we have seen, in some doubt and may be greater by about one-half than that used in the above example. It is also based on kaolinite and not on other clay minerals that may be present.

The heat required in practice is, of course, considerably greater than this theoretical value on account of the heat lost in the exit gases, in the clinker discharged from the cooler, in radiation and convection losses from the kiln, and, in the wet process, in the evaporation of water from the slurry. With a modern wet process kiln thermal efficiencies[2] are not much above 30 per cent, but with special heat recuperators values approaching 50 per cent are obtained in the semi-dry and dry processes and rather more with shaft kilns.

Cement burning in the rotary kiln

The investigation of the changes occurring during cement burning by experiments on full-size rotary kilns involves various difficulties in temperature measurement and sampling, but with the aid of the conclusions reached from small-scale experiments the results can be interpreted fairly clearly. The clinkering temperature, as measured by optical pyrometers, varies for different raw materials from 1300 to 1450° though occasionally it may fall outside these limits. The flame temperature is difficult to measure accurately, but the maximum temperature of combustion that can theoretically be reached with normal coals and a small excess of air is above 2000°. The actual flame temperature, which depends on the speed of combustion and other factors, has been ascribed various values up to 1700°.

The time of passage of a cement raw mix through a rotary kiln 200 feet long is about 2½ hours and about 6 hours for a 450 feet kiln; the time in the clinkering zone has been estimated to be 20 minutes or less, but it is questionable if the clinker is actually at its maximum temperature for more than a portion of even this short period. The temperature at which a clinker emerges from the kiln is normally between 1200° and 1400°.

The opinion seems to be fairly widely held that the actual formation of clinker takes place very rapidly in a kiln as the temperature rises from 1100–1200° to 1300–1350°. While clinker formation implies an increase in the liquid content sufficient to make the mix cohere into small balls, it does not, however, indicate that compound formation has necessarily proceeded with the same rapidity. Since clinkering is dependent on liquid formation, the minimum temperatures at which liquid forms in raw mixes is of importance. Hansen[3] has given the following data for the minimum eutectic melting temperature:

$3CaO.SiO_2$–$2CaO.SiO_2$–$3CaO.Al_2O_3$	1455°
$3CaO.SiO_2$–$2CaO.SiO_2$–$3CaO.Al_2O_3$–Na_2O	1430°
$3CaO.SiO_2$–$2CaO.SiO_2$–$3CaO.Al_2O_3$–Fe_2O_3	1340°
$3CaO.SiO_2$–$2CaO.SiO_2$–$3CaO.Al_2O_3$–MgO	1375°
$3CaO.SiO_2$–$2CaO.SiO_2$–$3CaO.Al_2O_3$–Fe_2O_3–MgO	1300°
$3CaO.SiO_2$–$2CaO.SiO_2$–$3CaO.Al_2O_3$–Na_2O–Fe_2O_3–MgO	1280°

[1] H. E. v. Gronow, 1936, loc. cit.
[2] P. Weber, loc. cit.; H. Gygi, *Sym., London 1952*, 750; R. J. Davies, ibid., 871.
[3] *J. Res. natn. Bur. Stand.* **4,** 55 (1930).

The lowest temperature of 1280° found in a mix containing ferric oxide, magnesia and soda in addition to the main lime, alumina and silica compounds is not likely to be lowered appreciably further by the addition of any other minor components and is in good agreement with various values of 1250–1280° reported on raw cement mixes. While almost all cement raw mixes are likely to show this same minimum temperature of liquid formation, the amount of liquid formed at this, and progressively higher, temperatures will vary much with the total proportion, and individual amounts, of the minor components. The clinkering temperatures of different mixes will be expected to vary considerably. This is found to be the case. The clinkering range represents a range of liquid contents varying from a minimum amount necessary to give a coherent clinker and a maximum amount where serious balling up commences to occur. In mixes in which the amount of liquid increases slowly with temperature the clinkering range may be fairly wide, but in mixes in which the rate of increase is rapid it will be small. Portland cements with a rather high iron content are examples of the latter type. Clinkers can of course be, and are, produced which in the usual sense of the term appear overburnt or underburnt. In a normal clinker the amount of liquid present at the clinkering stage is 20–30 per cent. This can be roughly estimated from the amount of interstitial matter visible in a thin section of a clinker.

The clinkering range will be a function not only of the amount of liquid formed but of the viscosity and surface tension of the liquid and of the manner in which these properties vary with temperature. This question has been discussed by Terrier[1] and some experimental work has been described by Endell and Hendrick.[2] These authors prepared a basic mix corresponding to a possible liquid composition in a cement comprising only CaO, Al_2O_3, Fe_2O_3 and SiO_2. Small additions of oxides, fluorides, etc. were made to the basic mix and the viscosity measured at various temperatures. It was concluded that the viscosity at 1400° was reduced by mixtures of up to 3 per cent in the following order:

$$Na_2O < CaO < MgO < Fe_2O_3 < MnO.$$

Viscosity increased markedly with increase of SiO_2 and to a much less extent with increase of Al_2O_3. Calcium fluoride in low concentrations (< 0·7 per cent) markedly decreased the viscosity but in larger amounts this substance acted as a mineraliser promoting crystallisation of the liquid. When it is used as a flux (see p. 156), therefore, the clinkering temperature is lowered and the liquid is initially more fluid, but the rapid onset of crystallisation stiffens the mass and the risk of formation of clinker rings is much reduced.

It has been shown[3] from studies on the quaternary system $CaO–Al_2O_3–SiO_2–Fe_2O_3$, and on the effect of the alkalis and magnesia on liquid formation, that the amount of liquid formed in a commercial raw mix at any given temperature can be calculated approximately from the following formulae. These

[1] P. Terrier, *Revue Matér. Constr. Trav. publ.* **585,** 188 (1964).
[2] K. Endell and G. Hendrickx, *Zement* **31,** 357, 416 (1942).
[3] F. M. Lea and T. W. Parker, *Building Research Tech. Pap.* No. 16 (1935). H.M.S.O., London.

formulae apply to any raw mix falling within the usual composition zone of Portland cements.

$$\text{Percentage liquid formed at } 1340^\circ = 6{\cdot}1y+a+b$$
$$\text{Percentage liquid formed at } 1400^\circ = 2{\cdot}95x+2{\cdot}2y+a+b$$
$$\text{Percentage liquid formed at } 1450^\circ = 3{\cdot}0x+2{\cdot}25y+a+b$$

where x, y and b are respectively the percentage contents of alumina, ferric oxide and alkalies in the clinker and a is the percentage of magnesia with the restriction that its maximum value is 2 per cent. This restriction arises from the limit of 5–6 per cent for the solubility of MgO in the clinker liquid corresponding, for a liquid content of, say, 30 per cent, to rather less than 2 per cent of the whole mix. It will be noticed that the amount of liquid formed does not vary with the lime-silica ratio since these do not enter into the equations. The total content of lime and silica does of course affect the amount of liquid formed since it alters the amount of the alumina, ferric oxide and minor components present. The formula for 1338° applies only to clinkers with a weight ratio of alumina to ferric oxide above 1·38. For Portland cements with an alumina:ferric oxide ratio below 1·38 the required formula is

$$\text{Percentage liquid formed at } 1338^\circ = 8{\cdot}5x-5{\cdot}22y+a+b$$

The way in which the liquid content varies with the Al_2O_3:Fe_2O_3 ratio of the clinker is illustrated in Table 14.

TABLE 14 Variation of percentage liquid content of clinker with Al_2O_3 : Fe_2O_3 ratio

Temperature	Al_2O_3/Fe_2O_3 ratio		
	2·0	1·25	0·64
1338°	18·3	21·1	0
Boundary Surface C_3S–C_3A or C_3S–C_4AF	23·5 (1365°)	22·2 (1339°)	20·2 (1348°)
1400°	24·3	23·6	22·4
1450°	24·8	24·0	22·9

The content of liquid formed at 1400° calculated from the formula given varies generally from about 20–30 per cent, though some cements fall outside this range. White cements for example which are deficient in fluxes may have liquid contents at 1400° as low as 15 per cent. Such cements require high burning temperatures and even then, though their total lime content is kept fairly low, the amount of free lime remaining is apt to be high.

Course of the reaction in the rotary kiln

Much interesting information has been obtained from the sampling of materials and the measurement of temperatures at successive points along rotary kilns. Following other earlier tests, comprehensive investigations were made by

Weber[1] on a wet-process rotary kiln 132 metres long, on a dry-process rotary kiln 40 metres long with cyclone suspension preheaters, and on a semi-dry Lepol kiln with travelling grate preheater. In all cases the fuel was pulverised coal and the raw materials limestone and marl. Samples were taken, and temperatures measured, through sampling holes in the rotary kilns or by means of a scoop in the front sintering zone. The results for the wet- and dry-process kilns are shown in Figs. 38 and 39.

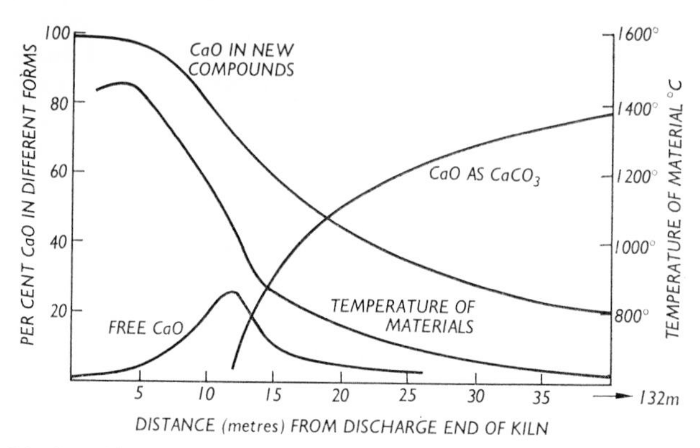

FIG. 38 Combination of a cement mix in a wet-process kiln (132 m).

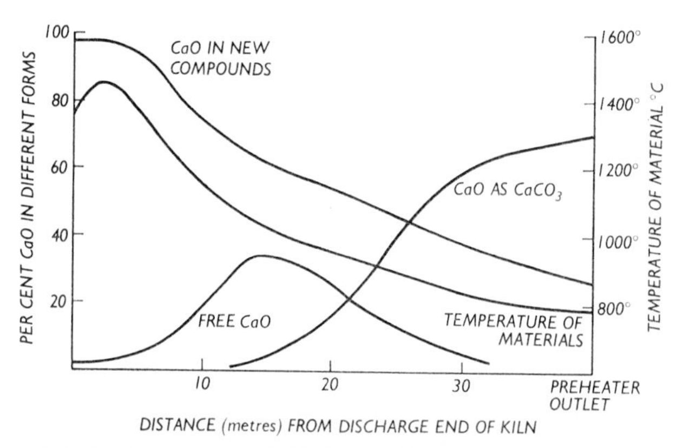

FIG. 39 Combination of a cement mix in a dry-process kiln (40 m) with preheater.

It must be noted that the temperatures measured on the materials are the average temperatures of the mass. The surface material which is heated very rapidly by radiation from the hot gases attains a temperature at least 200° higher than the general mass of the material and it is in this hotter surface material that the various reactions first start. The circulatory movement of the material in the rotary kiln caused this surface material to disappear into the

[1] P. Weber, loc. cit.

mass with the exposure of fresh surface material, but the rate of heating by radiation is so high that a substantial temperature difference between surface and interior of the kiln charge is maintained. The average temperature measured in the materials is not therefore an indication of the temperature at which the various reactions set in. This information has to be derived from laboratory experiments on burning materials at constant temperature as described earlier. Even allowing for this correction it will be evident from Figs. 38 and 39 that decomposition of calcium carbonate by reaction with clay sets in, as the laboratory experiments have indicated, at a temperature well below that of the dissociation temperature of 890°. This reaction in fact starts at about 600°, but free lime does not appear in the charge in any appreciable amount until about half of the calcium carbonate has reacted to form the less basic aluminates and dicalcium silicate. A rapid rise in free-lime content is apparent as the average temperature of the materials approaches 1000° for it is not until the temperature exceeds 1000° that the less basic calcium aluminates and ferrites formed initially are converted into $3CaO.Al_2O_3$ and $4CaO.Al_2O_3.Fe_2O_3$, and not until it exceeds 1250° that the major removal of free lime can take place by conversion of $2CaO.SiO_2$ to $3CaO.SiO_2$. The heat evolved from the exothermic reactions between the limestone and the clay must to a substantial extent be released in the calcining zone up to 1000°. The formation of liquid which occurs above about 1250° is endothermic and involves the absorption of heat. The various temperature zones in a kiln system can broadly be classified as:

	Materials temperature
Drying zone	up to 100°
Preheating zone	100°–550°
Calcining zone	550°–1000°
Sintering zone	1000°–1450°

In Fig. 38 for the wet-process kiln only the results over the last 40 metres, or about one-third, of the kiln are shown. The first third of the kiln was essentially a drying zone and not until the end of this did the average temperature of the materials exceed 100°. The second third of the kiln was largely a preheating zone, though, because of the excess surface temperatures, appreciable release of carbon dioxide occurred by reaction of the calcium carbonate with the clay. The clinker composition was:

	per cent
CaO	64·6
MgO	1·0
SiO_2	20·2
Al_2O_3	6·6
Fe_2O_3	3·8

From this it may be calculated that after the formation of $3CaO.Al_2O_3$, $4CaO.Al_2O_3.Fe_2O_3$ and $2CaO.SiO_2$ was complete some 15 per cent free CaO was left to combine with the $2CaO.SiO_2$ to form $3CaO.SiO_3$. The maximum free-lime content shown in Fig. 38 is about 25 per cent of the total CaO, or some 17 per cent of the clinker. This confirms that the reaction following the peak in the free CaO content in this case is essentially the formation of tricalcium silicate.

From Fig. 39 for the dry-process kiln it will be seen that the raw materials had already attained an average temperature of about 800° when they were discharged into the short rotary kiln. The whole length of the kiln in this case corresponds to the last third of the kiln length in the wet process. The clinker composition was:

	per cent
CaO	66·9
MgO	0·8
SiO_2	21·6
Al_2O_3	6·2
Fe_2O_3	2·5

A calculation of the amount of CaO required to form C_3A, C_4AF and C_2S shows that some 15 per cent is left to convert C_2S to C_3S. This is less than the 23 per cent, calculated on a clinker basis, or 34 per cent of the total CaO shown as the peak free-lime content, indicating that in this dry-process kiln the earlier reactions were still not complete at this point.

Though these data show clearly that the combination of calcium carbonate with clay precedes its direction dissociation, the extent of these early reactions must vary with the kind of raw material and the fineness to which it is ground. The reactions between limestone and marl, which already contains much of the calcium carbonate needed in an intimate association, must for instance occur more readily than in a mix of a dense limestone and shale. For the latter, more of the combination may occur after direct dissociation of the calcium carbonate has commenced.

Influence of fineness and composition of raw mix

The ease with which substantially complete combination of the lime can be effected is influenced not only by the mineralogical nature of the raw materials but also by the fineness to which the raw mix is ground and its composition and physical condition.[1] Hendrickx[2] particularly has contrasted the ease of burning of raw materials in which there is an intimate chemical homogeneity with those made from coarsely crystalline rocks.

Coarse particles of silica or calcite fail to react completely under commercial burning conditions. Heilmann[3] found that not more than 0·5 per cent of silica particles above 0·2 mm size, or nor more than 1 per cent between 0·09 and 0·2 mm should be present in raw mixes with a lime saturation factor (see p. 164) as high as 0·95, but that for a rather lower factor of 0·89 double these amounts may be permissible. Contents up to 5 per cent of pure calcite particles above 0·15 mm in size could be tolerated without serious effect on the ease of burning, while with impure siliceous limestones the proportion of coarse particles could be higher. The relative ease of burning of a mix of a fine-grained siliceous lime-

[1] G. Mussgnug, *Zement-Kalk-Gips* **6**, 46 (1953).
[2] *Revue Matér. Constr. Trav. publ.* **377/378**, 166; **381**, 254; **382**, 273 (1947).
[3] T. Heilmann, *Sym., London 1952*, 711.

stone and marl, compared with that of one made from a pure hard limestone and clay, both with a lime saturation factor of 0·91, is illustrated[1] in Figs. 40 and 41. In practice it is necessary to balance the relative costs of grinding a raw mix more finely and of burning at a higher temperature, allowing also in the wet process for the influence of fineness on the water content needed to form a slurry that can be readily pumped.[2]

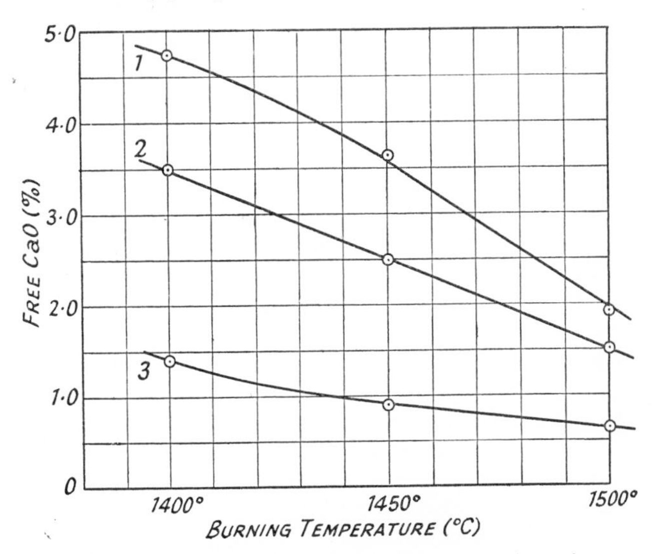

FIG. 40 Burning of pure hard limestone-clay mix.
Mix 1. 4·6 per cent. above 0·2 mm., 9·9 per cent. above 0·09 mm. size.
2. 1·5 ,, ,, ,, ,, ,, 4·1 ,, ,, ,, ,, ,, ,,
3. 0·6 ,, ,, ,, ,, ,, 2·3 ,, ,, ,, ,, ,, ,,

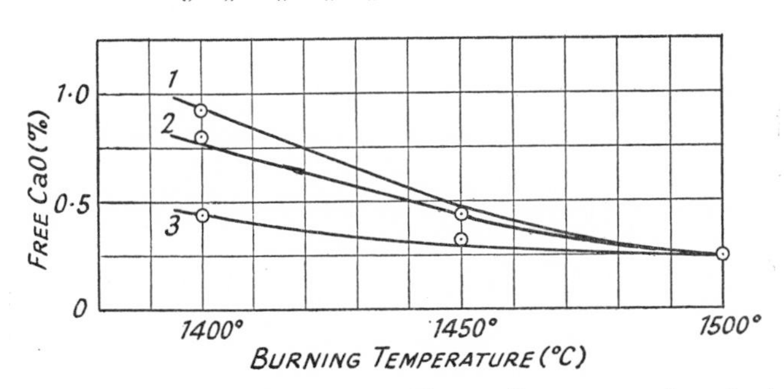

FIG. 41 Burning of fine-grained siliceous limestone and marl mix.
Mix 1. 11·5 per cent. above 0·09 mm. size.
2. 8.5 ,, ,, ,, ,, ,, ,,
3. 5·9 ,, ,, ,, ,, ,, ,,

These conclusions, though based on tests on a variety of raw materials, may need modification for other sources of calcareous aluminosilicate raw

[1] T. Heilmann, loc. cit.

[2] J. C. Gullen, *Sym., London 1952*, 746.

materials. Thus from studies[1] in the U.S.S.R. on the formation of $2CaO.SiO_2$ at 1200° with ground marble and quartz as the raw materials the maximum desirable size of the marble particles was set at 0·1 mm. Comprehensive investigations in Germany[2] showed a progressive reduction in the free-lime content of laboratory clinkers burnt at 1450° as the particle size of a limestone in mixes with clays was reduced below 0·1 mm. A maximum desirable size of 0·1 mm for grains of limestone or quartz was also found by Krämer.[3]

Alumina and iron oxides are, as we have seen, the main fluxes in cement burning. In the earlier stages of liquid formation, up to 1300°, say, the amount of liquid formed for mixes of $Al_2O_3:Fe_2O_3$ weight ratio above 1·38 increases with the iron oxide content, but for compositions with lower values of this ratio the liquid content increases as the ratio rises towards 1·38. At a clinkering temperature of, say, 1400°, however, rather more liquid is formed for each per cent of alumina than of ferric oxide. The latter, nevertheless, appears to have the greater influence on the speed of the solid reaction between lime and silica. It has been shown by Hedvall that the presence of foreign particles in the crystal lattice of materials increases their reactivity, and that, for example, crystobalite containing ferric oxide in solid solution reacts with calcium carbonate at a lower temperature than does the pure material. Alumina does not appear to have a comparable favourable influence on the solid reaction. Since in cement burning a considerable amount of the reactions occurs between the solids before any liquid is formed, this may account for the favourable influence on the speed of reaction often found on lowering the $Al_2O_3:Fe_2O_3$ ratio of a mix.[4] The replacement of alumina by the same percentage of iron oxide increases the potential tricalcium silicate content since proportionately less lime per unit of R_2O_3 is taken up in C_4AF than in C_3A. In high lime mixes, where little C_2S remains to combine with CaO as equilibrium is approached, the decrease thereby caused in the rate of removal of the uncombined lime must eventually more than offset the favourable influence that the iron oxide has exerted at an earlier stage in the burning, as has been shown by Bogue and Taylor.[5] If, however, the substitution is made in such a way that the lime saturation factor (p. 164) is maintained constant, no such risk of the cement becoming too high in lime content arises.

The reaction of CaO with $2CaO.SiO_2$ to form $3CaO.SiO_2$ depends on the lime dissolving in the clinker liquid and must therefore be dependent on the rate of solution. This is controlled by the size of the CaO particles which in turn is dependent on the size of the limestone particles in the raw mix. The time required for solution has been found to be proportional to the particle diameter and to depend exponentially on the reciprocal of the absolute temperature.[6]

[1] M. M. Sychev and M. A. Astakhova, *Tsement* 1962, 12.

[2] H. Lehmann, F. W. Locher and P. Thormann, *Tonind.-Ztg. keram. Rdsh.* **88**, 489, 534 (1964).

[3] H. Krämer, *Zement-Kalk-Gips* **10**, 305 (1957).

[4] K. Akiyama, *Cem. & Cem. Mf.* **6** (1), 17 (1932); S. Nagai and K. Akiyama, *J. Soc. chem. Ind., Japan* **35** (3), 65B, 118B (1932).

[5] R. H. Bogue, *The Chemistry of Portland Cement* (1955), p. 218.

[6] N. A. Toropov and P. F. Rumyantsev, *Zh. prikl. Khim.* **38**, 1614, 2115 (1965).

The times found for the solution of CaO grains in the clinker liquid are shown in Table 15.

TABLE 15 Rate of solution of CaO in clinker liquid

Temperature °C	Time (min) of solution of CaO grains Diameter of grains (mm)			
	0·1	0·05	0·025	0·01
1340°	115	59	25	12
1375°	28	14	6	4
1400°	15	5·5	3	1·5
1450°	5	2·3	1	0·5
1500°	1·8	0·7	—	—

These data, it was found, could be represented approximately by the equation

$$\log t = \log D/A + 0{\cdot}43E/RT$$

where t is the time, D the particle diameter, A a constant, T the absolute temperature and E the activation energy with a value of 146 kcal/mol. A similar form of relation was also found for the rate of solution of quartz grains[1] in the clinker liquid.

The silica modulus, that is the weight ratio of silica to alumina plus ferric oxide, has an important influence on the burning of Portland cement. When the sum of the $Al_2O_3+Fe_2O_3$ is low the amount of liquid formed at the clinkering temperature becomes insufficient to permit sufficiently rapid combination of the remaining CaO. The maximum upper limit of the silica modulus for most raw materials is about four, but in favourable cases, as with some siliceous limestones in which most of the quartz present is not above 0·005 mm size, it is possible[2] to burn commercially mixes with silica moduli between 6 and 12.

Clinker equilibrium and the cooling process

In the preceding pages we have traced the reactions occurring in cement burning up to the point where clinker is formed and compound formation practically completed. The condition which exists at the maximum temperature reached by the clinker, and the subsequent changes which occur on cooling, can now be examined in the light of our knowledge of the liquid-solid equilibria.

An enlarged diagram of the tricalcium silicate and neighbouring fields in the system $CaO–Al_2O_3–SiO_2$ is shown in Fig. 42. Most modern Portland cements lie in the higher lime portion of the Portland cement zone shown. A lime-alumina-silica mix of such composition forms a liquid and one, or two, solids at the clinkering temperature. Taking the clinkering temperature for lime-alumina-silica mixes as 1500°, a cement of composition M will at that temperature consist of solid C_3S and a liquid V, given by the point where the

[1] N. A. Toropov and P. F. Rumyantsev, *Zh. prikl. Khim.* **38,** 2113, 2115 (1965).
[2] K. Dyckerhoff, *Zement-Kalk-Gips* **11,** 196 (1958).

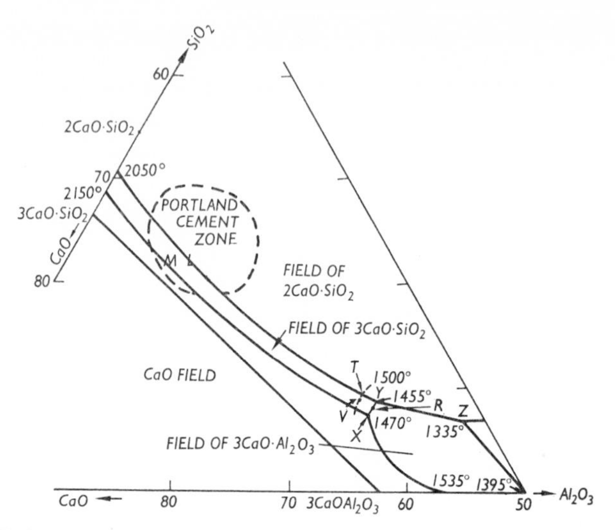

FIG. 42 The $3CaO.SiO_2$ field in system $CaO–Al_2O_3–SiO_2$.

line from C_3S to M produced cuts the 1500° isotherm, shown dotted. On cooling such a clinker, more C_3S will separate from the liquid, the composition of which will move along MV produced until the C_3S–C_3A boundary is met at R. Here C_3A will commence also to separate and the liquid composition will now follow the boundary, along RY, until the quintuple point Y is reached. At this point C_2S also separates and the last remaining liquid will finally solidify at this point.

A cement mix of rather lower lime content, such as represented by the point L, will at 1500° be composed of the solids C_3S and C_2S, and a liquid T lying at the point where the 1500° isotherm intersects the boundary between the primary phase fields of these two compounds. This liquid on cooling will deposit more of the two already existing solids and move down the boundary curve until the quintuple point Y is reached. Here solid C_3A also separates and the crystallisation of the liquid is completed.

These crystallisation paths are those which will be followed if the solid and liquid maintain equilibrium during cooling. In practice, however, it seems probable that cooling of a clinker is too rapid for equilibrium to be maintained and the liquid may either crystallise separately without regard to the solids which are already present or some of it may solidify as an uncrystallised glass. The meaning of this we may examine a little further. The liquid compositions V, R, T, Y all fall outside, on the low lime side, of the triangle formed by joining together the points representing the compositions C_3S, C_2S and C_3A. Such liquids *alone* therefore could never crystallise to give these three solids since the lime content is too low. The deficiency of lime in the liquid is made up by some of the solid tricalcium silicate redissolving in the liquid when the quintuple point Y is reached and dicalcium silicate crystallising out. If the time available is insufficient to allow this reaction to occur on cooling, the product must contain more tricalcium silicate than would be present if equilibrium were maintained. The residual liquid *Y* then crystallises to form C_2S, C_3A and $C_{12}A_7$. In this event the crystallisation path does not end at *Y* and the liquid does not completely solidify until the lower eutectic *Z* is reached. The proportion of C_3S in the product has thereby been increased at the expense of the content of C_2S and C_3A, and some $C_{12}A_7$ is formed in part replacement of the latter. The content of C_3S in the cooled clinker is similarly raised if the liquid *V*, for instance, failed to crystallise at all and formed a glass on cooling, or if crystallisation occurred as far as *Y*, but at this point the liquid was quenched to a glass. It is of interest to calculate the differences which will result in the composition of the cooled clinker.

The mix *M*, of composition CaO 69·5 per cent, SiO_2 20·5 per cent, Al_2O_3 10 per cent, when crystallised to yield its final equilibrium products will contain 60 per cent C_3S, 13·5 per cent C_2S, and 26·5 per cent C_3A. At 1500°, when the liquid composition is *V*, the mix is made up of 68 per cent C_3S and 32 per cent liquid.[1] At *R*, just as C_3A commences to crystallise, the mix consists of 69 per cent C_3S and 31 per cent liquid. When the quintuple point *Y* is just reached, but before any C_2S has actually separated, the mix consists of 70·5 per cent C_3S, 6·5 per cent C_3A and 23 per cent liquid. Hence by cooling rapidly from temperatures corresponding to *V*, just above *R*, or just above *Y*, we obtain respectively 68 per cent, 69 per cent, and 70·5 per cent C_3S in the clinker as compared with 60 per cent when the mix is completely and slowly crystallised to its equilibrium products. If we now examine cementing mixes containing ferric oxide in addition, a rather striking contrast is found in certain cases. Let us consider mixes containing a low ratio of Al_2O_3 to Fe_2O_3, taking a value for that ratio of 0·64 such that all the alumina and iron are combined in the compound C_4AF, and no C_3A is present. The system CaO–C_2S–C_4AF is shown in Fig. 43. Consider a cement composition CaO 65·4, SiO_2 22·7, Al_2O_3 4·6 and Fe_2O_3 7·3 represented by the point *E*. In order to keep conditions comparable with our previous calculation we will take a clinkering temperature such that the same amount of liquid, 32 per cent, is present as in our previous example. The required temperature is 1450°. At this temperature the mix is composed of solids C_3S and C_2S and a liquid of composition *F*, this being the point of intersection of the 1450° isotherm and the boundary between the primary phase fields of these two compounds. This liquid composition *F* is seen to lie just on the high lime side of the triangle C_3S–C_4AF–C_2S, and hence actually contains slightly more lime than is required to form these three compounds. If the mix of gross composition *E* is cooled from 1450° the liquid composition

[1] The method by which these values are derived is as follows. At *V* $\frac{\%\ \text{liquid}}{\%\ C_3S} = \frac{SM}{MV}$, where *S* designates the C_3S composition point. At *R*, similarly, $\frac{\%\ \text{liquid}}{\%\ C_3S} = \frac{SM}{MR}$. When *Y* is reached, but no C_2S yet crystallised, $\frac{\%\ \text{liquid}}{\%\ \text{solid}} = \frac{DM}{MY}$, where *D* represents the point of intersection of *YM* produced on the line joining the composition points C_3S and C_3A. The solids present are C_3S and C_3A and their relative proportions are $\frac{C_3S}{C_3A} = \frac{DA}{DS}$, where A and S represent the C_3A and C_3S composition points.

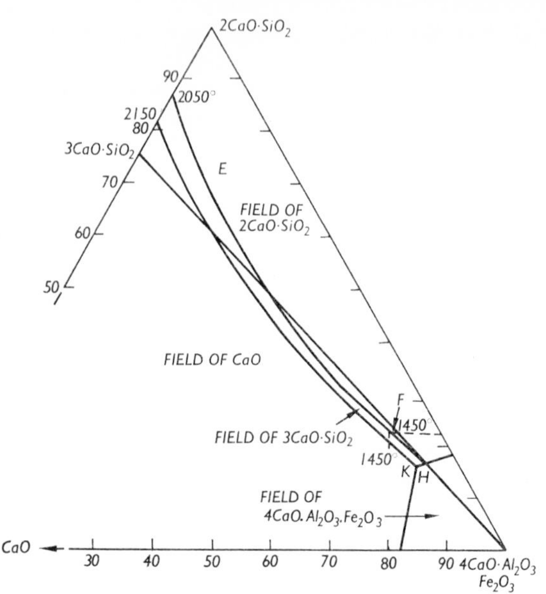

FIG. 43 The $3CaO.SiO_2$ field in system CaO–$2CaO.SiO_2$–$4CaO.Al_2O_3.Fe_2O_3$

moves from F to the quintuple point H, and at this point some solid C_2S redissolves, combined with the excess lime in the liquid and deposits out as C_3S, the whole mix going solid at this point to form C_3S, C_2S and C_4AF. If cooling is too rapid for equilibrium to be maintained between solid and liquid there will thus be a deficiency of C_3S and a small amount of free CaO will separate when the liquid finally solidifies. In this event final solidification takes place at K and not at H. It may be calculated that the cooled clinker will contain 52·9 per cent C_3S if equilibrium is completely maintained on cooling, 50·6 per cent if the liquid F fails to maintain equilibrium with the pre-existing solid and crystallises independently, and only 41·1 per cent if the liquid F fails to crystallise and is quenched to form a glass. It is seen that while in the previous case the C_3S content obtained by rapid cooling from the clinkering temperature exceeded the equilibrium value by 10 per cent, in the present case it is some 12 per cent lower if the liquid forms a glass and 2 per cent lower if it crystallises independently.

For ordinary Portland cements, which have an alumina-ferric oxide ratio intermediate between the two cases considered, the tricalcium silicate content obtained by rapid cooling from the clinkering temperature is higher than that obtained by slow cooling when the $Al_2O_3:Fe_2O_3$ ratio is moderate or high, and lower when the ratio is small. This is illustrated by the data[1] in Table 16 (p. 140), where the compound contents of four cements composed of lime, alumina, silica and ferric oxide are given. The values shown are those obtained as follows: (1) when the cement crystallises completely to yield its equilibrium products

[1] Cf. F. M. Lea and T. W. Parker, *Building Research Tech. Pap.* No. **16** (1935), H.M.S.O., London who have given methods whereby such values can be calculated. Also F. M. Lea, *Cement* **8**, 29 (1935).

and the compound content given by the Bogue calculation (see p. 115), (2) when the cement is cooled so rapidly from a clinkering temperature of 1450° that the liquid present at that temperature fails to crystallise at all and forms a glass, (3) when the cement is cooled from 1450° and the liquid crystallises independently of the solids already existing.

For the first case, where the $Al_2O_3:Fe_2O_3$ ratio is 2, the content of C_3S is slightly raised by rapid cooling; rather larger differences can be obtained when the $Al_2O_3:Fe_2O_3$ ratio is higher as in the second and third cases. In the fourth case where the $Al_2O_3:Fe_2O_3$ ratio is 0·67, the content of C_3S is decreased by rapid cooling and this is particularly evident when a glass is formed. The glass in this case has a high lime content. The contents of C_3A and C_4AF are reduced to zero in each case if the liquid is entirely quenched to a glass, since at the clinkering temperature all the alumina and ferric oxide are present in the liquid.

The compound content of cements in which the equilibrium is assumed to be frozen from the clinkering temperature can be calculated from the data given by Lea and Parker. The complete calculation is somewhat tedious, but, if a clinkering temperature of 1400° is assumed, approximate corrections for the deviation of the values from those given by the Bogue method (p. 115) can be obtained from the equations given below for cements of $Al_2O_3:Fe_2O_3$ weight ration between 0·9 and 6·1. These corrections give values of the compound content accurate to within ± 2 per cent for a clinkering temperature of 1400°. They are not appreciably affected by the use of higher clinkering temperatures, but may lead to rather larger errors for mixes clinkered at much lower temperatures. The following formulæ show the amounts to be added to the Bogue values for the two alternate assumptions that the liquid present at the clinkering temperature crystallises independently of any pre-existing solid, or fails to crystallise and forms a glass. The corrections depend only on the alumina and ferric oxide content of the cement. Let % content of Al_2O_3 in cement = X, % content of Fe_2O_3 in cement = Y.

1. Clinker liquid crystallises independently.

(*a*) Cements of $Al_2O_3:Fe_2O_3$ ratio between 0·9 and 1·7.
The Bogue values hold without correction.

(*b*) Cements of $Al_2O_3:Fe_2O_3$ ratio between 1·7 and 6·1.
The corrections to be added to the Bogue values are:

C_3S	+(1·8X−2·8Y)
C_2S	+(2·1Y−1·4X)
C_3A	+(2·5Y−1·6X)
$C_{12}A_7$	+(1·2X−1·8Y)
C_4AF	Nil.

The $C_{12}A_7$ content by the Bogue method is nil but, as explained earlier, it appears in the products of the frozen equilibrium.

TABLE 16 Effect of cooling conditions on compound content

Cement composition	$CaO = 68\% : SiO_2 = 23\% : Al_2O_3 = 6\% : Fe_2O_3 = 3\%$		
	Complete crystallisation	1450° liquid quenched to glass	1450° liquid crystallises independently
$3CaO.SiO_2$	57·5	59·6	59·6
$2CaO.SiO_2$	22·6	15·6	21·0
$3CaO.Al_2O_3$	10·8	0	9·8
$4CaO.Al_2O_3.Fe_2O_3$	9·1	0	9·1
Free CaO	0	0	0
$12CaO.7Al_2O_3$	0	0	0·5
Glass	0	24·8	0
Cement composition	$CaO = 68\% : SiO_2 = 23\% : Al_2O_3 = 7\% : Fe_2O_3 = 2\%$		
$3CaO.SiO_2$	52·2	59·4	59·4
$2CaO.SiO_2$	26·5	15·3	21·4
$3CaO.Al_2O_3$	15·2	0	9·7
$4CaO.Al_2O_3.Fe_2O_3$	6·1	0	5·9
Free CaO	0	0	0
$12CaO.7Al_2O_3$	0	0	3·8
Glass	0	25·5	0
Cement composition	$CaO = 66\% : SiO_2 = 24\% : Al_2O_3 = 7·5\% : Fe_2O_3 = 2·5\%$		
$3CaO.SiO_2$	32·4	38·5	38·5
$2CaO.SiO_2$	44·5	33·3	39·9
$3CaO.Al_2O_3$	15·7	0	10·6
$4CaO.Al_2O_3.Fe_2O_3$	7·6	0	7·6
Free CaO	0	0	0
$12CaO.7Al_2O_3$	0	0	3·5
Glass	0	28·1	0
Cement composition	$CaO = 67\% : SiO_2 = 23\% : Al_2O_3 = 4\% : Fe_2O_3 = 6\%$		
$3CaO.SiO_2$	62·7	55·1	61·3
$2CaO.SiO_2$	18·6	19·4	19·4
$3CaO.Al_2O_3$	0·5	0	0·5
$4CaO.Al_2O_3.Fe_2O_3$	18·2	0	18·2
Free CaO	0	0	0·6
$12CaO.7Al_2O_3$	0	0	0
Glass	0	25·5	0

2. Clinker liquid forms a glass.

No C_3A or C_4AF appear and in their place is formed a glass containing all the alumina and ferric oxide, and some lime and silica. The amount of this glass is given by the formula:

$$\text{Glass} \qquad +(2{\cdot}95X+2{\cdot}2Y)$$

The corrections to be added to the Bogue values for C_3S and C_2S are:

$$C_3S \qquad +(1{\cdot}8X-2{\cdot}8Y)$$
$$C_2S \qquad +(1{\cdot}9Y-2{\cdot}1X)$$

The corrections in this section apply to cements of Al_2O_3:Fe_2O_3 ratio between 0·9 and 6·1.

For most cements the correction to be added to the Bogue value is positive for C_3S and negative for C_2S.

There is a further point arising from the phase equilibrium diagrams which merits a little consideration since it forms the basis of a formula for the proportioning of cement raw mixes. Referring to Fig. 42, it will be observed that any cement composition lying on the high lime side of the line joining C_3S composition to the quintuple point X must at a clinkering temperature above 1470° consist of a liquid lying on the boundary curve between the fields of lime and tricalcium silicate.[1] These two solids are therefore present at the clinkering temperature. On cooling, the free lime should react with the liquid when the quintuple point X is reached and disappear. In practice, however, cooling is probably too rapid for this to occur and, however well burnt, cement of such composition must contain some free lime in the clinker. It is therefore considered that the line from C_3S to the point X marks the maximum lime content that can be combined in a cement consisting of lime, alumina and silica alone. A consideration of similar relationships in mixes containing ferric oxide in addition had led to the conclusion that for cements composed of the four oxides—lime, alumina, silica and ferric oxide—the plane joining this line to the compound C_4AF (see Fig. 21) represents a similar maximum limit to the amount of lime which can be combined by clinkering in the rotary kiln.

The full meaning of our conclusion in the previous chapter that Portland cement approximates to an equilibrium product *at its clinkering temperature* will now be apparent. Equilibrium is probably not maintained during cooling, with the result that the relative proportions of the different compounds differ from those calculated on the basis of complete equilibrium. From recent studies of the interstitial phase in Portland cement clinker it appears probable that the content of true glass is usually small and that much of the glass is a crypto-crystalline material so disordered in structure that it has a higher heat of solution than that of the fully ordered crystals. Its properties may correspondingly be expected to differ from that of true crystals. The heat-of-solution method for

[1] For such compositions a line drawn from a point on, say, the 1500° contour in the C_3S primary phase field through the original mix composition will, when produced, intersect the CaO–C_2S side of the triangle (Fig. 42) between the CaO and C_3S composition points, showing that CaO and C_3S must both be present in the solid at the clinkering temperature and the liquid composition lie on the CaO–C_3S boundary curve.

determination of glass content must measure not glass only but also crypto-crystalline material. For convenience we will still refer to this mixture of crypto-crystalline material and glass as glass. Dahl[1] used the heat-of-solution method as an estimate of glass content and assumed that crystallisation of clinker liquid proceeded normally down to the point at which freezing occurred, or alternatively that the liquid crystallised independently down to this point. Equations for the compound content were then derived appropriate to these two alternative cases on the basis of Lea and Parker's data for the quarternary system CaO–Al_2O_3–SiO_2–Fe_2O_3.

Factors influencing the compound content

We have seen in the previous section that the compound content as conventially calculated on the assumption of complete equilibrium requires correction to take account of the failure of the clinker to maintain equilibrium during cooling. Though the modified values just described give an improved picture of the make-up of Portland cement, the values they yield for the compound contents are still not true values. They still ignore the solid solution of Al_2O_3, Fe_2O_3 and MgO in tricalcium silicate, the variability in $A:F$ ratio of the iron compound, and the effect of the minor components.

Iron compound

The composition of the ferrite compound in different Portland cements can vary from about C_6A_2F to C_6AF_2 though the median value is fairly close to C_4AF. More alumina is taken up per unit of ferric oxide in C_6A_2F than in C_4AF thus reducing the amount of alumina available to form C_3A. For each 1 per cent Fe_2O_3 combined in C_6A_2F for example, instead of in C_4AF, the content of the ferrite compound will increase by 1·36 per cent, that of C_3A decrease by 1·70 per cent and 0·35 per cent less lime will be needed. This is available to convert more C_2S to C_3S decreasing the former by 1·08 per cent and increasing the latter by 1·43 per cent. Formation of C_6AF_2 instead of C_4AF would have the reverse effect, increasing the C_3A and decreasing the C_3S content. It is not possible to allow for these variable corrections in the calculation of compound contents and, to proceed further, recourse has to be made to X-ray diffraction methods as discussed later.

Alkalis

The K_2O and Na_2O present in clinker combine preferentially with sulphur trioxide to form a sodium-potassium sulphate solid solution (aphthitalite), but the content of the sulphur trioxide (0·1–0·5 per cent) is not normally sufficient to satisfy all the alkalis. The excess forms the compounds $KC_{23}S_{12}$ (actually a solid solution) and NC_8A_3 if complete crystallisation of the clinker liquid occurs, but unless cooling is slow they are likely to remain in the glass. Since, however, in both compounds alkali substitutes for lime, 1 molecule K_2O for CaO in every 12 molecules C_2S and 1 molecule Na_2O for CaO in every 3

[1] L. Dahl, *Bull. Res. Dev. Labs Portld Cem. Ass.*, No. 1, 1939; No. 59, 1956.

molecules C_3A, additional CaO is left free to convert C_2S to C_3S. In the majority of cements the content of K_2O exceeds that of Na_2O and the sulphate formed then tends to have a composition[1] around $K_{0\cdot75}Na_{0\cdot25}SO_4$. Thus, for example, in a clinker containing 0·5 per cent K_2O, 0·25 per cent Na_2O, 0·2 per cent SO_3, there will be formed 0·4 per cent of this compound, leaving about 0·3 per cent K_2O and 0·2 per cent Na_2O available to form the silicate and aluminate compounds. The 0·3 per cent K_2O will react with 6·6 per cent C_2S to form 6·7 per cent $KC_{23}S_{12}$ and will liberate 0·2 per cent CaO which will convert 0·5 per cent of the remaining C_2S to 0·7 per cent C_3S. The C_2S content will thus decrease by 7·1 per cent. Similarly the 0·2 per cent Na_2O will form 2·6 per cent NC_8A_3 decreasing the C_3A content by 2·6 per cent and, because of the additional CaO available, increasing the C_3S content by 0·7 per cent and decreasing that of C_2S by 0·6 per cent.

In this example the presence of alkalis only affects the C_3S content of a cement by the order of 1 or 2 per cent. The C_2S content is substantially affected but, if the properties of C_2S and $KC_{23}S_{12}$ are, as is probable, very similar, the C_2S content can be regarded as the sum of the two and this is affected by less than 1 per cent. The influence of potash on the relative content of the silicates depends, however, on the lime content of the clinker. The $KC_{23}S_{12}$ is not available for conversion into C_3S and the maximum amount of lime that can be combined in the silicate compounds is thus reduced. In the above example, the 6·7 per cent $KC_{23}S_{12}$ cannot be converted to C_3S whereas the 6·6 per cent of C_2S it replaced could have combined with 2·1 per cent CaO to form 8·8 per cent C_3S. The maximum possible C_3S content of the cement is thus substantially reduced and, if the amount of C_2S available is insufficient to react with the K_2O, this will decompose C_3S according to the equation

$$12(3CaO.SiO_2)+K_2O = K_2O.23CaO.12SiO_2+13CaO$$

From this it can be calculated that each 0·1 per cent K_2O available to react with the silicates will reduce the maximum possible C_3S content by 2·9 per cent and liberate 0·77 per cent free CaO. While the presence of potash can have an important effect on high lime cements it is not possible to calculate the amount of K_2O available for the above reaction with any precision since small amounts of it may be taken up in solid solution in other compounds.

Displacement of tricalcium silicate field by minor components

In the discussion earlier in this chapter on the clinker equilibrium and the cooling process (p. 135) we have seen that in the ternary system $CaO–Al_2O_3–SiO_2$ the liquid at the clinkering temperature is deficient in lime and that to maintain equilibrium during cooling it has to dissolve some solid C_3S and deposit out C_2S. Conversely in the system $CaO–Al_2O_3–Fe_2O_3$ the corresponding liquid contains an excess of lime and has to dissolve C_2S crystals and deposit out C_3S. In the quaternary system $CaO–Al_2O_3–SiO_2–Fe_2O_3$ either condition can arise, depending on the $Al_2O_3:Fe_2O_3$ ratio, and lead, if the equilibrium is frozen, to the deviations from the values calculated by the Bogue

[1] T. F. Newkirk, *Sym., London 1952*, 151; *Bur. Res. natn. Bur. Stand.* **47**, 349 (1951)

formula for the compound content given on page 139. These effects arise from the position of the C_3S primary phase field in the particular system concerned. Other components may be expected also to have some influence on the position of this field and so affect the relative amounts of C_3S and C_2S present at the clinkering temperature. Thus the results of Swayze show that the introduction of 5 per cent MgO into mixes of CaO, Al_2O_3, SiO_2, Fe_2O_3 causes a slight displacement of the C_3S primary phase field towards more acidic regions; and the same applies when Na_2O is added to the lime-alumina-silica mix. Both these additions will, therefore, tend to give a more basic solid and a more acidic liquid, i.e. to raise slightly the amount of C_3S present at the clinkering temperature. For a frozen equilibrium this would tend to increase somewhat the magnitude of the corrections to the Bogue values given on page 141. This division of the mix into a more basic solid and a more acidic liquid does not affect the compound content if the liquid reacts with the solid during cooling and so maintains equilibrium, as assumed in the Bogue formulae. As we have seen, however, this cooling reaction can only occur partially at best and with quick cooling is unlikely to occur at all. Since data are not yet available on all minor components to fix the position of the C_3S primary phase field in the system of some eight or nine components, the corrections for this position may be left, probably without undue error, to rest on the data obtained for the $CaO–Al_2O_3–SiO_2–Fe_2O_3$ system.

Solid solution. The content of the crystals of the silicates is influenced by solid solution of minor components. The alite in clinker, as we have seen earlier, is not pure C_3S but contains up to 4 per cent of alumina, ferric oxide and magnesia. The Mg^{2+} ion replaces Ca^{2+} thus making more lime available to convert C_2S to C_3S. The ferric oxide seems partly to replace lime and partly silica while part of the alumina replaces silica. On balance these effects would tend to leave more silica available for C_2S formation. Dicalcium silicate can also take up a few per cent of other oxides in solid solution thus raising the content of the crystals of this compound a little above the calculated value. There are consequential effects on the contents of the C_3A and ferrite compound but the content of these is predominantly influenced by the composition of the ferrite.

Deviation of compound content from calculated values

The effect of the various composition factors just discussed in causing deviations from the Bogue formula can only be qualitatively assessed and not calculated quantitatively as was done for the equilibrium between clinker and liquid. The broad qualitative effects are summarised in Table 18. When the actual content is raised above the Bogue value it is shown as +, where decreased as −, and were unchanged as 0. The uncertainties inherent in the calculation of compound content from chemical composition stimulated the use of microscopic and X-ray diffraction methods of estimation. Neither of these two methods is free from its own particular difficulties but with the development of the X-ray method this has now become the predominant one.

In measuring phases by microscopic methods the result is a volume percentage. Table 17 gives values of the densities used for converting the measured amounts to weight percentages.

TABLE 17 Densities of clinker compounds

C_3S	3·13	
C_2S	3·28	
C_4AF	3·77	
C_3A	3·00	dark interstitial
Glass	3·00	dark interstitial
MgO	3·58	
CaO	3·32	

There is usually no difficulty in microscopic measurements[1] in identifying C_3S and C_2S though the former may have reaction rims and the latter may be underestimated when very small secondary crystals are present. In slowly cooled or in quenched clinker the interstitial compounds can also be measured with some accuracy but at intermediate rates of cooling the scale of crystallisation of the interstitial matter may be too small for accurate work. With synthetic clinker containing only CaO, Al_2O_3, SiO_2 and Fe_2O_3, Parker and Nurse[2] obtained very good agreement with the Bogue formula when the clinkers were fully crystallised and with the Lea and Parker calculation when they were quenched. When alkalis and MgO were present the C_3S as measured rose above the calculated value.

TABLE 18 Effect of various factors on actual content of compounds as compared with values calculated from the Bogue formula

Factor	Effect on compound content				
	C_3S	C_2S	C_3A	C_4AF	MgO
Liquid quenched to glass:					
high A/F	+	−	disappears	disappears	disappears
low A/F	−	+	,,	,,	,,
Liquid crystallises independently:					
high A/F	+	−	−	0	0
low A/F	0	0	0	0	0
Iron phase C_6A_2F	+	−	−	+	0
C_6AF_2	−	+	+	−	0
High K_2O	+	−*	0	0	0
,, Na_2O	+	−	−†	0	0
C_3S field displacement by MgO	+	−	−	−	0
Probable solid solutions	?	?	−	−	−

* The potassium compound $K_2O.23\,CaO.12\,SiO_2$ is not included as C_2S
† The sodium compound $Na_2O.8\,CaO.3\,Al_2O_3$ is not included as C_3A

[1] L. S. Brown, *Proc. Am. Concr. Inst.* **44,** 877 (1948); *J. Res. Dev. Labs Portld Cem. Assoc.*, **1** (3), 23 (1959).
[2] T. W. Parker and R. W. Nurse, *J. Soc. chem. Ind., Lond.* **58,** 255 (1939).

The first major study on commercial clinker was made by Insley and his co-workers.[1] Following this G. W. Ward[2] measured 10 clinkers and compared the results with those obtained both from the Bogue formula and from a modified calculation using the result of a glass determination by heat of solution combined with the assumption that crystallisation had followed an equilibrium path down to the point at which this glass was formed. Each clinker was studied as it came from the plant and also when reheated and quickly and slowly cooled. The two calculations do not differ appreciably with regard to C_3S and C_2S and in general the observed values for these two components were from 0 to 10 per cent high and low respectively.

The C_3A content of the plant clinkers was always lower than the Bogue value, by up to 10 per cent in extreme cases, and to a lesser extent below the results from the modified calculation. Only in the case of the slowly cooled samples did the C_3A content approach the Bogue value. The content of C_4AF in plant clinkers approached the Bogue value for clinkers of $A:F$ ratio up to about two, but was low at higher ratios. The amount of glass determined microscopically tended to be lower than that given by the heat of solution method, but for most of the plant clinkers the discrepancy was only a few per cent. The free MgO found was less than the total present to an extent that increased with the rate of cooling, as would be expected from its solubility in the glass.

The difficulty of identification of the interstitial phases was brought out clearly in L. S. Brown's study of the 21 clinkers used in preparing cements for the Portland Cement Association's long-term study of concrete durability.[3] Glass, C_4AF and C_3A were identified microscopically but some interstitial material had to be classified as 'undifferentiated'. Again the measured C_3S was generally higher than calculated except for some clinkers of low $A:F$ ratio, and C_2S the reverse. The discrepancies for C_3A were reduced if 'dark prismatic' matter was counted with C_3A, but even then the measured values were low by amounts ranging from 1 to 8 per cent. The measured C_4AF tended to fall progressively more below the calculated value as the amount of glass increased. The microscopic measurement of the glass gave lower values, and sometimes substantially so, than the heat of solution method. Some C_3A, C_4AF and glass were, of course, lost in the microscopically undifferentiated fraction of the clinker. It will be evident that accurate microscopic measurements of the C_3A, C_4AF and glass are often difficult, and sometimes impossible.

Estimation of compound content by X-ray diffraction

The X-ray diffraction methods[4] for estimating the compound content of cements and the difficulties that arise have been discussed earlier (p. 104). Broadly they depend on the measurement of the intensities of suitable X-ray diffraction lines combined with a calibration from the single compounds. In the case of the ferrite compound the $A:F$ ratio can also be determined from the change in

[1] H. Insley, E. P. Flint, E. S. Newman and J. A. Swenson, *J. Res. natn. Bur. Stand.* **21,** 355 (1938).
[2] G. W. Ward, *J. Res. natn. Bur. Stand.*, **26,** 49 (1941).
[3] *Proc. Am. Concr. Inst.* **44,** 441 (1948).
[4] See *Sym. Analysis of Calcareous Materials: Monogr. Soc. chem. Ind., Lond.*, No. 18 (1964).

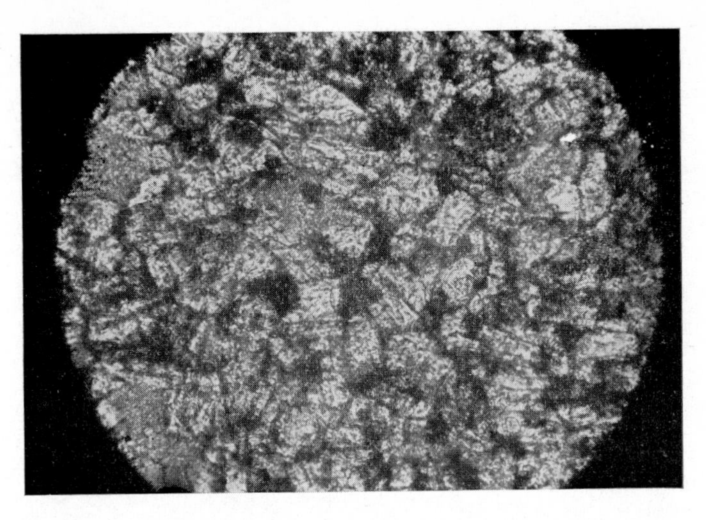

(i) Thin Section of Portland Cement Clinker Showing Prismatic Tricalcium Silicate (× 235)

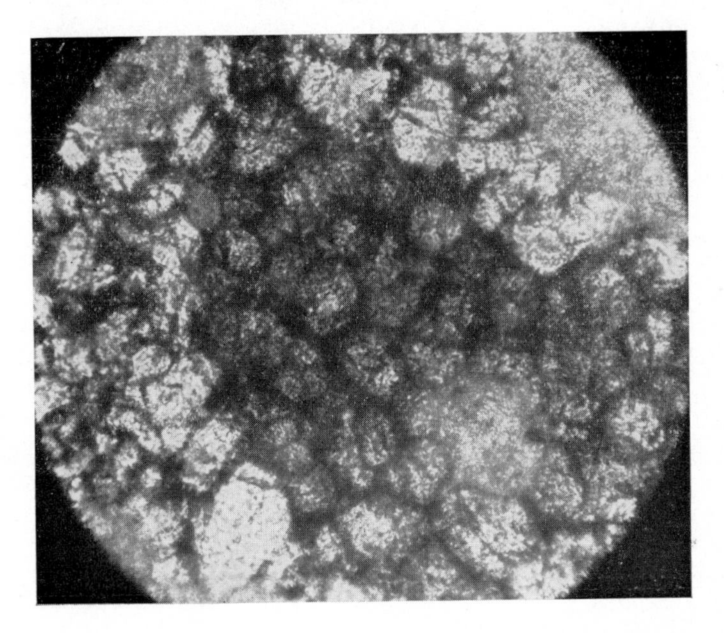

(ii) Thin Section of Portland Cement Clinker Showing Dicalcium Silicate in Centre and Tricalcium Silicate in Outer Part (× 235)

PLATE IV

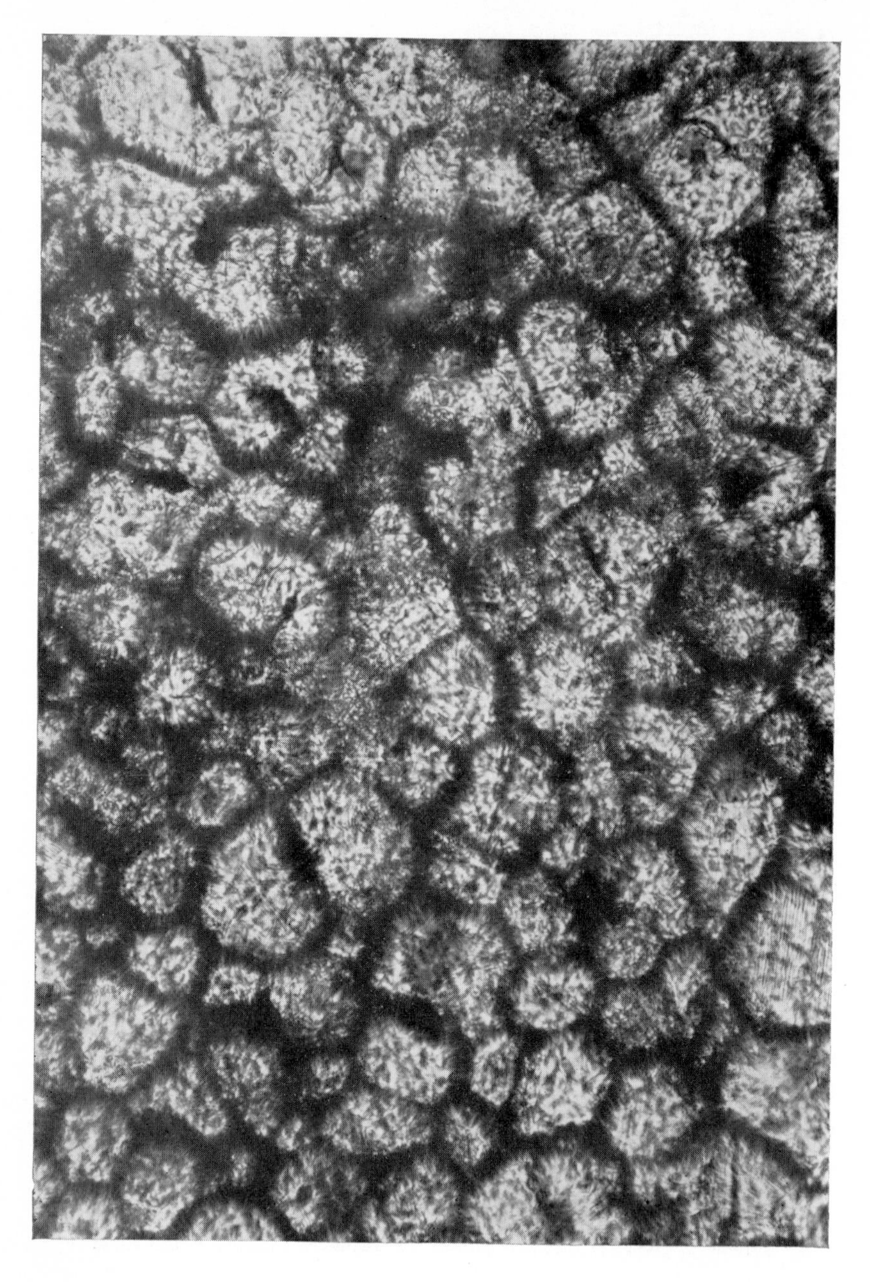

Thin Section of Portland Cement Clinker Showing Large Primary Crystals and Well-differentiated Interstitial Material (× 350)

spacing of a selected diffraction line. Estimations can be derived directly from the X-ray data or from a combination of X-ray data and chemical analysis of the contents of the individual oxides in the cement. These are known respectively as the 'X-ray method' and the 'combined method'. The latter is probably the more reliable in the present state of the techniques, though there are still difficulties to be overcome in the X-ray diffraction measurements.

A comparison of the compound contents as determined by the Bogue calculation, microscopic measurement, X-ray diffraction and 'combined' methods is shown in Table 19. No one of these values can be considered as a fully correct estimation but those obtained by the 'combined' method are probably closest to reality. From measurements by Kantro and his collaborators on over sixty cements the Bogue value for the ferrites varied from 4 per cent above the 'combined' value to 1 per cent below. On average the Bogue value was 0·5 per cent higher. For tricalcium aluminate the Bogue value varied from 6 per cent above to 2 per cent less than the X-ray method. On average the Bogue value was 3 per cent higher. The entry of alumina into the alite phase obviously reduces the amount available to form C_3A, but this amount is also influenced to an important degree by changes in the composition of the ferrite compound. For alite the 'combined' value varied from 21 per cent above the Bogue value to 8 per cent less, being on average 2·4 per cent higher. For dicalcium silicate the 'combined' value varied from 8 per cent above the Bogue value to 14 per cent less and on average was 1·8 per cent higher. From Table 19 it will be seen that the microscopic method tends to underestimate the quantity of ferrite and that for tricalcium aluminate the microscopic value is rather lower than the Bogue value, but above the X-ray or 'combined' value. For the silicates the relation between the microscopic and the other values is more erratic.

The content of C_3A in a cement is of particular importance in relation to its chemical resistance to sulphates. In the results quoted above the content obtained by either of the methods involving X-ray diffraction was lower than the calculated Bogue value. Some repeat[1] determinations with different X-ray techniques on the same cements did not show this effect so consistently and for half of the cements with less than 5 per cent (Bogue) C_3A content the X-ray value exceeded the Bogue value by up to 2 per cent. For some sulphate-resisting cements of low C_3A content Midgley[2] found X-ray diffraction values higher than the Bogue values. The ferrite and the C_3A contents (1) were directly estimated from the X-ray diffraction intensities and (2) the composition of the ferrite was determined from the X-ray spacing and by combining this with the Fe_2O_3 and Al_2O_3 contents, determined by analysis, the ferrite and C_3A contents were calculated. The results are shown in Table 20 from which it will be seen that there is little systematic difference between the Bogue value and the other two at the higher C_3A contents, but that at low C_3A contents the Bogue value is substantially lower than the others. The estimation of the C_3A content is thus still open to proportionately more uncertainty than that of the other three major compounds.

[1] R. L. Berger, G. J. C. Frohnsdorff, P. H. Harris and P. D. Johnson, *Spec. Rep. Highw. Res. Bd*, No. 90 (1966).
[2] H. G. Midgley, D. Rosaman and K. E. Fletcher, *Sym., Washington 1960*, 69.

TABLE 19 Compound content (%) of cements as determined by various methods

	Cement	11	12	15	17	18	23	25	31	33	41	42	51
Ferrite	Bogue	7·5	7·6	7·8	9·9	8·6	16·6	15·1	6·7	7·8	16·2	8·7	10·3
	Micr.	8·4	5·2	5·1	6·0	5·4	8·0	14·2	4·8	4·1	18·3	5·4	9·2
	X-ray	8·0	7·0	5·5	8·2	8·1	11·9	12·4	6·6	8·2	14·7	11·0	9·4
	Combined	7·1	7·6	7·1	8·6	9·7	14·2	13·4	7·3	7·4	15·0	9·5	9·0
	A/F Molec	0·9	11·00	0·84	0·76	1·23	0·74	0·80	1·17	0·92	0·86	1·16	0·76
C_3A	Bogue	11·8	12·4	11·9	10·0	12·1	3·7	4·7	10·6	10·4	4·1	3·1	3·5
	Micr.	10·8	11·3	10·1	9·1	12·3	2·3	3·9	9·7	9·6	3·7	2·9	2·6
	X-ray	8·3	7·8	9·7	8·8	7·0	3·8	1·4	7·4	7·1	1·0	1·4	1·2
	Combined	8·4	7·6	9·9	7·6	7·4	2·3	0·2	7·4	7·4	0·2	1·3	0·6
C_3S	Bogue	53·0	46·0	67·4	52·5	46·1	53·1	35·8	57·8	61·0	21·6	28·9	42·8
	Micr.	56·1	47·3	56·2	50·0	58·5	53·7	40·9	59·8	67·6	25·6	29·4	38·7
	X-ray	57·1	46·3	63·8	60·4	51·7	54·3	34·7	57·6	58·0	22·0	28·0	45·9
	Combined	56·8	50·2	66·7	57·8	49·3	55·0	37·6	55·6	55·0	23·4	29·1	45·5
C_2S	Bogue	19·4	26·7	7·1	21·7	26·3	21·3	37·8	14·6	11·4	49·4	53·3	37·7
	Micr.	12·4	22·2	6·0	22·0	16·4	21·8	35·7	6·8	4·0	50·0	52·8	43·5
	X-ray	20·6	26·1	10·0	21·3	29·4	22·0	38·6	19·5	20·6	49·9	52·5	40·0
	Combined	20·4	28·3	10·4	20·4	28·0	22·3	41·9	18·9	19·6	53·1	54·3	39·6

Data from D. L. Kantro, L. E. Copeland, C. H. Weise and S. Brunaver, *J. Res. Dev. Labs. Portld. Cem. Assoc.*, **6** (1) 20 (1964); S. Brunaver, L. E. Copeland, D. L. Kantro, C. H. Weise and E. G. Schulz, *Proc. Am. Soc. Test. Mater.*, **59**, 1091 (1959).

TABLE 20 Ferrite and tricalcium aluminate content (%) of cements as determined by various methods

	Cement	625	561	592	A.S.R.	540	539	P893
Ferrite	Bogue	9·2	9·0	5·6	19·6	19·3	19·0	18·0
	(i) X-ray	11·0	12·5	7·5	17·5	16·5	17·5	17·5
	(ii) Combined	9·9	9·7	6·1	17·1	16·6	17·0	15·2
C_3A	Bogue	12·9	10·9	9·7	0·9	0·9	0·7	2·5
	(i) X-ray	12·8	9·8	9·7	3·6	4·5	4·8	3·7
	(ii) Combined	12·1	10·0	9·1	4·0	4·7	3·0	5·8

Effect of rate of cooling on the properties of portland cement

The rate at which the cement clinker is cooled in manufacture and the proportion of the clinker liquid which is frozen to a glass has a significant effect on the properties of the resultant cement.[1] Close control of the cooling rate is not normally exercised and it is not surprising therefore to find that the glass content varies over a wide range, values from 2 to 21 per cent being found by Lerch[2] for commercial clinkers. The heat-of-solution method by which these values were determined usually gives higher results than are obtained from microscopic or X-ray measurement, but the order of these results is not in doubt. The majority of commercial clinkers show glass contents between 2 and 12 per cent though, as we have seen, some or all of this glass may actually be a very fine-textured devitrification product.

In laboratory tests increasing contents of glass have been found to make clinker rather less easy to grind, but in commercial manufacture it has sometimes been found that air quenching has a favourable influence on grinding.[3] This apparently variable behaviour may perhaps be explained by the observation[4] that the matrix of cement clinker is made more difficult to grind by rapid cooling from the burning temperature to 1250°, but that silicates become more easily ground. The setting time is not influenced in any consistent manner by the proportion of glass except that completely crystalline clinker, containing more than 10 per cent C_3A, tends to have a flash set. The rate of cooling, within the limits practicable commercially, seem to have no very significant effect on the strength up to 7 days, but fast rates tend to increase the strength somewhat at 28 days. Very slow cooling, such as is only practicable in the laboratory, reduces the strength at all ages. This is associated with the tendency to dust and the conversion of the calcium orthosilicate to the γ form. There is a little evidence that the drying shrinkage may be lower in glassy than crystalline cements but the available data are inadequate to support any general conclusion.

[1] W. Lerch and W. C. Taylor, *Concrete* (*U.S.A.*) **45,** 199, 217 (1937); T. W. Parker, *J. Soc. chem. Ind., Lond.* **58,** 203 (1939); G. W. Ward, *J. Res. natn. Bur. Stand.* **26,** 49 (1941).

[2] *J. Res. natn. Bur. Stand.* **20,** 77 (1938).

[3] O. Schwachheim, *Zement* **25,** 291 (1936); B. Nordberg, *Rock Prod.* **41** (7), 37 (1938); **42** (9), 39 (1939).

[4] G. Ackman and F. Keil, *Tonind.-Ztg. keram. Rdsh.* **81,** 1 (1957).

The heat of hydration, as would be expected, increases with the glass content for the same cement composition. No significant effect is found at early ages, but at 7 and 28 days there is a consistent increase with increasing glass contents. At the latter age Lerch[1] found differences between clinkers cooled slowly and quickly in a laboratory kiln varying from 3 to 16 cal/g for different cements. The commercial clinkers, as cooled in normal production, tended to have heats of hydration roughly intermediate between those of the slowly and quickly cooled samples, except that when the differences between these was large the plant clinker was closer to the slowly cooled product. Some investigators claim that the best cement is obtained by cooling the clinker slowly to 1250° and then rapidly.[2]

Much the most important effect of the rate of cooling of a clinker is on the soundness of the resultant cement and its resistance to the action of sulphate solutions.

The presence of crystals of MgO, periclase, in a clinker, has been found to lead to long-term unsoundness in the resultant cement (see p. 369) because of the expansion that accompanies its slow hydration, but, when present as part of the glass, it has no such undesirable effects. As we have seen earlier, the liquid present in the burning of clinker can dissolve about 5 per cent MgO which is equivalent to 1·5–2 per cent on the total cement composition. Rapid cooling of the clinker is, therefore, of much value in offsetting the adverse effects of high magnesia contents. This must be attributed primarily to the smaller size of the periclase crystals. Small grains of periclase hydrate more rapidly than larger ones and have less tendency to cause delayed expansion.[3] Larger contents of MgO can, therefore, be tolerated in clinkers that are quickly cooled than in those more slowly cooled.[4] With small magnesia contents, up to 1·5 or 2 per cent, the influence of the cooling rate is not important, but at the higher contents, up to 4 or 5 per cent, permitted under cement specifications, it can have a decisive influence on soundness.

The resistance of Portland cements to attack by sulphate solutions is known to be related in a general way to their calculated C_3A content, but there are many anomalies in this relationship. One explanation is to be found in Parker's observation[5] that sulphate resistance is much increased by rapid cooling and that when the aluminate is present in the glass it is much less susceptible to attack by sodium or magnesium sulphate. The relative rates of expansion of mortar rods in 5 per cent magnesium sulphate solution for a normal commercial clinker containing at complete crystallisation 11 per cent C_3A and the same clinker quenched as it came from the hot zone of the kiln are shown in Fig. 44. In cements of low C_3A content the reverse appears to hold,[6] crystalline C_4AF being more resistant than a high iron glass.

[1] *J. Res. natn. Bur. Stand.* **21,** 235 (1938).
[2] H. Lehmann, S. Traustel and P. J. Jacob, *Tonind.-Ztg. keram. Rdsh.* **86,** 316, 339 (1962).
[3] H. H. Vaughan, *Rock Prod.* **43** (4), 48 (1948); F. Gille, *Zement-Kalk-Gips* **5,** 142 (1952); F. Keil, *Rev. Mat. Constr.* **503/504,** 262 (1957).
[4] W. Lerch and W. C. Taylor, *Concrete* (*U.S.A.*) **45,** 199, 217 (1937); R. H. Bogue, *Portland Cement Association Fellowship*, Paper No. 55 (1949).
[5] *J. Soc. chem. Ind., Lond.* **58,** 203 (1939).
[6] R. H. Bogue, loc. cit.

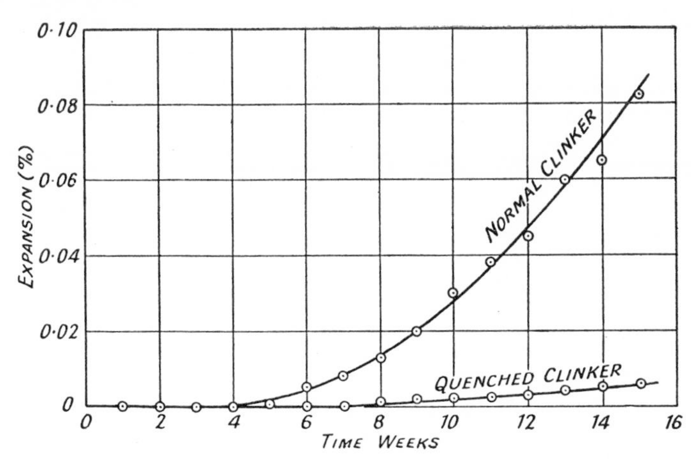

FIG. 44 Expansion of 5 × 1 cm rods of 1 : 3 mortar in 5 per cent $MgSO_4$ solution.

Clinker structure

The view has long been held, as, for example, by the French cement technologist Hendrickx, that it is necessary to break the crystals of the cement compounds in grinding and expose a fresh surface before they can hydrate rapidly. On this theory, a clinker in which the crystals of the silicates are small has been considered not to produce as good a cement as one composed of larger crystals. Little evidence seems to be available to support this opinion and indeed Sanada and Nishi[1] reached a reverse conclusion though they agreed that the alite crystals needed to be broken in grinding. Grzymek[2] also produced some evidence that the strength is higher with alite crystals of 15 μ than of 40 μ dimensions and with increasing diameter:thickness ratio. The higher the fineness of limestone in a raw mix the larger was this ratio. A more correct explanation may be that in clinkers with an Al_2O_3:Fe_2O_3 ratio above about 1·7 the alite crystals react with the liquid during cooling, unless this is rapid, and become coated with dicalcium silicate. Unless this is broken away during grinding it might be expected to slow down the hydration of the alite.

The opinion also exists fairly widely that a Portland cement can be overburnt. Thus, in the test[3] of the litre-weight of the clinker, which is often used in the control of cement burning, it has been found at various works that there is an optimum value and that harder burnt clinker giving a higher litre-weight does not give the highest strength. Some results of Dycherhoff and Nacken showed a lower strength for the extreme case of a cement melted, compared with one clinkered, as also did other laboratory tests at less extreme temperatures by Lerch and Taylor. These authors concluded, however, that the effect was probably to be attributed more to differences in cooling conditions than in the

[1] *J. Soc. chem. Ind. Japan* **39** (1), 9B (1936).
[2] J. Grzymek, *Silikattechnic* **6,** 296 (1955); **10,** 81 (1959); see also H. Schloemer, *Zement-Kalk-Gips* **12,** 102 (1959).
[3] W. Anselm, *Zement* **25,** 633 (1936); *Cement* **9,** 248 (1936).

burning temperature, and that at the higher temperature in the furnace used there was a longer cooling zone at temperatures conducive to more perfect crystallisation. Some work by Butt[1] on tricalcium silicate in which an optimum burning temperature of 1500° was found, with a reduced activity at higher temperatures, supports these conclusions. In contrast in other studies reported from the U.S.S.R. it is claimed that fusion at 2000° improves both strength[2] and soundness.[3] High contents of free CaO and MgO could be tolerated because they crystallised from the fused mix as very small grains uniformly distributed.

A suggestion made by Jander and Wuhrer[4] may well be apposite here. They point out that crystals with a deformed or strained lattice are more reactive than perfect crystals, and it may, therefore, be important to get small and faulty crystals in cement burning rather than larger and more fault-free crystals such as would tend to be produced by overburning.

For certain mix compositions of high lime content an undue increase in burning temperature can also result in free CaO becoming a phase in equilibrium with the clinker liquid. Though this should be recombined on cooling, the rate of solution of the lime crystals may be too slow for this to occur.

It is generally considered that reducing conditions in burning have an adverse influence on the cement and this is ascribed to the influence of ferrous oxide. The Fe^{2+} ion is taken up in solid solution in tricalcium silicate, replacing Ca^{++}, and promotes the decomposition of C_3S during cooling.[5] Ferrous oxide is also taken up in solid solution in $\beta\ C_2S$ and promotes its inversion to the γ form.[6] Reducing conditions are more apt to arise locally in a shaft kiln or on a sinter grate than in a rotary kiln. Ferrous oxide has a much stronger fluxing action than ferric oxide and the increased difficulty it has been reported to cause in the grinding of clinker may be due to a higher glass content.

Colour of Portland cement clinker

The normal greenish-black colour of Portland cement clinker may be considerably changed if reducing conditions exist locally at any point in a kiln, even though the general atmosphere is an oxidising one. The production of clinker with a black external shell and a yellow or brown interior has been ascribed to partial reduction of the iron followed by a subsequent reoxidation which has not penetrated beyond the surface. It should be noted in this connection that even in an oxidising atmosphere some dissociation of ferric oxide may occur at clinkering temperatures; reoxidation occurs during cooling, but may not necessarily be complete and FeO contents from a trace to 0·45 per cent have been reported. Overburning has been found to give a clinker that is black on the surface and deep yellow inside, and this has been ascribed to the increase in the ferrous oxide content.

[1] Yu. M. Butt and V. V. Timashov, *Tsement* **27** (2), 17 (1961).
[2] P. F. Konovalov and T. P. Kiseleva, *Proc. Conf. Chemistry and Technology of Cements, Moscow 1961.*
[3] L. Ya Gol'dshtein and S. D. Okorov, *Dokl. Akad. Nauk SSSR* **159** (2), 420 (1964).
[4] *Zement* **27,** 73 (1938).
[5] N. A. Toropov and B. V. Volkonskii, *Tsement* **26** (6), 17 (1960); E. Woermann, *Sym., Washington 1960*, I, 119; F. Trojer, *Zement-Kalk-Gips* **6,** 312 (1953).
[6] Y. Suzukawa and T. Sasaki, *Sym., Washington 1960*, I, 83.

The rate of cooling also affects the colour[1] and quenched clinkers tend to be more brownish-yellow owing to the decrease of the amount of crystalline iron compound present and its replacement by glass. The ferrite crystals darken as the amount of MgO in the cement increases but if rapid cooling occurs the glass is still yellowish brown. Dark brown colours can be produced, as Larmour[2] found in commercial production, by slow cooling in a reducing atmosphere from immediately below the clinkering temperature. The presence of manganese oxide also gives a brown colour to cement.

A quite different cause of a light-coloured centre to a clinker particle lies in underburning, for the cement mix changes from yellow to greenish-black as it passes from a lightly sintered mass into a clinker.

Sulphur compounds in cement burning

The clay or shale used in the manufacture of Portland cement usually contains a small proportion of sulphur either as pyrites, sulphates, or more rarely, as organic sulphur compounds. The other source of sulphur is the fuel. In the usual oxidising atmosphere of the kilns the sulphur compounds are oxidised finally to sulphur trioxide and leave the kiln distributed between the clinker, the exit gases, and that part of the dust which escapes and is not returned to the kiln. Within the kiln there is a sulphur cycle. The raw mix as it passes along a kiln and is heated absorbs sulphur compounds from the gas. The first reaction is with the alkalis, and particularly potassium, and later with calcium to form the sulphates. The quantity of sulphur compounds in the raw mix thus increases as it passes down the kiln and reaches a maximum at an average raw materials temperature of 700–800°. As the regions of high temperature are approached the alkali sulphates are vaporised and the calcium sulphate present is partly decomposed. A sulphur cycle is thus set up within the kiln. Obviously the total sulphur entering and leaving the kiln must balance. More than half the sulphur intake appears in the clinker, the SO_3 content of which usually ranges from 0·1 to 0·5 per cent, and the rest is lost in the escaping flue gases and in the small amount of dust that finally escapes with them or is discarded after collection. The kiln dust which is high in alkali sulphates and is collected in precipitators re-enters the cycle if it is returned to the kiln with the raw mix.

Detailed studies[3] have been made in Germany on the sulphur balance in various types of kiln. The amount of sulphur trioxide in the mix at the maximum point was about 1 per cent with a Lepol kiln, 2 per cent with a suspension preheater kiln and 3 per cent in a long wet-process kiln, compared with 0·5 per cent or less in the raw mix as fed to the kiln. While these values are not strictly comparable, as the coal used for firing contained respectively 1·5, 1·2 and 4·6 per cent S, they show the order of magnitude of the build-up in the sulphur cycle.

[1] A. J. Pool, *Rock Prod.* **34** (26), 50 (1931).
[2] H. McC. Larmour, ibid. **37** (7), 40 (1934).
[3] P. Weber, *Zement-Kalk-Gips*, special issue No. 9, English version (1963); S. Spring, *Tonind.-Ztg. keram. Rdsh.* **89,** 124 (1965); *Schriftenreihe der Zemintindustrie* **31** (1964); E. Vogel, *Silikattechnik* **9,** 361 (1958).

As we have seen elsewhere sulphur trioxide in cement clinker is taken up preferentially in the sodium-potassium sulphate solid solution. Any excess sulphate first combines to form $(2CaO.SiO_2)_2CaSO_4$ or $4CaO.3Al_2O_3.SO_3$ which may build up in clinker rings, but these compounds decompose again at clinkering temperatures. There is evidence[1] that in the presence of C_3A, though not of C_4AF, formation of C_3S is reduced by high SO_3 contents, but that this is counteracted by the presence of MgO. A combination of high SO_3 and low MgO in a clinker mix may therefore be detrimental to strength development. In cement compositions of normal MgO content the presence of a few per cent SO_3 has some accelerating effect on the formation of C_3S and it has been used, as in the U.S.S.R., as a mineraliser.

The alkalis in cement burning

The alkalis show a somewhat similar behaviour to sulphur compounds in cement burning and an alkali cycle is set up.[2] The alkalis are derived from the raw materials and from the ash of pulverised coal. Something like half of the alkalis are volatilised as the mix approaches the hot zone, starting at between 800° and 1000°, and are partly condensed again in those parts of the system where the temperature of the materials is below about 700°C. This condensation occurs in the grate preheater in a Lepol kiln and in the suspension preheater of a dry-process kiln. In order to avoid an excessive build up and recirculation of potash, it may be necessary to let part of the kiln gases bypass the preheater.[3] The content of potash in the kiln dust is high, as noted earlier, and if this is returned to the kiln it adds to the alkali load. The extent of volatilisation of alkalis varies with different raw materials and is higher for clays than for feldspars. Though potash is usually much more volatile than soda this seems not to be universally the case and some raw materials have been found to lose soda the more readily.[4] The extent of volatilisation is also influenced by the type of kiln. Thus in the tests by Weber the proportion of the alkalis in the raw materials and coal ash that appeared in the clinker was 64 per cent in a Lepol kiln, 80 per cent in a suspension preheater dry-process kiln, and 88 per cent in a long wet-process kiln. In tests by Goes and Keil the proportion of the total potash entering the kiln which was retained in the clinker was 81–88 per cent in Lepol kilns, with a drop to 56–66 per cent when the cyclone dust was not returned, and 67 to 90 per cent in long wet-process kilns. Tests by Woods[5] on older kilns with lower heat economy showed for four dry-process plants a retention in the clinker of 40–85 per cent of the Na_2O and 20–60 per cent of the K_2O in the raw mix. For three wet process plants the values were 70–90 per cent of the Na_2O and 50–80 per cent of the K_2O. There is a general tendency for the retention to be higher as the heat efficiency of kiln system increases.

In the first world war the recovery of potash from flue dust was practised in various countries, to meet a shortage, and, in the U.S.A., additions of feldspar

[1] W. Gutt and M. A. Smith, *Trans. Br. Ceram. Soc.* **67**, 487 (1968).
[2] P. Weber, loc. cit.; C. Goes and F. Keil, *Tonind.-Ztg. keram. Rdsh.* **84**, 125 (1960).
[3] H. Carlsen, *Rock Prod.* **69** (5), 87 (1966); C. Goes and F. Keil, loc. cit.
[4] M. Rustom, *Schweizer. Arch. angew. Wiss. Tech.* **22** (6), 197 (1956).
[5] H. Woods, *Rock Prod.* **45** (2), 66 (1942).

were made to the raw mix to increase the alkali content of the flue dust. Some interest in the recovery of potash also arose in the second world war, but, by that time, it was the desire to reduce the alkali content of the cement that had become dominant in the U.S.A. This had arisen from the discovery of the deleterious reaction of alkalis with certain types of aggregates (see p. 569) and the introduction of a limitation in the total permissible alkali content of the cement when reactive aggregates had to be used in concrete construction. Additions of calcium sulphate or fluoride have not been found to promote the loss of alkalis during burning, but that of calcium chloride is of some value, though of doubtful economy. Holden[1] found that the molecular sum of the Na_2O and K_2O released was proportional to the amount of calcium chloride added, but that the efficiency with which this is utilised is relatively low, about 32 per cent in a wet process, and 48 per cent in a dry process kiln on which tests were made. The alkali content of clinkers has also been reduced commercially by heat treatment[2] to a value below 0·6 per cent.

Phosphates in cement burning

It is well known in the cement industry that small amounts of P_2O_5 in the raw materials, as can occur for instance in some limestones, can lead to trouble. As explained in Chapter 4, P_2O_5 decomposes C_3S forming a series of solid solutions between C_2S and $3CaO.P_2O_5$. Nurse[3] found that for each 1 per cent P_2O_5 added to a cement mix the C_3S is lowered by 9·9 per cent and the C_2S (now a solid solution) raised by 10·9 per cent. With increasing amounts of phosphate the phosphatic dicalcium silicate phase can be stabilised in turn in the β, α' and α forms.[4] When so stabilised with phosphate the α form has no cementing properties and the α' very little but the β form has good cementing properties.[5] Provided the lime content is reduced so that free CaO is not formed, sound cements can be burnt from phosphatic raw materials but they develop strength more slowly because of the reduced C_3S:C_2S ratio. It is important therefore for such cements to so proportion the raw materials as to work as close as possible to the reduced lime saturation factor. Nurse has given an equation for calculating raw mix compositions in terms of the CaO, SiO_2, Al_2O_3, Fe_2O_3 and P_2O_5 contents which produce satisfactory cements with up to 2·0–2·5 per cent P_2O_5 content. There is evidence that if fluoride is present in the raw materials (e.g. as fluorapatite), or is added to the raw mix, then the tolerance for P_2O_5 is increased to some extent provided that the fluorine is retained in the burnt product.

Guye[6] has suggested that C_3S is formed in the early stages of burning of phosphatic mixes and is later decomposed. If this is correct a short burning time and rapid cooling would be advantageous when phosphate is present. Phosphatic raw materials have been used in a cement plant in Uganda and

[1] *Ind. Engng Chem.* **42** (2), 337 (1950).
[2] H. McC. Larmour, E. L. McMaster and W. Jacques, *Rock Prod.* **46** (7), 59 (1943).
[3] R. W. Nurse, *J. appl. Chem.* **2,** 708 (1952).
[4] R. W. Nurse, J. H. Welch and W. Gutt, *J. chem. Soc.*, 1077 (1959).
[5] J. H. Welch and W. Gutt, *Sym., Washington 1960*, 59.
[6] F. Guye, *Revue Matér. Constr. Trav. publ.* **460,** 7 (1954).

Steinour[1] has cited a plant in the U.S.A. producing a cement containing nearly 2 per cent P_2O_5. An account of cement manufacture from phosphatic raw materials has been given by Gutt.[2]

Use of fluxing agents in cement burning

An addition of calcium fluoride, or of various fluoride wastes, has sometimes been made to cement raw mixes to facilitate clinkering. Calcium fluoride lowers the temperature at which liquid is formed and thus reduces the clinkering temperature. The temperature of initial liquid formation is reduced by several hundred degrees and the clinkering temperature by 50–100° by the addition of 1–3 per cent calcium fluoride.[3]

The reduced clinkering temperature is an advantage with mixes high in lime, or otherwise difficult to burn, but the fluoride forms solid solutions with tricalcium silicate which give lower strengths[4] than pure C_3S. Tricalcium aluminate does not form[5] in the presence of calcium fluoride and is replaced by $C_{12}A_7$ making some additional lime available for formation of tricalcium silicate. The net effect of calcium fluoride is thus determined by the balance between the increase in C_3S formation and the decrease in its cementing properties. It also appears (see p. 78) that a calcium silicofluoride $(2CaO.SiO_2)_2CaF_2$, and another with the composition $(3CaO.SiO_2)_3CaF_2$, may be formed as intermediate compounds in the burning of mixes containing fluorides.[6] The dicalcium silicate compound has no cementing properties and those of the tricalcium silicate compound are poor.

The presence of calcium fluoride seriously increases the rate of deterioration of the refractory lining to the kiln but it has been claimed[7] that with an addition of 0·5 per cent, on the dry mix a stable coating is formed, though not with appreciably higher or lower contents. Difficulties with setting time and increased dusting of the clinker, as well as a reduction in early strength, have also occurred in some cases.

When fully burnt materials are compared the addition of calcium fluoride tends to cause some decrease in strength. However if cements burnt at the lower clinkering temperature made possible by the addition of calcium fluoride are compared then the strength is increased. Various workers[8] have found that calcium fluoride promotes the low temperature decomposition of C_3S below 1250° and therefore necessitates rapid cooling of the clinker. Studies of a

[1] H. H. Steinour, *Pit Quarry*, Sept. 1957, reproduced in *Bull. Res. Dev. Labs Portld Cem. Ass.*, No. 85 (1957).
[2] W. Gutt, *Sym., Tokyo 1968.*
[3] A. Guttmann and K. Biehl, *Zement* **13,** 48 (1921); H. Kuhl, ibid., 3; N. Nagy, *Rev. Matér. Constr. Trav. publ.* **638,** 428 (1968).
[4] J. H. Welch and W. Gutt, *Sym., Washington 1960*, I, 59.
[5] W. Eitel, *Zement* **30,** 17, 29 (1941).
[6] W. Gutt, *Sym., Tokyo 1968*; W. Gutt and G. J. Osborne, *Trans. Br. Ceram. Soc.* **67,** 125 (1968); M. Tanaka, G. Sudoh and S. Akaiwa, *Sym., Tokyo 1968.*
[7] P. F. Konovalev and F. R. Skue, *Tsement* **14** (5), 14 (1948); **18** (3), 14 (1952).
[8] E. P. Flint, *Rock Prod.* **42** (10), 40 (1939); W. Eitel, *Zement* **27,** 455, 469 (1938); J. H. Welch and W. Gutt, *Sym., Washington 1960*, I, 59.

wide range of fluxing agents have been made in the U.S.S.R.[1] The fluosilicates used to the extent of 0·2–0·3 per cent proved more efficient than calcium fluoride, with the sodium and magnesium fluosilicate somewhat better than the calcium fluosilicate. Cement mixes, which when burnt at 1300° without these additions contained large amounts of free CaO, produced cements of normal free CaO contents where they were added.

Coal ash in cement burning

The ash of the coal used for firing a rotary kiln combines with the raw mix during burning and thus causes some alteration in its composition. The ash consists mainly of silica, alumina and ferric oxide.

The effect of the absorption of the coal ash on the clinker composition is shown by the following data quoted by Newberry:

	Cement I			Cement II		
	Raw mix	Clinker calculated from mix	Clinker found	Raw mix	Clinker calculated from mix	Clinker found
SiO_2	14·33	22·18	22·96	13·50	22·02	22·33
Al_2O_3	4·32	6·68	6·78	3·43	5·60	5·53
Fe_2O_3	1·46	2·26	2·54	1·27	2·07	3·28
CaO	42·69	66·08	63·95	40·76	66·49	64·40
MgO and SO_3	1·81	2·80	2·94	3·27	3·82	3·61

It is seen that the coal ash causes a quite considerable change in the composition of the clinker. In practice the composition of the raw mix is adjusted so as to allow for this change. It is commonly found that about two-thirds of the ash is taken up by the clinker and about one-third passes out of the kiln with the gases.

The proportion of the ash which is absorbed into the cement mix depends partly on the type of kiln and the heat exchangers since, though some is absorbed in the high-temperature zone, some is carried up the kiln by the gases. The absorption may vary from about two-thirds to almost the whole.

Apart from its effect on the gross composition of the mix, the ash also causes a degree of heterogeneity in the clinker.[2] The melting point of the ash is commonly within the range 1050–1250° and thus in the coal flame it is molten. Part of the ash is absorbed by the clinker nodules in the burning zone by precipitation on to their surfaces where it reacts with the already formed clinker minerals. Tricalcium silicate is transformed into C_2S and the lime thus made available forms C_2S in the ash layer. The depth of this layer is generally not more than about $\frac{1}{2}$–1 mm.

[1] Various papers in Transactions of Leningrad Technological Institute, *Lensoviet*, No. 56 (1966) and *Sym., Moscow 1956*; S. D. Orokov et al., *Tsement* **30** (3), 6 (1964); see also E. P. Flint, loc. cit.

[2] T. Heilmann, *Sym., Washington 1960*, 87.

8 The Proportioning of Portland Cement

There was a gradual change in the composition of Portland cements over a long period, the content of lime rising steadily and that of silica falling somewhat. This was rendered possible by improved methods of manufacture which enabled the manufacturer to produce cements of higher lime content without attendant unsoundness. It resulted in an increase in the tricalcium silicate content of cements and a consequent more rapid development of strength. The limits of composition of Portland cements are approximately:

	Per cent
CaO	60–67
SiO_2	17–25
Al_2O_3	3–8
Fe_2O_3	0·5–6·0
MgO	0·1–5·5
Na_2O+K_2O	0·5–1·3
SO_3	1–3

Cements of many years ago sometimes had lime contents below 60 per cent, but modern cements almost always show values in excess of that figure.

The change in composition over the years may be illustrated by a comparison of the composition and Bogue compound contents of some cements manufactured between 1900 and 1910 and the corresponding values for modern cements.[1] Analyses of modern cements are shown in Tables 21 and 22. The average composition of sixteen cements of the 1900–1910 period is CaO 62·2, MgO 1·84, Al_2O_3 7·27, Fe_2O_3 3·48, SiO_2 21·6, SO_3 1·27. The corresponding average compound contents are C_4AF 10 per cent, C_3A 13 per cent, C_3S 32 per cent, C_2S 39 per cent, making no allowance for free lime for which no data are available. If we assume an average content of 1·5 per cent free lime, the average content of C_3S is about 25 per cent and that of C_2S 45 per cent in these 1900–1910 cements.

[1] Comparisons of cements manufactured in the U.S.A. over similar periods have been made by H. F. Gonnerman and W. Lerch, *Spec. Publ. Am. Soc. Test. Mater.*, No. 127 (1951). The same general trends are apparent.

TABLE 21 Composition of some British Portland cements

	CaO	MgO	Al_2O_3	Fe_2O_3	SiO_2	TiO_2	Na_2O	K_2O	SO_3	Free CaO	C_4AF	C_3A	C_3S	C_2S
Ordinary	65·6	0·70	4·31	2·55	23·73	0·24	0·31	0·66	1·00	1·0	8	7	47	32
	65·5	1·23	5·90	1·59	22·76	0·33	0·43	0·50	1·60	1·4	5	13	41	34
	64·4	0·89	5·36	3·27	21·19	0·34	0·36	0·58	2·53	1·9	10	9	45	27
	64·6	0·56	7·64	3·30	19·09	0·34	0·25	0·57	2·19	0·6	10	15	53	15
	65·5	0·97	6·85	2·30	20·54	0·35	0·16	0·76	1·54	2·0	7	14	48	22
	63·1	0·82	6·28	3·59	20·56	0·37	0·27	0·58	2·59	1·7	11	11	39	30
Rapid-hardening	64·5	1·28	5·19	2·91	20·66	0·30	0·08	0·70	2·66	2·0	9	9	50	21
	65·4	0·51	5·00	4·31	20·04	0·42	0·48	0·78	1·47	1·4	13	6	64	9
	63·0	1·46	6·07	2·67	20·21	0·33	0·12	0·94	2·10	1·5	8	12	46	23
	64·3	1·27	4·74	2·15	22·37	0·36	0·18	0·53	1·82	2·3	7	9	42	32
Sulphate-resisting	63·8	0·92	4·07	4·65	21·09	0·28	0·13	0·67	2·56	2·9	14	3	58	17
	64·5	0·89	3·13	5·23	22·14	0·21	0·18	0·45	2·08	1·5	16	0	54	22
Low-heat	61·8	1·69	4·60	2·07	25·08	0·25	0·19	0·77	2·57	0·7	6	9	17	59
	62·0	1·59	4·54	2·06	25·80	0·23	0·20	0·65	1·87	0·9	6	9	15	63

TABLE 22 Composition of some American Portland cements

ASTM type	CaO	MgO	Al_2O_3	Fe_2O_3	SiO_2	TiO_2	Na_2O	K_2O	SO_3	Free CaO	C_4AF	C_3A	C_3S	C_2S
I	63·8	3·7	5·6	2·4	20·7	0·23	0·21	0·51	1·6	0·4	7	11	55	18
	63·1	2·5	4·7	3·0	22·1	0·21	0·06	1·30	1·7	0·2	9	7	47	28
	65·8	1·1	4·7	2·1	22·2	0·30	0·04	0·19	1·6	1·6	6	9	54	23
	62·8	1·7	6·7	2·5	21·1	0·39	0·95	0·51	1·8	2·0	8	14	33	35
II	61·4	3·1	4·8	4·8	20·8	0·21	0·06	1·30	1·8	0·9	15	5	44	26
	64·9	1·9	4·0	2·1	24·0	0·23	0·23	0·55	1·7	1·5	6	7	41	38
III	65·6	1·4	5·2	2·5	20·0	0·27	0·21	0·44	2·3	1·8	8	10	63	10
	63·3	4·3	5·1	2·0	20·3	0·21	0·19	0·28	2·5	1·9	6	10	51	19
IV	59·6	3·0	4·6	5·0	22·9	0·23	0·06	1·19	1·3	0·4	15	4	25	47
	63·6	1·1	3·7	3·1	25·2	0·19	0·33	0·01	1·9	0·4	9	5	31	49
V	64·3	1·7	3·1	3·3	24·4	0·19	0·08	0·22	1·4	0·5	10	3	45	36
	64·2	2·5	1·9	1·3	26·1	0·12	0·10	0·15	2·0	1·8	4	3	35	48
	63·3†	1·2	3·3	4·7	23·1	—	0·08	0·37	1·7	—	14	1	49	30

† Corrected for free CaO.

The average values for modern ordinary Portland cements are about 45 per cent C_3S and 25 per cent C_2S. The sum of the contents of C_3A and C_4AF has decreased slightly in the modern cements.

Although the amounts of the various oxides present only vary between somewhat restricted limits, the calculated contents of the different compounds present in the cement vary much more widely. A relatively small change in the analytical composition of the cement alters the compound content and the properties quite considerably. An increase in the proportion of lime renders complete combination more difficult, but provided it is attained the strength of the cement at early ages in general is raised. When a certain limit, beyond which no more lime can be combined, is exceeded, however, the further increase of lime content must result in the presence of a considerable quantity of free lime in the clinker and the appearance of unsoundness in the cement.

The silica in a Portland cement forms with lime the essential cementing compounds C_3S and C_2S; any change in the silica content which increases the proportion of C_3S will affect the early strengths favourably, but at long ages the strength depends only on the sum of the two present. An increase in silica content at the expense of the content of alumina and ferric oxide will make the cement harder to clinker and its setting time easier to control. An increase of silica, at the expense of the lime content, reduces the rate of development of strength.

The alumina and ferric oxide contents in a cement need to be considered together, since although they are by no means equivalent to one another, their effects are closely interconnected. Cements with a high total alumina and ferric oxide content are easily clinkered and, unless carefully burnt, tend to cause ring formation in the kiln. In this respect the two oxides act somewhat similarly, but in most other respects they cannot be treated as similar. They form the compounds C_3A and C_4AF; the relative proportion of these two compounds depends on the ratio of alumina to ferric oxide present. Increase of alumina with no change, or with a reduction, in the ferric oxide content, hastens the setting of a cement and a point is eventually reached at which it becomes impossible to control the setting time adequately. The substitution of ferric oxide for alumina, or an increase in ferric oxide content, reduces the proportion of C_3A and increases that of C_4AF in the cement. The latter has less rapid-setting properties and hence, if the iron oxide content is raised along with the alumina, an increased alumina and ferric oxide content can be carried without setting troubles arising. This is well illustrated by some of the Portland cements manufactured in the early years of the present century and by the Kühl-type cements which contain about 14 per cent alumina and ferric oxide, of which about 6 per cent, or slightly more, is ferric oxide. The calculated content of C_3A in such a cement is about 10–12 per cent, a value close to the average content in normal Portland cements; all the excess alumina and ferric oxide is present in C_4AF, which rises to 18–19 per cent as compared with an average of 9 per cent in normal Portland cements. The maximum calculated content of C_3A normally found in Portland cement is about 18 per cent; higher values are likely to cause a rapid set which cannot be controlled. The content of C_4AF can be raised to 20 per cent without danger and the sum of this and the C_3A to about 30 per cent. These limits will depend somewhat on the fineness to which the cement is ground, but they appear to be fairly well applicable to cements as manufactured at the present time. In no ordinary Portland cement is the ratio of ferric oxide to alumina sufficient to convert all the C_3A to C_4AF, for this would require a ratio of $Fe_2O_3 : Al_2O_3$ of 1·56, but the sulphate-resisting Portland cements can approach close to this value. In the former German Erz cement excess iron was present, forming a solid solution of C_2F with C_4AF. This cement was rather slower setting than a normal Portland cement.

Alumina, whether present in the form of C_3A or C_4AF, probably makes little direct contribution to the strength of Portland cement but it accelerates the hydration of C_3S. The main role of alumina and ferric oxide in Portland cement seems to be the commercially most important one of reducing the clinkering temperature and rendering burning possible at temperatures which are economically attainable. In particular a high alumina and ferric oxide content renders it easier commercially to produce a cement with such a lime content that almost all the silica can be converted to tricalcium silicate without an undue content of free lime remaining in the clinker. The Kühl-type of cement represents the extreme case in the use of this method for the production of high $3CaO.SiO_2$ contents, but examples are to be found in ordinary Portland cements. The view which is often expressed that a high alumina, or alumina and ferric oxide, content in a cement is favourable to the production of high early strengths is to be considered therefore mainly as a tribute to their influence in cement burning rather than to the cementing qualities of these constituents. As will be seen in later chapters, cements of high C_3A content have a lowered resistance to seawater and certain other destructive agencies and a high heat evolution during setting. The substitution of C_3A by C_4AF decreases the heat evolution and tends to raise the resistance to attack by seawater.

Magnesia is a relatively unimportant constituent in most British Portland cements, averaging only around 1 per cent, but in many foreign cements it may rise to 3–4 per cent. As shown earlier, it has some value as a flux and this may facilitate combination and render clinkering easier. Specifications in all countries place a maximum limit, usually 4 or 5 per cent, on the permissible content of MgO because higher contents lead to long-term unsoundness. This arises from the very slow hydration, accompanied by expansion, of free MgO (periclase) crystals when these are above certain size. Unsoundness of this type has indeed proved a problem in some countries where the magnesium content of cements is around 3 or 4 per cent and this led to the introduction of the autoclave expansion test (p. 369). It has nevertheless been found that under some conditions of burning higher magnesia content can be tolerated. Thus Bates[1] obtained cements which were sound and had satisfactory strengths with MgO contents of up to 6·5 per cent and similar instances were quoted in a paper[2] summarising later U.S.A. investigations (see p. 369). It has also been claimed

[1] P. H. Bates, *Proc. Am. Soc. Test. Mater.* **27** (2), 324 (1927).

[2] H. F. Gonnerman, W. Lerch and T. M. Whiteside, *Bull. Res. Dev. Labs Portld Cem. Ass.*, No. 45 (1953).

that still higher contents can be tolerated in cements of high iron oxide content[1] or to which pozzolanas have been added.[2]

Minor constituents

Portland cements also contain various minor constituents in amounts varying from a trace to one or two per cent. Alkalis are always present in amounts varying from below 0·2 per cent Na_2O+K_2O up to about 1·3 per cent. Titanium oxide is commonly present to the extent of 0·1–0·4 per cent TiO_2 and may rise to 1 per cent. Manganese oxide is present to the extent of 1 per cent Mn_2O_3 or above in Portland cements made from a blastfurnace slag as raw material, but the content is usually less than 0·1 per cent in cements made from the normal raw materials. Sulphur trioxide in cement clinkers, before the gypsum is added during grinding, varies from about 0·1 to 0·5 per cent. Phosphorus pentoxide is found in amounts up to 2 per cent in cements made from phosphatic limestones, but for most cements its content is small, 0·2 per cent or less. Elements such as barium and strontium are usually only present in small amounts, below 0·05 per cent for BaO and 0·2 per cent for SrO, though in exceptional cases contents up to 0·2 per cent BaO and 0·5 per cent or more SrO have been reported. The content of chromium oxide, though small, is of particular significance in relation to the incidence of cement dermatitis.[3] In most Portland cements its content is below 0·01 per cent Cr_2O_3 but it may rise to 0·02 per cent. Analyses of minor components in a large number of U.S.A. cements have been reported.[4]

Proportioning formulae

The object of proportioning formulæ is to provide a means for calculating the maximum proportion of lime which can be made to combine with the acidic oxides during burning. Numerous formulæ have been proposed but the earlier ones were vitiated by inadequate knowledge of the compounds formed in cement clinker. The theoretical limit for the amount of lime that can be combined in the compounds $3CaO.SiO_2$, $3CaO.Al_2O_3$ and $4CaO.Al_2O_3.Fe_2O_3$ was used by Guttmann and Gille[5] to give the following formulæ (weight proportions) for the proportioning of Portland cement raw mixes.

$$CaO \gtreqless 2{\cdot}8SiO_2+1{\cdot}65Al_2O_3+0{\cdot}35Fe_2O_3$$

Manganic oxide (Mn_2O_3) when present was calculated as equivalent to ferric oxide.

An empirical formula, widely discussed in the German literature, was founded on the assumption that a cement in its most basic state contained the compounds

[1] P. P. Budnikov and Kh. S. Vorobev, *Silikattechnic* **9**, 158 (1958); *Tsement* **26** (1), 14 (1960); D. Nicoletti, *Cem. Lime Mf.* **29**, 1 (1956).
[2] J. Rosa, *Zement-Kalk-Gips* **18**, 460 (1965); A. J. Majumdar and S. S. Rehsi, *Mag. Concr. Res.*, **21**, 67, 141 (1969).
[3] C. D. Calman, *J. occup. Med.*, Jan. 1960; H. Pisters, *Zement-Kalk-Gips* **19**, 467 (1966); C. R. Denton, R. G. Keenan and D. J. Birmingham, *J. invest. Derm.* **23**, 189 (1954).
[4] D. G. Miller and P. W. Manson, *Tech. Bull. Minn. agric. Exp. Stn*, No. 194 (1951); R. L. Blaine, L. Bean and E. K. Hubbard, *Bur. Stand. Bldg Sci. Ser.*, No. 2, 33 (1965).
[5] *Zement* **18**, 571, 912 (1929).

$3CaO.SiO_2$, $2CaO.Al_2O_3$ and $2CaO.Fe_2O_3$. The maximum lime content is then given by:

$$CaO = 2{\cdot}8SiO_2 + 1{\cdot}1Al_2O_3 + 0{\cdot}7Fe_2O_3$$

The ratio of the lime content of the cement, after subtracting that present as calcium sulphate, to that calculated from this formulæ was termed the Lime Saturation value by Kühl.[1] It could approach unity as a limit. Forsen[2] proposed the formula:

$$CaO = 2{\cdot}8SiO_2 + 1{\cdot}4Fe_2O_3 + f(Al_2O_3 - 0{\cdot}64Fe_2O_3)$$

in which f varied from 1 to 1·65 for different raw materials. This was based on the assumption that at the burning temperature the maximum amount of lime was combined when the clinker consisted of $3CaO.SiO_2$ and a melt and that this equilibrium was frozen on cooling. The basicity of the aluminate contained in the melt was considered to vary in accordance with data derived from the known liquid compositions in the binary and ternary systems involved, data on the quaternary system not then being available. When $f = 1{\cdot}65$ the above formulæ corresponds to that of Guttmann and Gille and when $f = 1$ it approximates to the Kühl lime saturation value.

A formula which has been put forward by Hendrickx[3] for proportioning raw mixes is:

$$CaO = 2{\cdot}33SiO_2 + 1{\cdot}65Al_2O_3 + 1{\cdot}05Fe_2O_3$$

This formula is not intended to represent the maximum amount of lime that can be combined, but rather to define mixes which experience shows can be burnt economically taking account of the imperfect homogeneity of most raw materials. It is based on the assumption that such mixes have proportions corresponding to $2{\cdot}5CaO.SiO_2$, $3CaO.Al_2O_3$ and $3CaO.Fe_2O_3$. Essentially it is to be regarded as an empirical formula based on experience and, in fact, Hendrickx states that the lime content calculated may be increased for suitable raw materials by a factor determined from the results of burning tests.

It has been shown,[4] as mentioned earlier, from constitutional work on the system $CaO–Al_2O_3–SiO_2–Fe_2O_3$, that although the compounds $3CaO.SiO_2$, $3CaO.Al_2O_3$ and $4CaO.Al_2O_3.Fe_2O_3$ can be formed in a Portland cement mix, it is not possible in practice to carry such a high lime content. Mixes of such high lime contents have been found at the clinkering temperature to contain uncombined lime existing in equilibrium with the liquid formed. Although, theoretically, this solid lime could be redissolved in the liquid during cooling, in practice the cooling is probably too rapid for this to occur. It is necessary therefore to restrict the lime content to somewhat lower values. The following formula expresses the maximum lime content that can be present without free lime appearing at the clinkering temperature in equilibrium with the liquid present.

$$CaO = 2{\cdot}80SiO_2 + 1{\cdot}18Al_2O_3 + 0{\cdot}65Fe_2O_3$$

[1] H. Kühl, *Zement* **18,** 833 (1929); E. Spohn, ibid. **21,** 702 (1932).
[2] *Angew. Chem.* **47,** 162 (1934); *Zement* **24,** 17 (1935), translation in *Concrete* **45** (9), 231 (1937).
[3] *Chim. Ind.* **8,** 196 (1922); *Revue Matér. Constr. Trav. publ.* **377,** 166; **381,** 254; **382,** 273 (1947).
[4] F. M. Lea and T. W. Parker, *Tech. Pap. Bldg Res. D.S.I.R.*, No. 16 (1935).

This formula represents fairly well the maximum amount of lime found combined in any Portland cement. It is very similar to the lime saturation value of Kühl. As Dahl[1] has pointed out, this relation can also be expressed by the statement that the maximum combined lime content is reached when the content of $2CaO.SiO_2$ is less than 6/11 of the $3CaO.Al_2O_3$, when both are calculated by the Bogue formula for compound content, irrespective of the contents of $3CaO.SiO_2$ and $4CaO.Al_2O_3.Fe_2O_3$. Dahl[2] has also given methods of calculating the proportions of raw materials to control the content of particular compounds.

A refinement of the Lea and Parker formula for the maximum combined lime content to allow for the small amount of MgO combined in C_3S has been suggested[3] as follows:—

$$CaO+0.75MgO=2.8SiO_2+1.18Al_2O_3+0.65Fe_2O_3$$

The maximum value for MgO to be inserted in this formula is 2 per cent since any excess is present as free MgO.

A comparison of the various proportioning formulæ can best be based on the composition of rapid-hardening Portland cements, since in these the lime content is often carried as high as practicable. Such a comparison is made in Table 23

TABLE 23 Moduli of various rapid-hardening Portland cements

No.	I Kühl lime saturation factor	II Lea and Parker saturation factor	III Lea and Parker combination factor	IV Guttmann and Gille lime saturation factor	V Silica modulus	VI Iron modulus
1	0·94	0·94	0·92	0·91	2·59	1·96
2	1·00	1·00	0·99	0·98	1·88	1·42
3	0·90	0·90	0·88	0·86	1·80	2·09
4	1·04	1·03	1·00	1·00	1·97	2·37
5	1·02	1·02	0·97	0·99	2·62	2·18
6	1·01	1·02	0·98	0·99	2·36	1·48
7	0·98	0·98	0·92	0·94	2·43	1·76
8	1·04	1·03	0·99	1·00	2·11	2·16
9	1·01	0·99	0·96	0·96	2·88	1·98

I. Calculated as $CaO/(2{\cdot}8SiO_2+1{\cdot}1Al_2O_3+0{\cdot}7Fe_2O_3)$.
II. Calculated as $CaO/(2{\cdot}8SiO_2+1{\cdot}18Al_2O_3+0{\cdot}65Fe_2O_3)$.
III. Calculated as $(CaO-\text{Free } CaO)/(2{\cdot}8SiO_2+1{\cdot}18Al_2O_3+0{\cdot}65Fe_2O_3)$.
IV. Calculated as $CaO/(2{\cdot}8SiO_2+1{\cdot}65Al_2O_3+0{\cdot}35Fe_2O_3)$.
V. Calculated as $SiO_2/(Al_2O_3+Fe_2O_3)$.
VI. Calculated as Al_2O_3/Fe_2O_3.

for well-known makes of cement from England, a number of other European countries, and the U.S.A. The lime saturation factors represent in each case the

[1] *Sym., Stockholm 1938*, 138; *Rock Prod.* 1955 (5), 71, (6), 102, (7), 78; 1956 (8), 154.
[2] *Rock Prod.*, January–April 1947; August 1956.
[3] E. Spohn, E. Woermann and D. Knuefel, *Zement-Kalk-Gips*, **22**, 55, (1969)

amount of lime (corrected for the gypsum content) in the cement divided by the theoretical amount of lime calculated from the appropriate formula. Where this value exceeds unity the lime content of the cement is above that given by the formula for the theoretical maximum lime content. A lime combination factor is also calculated from the Lea and Parker formula. To obtain this factor the free lime present in the cement was subtracted from the total lime content and it thus represents the ratio of the lime actually combined to the theoretical maximum given by this formula. While it is clear that several of the cements are overlimed the maximum amount actually combined follows closely the value given by this formula.[1]

It will be noted that the lime moduli are based on the maximum amount of lime which can be combined under the optimum conditions of manufacture. Such a limit can only be approached closely, without free lime appearing in the clinker, when the raw materials are sufficiently finely ground and mixed, and the burning adequate, to allow complete combination in the kiln.

The formulæ take no account of minor components which, as we have seen in Chapter 7, must have some influence on the maximum amounts of lime that can be combined at the clinkering temperature. The influence of Na_2O is probably relatively slight, but K_2O, in excess of the amount combined in the sodium-potassium sulphate solid solution, appears to cause an appreciable reduction in the maximum combined lime content. Sufficient data are not available to permit of quantitative corrections being made and for practical purposes the formula of Lea and Parker above can be regarded as an adequate guide.

The relations between the components other than lime

The relations between the silica, alumina, and ferric oxide contents of a cement have been expressed by Kühl in the form of the following moduli in which the percentage weight ratios are used:

Silica modulus	$SiO_2/(Al_2O_3+Fe_2O_3)$
Iron modulus	Al_2O_3/Fe_2O_3

There are some objections to the use of a percentage weight factor in the silica modulus, since it suggests that the alumina and ferric oxide are equivalent, weight for weight. As far as their combining ratios with lime are concerned, this is far from being the case, though they are more nearly equivalent in their effects on clinkering temperatures. However, in this book we shall conform to the use of a percentage weight ratio for the silica and iron moduli.

The silica modulus has an average value of about 2·4–2·7, but drops to 1·7 in Portland cements with a high content of the sesquioxides, and to 1·2 in Kühl cements. It rarely exceeds 3 in grey Portland cements, but may rise to over 4 in white cements. With specially favourable raw materials it can rise still higher (see p. 135). The iron modulus varies from about 1 to 4 in grey cements and may even exceed 10 in white cements. In practice it is found that Portland cements of low silica modulus tend to have excellent early strengths, but show little progression with increasing age. Cements of high silica modulus may not give quite

[1] For other data supporting this conclusion, see F. M. Lea, *Sym., Stockholm 1938*, 128.

as high strengths at the shortest ages, but tend to show a better progression with age and higher ultimate strengths. This behaviour is illustrated in Fig. 45. The two cements shown there have the composition:

	CaO	SiO_2	Al_2O_3	Fe_2O_3	SO_3	Free CaO
No. 1	64·5	18·4	7·0	3·9	2·3	2·1
No. 2	64·7	20·9	5·9	2·6	1·9	2·4

the Bogue compound compositions being:

	$3CaO.SiO_2$	$2CaO.SiO_2$	$3CaO.Al_2O_3$	$4CaO.Al_2O_3.Fe_2O_3$
No. 1	55	11	12	12
No. 2	47	25	11	8

The silica moduli are (1) 1·69 and (2) 2·46.

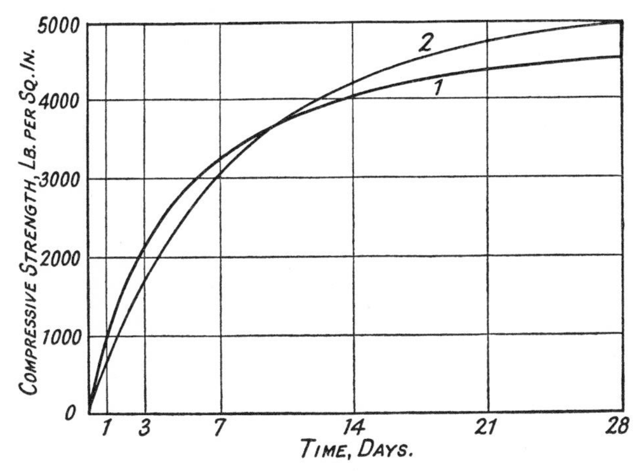

FIG. 45 Strength-age curves. 1. Silica modulus 1·69. 2. Silica modulus 2·46.

The limitation of cement composition in specifications

The specifications for Portland cement in many countries contain some clauses relating to the composition. Some limitation is often placed on the lime content with the object of ensuring that the cement shall not contain an excess of lime over that which can combine with the acidic oxides and so be liable to show unsoundness. It is true that a separate specific test for the soundness is always included, but it has usually been considered that an additional safeguard is provided by limiting the lime content. With the continued demand for cements of higher earlier strength it has been necessary for manufacturers to raise the lime content, and the specification limits have been progressively raised to enable this to be done.

In the British Standard for ordinary, rapid-hardening and sulphate-resisting Portland cement the ratio, calculated in percentage contents, is limited to a maximum

$$CaO/(2{\cdot}8SiO_2+1{\cdot}2Al_2O_3+0{\cdot}65Fe_2O_3)$$

of 1·02 and a minimum of 0·66. The lime content is first corrected for the amount combined with sulphur trioxide. This limit is derived from the Lea and Parker formula with the coefficients rounded off, and a small tolerance allowed by making the maximum value of the ratio 1·02 instead of unity. The ratio of the percentage of alumina to that of ferric oxide must not fall below 0·66. For low-heat Portland cement the maximum permitted lime content is lower and must not exceed $2{\cdot}4SiO_2+1{\cdot}2Al_2O_3+0{\cdot}65Fe_2O_3$.

Many countries place no limit on the lime content of Portland cement. The ASTM specification (C150–68) does not limit the lime content for ordinary (Type I) and rapid-hardening (Type III) Portland cement, but for Modified (Type II), low heat (Type IV) and sulphate-resisting cement (Type V) there are composition requirements as set out below:

	Type II*	Type IV	Type V
SiO_2 min. per cent	21·0	—	—
Al_2O_3 max. per cent	6·0	—	†
Fe_2O_3 max. per cent	6·0	6·5	†
$3CaO.SiO_2$ max. per cent‡	—	35	—
$2CaO.SiO_2$ min. per cent‡	—	40	—
$3CaO.Al_2O_3$ max. per cent‡	8·0	7	5
SO_3 max. per cent	2·5	2·3	2·3

* Where moderate heat of hydration is required the sum of $3CaO.SiO_2$ plus $2CaO.SiO_2$ is limited to 58 per cent.

† $3CaO.Al_2O_3 \ngtr 5$ per cent; $4CaO.Al_2O_3.Fe_2O_3$ plus twice the $3CaO.Al_2O_3 \ngtr 20$ per cent.

‡ Calculated from the usual Bogue formula.

Most countries place a limit varying from 3 to 5 per cent on the magnesia content because of the risk of unsoundness. In the British Standard the limit is 4 per cent and in the ASTM specification 5 per cent.

The maximum content of gypsum which may be added to a cement during grinding to control the setting time is limited in BS 12:1958 for ordinary and rapid-hardening Portland cement, and in BS 1370:1958 for low-heat Portland cement, to 2·5 per cent SO_3 when the C_3A content is 7 per cent or less, and 3·0 per cent when it exceeds 7 per cent. For sulphate-resisting Portland cement (BS 4027:1966) the C_3A content is limited to a maximum of 3·5 per cent and the SO_3 to 2·5 per cent. The ASTM specification limits for SO_3 are:

	Type I	Type II	Type III	Type IV	Type V
$C_3A \leqq 8$ per cent	2·5	2·5	3·0	2·3	2·3
$C_3A > 8$ per cent	3·0	—	4·0	—	—

There is also an ASTM method (C265–64) for determining the amount of water-soluble sulphur trioxide in a cement paste at 24 hours after mixing and for the determination of the optimum SO_3 content for highest strength. Neither of these tests is a specification requirement.

Some specifications limit the content of matter insoluble in dilute acid, e.g. to 1·5 per cent in the British, 0·75 per cent in the ASTM, and up to 3 per cent in other national specifications. The loss on ignition is also usually controlled, the maximum permitted under the British specifications being 3 per cent for cements in temperate climates and 4 per cent in tropical climates. The ASTM limit is 3 per cent, except for Type IV where it is 2·5 per cent. Limits in other countries, e.g. Germany, go up to 5 per cent.

Rapid-hardening Portland cement

Rapid-hardening Portland cement, or high-early-strength Portland as it is sometimes known, represents a development of ordinary Portland cement, and between them no rigid dividing-line can be drawn. The development of rapid-hardening Portland cements started in Austria in 1912–1913 at the Vorarlberg works, and a little later was taken up by the Holderbank Company in Switzerland. Their production was commenced in France by one cement company in 1916, but only became general after 1918, as was also the case in England.

The production of rapid-hardening Portland cements has involved a number of changes in manufacture, such as finer grinding and more perfect mixing of the raw materials, the adjustment of composition, usually so as to give rather higher lime contents, and the finer grinding of the cement clinker. Each of these factors has played a part. Without improved methods of preparation of the raw materials it would not have been possible to carry high-lime ratios in the mix and still obtain the required degree of combination on burning. The finer grinding of the clinker has caused the development at earlier ages of the strength which was latent in the more coarsely ground cement.

The relative extent to which emphasis is laid on these various factors will depend on the plant conditions. Thus a plant which is favoured by a raw material of the nature of a marl approaching almost to cement composition may find it possible to burn a high-lime mix with little trouble. A plant using a hard limestone and shale will need to grind these very fine in the raw mix and to burn rather harder in order to obtain the necessary degree of combination. A favourable chemical composition will render it possible to obtain the same high strength at early ages with a relatively coarsely ground cement as is only obtained in another cement of less favourable composition by much finer grinding of the clinker. On the average the lime content of rapid-hardening Portland cements tends to be slightly higher, and the silica content slightly lower, than that of ordinary Portland cement, but there are many exceptions to this generalisation.

A double-burning process has occasionally been used, and still is for instance in Italy, to enable the combined lime content of rapid-hardening Portland cement to be raised close to the theoretical limit without leaving much free lime in the product. A normal cement is first burnt, the clinker ground, mixed with a little more lime, and reburnt. A very rapid rate of hardening over the first few days, and particularly the first day, can also be secured by extremely fine grinding of the cement. Such a cement made in England with a specific surface of over 7000 cm^2/g gives a strength in 1 : 2 : 4 concrete, w/c 0·6, of over 3000 lb/in^2 (211 kg/cm^2) at 1 day and correspondingly higher strengths at lower water-cement ratios. Because of the fineness the amount of gypsum added to control the set may be above that permitted under BS 12 for Portland cement. Cements with a specially rapid rate of hardening over the first day are also made in some countries, particularly for concreting in cold weather, by intergrinding calcium chloride with the cement.

Though the minimum strength requirements laid down in specifications differ, no definite dividing line can be drawn between ordinary and rapid-hardening Portland cement and the two classes overlap as may be seen from Fig. 46. The

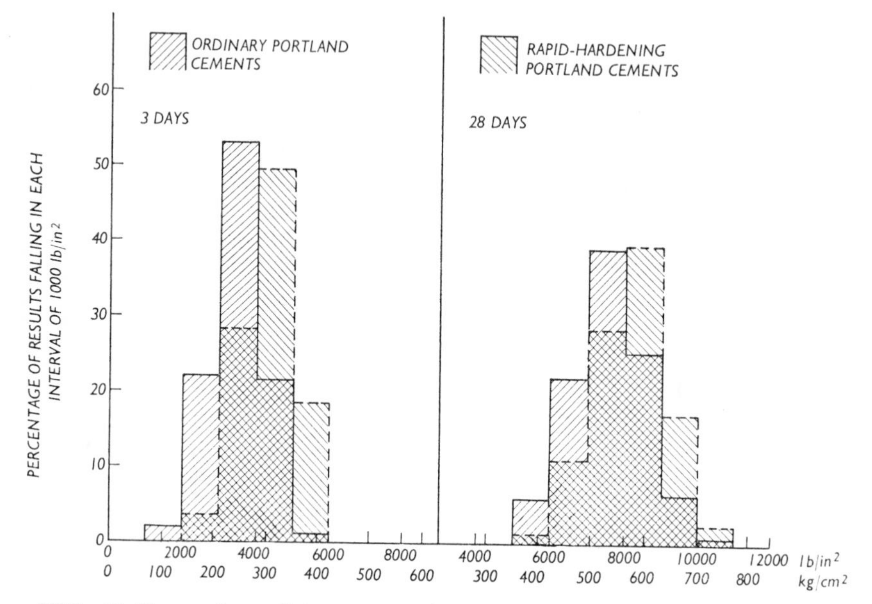

FIG. 46 Comparison of the compressive strengths of vibrated mortar cubes made with ordinary and rapid-hardening Portland cements.

histograms, which are based on strength tests on a large number of cements, show for each age the proportion of cements with strengths falling in successive strength intervals.[1]

Other types of Portland cement

Sulphate-resisting Portland cement differs primarily from ordinary cement in having a low calculated content of tricalcium aluminate. Ferrari cement also falls into this class. This may be achieved either by using raw materials of low-alumina content or by adding iron oxide so as to produce a low ratio of $Al_2O_3 : Fe_2O_3$, i.e. about unity or less. Except for the former type of Erz cement

[1] Strength distribution curves are given by A. J. Newman, *Proceedings Symposium on Concrete Quality*, p. 53, Cement and Concrete Association, London (1966); for data on U.S.A. cements see S. Walker and D. L. Bloem, *Proc. Am. Soc. Test. Mater.* **58**, 1009 (1958).

it is not usual to drop this ratio below 0·64, corresponding to the compound $4CaO.Al_2O_3.Fe_2O_3$. The strength requirements for sulphate-resisting Portland cement are the same as for ordinary Portland cement in the British specifications and slightly lower in the ASTM specifications.

Low-heat Portland cement was developed in the U.S.A. soon after 1930 for use in mass concrete construction, such as dams, where the temperature rise caused by the heat evolved on hydration can become excessively large. Its manufacture spread later to many countries, though with the introduction of other methods of cooling its use has now declined. A low-heat evolution is achieved by reducing the contents of $3CaO.SiO_2$ and $3CaO.Al_2O_3$ which are the compounds evolving the most heat on hydration. Control is exercised in the British Standard 1370:1958 for low-heat Portland cement by limiting the permissible heat of hydration to 60 and 70 calories per gram respectively at 7 and 28 days, and by a reduction in the permissible maximum lime content. In the earlier American specification both the heat of hydration and the calculated compound content were controlled, but the former was subsequently dropped and reliance placed solely on the latter as indicated elsewhere (p. 175). The strength of low-heat cement at 7 and 28 days is less than that of ordinary cement but the ultimate strength is not reduced.

A form of cement with a heat evolution intermediate between that of low-heat and the average ordinary Portland cement, but with a rate of strength development nearly comparable to the latter, was introduced in the U.S.A. about 1934, as a compromise to avoid some difficulties that had arisen in concreting with low-heat cement in cold weather.[1] Later this cement, first called 'modified' cement, and subsequently Type II cement, was adopted for some other types of constructional work on account of its supposed superior durability to ordinary Portland cement. It is defined in the ASTM specification as for use for concrete exposed to moderate sulphate action or where moderate heat of hydration is required. The method of control used in the specification is based on a limitation of composition and compound content (see p. 168).

Properties and compound content of cements

The relation of the properties of cements to their composition is an old problem, but for long only broad deductions, such as that the rate of strength development tended to increase with lime content or that resistance to sulphate attack increased as the alumina : ferric oxide ratio decreased, were possible. With advance in our knowledge of the constitution of cements, and of the compounds present, a fresh approach became possible. The calculation of the contents of the four major compounds present, erroneous to some extent as are the values derived, provides a basis for the analysis of properties in terms of the content of these compounds. The latter does not provide all the data required for the purpose, since the properties of a cement depend also on the fineness to which it is ground and to some extent on the amount of gypsum added. Uncertainties also still

[1] M. Swayze, *Sym., London 1952*, 790, has reviewed the development of this and other types of Portland cement in the U.S.A.; also H. F. Gonnerman, *Bull. Res. Dev. Labs Portld Cem. Ass.*, No. 39 (1952); No. 93 (1958).

remain as to the influence on the properties of the major compounds of minor components taken up in solid solution and of the effect of the latter on the actual contents of the former. For comparative purposes the effect of some variables can be eliminated, or much reduced, by using data on cements burnt under similar conditions and ground to the same fineness, but other variables cannot be controlled in this way.

A comparison[1] of the strengths developed in a series of laboratory-made cements composed of CaO, SiO_2, Al_2O_3, Fe_2O_3 and MgO all ground to the same fineness and with the same amount of gypsum added indicated that strength is primarily a function of the contents of $3CaO.SiO_2$ and $2CaO.SiO_2$. The strength at ages up to 28 days was found to be a function of the content of C_3S while the increase in strength between 28 days and 6 months was roughly linearly related to the content of C_2S. With a more finely ground cement the dicalcium silicate probably begins to produce its effect somewhat earlier. Typical curves for the development of strength in cements with (1) 70 per cent C_3S, 10 per cent C_2S, and (2) 30 per cent C_3S, 50 per cent C_2S are shown in Fig. 47.

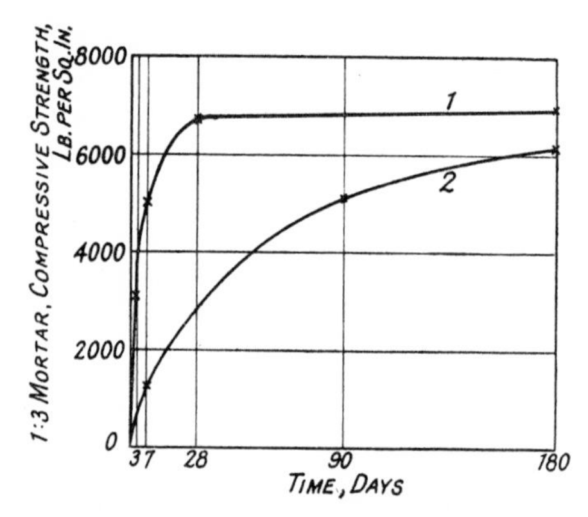

FIG. 47 Characteristic strength-age curves (Woods, Steinour and Starke).
1. 70 per cent. $3CaO . SiO_2$; 10 per cent. $2CaO . SiO_2$.
2. 30 per cent. $3CaO . SiO_2$; 50 per cent. $2CaO . SiO_2$.

The cements rich in tricalcium silicate showed high strength at early ages, while those which were low in this compound showed much lower strengths at early ages, but a progressive increase with age, so that at six months the difference in the strengths was relatively small. At twelve months the strengths of the two groups of cements were about equal and it seemed likely that the ultimate strength of the group high in dicalcium silicate would be the greater. Figure 48 shows the relation obtained between the content of C_3S in the cement and the strength at three and seven days, while in Fig. 49 the increase in strength between twenty-eight days and one year is plotted against the content of C_2S.

Attempts have been made, by analysis of data on a series of cements, to derive factors representing the contribution of each of the four major cement compounds

[1] H. Woods, H. R. Starke and H. H. Steinour, *Engng News Rec.* **109**, 404, 435 (1932); **110**, 431 (1933).

to a particular property at a given age and under given conditions. In this analysis the assumption is made that the contributions made by the different compounds

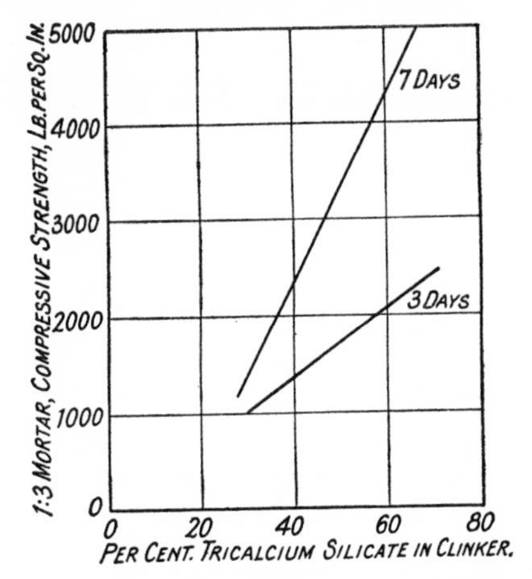

FIG. 48 Strength and tricalcium silicate content (Woods, Steinour and Starke).

are additive, i.e. that a property of a cement may be expressed quantitatively by the equation:

$$P = aA + bB + cC + dD$$

where P represents the numerical value of the property, e.g. strength, at a given

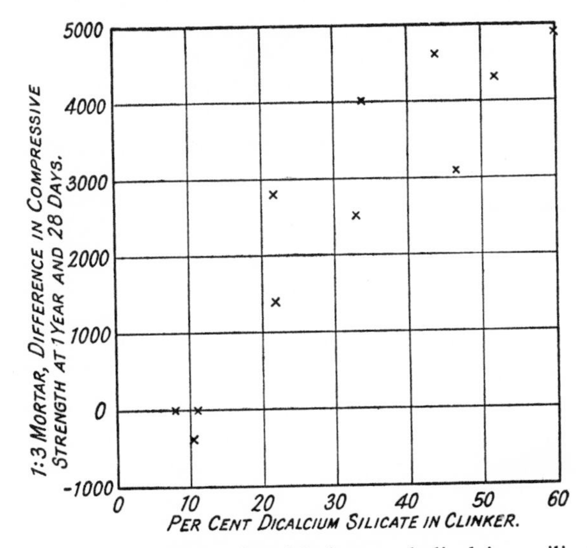

FIG. 49 Increase in strength after 28 days and dicalcium silicate content (Woods, Steinour and Starke).

age and under given conditions; *A*, *B*, *C* and *D* are the percentage contents of $3CaO.SiO_2$, $2CaO.SiO_2$, $3CaO.Al_2O_3$ and $4CaO.Al_2O_3.Fe_2O_3$ present in the cement and *a*, *b*, *c* and *d* coefficients representing the contribution of 1 per cent of the corresponding compound to the property considered. The use of such an equation also assumes that the contribution of a unit weight of any given compound to a particular property under the conditions postulated remains unchanged, i.e. that the coefficients *a*, *b*, *c* and *d* are constants. Errors in the calculated amount of the compounds present do not invalidate this method of analysis, but they will be reflected in the degree of uncertainty attaching to the coefficients deduced.

The assumption of an additive relationship, which is fundamental to this method of analysis, will bear a little more discussion.[1] The heat of hydration of a cement is clearly an additive property except in so far as changes in the relative proportions of different constituents present may influence their rates of hydration. The extent to which strength and shrinkage can be treated as additive properties is much more uncertain. Both are more complex properties than the heat of hydration and may depend not only on the relative amounts and specific individual properties of the compounds concerned, but also on the structure of the set cement mass as a whole and the way in which it is first formed during setting.

This method of approach has, in fact, been more successful when applied to the heat of hydration, as will be seen in Chapter 10, than when applied to strength or shrinkage data. The relative contributions to strength of the four cement compounds derived from different sets of data are very variable and it cannot be claimed that they do more than confirm the conclusion that $3CaO.SiO_2$ makes the major part of its contribution to strength in the first 28 days and $2CaO.SiO_2$ from 28 days onwards. The results for $3CaO.Al_2O_3$ and $4CaO.Al_2O_3.Fe_2O_3$ are very erratic and this is not surprising since the content of these compounds calculated by the Bogue formula can be wide of the truth. The variation in the Al_2O_3/Fe_2O_3 ratio of the iron compound, and in the extent to which the clinker liquid has fully crystallised on cooling, account for these errors. All that can be said is that the available data indicate that these two compounds and glass contribute to early strength, but under damp storage may actually cause retrogression in strength at long ages.

The coefficients derived for the relative contributions of the different compounds to drying shrinkage are not at all satisfactory. There is some indication that cements with a high content of $2CaO.SiO_2$ show somewhat higher shrinkage values than those with a high content of $3CaO.SiO_2$. There is no consistent relation with the content of $3CaO.Al_2O_3$, or $4CaO.Al_2O_3.Fe_2O_3$, and their effect seems to depend much on the gypsum content of the cement. There is evidence[2] to show that the amount of gypsum required for minimum shrinkage increases with the content of $3CaO.Al_2O_3$, and of alkalis, in the cement, and that with amounts below the optimum the shrinkage increases considerably

[1] F. M. Lea, *J. Soc. chem. Ind., Lond.* **54**, 522 (1935).
[2] W. Lerch, *Proc. Am. Soc. Test. Mater.* **46**, 1252 (1946); G. Pickett, *Proc. Am. Concr. Inst.* **44**, 149 (1948); G. Haegermann, *Zement* **28**, 599, 609 (1939); H. Roper, *Sym., Tokyo 1968.*

(see p. 308). It has often been held that the replacement of $3CaO.Al_2O_3$ by $4CaO.Al_2O_3.Fe_2O_3$ tends to reduce shrinkage, but the evidence is too uncertain for valid conclusions to be drawn.

The influence of minor components, e.g. the alkalis and trace elements present in amounts mostly below 0·1 per cent, on cement properties has been examined statistically for nearly 200 cements tested at the U.S. Bureau of Standards.[1] Their influence on properties, though statistically significant in many instances, was small compared with that of the major cement components.

Use of compound content in cement specifications

The use of calculated compound contents as a method of control of the properties of cements originated in some of the specifications for low-heat cement drawn up in the U.S.A. during the years 1932–1934.[2] It later spread there into specifications for 'modified' (ASTM Type II) and sulphate-resisting cement and has since been adopted in other countries for such special types of Portland cement. As a method of limiting the chemical composition it has advantages over the more direct limitation of the contents, or ratios, of the individual oxides. Its use to control properties does, however, presuppose that the relationship between the calculated compound content and cement properties is sufficiently well established for the method not to be unduly restrictive on the manufacturer, i.e. it infers that the particular properties required can, broadly, only be obtained in a Portland cement conforming to the particular compound content requirements specified. In some cases the compound content requirement is used to reinforce direct physical tests, in others as a substitute for them. Thus the earlier U.S.A. specifications for low-heat cements required the measurement of heat of hydration as well as conformity to the compound content requirements, but later the direct physical test was dropped and reliance placed solely on the latter. Limitations based on the composition of a cement, or the compound contents derived from them, have the advantage of speed, since direct physical tests have to be carried out at ages up to 28 days. In this instance, however, the method can only do very rough justice, for the heat of hydration calculated from the compound content (p. 297) can vary up to 10 cal/g from the measured value,[3] a serious discrepancy. For any one cement works a relationship can undoubtedly be established that is closer than this and the calculated compound content used as rapid check in production. No attempt has been made to use compound content as a substitute for strength tests for here, as we have seen, the discrepancies become much larger. For resistance to chemical attack by sulphate solutions, it is shown elsewhere (p. 640) that a limitation of the calculated content of $3CaO.Al_2O_3$ to about 5 per cent ensures a relatively high resistance to sulphate attack, but the converse does not hold and some cements of higher content can have a good resistance. This, as we have seen, is connected with the formation of glass from

[1] R. L. Blaine *et al.*, *Bur. Stand. Bldg Sci. Ser.*, No. 2 (1965); No. 5 (1966); No. 8 (1968); *Sym., Tokyo 1968.*

[2] See F. M. Lea, *J. Inst. Civ. Engng*, 1936–1937 (4), 217.

[3] *Boulder Canyon Project, Final Reports*, Part VII, Bull. 2, p. 364, Bureau of Reclamation, Denver, U.S.A., 1949; G. J. Verbeck and C. W. Foster, *Proc. Am. Soc. Test. Mater.* **50,** 1235 (1950); G. J. Verbeck, *Sym., Washington 1960*, 453.

the clinker liquid or the way in which it forms a sub-microscopic crystalline mass. For control of either heat of hydration or sulphate resistance the use of compound content limitations is to be regarded, therefore, as an expedient, to be justified on grounds of convenience or speed, and not as the infallible result of well-established laws. Calculated compound contents are valuable; they afford a better descriptive picture of the nature of a cement than does the analytical composition expressed in terms of oxides, but they must not be regarded as precision tools and their imperfections overlooked.

9 The Hydration of Portland Cement

Portland cement is composed of a number of compounds the reaction of which with water leads to the setting and hardening of the cement. It is convenient therefore, before passing on to consider the setting and hardening processes, to study first the reaction of these various compounds with water and the nature and properties of their hydration products.

All the compounds present in Portland cement clinker are anhydrous, but when brought into contact with water they are all attacked or decomposed, forming hydrated compounds. Supersaturated and unstable solutions are formed temporarily, but these gradually deposit their excess solids and tend to come into equilibrium with the hydrated compounds produced. The rate of attack, and the degree of temporary supersaturation of the solutions, are determined by the physical state of the cement compounds as well as by their chemical nature. Since the original anhydrous compounds cannot exist in equilibrium with aqueous solutions the ultimate result of the action of water must be complete hydration. We cannot speak of the solubility of, for example, tricalcium or dicalcium silicate in water since it is not a definite physical quantity; there is no aqueous solution in which these compounds can be placed and remain in equilibrium, or from which they would separate on evaporation. The hydrated compounds, on the other hand, can exist permanently in contact with certain solutions, though in many cases they form what are known as incongruent solutions. This will be seen later to be the cause of the complete decomposition of the hydration products of cement when they are continually extracted with water, and to be one of the important factors in deterioration, under certain conditions, of concrete structures.

The nature of the chemical action of water on cement compounds may perhaps be more clearly illustrated if we anticipate some of the later discussion and consider the hydration of $3CaO.SiO_2$. The hydration of this compound may be represented as

$$3CaO.SiO_2 + \text{water} \rightarrow Ca(OH)_2 + xCaO.ySiO_2.aq$$

This action does not stop when the solution is saturated with calcium hydroxide, but hydration continues and the further lime liberated by the reaction is deposited as crystals of calcium hydroxide. The hydrated silicate formed remains stable in contact with the saturated lime solution, but if it is placed in water it undergoes hydrolysis, liberating some lime into solution until the concentration is raised to the value required to stabilise it. Continued extraction with water of the hydrated calcium silicate eventually leaves a solid composed only of hydrated silica, all the lime having been dissolved and removed whilst very little of the silica is dissolved.[1] The hydrated calcium silicate is an example of a hydrated compound which forms an incongruent solution; the compound is partially decomposed when it passes into solution, the quantity of lime dissolved being in excess of that of silica. Equilibrium is only attained when a less basic hydrated calcium silicate is also present in the solid and the concentration of lime in solution has attained the value required to render the remaining more basic silicate stable. The two solids and the solution form, at constant temperature, an invariant point in the ternary system lime-silica-water. If more lime is added to the solution it combines with the less basic silicate, while if more water is added further decomposition of the more basic silicate occurs. In both cases the lime concentration in solution eventually returns to the original value though, owing to the slowness with which equilibrium is reached, this may require a long time. On continued addition of water all the more basic silicate is eventually decomposed and the less basic silicate remains. This in turn decomposes if more water is added, liberating lime, and hydrated silica is formed.

The hydration products of cement are all compounds of relatively low solubility; were it not so, mortars and concretes would not remain stable in contact with water, but would rapidly suffer attack.

Two extreme mechanisms of cement hydration can be visualised. In the first 'through-solution' mechanism, the cement compounds may dissolve to produce ions in solution and these then combine to precipitate the hydrated products. On the other hand, in view of the very low concentrations of alumina and silica observed in the liquid phase very shortly after mixing cement with water, Hansen[2] considered that this mechanism would operate too slowly to account for the observed rates of reaction, and he therefore suggested that the hydration reactions may take place without the cement compounds going into solution and by a direct 'topochemical' or 'solid-state' reaction.

A definite relationship between the structures and crystallographic orientation of the hydrated and anhydrous compounds may be expected from a solid-state reaction. Such a relationship has been observed in the hydration[3] of C_3S and βC_2S and also in the hydration of the minerals larnite (βC_2S) and bredigite ($\alpha' C_2S$) in nature.[4] However, in other cases, a 'through-solution' mechanism is indicated by the observations.[5] Such a mechanism seems also to be clearly involved in many other hydration reactions, despite the extremely low concentrations of alumina and silica that prevail in saturated or supersaturated lime solu-

[1] W. Lerch and R. H. Bogue, *J. Phys. Chem.* **31,** 1627 (1927).

[2] *Sym., London 1952,* 318; *Spec. tech. Publ. Am. Soc. Test. Mater.*, No. 266, 3 (1960); *Mater. Res. Stand.* **2,** 490 (1962). M. W. Grutzeck and D. M. Roy, *Nature, Lond.* **223,** 495 (1969).

[3] R. Nacken, *Zement* **24,** 183 (1935); *Zement-Kalk-Gips* **6,** 69 (1953); F. Trojer, *Zem. Bet.*, 1964 (29), 1; H. Funk, *Sym., Washington 1960,* 291; H. Funk and B. Fahlke, *Z. anorg. allg. Chem.* **334,** 99 (1964).

[4] J. D. C. McConnell, *Min. Mag.* **30,** 672 (1955); J. V. P. Long and J. D. C. McConnell, ibid. **32,** 117 (1959).

[5] H. Funk, loc. cit.; S. A. Greenberg and T. N. Chang, *J. phys. Chem.* **69,** 553, 2489 (1965).

tions. Such reactions include the formation of hydrated products containing both silica and alumina during the reaction of C_3S and C_3A mixtures in water, the production of high-sulphate sulphoaluminate from alumina-containing C_3S, the conversion of hexagonal-plate calcium aluminate hydrates into the cubic C_3AH_6, and the transformation of initially formed high-sulphate sulphoaluminate into the low-sulphate sulphoaluminate or its related solid solution. It seems probable, therefore, that both 'through-solution' and 'solid-state' types of mechanism occur during the course of the reaction between cement and water, and while the former mechanism may predominate in the early stages of hydration the latter mechanism may also operate and perhaps more especially during the later stages when diffusion has become more difficult.

HYDRATION OF THE CEMENT COMPOUNDS

The calcium silicates

During the course of the reaction of tri- or di-calcium silicate with water, calcium hydroxide is split off and a calcium silicate hydrate gel of lower basicity formed. It is necessary for the determination of the CaO : SiO_2 ratio of this gel to estimate the content of free calcium hydroxide present. A determination by X-ray methods gives a lower value[1] than that obtained by a modification of the Franke extraction method (see p. 247). This is attributed to the presence of part of the calcium hydroxide in an amorphous form so the results obtained by the latter method are used in the calculation of the composition of the calcium silicate hydrate gels. The evidence for the correctness of this procedure is strong, though doubts have been expressed as to whether in the solvent extraction method some loosely bound lime might not be removed from the gel.[2] The composition of the calcium silicate hydrate gels produced during the hydration of the anhydrous silicates changes during the period of the reaction and it also varies with the water : solid ratio of the mix and the temperature. The reaction product may be denoted as a *C-S-H* gel without implying any particular composition. More specific forms of the gel (see p. 189) are denoted as *C-S-H* (I) (or *CSH* (I)), consisting of poorly crystallised foils or platelets with a tobermorite-like structure and CaO : SiO_2 molar ratio of 0·8 to 1·5 and *C-S-H* (II) (or *CSH* (II)) with a fibrous structure and a molar ratio of 1·5 to 2. The *C-S-H* gels obtained in the hydration of C_3S or C_2S are very poorly crystallised products showing only a few of the X-ray diffraction lines of well crystallised tobermorites. They are commonly described as 'tobermorite gels' but in view of the difficulty of precise identification the term *C-S-H* gels seems preferable.

The *C-S-H* gels show no distinct structure under the microscope and only a mean refractive index of 1·5 to 1·55, increasing with age can be obtained.[3]

Tricalcium silicate. When finely ground C_3S is mixed with water, hydration commences quickly and both lime and silica pass into solution initially in the

[1] S. Brunauer, D. L. Kantro and L. E. Copeland, *J. Am. chem. Soc.* **80,** 761 (1958); S. Brunauer and S. A. Greenberg, *Sym., Washington 1960*, 135.
[2] E. R. Buckle and H. F. W. Taylor, *J. appl. Chem.* **9,** 163 (1959).
[3] H. F. W. Taylor, *Prog. ceram. Sci.* **1,** 89 (1961).

same molecular ratio 3 : 1 as in the anhydrous compound. The concentration of lime in solution increases steadily while that of silica rapidly decreases. Crystals of calcium hydroxide soon appear together with a gelatinous or nearly amorphous hydrated calcium silicate. Complete hydration cannot be obtained under periods of a year or more unless the tricalcium silicate is very finely divided and the mix reground at intervals to expose fresh surfaces to the water. Otherwise the product obtained shows unattacked cores of tricalcium silicate surrounded by a layer of hydrated silicate which, being relatively impervious to water, renders further attack slow. At complete hydration the reaction can approximately be represented by the equation:

$$2(3CaO.SiO_2)+6H_2O = 3CaO.2SiO_2.3H_2O+3Ca(OH)_2$$

but this simple equation does not bring out the complexities of the reaction.[1] The immediate product formed in pastes on mixing has a $CaO : SiO_2$ ratio near to 3. This forms as a coating on the C_3S surfaces and retards the reaction. After a few hours, dissolution or splitting off of this initial product results in an acceleration of the hydration and the formation as the second product of a *C-S-H* gel of lower $CaO : SiO_2$ ratio, 1·5 or less. This is followed by formation of a third stable product. Evidence obtained by electron microscopy suggests that the second product may be a poorly crystallised *CSH* (I) and the third *CSH* (II) or a closely related product. The completely hydrated silicate, which may well be a heterogeneous material containing both these latter products, has a $CaO : SiO_2$ ratio[2] of about 1·4 to 1·6. This ratio increases somewhat as the water : solid ratio of the mix is decreased.

The hydration reaction is thus more correctly represented in a generalised form by the equation

$$C_3S+(2{\cdot}5+n)H = C_{1\cdot5+m}SH_{1+m+n}+(1{\cdot}5-m)C\,H$$

where *CH* represents $Ca(OH)_2$ and *H* represents the water retained in drying to equilibrium with the vapour pressure of ice at $-78{\cdot}5°$ C (5×10^{-4} mm).

Complete hydration of tricalcium silicate can be obtained in one or two days by grinding with an excess of water in a small steel ball mill. Under these conditions the hydration products are continuously removed from the surface of the C_3S grains so that the reaction continually proceeds at fresh surfaces and is not dependent on the diffusion of water through a surface coating. The first reaction product[3] obtained in this way is an unstable amorphous *C-S-H* gel with $CaO : SiO_2$ ratio 1·5 which converts within about seven days into a well-defined compound identical with the natural mineral afwillite, $3CaO.2SiO_2.3H_2O$. Why the initial *C-S-H* gel should convert to afwillite rather than to the tobermorite-like gel *CSH* (I) is not clear, nor are the relative stabilities of *CSH* (I) and afwillite.

Another method for obtaining complete hydration quickly is to allow tricalcium silicate to harden for twenty-four hours in water and then heat in an

[1] H. F. W. Taylor, *Sym., Tokyo 1968*, has summarised the many relevant studies.
[2] D. L. Kantro, S. Brunauer and C. H. Weise, *Adv. Chem. Ser.* **33**, 199 (1961); *Spec. Rep. Highw. Res. Bd*, No. 90, 309 (1966); F. W. Locher, ibid., 300; *Zement-Kalk-Gips* **20**, 402 (1967); J. G. M. de Jong, H. N. Stein and J. M. Stevels, *J. appl. Chem.* **17**, 246 (1967).
[3] D. L. Kantro, S. Brunauer and C. H. Weise, *J. Colloid Sci.* **14**, 363 (1959).

autoclave under steam pressure. The hydration products, as discussed later (p. 196) are not the same as those formed at ordinary temperatures.

Dicalcium silicate. There are four main polymorphic forms of the dicalcium silicate, γ, β, α' and α of which only the β or occasionally the α' or γ form occurs in Portland cement. The β form is only slowly attacked by water and even after some weeks the original crystals show under the microscope only a surface coating of an amorphous hydrated silicate, the thickness of which slowly increases with the passage of time. A hardened paste more than four years old has been found still to contain about 15 per cent unhydrated C_2S. Although the reaction is considerably slower than with C_3S and substantially less calcium hydroxide is produced, and is not detectable microscopically until after many weeks, the *C-S-H* gels produced are of the same type as those from C_3S. There are, however, differences in the course of the two reactions. With C_2S pastes the initial product, formed as a surface coating, has a $CaO:SiO_2$ ratio close to 2. Within twelve hours or so this converts into a low-lime product related to *CSH* (I) and the molar ratio drops to a minimum value between 1·1 and 1·2. A stable final product related to *CSH* (II) is then gradually formed and the molar ratio gradually increases reaching a final value at 25° of 1·65 to 1·8 after a year or so. At 5° the ultimate value is rather lower, about 1·55. The molar ratio increases as the water : solid ratio of the original paste decreases.

When dicalcium silicate is hydrated in excess water in a ball mill, complete hydration has been obtained in a period of 46 days.[1] In contrast to C_3S which under these conditions gives afwillite the C_2S gives *C-S-H* products similar to those obtained by hydration in pastes, but with $CaO:SiO_2$ ratio of 1·5. The reaction equation for this completely hydrated ball-mill product can, therefore, be represented by the equation

$$2(2CaO.SiO_2)+4H_2O = 3CaO.2SiO_2.3H_2O+Ca(OH)_2$$

This equation has to be modified to apply to hydration in pastes where the $CaO:SiO_2$ ratio varies with age and other factors. For the later stages of hydration in a paste with water : solid ratio of 0·7 at 25°, which gives a *C-S-H* gel with a $CaO:SiO_2$ ratio of about 1·65, the equation becomes

$$2(2CaO.SiO_2)+4H_2O = Ca_{3\cdot3}Si_2O_{7\cdot3}.3\cdot3H_2O+0\cdot7Ca(OH)_2$$

A more generalised equation applicable at any age or water : solid ratio is

$$C_2S+(1\cdot5+n)H = C_{1\cdot5+m}\,S\,H_{1+m+n}+(0\cdot5-m)C\,H$$

The rate of hydration and the strength developed by βC_2S depends on the nature of the stabilisers (see p. 242). Some of the lime in dicalcium silicate can be replaced by K_2O to give $K_2O.23CaO.12SiO_2$. This seems to be more readily hydrated and some calcium hydroxide crystals appear within a day or two, along with a hydrated silicate. At ordinary temperatures γC_2S is attacked by water more slowly than the β form but eventually a hydrated silicate is formed.

[1] S. Brunauer, D. L. Kantro and L. E. Copeland, *J. Am. chem. Soc.* **80,** 761 (1958).

Tricalcium aluminate

Finely ground C_3A reacts very rapidly with water, though apparently less so in a saturated lime solution. In the presence of excess water a plentiful formation of hexagonal plate crystals is observed, and these when seen on edge appear as small birefringent 'needles' which often form clusters radiating from a centre. A photomicrograph of the hexagonal-plate crystals from C_3A and water is shown in Plate V (i). Some plates on edge lie across the hexagonal plate groups and appear as 'needles'. These crystals begin to form within a few minutes and increase rapidly in size and amount. They consist of a mixture of the hydrates $4CaO.Al_2O_3.19H_2O$ and $2CaO.Al_2O_3.8H_2O$ or a closely-related solid-solution of limiting composition $C_{2\cdot4}A.10\cdot2H_2O$. No calcium hydroxide or hydrated alumina is precipitated but, with a limited amount of water present, concentrations of about 0·6 g CaO per litre and 0·1 g Al_2O_3 per litre are obtained in solution at ordinary temperatures. The hexagonal-plate hydrates are however metastable at these temperatures, and although they can persist for some time they eventually transform into the less soluble and more stable isometric compound $3CaO.Al_2O_3.6H_2O$. As described later, this hydrate may crystallise in a number of different forms all belonging to the cubic crystal system. A photomicrograph of C_3AH_6 crystallised as hexakis octahedra is shown in Plate V (ii).

The hydration of C_3A under steam pressure, or even in water at 50°, probably results in the direct formation of C_3AH_6 without the intermediate formation of the hexagonal-plate hydrates. Similarly, C_3AH_6 can also be formed very rapidly when C_3A is mixed with a limited amount of water at normal temperatures to form a plastic mass, since there is considerable heat evolution and the temperature of the material rises, favouring the formation of C_3AH_6. Complete hydration in such a mass is only reached slowly owing to the protection afforded to the unhydrated cores by the cubic hydrate surrounding them. With larger amounts of water the hexagonal-plate hydrates produced initially can also form on the C_3A grains and moderate the reaction by limiting the supply of water to the C_3A. At a later stage, perhaps after only a few hours, the transformation to the cubic C_3AH_6 occurs and the reaction then accelerates owing to the removal of the hexagonal hydrates from the surface of the C_3A grains.[1]

Some of the lime in C_3A may be replaced to give the compound $Na_2O.8CaO.3Al_2O_3$ and, while the presence of alkali may have a slight accelerating effect on the rate of reaction, the hydration products remain the same. The rate of release of alkali into solution can be used to obtain a measure of the rate of hydration of this aluminate phase. In saturated lime solution both compounds form the hydrated tetracalcium aluminate C_4AH_{19} which tends to change, the more rapidly with rising temperature, to the cubic C_3AH_6.

Tetracalcium aluminoferrite

Although the ferrite phase in Portland cement is not necessarily C_4AF as such, but a solid solution of composition lying between C_6AF_2 and C_6A_2F, the hydra-

[1] R. F. Feldman and V. S. Ramachandran, *J. Am. Ceram. Soc.* **49**, 268 (1966); K. Murakami, H. Tanaka and Y. Nakura, *Chemy Ind.*, 1968, 1769.

tion behaviour of the ferrite phase is typified by that of the C_4AF composition. The rate of reaction of the calcium aluminoferrites with water increases with the proportion of alumina in the aluminoferrite, and C_4AF therefore though reacting quickly does so less rapidly than C_3A. When mixed with water, hexagonal-plate crystals are rapidly formed, as in the photomicrograph shown in Plate VI (i). The hexagonal-plate crystals occur mostly as foliated masses. The residual unhydrated C_4AF and a probable hydrated iron oxide or amorphous αFe_2O_3 (hematite) deposited around it form the dark mass.

Calcium hydroxide is not precipitated during the hydration of C_4AF, but the course of hydration[1] in excess water resembles that of anhydrous CA or $C_{12}A_7$ (see p. 205) in that a supersaturated calcium aluminate solution is formed, but one in which the molar ratio of $CaO : Al_2O_3$ is between about 2 : 1 and 3 : 1. Only extremely small amounts of ferric oxide are present in this supersaturated solution which can be expected to precipitate C_2AH_8 or a closely-related solid solution. Both Brocard and Carlson obtained X-ray evidence for the formation of this hexagonal-plate phase in pastes or aqueous suspensions of C_4AF in water at 1°–15°.

Chatterji and Jeffery[2] also observed the additional formation of C_4A.aq in pastes with a water : solid ratio of 0·6, as well as the probable incorporation of some ferric oxide to give a hexagonal-plate-phase solid solution. At temperatures above about 15° the hexagonal-plate phase readily converts into a cubic C_3AH_6–C_3FH_6 solid solution which has a raised refractive index and an increased lattice parameter of the cubic unit cell compared with those of C_3AH_6. While some iron oxide is present in this cubic phase, some hydrated iron oxide or amorphous αFe_2O_3 (hematite) is also produced.

In saturated lime-water, or in the presence of excess lime, the reaction of C_4AF is less rapid and rather different in that hydrated iron oxide or αFe_2O_3 is not produced and the reaction products are white.[3] Moreover, the hexagonal-plate phase formed is now a solid solution of the tetracalcium compounds C_4A.aq and C_4F.aq. This is metastable with respect to the cubic C_3AH_6–C_3FH_6 solid solution and transformation to the latter occurs more and more rapidly with increasing temperature above about 15°, with liberation of calcium hydroxide.

Effect of gypsum. In the presence of gypsum the hydration products of C_3S and βC_2S are slightly modified in that some sulphate can enter the structure of the calcium silicate hydrate gel and change its morphology. The manner of hydration of C_3A is, however, altered considerably, and if this compound is placed in a calcium sulphate solution on a microscope slide the rapid formation of extremely fine needles is observed. These are needles of a calcium sulpho-aluminate hydrate $C_3A.3CaSO_4.31H_2O$, and are characterised by a negative elongation in contrast to the positive elongation of the hydrated calcium aluminate 'needles'. A photomicrograph of needles of $C_3A.3CaSO_4.31H_2O$ is shown in

[1] E. P. Flint, H. F. McMurdie and L. S. Wells, *J. Res. natn. Bur. Stand.* **26,** 13 (1941); M. J. Brocard, *Annls Inst. Bâtiment*, New Series No. 12 (1948); E. T. Carlson, *J. Res. natn. Bur. Stand.* **68A,** 453 (1964).
[2] *J. Am. Ceram. Soc.* **45,** 536 (1962).
[3] G. Malquori and V. Cirilli, *Sym., London 1952*, 321; E. T. Carlson, *Bur. Stand. Bldg Sci. Ser.*, No. 6 (1966); H. E. Schwiete and T. Iwai, *Zement-Kalk-Gips* **17,** 379 (1964).

Plate VI (ii). In the case of C_4AF, hydration in the presence of gypsum and lime yields a solid solution of the sulphoaluminate with an analogous sulphoferrite hydrate. There is some evidence that with smaller amounts of gypsum the trisulphate solid solutions formed first later convert to the monosulphate solid solutions.[1]

Effect of other salts. When C_3A is treated with $CaCl_2$ solution a hydrated calcium chloroaluminate $C_3A.CaCl_2.10H_2O$ is produced, and this may form as a reaction-inhibiting coating on the particles and reduce the rate of hydration. On the other hand,[2] the rate of hydration of both C_3S and βC_2S is increased in pastes containing 2 per cent $CaCl_2$. Similar effects can be produced, usually to a lesser degree, by $MgCl_2$, NaCl and LiCl, as well as by the corresponding bromides and nitrates, but zinc salts have a retarding action which is probably associated with the formation of a protective coating of hydrous zinc hydroxide or related oxycompounds on the surface of the anhydrous compounds.[3]

The less-basic calcium aluminates

Though not of importance in connection with Portland cement, it is convenient to describe here the hydration of calcium aluminates less basic than C_3A. The compounds $C_{12}A_7$ and CA react rapidly with excess water to produce highly-supersaturated calcium aluminate solutions followed by the formation of hexagonal plates and 'needles' of hydrated calcium aluminates together with hydrated alumina in an ill-defined form. What was previously thought to be a definite compound of composition $3CaO.5Al_2O_3$ is really a mixture of CA and CA_2. The latter compound probably has little hydraulic value, since Buttler and Taylor[4] found that reaction with water was very slow and incomplete and Nurse found that little strength was developed in 7 days.

The products obtained by hydration of the anhydrous calcium aluminates are dependent upon the temperature, and from the work of many investigators it is possible to anticipate later discussion and to indicate schematically a probable hydration mechanism as follows:

$$2C_3A + aq \longrightarrow C_2A.aq + C_4A.aq$$

$$C_{12}A_7 + aq \xrightarrow{5^\circ} 3CA.aq + 4C_2A.aq + Ca(OH)_2$$

$$\xrightarrow{\text{Rising temp.}} 6C_2A.aq + Al_2O_3.aq$$

$$2CA + aq \xrightarrow{\text{less than about } 15\text{–}20^\circ} 2CA.aq$$

$$\xrightarrow{\text{Rising temp.}} C_2A.aq + Al_2O_3.aq$$

$$2CA_2 + aq \xrightarrow{\text{less than about } 15\text{–}20^\circ} 2CA.aq + 2Al_2O_3.aq$$

$$\xrightarrow{\text{Rising temp.}} C_2A.aq + 3Al_2O_3.aq$$

$$\left.\right\} \xrightarrow{\text{Rising temp.}} C_3A.6H_2O\,(\text{cubic}) + \underset{(\text{Gibbsite})}{Al_2O_3.3H_2O}$$

[1] N. Tenoutasse, *Revue Matér. Constr. Trav. publ.* **604,** 18; **614,** 452 (1966); W. L. de Keyser and N. Tenoutasse, *Sym., Tokyo 1968.*
[2] N. Tenoutasse, *Sym., Tokyo 1968.*
[3] G. C. Edwards and R. L. Angstadt, *J. appl. Chem.* **16,** 166 (1966).
[4] *J. appl. Chem.* **9,** 616 (1959).

The general tendency at lower temperatures is first to produce a hydrated monocalcium aluminate, but with rising temperature a hydrated dicalcium aluminate is formed. It seems probable that the tendency towards formation of the CA.aq decreases as the CaO/Al_2O_3 ratio of the anhydrous aluminate increases. In the case of C_3A the CA.aq is not formed at all and instead a mixture of C_2A.aq and C_4A.aq is produced. The cubic C_3AH_6 becomes the stable hydrate at temperatures above about 15°–20°, and the other hydrates convert into it the more rapidly with rising temperature or increasing CaO/Al_2O_3 ratio in the solid and the solution.

THE HYDRATED CEMENT COMPOUNDS

Calcium hydroxide

Calcium hydroxide crystallises in the hexagonal system as plates or short hexagonal prisms with perfect cleavage parallel to the face (0001). Its refractive indices are ω 1·574, ε 1·545 and it is uniaxial negative. Its density is 2·30. The dissociation pressure[1] attains 760 mm at 512°. The solubility of calcium hydroxide in water decreases with rise in temperature though the recorded data are not in good agreement. The values given below, obtained by Bassett,[2] and closely confirmed by Hedin[3] and by Bates,[4] seem the most reliable.

Temperature °C	0	10	18	25	50	75	100
CaO g per litre	1·30	1·25	1·20	1·13	0·92	0·72	0·52

The solubility as a function of temperature between 297° and 330° K (24–57° C) was expressed by Hedin as C(g CaO per litre) $= 3{\cdot}947 - 0{\cdot}0094T$. The solubility product[5] at 25° is $K = a_{Ca^{++}}\, a^2_{OH^-} = 9{\cdot}10 \times 10^{-6}$ where a is the activity of the ions. Calcium hydroxide crystals of small particle size give apparently higher solubilities because the supersaturated solution crystallises only very slowly to form larger crystals.

The pH values and specific conductivities of the solutions are:

CaO g per litre	1·20	1·00	0·70	0·40	0·20	0·10	0·05	0·01
pH at 25°[6]	12·55	12·45	12·30	12·05	11·75	11·47	11·15	—
pH at 30°[7]	12·33	12·25	12·11	11·87	11·57	—	10·99	—
$k \times 10^3$ at 25°[8]	8·45	7·2	5·2	3·05	1·6	0·8	—	—

A saturated solution of calcium hydroxide, recommended by Bates as a highly alkaline pH standard, is assigned a pH of 12·45 at 25° and 1·14 g CaO per litre.

[1] P. E. Halstead and A. E. Moore, *J. chem. Soc.*, 1957, 3873.
[2] H. Bassett, *J. chem. Soc.*, 1934, 1270.
[3] R. Hedin, *Proc. Swed. Cem. Concr. Res. Inst.*, No. 27 (1955).
[4] R. G. Bates, V. E. Bower and E. R. Smith, *J. Res. natn. Bur. Stand.* **56**, 305 (1956).
[5] A. Greenberg and L. E. Copeland, *J. phys. Chem.* **64**, 1057 (1960).
[6] F. M. Lea and G. E. Bessey, *J. chem. Soc.*, 1937, 1612.
[7] E. P. Flint and L. S. Wells, *J. Res. natn. Bur. Stand.* **11**, 163 (1933).
[8] G. Ringqvist, *Proc. Swed. Cem. Concr. Res. Inst.*, No. 19 (1952). This author gives specific conductivity data from 0° to 100°C.

The solubility of calcium hydroxide is much reduced in the presence of alkali hydroxides. Values at 20° are shown below:

KOH g per litre	0·4	1·6	2·5	5·0	8·0	20·0	
CaO g per litre	1·05	0·73	0·57	0·34	0·22	0·10	Hedin[1]
	1·06	0·75	0·57	0·32	0·20	0·11	Fratini[2]

NaOH g per litre	0·4	1·6	2·5	5·0	8·0	20·0	
CaO g per litre	0·96	0·49	0·36	0·18	0·09	0·02	Hedin[1]
	1·04	0·61	0·44	0·24	0·15	0·05	Fratini[2]

The values fall on a smooth curve when plotted against the molar concentration of NaOH or KOH, so the solubility is practically the same in NaOH and KOH solution of equivalent concentration.

Calcium sulphate

Calcium sulphate crystallises under normal conditions as the dihydrate $CaSO_4.2H_2O$ in the form of needles or prisms belonging to the monoclinic system. Refractive indices α 1·520, β 1·523, γ 1·530, biaxial positive 2V = 58°. Density 2·32.

The solubility increases to a maximum at about 40° and then decreases again.

Temperature	0	10	15	20	25	30	40	50	75	100
$CaSO_4$ g per litre	1·76	1·93	1·99	2·05	2·09	2·105	2·11	2·07	1·89	1·68

The solubility of calcium sulphate is decreased by the presence of lime, and that of lime is decreased slightly by the presence of increasing amounts of calcium sulphate in solution. The solubility curves[3] are shown in Fig. 50.

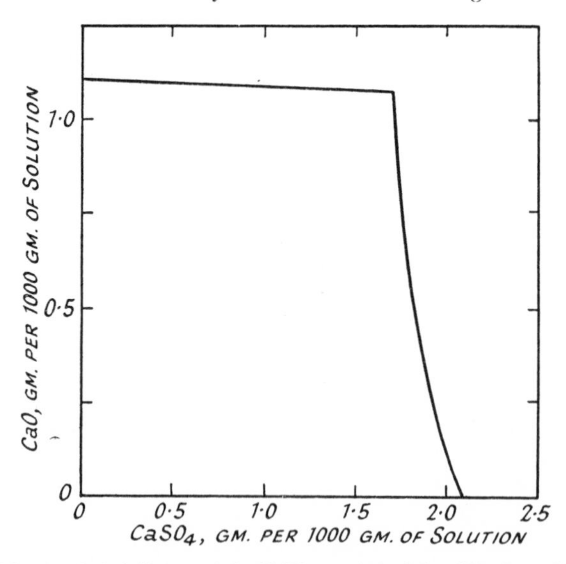

FIG. 50 Solubilities of $Ca(OH)_2$ and $CaSO_4.2H_2O$ at 25°.

[1] *Proc. Swed. Cem. Concr. Res. Inst.*, No. 3 (1945).
[2] *Annali Chim. appl.* **39,** 616 (1949).
[3] F. E. Jones, revised data.

The hydrated calcium silicates

The hydrated calcium silicates form a group of compounds the study and identification of which is difficult, and our knowledge of them is as yet incomplete. Many different hydrated calcium silicates occur in nature, though most of them are rare.[1] Studies on the naturally-formed crystals have helped considerably in elucidating the characteristics of hydrated calcium silicates. Some physical properties of various natural minerals and synthetic compounds are given in Tables 24 and 44 (p. 335). The temperatures of formation indicate the conditions, with saturated steam pressure, under which a compound can most usually be obtained and not the temperatures at which it is a stable phase. The precise composition of some of these compounds is uncertain, and this may be mainly due to difficulties in differentiating between combined and 'free' lime and silica, as well as in distinguishing adsorbed water from water of constitution. Contamination with atmospheric carbon dioxide may also play some part. Of the compounds listed in Table 24, tricalcium silicate hydrate ($C_6S_2H_3$), calciochondrodite (C_5S_2H), C_2SH (A) and 'Z-phase' (CS_2H_2) are not known as natural minerals, while the minerals nekoite and okenite have not yet been synthesised. Almost all the reported hydrated calcium silicates are prepared hydrothermally in the presence of water above 100° at pressures exceeding atmospheric, and only the poorly crystallised tobermorite-like compounds CSH (B) and C_2SH_2 (now usually called C-S-H (I) and C-S-H (II) respectively) are obtained at ordinary temperatures. The latter two hydrates appear to be indefinitely persistent, but when formed on hydration of Portland cement they may be metastable with respect to afwillite ($C_3S_2H_3$), since, as an exception to the general behaviour, afwillite is eventually formed when C_3S is ball-milled with water at room temperature.

It was for long uncertain whether any crystalline hydrated calcium silicate was produced by the action of water at ordinary temperatures on the anhydrous calcium silicates, or by the action of lime on silica gel. That such crystals are formed in sub-microscopic size has been demonstrated by X-ray and electron microscopy. Though many observers[2] have failed to find them under the optical microscope, Keisermann,[3] and later Bessey,[4] found fine needle crystals in the products of the action of water on dicalcium silicate at long ages, while Pulfrich and Linck[5] obtained from tri- and di-calcium silicate, and by boiling silica gel and lime, needles of a hydrated calcium silicate of low birefringence with refractive indices α 1·498, γ 1·500.

An amorphous hydrated calcium monosilicate of composition $CaO.SiO_2.aq$ was observed by Le Chatelier. It can be obtained by the action of lime-water on silica gel and it can absorb additional lime. It is unstable in contact with water and dissolves incongruently, liberating lime into solution until a concentration of about 0·05–0·10 g CaO per litre, and a pH value of 10·5–11 is reached.[6]

[1] L. Heller and H. F. W. Taylor, *Crystallographic Data for the Calcium Silicates*, H.M.S.O., London, 1956; H. F. W. Taylor, *Sym., Washington 1960*, 167, and *Sym., Tokyo 1968*.
[2] H. H. Steinour, *Chem. Rev.* **40,** 391 (1947), has reviewed the various studies mentioned, and others published up to 1947.
[3] *Kolloidchem. Beih.* **1,** 423 (1909–1910).
[4] *Sym., Stockholm 1938*, 178.
[5] *Kolloidzeitschrift* **34,** 117 (1924).
[6] See G. E. Bessey, *Sym., Stockholm 1938*, 206; H. H. Steinour, loc. cit.

TABLE 24 Physical properties of calcium silicate hydrates

Compound or mineral[1]	Composition[2]	Temperature of formation °C	Density (g/cm^3)	Refractive indices α	Refractive indices γ	Characteristic X-ray powder spacings (Å)			
Tricalcium silicate hydrate	$C_6S_2H_3$	150–500	2·61	1·593	1·597	8·6	3·28	3·03	2·89
Calciochondrodite	C_5S_2H	250–800	2·84	Mean 1·630		5·42	3·79	3·31	3·02
C_2SH (A)	C_2SH	100–200	2·8	1·614	1·633	4·22	3·90	3·54	3·27
C_2SH (B) (Hillebrandite)	C_2SH	140–350	2·66	1·605	1·612	4·74	3·51	3·32	3·00
C_2SH (C)	C_2SH [3]	160–300	2·67	Mean 1·630		3·03	2·83	2·69	2·49
C_2SH (D) [4]	C_6S_3H	350–800	2·98	1·650	1·664	3·06	2·89	2·86	2·82
Afwillite	$C_3S_2H_3$	100–160	2·63	1·617	1·634	6·46	5·74	4.73	3·19
Foshagite	C_4S_3H	300–500	2·7	1·594	1·598	10·	6·8	4·95	3·37
Xonotlite	C_5S_5H	150–400	2·7	1·583	1·592	3·65	3·23	3·07	2·04
CSH (A)	CSH [5]	150–300		Mean 1·603		3·21	3·01	2·23	1·89
C_2SH_2 [6] (or C-S-H II)	$C_{1.5-2.0}S$.aq	< 100				10·6–9·8	3·07	2·85	2·80
CSH (B) [6] (or C-S-H I)	$C_{0.8-1.5}S$.aq	< 100				14–9	3·07	2·80	1·83
Crystalline tobermorites	$C_5S_6H_9$	60 (?)	2·2	Mean 1·550 ?		14·	5·53	3·25	3·07
	$C_5S_6H_5$	110–140	2·44	1·570	1·575	11·3	3·07	2·97	2·80
	C_5S_6H	250–450	2·7	1·600	1·605	9·3	3·59	3·03	2·78
	C_5S_6	450–650				9·7			
	?					12·6	3·07	2·80	1·83
	?	—				10·0	3·05	2·93	2·80
Gyrolite	$C_2S_3H_2$	120–200	2·39	1·536	1·549	22·	11·	4·20	3·12
Truscottite	$C_6S_{10}H_3$	200–300	2·36–2·48	Mean 1·550		19·	9·4	4·13	3·14
Nekoite	$C_3S_6H_8$	—	2·21	Mean 1·535		9·3	3·36	2·82	2·47
Okenite	$C_3S_6H_6$	—	2·33	1·530	1·541	21·	8·80	3·56	3·07
'Z-Phase'	CS_2H_2	140–240				15·0	8·35	5·07	3·03

[1] Bogue's nomenclature.
[2] Approximate in some cases.
[3] Probably a mixture of calciochondrodite (C_5S_2H) and a phase related to kilchoanite (C_3S_2).
[4] Also known as Phase Y and occurs as the mineral dellaite.
[5] Contains carbonate and is now known to correspond to scawtite $7CaO.6SiO_2.CO_2.\ 2H_2O$.
[6] Similar to tobermorite but poorly crystallised. Gels, or near amorphous samples, give weak X-ray patterns of one to three *hk* lines

Numerous investigations have been made on the system $CaO-SiO_2-H_2O$. It is convenient to represent the data by plotting the molar $CaO:SiO_2$ ratio in the solid phase against the CaO concentration of the solution. The curve obtained by Taylor[1] at 17° together with those observed at 25° by Kalousek[2] and by Greenberg and Chang[3] are shown in Fig. 51. There have been various

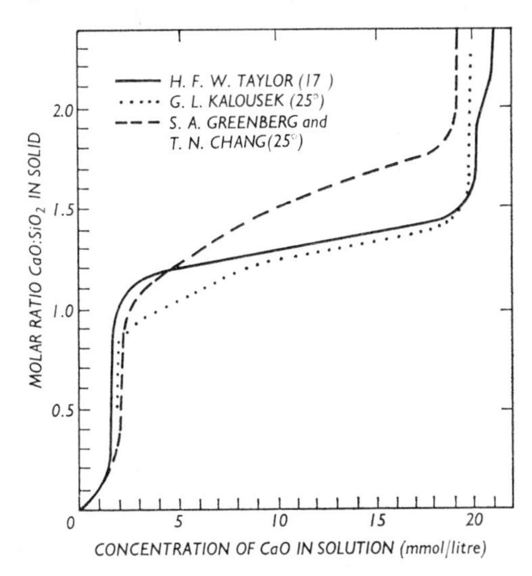

FIG. 51 Equilibria between hydrous calcium silicates and solutions.

suggestions[4] that a hydrated calcium silicate less basic than the monosilicate exists, but the vertical portion of the curves indicates that there is an invariant point between hydrated silica and a monocalcium silicate hydrate at a lime solution concentration of about 1–2 millimoles per litre (0·06–0·11 g CaO per litre). At this lime concentration and below, the concentration of silica in solution can rise to about 0·3 g SiO_2 per litre, but extremely small amounts of silica (< 0·006 SiO_2 per litre) occur in solution at higher lime concentrations when the $CaO:SiO_2$ ratio of the solid is near unity, or above. According to Taylor, X-ray examination of the solid products with $CaO:SiO_2$ ratios varying from 1·0 to 1·5 showed a considerable degree of crystallinity and no significant change in the diffraction pattern, indicating that the basic structure remained the same. This hydrate of composition varying from $CaO.SiO_2.aq$ to $3CaO.2SiO_2.aq$ was called calcium silicate hydrate I, now usually abbreviated to *C-S-H* (I). Bogue designated it as *CSH* (B). From the results of Kalousek, the range of composition of this *C-S-H* (I) hydrate may extend down to a minimum $CaO:SiO_2$ ratio of 0·8. In addition to the reaction of calcium hydroxide solutions with silica gel,

[1] *J. chem. Soc.*, 1950, 3682. For a general review see S. Brunauer and S. A. Greenberg, *Sym., Washington 1960*, 135.
[2] *Sym., London 1952*, 296.
[3] *J. phys. Chem.* **69**, 182 (1965).
[4] e.g. K. G. Krasilnikov, *Sym., Moscow 1956*, 351; also H. H. Steinour, *Chem. Rev.* **40**, 391 (1947) for a review of earlier work.

this compound is also obtained by the repeated extraction of C_3S or βC_2S with water, or by precipitation of calcium nitrate and sodium silicate solutions, or by reaction[1] of ethyl orthosilicate with lime solutions. In suspensions under hydrothermal conditions it can be formed as an intermediate product which rapidly recrystallises to give a tobermorite compound with a basal spacing of 11·3 Å.

Later studies have stressed the close resemblance of the X-ray patterns of *C-S-H* (I) to those of the crystalline tobermorite minerals, but the patterns of the former show fewer spacings and a broad basal spacing which may even be absent in poorly crystallised samples. The basal spacing in X-ray patterns of *C-S-H* (I) depends on both the $CaO:SiO_2$ ratio and the $H_2O:SiO_2$ ratio, and possibly on other ill-defined factors. For samples dried by washing with alcohol and ether, the basal spacing falls from 13–14 Å for $CaO:SiO_2 = 0{\cdot}8$ to a value of 12–12·5 Å at a ratio of about 1, and then to about 10 Å for $CaO:SiO_2 = 1{\cdot}5$. The $CaO:SiO_2$ ratio of *C-S-H* (I) is generally higher than that of the tobermorite minerals, and the reasons for this and for the varying $CaO:SiO_2$ ratio from about 0·8 to 1·5 are discussed in Chapter 11. The incorporation of extra calcium ions in the crystal lattice may occur, or the replacement of silicate by hydroxyl ions. There may also be a condensation of SiO_4 anions to give Si_2O_7 or polysilicate anions associated with expulsion of Ca^{2+} ions from the structure.

Greenberg and Chang concluded that at $CaO:SiO_2$ ratios of 0·14 to 1 the solids consisted of mixtures of CaH_2SiO_4 (or $CaO.SiO_2.H_2O$) and silica gel, and at ratios of 1 to 1·75 the solid was regarded as a solid solution of composition $CaH_2SiO_4.nCa(OH)_2$. From determinations of pH and calcium and silicic acid concentrations in solutions, they also derived values for the solubility product

$$K_{Sp1} = a_{Ca^{++}} \times a_{H_2SiO_4^{--}}$$

and obtained an average value of $10^{-7{\cdot}2}$ over a range of $CaO:SiO_2$ ratios in the solid from 0·2 to 1·8.

The sharp rise in Fig. 51 in the lime : silica ratio of the solid phase at lime solution concentrations near saturation could be attributed to an invariant point between *C-S-H* (I) and $Ca(OH)_2$ or to the formation of a more basic calcium silicate hydrate. Many investigators have indicated the existence of a compound $2CaO.SiO_2.aq$ and its formation from lime and silica gel, or from C_3S, provided that the lime solution concentration was close to saturation. In conformity with these earlier observations, the X-ray pattern of solids of $CaO:SiO_2$ ratio of about 2 was found by Taylor to be somewhat different from that of *C-S-H* (I), and not to show the presence of $Ca(OH)_2$, thus indicating the formation of a dicalcium silicate hydrate which was termed calcium silicate hydrate II, or C_2SH (II). Bogue later designated it as C_2SH_2, but the term now favoured is *C-S-H* (II).

In view of the variability of the $CaO:SiO_2$ ratio found in different preparations of this material, the term *C-S-H* (II) is now taken to denote a semi-crystalline material with ratios from 1·5 to 2·0. It has been obtained[2] in aqueous suspensions

[1] A. Grudemo, *Proc. Swed. Cem. Concr. Res. Inst.*, No. 26 (1955).

[2] H. F. W. Taylor, loc. cit.; G. L. Kalousek and A. F. Prebus, *J. Am. Ceram. Soc.* **41**, 124 (1958); H. Funk, *Z. anorg. allg. Chem.* **291**, 276 (1957); *Silikattechnik* **11**, 375 (1960); J. A. Gard, J. W. Howison and H. F. W. Taylor, *Mag. Concr. Res.* **11** (33), 151 (1959).

of C_3S or C_2S at room temperature, by the action of steam at 100° on C_2S, by hydrothermal treatment of mixtures of silica, or *C-S-H* (I), with $Ca(OH)_2$ at 100°–200°, and by precipitation from sodium silicate solution. It has also been reported[1] that an almost pure precipitate of $2CaO.SiO_2.H_2O$ with an X-ray pattern agreeing with *C-S-H* (II) is obtained by treating amorphous hydrated silica with calcium glycerate solution at 180°. As we have seen earlier, the CaO : SiO_2 ratio of the hydration products in pastes of C_3S and C_2S also fall between 1·5 and 2·0, but the poorly crystallised material formed shows some differences in morphology from *C-S-H* (II) so that it is not certain that it is identical with this compound. Under the electron microscope[2] *C-S-H* (II) appears as fibres, or foils of a fibrous texture, as compared with the crumpled foils generally observed with *C-S-H* (I). The X-ray pattern is very similar to, but can be distinguished from, that of *C-S-H* (I).

In view of these results, the conclusion of some earlier investigators that the most basic hydrated calcium silicate formed at ordinary temperatures is $3CaO.2SiO_2.aq$ must now be revised, though there is still some uncertainty regarding the CaO : SiO_2 ratio of the hydrated calcium silicate in contact with a saturated lime solution. It is generally concluded that over a range of lime concentrations from about 0·05 g CaO per litre up to close to saturation the *C-S-H* (I) is formed with a composition varying over the range $CaO_{0\cdot8-1\cdot5}SiO_2.aq$ while in the region of lime saturation the *C-S-H* (II) with a composition $CaO_{1\cdot5-2\cdot0}SiO_2.aq$ is formed. Since *C-S-H* (I) and (II) are so closely related in their structure and properties, their distinction is not easy and almost impossible at CaO : SiO_2 ratios near 1·5. It is evident that the poorly crystallised calcium silicate hydrates formed at ordinary temperatures probably comprise a continuous range of phases with varying CaO : SiO_2 content, water content, basal spacing, degree of crystallinity and other properties.

As well as the CaO : SiO_2 ratio, the water content of the calcium silicate hydrates formed at room temperature is also difficult to define. They have a very large specific surface which is hydrophilic and readily adsorbs water, and this adsorbed water cannot be completely separated from water of constitution. Various attempts have been made to determine the water content by the study of isotherms and isobars. Taylor[3] found that the water-content/temperature curve on isobaric dehydration of *C-S-H* (I) at a water vapour pressure of 6 mm was independent of the CaO : SiO_2 ratio of the compound for values from 1 to 1·5, as is illustrated in Fig. 52. An isotherm showing the loss at different vapour pressures at room temperatures for material with a CaO : SiO_2 ratio of unity is shown in Fig. 53. This indicates that water in excess of one molecule can be removed, or replaced, by exposure to varying vapour pressures and that the maximum content is 2 to 2·5 H_2O at a water vapour pressure of 15 mm. It is evident that at ordinary temperatures no hydrate with less than one molecule of H_2O will be formed and that in contact with aqueous solution the hydrate must

[1] N. A. Toropov, A. I. Borisenko and P. V. Shirakova, *Chem. Abstr.* **48,** 9247 (1954); F. D. Tamás, *Silikattechnik* **11,** 378 (1960).
[2] A. Grudemo, *Proc. Swed. Cem. Concr. Res. Inst.*, No. 26 (1955); *Sym., Washington 1960*, 615; H. Funk, *Z. anorg. allg. Chem.* **291,** 276 (1967); L. E. Copeland and E. G. Schulz, *J. Res. Dev. Labs Portld Cem. Ass.* **4** (1), 2 (1962).
[3] *J. chem. Soc.*, 163 (1953).

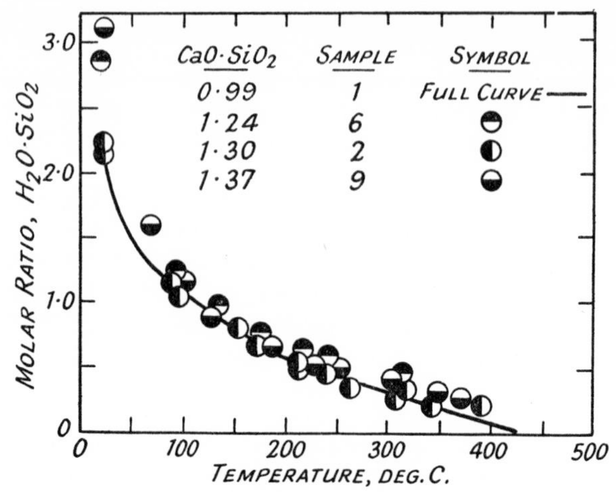

FIG. 52 Dehydration isobars for preparation of calcium silicate hydrate (I) with CaO : SiO_2 ratio varying between 1 and 1·5.

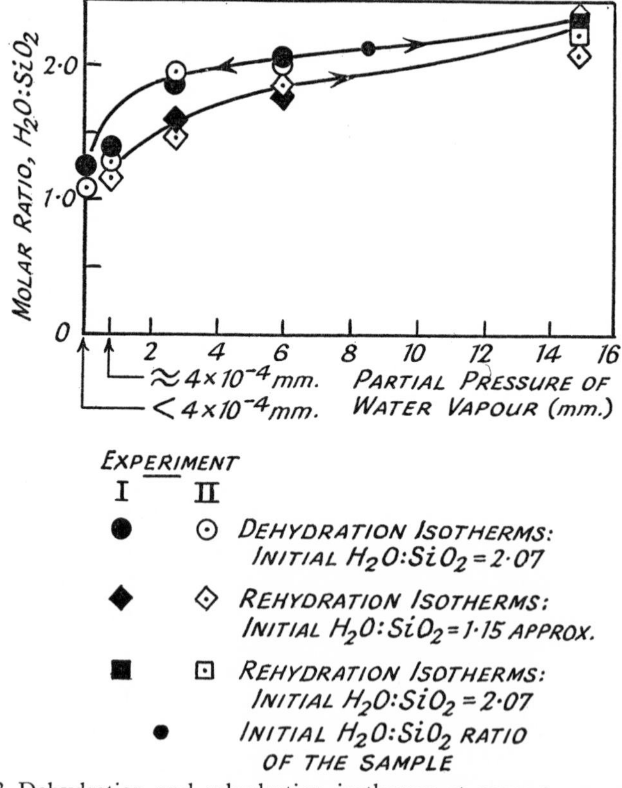

FIG. 53 Dehydration and rehydration isotherms at room temperature for calcium silicate hydrate (I) CaO : SiO_2 = 0·99.

contain at least 2 to 2·5 H_2O. Similarly, for ill-crystallised calcium silicate hydrate with a CaO : SiO_2 ratio of 1·5, H_2O : SiO_2 values of about 1·4 and 1·0 have been obtained[1] for samples in equilibrium at room temperature with the water vapour pressure over $Mg(ClO_4)_2.2H_2O$–$Mg(ClO_4)_2.4H_2O$ (0·008 mm) and ice at −78° (0.0005 mm) respectively. Using the latter method of drying, the data also indicated that an increase in the CaO : SiO_2 ratio above 1·5 is accompanied by a like increase in the H_2O : SiO_2 ratio, the values gradually increasing from about 1 to 1·3 at a CaO : SiO_2 ratio of 1·75.

The X-ray pattern of *C-S-H* (I) varies with the water content, and on heating a preparation with a CaO : SiO_2 ratio of 0·95 at a water vapour pressure of 6 mm, Taylor observed that the initial basal spacing of 12·3 Å fell to 10·4 Å at 120° and 9·3 Å at 240°–450°. A partial but not complete reversal of the lattice shrinkage was also indicated, since on rehydrating the sample heated at 120° over water at room temperature the basal spacing became diffuse, with a maximum at about 11 Å. The interpretation of the dehydration behaviour was much helped by a comparison with the calcium silicate hydrate occurring as intergrowths in the natural mineral crestmoreite (or riversideite).[2] Because of its higher degree of crystallinity more definite data could be obtained as follows:

Temperature of heating at a water vapour pressure of 6 mm.	< 20°	100°	250°
Water content, moles H_2O per mole SiO_2	ca. 2·5	1·0	0·5
Interlayer spacing *c*-axis. Å	14 ± 0·4	10·4 ± 0·4	9·3 ± 0·3

It appears that one molecule of H_2O is essential to the structure and that the *C-S-H* (I) has a layer structure, and that additional molecules of water can enter between the layers increasing their spacing. In synthetic preparations of *C-S-H* (I) there is, Taylor suggests, an incomplete equilibrium and different hydrates may exist interstratified at random in layers normal to the *c*-axis.

The *C-S-H* (I) and *C-S-H* (II) hydrates are apparently both very closely related to the natural mineral tobermorite, which has the composition 5CaO. $6SiO_2.5H_2O$.[3] The terms 'ill-crystallised tobermorite' or 'tobermorite gel' are often used to include these two compounds in their various states of hydration. Several varieties of the tobermorite minerals are known, and these are classified as the 14, 12·6, 11·3, 10·0 and 9·3 Å tobermorites. Samples from different sources may show slightly different characteristics on heating. A composition approximating to $C_5S_6H_9$, with small amounts of B_2O_3 and CO_2, has been attributed[4] to a 14 Å tobermorite mineral. On heating in dry air (CO_2-free), this mineral lost water at 55° ± 5° to give a 11·3 Å tobermorite of approximate composition $C_5S_6H_5$, and at 250–450° a C_5S_6H composition with a 9·3 Å basal spacing. At 450–650° the remaining water and CO_2 are lost and the basal spacing increases to 9·7 Å, and above 730° wollastonite (*β-CS*) is formed.

[1] S. Brunauer, D. L. Kantro and L. E. Copeland, *J. Am. chem. Soc.* **80,** 761 (1958).
[2] H. F. W. Taylor, *Mineralog. Mag.* **30,** 155 (1953).
[3] G. F. Claringbull and M. H. Hey, *Mineralog. Mag.* **29,** 960 (1952); see also J. D. C. McConnell, ibid. **30,** 293 (1954).
[4] V. C. Farmer, J. Jeevaratnam, K. Speakman and H. F. W. Taylor, *Spec. Rep. Highw. Res. Bd,* No. 90, 291 (1966).

The poorly-crystalline or near-amorphous calcium silicate hydrates formed in Portland cement pastes at ordinary temperatures are of even more indefinite nature than the *C-S-H* (I) and *C-S-H* (II) hydrates discussed above. They comprise a range of phases of differing compositions, depending on such factors as degree of hydration, temperature and water : cement ratio. The position is further complicated by the entry into the structure of the hydrates of other ions present in cement, such as sulphate, aluminium, iron, magnesium and alkali ions. Taylor[1] has suggested that the hydrated calcium silicate formed may be more closely related in structure to the alkali-containing mineral jennite ($8CaO.5SiO_2.Na_2O.11H_2O$) than to tobermorite.

The influence of alkalis on the formation of hydrated calcium silicates at room temperature has been studied by Kalousek[2] in a partial study of the system $Na_2O–CaO–SiO_2–H_2O$. Only calcium hydroxide and a four-component gel of variable composition appeared as solid phases. In the presence of solid $Ca(OH)_2$ the composition of the gel varied from $0{\cdot}003\ Na_2O.2CaO.SiO_2.aq$ in solutions of 0·2 g Na_2O per litre to about $0{\cdot}25\ Na_2O.CaO.SiO_2.aq$ at 20 g Na_2O per litre and upwards to 101 g. The reduction in the CaO : SiO_2 ratio of the gel with increasing soda in solution is to be attributed to the lowering of the solubility of calcium hydroxide. It is also evident that the gel can take up an appreciable amount of alkali from the aqueous phase. Kalousek also indicated that the equilibria depended strongly on the nature of the solid phase, and that slow changes in crystallinity over a long period of time could produce very different equilibrium points. This indication has been confirmed by later studies[3] in which various cements, anhydrous calcium silicates and synthetic calcium silicate hydrates were treated for 6 months with alkali solution of concentration 0·125–2 molar with respect to NaOH or KOH. All the solid phases were found to develop some degree of crystallinity as revealed by the presence of basal spacings in the range 9·6–12·4 Å in the X-ray diffraction patterns. The alkali contents of the solid phases were not more than about one-fifth of those observed by Kalousek. It seems, therefore, that alkalis can only be incorporated to a minor extent in *CSH* gels.

The incorporation of Al ions in the structure of 11 Å tobermorite prepared hydrothermally was first shown by Kalousek[4] and later confirmed by other investigators.[5] It was established that up to 15 per cent of the Si^{4+} in the structure could be replaced by Al^{3+} resulting in an increase in the basal spacing from 11·18 Å to 11·45 Å for maximum substitution. At the same time, slight changes occurred in the infra-red spectrum and DTA thermogram. Preparations of 11 Å tobermorite in the presence of MgO or of hydrated iron oxide suggested that both Fe^{3+} and Mg^{2+} could enter the structure of 11 Å tobermorite.

In the case of the ill-crystallised calcium silicate hydrates formed in cement pastes, it has been established that substitution reactions occur to an appreciable extent. Quantitative DTA investigations[6] have indicated that not all of the sulphate in cement paste can be accounted for by the amounts of gypsum and

[1] *Sym., Tokyo 1968.*
[2] *J. Res. natn. Bur. Stand.* **32,** 285 (1944).
[3] P. Seligmann and N. R. Greening, *Sym., Tokyo 1968.*
[4] *J. Am. Ceram. Soc.* **40,** 74 (1957).
[5] S. Diamond, J. L. White and W. L. Dolch, *Am. Miner.* **51,** 388 (1966).
[6] G. L. Kalousek, *Mater. Res. Stand.* **5,** 292 (1965).

sulphoaluminate phases present, and it was suggested that the 'missing sulphate' was incorporated in the lattice of the calcium silicate hydrate gel. Similarly, Smolczyk[1] observed that the calcium aluminate hydrates could not account for all the alumina present in blastfurnace slag cement pastes, and it was concluded that the calcium silicate hydrate phase retained the missing alumina. Other studies[2] on the reaction of 'tobermorite gel' with aluminates, ferrites and sulphates have established that Al^{3+}, Fe^{3+} and SO_4^{2-} can all enter the structure of the calcium silicate hydrate gel formed in C_3S and βC_2S pastes. The maximum amount of Al^{3+} that can be incorporated is about 1 atom of Al per 6 atoms of Si. This amount is comparable to that found with the crystalline tobermorite, but with the gel material it is suggested that Al^{3+} can enter the structure in two ways, firstly $Al^{3+}+H^+$ replacing Si^{4+} and secondly $2Al^{3+}$ replacing $3Ca^{2+}$. For iron-substituted gels prepared from C_3S and C_2F pastes it seems that $2Fe^{3+}$ replaces Ca^{2+} and Si^{4+} and the upper limit of solid solution is 1 atom of Fe per 6 atoms of Si. Similarly, the composition of the products from hydrated mixtures of C_3S and gypsum suggests that S^{6+} replaces Si^{4+}. There is also an uptake of 1·5 moles of lime of which 1 mole enters the structure to maintain change balance with the additional lime entering as $Ca(OH)_2$.

While these values indicate the maximum substitution possible there is some evidence[3] that the calcium silicate hydrate phase finally formed in set cement contains considerable alumina but little ferric oxide or sulphate ion though the latter is taken up initially. The introduction of alumina seems to cause significant reduction in shrinkage and all these substitution reactions produce morphological changes in the gels when examined by electron microscopy. There is need for more study of these substitutions and their effects.

Nomenclature of the hydrated calcium silicates

The nomenclature of the hydrated calcium silicates used by successive investigators has differed and is confusing, particularly when we come to consider the products obtained in high pressure steam. Bogue[4] has suggested the revised system of nomenclature shown below:

Composition	Nomenclature		
	Thorvaldson	Taylor, Taylor and Bessey	Bogue revision
$C_2SH_{0\cdot9-1\cdot25}$	C_2S hydrate I	C_2S α hydrate	C_2SH(A)
$C_2SH_{1\cdot1-1\cdot5}$	C_2S hydrate II	C_2S β hydrate	C_2SH(B)
$C_2SH_{0\cdot3-1\cdot0}$	C_2S hydrate III	C_2S γ hydrate	C_2SH(C)
$C_2SH_{0\cdot67}$		C_2S δ hydrate	C_2SH(D)
$C_{1\cdot5-2\cdot0}SH_2$		C_2SH(II), CSH(II)	C_2SH_2
$CSH_{1\cdot1}$		Flint's CSH	CSH(A)
$C_{0\cdot8-1\cdot5}SH_{1\cdot0-2\cdot5}$		CSH(I)	CSH(B)

[1] *Zement-Kalk-Gips* **18,** 238 (1965).
[2] L. E. Copeland, E. Bodor, T. N. Chang and C. H. Weise, *J. Res. Dev. Labs Portld Cem. Ass.* **9,** 61 (1967).
[3] P. Seligmann and N. R. Greening, *Sym., Tokyo 1968.*
[4] R. H. Bogue, *Mag. Concr. Res.,* No. 14, 87 (1953).

For convenience we will use the Bogue nomenclature in discussing the hydrothermal products. It should be noted, however, that the compound given as C_2SH_2 by Bogue has a variable composition $C_{1\cdot5-2\cdot0}S$.aq and it is now designated as *C-S-H* (II) to show that no particular composition is implied. Similarly, the *CSH* (B) phase of variable composition $C_{0\cdot8-1\cdot5}.S$.aq is designated as *C-S-H* (I). We shall retain these designations for these two lower temperature products. It has also been established in later studies that the *CSH* (A) compound is a quaternary carbonate-containing hydrate $C_7.S_6.CO_2.2H_2O$ (scawtite), and that C_2SH (C) is probably a mixture of calciochondrodite and kilchoanite (C_3S_2) or a related phase. Kilchoanite is substantially identical with a 'phase Z' found by Roy.[1] This is quite different from the 'Z phase' mentioned later.

Hydrothermal reactions of silicates

The hydrated calcium silicates which can arise in reactions above room temperature have been examined by many authors.[2] As shown in Table 24, a large number of compounds with different CaO : SiO_2 ratios and varying water content have been obtained. In recent years much effort has been given to the X-ray examination of both natural and synthetic products, and the identity of the former with several of the latter has been established. Such work contributes to our knowledge of behaviour in the system lime–water–silica at elevated temperatures and helps both to throw light on behaviour at normal temperatures and also under conditions of steam-curing. The general complexity of the relations over a wide range of temperature and the practical difficulties involved enable only rather general indications of behaviour to be given. Experimental procedure has been either to maintain the solid phases in contact with water (up to the critical temperature 374°, 218 atm) or to keep solid and water separate, apart from water present in an initial paste, so that the solid phase was treated with water-vapour. It has been suggested that the latter procedure may tend to give hydrated products richer in lime and poorer in water.

This is an indication of the general tendency that exists in this system for metastable products to be formed, some of which transform only very slowly. Thus Peppler[3] in an investigation of the equilibrium relations of the system at 180°, found that almost all the known hydrated calcium silicates can appear as transitory or metastable phases, and that, in general, the initial compounds tended to have the same CaO : SiO_2 ratio as the original mix. Final equilibria are obtained only very slowly and may take months at 100–200° and weeks at higher temperatures. The approach to equilibrium can proceed through the formation of more than one intermediate phase[4] and the sequence of these phases may be markedly affected by variations in experimental conditions, such as the use of pastes or aqueous suspensions. For practical purposes, therefore, we must pay

[1] D. M. Roy, *Am. Miner.* **43,** 1009 (1958); *J. Am. Ceram. Soc.* **41,** 293 (1958).
[2] For general reviews, see T. Thorvaldson, *Sym., Stockholm 1938*, 246; H. W. F. Taylor and G. E. Bessey, *Mag. Concr. Res.*, No. 4, 15 (1950); various contributions to *Sym., London 1952*, *Washington 1960*, *Tokyo 1968* and to *Sym. Autoclaved Calcium Silicate Building Products, Soc. Chem. Ind. London 1965.*
[3] *J. Res. natn. Bur. Stand.* **54,** 205 (1955).
[4] L. Heller and H. F. W. Taylor, *J. chem. Soc.*, 1952, 2535; G. L. Kalousek, *Proc. Am. Concr. Inst.* **51,** 989 (1955); G. O. Assarsson, *J. phys. Chem.* **62,** 223 (1958).

primary attention to the products formed in relatively short periods of hydrothermal treatment. For temperatures and times of autoclaving up to 200° and 24 hours respectively the phases most likely to be obtained from lime and silica are *C-S-H* (I) and (II), 11 Å tobermorite and dicalcium silicate hydrate A(C_2SH(A)), together with xonotlite and gyrolite at the higher end of this temperature range. Afwillite, 14 Å tobermorite and hillebrandite all form too slowly to be significant under these conditions. With the calcium silicates, or Portland cement-silica starting materials, the phases formed most readily are generally similar except that with C_3S the hydrate $C_6S_2H_3$ can also be formed within a few hours. The ultimate stable phases have been discussed by Taylor[1] and also deduced[2] from thermodynamic calculations of the free-energy changes in the formation and transformation of the various hydrated silicates.

A tricalcium silicate hydrate of composition $3CaO.SiO_2.1\frac{1}{2}H_2O$, or $6CaO.2SiO_2.3H_2O$, has been obtained[3] under steam pressure at temperatures ranging from about 150° to 500° from C_3S or other starting materials of appropriate CaO : SiO_2 ratio. It can also be formed at lower temperatures in C_3S pastes where transport of material is difficult. The H_2O : SiO_2 ratio is probably 1·5, and not 2·0 as was supposed in early studies, and it dehydrates in air at 420–550° to give a modified form of γC_2S. The ionic constitution is indicated to be $Ca_6(Si_2O_7)(OH)_6$. On treating suspensions of C_3S, or of other starting materials with a CaO : SiO_2 ratio of 3, below about 180° the C_2SH (A) hydrate is produced together with free lime. The temperature of the transition to $C_6S_2H_3$ is rather uncertain for Heller and Taylor obtained C_2SH (A) from C_3S at temperatures from 120° to 200°. At lower temperatures the product was the normal low-temperature *C-S-H* (I) and/or *C-S-H* (II). The picture presented is thus that hydration of C_3S leads first to *C-S-H* (I) or *C-S-H* (II) at temperatures up to about 110° and that as the temperature (and pressure) is raised further these are found, if at all, only as an intermediate stage in the formation of C_2SH (A) at 120° upwards and of $C_6S_2H_3$ at 150° upwards. Even this series probably fails to represent the true equilibrium relations, for Peppler gives C_2SH (B) as the final stable phase at 180°, and it seems probable that $C_6S_2H_3$ is not really stable below 200°. At considerably higher temperatures the $C_6S_2H_3$ becomes unstable with respect to a compound of composition $5CaO.2SiO_2.H_2O$. This compound has been obtained[4] hydrothermally at 400–800°, but it can also apparently be formed at saturated steam pressures at temperatures at least as low as 250°. Buckle and Taylor[5] prepared C_5S_2H from C_3S at 600–700°, and showed it to be analogous to the mineral chondrodite $(Mg, Fe)_5(SiO_4)_2(OH, F)_2$. The name calcio-chondrodite was, therefore, suggested for C_5S_2H, which can be written as $Ca_5(SiO_4)_2(OH)_2$.

[1] H. F. W. Taylor, *Sym., Tokyo 1968*.
[2] O. P. Mchedlov-Petrosyan and W. I. Babushkin, *Sym., Washington 1960*, 533; *Silikattechnik* **10**, 605 (1959).
[3] N. B. Keevil and T. Thorvaldson, *Can. J. Res.* **B14**, 20 (1936); G. E. Bessey, *Sym., Stockholm 1938*, 178; E. P. Flint, H. F. McMurdie and L. S. Wells, *J. Res. natn. Bur. Stand.* **21**, 617 (1938); E. R. Buckle, J. A. Gard and H. F. W. Taylor, *J. chem. Soc.* 1958, 1351.
[4] D. M. Roy, *Am. Miner.* **43**, 1009 (1958).
[5] Ibid. **43**, 818 (1958).

The C_2SH (A) compound referred to above was first found by Thorvaldson and Shelton[1] in steam-cured Portland cement mortars at about 150°. Other investigators have confirmed its formation as a distinct species at temperatures between 100° and 200° and from a wide range of starting materials. Vigfusson, Bates and Thorvaldson[2] prepared crystals as large as 0·5 mm × 0·15 mm by the action of saturated limewater on plates of quartz or fused silica at 170°. C_2SH (A) is most readily obtained pure by treatment of βC_2S for about 14 days at 140–160°. On dehydration in air at about 450° βC_2S is formed, and the constitutional formula Ca_2 $(HSiO_4)$ (OH) has been indicated by crystal structure determinations.[3] The formation of C_2SH (A) is much assisted by the presence of seed crystals[4] as well as by the presence of excess calcium hydroxide in reaction mixtures treated at temperatures up to about 180°. At somewhat higher temperatures, C_2SH (B) is obtained from βC_2S and from mixtures of lime and silica, while at still higher temperatures a third form C_2SH (C) has been reported. The general order of the production of the various hydrates from materials of 2CaO : SiO_2 ratio may be illustrated by the results obtained by Heller and Taylor on heating with water mixtures of lime and silica gel in the molar ratio 2 : 1, and β and γC_2S, at 100–200°. With increasing temperature there appeared in succession *C-S-H* (I), afwillite $(3CaO.2SiO_2.3H_2O)$, C_2SH (A), C_2SH (B) and finally C_2SH (C). The afwillite appeared at temperatures as low as 110°, in general agreement with the observation of Bessey that a gel of dicalcium silicate composition treated with excess water at 98° gave afwillite and some $Ca(OH)_2$. The C_2SH (B) hydrate of possible ionic constitution $Ca_2(SiO_3)$ $(OH)_2$ is substantially identical to the natural mineral hillebrandite, but the nature of C_2SH (C) is less certain. The latter compound is formed most readily by hydrothermal treatment of γC_2S at 160–300°, and this may be attributed to a similarity in structure. It has been suggested[5] that a variable CaO : SiO_2 ratio may arise with C_2SH (C), as well as with C_2SH (A) and C_2SH (B). Taylor[6] has indicated that C_2SH (C) is probably a mixture of calciochondrodite (C_5S_2H) and a hydrous variety of kilchoanite (C_3S_2), and later work[7] has confirmed that calciochondrodite is a major constituent of C_2SH (C) preparations.

The remaining dicalcium silicate hydrate, designated as C_2SH (D), is a high-temperature phase of composition $6CaO.3SiO_2.H_2O$ which is readily prepared[8] from a range of starting materials under steam pressure at 350–800°. It has also been found as a natural mineral and named dellaite.[9] On heating in air, it dehydrates at 640–700° to give βC_2S, and the possible constitution $Ca_6(SiO_4)$ (Si_2O_7) $(OH)_2$ is indicated. Another high-temperature product which is readily formed from various starting materials at 300–500° is the compound correspond-

[1] *Can. J. Res.* **1,** 148 (1929).
[2] Ibid. **2,** 520 (1934).
[3] L. Heller, *Acta crystallogr.* **5,** 724 (1952).
[4] H. Funk, *Z. anorg. allg. Chem.* **297,** 103 (1958).
[5] G. L. Kalousek, J. S. Logiuduce and V. H. Dodson, *J. Am. Ceram. Soc.* **37,** 7 (1954).
[6] *Sym., Washington 1960,* 167.
[7] J. M. Bennett, J. A. Gard, K. Speakman and H. F. W. Taylor, *Nature, Lond.* **209,** 1127 (1966); *J. chem. Soc.* **A,** 1052 (1967).
[8] L. S. Dent Glasser, H. Funk, W. Hilmer and H. F. W. Taylor, *J. appl. Chem.* **11,** 186 (1961).
[9] S. O. Agrell, *Mineralog. Mag.* **34,** 1 (1965).

ing to the mineral foshagite. In early studies this was given the composition $5CaO.3SiO_2.3H_2O$, but later work[1] has established the composition $4CaO.3SiO_2.H_2O$ and the constitutional formula $Ca_4(Si_3O_9)(OH)_2$.

With starting materials of CaO : SiO_2 ratio near 1·5, earlier investigators found that xonotlite ($5CaO.5SiO_2.H_2O$) was formed above about 180°. According to Heller and Taylor[2] a mixture of lime and silica gel in this ratio gives *C-S-H* (I) at temperatures up to 130°, and this hydrate also appears as an initial product up to 180°. With longer periods of steam treatment afwillite ($3CaO.2SiO_2.3H_2O$) is formed at 140–160°, and xonotlite and hillebrandite at 180–200°. The latter two compounds probably remain up to about 300°, but at higher temperatures foshagite may be formed. It appears that afwillite is a stable phase in the system $CaO–SiO_2–H_2O$ at 100–160°, if not also at lower temperatures, and the intermediate phases *C-S-H* (I) and C_2SH (A) which often occur are probably metastable with respect to afwillite, though once formed they are highly persistent. X-ray and infra-red studies[3] on afwillite have indicated the constitutional formula $Ca_3(HSiO_4)_2.2H_2O$.

Xonotlite occurs as a natural mineral in many localities, and it is readily formed when reaction mixtures of 1 : 1 CaO : SiO_2 ratio are treated hydrothermally at 150–400°. There appears to be some uncertainty as to its precise composition, and while many preparations give a composition of $5CaO.5SiO_2.H_2O$ the formula $6CaO.6SiO_2.H_2O$ or $Ca_6(Si_6O_{17})(OH)_2$ is indicated by X-ray structure determination.[4] The formation of xonotlite proceeds through the intermediate stages of *C-S-H* (II), *C-S-H* (I) and tobermorite.[5] The 11·3 Å tobermorite appears to persist indefinitely under hydrothermal conditions at 110–140°, and it has been shown to be a major constituent of the binding material in many autoclaved cement-silica or lime-silica products. At temperatures below about 100° the less well-crystallised *C-S-H* (I) occurs, though the formation of 14 Å tobermorite in lime-silica slurries at 60° has been reported.[6] Another phase which was considered to be $CaO.SiO_2.H_2O$ was obtained by Flint, McMurdie and Wells from lime-water and silica gel at 150–225°. This compound has been designated *CSH* (A), and many other investigators have also reported its formation at 140–300°. The X-ray pattern of *CSH* (A) is very similar to that of the natural mineral scawtite which is considered to have the composition $7CaO.6SiO_2.CO_2.2H_2O$, though the carbonate content may be variable. It has been suggested[7] that *CSH* (A) is in fact identical with scawtite. The view that *CSH* (A) is a quaternary carbonate-containing phase arising by inadvertent carbonation is supported by later work[8] in which scawtite was synthesised by hydrothermal treatment of a mixture of $Ca(OH)_2$ and silicic acid together with excess water containing dissolved CO_2. It appears that scawtite is formed mainly in place of

[1] J. A. Gard and H. F. W. Taylor, *Am. Miner.* **43,** 1 (1958); *Acta crystallogr.* **13,** 785 (1960).
[2] *J. chem. Soc.*, 1952, 1018.
[3] H. D. Megaw, *Acta crystallogr.* **5,** 477 (1952); **9,** 26 (1956).
[4] Kh. S. Mamedov and N. V. Belov, *Dokl. Akad. Nauk SSSR* **104,** 615 (1955).
[5] G. O. Assarsson, *J. phys. Chem.* **62,** 223 (1958); L. Heller and H. F. W. Taylor, *J. chem. Soc.*, 1951, 2397.
[6] G. L. Kalousek and R. Roy, *J. Am. Ceram. Soc.* **40,** 236 (1957).
[7] D. A. Buckner, D. M. Roy and R. Roy, *Am. J. Sci.* **258,** 132 (1960).
[8] R. I. Harker, *Mineralog. Mag.* **34,** 232 (1965).

xonotlite, since below 140° scawtite is unstable relative to 11·3 Å tobermorite, while above 300° it is unstable relative to xonotlite and calcite.

A compound corresponding to the mineral gyrolite ($2CaO.3SiO_2.2H_2O$) was first synthesised by Wells, Clarke and McMurdie from glasses and gels having CaO : SiO_2 ratios of 0·5–0·66 at 150° and above. Its formation has since been confirmed in many other studies, though some controversy exists with regard to the precise composition, especially the water content. Various compositions have been suggested for gyrolite, but Harker[1] concluded that the generally accepted formula $C_2S_3H_2$ showed the best agreement with available evidence. Gyrolite appears to be a stable phase under saturated steam conditions between 120° and 200°. At higher temperatures up to over 300° another compound corresponding to the mineral truscottite is formed. The exact composition of truscottite is uncertain and the composition may be variable. Its ideal formula has been indicated[2] to be within the range C_2S_4H–$C_6S_{10}H_3$. The mineral referred to as reyerite closely resembles truscottite, but its X-ray pattern is slightly different and it seems that, unlike truscottite, the mineral reyerite contains additional essential components (Na and/or K) other than CaO, SiO_2 and H_2O. Another product which is probably closely related in structure to gyrolite has been described by Assarsson[3] and designated as Z-phase with an approximate composition $CaO.2SiO_2.2H_2O$ or $CaO.2SiO_2.3H_2O$. It was obtained by autoclaving lime-silica mixtures at 140–240° and considered to be an intermediate phase in the formation of gyrolite. The X-ray pattern of Z-phase is distinct from those of *C-S-H* (I), gyrolite and truscottite, and is characterised by a longest spacing of 15 Å. Later studies[4] have confirmed the existence of Z-phase as a distinct species, but its composition still appears to be uncertain. A composition near to $CaO.2SiO_2.2H_2O$ has also been assigned to the minerals okenite and nekoite described by Gard and Taylor.[5] These two minerals have not yet been synthesised.

Though present knowledge allows of no great precision of detail, a broad picture of the various hydrated calcium silicate phases apparently stable at various temperatures is shown in Table 25. As we have seen, intermediate phases may also appear but these are not shown. It should also be noted that the temperature ranges in which various compounds are formed may be altered at pressures above that of saturated steam. For example, afwillite is apparently stable above 200° at high pressures, and 11·3 Å tobermorite remains stable relative to xonotlite until about 285° at pressures of 800–1700 kg/cm^2. Many studies[6] have been made on the CaO–SiO_2–H_2O system at elevated temperatures and pressures, and these have shown that increasing the pressure beyond about 300 kg/cm^2 usually has only a minor effect on the stability relations of various compounds.

[1] *J. Am. Ceram. Soc.* **47,** 521 (1964).
[2] H. F. W. Taylor, *Sym., Tokyo 1968.*
[3] *Sym., Washington 1960*, 190.
[4] H. Funk, *Z. anorg. allg. Chem.* **313,** 1 (1961); R. I. Harker, *J. Am. Ceram. Soc.* **47,** 521 (1964).
[5] *Mineralog. Mag.* **31,** 5 (1956).
[6] D. M. Roy and R. I. Harker, *Sym., Washington 1960*, 196; D. A. Buckner, D. M. Roy and R. Roy, *Am. J. Sci.* **258,** 132 (1960); R. I. Harker, D. M. Roy and O. F. Tuttle, *J. Am. Ceram. Soc.* **45,** 471 (1962); C. W. F. T. Pistorius, *Am. J. Sci.* **261,** 79 (1963); R. I. Harker, *J. Am. Ceram. Soc.* **47,** 521 (1964); D. M. Roy and A. M. Johnson, *Sym. Autoclaved Calcium Silicate Building Products, Soc. Chem. Ind. London 1965,* 114.

TABLE 25 Apparently stable phases formed from lime-silica mixes in the presence of water at saturated steam pressures*

Molar ratio of initial mix $CaO : SiO_2$	Temperature 0	50	100	150	200	250	300	350
						← Calciochondrodite →		
3 : 1	← *C-S-H* (II) + $Ca(OH)_2$ →		← C_2SH (A) + $Ca(OH)_2$ →		← C_2SH(B) + $Ca(OH)_2$ →	← Tricalcium silicate hydrate →		
			← C_2SH (A) →			← C_2SH (C) →		
2 : 1	← *C-S-H* (II) →		← Afwillite + $Ca(OH)_2$ →		← C_2SH (B) →			
							← Foshagite →	
3 : 2	← *C-S-H* (I) →		← Afwillite →		← Xonotlite + C_2SH (B) →			
1 : 1	← *C-S-H* (I) →		← 11 Å Tobermorite →		← Xonotlite →			
2 : 3	← *C-S-H* (I) + SiO_2 →		← Gyrolite →			← Truscottite + Xonotlite →		

* Bogue nomenclature above 100°.

Structure of the hydrated silicates

We shall consider in a later chapter the structure of cement compounds, but it may be noted here that a group of hydrated silicates structurally related to the anhydrous compound wollastonite (β-CS) may be distinguished. Into this group there fall Taylor's C-S-H (I) and C-S-H (II), the minerals okenite and nekoite, and the synthetic and natural forms of tobermorite, hillebrandite, xonotlite and foshagite. These compounds seem to contain metasilicate $(SiO_3)^{2-}$ chains which are linked so as to repeat every third tetrahedron and to give a unit repeat distance of 7·3 Å along the *b*-axis, which in varying degrees is a direction of fibrous or prismatic growth. Of the remaining hydrated silicates afwillite and C_2SH (A) have a structure containing $(HSiO_4)^{3-}$ tetrahedra, while gyrolite, truscottite and 'Z-phase' form hexagonal or pseudohexagonal crystals with an *a*-axis of about 9·7 Å and possibly contain Si–O sheets of empirical composition $(Si_2O_5)^{2-}$. Calciochondrodite is closely related structurally to γC_2S and contains $(SiO_4)^{4-}$ tetrahedra. The $(SiO_4)^{4-}$ group together with a $(Si_2O_7)^{6-}$ group may occur in C_2SH (D), while only the $(Si_2O_7)^{6-}$ seems to be present in the tricalcium silicate hydrate ($C_6S_2H_3$).

The layer type structure of C-S-H (I) and the tobermorite minerals leads to the changing lattice spacing with changing water content, whereas the other hydrated silicates show no such variation in lattice dimension. The possible significance of this in relation to moisture movement and drying shrinkage of set cement we shall discuss later.

Hydrated magnesium silicates

There are a number of natural minerals,[1] including serpentine $3MgO.2SiO_2.2H_2O$, which occurs as the fibrous chrysotile or the platy antigorite, talc $3MgO.4SiO_2.H_2O$, and meerschaum $2MgO.3SiO_2.4H_2O$. The two former minerals have been prepared hydrothermally from the oxides.[2] There is evidence (see p. 346) that hydrated magnesium silicates may be formed at room temperature from the action of magnesium sulphate solutions on set cement. Using this method, Cole and Hueber[3] obtained a preparation which was considered to consist largely of a magnesium silicate hydrate of composition $4MgO.SiO_2.11H_2O$. The X-ray pattern of this supposed hydrate could be indexed on a hexagonal unit cell, $a = 8{\cdot}95$ Å, $c = 8{\cdot}26$ Å. On heating, the water content fell to 9·1 H_2O at 110–170° and 5·1 H_2O at 310°, and the longest basal spacing decreased from the initial 8·26 Å to 8·03 Å at 310°. The basal spacing disappeared at 350° and the presence of periclase (MgO) was then indicated in the X-ray pattern. However, attempts to synthesise the $4MgO.SiO_2.11H_2O$ compound by reaction of MgO or $MgCl_2$ with sodium silicate solution or silica sol at 37° or higher temperatures were unsuccessful, and a poorly-crystallised hydrous magnesium silicate of

[1] N. L. Bowen and O. F. Tuttle, *Bull. Geol. Soc. Amer.* **60**, 439 (1949).
[2] E. T. Carlson, R. B. Peppler and L. S. Wells, *J. Res. natn. Bur. Stand* **51**, 179 (1953); G. L. Kalousek and D. Mui, *J. Am. Ceram. Soc.* **37**, 38 (1954).
[3] W. F. Cole and H. V. Heuber, *Silic. ind.* **22**, 75 (1957).

composition $MgO.SiO_2.1{\cdot}9H_2O$ was obtained instead. The latter preparations gave X-ray patterns with *c*-axis spacings of antigorite together with broad *hk* two-dimensional reflections, and water was lost very readily on heating, almost all being removed slightly above 100°. It appears that the $MgO.SiO_2.aq$ gel may be similar to *C-S-H* (I) or *C-S-H* (II), but the $4MgO.SiO_2.aq$ compound is not analogous to any calcium silicate hydrate, and since it is only obtained in the presence of other ions or compounds its identity must be regarded as uncertain. The formation of hydrated 2 : 1, 1 : 1 and 2 : 3 magnesium silicate hydrates by precipitation from magnesium chloride solutions with silica hydroxide[1] and of the 2 : 3 compound from silica gel and magnesium hydroxide in aqueous suspension[2] has also been claimed.

Steam-cured cement

The hydrated silicates formed on curing Portland cement products in steam at 100°[3] are similar to those obtained from lime-silica mixes or C_3S and C_2S but the CaO : SiO_2 ratio of the *C-S-H* gel is increased. Some slight transformation of *C-S-H* gels into crystalline $C_6S_2H_3$ and hillebrandite (C_2SH (B)) may occur. There is evidence that increased amounts of alumina and sulphur trioxide present in the cement pass into the *C-S-H* gel.[4] Ettringite is decomposed and, if there is any separate formation of alumina compounds, it is as the monosulphate and tetracalcium aluminate hydrate, or their solid solution, as transient phases. There is no evidence of hydrogarnet formation.

When cured in high-pressure steam at 150° to 200° for the normal periods of six to twelve hours, the main hydrated silicates formed from Portland cement[5] are poorly-crystallised *C-S-H* gels or tobermorite, $C_6S_2H_3$ and C_2SH (A). The formation of C_2SH (A) is unfavourable to strength development. If, as is common for autoclave curing of Portland cement products, a proportion of finely ground quartz is substituted for cement, the formation of $C_6S_2H_3$ no longer occurs. With small replacements of 10 per cent, or possibly up to 20 per cent, there is increased formation of C_2SH (A) and a loss in strength. Replacement of 30 to 40 per cent of the cement by quartz usually gives optimum strengths and favours the formation of the 11 Å tobermorite. Xonotlite may also form above 175° with the higher silica additions and gyrolite has on occasions been detected. It is doubtful if any hydrogarnet is formed, at least in the 6–12 hours autoclave treatment used commercially. No hydrated calcium aluminate or sulphoaluminate phases have been detected.

[1] J. Wiegmann and C. H. Horte, *Silikattechnik* **11,** 380 (1960).

[2] N. Hast, *Ark. Kemi* **9,** 343 (1956).

[3] G. N. Idorn, *Sym., Tokyo 1968.*

[4] G. L. Kalousek, *Mater. Res. Stand.* **5,** 292 (1965).

[5] G. L. Kalousek, *Sym., Tokyo 1968*; G. L. Kalousek and J. E. Kopanda, *J. Mater.* **3,** 304 (1968); A. Aitken and H. F. W. Taylor, *Sym., Washington 1960*, 285; A. V. Volzhensky, *R.I.L.E.M. Sym. Problems of Accelerated Hardening of Concrete, Moscow 1964*, has reviewed U.S.S.R. work.

Hydrated calcium aluminates

The hydrated calcium aluminates have been the subject of numerous investigations[1] but there is still uncertainty about some of the reported compounds, a list of which is given in Table 26. Difficulties arise in the study of these compounds because of such complicating factors as the susceptibility of the hydrates to attack

TABLE 26 Physical properties of calcium aluminate hydrates

Compound	Crystal form	Density (g/cm³)	Refractive indices ω	ε	Characteristic X-ray powder spacings (Å)
C_6AH_{33}*	Needles		1.475	1·466	
C_5AH_{34}*	Hex. prisms		1·487	1·480	
α_1-C_4AH_{19}	Hex. plates	1·79	1·500	1·485	10·7, 5·35, 4·24, 3·93
α_2-C_4AH_{19}	,, ,,	1·81	1·500	1·485	10·7, 5·35, 4·10, 3·66
α-C_4AH_{13}*	,, ,,		1·539	1·514	8·2, 4·10, 3·88, 2·88
C_4AH_{13}†	,, ,,		—	—	8·2, 4·10
β-C_4AH_{13}	,, ,,	2·02	1·539	1·514	7·9, 3·95, 2·88, 2·86
C_4AH_{12}†	,, ,,		—	—	7·9, 3·95
C_4AH_{11}	,, ,,	2·08	1·539	1·524	7·4, 3·90, 3·70, 2·87
C_4AH_7	,, ,,	2·28	1·555	1·544	ca. 6 (or 7·2–7·4?)
C_3AH_{18-21}†	Needles		1·495 (γ)	1·479 (α)	
C_3AH_{10-12}*	Hex. plates				7·6, 3·80
C_3AH_6	Cubic forms	2·52	1·605		5·13, 4·45, 3·36, 3·14
α_1-C_2AH_8	Hex. plates	1·95	1·520	1·505	10·7, 5·36, 4·10, 3·96
α_2-C_2AH_8	,, ,,	1·95	1·520	1·505	10·7, 5·36, 2·89, 2.87
$C_2AH_{7·5}$	,, ,,	1·98	1·520	1·505	10·6, 5·30, 3·53, 2·86
β-C_2AH_8	,, ,,	1·98	1·520	1·505	10·5, 5.24, 3·49, 2·87
C_2AH_5	,, ,,	2·09	1·534	1·524	8·7, 4·34, 3·18, 2·87
C_2AH_4†	,, ,,	2·27	1·565	1·559	7·4, 3·72, 2·87
CAH_{7-10}	Ill-defined	1·70–1·74	Mean 1·48		14·3, 7·16, 3·72, 3·56
C_2AH_3?†	—		1·576	1·589	
$C_4A_3H_3$	Orthorhombic plates	2·71	Mean 1·627		3·61, 3·27, 2·80

* Probably a quaternary hydrate containing carbonate.
† Existence doubtful.

[1] The more recent investigations are F. E. Jones and M. H. Roberts, *Building Research Current Papers, Research Series 1* (1962); R. Alègre, *Revue Matér. Constr. Trav. publ.* **566,** 301 (1962); F. Lavanant, *Revue Matér. Constr. Trav. publ.* **592,** 1; **593,** 76; **595,** 193; **596,** 251; **597,** 298 (1965); W. Dosch and H. zur Strassen, *Zement-Kalk-Gips* **18,** 233 (1965). The earlier literature on these and the corresponding ferrite hydrates can be traced through the reviews by H. H. Steinour, *Bull. Res. Dev. Labs Portld Cem. Ass.*, No. 34 (1951); F. E. Jones, *Sym., Washington 1960*, 205; and H. E. Schwiete and U. Ludwig, *Sym., Tokyo 1968*.

by atmospheric carbon dioxide and the occurrence of some compounds as polymorphs and polytypes as well as in different hydration states. It is likely that some confusion and misinterpretation in identification of the various hydrates examined in past studies has occurred from lack of precise control of water content of solid phases, and especially from the effects of carbon dioxide contamination. These two factors may well account for the existing uncertainties and conflicting views in the literature with regard to the composition and identity of some compounds, as well as to their stability relations in the system CaO–Al_2O_3–H_2O. In particular, carbonation has been responsible for erroneous interpretation in the past, as for example in the case of the hexagonal-plate compound of reported formula $C_3A.10$–$12\ H_2O$ which on the basis of many investigations[1] is now accepted really to be the quaternary carbonate hydrate $C_3A.CaCO_3.11H_2O$. It is also probable that the reported compounds C_6AH_{33}, C_5AH_{34}, and α-C_4AH_{13} are actually quaternary hydrates containing carbonate, as will be discussed later.

The hydrated calcium aluminates generally crystallise more readily than do the hydrated calcium silicates, and different forms with varying optical character have been reported for the compounds shown in Table 26. The isotropic C_3AH_6 crystallises in a variety of cubic forms such as cubes, rhombic dodecahedra, octahedra, trapezohedra, hexakisoctahedra and combinations of these forms, and the interconversion between these crystalline habits under the influence of various factors, including temperature, CaO : Al_2O_3 ratio in mother-liquor, pH, etc., has been studied by Sersale.[2] Apart from the doubtful hydrate C_3AH_{18-21} and the hydrothermal compounds C_2AH_3 and $C_4A_3H_3$, the remaining compounds are apparently uniaxial negative. The monocalcium aluminate hydrate usually occurs in an ill-defined form, but the formation of hexagonal prisms has been reported.[3] The doubtful C_3AH_{18-21}, C_6AH_{33} and C_5AH_{34} compounds give needles or prisms showing negative elongation, while the C_4A and C_2A hydrates crystallise as very thin hexagonal-shaped plates which on edge appear as 'needles' with positive elongation. In general, both the density and the refractive indices gradually increase as water is removed from a particular compound. The close similarities in the optical and morphological properties of many of the hydrates make identification by the microscopical method rather difficult, and X-ray methods are generally more suitable. Provided that adequate care is taken to prevent carbonation and to control the water content of solid phases, X-ray examination is especially useful in characterising particular hydrates, each compound giving characteristic reflections in the X-ray powder patterns as shown in Table 26.

When monocalcium aluminate, or $C_{12}A_7$, or alkali-free high-alumina cement is shaken with excess water, a supersaturated solution is first obtained in which the molecular ratio of CaO : Al_2O_3 approaches unity, but contains some excess of lime which increases with time (Fig. 54). If the solution is allowed to stand at

[1] R. Turriziani and G. Schippa, *Ricerca scient.* **26,** 2792 (1965); M. H. Roberts, *J. appl. Chem.* **7,** 453 (1957) and *Sym., Washington 1960,* 1033; F. E. Jones and M. H. Roberts, loc. cit.; R. Rabot and M. T. Mounier, *Revue Matér. Constr. Trav. publ.* **554,** 449 (1961).

[2] R. Sersale, *Ricerca scient.* **27,** 777 (1957).

[3] J. Farran, *Revue Matér. Constr. Trav. publ.* **155** (1956); E. T. Carlson, *J. Res. natn. Bur. Stand.* **59,** 107 (1957).

room temperature it deposits alumina and hydrated calcium aluminate, mostly the dicalcium aluminate. The addition of lime solution, or solid CaO or $Ca(OH)_2$, to the supersaturated monocalcium aluminate solution results in the immediate precipitation of hydrated calcium aluminates, and this method has been much used in studying the nature of the compounds formed in this system. Other less satisfactory methods of preparation which have also been used include reaction of lime solution with aluminium or alumina gel, treatment of alkali aluminate solution with lime and calcium salt solutions, and hydration of anhydrous calcium aluminates.

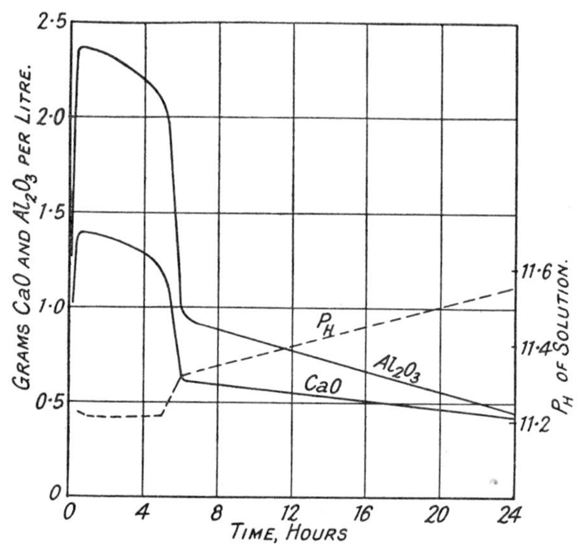

FIG. 54 Composition and pH of solutions obtained on shaking 50 g $CaO.Al_2O_3$ with 1 litre of water.

The monocalcium aluminate hydrate was first obtained by Assarsson[1] in the form of a 'gel' by the action of water on anhydrous calcium aluminates and high-alumina cements at temperatures below room temperature. It can be prepared by precipitation at temperatures around 1° from metastable monocalcium aluminate solution. At higher temperatures there is an increasing tendency for the formation of dicalcium aluminate hydrate and hydrated alumina, and the monocalcium aluminate hydrate does not ordinarily form in the aqueous system at 25°. On the basis of phase-equilibrium studies, Jones and Roberts concluded that there is a transition temperature at about 22°, CAH_{10} being relatively stable below, and C_2AH_8 relatively stable above, that temperature. However, the behaviour in the presence of the limited amount of water in pastes may be somewhat different, and CAH_{10} may be formed in high-alumina cement pastes at

[1] G. Assarsson, *Sver. geol. Unders. Afh.*, Ser. C, No. 399, Arsbok 30 (1936).

about 25°, though apparently not at 30° and above. Many investigations[1] have shown that CAH_{10} is the main product of hydration of high-alumina cement under normal conditions.

Various water contents for samples of monocalcium aluminate hydrate dried under different conditions have been reported.[2] The hydrate contains $10H_2O$ when dried at room temperature at relative humidities in the range 79–98 per cent, and approximately $7H_2O$ on drying at 45 per cent relative humidity or over solid NaOH or $CaCl_2$. This loss of $3H_2O$ from the fully hydrated CAH_{10} does not appear to have any significant effect on the X-ray diffraction pattern. More intensive drying over P_2O_5 or at 100–105° results in a decrease in the water content to about $2{\cdot}5H_2O$ and in some deterioration of the X-ray pattern, with diminished intensity of the reflections or even in complete disappearance of the pattern. The thermal dehydration curve (Fig. 55) gives no steps indicative of the

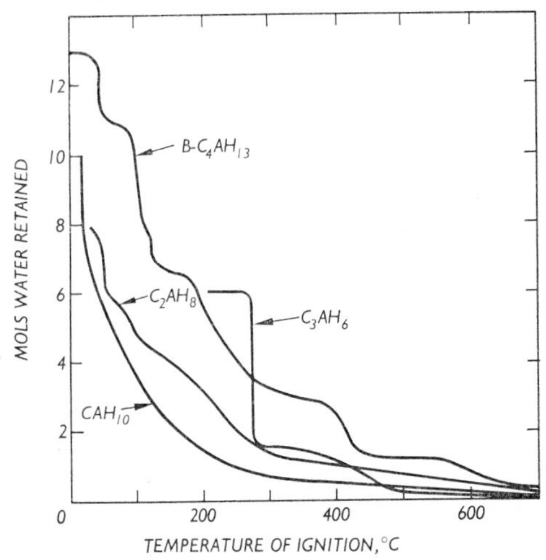

FIG. 55 The loss of water from hydrated calcium aluminates on ignition.

formation of other definite hydration states, but shows only progressive loss of water until almost all is removed at about 600°. Infra-red investigations[3] make it seem likely that CAH_{10} contains $8H_2O$ as molecular water which may be adsorbed and/or zeolitic, and $2H_2O$ as 'hydroxyl water'. The crystal structure of CAH_{10} is unknown, and while many of the reflections in the X-ray powder pattern can be indexed on the basis of the hexagonal cell with $a = 9{\cdot}45$ Å, $c = 14{\cdot}6$ Å given by Brocard,[4] this cell does not appear to account for all the spacings.

[1] L. S. Wells and E. T. Carlson, *J. Res. natn. Bur. Stand.* **57,** 335 (1956); S. J. Schneider, *J. Am. Ceram. Soc.* **42,** 184 (1959); H. Lehmann and K. J. Leers, *Tonind.-Ztg. keram. Rdsh.* **87,** 29 (1963); H. E. Schwiete, U. Ludwig and P. Muller, *Silikattechnik* **16,** 103, 146 (1965); H. G. Midgley, *Trans. Br. Ceram. Soc.* **66,** 161 (1967).

[2] P. Longuet, *Sym., London 1952,* 328; E. T. Carlson, *J. Res. natn. Bur. Stand.* **59,** 107 (1957).

[3] G. Emschwiller, L. Henry, R. Rabot and M. T. Mounier, *J. Chim. phys.* **59,** 419 (1962); G. Emschwiller, L. Henry, C. Troyanowski and J. Volant, *C.r. hebd. Séanc. Acad. Sci., Paris* **259,** 1329 (1964).

[4] J. Brocard, *Annls Inst. Bâtiment,* New Series No. 12 (1948).

The dicalcium aluminate hydrate has been variously assigned a water content of 7–9H_2O, but the fully hydrated compound is generally considered to contain 8H_2O as essential to the structure. It can be prepared conveniently from high-alumina cement (see p. 502), or by adding CaO or lime solution to a supersaturated monocalcium aluminate solution so that the CaO : Al_2O_3 molar ratio is near to 2. Jones and Roberts established the existence of three modifications of C_2AH_8, designated α_1, α_2 and β. The α_1-and-α_2 forms are polytypes,[1] giving X-ray patterns with the same basal spacings and only differing slightly in the positions of non-basal reflections, and the β-form is a polymorph which is characterised in the X-ray powder patterns by a reduced longest basal spacing of 10·5 Å. The β-C_2AH_8 precipitates first from solutions of high-alumina content as an unstable layer structure, and gradually converts in contact with solution into the normal α_1-C_2AH_8. The α_2-C_2AH_8 has so far not been found in the aqueous system, but is obtained by dehydrating the α_1-form to a 5H_2O hydrate followed by rehydration in a damp atmosphere. It gives a simplified X-ray pattern, and unlike the patterns from the α_1- and β-forms this pattern can be satisfactorily indexed on a hexagonal unit cell of dimensions $a = 5{\cdot}80$ Å, $c = 21{\cdot}59$ Å.

In contrast to the monocalcium aluminate hydrate, distinct lower hydrates with different X-ray patterns are produced by dehydrating C_2AH_8 under various drying conditions. According to Roberts,[2] $C_2AH_{7\cdot5}$, C_2AH_5 and C_2AH_4 are formed successively, and these give respective longest basal spacings of 10·6 Å, 8·7 Å and 7·4 Å. Similar results have also been reported by other investigators,[3] but slightly higher water contents of 9H_2O, 8H_2O, 6H_2O and 5H_2O were indicated. The 7·5H_2O hydrate is obtained by drying at 34 per cent relative humidity, and the 5H_2O (8·7 Å) hydrate at 12 per cent relative humidity, or over anhydrous $CaCl_2$ or P_2O_5, or by heating at 102°. These two hydrates are very readily rehydrated to C_2AH_8 on exposure to air at 81 per cent relative humidity, but the 4H_2O (7·4 Å) hydrate, obtained by heating at 120°, does not rehydrate at this humidity. Infra-red studies by Emschwiller and his co-workers have indicated that the 5H_2O hydrate, obtained by drying over P_2O_5 at room temperature, contains only 'hydroxyl water'. The constitution and indeed the existence of the alleged 4H_2O hydrate is uncertain. The thermal dehydration curve for C_2AH_8 given in Fig. 55 shows only slight modifications of slope which agree roughly with the formation of the lower hydrates discussed above, but more distinct steps at temperatures below 100° have been observed by Lavanant. The remaining water is almost all lost at about 600°, and at higher temperatures $C_{12}A_7$ and CaO are formed.[4] Some variation occurs in the position and relative intensity of the endothermic peaks in the reported DTA thermograms of C_2AH_8 preparations.[5]

Three tricalcium aluminate hydrates of different crystal form and structure have been reported, but of these only the cubic C_3AH_6 is well established. The

[1] Polymorphs differing only by minor variations in the way in which successive layers of the crystal structure are stacked.
[2] M. H. Roberts, *J. appl. Chem.* **7**, 543 (1957).
[3] E. T. Carlson, *J. Res. natn. Bur. Stand.* **61**, 1 (1958); F. Lavanant, loc. cit.
[4] G. Maekawa, *J. Soc. chem. Ind. Japan* **46**, 750 (1943).
[5] M. Rey, *Silic. ind.* **22**, 533 (1957); J. F. Young, *Mag. Concr. Res.* **14**, 137 (1962) and **16**, 231 (1964); R. F. Feldman and V. S. Ramachandran, *J. Am. Ceram. Soc.* **49**, 268 (1966).

alleged needle-compound C_3AH_{18-21} was first obtained by Travers and Sehnoutka[1] and later by Mylius[2] in precipitation experiments involving potassium aluminate solutions and the presence of alkali may be of some significance. It seems unlikely that such a compound occurs in the $CaO–Al_2O_3–H_2O$ system. Only optical evidence of the formation of C_3AH_{18-21} appears to be available, and in the absence of X-ray data its existence must be regarded as uncertain. There can now be little doubt that confusion and misinterpretation in identification has arisen in the case of the alleged hexagonal-plate compound of formula C_3AH_{10-12}. It is very probable that inadvertent carbonation occurred in preparations believed in previous studies to have this composition, with the result that these actually consisted of the quaternary carbonate hydrate $C_3A.CaCO_3.11H_2O$. In the absence of carbonate contamination a hexagonal-plate phase of this composition consists of a mixture of the dicalcium and tetracalcium aluminate hydrates.

The cubic compound C_3AH_6 is probably the only stable hydrated calcium aluminate in the system $CaO–Al_2O_3–H_2O$ at temperatures from 20° to 225°. It is formed slowly at room temperature by conversion from the metastable mono-, di- and tetra-calcium aluminate hydrates, probably by a 'through-solution mechanism' involving the crystallisation of C_3AH_6 from solution and the dissolution of CAH_{10}, C_2AH_8 or C_4AH_{19} to maintain the solution composition. This transformation process becomes increasingly rapid with increasing temperature or pH. The C_3AH_6 can be prepared from anhydrous C_3A by boiling or heating under steam pressure, or by autoclaving a mix of lime and alumina at 150°. It loses no water when dried over P_2O_5, nor at 105°, but as shown in Fig. 55 the water content drops sharply at about 275° to give a composition $C_3AH_{1.5}$ with a reduced refractive index of 1·543 and an X-ray pattern closely resembling that of $C_{12}A_7$. On continued dehydration above 275°, decomposition[3] occurs with liberation of CaO and between 550° and 950° it is quantitatively transformed to CaO and $C_{12}A_7$. At about 1050° recombination of lime to form C_3A occurs and this gives an easier method of preparing this compound than by burning a mix of lime and alumina. The DTA thermograms of C_3AH_6 agree with the water-loss data and show an intense endotherm at 315–360° and a weaker endotherm at 500–550°. The infra-red spectra for C_3AH_6, obtained by Emschwiller and his co-workers and by Majumdar and Roy,[4] indicate the absence of free or molecular water and the hydrate therefore contains only 'hydroxyl water'. Crystal structure studies[5] have shown that C_3AH_6 is body-centred cubic, with a = 12·56–12·58 Å, space group Ia_3d and 8 formula units in the unit cell (see p. 333).

In addition to C_3AH_6 which remains stable up to about 225°, two other hydrated calcium aluminates have been reported to be formed under hydrothermal conditions. According to Schneider and Thorvaldson[6] a hexagonal

[1] *Annali Chim.* **13,** 253 (1930).
[2] C. R. W. Mylius, *Acta Acad. abo.*, Math. et Phys. **7,** No. 3 (1933).
[3] T. Thorvaldson and W. G. Schneider, *Can. J. Res.*, **B19,** 109, 123 (1941).
[4] A. J. Majumdar and R. Roy, *J. Am. Ceram. Soc.* **39,** 434 (1956).
[5] T. Thorvaldson, N. S. Grace and V. A. Vigfusson, *Can. J. Res.* **1,** 201 (1929); E. Brandenburger, *Schweizer. Arch. angew. Wiss. Tech.* **2,** 45 (1936); R. Weiss, D. Grandjean and J. L. Pavin, *Acta crystallogr.* **17,** 1329 (1964).
[6] W. G. Schneider and T. Thorvaldson, *Can. J. Res.* **B 21,** 34, 65 (1943).

ternary hydrate of $CaO : Al_2O_3$ ratio less than 3 : 1 can be formed on autoclaving anhydrous calcium aluminates at 105–150°. Spangenburg[1] later suggested that this phase possibly had a composition C_2AH_3. This alleged C_2AH_3 hydrate has, however, not been further observed in later hydrothermal studies by other investigators and its existence remains uncertain. The other hydrothermal hydrate $C_4A_3H_3$ is now well established.[2] It can be prepared by treating C_3A, or stoichiometric mixtures of $Ca(OH)_2$ and alumina gel or of $Ca(OH)_2$ and CA, in saturated steam at 250–350°. On heating in dry air, $C_4A_3H_3$ starts to lose water at about 500°, and at higher temperatures $C_{12}A_7$ forms. Under hydrothermal conditions with high water-vapour pressures and temperatures, the decomposition products are $C_{12}A_7$, CA_2 and water.[3] X-ray data for $C_4A_3H_3$ have been given by Percival and Taylor[4] and an orthorhombic unit cell containing 4 formula units and with $a = 12{\cdot}78$ Å, $b = 12{\cdot}42$ Å, $c = 8{\cdot}90$ Å was obtained. The ionic constitution is probably $Ca_4Al_6O_{10}(OH)_6$ or, as suggested by Spangenburg, $4CaO.6AlO(OH)$.

A tetracalcium aluminate hydrate, crystallising in small hexagonal plates, can be prepared at 25° or below by adding CaO or lime solution to a supersaturated monocalcium aluminate solution; sufficient lime must be added to combine with all, or almost all, of the Al_2O_3 as the C_4A hydrate and to give under metastable equilibrium conditions a final lime concentration in solution within the range 0·7 to 1·1 g CaO per litre. Using similar methods, this hydrate was first prepared by Lafuma[5] and later by Wells[6] and others. In all of the work before 1957, in which 'dried' solids were examined, the hydrate precipitated from solution was considered to be C_4AH_{13}, but it was then established[7] that the phase which exists in the aqueous system in contact with solution at 25° is actually C_4AH_{19} and this changes to the $13H_2O$ composition on drying under suitable conditions. This has since been confirmed by many other studies, and it is now widely accepted that when hydrated tetracalcium aluminate is formed at temperatures in the range 1° to about 50° it is present as the $19H_2O$ hydrate in contact with aqueous solutions. At some temperature above 50°, C_4AH_{13} may occur in the aqueous system.

The fully-hydrated C_4AH_{19} occurs in α_1- and α_2-modifications,[8] similar to α_1- and α_2-C_2AH_8. The α_1-C_4AH_{19} is considered to possess an unstable or 'disordered' layer structure which forms initially in the aqueous system, and it readily converts into the more stable α_2-C_4AH_{19}. According to Aruja[9] these two forms are polytypes with structures differing only in the way in which successive layers are stacked, and the α_1-form has a hexagonal unit cell of dimensions $a = 5{\cdot}77$ Å, $c = 64{\cdot}08$ Å, while the α_2-form gives $a = 5{\cdot}77$ Å, $c = 21{\cdot}37$ Å.

Dehydration of C_4AH_{19} occurs very readily on washing with alcohol or acetone, or on drying at room temperature at relative humidities below about

[1] K. Spangenburg, *TagBer. Zemind.* **4,** 102 (1951).
[2] H. Johnson and T. Thorvaldson, *Can. J. Res.* **B 21,** 236 (1943).
[3] C. W. F. T. Pistorius, *Am. J. Sci.* **260,** 221 (1962); A. J. Majumdar and R. Roy, loc. cit.
[4] A. Percival and H. F. W. Taylor, *Acta crystallogr.* **14,** 324 (1961).
[5] H. Lafuma, Thesis, Paris (Libraire Viubert) (1925).
[6] L. S. Wells, *J. Res. natn. Bur. Stand.* **1,** 951 (1928).
[7] M. H. Roberts, loc. cit.
[8] F. E. Jones and M. H. Roberts, loc. cit.
[9] *Acta crystallogr.* **14,** 1213 (1961).

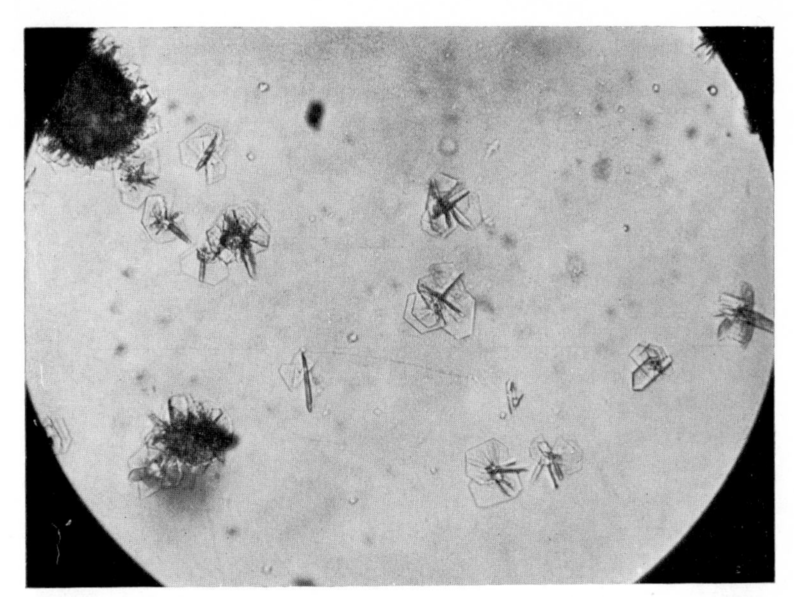

(i) HYDRATED PLATES AND NEEDLES FROM TRICALCIUM ALUMINATE (× 260)

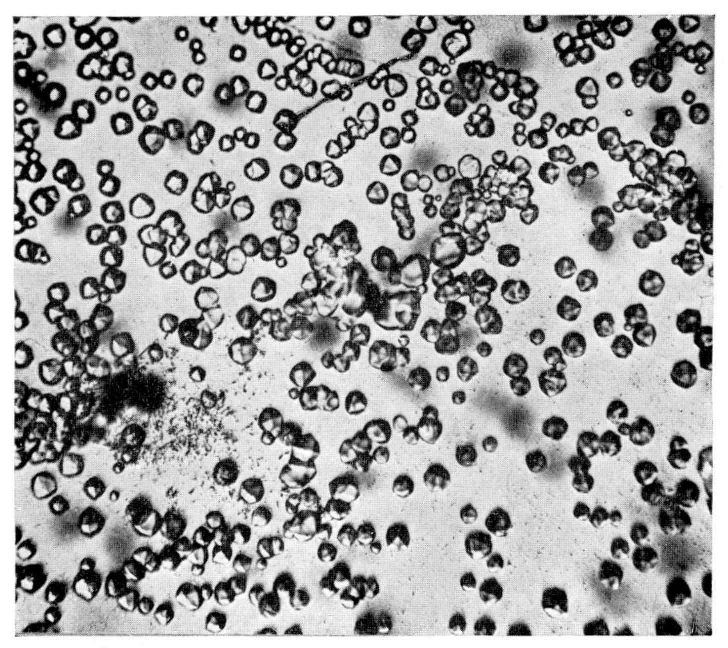

(ii) $3CaO.Al_2O_3.6H_2O$ IN FORM OF HEXAKIS OCTAHEDRA (× 260)

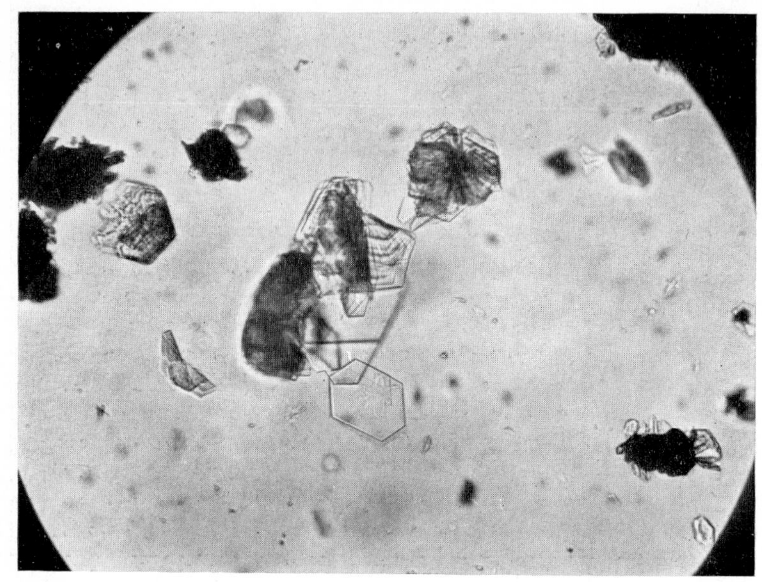

(i) HYDRATED $4CaO.Al_2O_3.Fe_2O_3$ SHOWING PLATES OF HYDRATED ALUMINATE AND AMORPHOUS IRON COMPOUND (× 260)

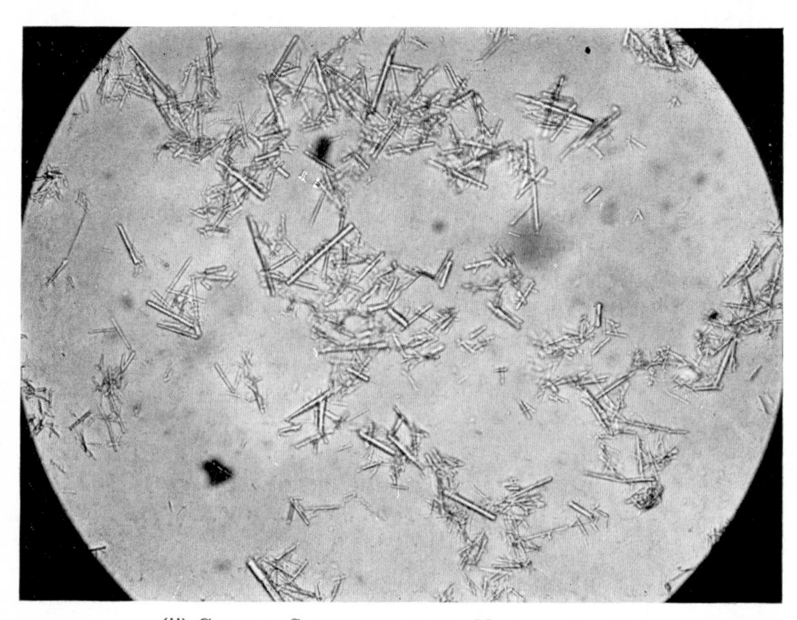

(ii) CALCIUM SULPHOALUMINATE NEEDLES (× 260)

88 per cent, and, depending upon the actual drying conditions, various lower hydrates not occurring in the aqueous system at ordinary temperatures are formed. Several lower hydrates have been described in the literature, and some uncertainty and controversy exists with regard to their composition and identity. Many of these studies were undoubtedly affected by failure to control the water content of solid phases and/or by inadvertent carbonation.

Assarsson concluded that α- and β-polymorphs of C_4AH_{12-14} existed, but later work by Jones and Roberts and other authors[1] has led to the conclusion that there is only a single $13H_2O$ hydrate (β-C_4AH_{13}) with a basal spacing of 7·9 Å existing at relative humidities between 12 and 81 per cent, and that the supposed α-C_4AH_{13} with a basal spacing of 8·2 Å is actually a related quaternary hydrate of composition $4CaO.Al_2O_3.\frac{1}{2}CO_2.12H_2O$.

Another quaternary carbonate compound closely related in structure to C_4AH_{13} occurs naturally as the mineral hydrocalumite. This mineral was originally studied by Tilley, Megaw and Hey[2] and later by Buttler, Dent, Glasser and Taylor[3] and the composition was reported as $Ca_{16}Al_8(OH)_{54}(CO_3)$ $21H_2O$ or $C_4A.\frac{1}{4}CO_2.12H_2O$. The latter authors obtained a monoclinic unit cell of dimensions a = 9·9 Å, b = 11·4 Å, c = 16·8 Å, β = 111° for hydrocalumite, but a pseudohexagonal cell with a = 5·7 Å, c = 7·86 Å, almost identical to the values obtained for βC_4AH_{13}, was also indicated. The X-ray powder pattern apparently shows basal spacings of 7·86 Å and 3·93 Å, which may be compared with those of 7·9 Å and 3·95 Å for C_4AH_{13}, 8·2 Å and 4·10 Å for $C_4A.\frac{1}{2}CO_2.12H_2O$, and 7·6 Å and 3·80 Å for $C_4A.CO_2.11H_2O$ ($C_3A.CaCO_3.11H_2O$). The precise relationship of the latter compounds and hydrocalumite is not clear as there may be some doubt as to the exact carbonate content of the crystals of hydrocalumite examined by X-rays. It may well be a stage in a solid-solution series $C_3A.Ca(OH)_2.12H_2O$–$C_3A.CaCO_3.11H_2O$ (see later). The density of Tilley's hydrocalumite was 2·15 and the refractive indices α = 1·535, γ = 1·557.

The thermal dehydration curve of β-C_4AH_{13} (Fig. 55) shows definite steps indicating the formation of $11H_2O$ and $7H_2O$ hydrates, and partial dehydroxylation occurs between 150° and 300° to give $C_4A_3H_3$ and $Ca(OH)_2$. The DTA thermograms of C_4AH_{13} reported by several investigators are very similar to those of C_2AH_8. The C_4AH_{11} hydrate can also be obtained by drying C_4AH_{13} over anhydrous $CaCl_2$ or solid NaOH at room temperature, and it may even be formed at 11 per cent relative humidity. It is generally accepted to give an X-ray powder pattern with a longest basal spacing of 7·4 Å. Further dehydration of C_4AH_{11} over P_2O_5 at room temperature or by heating at 120° gives a composition C_4AH_7. According to the infra-red data obtained by Emschwiller and his co-workers, this $7H_2O$ hydrate contains only 'hydroxyl' water, while the higher hydrates also include molecular water. Both the $7H_2O$ and $11H_2O$ hydrates readily rehydrate to C_4AH_{13} in air at 81 per cent relative humidity, and while C_4AH_{13} becomes further hydrated to C_4AH_{19} in nearly saturated lime solution,

[1] M. H. Roberts, *Sym., Tokyo 1968*; P. Seligmann and N. R. Greening, *J. Res. Dev. Labs Portld Cem. Ass.* **4** (2) 2 (1962) and *Highw. Res. Rec.*, No. 62, 89 (1964). For contrary views see Alègre, Lavanant, and Dosch and zur Strassen (loc. cit.).
[2] *Mineralog. Mag.* **23,** 607 (1934).
[3] *J. Am. Ceram. Soc.* **42,** 121 (1959).

it apparently only rehydrates slowly in air at 99 per cent relative humidity, possibly because of complications from carbonation and the consequent formation of quaternary carbonate hydrates.

It now seems unlikely that any hydrated calcium aluminates more basic than the tetracalcium compound exists. A compound of supposed formula C_5AH_{34} was reported by Assarsson[1] and a C_6AH_{33} hydrate by Flint and Wells.[2] Later work has failed to confirm the existence of such compounds which were probably carboaluminate hydrates.

Structure of the hydrated aluminates

Of the various hydrated calcium aluminates described above, only in the case of C_3AH_6 with the ionic constitution $Ca_3[Al(OH)_6]_2$ is the crystal structure known (see p. 333). Several hypothetical structures have been suggested for the remaining hydrates, and such structural considerations are of importance in establishing what calcium aluminate hydrates, as well as related quaternary hydrates or solid solutions, are practically possible. These hydrates evidently have layer structures, and Brandenburger[3] suggested that they may be expressed by the general formula $m\mathrm{Ca(OH)_2}.n.\mathrm{Al(OH)_3}.p\mathrm{H_2O}$, where $p\mathrm{H_2O}$ represents the water lost on drying over P_2O_5 or by heating at about 100°. A similar arrangement, but with $Al(OH)_6$ or $Al(OH)_4$ groupings in particular cases, has also been indicated by Barret and Grange.[4] These representations agree with the water-loss and infra-red data for both the di- and tetra-calcium aluminate hydrates, the C_2AH_8 being indicated to have the constitution $2Ca(OH)_2.2Al(OH)_3.3H_2O$ while C_4AH_{13} gives $4Ca(OH)_2.2Al(OH)_3.6H_2O$ or $2Ca(OH)_2.Al(OH)_3.3H_2O$. The mono-calcium aluminate hydrate, which gives an X-ray pattern markedly different from those of C_2A.aq and C_4A.aq does not appear to fit into this representation, and a constitution such as $Ca(OH)_2.2AlO(OH).8H_2O$ may be possible for CAH_{10}. Similar structures in which Al is linked with O and OH as in the boehmite (2AlO(OH)) structure may also occur in the hydrothermal compounds C_2AH_3 (?) and $C_4A_3H_3$. Brandenburger also suggested that the structure of the hexagonal-plate hydrates is based on alternate layers of $Ca(OH)_2$ and $Al(OH)_3$, a view supported by the work of Tilley, Megaw and Hey on the natural mineral hydrocalumite which is closely related to C_4AH_{13}. For various reasons, Buttler, Dent, Glasser and Taylor considered this arrangement unsatisfactory and suggested instead that the C_4AH_{13} structure may be based on layers of octahedrally co-ordinated cations $Ca_2Al(OH)_6^+$, between which are placed additional hydroxyl ions and water molecules, and the formula may be written as $2[Ca_2Al(OH)_6.OH.3H_2O]$. The higher hydrate C_4AH_{19} is presumably derived from this by the incorporation of an extra layer of water molecules. The C_2AH_8 may well have a very similar structure, possibly represented by the formula $Ca_2Al(OH)_6Al(OH)_4.3H_2O$. It is also presumed possible for other anions to replace the OH groups outside the octahedral layer to give the related quaternary

[1] loc. cit.
[2] *J. Res. natn. Bur. Stand.* **33,** 471 (1944).
[3] *Schweiz. miner. petrogr. Mitt.* **13,** 569 (1933); *Schweizer. Arch. angew. Wiss. Tech.* **2,** 45 (1936).
[4] *C.r. hebd. Séanc. Acad. Sci., Paris* **257,** 2492 (1963).

basic salt hydrates of the hexagonal-plate type. Depending on the size of the anion and the number of water molecules displaced by it, such replacement can either increase or decrease the thickness of the elementary layer, as given by the longest basal spacing in the X-ray powder pattern, but the value of the *a*-axis appears to remain at about 5·7 Å for all hydrates of the hexagonal-plate type. Later studies[1] of the crystal structures of calcium aluminate hydrates and derived basic salts appear to confirm the structures proposed by Buttler, Dent Glasser and Taylor, but alternative hypotheses based on analogy with the clay minerals have also been advanced by other investigators.[2]

The system $CaO–Al_2O_3–H_2O$

The stability relations and equilibria in the $CaO–Al_2O_3–H_2O$ system at temperatures from 1° to 1000° have received much attention and in recent years considerable advances have been made as a result of improved techniques for the X-ray examination and control of water content of solid phases, as well as for the prevention of carbonate contamination. At temperatures up to 90° the equilibria are characterised by the existence of various metastable phases giving solubility curves lying above those of the stable phases C_3AH_6, $Ca(OH)_2$ and γ-Al_2O_3. $3H_2O$, and difficulties have especially occurred in interpreting the metastable equilibria.

The many successive investigations that have been made have led to a progressive refinement in the phase relations, and rather than detail these we may base the present discussion on the most recent comprehensive study of Jones and Roberts of the system at 25°. The phase diagram shown in Fig. 56 is of the type found by numerous investigators. The curve *T′TYV* gives the compositions of solutions in metastable equilibrium with ternary hexagonal-plate hydrates. The position of this curve is approximately coincident with the experimental curves obtained previously by other investigators, but differences occur on comparison of the variation in $CaO : Al_2O_3$ ratio of the solids along the various curves. Such a comparison is made in Fig. 57, in which the molar ratio of CaO to Al_2O_3 in the solids is plotted against both the CaO and Al_2O_3 concentration in solution. This plot for the results of Jones and Roberts clearly shows that between *V* and *Y* the solid has a $CaO : Al_2O_3$ ratio very close to 4, so that *VY* is the metastable solubility curve of C_4AH_{19} apparently relating to both the α_1 and α_2 modifications; that *V* is a metastable invariant point C_4AH_{19}–$Ca(OH)_2$; and that *Y* is another invariant point at which the $CaO : Al_2O_3$ ratio may vary from nearly 4 to 2·4, followed subsequently by a gradual fall from 2·4 to 2·0 as the solution concentrations decrease with respect to CaO and increase with respect to Al_2O_3. It was concluded that the second solid phase formed at the invariant point *Y* with C_4AH_{19} is a solid solution of α_1 or α_2 C_4AH_{19} and α_1 C_2AH_8 with the probable composition $C_{2\cdot4}A.10\cdot2H_2O$, and that along the solubility curve *YT* the solid solution gradually changes in composition until metastable α_1 C_2AH_8 forms near *T*.

[1] W. Dosch, *Neues Jb. Miner. Abh.* **106,** 200 (1967); S. J. Ahmed and H. F. W. Taylor, *Nature, Lond.* **215,** 622 (1967); S. J. Ahmed, L. S. Dent Glasser and H. F. W. Taylor, *Sym., Tokyo 1968*; H. J. Kuzel, *Sym., Tokyo 1968.*
[2] W. Feitknecht and H. W. Buser, *Helv. chim. Acta* **32,** 2298 (1949) and **34,** 128 (1951); A. Grudemo, *Sym., Washington 1960*, 615.

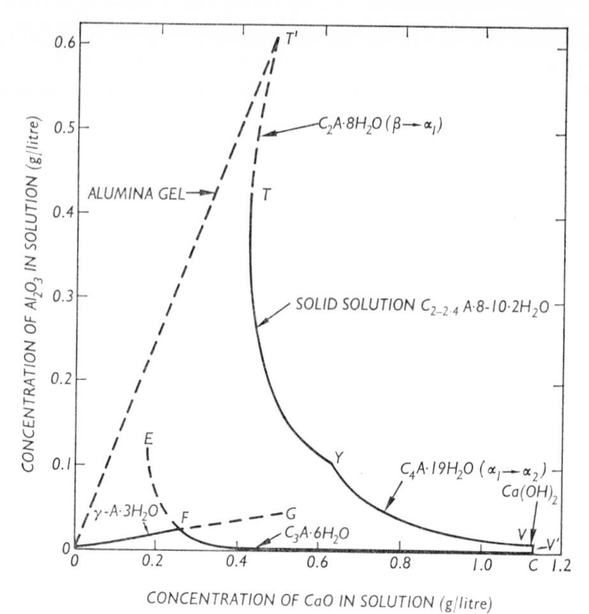

FIG. 56 System $CaO–Al_2O_3–H_2O$ at 25° (Jones and Roberts).

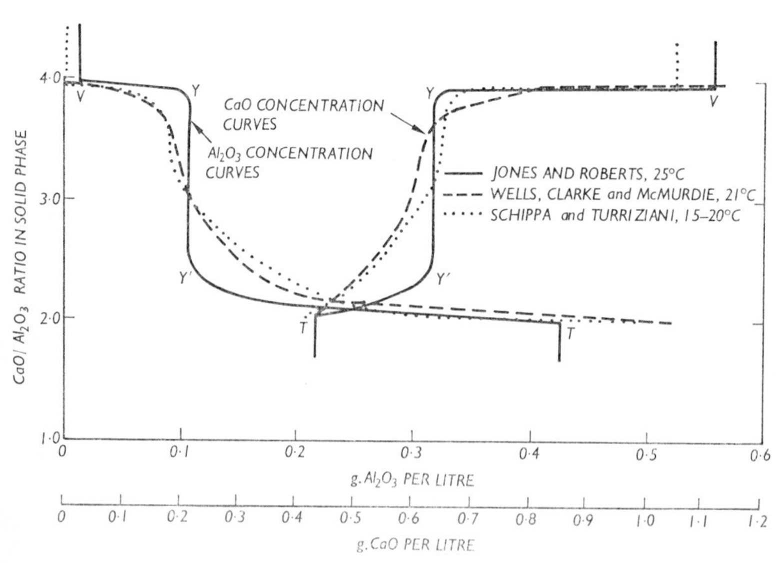

FIG. 57 Variation of $CaO : Al_2O_3$ ratio in solid phase with metastable equilibrium solution composition.

However, in view of the difficulty in differentiating between C_2AH_8 and members of this supposed solid-solution series either optically or by X-rays, and of avoiding any traces of carbonation, it is difficult to obtain conclusive evidence. The results of earlier investigations[1] show a more gradual change in the CaO : Al_2O_3 ratio along the curve TYV which was attributed by the various authors to the existence of two closely overlapping solubility curves for C_4AH_{13} and C_2AH_8. This, however, was before it was recognised that the former hydrate existed in the aqueous system as C_4AH_{19}, and lack of precise control of water content of solids, probably together with inadvertent carbonation by atmospheric CO_2, must have led to confusion in the interpretation of the optical and X-ray data obtained for various solids.

The α_1-C_2AH_8, or another less stable β-C_2AH_8 which slowly converts into α_1-C_2AH_8, may be formed in equilibrium with solutions represented by a prolongation of YT to T' (Fig. 56). The curve OT' represents the solubility of alumina gel in its most unstable condition, and the position of this curve varies as the nature of the hydrated alumina changes on 'ageing' in the sequence, highly gelatinous hydrous alumina → γ-$Al_2O_3.H_2O$ (boehmite) → α-$Al_2O_3.3H_2O$ (bayerite) → γ-$Al_2O_3.3H_2O$ (gibbsite). It follows, therefore, that in region $T'T$ and to a decreasing extent along the curve $T'TY$ precipitation of hydrated alumina can occur with C_2AH_8 or solid solution, giving a series of unstable 'invariant points' for C_2AH_8, or solid solution, and hydrated alumina. Jones and Roberts observed that this unstable invariant point had fallen to a solution composition of 0·49 g CaO and 0·19 g Al_2O_3 per litre at 30 days, but theoretically this unstable 'invariant point' curve may proceed to the point where a metastable prolongation of the stable gibbsite curve OF cuts the C_4AH_{19} curve VY to give an invariant point gibbsite—C_4AH_{19}.

With regard to the stable equilibria, these are represented in Fig. 56 by the curves OF for crystalline γ-$Al_2O_3.3H_2O$ (gibbsite), FV' for C_3AH_6 and $V'C$ for $Ca(OH)_2$ with the stable invariant points γ-$Al_2O_3.3H_2O$–C_3AH_6 at F and $Ca(OH)_2$–C_3AH_6 at V'; the solubility of crystalline $Ca(OH)_2$ (1·11 g CaO per litre) is given by the point C. Various values have been given by different authors for the lime concentration at the invariant point gibbsite—C_3AH_6 ranging from about 0·25 to 0·35 g CaO per litre, with an Al_2O_3 concentration around 0·02 g per litre. Though C_3AH_6 is thus incongruently soluble, and the prolongation FE of the stable C_3AH_6 curve $V'F$ is unstable with respect to crystalline gibbsite, it is possible to obtain a persistent metastable congruent solution of C_3AH_6 because of the higher solubility of alumina gel and its reluctance to crystallise as gibbsite. Many authors have reported congruent solution of C_3AH_6, corresponding to 0·19 to 0·32 g C_3A per litre at temperatures in the range 15–40°.

The effect of temperature on the phase diagram is shown in Fig. 58 in which the stable solubility curves for γ-$Al_2O_3.3H_2O$ and $C_3A.6H_2O$ are omitted. With decreasing temperature the solubility curves are displaced to lower lime and alumina concentrations, and an important feature at lower temperatures is the formation of CAH_{10}. The transition temperature of $C_2AH_8 \rightarrow CAH_{10}$ is about

[1] L. S. Wells, W. F. Clarke and H. F. McMurdie, *J. Res. natn. Bur. Stand.* **30,** 367 (1943); J. D'Ans and H. Eick, *Zement-Kalk-Gips* **6,** 197 (1953); G. Schippa and R. Turriziani, *Ricerca scient.* **27,** 3654 (1957).

22° according to Jones and Roberts. The latter hydrate is precipitated preferentially from highly concentrated monocalcium aluminate solutions, and this behaviour is consistent with the observation that at temperatures even up to about 25° CAH_{10} is the main hydration product in pastes of high-alumina cement and $CaO.Al_2O_3$. The CAH_{10} solubility curve ZT extends from an unstable invariant point Z with hydrated alumina to another invariant point T with the C_2AH_8–C_4AH_{19} solid solution. The position of this curve either at 5° or at 21° is difficult to fix accurately because of the unstable nature of the equilibria, and it may also be influenced by crystal size. This probably accounts for the different positions assigned to the curve at 21° and 5° by Taylor and his co-workers[1] who at 5° found an invariant point CAH_{10}–C_4AH_{19} instead of CAH_{10} 'solid solution' as indicated by Jones and supported by the results of Carlson[2] at 1°.

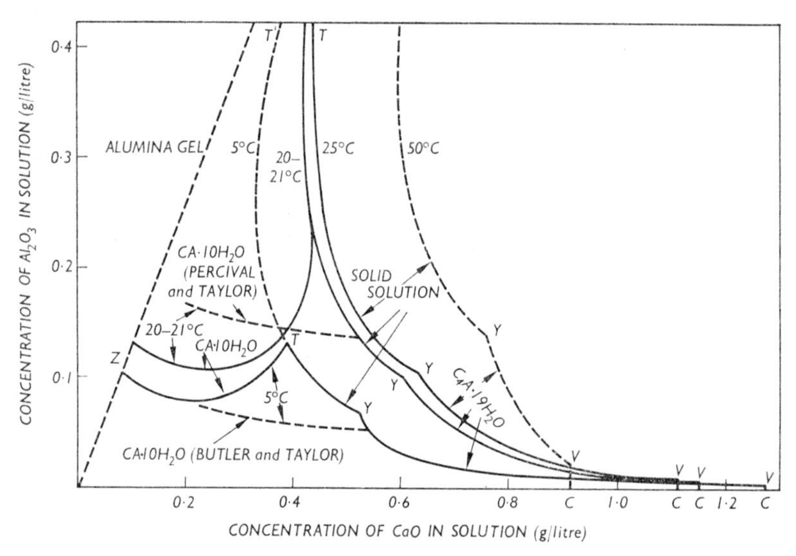

FIG. 58 System CaO–Al_2O_3–H_2O at various temperatures (*Jones, 1960*).

Both Jones and Buttler and Taylor inferred that the stable solid phases at 1–5° remain C_3AH_6, gibbsite and $Ca(OH)_2$, but later work[3] has shown that C_3AH_6 is no longer a stable phase at 5° or lower and that it transforms into C_2AH_8 or C_4AH_{19} according to the lime concentration in the solution. This transformation tends to be sluggish and may be interfered with by small amounts of carbonation. This may account for some disagreement between various investigators. Thus Carlson placed the temperature of the transformation

[1] A. Percival, F. G. Buttler and H. F. W. Taylor, *Sym., Washington 1960*, 277; F. G. Buttler and H. F. W. Taylor, *J. chem. Soc.*, 1958, 2103.

[2] *J. Res. natn. Bur. Stand.* **61,** 1 (1958).

[3] P. Seligmann and N. R. Greening, *J. Res. Dev. Labs Portld Cem. Ass.* **4** (2), 2 (1962); E. T. Carlson, *Bur. Stand. Bldg. Sci. Ser.*, No. 6 (1966); J. H. P. van Aardt and S. Visser, *Cem. Lime Mf.* **40,** 7 (1957); M. H. Roberts, *Sym., Tokyo 1968*.

$C_4AH_{19} \rightleftarrows C_3AH_6+CH$ at 15° or even higher, while Buttler and Taylor had earlier concluded that C_3AH_6 dissolved congruently at 5°.

In the presence of alkalis, as is the case in the liquid phase in cement pastes, the equilibria in the system $CaO–Al_2O_3–H_2O$ will be modified considerably because the solubility of $Ca(OH)_2$ is much reduced in NaOH or KOH solution while that of hydrated alumina is increased. Jones[1] has shown that with 1 per cent KOH or NaOH at 25° the alumina gel curve is virtually absent and the C_3AH_6 solubility curve becomes eventually practically coincident with the Al_2O_3 axis at concentrations from about 0·3 up to 6·6 g Al_2O_3 per litre. A similar displacement of the hexagonal-plate solubility curve *TYV* (Fig. 56) occurs progressively at 25° with increasing NaOH concentration[2] and, for example, the invariant point $C_4AH_{19}–Ca(OH)_2$ gives solution compositions at 25° as follows:

g NaOH per litre	0	1	4	10	20	40
g CaO ,, ,,	1·11	0·79	0·27	0·09	0·04	0·02
g Al_2O_3 ,, ,,	0·01	0·02	0·02	0·04	0·11	0·41

The metastable equilibria in the system $CaO–Al_2O_3–H_2O$ become increasingly difficult to study at temperatures above 25° owing to the rapid transformation of the hexagonal-plate hydrates into the cubic hexahydrate, and the solution data obtained by Peppler and Wells[3] at 50° and 90° are considered to represent a metastable region rather than to define a metastable equilibrium. This and other work[4] has shown that the general form of the stable solubility curves at these temperatures is similar to those at 20–25°. The invariant point between C_3AH_6 and crystalline $Al_2O_3.3H_2O$ remains at a solution concentration of about 0·3 g CaO per litre but the alumina concentration rises to 0·11 g per litre at 90°, or even higher to 0·18 g per litre on the basis of the data of Wells, Clarke and McMurdie. Under hydrothermal conditions, the highest alumina concentrations occur at this invariant point with a maximum value of 0·3 g per litre at 120°. With increasing temperature the solubility of $Ca(OH)_2$ decreases steadily to 0·037 g CaO per litre at 250°, and only similar very small amounts of alumina occur in solution at this temperature. Gibbsite remains the stable form of hydrated alumina up to 150°, and above this temperature is replaced by boemhite ($Al_2O_3.H_2O$) or its polymorph diaspore. The cubic C_3AH_6 is still the only stable ternary compound up to about 225° and above this temperature it becomes metastable with respect to $C_4A_3H_3$.

Hydrated magnesium aluminates

The existence of different magnesium aluminate hydrates has been indicated, and some of these are apparently isomorphous with the corresponding calcium aluminate hydrates, though not as well crystallised. A tetramagnesium aluminate hydrate was prepared by Cole and Hueber[5] and also a dimagnesium compound. The latter preparation with a probable composition $2MgO.Al_2O_3.9{\cdot}9H_2O$, but

[1] *J. phys. Chem.* **48,** 356 and 379 (1944).
[2] M. H. Roberts, private communication.
[3] *J. Res. natn. Bur. Stand.* **52,** 75 (1954).
[4] L. S. Wells, W. F. Clarke and H. F. McMurdie, loc. cit.
[5] *Silic. ind.* **22,** 75 (1957).

also containing $MgSO_4$, gives an X-ray pattern similar to that of C_2AH_8 and shows a longest basal spacing of 10·6 Å. Its water content is reduced to $5{\cdot}7H_2O$ on drying at 110°. The X-ray powder pattern of $4MgO.Al_2O_3.12{\cdot}6H_2O$ shows a longest basal spacing of 7·6 Å, and this remains unchanged after drying at 60° with the loss of $0{\cdot}55H_2O$. On heating to 150° the basal spacing decreases to 6·6 Å, and its water content is reduced to $7{\cdot}4H_2O$, suggesting the structural formula $Mg_4Al_2(OH)_{14}$. Other compounds of uncertain composition with basal spacings of 7·95 Å[1] and 8·2 Å[2] have also been reported.

In view of the impure nature of the various preparations and the possibility of the formation of quaternary hydrates, the precise compositions of all the different hydrated magnesium aluminates described above must be considered uncertain.

Hydrated calcium ferrites

Compounds analogous to certain of the hydrated calcium aluminates are known, and the equilibria in the system $CaO–Fe_2O_3–H_2O$ probably are similar to those in the $CaO–Al_2O_3–H_2O$ system. The iron-containing hydrates have, however, very low solubilities, and experimental difficulties arise from this and from the fact that metastable calcium ferrite solutions cannot be obtained. Methods of preparation involving the reaction of ferric salt solutions with lime or alkali solutions are generally unsatisfactory because the salt anion is easily incorporated in the solid to give a quaternary hydrate and there seems little doubt that the solids obtained in many early studies[3] were contaminated with chloride and/or silica from attack of lime or alkali solution on the glass vessels containing the reactants. The remaining methods involve either direct hydration of anhydrous calcium ferrites or reaction between ferric hydroxide gel and lime solution, and these give slow and usually incomplete reaction. The pure hydrates are white, and contamination with hydrated iron oxide is shown by a brown colour.

A cubic C_3FH_6 isomorphous with and forming a complete solid solution series with C_3AH_6 has been considered to exist.[4] Preparations had a density of 2·77, a refractive index of 1·71–1·72 and an X-ray pattern very similar to C_3AH_6 but with slightly increased spacings and a cubic unit cell with a = 12·76 Å. Later workers[5] have not been able to prepare pure C_3FH_6 and it has been concluded that the C_3FH_6 structure is not stable without some silica in solid solution to give a member of the hydrogarnet series $C_3FH_6–C_3F.3SiO_2$. However, the possibility still remains that an unstable pure C_3FH_6 phase may be formed exceptionally under special conditions.

[1] M. M. Mortland and M. C. Gastuche, *C.r. hebd. Séanc. Acad. Sci., Paris* **255,** 2131 (1962).
[2] S. Diamond, *Nature, Lond.* **204,** 183 (1964).
[3] A. Eiger, *Sym., Stockholm 1938*, 173; G. Malquori and V. Cirilli, *Ricerca scient.* **18,** 316 (1940); H. Hoffmann, *Uber Calcium Ferrithydrate*, Zementwerlag, Berlin, 1935 and *Zement* **25,** 113, 130 (1936); see also F. E. Jones, *Sym., Washington 1960*, 229 for review of work to 1960.
[4] A. Eiger, *Revue. Matér. Constr. Trav. publ.* **334,** 141 (1937); A. Burdese and S. Gallo, *Annali Chim.* **42,** 349 (1952).
[5] H. zur Strassen and C. H. Schmitt, *Sym., Washington 1960*, 243; R. Turriziani and P. Verdis, *Annali Chim.* **55,** 1141 (1965).

Hexagonal-plate hydrates also exist, but only in the case of the tetracalcium ferrite hydrates is the evidence definite. The water contents of these hydrates have not been precisely determined owing to difficulties in obtaining pure preparations, and these are largely inferred by analogy with the corresponding calcium aluminate hydrates. The occurrence of C_4FH_{19}, C_4FH_{13} and C_4FH_{11} is indicated, and these are characterised in their X-ray patterns by respective longest basal spacings of 10·7 Å, 8·0 Å and 7·6 Å, which are almost the same, or very slightly larger, than those of the corresponding tetracalcium aluminate hydrates. In contrast to C_4AH_{19}, the C_4FH_{19} is unstable in solution at 25°, but it is formed at 1° and 15°. At 25° and higher temperatures up to 45°, C_4FH_{13} is present in contact with solution and this hydrate decomposes in solution at 60° or above to give $Ca(OH)_2$ and α-Fe_2O_3 (hematite). According to Hoffmann, the tetracalcium ferrite hydrate decomposes in solutions containing less than 1·06 g CaO per litre to give a less basic hydrate which in turn hydrolyses to ferric hydroxide below 0·64 g CaO per litre, but these stability relations must be considered uncertain.

A complete solid-solution series exists between the tetracalcium aluminate and ferrite hydrates, and this is metastable with respect to the cubic C_3AH_6–C_3FH_6 series. Thus the hydration of various members of the anhydrous calcium aluminoferrite solid-solution series, ranging in composition from C_6A_2F to $C_{12}AF_5$, in the presence of $Ca(OH)_2$ at temperatures below 15° gives the hexagonal-plate solid solutions, while at higher temperatures the same solid solutions appear to be formed initially and then to convert to the cubic hexahydrate series.[1] In the paste hydration of these compounds in a limited amount of water, C_2AH_8 is also formed at the lower temperatures from C_4AF and C_6A_2F. It is also significant that the hydration of C_2F in water or lime solution at temperatures up to 45° gives products containing the C_4F.aq hydrate but not the C_3FH_6 hydrate. This observation indicates that less basic hexagonal-plate hydrates probably do not occur, and also appears to support the conclusion that C_3FH_6 does not exist and that solids near this composition must contain a little Al_2O_3 or SiO_2 in solid solution. Both Carlson and Schwiete and Iwai[2] have established that the crystal lattice parameter, a, of the cubic C_3AH_6–C_3FH_6 series increases linearly with increasing Fe_2O_3 content. While this is probably also true for the unit cell of the hexagonal-plate series, the X-ray data for these solid solutions appear to be rather conflicting. For reaction products prepared at 1° and containing the $19H_2O$ hydrate, Carlson concluded that the basal spacings are very slightly affected, or unchanged, as Fe_2O_3 replaces Al_2O_3.

The evidence for hexagonal hydrated calcium ferrites less basic than the tetracalcium ferrite hydrate is not clear. Mather and Thorvaldson[3] considered that $C_2F.2H_2O$ is formed when C_2F is hydrated in saturated steam at 150°, and this later converted into $CF.H_2O$ which then decomposed to ferric oxide. With C_4AF it is similarly indicated that at 150° the first product is C_3AH_6 and $CF.H_2O$ and that the latter is slowly decomposed on prolonged treatment. Bogue

[1] E. T. Carlson, *J. Res. natn. Bur. Stand.* **68A**, 453 (1964); *Bur. Stand. Bldg Sci. Ser.*, No. 6 (1966).
[2] *Zement-Kalk-Gips* **17,** 379 (1964).
[3] *Can. J. Res. B.* **15,** 331 (1937).

and Lerch,[1] Yamauchi[2] and Budnikov[3] concluded that a $C_2F.5H_2O$ was formed on hydration of C_2F at ordinary temperature, and the first authors that a hydrated monocalcium ferrite was produced on hydration of C_4AF. These observations were not confirmed in the later work by Carlson, and no satisfactory X-ray evidence has yet been produced establishing the existence of such hydrates. Chatterji and Majumdar[4] have indicated that treatment of the compounds C_2F, $C_2F_{0{\cdot}7}\ A_{0{\cdot}3}$ and $C_2F_{0{\cdot}3}A_{0{\cdot}7}$ with water produces in addition to the tetracalcium hydrates a C_2F.aq hydrate, or a $C_2(AF)$.aq solid solution, which gives an X-ray pattern with basal spacings of 10·8 Å and 5·3 Å, but these spacings may really arise from C_4FH_{19} and/or C_2AH_8.

The hydrogarnet solid solutions

We have seen that at room temperatures solid solutions occur between the cubic tricalcium aluminate and ferrite hydrates. They are part of a wider series[5] with grossularite garnet $3CaO.Al_2O_3.3SiO_2$ and andradite $3CaO.Fe_2O_3.3SiO_2$. These solid solutions can be prepared by treatment in saturated steam at 150–250° of $CaO–Al_2O_3–SiO_2–Fe_2O_3$ glasses of appropriate compositions representative, for example, of the composition of the liquid phase formed in the burning of Portland cement. In addition, their limited formation occurs at atmospheric pressure and room temperature, or slightly above, as for example in the reaction of C_2F or ferric hydroxide gel with lime solution in the presence of reactive silica or calcium silicates, in the hydration of C_3A with added C_3S or reactive silica, and in the treatment of dehydrated kaolin with lime solution at 50°.

The relation can be represented as follows:

$$3CaO.Al_2O_3.6H_2O \quad 3CaO.Al_2O_3.3SiO_2$$
$$3CaO.Fe_2O_3.6H_2O \quad 3CaO.Fe_2O_3.3SiO_2$$

Each member is capable of forming a solid solution with the other three, Fe_2O_3 and Al_2O_3 being mutually replaceable and $1SiO_2$ substituting for $2H_2O$, but a certain minimum amount of SiO_2 may be necessary to stabilise the cubic structure.

The solid solutions containing silica and/or iron oxide are much more resistant to attack by sulphate solutions than is C_3AH_6.

Calcium aluminate and ferrite complex salts

A wide range of organic and inorganic anions will combine with lime, alumina and water under suitable conditions to form quaternary hydrates.[6] The com-

[1] *Ind. Engng Chem.* **26,** 837 (1934).
[2] *J. Ceram. Ass. Japan* **45,** 277 (1937); *Zement* **26,** 830 (1937).
[3] P. P. Budnikov and V. S. Gorshkov, *Zh. prikl. Khim.* **33,** 1246, 1960.
[4] *Indian Concr. J.* **40,** 152 (1966).
[5] E. P. Flint, H. F. McMurdie and L. S. Wells, *J. Res. natn. Bur. Stand.* **26,** 13 (1941).
[6] F. E. Jones, *Sym., Stockholm 1938*, 231; W. Feitknecht, *Helv. chim. Acta* **25,** 106 (1942), **32,** 2298 (1949), **34,** 119 (1951); G. Malquori and V. Cirilli, *Sym., London 1952*, 321; J. H. P. van Aardt, *Sym., Washington 1960*, 835; H. J. Kuzel, *Sym., Tokyo 1968*.

pounds may be assigned to two series of empirical formulæ:

I		II
$C_3A.3CaX.mH_2O$		$C_3A.CaX.nH_2O$
or	and	or
$C_3A.3CaY_2.mH_2O$		$C_3A.CaY_2.nH_2O$

where X = divalent anion: CO_3, SO_3, SO_4, S_2O_3, CrO_4, SeO_4, WO_4
Y = monovalent anion: OH, Cl, Br, I, NO_3, NO_2, MnO_4, ClO_3, IO_3, BrO_3, SiO_3, HCO_2, CH_3CO_2, $C_2H_5CO_2$, $C_6H_5CO_2$, $C_6H_4NO_2CO_2$.

The water contents in the fully hydrated state are usually about m = 30–32 H_2O, n = 10–12 H_2O, and the compounds dissolve incongruently in water. The methods of preparation are similar to those of the calcium aluminate hydrates, and while compounds of both series can be obtained with some anions, other anions appear to give only the series II hydrates. Furthermore, in many cases Al_2O_3 can be replaced by Fe_2O_3 to give the corresponding ferrite complex salts and similar substitution by Cr_2O_3 or Ga_2O_3 has been reported for the sulphate and chloro compounds. Of the numerous calcium aluminate and ferrite complex salts, those containing sulphate, carbonate, chloride or silicate are of interest to the chemistry of cement, and a list of these compounds is given in Table 27.

TABLE 27 Physical properties of calcium aluminate and ferrite complex salts

Compound	Crystal form	Density (g/cm³)	Refractive indices		Characteristic X-ray powder spacings (Å)
			ω	ε	
$C_3A.3CaSO_4.31H_2O$	Needles	1·73	1·464	1·458	9·7, 5·61, 4·70, 3·88
$C_3A.3CaSO_4.8H_2O$	,,		1·535?		ca. 7·3 (poor pattern)
$C_3A.CaSO_4.15H_2O$*	Hex. plates				10·3, 5·15, 4·03, 2·87
$C_3A.CaSO_4.12H_2O$*	,, ,,				9·6, 4·78, 3·75, 2·87
$C_3A.CaSO_4.12H_2O$	,, ,,	1·99	1·504	1·488	9·0, 4·48, 4·00, 2·87
$C_3A.CaSO_4.10H_2O$	,, ,,				8·2
$C_3A.CaSO_4.7H_2O$*	,, ,,				8·0
$C_3F.3CaSO_4.31H_2O$	Needles		1·486	1·492	9·8, 5·62, 4·81, 3·95
$C_3F.CaSO_4.15H_2O$*	Hex. plates				10·3, 5·15, 3·44, 2·95
$C_3F.CaSO_4.12H_2O$	,, ,,		1·564	1·539	9·0, 4·45, 4·05, 2·94

* Water content uncertain.

TABLE 27—(*contd.*)

$C_3A.3CaCO_3.30H_2O$	Needles		1·480–1·490	1·456–1·470	9·4, 5·43, 4·62, 3·80
$C_3A.CaCO_3.11H_2O$	Hex. plates	2·14	1·554	1·532	7·6, 3·80, 2·86
$C_3A.CaCO_3.6H_2O$	,, ,,				7·1 (or 6·3?)
$C_4A.\frac{1}{2}CO_2.12H_2O$	,, ,,		1·539	1·514	8·2, 4·10, 3·88, 2·88
$C_4A.\frac{1}{2}CO_2.10H_2O$	,, ,,				7·7, 3·85, 3·78, 2·87
$C_4A.\frac{1}{2}CO_2.8H_2O$*	,, ,,				7·3
$C_4A.\frac{1}{2}CO_2.7H_2O$	,, ,,				6·1
$C_3F.CaCO_3.12H_2O$	,, ,,		1·592	1·573	8·0, 3·99, 3·89, 2·95
$C_3A.3CaCl_2.30H_2O$	Needles	1·69			10·2, 5·06, 4·69, 3·66
α-$C_3A.CaCl_2.10H_2O$	Hex. plates	2·03			7·9, 3·94, 3·85, 2·87
β-$C_3A.CaCl_2.10H_2O$	,, ,,	2·12	1·550	1·535	7·8, 3·91, 3·80, 2·87
$C_3A.CaCl_2.6H_2O$	,, ,,				6·9
$C_3F.CaCl_2.10H_2O$	,, ,,		1·607	1·597	7·8, 3·89, 3·83, 2·93
$C_3A.3CaSiO_3.32H_2O$†	Needles		1·487	1·479	
$C_3A.CaSiO_3.12H_2O$†	Hex. plates		1·538	1·523	
$C_2A.SiO_2.8H_2O$	,, ,,	1·89	1·512		12·6, 6·3, 4·18, 2·87
$C_2A.SiO_2.H_2O$	Cubic forms		1·628		3·61, 2·79, 2·55, 2·36

* Water content uncertain.
† Existence doubtful.

The quaternary hydrates appear all to be hexagonal or pseudohexagonal and to crystallise either as needles in the tri- or 'high' forms of series I or as hexagonal plates and spherulites in the mono- or 'low' forms of series II. The hydrated tetracalcium aluminate hydrate may be written as $C_3A.Ca(OH)_2.nH_2O$, and can be regarded as belonging to the $C_3A.CaY_2.nH_2O$ series. A close structural relationship evidently exists between this compound and the quaternary hydrates of the mono-forms. The value of the *a*-axis in the pseudohexagonal crystal lattice structures appears to be near 5·75 Å for all compounds of this series, while the *c*-axis varies depending upon the space requirements of the anion and the water present. The *a*-axis value for the Fe_2O_3-containing hydrates of the mono-forms is about 5·9 Å, and the *c*-axis values are very near to those of the corresponding aluminate hydrates. Because of the close resemblance between the various hydrates, there is the possibility of the formation of complex solid-solution series in the presence of two or more anions and of partial replacement of Al_2O_3 by Fe_2O_3. Similar solid-solution formation can also occur in the tri-forms, but solid solutions do not arise between the tri- and mono-forms because their crystal structures are different.

In addition to the above quaternary hydrates, a large number of adsorption complexes, similar to those obtained with layer-lattice clay minerals, have been prepared by Dosch[1] by treating C_4AH_{11} at room temperatures with a variety of organic compounds, including alcohols, mercaptans, sugars, amines, aldehydes and carboxylic acids. The complexes with primary alcohols give X-ray patterns with well-defined basal spacings, and the latter increase with increasing number of carbon atoms in the alcohol. Values for the longest basal spacing range from 12·9 Å for the CH_3OH complex to 36·2 Å for the $CH_3(CH_2)_9OH$ complex, and the carbon chains appear to be arranged perpendicular to the lattice layers of C_4A.aq. Both a perpendicular and parallel type of arrangement can occur with other organic compounds and the X-ray patterns can vary widely.

Calcium sulphoaluminates

Both 'high' and 'low' forms of calcium sulphoaluminate hydrate are produced by the action of calcium sulphate on calcium aluminate solutions. The compound $C_3A.3CaSO_4.31H_2O$ was first observed in 1890 by Candlot[2] and it occurs naturally as the mineral ettringite. It is formed during the destructive attack of sulphate solutions on cement mortars and concretes and has for this reason, sometimes been called the 'cement bacillus'. The low-sulphate form has been known since 1929.[3]

Compound	Crystal system	Crystal form	Refractive indices ω	ε
$3CaO.Al_2O_3.3CaSO_4.30\text{–}32H_2O$	Pseudo-hexagonal Uniaxial Negative	Needles Negative elongation	1·464	1·458
$3CaO.Al_2O_3.CaSO_4.12H_2O$	Pseudo-hexagonal Uniaxial Negative	Hexagonal plates and 'needles' (plates on edge) Positive elongation	1·504	1·488

The density of the high-sulphate form is 1·73 at 25° and of the low-sulphate form 1·99 at 20°.

The high-sulphate form can be prepared by slowly adding a volume of nearly saturated lime solution with stirring to an equal volume of a solution containing the appropriate amount of aluminium sulphate and calcium sulphate to satisfy the equation

$$Al_2(SO_4)_3.aq+6CaO \rightarrow 3CaO.Al_2O_3.3CaSO_4.31H_2O$$

and the condition that the final solution will contain at least 0·215 g $CaSO_4$,

[1] *Neues Jb. Miner. Abh.* **106,** 200 (1967).
[2] *Bull. Soc. Encour. Ind. natn.* ,682 (1890).
[3] W. Lerch, F. W. Ashton and R. H. Bogue, *J. Res. natn. Bur. Stand.* **2,** 715 (1929).

0·043 g CaO and 0·035 g Al_2O_3 per litre. Other methods of preparation include the direct hydration[1] of a mixture of $4CaO.3Al_2O_3.SO_3$ with CaO and $CaSO_4$ of overall composition $6CaO.Al_2O_3.3SO_3$, the addition of CaO to ammonium alum solution and the treatment of monocalcium aluminate solution with CaO and $CaSO_4.2H_2O$. In the latter two methods it is again necessary to have excess lime and calcium sulphate in solution at the end of the reaction, because $C_3A.3CaSO_4.31H_2O$ is incongruently soluble in water and decomposes slightly to give alumina gel as a second solid phase and to form a solution of the above concentration at 25° with a pH value of 10·80. If $C_3A.3CaSO_4.31H_2O$ is dissolved in a solution of lime or calcium sulphate, the equilibrium values required to stabilise the compound are altered. Thus in a lime solution containing 0·53 g CaO per litre, it dissolves to give a solution containing only about 0·01 g SO_3 and 0·008 g Al_2O_3 per litre. In a solution of 0·68 g $CaSO_4$ per litre the solubility is about 0·089 g SO_3, 0·02 g Al_2O_3, and 0·02 g CaO per litre. From the equilibrium data for the quaternary system at 25° it appears that the compound $C_3A.3CaSO_4.31H_2O$, although decomposed by very dilute lime solutions, becomes congruently soluble at 0·027 g CaO per litre, remains stable up to a lime concentration of 0·15 g CaO per litre, and at higher concentrations should tend to form cubic C_3AH_6 as a second solid phase. In the absence of this change the solubility (partly metastable) of $C_3A.3CaSO_4.31H_2O$ at 25° can be calculated as:

g CaO per litre	0·056	0·112	0·168	0·224	0·67	1·08
g $C_3A.3CaSO_4.31H_2O$ per litre	0·255	0·165	0·115	0·080	0·030	0·024

The solubility increases with temperature, and $C_3A.3CaSO_4.31H_2O$ remains a stable phase in solution at 90°, but at higher temperature it decomposes with formation of the monosulphoaluminate hydrate and gypsum.[2]

In aqueous solutions of Na_2SO_4, NaCl, or $CaCl_2$ the solubility is slightly increased, as compared with that in water, but apart from the slight decomposition owing to the formation of incongruent solutions, the compound remains stable. It is decomposed by solutions of magnesium sulphate with formation of gypsum, aluminium hydroxide and magnesium hydroxide. The solubility of magnesium hydroxide is very low and the pH value of its saturated solution is only about 10·5. This is less than the pH of the solution which $C_3A.3CaSO_4.31H_2O$ forms when it dissolves in water, and progressive decomposition, therefore, occurs. It is also decomposed by alkali carbonate solutions, and by carbon dioxide.

The heat of formation[3] at 25° of the hydrated trisulphoaluminate from $3CaO.Al_2O_3$, $CaSO_4.2H_2O$, and water is 47 kcal per mole or 195 cal per g SO_3. For the monosulphate the corresponding values are 15 kcal per mole and 187 cal per g SO_3. The 31 moles of water are retained until a vapour pressure below 0·7 mm at 20° is reached and reduced to $26H_2O$ and $18H_2O$ on drying at room temperature over anhydrous $CaCl_2$ and solid NaOH, respectively, but the X-ray pattern is apparently unchanged under these conditions. More intensive drying over

[1] P. K. Mehta and A. Klein, *J. Am. Ceram. Soc.* **48,** 435 (1965).
[2] Yu. M. Butt, A. A. Maier and B. G. Varshal, *Dokl. Akad. Nauk SSSR* **136,** 398 (1961); W. Lieber, *Zement-Kalk-Gips* **16,** 364 (1963).
[3] H. A. Berman and E. S. Newman, *J. Res. natn. Bur. Stand.* **67A,** 1 (1963); *Sym., Washington 1960*, 247.

P_2O_5 results in dehydration to an octahydrate and the X-ray pattern deteriorates and notably the longest spacing changes from 9·7 Å to about 7·3 Å. Thermal dehydration leaves 7–8H_2O at 105–110°, 4–6H_2O at 145° and 2–3H_2O at 200°. Ignition at 1000° produces an anhydrous calcium sulphoaluminate compound of composition $4CaO.3Al_2O_3.SO_3$, together with $CaSO_4$ (anhydrite) and CaO. The DTA thermogram of $C_3A.3CaSO_4.31H_2O$ is characterised by an endotherm at low temperatures with peak temperatures, as determined by different investigators, ranging from 110° to 180°. Infra-red absorption curves for the high-sulphate form, and the low-sulphate form also, have been reported by Schwiete and Niel.[1] Early crystal structure studies[2] established that $C_3A.3CaSO_4.31H_2O$ was hexagonal (see p. 334) but later work[3] indicates that the symmetry is really trigonal and that the apparent hexagonal symmetry must be attributed to twinning.

The low-sulphate form $C_3A.CaSO_4.12H_2O$ is difficult to obtain pure since it enters into solid solution with C_4AH_{13} and has no separate field of existence at ordinary temperatures. It can be prepared at 25° by treating a monocalcium aluminate solution with lime-calcium sulphate solution, or with solid CaO and $CaSO_4.2H_2O$, in such amounts that the $CaSO_4 : Al_2O_3$ molar ratio of the initial mixture is near unity and the final lime concentration in solution is about 1 g CaO per litre. If insufficient CaSO is present the solid solution with C_4AH_{13} is obtained, while if the $CaSO_4 : Al_2O_3$ ratio exceeds 1 mixtures of $C_3A.CaSO_4.12H_2O$ and $C_3A.3CaSO_4.31H_2O$ are formed. Berman[4] has described techniques for the preparation of carbonate-free $C_3A.CaSO_4.12H_2O$ from lime and aluminium sulphate solutions.

As shown in Table 27 the monosulphoaluminate hydrate exists in various forms giving different X-ray patterns.[5] Three of these modifications occur in contact with solution, and these are characterised in the X-ray powder patterns by respective longest basal spacings of 10·3 Å, 9·6 Å and 9·0 Å. It is widely accepted that the 9·0 Å form corresponds to $C_3A.CaSO_4.12H_2O$, but the water content of the other two forms and the relationship between them has not been conclusively established.[6] Their formation appears to depend both on the temperature and the composition of the reaction mixture. The 10·3 Å form may be a 15H_2O hydrate more stable below 25° in contact with solution.[7] The 9·6 Å form is probably a polymorph of the 9·0 Å, 12H_2O, hydrate though higher water contents for it have been reported.[8] The 10·3 Å form, like C_4AH_{19}, readily loses water at relative humidities near to 90 per cent and dehydrates to the 9·6 Å or 9·0 Å form, or both. Below about 33 per cent relative humidity the 8·2 Å, 10H_2O hydrate begins to form. This hydrate is obtained on drying at room temperature

[1] *Zement-Kalk-Gips* **18,** 157 (1965).
[2] F. A. Bannister, M. H. Hey and J. D. Bernal, *Mineralog. Mag.* **24,** 324 (1936).
[3] A. Moore and H. F. W. Taylor, *Nature, Lond.* **218,** 1048 (1968).
[4] *J. Res. natn. Bur. Stand.* **69A,** 45 (1965).
[5] R. Turriziani and G. Schippa, *Ricerca scient.* **24,** 2356 (1954); **25,** 2894 (1955); M. H. Roberts, *J. appl. Chem.* **7,** 543 (1957).
[6] P. Seligmann and N. R. Greening, *Sym., Tokyo 1968.*
[7] *Building Research 1965*, p. 72.
[8] W. Dosch and H. zur Strassen, *Zement-Kalk-Gips* **20,** 392 (1967); R. Turriziani and G. Schippa, loc. cit.; P. K. Mehta and A. Klein, *Spec. Rep. Highw. Res. Bd*, No. 90, 328 (1966).

over anhydrous $CaCl_2$ or solid NaOH, or by heating at 50°. More intensive drying over P_2O_5 results in dehydration to a 7–8H_2O hydrate giving a longest basal spacing of about 8 Å. These changes in the state of hydration are largely reversible on storage in a damp atmosphere, but the 9·6 Å and 10·3 Å forms are not always reformed.

On heating $C_3A.CaSO_4.12H_2O$ at 105–110° the water content is reduced to 6–8H_2O, and $4CaO.3Al_2O_3.SO_3$ together with $CaSO_4$ and CaO are formed on ignition at 1000°. The DTA thermogram of $C_3A.CaSO_4.12H_2O$ resembles that of C_4AH_{13}, and gives a strong endotherm in the region of 200°. X-ray studies[1] on hydrothermally prepared single crystals of $C_3A.CaSO_4.12H_2O$ have established a pseudohexagonal unit cell with $a = 5{\cdot}76$ Å, $c = 26{\cdot}79$ Å. This unit cell will contain $1\frac{1}{2}$ formula units, and electron diffraction evidence[2] shows the true a-axis to be double the above value.

The existence has been postulated of a continuous solid solution series between C_4AH_{13} and $C_3A.CaSO_4.12H_2O$ on the basis of variations in the refractive indices, but X-ray studies[3] have clearly established that it exists only over a restricted composition range. Thus Roberts established that a solid-solution series is formed with $CaSO_4 : Al_2O_3$ ratios from near 1 to 0·5, the longest basal spacing in the X-ray patterns decreasing linearly within this range from 8·96 Å to 8·77 Å, but the a-axis of the pseudohexagonal unit cell remaining constant at 5·75 Å. The limiting solid-solution composition $C_3A.\frac{1}{2}Ca(OH)_2.\frac{1}{2}CaSO_4.12H_2O$ gives the basal spacing of 8·77 Å. If slight carbonation occurs there is the possibility of the formation of quinary solid solutions containing carbonate and sulphate. The basal spacings in the region 8·3–8·5 Å which are sometimes observed in the X-ray patterns may arise from such a solid solution.

THE SYSTEM $CaO–Al_2O_3–CaSO_4–H_2O$

One, or both, of the calcium sulphoaluminates is formed during the hydration of cement by the reaction of gypsum with tricalcium aluminate and there has been much controversy concerning this. The basic system we have to consider is the quaternary system $CaO–Al_2O_3–CaSO_4–H_2O$ which can be represented as a reciprocal salt pair involving the reaction:

$$(Ca(OH)_2)_3 + Al_2(SO_4)_3.aq \rightleftarrows Al_2O_3.aq + (CaSO_4)_3.aq$$

The equilibrium in this system at 25°, and in the corresponding quinary systems with alkali hydroxides, has been studied by a number of investigators. Data for a quaternary system of this type can be represented in a tetrahedral space figure with H_2O at the apex and the four solid components of the salt pair at the four corners of the base. The data are plotted in one plane by the Jänecke projection in which a composition point in the space figure is projected on to the base by a

[1] J. H. Kuzel, *Neues Jb. Miner. Mh.*, No. 7, 193 (1965); R. Allmann, ibid., No. 5, 140 (1968).
[2] A. Grudemo, *Sym., Washington 1960*, 615.
[3] P. Seligmann and N. R. Greening, *Highw. Res. Rec.*, No. 62, 89 (1964); M. H. Roberts, *Sym., Tokyo 1968*.

line drawn from the water apex through the composition in question on to the base. Points so obtained on the base show the molal proportions of the four salt components of any solution, but not the water content of the solution. Each solid phase gives a field on the diagram over which it coexists with solution, two solid phases occur along the curves and at the intersection of these curves are invariant points at which three solid phases coexist with solution.

The equilibria in this system have been studied by Jones[1] at 25° and D'Ans and Eick[2] at 20° with results that are in broad agreement. The phase diagram as found by Jones is shown in Fig. 59. It will be seen that the only quaternary com-

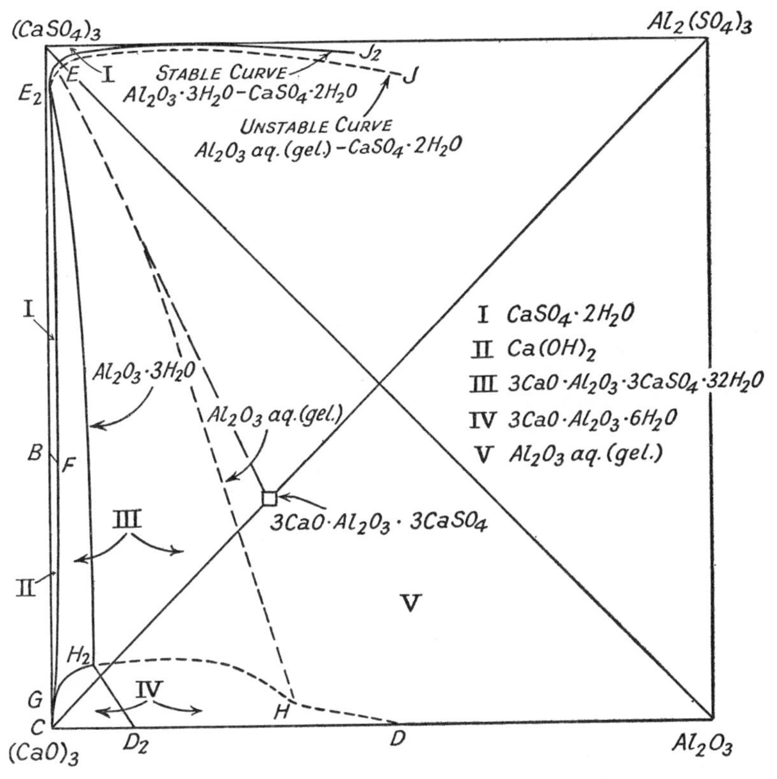

FIG. 59 System $CaO–Al_2O_3–CaSO_4–H_2O$.

pound appearing as a stable phase is the high-sulphate form of calcium sulphoaluminate. The extent of the composition field of the solutions in which this appears as a stable phase varies with the form of alumina. The boundary curve between the high-sulphate form and alumina gel is EH, while with crystalline $Al_2O_3.3H_2O$ (gibbsite) it is represented by E_2H_2; E and E_2 are the corresponding invariant points at which three solid phases, gypsum, high-sulphate calcium

[1] *Trans. Faraday Soc.* **35,** 1484 (1939); *J. phys. Chem.* **48,** 311 (1944).
[2] J. D'Ans and H. Eick, *Zement-Kalk-Gips* **9,** 203 (1953).

sulphoaluminate and hydrated alumina are in equilibrium. The points H and H_2 at the other end of the boundary represent the invariant points at which hydrated alumina, sulphoaluminate and $3CaO.Al_2O_3.6H_2O$ coexist. It was found, however, by Jones, that $C_3A.6H_2O$ only forms with great difficulty, and that in its place there appear metastable solid solutions of $3CaO.Al_2O_3.CaSO_4.aq$ and $4CaO.Al_2O_3.aq$ which persist indefinitely. In this condition the boundary of the sulphoaluminate field extends close to the $CaO–Al_2O_3$ line, down in fact to concentrations of $CaSO_4$ in solutions of about 0·004 g $CaSO_4$ per litre. The solid solution that appears along this boundary approximates to the low sulphate compound, $3CaO.Al_2O_3.CaSO_4.aq$, but as the $CaSO_4$ content of the solution decreases still further, the content of $4CaO.Al_2O_3.aq$ in it increases until finally when the $CaO–Al_2O_3$ boundary is reached the solid necessarily becomes pure aluminate. There was also evidence that, towards the CaO corner, the lime content of the solid solution increased, indicating that additional $Ca(OH)_2$ was taken up. On the other hand, towards the boundary between the solid solution and the alumina fields the lime content of the solid drops, suggesting that '$3CaO.Al_2O_3.12H_2O$' or $2CaO.Al_2O_3.8H_2O$ is entering in place of the tetracalcium aluminate.

Later work[1] has provided evidence that the $C_4AH_{13}–C_3A.CaSO_4.12H_2O$ solid solutions can in fact occur as stable phases and that there is a stable invariant point in the region IV close to the $(CaO)_3–Al_2O_3$ boundary at which $Ca(OH)_2$, C_3AH_6 and the lowest sulphate form $C_3A.\frac{1}{2}Ca(OH)_2.\frac{1}{2}CaSO_4.12H_2O$ of the solid solution series are in equilibrium.

There remains scope for further studies of the solid solution equilibria by X-ray methods for in the work both of Jones and D'Ans and Eick the identification of the solids was based solely on microscopic examination.

In connection with solid-solution formation, Kalousek[2] has suggested that there are five interconnected solid-solution series as follows:

$$\begin{array}{ccc}
6CaO.Al_2O_3.xH_2O & \text{—— 1 ——} & 6CaO.Al_2O_3.3SO_3.31H_2O \\
| & & | \\
5 & & 2 \\
| & & | \\
4CaO.Al_2O_3.13H_2O & \text{—— 4 ——} & 4CaO.Al_2O_3.SO_3.12H_2O \\
 & & | \\
 & & 3 \\
 & & | \\
 & & 3CaO.Al_2O_3.12H_2O
\end{array}$$

Of these various possibilities, solid solutions of series 2, 3 and 5 can now be dismissed on the basis of more recent evidence, and only the occurrence of solid solutions of series 4 are well established. The existence of solid solutions of series 1 appears to be doubtful because there is no satisfactory evidence for the formation of C_6A, xH_2O.

The introduction of alkalis to the extent of 1 per cent NaOH or KOH into the quaternary system, corresponding roughly to the content of alkali present in

[1] P. Seligmann and N. R. Greening, loc. cit.
[2] Thesis, Univ. Maryland, 1941; *J. phys. Chem.* **49**, 405 (1945).

the liquid phase in a setting cement paste, modifies considerably the solution compositions but does not alter the general solid phase relations.[1] This was largely confirmed by Kalousek but in contrast to Jones, he concluded that the monosulphate solid solution is a stable phase in the quinary system containing 1 per cent or more NaOH. The latter conclusion agrees with more recent findings on the quaternary system, but it may also be connected with the use of higher NaOH concentrations up to 2·8 per cent or 0·7 N in Kalousek's work, since decomposition of $C_3A.3CaSO_4.31H_2O$ with the formation of $C_3A.CaSO_4.12H_2O$ has been observed[2] to occur when the high-sulphate compound is treated with 0·5–1·0 N NaOH solutions at 25°, though it is unaffected in 0·1–0·25 N NaOH solutions.

The existence of a quinary hydrate with a variable composition, but approximating to $4CaO.0{\cdot}9Al_2O_3.1{\cdot}1SO_3.0{\cdot}5Na_2O.16H_2O$, has also been indicated[3] in the quinary system with NaOH concentrations of about 4 per cent or above, but the evidence cannot be regarded as conclusive. It appears rather improbable that the supposed Na_2O-containing quinary hydrate will occur in hydrated cement paste under normal conditions, owing to the lower alkali concentrations that usually prevail.

The deduction to be drawn from this work on the quaternary and quinary systems is that in a saturated gypsum and lime solution, tricalcium aluminate first forms the high-sulphate form of calcium sulphoaluminate. The course of the reaction in saturated lime solution then depends on the relative amounts of C_3A and calcium sulphate. If, and when, the gypsum is used up the low-sulphate phase or its solid solution with C_4AH_{13} will commence to appear. Eventually, the high-sulphate form may all change over completely to the solid solution $C_3A.\frac{1}{2}Ca(OH)_2.\frac{1}{2}CaSO_4.12H_2O$ and any further hydration of C_3A produces C_4AH_{19} or C_3AH_6. A somewhat different behaviour would be expected if the gypsum does not readily enter into solution or if local variations of sulphate concentration occur adjacent to the surface of the C_3A. Under such conditions the low-sulphate solid solution, or even hydrated calcium aluminates, could form first and then be converted to the $C_3A.3CaSO.31H_2O$ as the $CaSO_4$ concentration increased, the subsequent course being as before. It is evident that, while the knowledge of the equilibria in the quaternary and quinary systems that we have discussed is very helpful in indicating possible courses of the reaction of C_3A, it does not provide decisive evidence as to what happens in a setting cement with a limited amount of water where there may be restricted mobility of ions in solution.

Calcium sulphoferrites

Two series of calcium sulphoferrites exist analogous to the sulphoaluminates. The high-sulphate sulphoferrite $C_3F.3CaSO_4.32H_2O$ was prepared by Jones[4] as fine needles of mean refractive index 1·490 by shaking a solution of iron alum

[1] F. E. Jones, *J. phys. Chem.* **48,** 356, 379 (1944); **49,** 344 (1945).
[2] M. H. Roberts, private communication.
[3] W. Dosch and H. zur Strassen, *Zement-Kalk-Gips* **20,** 392 (1967); P. Seligmann and N. R. Greening, *Sym., Tokyo 1968.*
[4] *J. phys. Chem.* **49,** 344 (1945).

$(NH_4)_2SO_4.Fe_2(SO_4)_3.24H_2O$ with lime for 2 days. It can also be obtained by hydration of C_2F in lime-calcium sulphate solution.[1] The compound forms solid solutions[2] with $C_3A.3CaSO_4.31H_2O$ up to a limiting ratio of about 3 mole Fe_2O_3 to 1 mole Al_2O_3 when the mean refractive index is 1·48. Higher contents of Fe_2O_3 cannot be obtained in the solid solution.

The low-sulphate forms $3CaO.Fe_2O_3.CaSO_4.aq$ occurs as needle crystals and as hexagonal plates and spherulites. Hydrates with 13–14H_2O and 11–12 H_2O are obtained on drying at 94 per cent relative humidity and 55 per cent respectively.[3] Their X-ray patterns show longest basal spacings of 10·3 Å and 9·0 Å respectively. The 13–14H_2O hydrate is the phase existing in contact with solution at 18–25°. Unlike the monosulphoaluminate hydrate, the monosulphoferrite hydrate can occur in contact with saturated lime-calcium sulphate solution at room temperature with no great tendency to change to the trisulphoferrite hydrate. The compound retains 10 moles H_2O over $CaCl_2$, 8 moles over P_2O_5 and 6 moles at 110°. Complete dehydration at 1000° produces a mixture of C_2F, CaO and $CaSO_4$. Some evidence[4] has been advanced for the formation of solid solutions between $C_3F.CaSO_4.aq$ and $C_3A.CaSO_4.aq$ as well as between C_4F aq and $C_3F.CaSO_4.aq$ but their extent and the conditions under which they are formed remains an open question.

Calcium carboaluminates

The existence of carboaluminates is important because of the possibility of their formation during the hydration of cement containing carbonate impurities, or by the action of atmospheric carbon dioxide on hydrated cement compounds. Both 'high' and 'low' forms were first observed by Bessey.[5]

The 'high' form, $3CaO.Al_2O_3.3CaCO_3.30H_2O$, occurs as needles with refractive indices and X-ray patterns very similar to $3CaO.Al_2O_3.3CaSO_4.31H_2O$. The water content is reduced to $26H_2O$ and $12H_2O$ on drying at relative humidities of 10 and 5 per cent. The DTA curve shows an endotherm at 145–165° arising from loss of water and at 860–925° caused by liberation of carbon dioxide. There is no evidence of the formation of a solid solution with $3CaO.Al_2O_3.3CaSO_4.31H_2O$. This 'high' carbonate compound can be formed by the addition of sodium or ammonium carbonate solutions to a calcium monoaluminate or potassium aluminate solution in the presence of a lime-sucrose solution.[6]

The 'low' form, $3CaO.Al_2O_3.CaCO_3.11H_2O$, crystallises in hexagonal plates, $\omega = 1{\cdot}554$, $\varepsilon = 1{\cdot}532$. This water content is retained on drying at relative humidities down to 33 per cent relative humidity, and the X-ray pattern shows a longest basal spacing of 7·6 Å. A hexagonal unit cell with $a = 8{\cdot}72$ Å, $c = 7{\cdot}57$ Å

[1] R. H. Bogue and W. Lerch, *Ind. Engng Chem.* **26,** 837 (1934); P. P. Budnikov and V. S. Gorshkov, *Dokl. Akad. Nauk SSSR* **126,** 337 (1959).
[2] G. Malquori and V. Cirilli, *Sym., London 1952*, 321; *Ricerca Scient.* **14,** 78 (1943); V. Cirilli, ibid. **14,** 27 (1943).
[3] G. Schippa, *Ricerca scient.* **28,** 2334 (1958).
[4] H. E. Schwiete and U. Ludwig, *Sym., Tokyo 1968*; H. J. Kuzel, *Zement-Kalk-Gips* **21,** 493 (1968).
[5] G. E. Bessey, *Sym., Stockholm 1938*, 186, 233, 234.
[6] E. T. Carlson and H. A. Berman, *J. Res. natn. Bur. Stand.* **64A,** 333 (1960); H. A. Berman, ibid. **69A,** 407 (1965).

was proposed by Carlson and Berman, but in accordance with similar quaternary hydrates an *a*-axis of about 5·7 Å would appear more likely. Water is gradually lost on heating above 100° and the basal spacing diminishes.[1,2] The DTA curves are characterised by endotherms at about 230° and 900°. The 'low' carbonate compound can be prepared by mixing solutions of calcium aluminate with calcium hydroxide and sodium carbonate solution in such proportions that the $CO_3 : Al_2O_3$ molar ratio is near unity and the final CaO concentration in solution is about 0·4 g per litre. This compound is also formed when atmospheric carbon dioxide reacts with calcium aluminate solutions or hydrates but on continued exposure it decomposes to $CaCO_3$ and hydrated alumina. It is also formed in suspensions or pastes of C_3A, or C_4AF, or cement in the presence of calcium or magnesium carbonates.[3] The monocarboaluminate forms solid solutions[4] with C_4AH_{13} varying in composition from $C_4A.\frac{1}{2}CO_2.12H_2O$ to $C_3A.CaCO_3.11H_2O$ with basal spacings changing from 8·2 Å to 7·6 Å.

Studies[4] on the equilibria in the system $CaO–Al_2O_3–CaCO_3–H_2O$ at 25° have shown that in the presence of $C_4A.\frac{1}{2}CO_2.12H_2O$, or similar solid solution compositions, the solution concentrations fall on a curve approximately parallel to the solubility curve *YT* in the ternary system $CaO–Al_2O_3–H_2O$ (Fig. 56) and displaced to slightly lower CaO and Al_2O_3 concentrations. This curve may proceed to values of 0·39 g CaO per litre, 0·32 g Al_2O_3 per litre. When hydrated alumina and calcite are formed together with $C_3A.CaCO_3.11H_2O$ and $C_4A.\frac{1}{2}CO_2.12H_2O$ the solution concentration decreases to about 0·2 g CaO per litre, 0·08–0·10 g Al_2O_3 per litre. A similar solution composition is obtained by treating $C_3A.CaCO_3.11H_2O$ with water, while in lime solutions of increasing concentration the Al_2O_3 concentration decreases to practically nil at about 0·3 g CaO per litre and remains barely detectable at higher lime concentrations up to saturation. The solubility curve of $C_3A.CaCO_3.11H_2O$ is, therefore, very similar to that of C_3AH_6, but since the latter hydrate reacts with CO_2 it should be more soluble than $C_3A.CaCO_3.11H_2O$. The carboaluminate hydrates have, however, a higher solubility than the sulphoaluminate hydrates, since $C_3A.3CaSO_4.31H_2O$ is formed when $C_3A.CaCO_3.11H_2O$ and $C_4A.\frac{1}{2}CO_2.12H_2O$ are treated with sulphate solutions.[5]

Calcium carboferrites

Various possibilities exist for the formation of different calcium carboferrite hydrates analogous to the carboaluminate hydrates, but little information is available. A compound similar to the low-carbonate carboaluminate has been prepared[6] which has a composition after drying at 81 per cent relative humidity approximating to $C_3F.CaCO_3.12H_2O$. The X-ray pattern closely resembled that of $C_4F.H_{13}$ and showed a longest basal spacing of about 8 Å, somewhat

[1] W. Dosch and H. zur Strassen, *Zement-Kalk-Gips* **18,** 233 (1965).
[2] R. Turriziani and G. Schippa, *Ricerca scient.* **26,** 2792 (1956).
[3] P. P. Budnikov, V. M. Kolbasov and A. S. Panteelev, *Tsement* **27** (5), 1 (1961); T. Manabe, N. Kawada and M. Nishiyama, *Jap. Cem. engng Ass.*: Rev. 15th gen. Meet. 1961, 48; E. Spohn and W. Lieber, *Zement-Kalk-Gips* **18,** 483 (1965).
[4] M. H. Roberts, *Sym., Tokyo 1968.*
[5] T. Manabe, N. Kawada and M. Nishiyama, loc. cit.; *Building Research 1964,* p. 55.
[6] M. H. Roberts, private communication.

greater than that of $C_3A.CaCO_3.11H_2O$. It is likely that solid solutions between C_4FH_{13} and $C_3F.CaCO_3.12H_2O$ occur and there is some evidence[1] for the formation of a hemicarboferrite $C_4F.\frac{1}{2}CO_2.12H_2O$. Solid solutions may also well occur between the carboferrites and carboaluminates.

Calcium chloroaluminates and chloroferrites

The chloroaluminates and chloroferrites are of importance because of their possible formation when chloride-containing admixtures are used to accelerate the initial rate of reaction between cement and water.

The low chloride chloroaluminate $C_3A.CaCl_2.10H_2O$ can be prepared by mixing aluminium chloride and lime solutions, or by reaction of $CaCl_2$ with lime solutions and calcium aluminate solutions, or of mixed $CaCl_2$–$AlCl_3$ solutions with NaOH solution. Two polymorphs exist, a monoclinic α-$C_3A.CaCl_2.10H_2O$ stable below 28° and a trigonal β-$C_3A.CaCl_2.10H_2O$ formed at higher temperatures.[2] The longest basal spacings are 7·9 Å and 7·8 Å respectively. The X-ray pattern and the water content appear to be unchanged on drying over $CaCl_2$ but over P_2O_5, and at 110°, a $6H_2O$ hydrate with a longest basal spacing of 6·9 Å is obtained.[3] The compound is incongruently soluble in water, $CaSO_4$ solution, and lime and alkali solution.[4] It is stable in alkaline solutions containing chloride at a concentration equivalent to about 30 g $CaCl_2$ per litre while at lower chloride concentrations solid solutions with C_4AH_{13} are formed, with basal spacings in the range 8·0–7·9 Å. These appear to be limited to a restricted range of compositions similar to the $C_3A.\frac{1}{2}Ca(OH)_2.\frac{1}{2}CaSO_4.12H_2O$–$C_3A.CaSO_4.12H_2O$ series discussed earlier.[5] There is evidence that a trigonal hydrate $C_3A.\frac{1}{2}CaCl_2.\frac{1}{2}CaSO_4.12H_2O$ exists with a density of 2·2 and a longest basal spacing of 8·4 Å. Other solid-solution compositions containing chloride and sulphate may possibly exist. No information is available on the possible existence of quinary solid solutions containing chloride and carbonate.

The high chloride chloroaluminate $C_3A.3CaCl_2.30H_2O$ can be obtained[6] by hydration of C_3A in a 21–23 per cent $CaCl_2$ solution at −10°. It can also be formed at 20° but it is unstable at this temperature and changes into $C_3A.CaCl_2.10H_2O$ in saturated lime-water or in 1·5–10 per cent $CaCl_2$ solution. The formation of $C_3A.3CaCl_2.30H_2O$ appears to depend both on temperature and calcium chloride concentration and it seems very improbable that it could be formed under normal conditions in hydrated cement. Attempts to use large calcium chloride additions for concreting at below zero temperatures in the U.S.S.R. have shown that the formation of the high-chloride compound has a destructive effect.[7]

[1] P. Seligmann and N. Greening, *Sym., Tokyo 1968.*
[2] H. J. Kuzel, *Neues Jb. Miner. Mh.* No. 7, 193 (1966); *Sym., Tokyo 1968.*
[3] W. Feitknecht and H. W. Buser, *Helv. chim. Acta* **34,** 128 (1951).
[4] M. H. Roberts, *Mag. Concr. Res.* **42,** 143 (1962).
[5] M. H. Roberts, *Sym., Tokyo 1968.*
[6] N. N. Serb-Serbina, Yu. A. Savvina and V. S. Zhurina, *Dokl. Akad. Nauk SSSR* **3** (3), 659 (1956); H. E. Schwiete and U. Ludwig, *Sym., Tokyo 1968. Zement-Kalk-Gips* **22,** 225 (1969).
[7] Yu. A. Savvina and N. N. Serb-Serbina, *Stroit. Prom.*, 1956 (9), 31.

The formation of $C_3F.CaCl_2.10H_2O$ and limited solid solutions $C_3A.CaCl_2.10H_2O$–$C_3F.CaCl_2.10H_2O$ has been reported.[1] The monochloroferrite hydrate dissolves incongruently in water and various solutions in a similar manner to $C_3A.CaCl_2.10H_2O$ and these two compounds have very similar properties, though the formation of α and β polymorphs of $C_3F.CaCl_2.10H_2O$ has not been observed. Schwiete, Ludwig and Albeck[2] state that $C_3F.3CaCl_2.30H_2O$ can be obtained from mixtures of C_2F, CaO and $CaCl_2$ at $-10°$.

Calcium silicoaluminates and silicoferrites

Apart from the silica-containing hydrogarnets belonging to the series C_3AH_6–C_3AS_3, the only well-established hydrated calcium silicoaluminate is the compound $2CaO.Al_2O_3.SiO_2.8H_2O$ which is sometimes called 'gehlenite hydrate'. This compound was first observed by Stratling[3] who obtained it by the reaction of burnt kaolin and lime-water, by shaking C_3A with a silica sol and lime-water, and by long period reaction of C_3S and C_3A in water. Later work[4] has confirmed the existence of $C_2AS.8H_2O$, and has shown that this compound is also formed by the reaction of lime solutions with C_2AS glass, or alumina-silica gels, or some natural pozzolanas and clay minerals. It is difficult to obtain pure preparations of $C_2AS.8H_2O$, and these are usually contaminated with calcium silicate hydrate, hydrogarnet and possibly carboaluminate hydrates. The compound crystallises in weakly birefringent plates, with a mean refractive index of 1·512 and crystal structure studies[5] have established a pseudohexagonal unit cell with $a = 5{\cdot}73$ Å, $c = 12{\cdot}60$ Å, indicating a close structural relationship with the hexagonal-plate calcium aluminate hydrates. It dissolves incongruently in water to give a solution containing 0·01 g SiO_2, 0·013 g Al_2O_3, 0·08 g CaO per litre, and is unstable in saturated lime solution at room temperature, or in weaker lime solutions at 50°, changing into a silica-containing hydrogarnet. It is decomposed by $MgSO_4$ and lime-gypsum solutions but is apparently stable in saturated $CaSO_4$ solution and in 0·15 M $NaSO_4$ solution.[6] The DTA curves show an endotherm at about 210° and an exotherm at 940° due to the formation of C_2AS. Loss of water on heating starts at about 50°, becomes intensified at about 150° and practically all water is lost at 350–400°.

On hydrothermal treatment[7] $C_2AS.8H_2O$ decomposes into boehmite and a hydrogarnet at 250° and a new cubic phase approximating to $C_2AS.H_2O$ appears at 350°. This has a body-centred cubic lattice with $a = 8{\cdot}33$ Å and a

[1] G. Malquori and E. Caruso, *Atti del X° Congr. int. di Chim., Rome,* ii, 713 (1938); G. Malquori and V. Cirilli, *Ricerca scient.* **14,** 78 (1943); A. Burdese, ibid. **22,** 1447 (1952); H. J. Kuzel, loc. cit.

[2] *Naturwissenschaften* **55,** 179 (1968).

[3] W. Stratling, *Zement* **29,** 311 (1940); W. Stratling and H. zur Strassen, *Z. anorg. allg. Chem.* **245,** 257 (1940).

[4] R. Turriziani, *Ricerca scient.* **24,** 1709 (1954); N. Fratini and R. Turriziani, ibid. 1654; R. Turriziani and G. Schippa, ibid., 2645; C. H. Schmitt, *Sym., Washington 1960,* 244; A. Ariizumi, *Sym., Tokyo 1968.*

[5] H. zur Strassen, *Sym., Washington 1960,* 244; R. Sersale, P. G. Orsini and R. Aiello, *Atti Accad. naz. Lincei* **34,** 274 (1963); P. G. Orsini and R. Aiello, *Ricerca scient. Rend.* **8A,** 35 (1965).

[6] F. W. Locher, *Sym., Washington 1960,* 267.

[7] E. T. Carlson, *J. Res. natn. Bur. Stand.* **68A,** 449 (1964).

refractive index of 1·628. The optical properties and the X-ray pattern of this cubic hydrate are very similar to those of the orthorhombic hydrated aluminate $C_4A_3H_3$ and some confusion between them may have arisen in earlier hydrothermal studies.

The existence of two other silicoaluminates[1] $C_3A.CaSiO_3.12H_2O$ and $C_3A.3CaSiO_3.32H_2O$ has also been claimed, but later work has thrown much doubt on the former[2] while the latter[3] is probably a quinary solid solution containing both CO_2 and SiO_2.

Little information is available on the possible formation of calcium silicoferrite hydrates. A limited substitution of Fe_2O_3 for Al_2O_3 occurs in $C_2AS.8H_2O$ to give solid solutions up to $C_2(A_{0\cdot7}F_{0\cdot3}).8H_2O$ with a refractive index raised to 1·525. There does not appear to be any compound $C_2FS.8H_2O$.

The solid solutions between the hydrated compounds

In view of the variety of the solid solutions formed amongst the hydrated aluminates, ferrites and the complex salts they are summarised for easy reference in Table 28. In addition to these known solid solutions it must be expected that

TABLE 28 Solid solutions between the hydrated compounds

End members forming the solid solutions	Limits of the solid solution series
C_4A.aq, C_4F.aq	Complete solid solution series
$C_3A.3CaSO_4$.aq, $C_3F.3CaSO_4$.aq	From $C_3A.3CaSO_4$.aq to $C_3(A_{0\cdot25}F_{0\cdot75})\,3CaSO_4$.aq
$C_3A.CaSO_4$.aq, C_4A.aq	From $C_3A.CaSO_4$.aq to $C_3A.\frac{1}{2}Ca(OH)_2.\frac{1}{2}CaSO_4$.aq
$C_3A.CaSO_4$.aq, $C_3F.CaSO_4$.aq	Solid solutions, extent limited
$C_3F.CaSO_4$.aq, C_4F.aq	Solid solutions, extent limited
$C_3A.3CaCO_3$.aq, $C_3A.3CaSO_4$.aq	No solid solution
$C_3A.CaCO_3$.aq, C_4A.aq	From $C_3A.CaCO_3$.aq to $C_3A.\frac{1}{2}Ca(OH)_2.\frac{1}{2}CaCO_3$.aq
$C_3F.CaCO_3$.aq, C_4F.aq	Probable from $C_3F.CaCO_3$.aq to $C_3F.\frac{1}{2}CaOH_2.\frac{1}{2}CaCO_3$.aq but evidence only for the hemicarbonate
Ca carboaluminates and carboferrites	Solid solution may well occur
$C_3A.3CaCO_3$.aq, $C_3A.3CaSiO_3$.aq	Solid solution probable but the pure end silicate compound does not exist
$C_3A.CaCl_2$.aq, C_4A.aq	Restricted range of solid solution, perhaps from $C_3A.CaCl_2$.aq to $C_3A.\frac{1}{2}Ca(OH)_2.\frac{1}{2}CaCl_2$.aq
$C_3A.CaCl_2$.aq, $C_3F.CaCl_2$.aq	Limited solid solutions
$C_3A.CaSO_4$.aq, $C_3A.CaCO_3$.aq, C_4A.aq	Probable solid solutions of type $C_3A.xCa(OH)_2.yCaCO_3.(1-x-y)CaSO_4$.aq
$C_3A.CaSO_4$.aq, $C_3A.CaCl_2$. aq	A phase $C_3A.\frac{1}{2}CaCl_2.\frac{1}{2}CaSO_4$.aq known

[1] E. P. Flint and L. S. Wells, *J. Res. natn. Bur. Stand.* **33,** 471 (1944).
[2] R. Turriziani, *Ricerca scient.* **24,** 1709 (1954); E. T. Carlson, *Sym., Washington 1960,* 244; H. zur Strassen and C. H. Schmitt, ibid., 243.
[3] E. T. Carlson and H. A. Berman, *J. Res. natn. Bur. Stand.* **64A,** 333 (1960).

more complex ones may exist which in their most general form could be expressed as $3CaO(Al_2O_3, Fe_2O_3)\ Ca(SiO_3, (OH)_2, SO_4)$ or combinations of some of these anions.

PORTLAND CEMENT

We have discussed the reactions of the individual cement compounds with water and the nature of the hydrated compounds that exist in the complex aqueous systems involved. Our problem now is to integrate this knowledge into a picture of the reactions of cement itself.

The reaction of cement with water is in the first instance a reaction of the individual constituents. The C_3A and ferrite phases react quickly at first, as also does the C_3S, but the β-C_2S reacts with water more slowly. The gypsum in cement also commences to dissolve rapidly, and the alumina compounds crystallise out as hydrated calcium aluminate and/or sulphoaluminate. The formation of plates or needles of these hydrates may be observed when a cement, or a ground clinker, is treated with excess water on a microscope slide, though they are not observed equally easily with all clinkers and cements. The occurrence of calcium hydroxide plates and hexagonal prisms can also be observed, together with an apparently amorphous hydrated silicate covering the original grains.

Within a very short period of time, if not almost immediately, after a cement is mixed with water, the liquid phase is essentially a solution of the hydroxides and sulphates of calcium, sodium and potassium, tending towards the following equilibrium:[1]

$$CaSO_4 + 2MOH \rightleftarrows M_2SO_4 + Ca(OH)_2$$

where M is either K or Na or both. Only extremely small amounts of Al_2O_3, SiO_2, and other oxides have been found to be present in the aqueous phase. The solution composition with respect to calcium, sulphate and hydroxyl ions may change rapidly during the first several minutes, but subsequently it usually remains relatively constant for a period of several hours, the actual composition depending upon the particular cement and water : cement ratio. At this time, both $CaSO_4.2H_2O$ and $Ca(OH)_2$ are present in the solid, and the solution is saturated, or even supersaturated, with respect to these compounds. Whilst solid $Ca(OH)_2$ persists in hydrated Portland cement at all ages, the $CaSO_4.2H_2O$ is rapidly used up as an increasing amount of the almost insoluble hydrated calcium sulphoaluminate is formed, and eventually the sulphate is almost completely removed from solution, usually between about 10 and 24 hours after gauging the cement with water. Thereafter, the liquid phase in cement paste becomes a NaOH and/or KOH solution which is saturated, or perhaps supersaturated, with respect to calcium hydroxide, and in general the lime concentration continues to decrease gradually with slow further release of alkali from the cement. As we have seen, the alkalis exist in cement both as the sulphates which pass into

[1] S. A. Greenberg and V. S. Mehra, *Sym., Washington 1960*, 378; A. Rio, A. Celani and M. Collepardi, *Industria ital. Cem.* **35**, 275 (1965); C. D. Lawrence, *Spec. Rep. Highw. Res. Bd*, No. 90, 378 (1966); P. P. Budnikov and M. I. Strelkov, ibid., 447; M. H. Roberts, *R.I.L.E.M Sym. Admixtures for Mortars and Concrete, Brussels 1967*, *Bldg Res. Stn curr. Pap.*, 61/68.

solution rapidly and as components of the alumina and silica compounds from which they effectively dissolve as hydroxides. Probably arising from these different modes of combination of the alkalis, the total amount obtained in solution during the first few hours may vary from a few per cent to some 70 per cent of the alkalis present in different cements (see p. 548). Depending upon the alkali content of the cement and the water-cement ratio, concentrations of alkali hydroxide up to about 1 per cent, or approximately 0·2–0·25 N, may occur in the solution in a cement paste. The reduced solubility of lime in the presence of alkali hydroxides, as well as the possible increased solubility of alumina and silica, may have some influence on the nature of the cement hydration products, but there is not at present sufficient information on the influence of alkalis on the cement hydration reactions.

Some controversy has existed as to whether the immediate reaction of the C_3A with gypsum during the hydration of cement leads to the formation of the low-sulphate sulphoaluminate (or its solid solution with tetracalcium aluminate hydrate) or the high-sulphate compound. It is now generally accepted that the high-sulphate compound is usually formed first. However, this may well depend on the availability of calcium sulphate in the solution adjacent to the C_3A. In cement pastes of low water-cement ratio there may be a restricted mobility of ions in solution and, if local variations in solution concentration occur so that insufficient calcium sulphate is available in the solution near to the hydrating C_3A, then the low-sulphate sulphoaluminate will tend to form temporarily at an early age. On the other hand, if saturated solutions are established throughout the cement paste, it must be expected that the high-sulphate sulphoaluminate will form and persist until almost all the sulphate has been removed from the solution. At this point, and with further hydration of the C_3A, the high-sulphate sulphoaluminate will start to convert to the low-sulphate sulphoaluminate or to the hexagonal-plate solid solution of $C_3A.CaSO_4.12H_2O$ and C_4AH_{13}. Once the limiting solid-solution composition is established any remaining C_3A will hydrate to produce C_4AH_{19} which may then convert, depending on the temperature, into the cubic hydrate C_3AH_6 or by taking up silica to form a hydrogarnet.

The C_4AF which reacts less instantaneously than the C_3A may be expected to combine initially with gypsum and lime to form a solid solution of the high-sulphate sulphoaluminate and sulphoferrite. As with C_3A, this phase converts later, when the sulphate is exhausted, into a low-sulphate aluminoferrite solid solution and/or a more complex solid-solution phase in which sulphate ion is replaced by hydroxyl ion. The only exception to this behaviour occurs in cements of low C_3A content in which the $C_3(AF).3CaSO_4.aq$ solid solution seems to persist indefinitely.

Some of the Al_2O_3, Fe_2O_3 and SO_3 is taken up initially in the *C-S-H* gel but only the alumina seems to be retained in significant quantity indefinitely. Some Al_2O_3 as well as Fe_2O_3 and SiO_2 enters into a hydrogarnet phase. The rest of the alumina and ferric oxide from the C_3A and C_4AF, together with the sulphur trioxide, may be expected ultimately to be present in the hexagonal-plate solid solution of the $C_3A.CaSO_4.12H_2O$–C_4AH_{13} and $C_3(AF).CaSO_4.aq$ types. There is also the possibility of the SiO_3 group entering into these solid solutions.

If some carbonation of the set cement occurs one of the reactions involved is the decomposition of the low-sulphate sulphoaluminate to give $C_3A.\frac{1}{2}Ca(OH)_2.\frac{1}{2}CaCO_3.aq$ and the high-sulphate sulphoaluminate. This may account for the presence sometimes reported of ettringite in set ordinary Portland cement at long ages.

The initial hydration products of C_3S and β-C_2S have a composition close to C_3S.aq and C_2S.aq but within a matter of hours these change to a calcium silicate hydrate gel with a $CaO:SiO_2$ ratio in the region of 1·5. The final composition of this gel varies from about 1·5 to 1·8 with the cement, the water-cement ratio and the temperature. It is also influenced by the amount of other oxides remaining in the gel. It is still an open question whether the final stable product is afwillite $3CaO.2SiO_2.3H_2O$.

A schematic representation of the reactions during the hydration of Portland cement at ordinary temperatures is set out below:

HYDRATION OF PORTLAND CEMENT

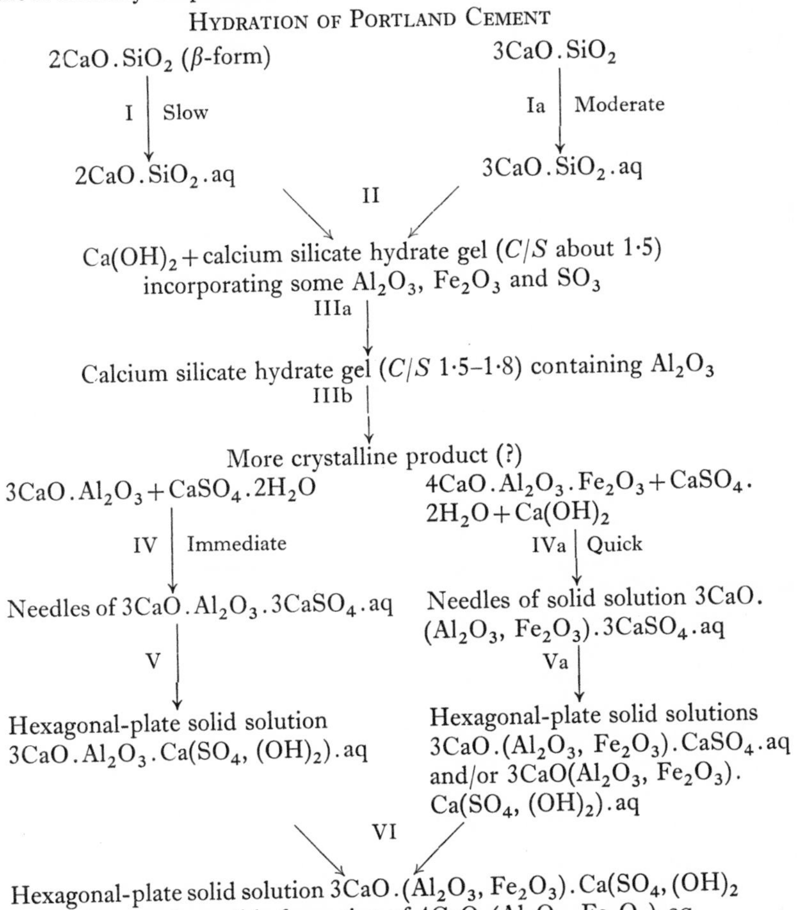

There are still uncertainties with regard to this scheme of the reactions. Thus we have no very direct evidence of stage VI with respect to incorporation of the SiO_3 group in the hexagonal-plate solid solutions, and this is inferred from general knowledge of the phases formed in various separate aqueous systems. The hydrogarnet phase may appear at an earlier stage. The extent of stage IIIb is also uncertain. The stages I, Ia, IV and IVa all belong to the very early reaction period, and stages V and Va will not be expected to commence until all the sulphate is used up at some time up to about 24 hours, or a little longer. In a complex system such as this some overlapping of the stages must obviously occur, and the later stages IIIb and VI may be very slow reactions. Furthermore, some interference and complex interactions can occur by carbonation, either arising from carbonate in the cement or from exposure to atmospheric carbon dioxide. For example, if hydrated calcium carboaluminates are thereby formed at an early stage of the reactions, then the transformation from high-sulphate to low-sulphate compounds in stages V and Va may be retarded, or perhaps even prevented. Alternatively, as noted earlier, the high-sulphate compounds can be re-formed if carbonation of the low-sulphate compounds and the formation of carboaluminates takes place at a later stage in the reactions.

Microscopic structure of set Portland cement

Set Portland cement can be examined microscopically by powdering the material or preparing thin sections or polished surfaces, but identification of the hydrated materials present is difficult. In a powdered set cement, calcium hydroxide crystals can be detected and, occasionally, calcium sulphoaluminate, but the remainder cannot be differentiated.

Staining tests have been found of some value and stains may also be useful for improving the contrast between hydrated and unhydrated material. Perhaps the most useful is naphthol green B (the ferrous sodium salt of nitroso-β-naphthol-β-monosulphonic acid) which stains all the hydrated lime compounds green except for the larger crystals of calcium hydroxide. The dye may be dissolved in the mixing water, or thin sections immersed for 12 hours in an alcoholic solution of the dye, containing a few per cent water, after fine grinding the first side of the specimen.[1] An alcoholic solution of patent blue has also been used in a similar manner on polished surfaces.[2] Alternatively, etching agents have been used such as a 1 per cent solution of borax in water or dilute nitric acid in alcohol. The former attacks calcium hydroxide crystals and unhydrated C_3S, and the latter unhydrated C_2S as well.[3]

Following earlier work by various authors thin sections have been studied by Parker and Hirst, and Brown and Carlson,[4] and polished surfaces using reflected light by Tavasci,[5] Brownmiller and Terrier and Moreau. In thin sections only three main constituents can normally be observed, unhydrated grains, calcium hydroxide crystals and the gel groundmass of hydrated products. The gel tends to occupy the whole of the space, apart from the calcium hydroxide, between the

[1] T. W. Parker and P. Hirst, *Cem. and Cem. Mf.* **8,** 235 (1935).
[2] L. T. Brownmiller, *Proc. Am. Concr. Inst.* **39,** 193 (1943).
[3] P. Terrier and M. Moreau, *Revue Matér. Constr. Trav. publ.* **584,** 129 (1964).
[4] *Proc. Am. Soc. Test. Mater.* **36,** II, 332 (1936).
[5] *Chim. Ind.* **21,** 656 (1939); *Zement* **30,** 43, 56 (1941).

original cement grains, whatever the original water : cement ratio, indicating that its porosity must vary correspondingly. It is possible by very careful polishing to prepare thin sections less than 5 μ thick and in these the reaction rims round the clinker grains can be seen. Brownmiller noted that many of the cement particles were composite and that hydration proceeded by a gradual penetration into the particle from the surface. Except at, or near, the surface, therefore, the hydration of the coarser particles may not be very selective even though the hydration rates of the various cement compounds differ. Some of the interstitial material in the cement grains seemed to hydrate more slowly than the silicate crystals. A content of 10–15 per cent of $Ca(OH)_2$ crystals was observed after 28 days.

As we shall see in the next chapter the use of X-rays and of the electron microscope has enabled various crystalline constituents which cannot be seen under the microscope to be identified.

Rate of hydration of Portland cement

The rate of hydration of Portland cement and of the individual cement compounds may be determined directly by microscopic or X-ray analysis or indirectly by the measurement of some property which is a function of the degree of hydration. Properties such as the amount of combined water, the heat of hydration, the density and the free calcium hydroxide content have been used. The indirect methods can be useful in the study of single compounds or to obtain an average rate of hydration for a cement as a whole, but they do not lend themselves without ambiguity to deductions about the rate of hydration of the individual phases in a cement. These methods involve the implicit assumption that the reaction products remain the same throughout the hydration. This is not strictly true. For example, as we have seen earlier the $CaO : SiO_2$ ratio of the calcium silicate hydrate produced from C_3S or C_2S changes as hydration proceeds. Only the X-ray diffraction method permits of the direct measurement of the amount of each compound remaining after any particular time of hydration, but its precision is not as good as would be desired.

The rate at which the hydration proceeds into the grains was measured microscopically by Anderegg and Hubbell.[1] Their results giving the depth in microns (0·001 mm) to which hydration of grains of 25–30 μ diameter proceeds in various times at 20–25° are shown in Table 29.

TABLE 29 Depth of hydration (microns)

	3 hours	1 day	3 days	7 days	28 days	5 months
$3CaO.Al_2O_3$	4·35	—	5·68	—	5·66	—
$3CaO.SiO_2$	1·68	2·25	—	4·32	4·44	—
$2CaO.SiO_2$ (β)	—	0·28	—	0·62	0·83	3·5
Portland cement A	—	0·43	—	2·60	5·37	8·9
Portland cement B	—	0·47	—	1·71	3·54	6·1
Portland cement C*	—	0·65	1·20	1·80	3·60	—

* Steinherz calculated from heats of hydration.

[1] *Proc. Am. Soc. Test. Mater.* **29** (II), 554 (1929); **30** (II), 572 (1930).

Various authors[1] have also calculated the depth of hydration from estimations of the heat of hydration, the combined water, or the change in density, using particles of a limited size range, and obtained results of the same general order. Thus Butt, calculating from the water combined by particles of 30–55 μ diameter, obtained the values given in Table 30.

TABLE 30 Depth of hydration (microns)

	3 days	7 days	28 days	6 months
$3CaO.Al_2O_3$	10·7	10·4	11·2	15·0
$4CaO.Al_2O_3.Fe_2O_3$	7·7	8·0	8·4	13·2
$3CaO.SiO_2$	3·5	4·7	7·9	15·0
$2CaO.SiO_2$	0·6	0·9	1·0	2·7

Higher rates of penetration for Portland cement, 7·4 μ at 1 day and 20 μ in 7 days, were found by Brownmiller[2] from photographs of polished sections by reflected light, and also by L'Hopitalier and James.[3] Tricalcium aluminate hydrates the most rapidly, followed by the iron compound C_4AF, then tricalcium silicate with dicalcium silicate much the slowest. The Portland cements show intermediate values.

The depth of hydration has also been calculated by Tsumura[4] from X-ray measurements for the initial hydration period during which the reaction was treated as a surface reaction with the depth of penetration proportional to time. The results shown in Table 31 illustrate the rapid reaction of C_3A and C_4AF over the first 10 to 20 minutes. These initial surface reactions are followed by a slower reaction, the rate of which is controlled by diffusion of water through the hydrated layer.

TABLE 31

	C_3A	C_4AF	β-C_2S	C_3S	Alite
Hydration time (hours)	0·18	0·35	4·32	15	20
Depth of hydration at end of surface reaction (microns)	1·3	0·65	0·35	0·4	0·8

Many investigators have studied the relative rates of hydration of the cement compounds by measurement of the amounts of water combined in pastes at various ages, this usually being defined by the content of water, expressed as a percentage of the anhydrous material, retained on drying at about 100°. The data obtained by Bogue and Lerch[5] are shown in Table 32.

[1] A. R. Steinherz, *Revue Matér. Constr. Trav. publ.* **509,** 48 (1958); **536,** 119 (1960); A. Eiger, ibid. **335,** 161 (1937); V. M. Butt, *Zh. prikl. Khim.* **22** (3), 223 (1949).
[2] L. T. Brownmiller, *Proc. Am. Concr. Inst.* **39,** 193 (1943).
[3] P. L'Hopitalier and L. M. James, *Publs tech. Cent. Etud. Réch. Ind. Liants hydraul.*, No. 16 (1948).
[4] I. S. Tsumura, *Zement-Kalk-Gips* **19** (11), 511 (1966).
[5] R. H. Bogue and W. Lerch, *Ind. Engng Chem.* **26,** 837 (1934).

TABLE 32 Per cent water retained after heating to 100°

Compound	Percentage of gypsum added	1 day	7 days	28 days	6 months	2 years
$3CaO.SiO_2$	0	8·6	12·0	14·6	15·0	18·6
$3CaO.SiO_2$	5	11·7	12·9	15·0	15·1	18·9
β-$2CaO.SiO_2$	0	0·8	1·7	3·4	10·8	12·1
β-$2CaO.SiO_2$	5	1·2	1·8	3·8	8·7	11·6
$3CaO.Al_2O_3$	0	30·2	32·4	34·9	36·2	38·5
$3CaO.Al_2O_3$	15	25·8	29·3	29·6	32·3	39·0
$4CaO.Al_2O_3.Fe_2O_3$	0	25·7	27·1	27·9	28·8	30·1
$4CaO.Al_2O_3.Fe_2O_3$	15	22·0	24·0	25·6	30·0	31·9

As will be seen from these data the presence of gypsum increases somewhat the amount of hydration of tricalcium silicate at one day and Jacquemin[1] found a similar acceleration with a Portland cement. In either case the effect at 28 days was negligible.

A comparison of the degrees of hydration at room temperatures derived from three investigations[2] of combined water content for the silicate compounds without gypsum is shown in Figs. 60 and 61, together with two sets of similar

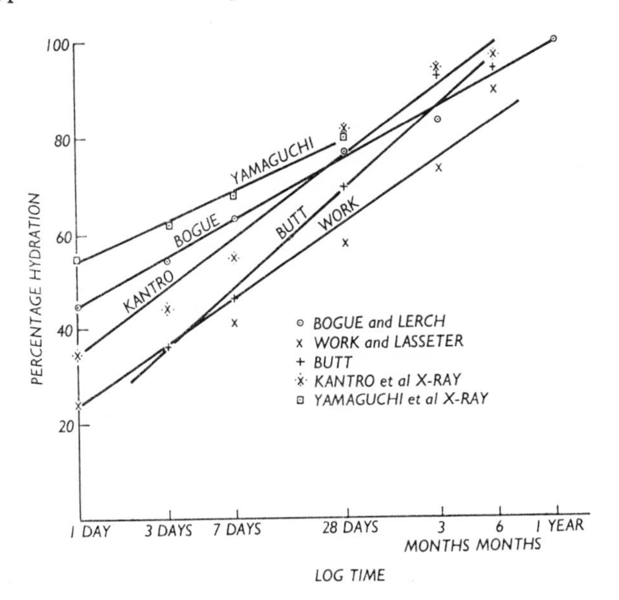

FIG. 60 Rates of hydration of $3CaO.SiO_2$ pastes.

[1] R. Jacquemin, *Bull. Cent. Etud. Rech. Essais scient. Genie civ., Liége* **11,** 25 (1947).

[2] Data of Bogue and Lerch, loc. cit.; T. L. Work and F. P. Lassiter, *Concrete* **38** (3) 81, (4) 89, (5) 79 (1931), as recalculated by S. Brunauer and S. A. Greenberg, *Sym., Washington 1960*, 135; and V. M. Butt (loc. cit.).

data obtained by X-ray diffraction measurements[1] of the unhydrated material remaining at various ages.

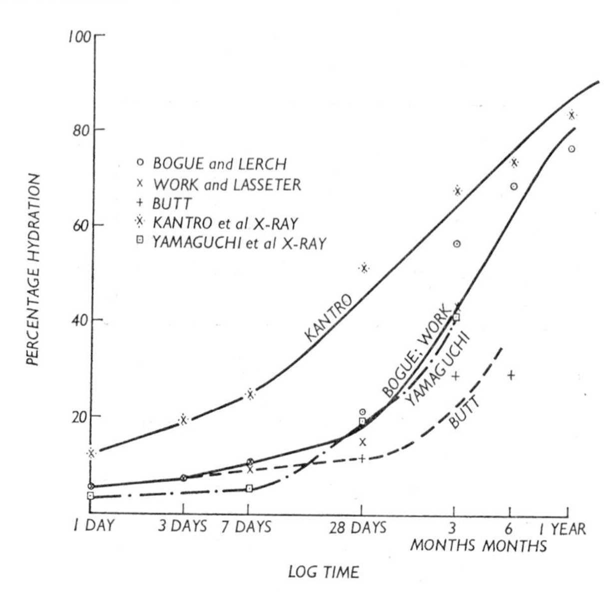

FIG. 61 Rates of hydration of $2CaO.SiO_2$ pastes.

While combined water contents do not afford a strict comparison between the degrees of hydration of the different compounds they do place them in the same order of reactivity as the microscopic measurements or heat of hydration data (p. 295). The X-ray diffraction measurements on the rate of hydration of $3CaO.SiO_2$ give results in the same general range, but there are wider differences for $2CaO.SiO_2$ for which the rates are likely to be affected by the stabiliser used to keep the $2CaO.SiO_2$ in the β form.

A close agreement between these various data is hardly to be expected not only because of differences in fineness of the samples tested and other variables, but also on account of the assumptions implicit in the methods. Indeed the data available in the literature as a whole show still wider variations in the figures obtained for percentage hydration at various ages.[2] For C_3S the figures vary from 15 to 55 per cent hydration at one day, 35 to 60 per cent at 3 days and 40 to 70 per cent at 7 days. The range for 2CaO.SiO is from negligible to 20 per cent at 3 days, 8 to 25 per cent at 7 days, 11 to 50 per cent at 1 month, and 20 to 75 per cent at 3 months.

The rate of hydration of the individual compounds in commercial cements can be derived from heat-of-hydration data (p. 295) or determined directly by X-ray

[1] D. L. Kantro, S. Brunauer and C. H. Weise, *Adv. Chem. Ser.* **33**, 199 (1961); G. Yagamuchi et al., *Sym., Washington 1960*, 135.

[2] Summarised in part by Yu. M. Butt, V. M. Kolbasov and V. V. Timashev, *Sym., Tokyo 1968* and S. Brunauer and S. A. Greenberg (loc. cit.).

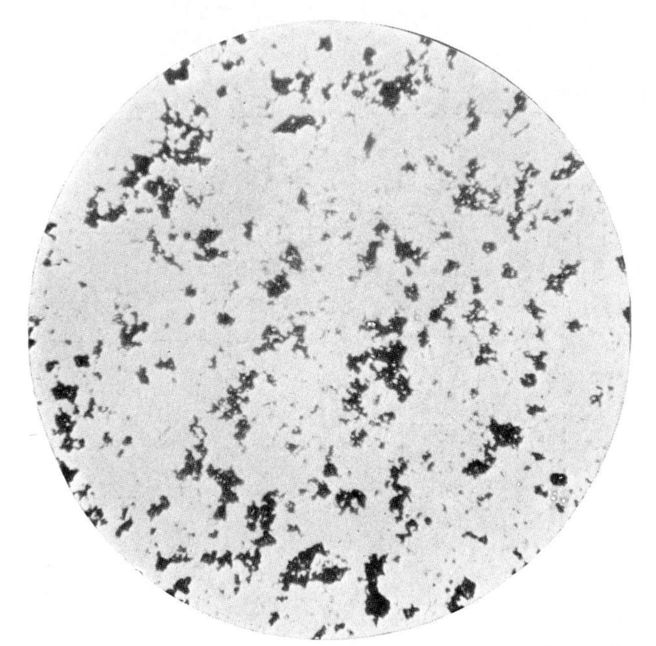

(i) PORTLAND CEMENT IMMEDIATELY AFTER MIXING WITH WATER (× 70)

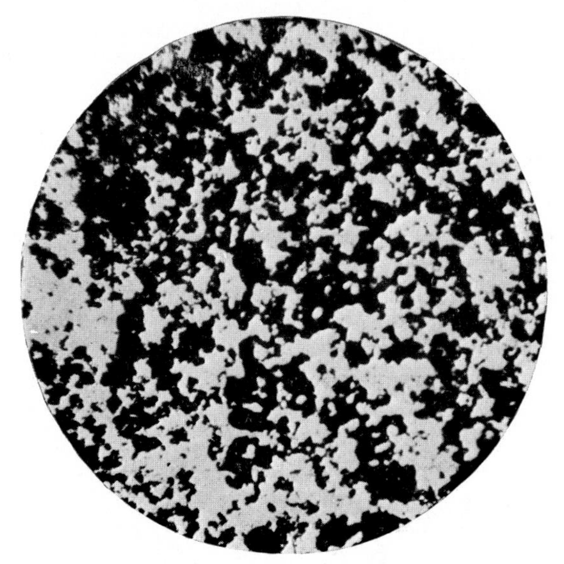

(ii) PORTLAND CEMENT 24 HOURS AFTER MIXING WITH WATER (× 70)

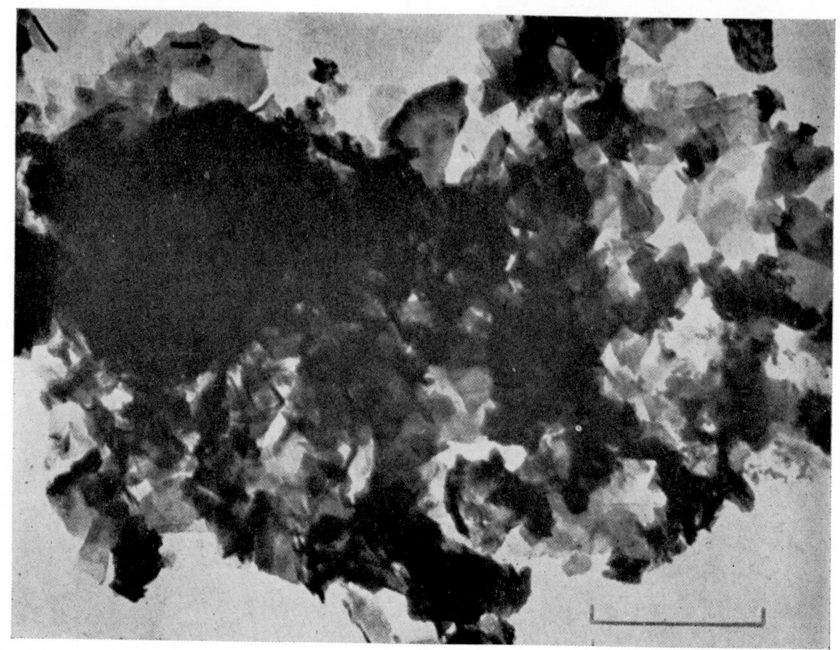

(i) ELECTRON MICROGRAPH OF $CaO.SiO_2.2.5H_2O$ (CSH(I)) ($\times$ 10,000)

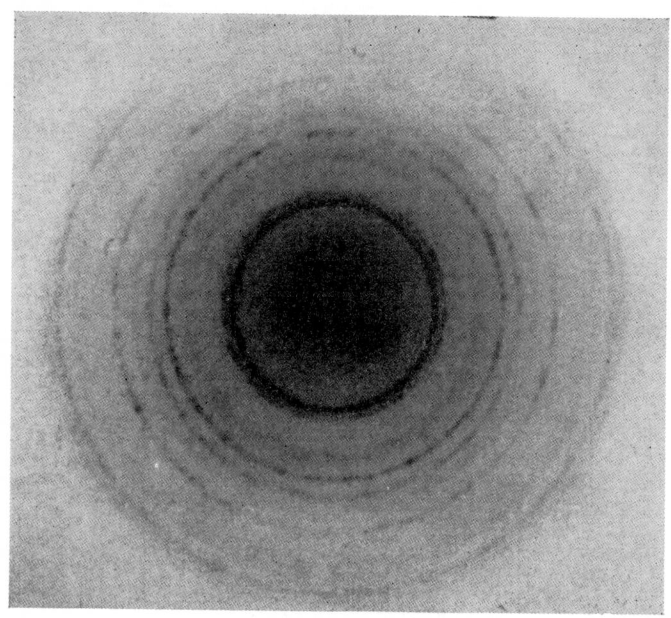

(ii) ELECTRON DIFFRACTION PATTERN OF CRYSTALS IN (i). GRUDEMO

diffraction measurements. Results[1] on pastes (w/c 0·40) of an ASTM Type I cement by X-ray methods for the four major mineral constituents are shown in Fig. 62. The alite and belite show a more rapid rate of hydration than the

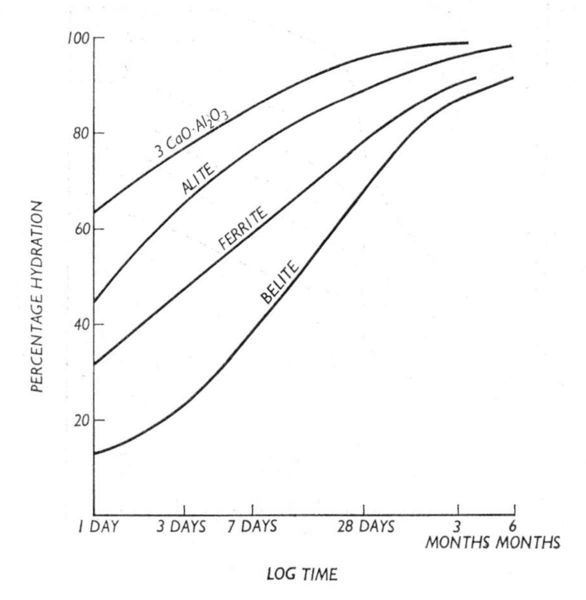

FIG. 62 Rates of hydration of compounds in a Portland cement.

$3CaO.SiO_2$ and $2CaO.SiO_2$ (Figs. 60 and 61) tested by Kantro, using the same method. Hurley and Angstadt[2] found very similar rates of hydration for alite in a Type I cement but no conclusive evidence for any appreciable hydration of belite up to 3 days, again illustrating the uncertainty attaching to the hydration rate of this compound. Significantly different curves were found by Copeland and Kantro with a Type IV cement. In particular, the ferrite compound, with its lower $Al_2O_3 : Fe_2O_3$ ratio, hydrated more slowly than in a Type I cement reaching only about 35 per cent hydration in 28 days and 60 per cent at 3 months. The rate of hydration increases with the water-cement ratio[3] and with the fineness of the cement.

The form of the curve for loss of water on heating and the progress of hydration is illustrated in Figs. 63 and 64 which are based on data[4] on set cement pastes (w/c 0·22) stored in lime water for various periods after gauging. The curves in Fig. 63 show the weight loss (as a percentage of the 110° weight) when the ground set cement, dried at 110°, is heated at various temperatures in a slow stream of carbon dioxide-free air dried over P_2O_5. Results for a number of

[1] L. E. Copeland and D. L. Kantro, *Sym., Washington 1960*, 443. Broadly similar results were also reported by Yamaguchi, ibid., 495.
[2] F. R. Hurley and R. L. Angstadt, *J. appl. Chem.* **16** (5), 162 (1966).
[3] J. H. Taplin, *Aust. J. appl. Sci.* **10**, 329 (1959); *Sym., Washington 1960*, 465.
[4] F. M. Lea and F. E. Jones, *J. Soc. chem. Ind., Lond.* **54**, 63T (1935). See also H. F. W. Taylor, *Proc. 27th Congr. ind. Chem., Brussels 1954*, **3**, 63.

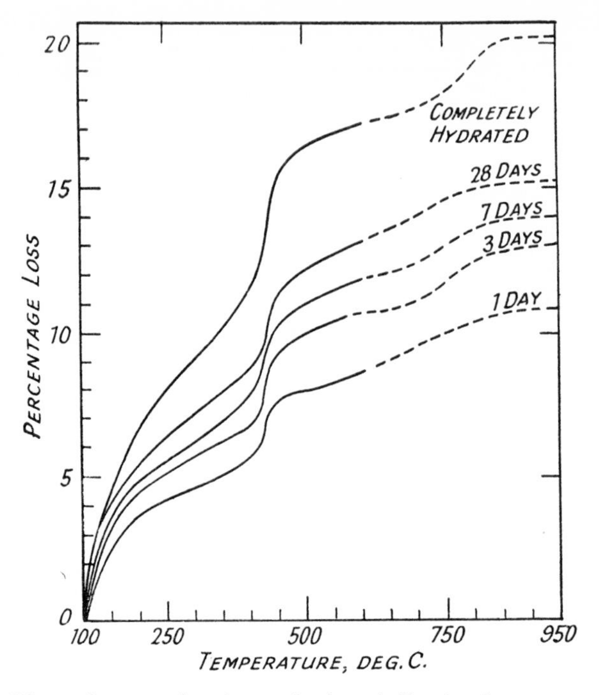

FIG. 63 Water lost on heating a hydrated Portland cement **at various** temperatures.

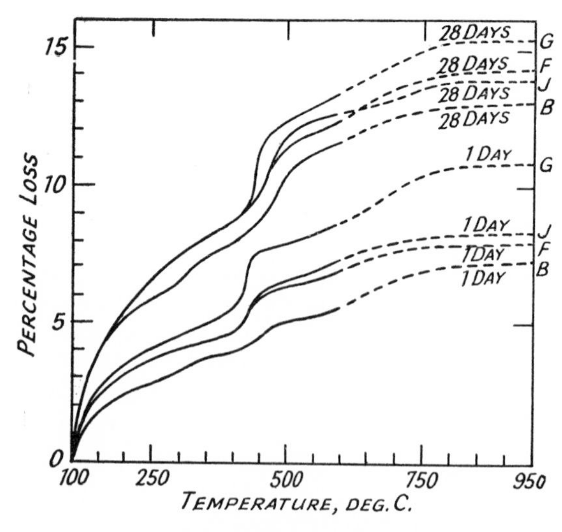

FIG. 64 Water lost on heating various hydrated Portland cements (aged 1 and 28 days) at various temperatures.

cements varying in composition and fineness are shown in Fig. 64. The amount of water retained at 110° rises as the proportion of water used in gauging the cement increases, but it approaches a limit at a w/c value of about 0·25 at 1 day, 0·35 at 3 days, 0·5 at 7 days, and progressively higher values as the age increases further. Combined water contents can also be determined by the measurement of non-evaporable water (p. 271). In a completely hydrated cement the amount of water retained at 110° is 20–25 per cent by weight of the anhydrous cement.

The rate of development of strength of Portland cements at the earlier ages increases with temperature and this is associated with an increased rate of hydration. Curves for combined water content (measured by the loss on drying between 125° and 540°) of an ordinary (Type I) cement are shown[1] in Fig. 65.

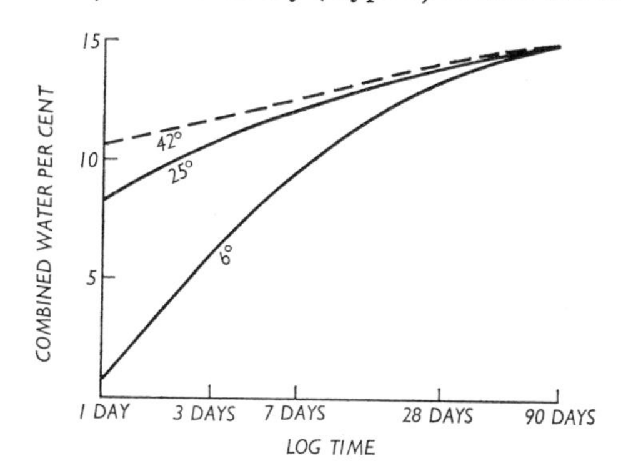

FIG. 65 Effect of temperature on rate of hydration of Portland cement.

Broadly similar curves were obtained by Carlson and Forbrich[2] from measurements of the effect of temperature on heats of hydration. Marked differences appear in the response to temperature of the silicate compounds, as may be seen from the curves[3] in Fig. 66, which were based on X-ray determinations of the unhydrated silicates. The hydration of tricalcium silicate is influenced little by temperature at early ages. In contrast dicalcium silicate shows a large effect of temperature on hydration rate, tailing off after some months. Correspondingly, Taplin found that with increasing dicalcium silicate content in cements the accelerating influence of temperature was extended to longer ages. The tendency for both silicates to show an actual decrease in hydration at long ages as the temperature is raised affords at least a partial reason for the failure of steam-cured concretes to reach as high an ultimate strength as those cured at normal temperatures. While no complete explanation is yet available of these effects of temperature on hydration rates there is evidence[4] to suggest that it arises from a change in the lime-silica ratio of the hydration product as hydration proceeds (see p. 180), and in the thickness and permeability of the gel coatings that are

[1] J. H. Taplin, *Aust. J. appl. Sci.* **13** (2), 164 (1962).
[2] R. W. Carlson and L. R. Forbrich, *Ind. Engng Chem. analyt.* **10** (7), 382 (1938).
[3] D. L. Kantro, S. Brunauer and C. H. Weise, *Adv. Chem. Ser.* **33,** 199 (1961); *J. phys. Chem.* **66** (10), 1804 (1962).
[4] Kantro et al., loc. cit. (1962).

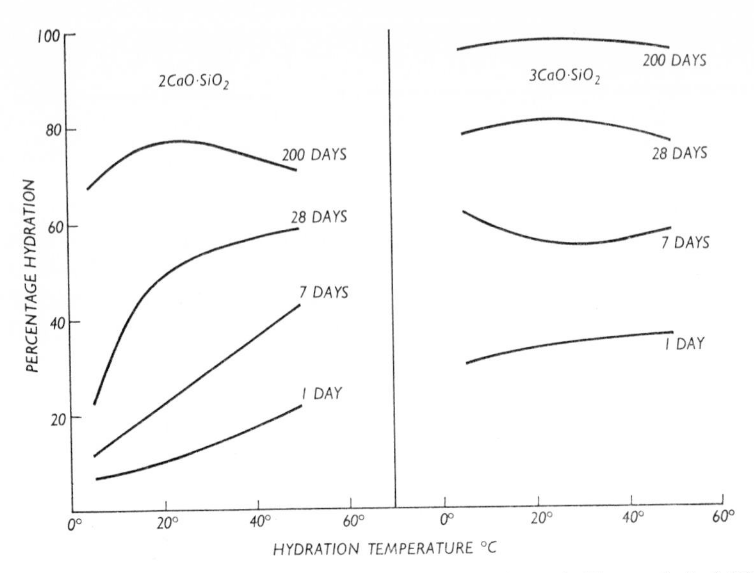

FIG. 66 Effect of temperature on rate of hydration of silicates (w/c 0·70).

formed. However not all the data available are consistent. Thus Schwiete and Muller-Hesse[1] found positive temperature coefficients up to 65° for C_3S and C_2S at degrees of hydration up to 90 per cent. Investigators[2] in the U.S.S.R. have also reported progressive increases in hydration rate for both the individual cement compounds and Portland cement up to 90° with a positive temperature coefficient of some 10–20 per cent for each 10° rise in temperature.

There have been numerous suggestions for reaction-rate equations to represent the hydration-time relations. Zur Strassen[3] has examined two primitive assumptions that could be made. The first is that diffusion through the coating of reaction products is so rapid that the reaction rate is controlled by the solid-water reaction. If the thickness of the hydrated layer is called y, then

$$dy/dt = k \quad \text{or} \quad y = kt \tag{I}$$

The second assumption is that the solid-liquid reaction is so fast that the rate is controlled by diffusion of water through the hydrated layer. Assuming this diffusion rate to be inversely proportional to the thickness of the layer we can write

$$dy/dt = (1/y)\,k \quad \text{or} \quad y^2 = 2kt$$

$$\text{and} \quad y = k't^{\frac{1}{2}} \tag{II}$$

Zur Strassen found that the hydration of $2CaO.SiO_2$ followed equation I up to

[1] H. E. Schweite and H. Muller-Hesse, *Zem. Bet.* **16,** 25 (1959).
[2] P. P. Budnikov et al., *Dokl. Akad. Nauk SSSR* **16,** 25 (1959); **148** (1), 91 (1963); *RILEM International Conference on Accelerated Hardening of Concrete, Moscow 1964*: T. M. Berkovitch, *Dokl. Akad. Nauk SSSR* **149** (5), 1127 (1963); *RILEM Conference on Accelerated Hardening of Concrete, Moscow 1964.*
[3] H. zur Strassen, *Zem. Bet.* **16,** 32 (1959).

about 120 days and that of $3CaO.SiO_2$ equation II up to about 30 days, but Taplin[1] was able to fit data on the dicalcium silicate also to equation II. Tsumura[2] from X-ray data also found that the hydration of the individual cement compounds could be divided into two stages following equations I and II respectively. The time of changeover from a surface reaction rate was found to be under an hour for C_3A and C_4AF, 15 hours for C_3S and 18 days for C_2S. These times are much less than those obtained by Zur Strassen.

Berkovitch[3] has also treated the hydration of cement after the first few hours as an heterogeneous diffusion process and expressed this by a first-order equation

$$dC/dt = -kC$$

where C represents the amount of unhydrated material present. This gives

$$-\log_{10} C = 0{\cdot}4343\ kt + \text{constant}$$

Plotting log C against t gave two straight lines of different slope with a changeover point at 1 to 3 days when hydration had proceeded 30 to 50 per cent. The rate constant before the changeover was 10–15 times higher than the subsequent rate and this was attributed to a change in the structure and adherence of the gel coatings. An empirical equation fitted to the data by Budnikov[4] is

$$L = k \log t - B$$

where L is the extent of hydration and B is a constant defined as the initial induction period of a few hours before the main hydration starts. This equation was found to give a linear relation up to 180 days for the tri- and di-calcium silicates. Various other equations have also been proposed.[5]

Perhaps the only valid conclusions that can be drawn at present is that the precision of the available data is not sufficient to afford any critical test of these various relations.

Estimation of calcium hydroxide in set Portland cement

The qualitative determination of the calcium hydroxide in set cements presents no difficulties, but the quantitative determination of the amount present is more uncertain. The solvent-extraction methods as used for the estimation of free lime in cements (see p. 108) are not satisfactory, because it is difficult to ensure that the calcium hydroxide is dissolved completely without at the same time removing some lime from the hydrated cement compounds, especially from the sulphoaluminates. Of the various solvents available, the acetoacetic ester–*iso*butyl alcohol mixture of Franke seemed the least open to objection, for Assarsson[6] found that this solvent only attacked hydrated calcium aluminates and sulphoaluminates to a small extent. It also had only a slight action on hydrated calcium silicate prepared in high-pressure steam and this was confirmed by Kalousek.[7]

[1] J. H. Taplin, *Sym., Washington 1960*, 263.
[2] T. Tsumura, *Zement-Kalk-Gips* **19** (11), 511 (1966).
[3] T. M. Berkovitch et al., loc. cit.
[4] Loc. cit.
[5] See Yu. M. Butt, V. M. Kolbasov and V. V. Timashev, *Sym., Tokyo. 1968*
[6] *Zement-Kalk-Gips* **7,** 167 (1954).
[7] *Sym., London 1952*, 302.

There is, however, evidence that it has more action on the hydrated silicates produced at ordinary temperatures for Kalousek[1] observed that while the hydrate *CSH* (I) with a CaO : SiO_2 ratio up to 1·33 was unattacked there was some action on the hydrates of higher lime content. The method can be modified by the addition of 5 ml ethyl ether to the 20 ml *iso*butyl alcohol–3 ml acetoacetic ester solvent. This reduces the boiling point and the extraction can be carried out at 70°, so reducing the risk of attack on the hydrated cement compounds. Even so, Brunauer and his co-workers[2] found that a single extraction did not remove all the free calcium hydroxide and with successive extraction periods of three hours a progressive attack occurred on samples of *CSH* (I) with a CaO : SiO_2 ratio of 1·5, though there was no action on afwillite. On the basis of the rate of removal of the free lime being much greater than that of the combined lime, a multiple extraction technique was then developed to correct for the decomposition of the hydrated calcium silicates.

In later work[3] other modifications of the acetoacetic ester–*iso*butyl alcohol extraction method were examined for the determination of free lime in pastes of calcium silicates and set cements, and procedures were developed involving either variation in the solvent to sample ratio or variation in the time of extraction of separate samples. In both cases, extrapolation of the data to zero solvent volume or zero extraction time, respectively, was used to obtain a measure of the free calcium hydroxide, and on a given sample practically the same result was obtained with either procedure. Additional refinements to the time variation method of extraction, resulting in increased precision of individual determinations, have been described by Brunauer, Kantro and Weise.[4] In this revised procedure, 0·2 g of the dried solid is refluxed with 100 ml of a 3 to 10 (by volume) mixture of acetoacetic ester and *iso*butyl alcohol in the presence of 0·25 ml of a solution of 1 g NaOH in 100 ml of absolute ethyl alcohol. Extractions are carried out on separate samples for four different periods up to about 2 hours, and blank determination is made for each set to correct for the added alkali. After extraction for the required time, the mixture is cooled and rapidly filtered on a sintered glass funnel. The residue is washed with 50 ml *iso*butyl alcohol and the solution is titrated with a standard 0·2 N solution of perchloric acid in *iso*butyl alcohol in the presence of methylene blue–thymol blue screened indicator until the colour changes from light green to dark blue.

The calcium hydroxide content of various hydrated cement products has also been determined by X-ray quantitative analysis.[5] This method usually gives lower values for the free calcium hydroxide than those obtained by solvent extraction, and the discrepancy may be mainly attributed to the presence of some non-crystalline, or X-ray amorphous, calcium hydroxide. Thus, Brunauer,

[1] *Proc. Am. Concr. Inst.* **51,** 989 (1955).
[2] S. Brunauer, L. E. Copeland and R. H. Bragg, *J. phys. Chem.* **60,** 112, 116 (1956); S. Brunauer, D. L. Kantro and E. E. Pressler, *Analyt. Chem.* **28,** 896 (1956).
[3] E. E. Pressler, S. Brunauer, D. L. Kantro and C. H. Weise, *Analyt. Chem.* **33,** 877 (1961).
[4] *Spec. Rep. Highw. Res. Bd*, No. 90, 324 (1966).
[5] S. Brunauer, D. L. Kantro and L. E. Copeland, *J. Am. chem. Soc.* **80,** 761 (1958); N. Fratini and G. Schippa, *Ricerca scient.* **30,** 2030 (1960); E. E. Pressler, S. Brunauer, D. L. Kantro and C. H. Weise, loc. cit; H. Lehmann, F. W. Locher and D. Prussog, *Tonind.-Ztg. keram. Rdsh.* **94,** 230 (1970).

Kantro and Copeland found that after a preliminary extraction with a solvent which would be expected to remove any amorphous calcium hydroxide the X-ray method then gave results in agreement with those obtained by extraction. Further support for the occurrence of amorphous calcium hydroxide is provided by the higher solvent-extraction values as compared with those of X-ray analysis for preparations of afwillite obtained by hydration of C_3S in a ball mill.[1] Indeed, it is very probable that in many hydrated materials the calcium hydroxide will appear partly in a crystalline form and partly in a finely-divided or amorphous form. Since these two forms will be expected to show a different behaviour on heating, there is, therefore, some doubt with regard to the values for uncombined lime which can be obtained by calorimetric determination,[2] dynamic differential calorimetric determination,[3] thermogravimetric determination[4] and differential thermal analysis.[5] Some corrections in these methods may also be required for recombination of lime with hydrated cement compounds during heating.

[1] S. Brunauer, D. L. Kantro and C. H. Weise, *J. Colloid Sci.* **14,** 363 (1959).
[2] G. E. Bessey, *Building Research Tech. Pap. D.S.I.R.*, No. 9 (1930); *Special Report,* No. 17 (1931).
[3] H. E. Schwiete and G. Ziegler, *Ber. dt. keram. Ges.* **35,** 193 (1958); H. G. Kurczyk and H. E. Schwiete, *Tonind.-Ztg. keram. Rdsh.* **84,** 585 (1960).
[4] A. van Bemst, *Industr. Chim. Belge* **20,** 67 (1955); R. Sersale, *Industria ital. Cem.* **25,** 237 (1955); J. A. Forrester, *Monogr. Soc. chem. Ind.*, No. 18, 407 (1964).
[5] H. G. Midgley, *Sym., Washington 1960*, 479; G. L. Kalousek, *Mater. Res. Stand.* **5,** 292 (1965).

10 The Setting and Hardening of Portland Cement

The practical utility of all calcareous cements depends on the power which they possess, when mixed with water, of forming a coherent mass which in course of time becomes hard and mechanically resistant. Cements are employed in the form of a powder, and, when mixed with an appropriate quantity of water and some suitable aggregate, yield at first a plastic mass, capable of being spread over a surface as in the construction of brickwork and masonry, or of filling and taking the shape of some confined space as in concrete work. The passage to the hard condition in cements takes place in two stages. In the first, the mass loses its plasticity and becomes more or less friable, so that if it is moulded or remixed with water the plasticity is not restored, or is restored only partially. In the second, consolidation takes place, the mass increasing in hardness until a stony texture is finally obtained. In the case of hydraulic cements the change is accompanied by the almost complete disappearance of the permeability for water. These two stages are distinguished as the processes of setting and hardening. In certain cementing materials such as plaster no such division is observed and the setting and hardening become practically a single process.

The mechanism of these changes is not the same in all cementing materials. The production of a coherent mass from a powdered solid and water may take place in several different ways, such as the following:

1. The crystallisation of a substance from a supersaturated solution producing a mass of interlaced crystals.
2. The formation of a semi-solid gel.
3. A chemical reaction between two or more substances in presence of water giving rise to either crystalline or colloidal products.
4. The transformation of a metastable compound to a more stable form.

Plaster

The setting of gypsum plaster affords an excellent example of a set due to crystallisation from a supersaturated solution. Lavoisier in 1765 showed that the setting of plaster of Paris was due to the recombination of the calcium sulphate with the water of crystallisation of which it had been deprived by heat, the hydrated salt then forming a confused mass of fine needle-shaped crystals, the intergrowth of which gave strength to the hardened mass. When lightly calcined, as in the ordinary open-kettle process, gypsum, which has the composition $CaSO_4.2H_2O$, loses three-quarters of its water of crystallisation, forming the hemihydrate $CaSO_4.\frac{1}{2}H_2O$. The hemihydrate has about five times the solubility of gypsum and, when mixed with water, forms a supersaturated solution which tends to deposit crystals of the dihydrate; the deposition begins at some point and the series of changes repeats itself, the hemihydrate dissolving, becoming further hydrated, and then depositing crystals of dihydrate. In the ordinary use of plaster the quantity of water used is so small that the particles are in close contact and the crystals produced form a confused and dense mass, the strength of which is due to the interlocking of the radiating needle-growths. The crystal growth is, however, so fine that under the microscope definite crystals cannot be observed, at least for some weeks, in a hardened mass. If, instead of water, dilute alcohol is used to gauge the plaster, the setting is retarded and the crystals develop better and can be observed in the set mass. Crystals of appreciable size can also be obtained by mixing plaster with an excess of water.

The formation of needle crystals can also be observed on a microscope slide in plaster mixed with water. Under such conditions where the proportion of liquid to solid is large the crystals do not become interlocked and the gypsum formed is incoherent. The presence of a strictly limited quantity of water, such that the crystal nuclei are formed in close contact with one another, is necessary for hardening. The quantity of water generally used in mixing is only capable of dissolving about one-thousandth part of the calcium sulphate with which it is in contact, but by the process of dissolving the hemihydrate and depositing the dihydrate it is able to effect the conversion of the whole.

Taking the specific gravity of the hemihydrate as 2·75 and the dihydrate as 2·32, it can be calculated that the setting is accompanied by a contraction of about 7 per cent of the initial volume of plaster and water. This seems at first sight contrary to experience, as the utility of plaster in the preparation of casts is known to depend on its power of filling a mould by expansion and of taking a sharp impression of its surface. But it may well happen owing to crystal thrust that, though the total volume of solid and water present decreases, the effective space taken up by the moist mass of interlocking crystals is actually increased while cavities are left in the interior. The overall expansion of a plaster mass on setting is about 0·5 per cent. It can be modified by the presence of retarders or accelerators.

Lime mortars

The hardening of a high-calcium lime mortar is totally different from that of plaster. The hydration of quick lime is completed in the process of slaking before the mortar is mixed. When the reaction $CaO+H_2O = Ca(OH)_2$ takes place, the quantity of water present is only sufficient to dissolve a small part of the lime and not to allow of crystallisation. Crystals of calcium hydroxide are indeed found in many specimens of hardened mortar, but they owe their formation to subsequent solution in water and redeposition and are not a primary product of hydration. During slaking we must imagine the lime to form a solution with water which is highly supersaturated and has only a momentary existence. Under such conditions the product of the reaction takes a colloidal or amorphous form. When allowed to dry, hardening of the mortar occurs, but relatively little development of strength is found. On exposure to air for any length of time carbon dioxide is absorbed and a part of the calcium hydroxide is converted to calcium carbonate.

This reaction confers considerable strength on lime mortars. This is illustrated by the following data[1] on the strengths developed by lime mortars stored in carbon dioxide-free air and in carbon dioxide respectively.

Tensile strength 1 : 3 (weight) hydrated lime–sand mortars

Relative humidity		100 per cent		50 per cent	
Storage		Air	CO_2	Air	CO_2
Tensile strength Lb/in^2	7 days	36	42	22	169
	28 days	48	56	19	167

The presence of some moisture is essential to the carbonation process, as carbon dioxide is without action on dry lime. When, however, the pores of the mortar are full of water, as in a mortar kept in saturated air, carbonation is slow; it proceeds most readily in mortars from which a large proportion of the water has evaporated, as for instance by storage in atmospheres of reduced humidity. The calcium carbonate formed is crystalline and the interlacing of the slowly growing crystals binds the whole mortar into a hard and coherent mass.[2] This process is confined to the outer layers of a mass of mortar, as it produces an impervious coating hindering the access of further carbon dioxide to the interior. Mortar taken from buildings many hundred years old, if uninjured, is found to consist mainly of calcium hydroxide, only the external portion having been converted to carbonate.

An example of two materials in solution which when mixed together under suitable conditions will produce a coherent mass is afforded by the sodium silicate–calcium chloride cementation process. This is applied to the consolidation *in situ* of sands and gravels by forcing these two solutions alternately into the mass. The reaction products are sodium chloride and a solid, gelatinous mass of hydrated calcium silicate and silica gel which form the cementing agent.

An example of the formation, by the chemical interaction of two substances, of a crystalline compound leading to the production of a hard mass, is to be found in magnesium oxychloride cement.

An example of setting produced by an inversion of one crystalline form to another is provided by calcium carbonate of which two forms, calcite and aragonite, exist. Both have very low solubilities in water but that of aragonite, the unstable form, is slightly the greater. A paste of very finely ground aragonite and water has been found to set very slowly as conversion to calcite occurs.[3]

[1] G. E. Bessey, *J. Soc. chem. Ind., Lond.* **52,** 287T (1933).
[2] P. Ney, *Zement-Kalk-Gips* **20,** 429 (1967).
[3] N. Ilchenko and H. Lafuma, *Chem. et Ind.* **38** (3), 438 (1937).

Theories relating to Portland cement

The setting and hardening of Portland cement is far more complex than that of lime or plaster. This complexity arises not only from the presence of several different compounds all of which undergo hydration, but also from the physical nature of their hydration products. Various theories have been advanced to explain the hardening of cements.[1] The first and oldest is the crystalline theory put forward in 1882 by Le Chatelier[2] which ascribes the development of cementing action to the passage of the anhydrous cement compounds into solution and the precipitation of the hydration products as interlocking crystals as occurs, for instance, in the setting of gypsum plasters. In the colloidal theory first put forward by W. Michaelis, Sr.,[3] cohesion was considered to result from the precipitation of a colloidal gelatinous mass which hardened as it lost water by external drying or by 'inner suction' caused by hydration of the inner unhydrated cores of the grains.

Many insoluble substances when precipitated from excessively supersaturated solutions give a product which, at least initially, takes a colloidal form. Thus barium sulphate, when precipitated by the slow mixing of dilute solutions of barium chloride and sulphuric acid, forms a finely crystalline product, but if strong solutions are rapidly mixed a gelatinous material is obtained. A truly amorphous material is one in which the molecules of the solid are not orientated in the regular manner that typifies the crystalline state. Colloidal behaviour, however, depends essentially on the state of division of a material and its existence in such a finely divided form that it has a very large surface area. It does not necessarily presuppose a non-regularity of the internal structure of the ultimate particles. Many gels are, in fact, built up of crystalline elementary constituents.

Gelatinous bodies are divided into elastic, or non-rigid, and rigid gels. The former, such as gelatine, swell on the addition of water and finally pass into solution, while on drying the gel is reformed. The rigid gels, of which silica gel is a typical example, do not show this reversible behaviour and once formed will not disperse again to a solution on the addition of water. Set cement on the colloid theory is somewhat analogous to silica gel which indeed reproduces some of its properties on an exaggerated scale.

A silica gel freshly prepared by adding hydrochloric acid to sodium silicate solution is unstable and begins to shrink, expelling water from the mass. During this process, which is known as syneresis, free liquid enmeshed in the gel is spontaneously expelled. Syneresis comes to an end when the water content of the

[1] For a review of much of the older literature, see R. H. Bogue, *Rock Prod.*, May to September 1928, and H. Kühl, *Cement Chemistry in Theory and Practice*, translated by J. W. Christelow, Concrete Publications, London.

[2] *Compt. Rend.* **94**, 13 (1882); *Experimental Researches on the Constitution of Hydraulic Mortars*, translated by J. L. Mack, McGraw, New York, 1905; *Trans. Faraday Soc.* **14**, 10 (1919).

[3] *Chemikerzeitung* **17**, 982 (1893); *Prot. Ver. deut. Portland Cement Fabrikanten* **30**, 199 (1907); **32**, 206, 243 (1909); translations by W. Michaelis, Jr., published by *Cem. Engng News*, Chicago, 'The Hardening Process of Hydraulic Cements,' and 'The Hardening of Cement under Water,' 1909.

gel is still very high, about 90 per cent. If the gel is then allowed to dry it continues to shrink until a point, known as the opacity point, at which the clear gel begins to come opaque, is reached. The path to this point is entirely irreversible and represents a continuation of the coagulation process. Beyond the opacity point the gel ceases to shrink appreciably with further removal of water, the capillary pores begin to empty, and the vapour pressure to fall. This further dehydration path beyond the opacity point is, except for a hysteresis effect, a reversible one and can be retraced by exposure to a moist atmosphere. It is this stage of the dehydration of silica gel, which is a non-crystalline material built up of silicon-oxygen tetrahedra linked in a random manner, that presents many similarities to the behaviour of set cement.

The theories of Le Chatelier and Michaelis were developed later by Baikov[1] who distinguished three periods in the setting of Portland cement, first a solution stage, second a colloidal stage and third a crystallisation stage. The colloidal stage was attributed to direct formation, followed by dispersion, of solid hydrated reaction products without intermediate solution. More recent work by Rebinder and many other U.S.S.R. investigators[2] has modified and extended this theory.

Rebinder's approach to the mechanism of the setting and hardening of cement stems from his general theory[3] of coagulation in suspension of a finely divided solid dispersed in a liquid. Two types of three-dimensional networks can be formed, a coagulation network and a crystalline network often referred to as a 'condensation-crystallisation' network.

In a coagulation network collisions occurring between colloidal particles brought into contact by Brownian movement lead to cohesion at favourable points of contact. Very thin films of the liquid medium remain along the microsurfaces of contact much weakening the molecular interaction and cohesion which arises from physical adsorption forces—the van der Waals' forces—which are much weaker than most chemical bonds. The coagulation centres are relatively few and localised at corners and edges of particles since it is in regions of maximum curvature that the liquid layers are most easily displaced. The network formed throughout the suspension—or paste—is weak and possesses marked plasticity. It is readily disrupted but capable of reforming. A characteristic of such networks is therefore that they exhibit thixotropic properties. The plastic strength of the network is determined by the number of contacts which in turn depends on the number of colloidal particles per unit volume.

A crystalline network is formed when the contacts between microcrystals are points of direct junction with no intervening liquid film and the cohesive forces are strong chemical valency bonds. Silica gel in which silicon-oxygen tetrahedra are linked in a random manner is an example of a three-dimensional 'condensed-

[1] A. A. Baikov, *C.r. hebd. Séanc. Acad. Sci., Paris*, 1926, **182,** 128 and numerous U.S.S.R. papers.

[2] *Sym., Moscow 1956*, P. A. Rebinder, 125; E. E. Segalova and E. S. Soloveva, 138; S. D. Okorov, 173; M. I. Strelkov, 183; E. E. Segalova, P. A. Rebinder and U. I. Lukyanova, *Vest. mosk. ges. Univ.* **9** (2), *Ser. Phys-Maths and Nat. Sci.*, No. 1, 17 (1954).

[3] P. A. Rebinder, *Discuss. Faraday Soc.*,1954, No. 18, 151; P. Rebinder and E. Segalova, *Proc. int. Congr. Surface Activity, London 1957*, **3,** 492; P. Rebinder, *Izv. Akad. Nauk SSSR*, Otdel. Khim. Nauk, 1957, 1284.

crystallisational' structure and is a typical rigid gel. Crystalline networks are formed through crystallisation from a supersaturated solution giving a conglomerate of fine crystals. The strength of the network is determined by the number per unit volume of the crystals and their size and strength and is of a higher order than that of a coagulational network. The structure is irreversible and if broken up does not reform as such, unless the solution remains supersaturated and further crystallisation can occur, in contrast to the coagulation network which can reform by forces of physical adsorption. A broken-up crystalline network can form a coagulation network by the operation of these adsorption forces but with a much lower strength.

The stability of the cement suspension, the duration of the period before the onset of the formation of coagulation or crystallisation networks, and the conditions under which these are formed provide the basis of the theories advanced by Rebinder and other U.S.S.R. workers to explain not only the setting and hardening of cements but also the effect of additions, such as surface active agents or electrolytes, of more prolonged mixing or mechanical action such as vibration, and of other factors that influence the setting and hardening processes.

In the earlier form of these concepts emphasis was laid also on the effect of the liquid media, water, in causing dispersion, or peptisation, of the anhydrous cement particles by adsorption at crystal defects and microcracks where the cohesive forces in the solid are low. This 'colloidisation' of the anhydrous cement was considered to be reinforced by the initial chemical hydration at similar weak points in the anhydrous cement minerals, creating internal stresses that caused rupture. Less emphasis has been laid in the more recent U.S.S.R. work on this dispersion of the anhydrous cement by water, as distinct from that of the hydration products, as a factor in producing the initial colloidal dispersion when cement and water are mixed.

Setting of cement is considered to consist of a number of stages though these may overlap in time. The various processes are to be found in their least complicated form in simple pastes of tricalcium aluminate with or without gypsum or surface active agents. The theory, as originally developed for this single mineral type of cement, has been somewhat modified in detail in relation to Portland cement.

When Portland cement and water are mixed there is an immediate rapid reaction with the formation of a supersaturated solution but the reaction rapidly slows down and this is ascribed to the formation of a film of micro-crystalline or gel-like calcium sulphoaluminate around the cement particles. There follows a period of slow reaction, termed the induction period, during which the amount of hydration products gradually builds up with time and slowly increases the plastic viscosity of the paste. The process of structure formation begins immediately the cement and water are mixed and the growth of plastic strength is divided into two stages.[1]

In the first stage a coagulational structure is formed characterised by the presence of a three-dimensional network formed by disordered coupling of the

[1] L. A. Si'lchenko, N. V. Mikhailov and P. A. Rebinder, *Dokl. Akad. Nauk SSSR* **162,** 1342 (1965).

finest particle in the disperse phase through thin layers of the dispersion medium. In this stage crystallisation occurs only in the form of individual crystallites, mainly of calcium hydroxide and sulphoaluminate. The coagulational or coagulational-crystallisation networks have low strength. The slow rise in plastic strength comes to an end at a certain critical time and is followed by a rapid growth in strength. In this second stage there is more intense crystal formation accompanied by crystal bonding forming a strong crystalline network. Here the cement loses its plasticity and takes on the properties of a solid. In this period the structure becomes predominately a condensed-crystallisation network. The subsequent development of strength comes from the growth of a hydrated calcium silicate crystalline structure as the crystals are formed from a supersaturated solution. The critical time, though somewhat arbitrary, represents the point in time at which any further mechanical deformation of the setting mix becomes detrimental to its ultimate strength.

This statement needs more explanation to bring out its full meaning. The colloidal suspension formed initially is considered to stem primarily from the aluminate compounds with the silicates at this stage remaining relatively inert. It may also contain dispersed particles of the cement of colloidal dimensions, these being formed by the dispersion mechanism mentioned earlier. The tricalcium aluminate is viewed as having the greatest capacity for such an adsorptive and chemical peptisation.[1] There is a rapid increase in the surface area of C_3A, some ten-fold after 30 minutes hydration,[2] but this could alternatively be attributed to the large specific surface of the hydration products. Only one or two per cent hydration, which occurs within a few minutes, would be sufficient to produce a ten-fold increase in surface area. The dispersion of grains by a liquid medium is well known and it may occur over a longer period of time and afford an explanation of the S-shaped hydration curve shown in Fig. 61 for dicalcium silicate, and perhaps during the first day for $3CaO.SiO_2$ as found by Forbrick.[3] An acceleration of this dispersion process by the introduction of surface active agents seems, as we shall see later (p. 258), to help explain the mechanism of their action.

The process of structure formation stems from the instability of the colloidal cement-water dispersion which coagulates to give a weak and thixotropic coagulation network. Because of the supersaturation in the cement solution, this is rapidly followed by the formation of a loose crystalline calcium sulphoaluminate network[4] which is non-thixotropic and irreversible. If it is broken down mechanically in the mixing process the very small separate crystals cohere again to form a coagulational network. This probably accounts for the reduction in slump, or consistence, of a concrete or cement mix as mixing is continued over long periods. Vibration of a cement paste is held to increase the quantity of colloidal fraction per unit volume, to increase the strength of the coagulation structure, and to raise the mobility proportionally to the frequency of vibration.[5]

[1] A. M. Smirnova and P. A. Rebinder, *Dokl. Akad. Nauk SSSR* **96,** 107 (1954).
[2] A. M. Smirnova, N. G. Zaitseva and P. A. Rebinder, *Kolloid. Zh.* **18,** 93 (1956).
[3] L. R. Forbrick, *Proc. Am. Concr. Inst.* **37,** 161 (1940).
[4] Du. Yu-Zho and E. E. Segalova, *Zh. prikl. Khim.* **34,** 521 (1961).
[5] Kalmykova and Mikhailov, *Dokl. Akad. Nauk SSSR* **152,** 389 (1963).

The clear-cut distinction between the formation of these coagulational and crystalline networks drawn in the earlier U.S.S.R. papers seems no longer to be maintained and they both seem more to be regarded as part of a more-or-less overlapping process. The essential point is that there is a slow rise in plastic viscosity as hydration products build up in the system and loose bonds are formed and that the amount of water required to give a paste of given workability (plastic strength) is greater for the coagulational network than for the initial dispersion. A delay in the formation of the coagulational network until placing is finished thus favours a lower water-cement ratio. The loose coagulational-crystalline network of calcium sulphoaluminate hydroxide is not thought to contribute to ultimate strength, but to form the framework in which the build-up of the subsequent calcium hydrosilicate crystalline structure takes place. It is visualised as forming a sort of template and, since it is a loose framework, to be a source of weakness rather than strength in the hardened cement. Its destruction can therefore be advantageous.[1] This may explain the beneficial effects of re-vibration of concrete after an interval.

Setting is identified with a certain stage in the development of the coagulational-crystallisation network, initially weak and thixotropic, while the process of hardening denotes the development of the much stronger irreversible, crystalline structures.

This general theory follows that of Le Chatelier in ascribing hardening to crystallisation, but the introduction of the concept of a preceding colloidal stage has the merit of providing explanations of a variety of phenomena associated with the gauging and setting of cements.

Retarders, such as gypsum, and surface active agents, such as calcium ligno-sulphonate, influence the rate of formation of the coagulational structure and the speed, and form, of the crystallisation of the aluminate hydration products. A cement without gypsum quickly forms a coagulated structure which is stiffened by rapid crystallisation of hydrated calcium aluminate from the supersaturated solution resulting in quick setting. When gypsum is added the calcium sulpho-aluminate coating formed on the cement particles acts as a more hydrophilic colloidal stabiliser than does hydrated calcium aluminate, delaying the coagula-tional process;[2] it leads to the formation later of a weak ettringite network, instead of a stronger one of hydrated calcium aluminate.[3] The ettringite network is weak according to the U.S.S.R. workers because crystallisation takes place from a more highly supersaturated solution resulting in high crystallisation pressures which strain or disrupt the network.[4]

The action of surface active agents is a complex one. The typical agent studied in much of the work in the U.S.S.R. has been calcium lignosulphonate. This is

[1] O. P. Mchedlov-Petrosyan and A. G. Bundakov, *Silikattechnik* **12,** 292, 338 (1961); *R.I.L.E.M. Sym. Problems of Accelerated Hardening of Concrete, Moscow 1964.*

[2] E. E. Segalova and E. S. Solovyeva, *Sym., Moscow 1956*, 138.

[3] Du. Yu-Zho and E. E. Segalova, loc. cit.

[4] T. K. Brutskus and E. E. Segalova, *Kolloid. Zh.* **26,** 8, 284 (1964); *Zh. prikl. Khim.* **37,** 1124 (1964).

adsorbed on the surface of the cement grains retarding their hydration and, because of its hydrophilic nature, increasing the stability of the suspension, delaying the formation of a coagulation network, and decreasing the initial plastic viscosity. This last effect accounts for the reduction found in the water-cement ratio required to obtain a given plasticity. Opposed to these effects is a dispersing action* on the aluminate minerals in the cement which increases their available surface and the rate of hydration and thus the concentration of colloidal products and the plastic velocity of the cement paste. The stabilising and dispersing effects need to be balanced since an increase in dispersion on its own would lead to more rapid structure formation while excessive stabilisation, and slowing up of hydration by the adsorbed surface films, unduly delays setting and strength development. This balance is determined by the tricalcium aluminate and gypsum contents of the cement and the quantity of the agent that is added. It accounts for the variability in the effect of calcium lignosulphonate on setting time observed with different Portland cements.

The calcium lignosulphonate is also adsorbed on the aluminate hydration products, stopping the crystal nuclei acting as crystallisation centres and the growth at too early a stage of a crystallisation network. Once all the surface active agent has been adsorbed and removed from solution, crystallisation and structure formation proceed rapidly and with increased strength because of the larger number of nuclei available for growth.[1]

The addition of salts, such as chlorides, to a cement mix is held not to affect the coagulation process but to influence the speed of crystallisation because of their effects on the solubility of the hydration products and the degree of supersaturation in the solution. This general subject of the influence of retarders and electrolytes is discussed in more detail elsewhere (p. 302).

In these theories primacy of place is given to the aluminate compound as determining the rheological properties of the cement paste and the setting, with the silicates responsible for the subsequent hardening and development of strength. However, tricalcium silicate must undoubtedly play a part in the setting phenomena. Its hydration is rapid and the amount hydrated in the initial stages is not much lower than that of tricalcium aluminate in an ordinary gypsum-retarded Portland cement. On its own C_3S has a setting time similar to that of Portland cement and it has a rapid initial reaction with water, producing a solution supersaturated with respect to the silicate ions and $Ca(OH)_2$ within a few minutes. Within half an hour some 5 per cent of C_3S is hydrated and the thickness[2] of the colloidal hydrated layer surrounding the grains is about 250 Å. Other work,[3] in the U.S.S.R., has also shown that C_3S contributes to structure

[1] O. I. Lukyanova, E. E. Segalova and P. A. Rebinder, *Kolloid. Zh.* **19,** 82, 459 (1957); *Dokl. Akad. Nauk SSSR* **117,** 1034 (1957); E. E. Segalova, R. B. Sarkisyan and P. A. Rebinder, *Kolloid. Zh.* **20,** 611 (1958); B. Blank, D. R. Rossington and L. A. Weinland, *J. Am. Ceram. Soc.* **46,** 395 (1963).

[2] D. L. Kantro, S. Brunauer and C. H. Weise, *J. phys. Chem.* **66,** 1804 (1962).

[3] E. P. Andreeva and E. E. Segalova, *Kolloid. Zh.* **22,** 503 (English Trans.) (1960).

* Large according to earlier Rebinder views; about 25–50 per cent according to F. E. Ensberger and W. G. France, *Ind. Engng Chem.* **37,** 598 (1965).

formation in the early period after mixing. A C_3S paste also shows thixotropic properties characteristic of a coagulational network.[1] It may be noted in passing that the rheological properties of Portland cement pastes are variable (see p. 364).

The distinction drawn in the Rebinder theories between coagulational and crystalline networks must depend more on the nature of the bonds involved than on the degree of crystallinity. Thus tobermorite in a hardened cement paste is poorly crystallised while at the earliest stages the gel coatings of the silicate have the same basic layered crystal structure. Both have a very high specific surface of colloidal dimensions, 100–400 m^2/g. In this sense, as we have noted earlier, the distinction between gel and crystal or highly dispersed solids can become a matter of semantics, but the difference between van der Waal and chemical bonds is a real one. There is further some evidence that in hardened cement paste the tobermorite slowly undergoes with age a condensation-type polymerisation by linking of the silicon-oxygen tetrahedra through O–Ca–O atom bridges forming an inorganic polysilicate polymer.[2]

To summarise these theories, there is an initial rapid reaction over the first few minutes followed, in a correctly retarded cement, by an induction period of slow reaction usually lasting from a half to two hours. During this induction period coatings of hydration products form over the cement grains and there is a slow build-up of hydration products of colloidal dimensions in the paste. The more rapid reaction that follows the induction period is ascribed to the break-up of the coatings and setting to the coagulation of the paste. The rupture of the coatings may, it has been suggested by Powers,[3] be due to osmotic pressure arising from a difference between the concentration of the ions in the solution immediately in contact with the cement grains and the solution outside the gel coating. This coating is taken to act as a semi-permeable membrane with increasing efficiency as it becomes thicker and the dormant period to be the time required for the osmotic pressure to increase sufficiently to rupture the coating. Some evidence in support of this is to be found in Steinour's observation[4] that a saturated cement paste undergoes a transient expansion at the end of the dormant period, suggesting that the cement grains were being forced apart by the displacement of the coatings at points where the grains are in contact. The dormant period is followed by a more rapid chemical reaction leading to initial and final set, which are arbitrary defined degrees of firmness. The subsequent hardening and strength development is ascribed to new crystal formation and recrystallisation of the hydrated silicates from a supersaturated solution and the filling of the spaces in the skeleton framework by these hydration products. The space structure of the cement gel as first formed is an expanded and unstable one with an inherent tendency to change to a more stable form. This change is accompanied by a diminution in surface area,[5] i.e. a growth of larger gel particles at the expense

[1] M. Ish-Shalom and S. A. Greenberg, *Sym., Washington 1960*, 731.

[2] C. W. Lentz, *Spec. Rep. Highw. Res. Bd*, No. 90, 269 (1966); S. Brunauer and S. A. Greenberg, *Sym., Washington 1960*, I, 143; O. P. Mchedlov-Petrosyan and W. I. Babushkin, ibid., 533; H. Funk, *Sym., Tokyo 1968*.

[3] T. C. Powers, *J. Res. Dev. Labs Portld Cem. Ass.* **3** (1), 47 (1961).

[4] See T. C. Powers, loc. cit.

[5] L. A. Tomes, C. A. Hunt and R. L. Blaine, *J. Res. natn. Bur. Stand.* **59**, 357 (1953); C. A. Hunt, L. A. Tomes and R. L. Blaine, ibid. **64A**, 163 (1960).

of smaller ones, and a shrinkage of the network,[1] a behaviour similar to that which silica gel shows on a larger scale. This process occurs on ageing to a limited extent but it is much accentuated by drying, thus accounting for the drying shrinkage (see p. 279). Pressure should have a similar effect and this must at least be one of the factors involved in creep of concrete (see p. 286).

The morphology of the set cement compounds

We have seen in a previous chapter that when a set Portland cement, gauged with only sufficient water to form a plastic mass, is examined microscopically, the hydration products appear as an undifferentiated amorphous mass apart from crystals of calcium hydroxide and occasionally small amounts of the hydrated hexagonal or cubic aluminate and of sulphoaluminate. The sharp, angular forms of the original cement grains disappear, the particles becoming surrounded by a zone of semi-transparent gelatinous material which slowly increases in bulk. Plate 7 (i) shows a Portland cement immediately after immersion in water, and Plate 7 (ii) the same specimen after 24 hours. The enlargement of the grains by reaction with the water to form a gelatinous mass is clear.

The use of X-ray methods has proved invaluable in showing the crystalline nature of the hydrated calcium silicate gels, but their application to set cement pastes at first did little more than confirm what had already been revealed by microscopic method. Strätling[2] was able to show that the gelatinous hydration product obtained when cement was mixed with a large excess of water gave an X-ray pattern similar to that of a hydrated calcium silicate gel, but conclusions so obtained are not necessarily applicable to pastes of low water content. Nurse and Taylor,[3] from X-ray examination of different size fractions separated from the ground cement paste by treatment with a super-centrifuge, identified the patterns of calcium sulphoaluminate, hydrated calcium aluminate (this might be the hexagonal solid solution with $3CaO.Al_2O_3.CaSO_4.12H_2O$), and hydrated calcium silicate, *CSH* (I), thus providing evidence of the essentially crystalline nature of the gelatinous material.

Further valuable evidence has come from studies with the electron microscope though there are difficulties in interpretation. In order to resolve the structure of a powdered material in the electron microscope, it is necessary to choose a particle size such that the thickest particles are just penetrated by the electron beam and at the same time to maintain such a degree of dispersion that the ultimate particles are separated one from the other. With a material such as set cement this is difficult, and further the thin dispersion of fine material is so rapidly affected by atmospheric carbon dioxide that what is frequently obtained is a mass of calcium carbonate retaining some of the details of the original structure. Further difficulties arise from the dehydration of hydrates in the high vacuum of the electron microscope and under the actual bombardment by the

[1] W. Czernin, *Zement-Kalk-Gips* **9,** 525 (1956).
[2] W. Strätling, *Ber. dt. keram. Ges.* **20,** 522 (1939).
[3] R. W. Nurse and F. W. Taylor, *Sym., London 1952,* 311.

electron stream. It is also important when possible to check the identity of the particles actually photographed by obtaining their electron diffraction pattern at the same time. When all these factors are considered it is not surprising that there has been some disagreement between workers in this field as to the interpretation of results.

The earliest studies on cement compounds were made by Eitel[1] who obtained electron micrographs showing the formation from tricalcium aluminate of hexagonal plates up to about a micron in length and 0·1 μ in thickness; and also, in an *iso*butyl alcohol-water medium, of icositetrahedra of $3CaO.Al_2O_3.6H_2O$ appearing almost as spheres of about 0·1 μ diameter. Later work by McMurdie has shown that C_3A, C_4AF and C_8NA_3 all form hexagonal plate hydrates up to one or two microns across. Eitel also noted the formation from tricalcium silicate of small hemispheres of calcium hydroxide, now considered to be amorphous, and needle crystals of hydrated calcium silicate.

Knowledge of the morphology of the hydrated silicates was much advanced by the work of Grudemo in Sweden. Grudemo[2] studied by electron microscopy and diffraction, and X-ray diffraction, Taylor's preparations of calcium silicate hydrates, *CSH* (I) and *CSH* (II), and various other products obtained by the hydration of tricalcium silicate, sometimes with ultrasonic treatment to accelerate hydration, and by the reaction of lime and silica gel and other means.

An electron micrograph of a sample of Taylor's calcium silicate hydrate *CSH* (I) prepared from lime and silica gel, is shown in Plate 8 (i) and its electron diffraction pattern which identified it in Plate 8 (ii). The particles consist of very thin flexible flakes or foils sometimes distorted and rolled at the edges, possibly owing to partial dehydration in the vacuum of the electron microscope. The thinnest flakes are probably of single crystal thickness (about 10 Å) and others are composed of stacks of superimposed single crystals up to perhaps ten units. The X-ray evidence similarly suggests, according to Bernal, that the crystallites have dimensions of the order of 50–200 Å. In saturated lime solutions, when the composition of the *CSH* (I) approximates to $1{\cdot}5CaO.SiO_2.aq$ the growth of fibrous or needle-like crystals seemed to be promoted, and may be caused by a degeneration of the sheets into lath-like structures and by twisting and rolling of the crystal sheets. Samples of *CSH* (I) obtained by hydration of tricalcium silicate with ultrasonic treatment appeared as clusters of the very thin wrinkled foils that are characteristic of this compound.

The more basic calcium silicate hydrate II of Taylor (*CSH* (II)) prepared by hydration of tricalcium silicate, and with an approximate composition of $2CaO.SiO_2.4H_2O$, gave another characteristic type of electron micrograph illustrated in Plate 9 (i). This compound appears as bundles of rod-like or lath-like thin particles and the broom-shaped structure at the ends of the bundles is very typical. The crystalline units of these aggregates are very distorted. With vigorous ultrasonic treatment of the *CSH* (II) in saturated lime solution, thin fibres protruded from the surfaces of the aggregates of crystals filling the whole space

[1] W. Eitel, *Z. angew. Chem.* **54,** 185 (1941); *Zement* **28,** 693 (1939); **31,** 489 (1942).

[2] A. Grudemo, *Proc. Swed. Cem. Concr. Res. Inst.*, No. 26 (1955); *Sym., London 1952*, 247; *K. tech. Högsk. Handl.*, No. 242 (1965); *Sym., Washington 1960*, 615.

between them with a fine network. Similar effects were also sometimes observed on ordinary shaking of $3CaO.SiO_2$ in saturated lime solution. The main product found by Grudemo in such shaking tests, however, was a conglomerate of rounded particles of sizes between 100 and 1000 Å, and some foils of *CSH* (I), but in none of these tests was hydration complete.

Plate 9 (ii) shows fibrous crystals formed on the surface of a hydrating tricalcium silicate particle and these are thought to be similar to the *CSH* (II) shown in Plate 9 (i). Thin plates of *CSH* (I), similar to Plate 8 (I), are also formed but the conditions which lead to the alternative formation of these two forms are not clear. Further work[1] has provided general confirmation of Grudemo's observations and shown that with Portland cement, clusters of acicular particles can be observed at about the time of initial set and that as time progresses the clusters become larger and denser. Studies on replicas and thin sections of hardened cement pastes and concrete have indicated that the silicate phases are present either as masses of plates or of splines. Both with C_3S and Portland cement the formation of plates of $Ca(OH)_2$, as well as the amorphous form, is found at an early stage. An electron micrograph replica of a fracture surface of a set cement composed of 75 per cent alite, 25 per cent dicalcium silicate with a water-cement ratio of 1·0 is shown in Plate 10. Fibrous *CSH* crystals growing out from the compact centre of the grains can be seen. In the voids between the grains, which are lines of weakness, less hydrate material is visible and takes the form of crumpled foils of *CSH*. Both the fibrous and the crumpled foils grow best in mixes of high water : cement ratio giving unconfined areas for growth.

The development of the scanning electron microscope with its greater depth of focus has thrown further light on the morphology of the hydrated cement compounds. Plate 11 (i) of a set cement mortar water-cured for forty days shows the characteristic calcium silicate hydrate fibres, probably *CSH* (II), some material platey in habit seen on edge, and some amorphous material. These may be *CSH* (I) and amorphous $Ca(OH)_2$. The fibres are shown at a higher magnification in Plate 11 (ii) and it will be noted that they lack well-developed crystal faces. The crystallisation of plates of $Ca(OH)_2$ in a set cement at 7 days is illustrated in Plate 12 which shows that some interlocking takes place. The very fine needle crystals are probably ettringite.

It will be evident that electron microscopy and diffraction have a very important part to play in increasing our understanding of the hydration process and the ultimate structure of set cement, but that the techniques employed influence the results. Thus supersonic treatment to aid hydration or dispersion, or ball-milling to aid hydration, may influence the morphology of the products.

We can now see that the ultimate particles in set cement have an ordered crystalline structure but that their dimensions, of the order of 100 Å (1×10^{-6} cm), are similar to those of the colloidal particles that make up such well-known gels as silica and gelatine. The surface area of the set mass is thus very large and it is this which determines many of its physical properties. Thus the volume change on wetting and drying, and the relation between the vapour pressure and

[1] L. E. Copeland and E. G. Schultz, *J. Res. Dev. Labs Portld Cem. Ass.* **4** (1), 2 (1962). H. G. Midgley, *Proc. Br. Ceram. Soc.*, 1969 (13) 89.

water content, are characteristic of the behaviour of a rigid gel. In this sense we can regard the older rival views of Le Chatelier and Michaelis as each representing one facet of the properties of set cement, the former stressing its crystalline nature and the latter its high surface area.

The bonding action

When we turn to the actual mechanism of the bonding action there is still room for divergence of opinion and we have yet to reach any firm conclusion.

The development of strength, as we have seen, is determined by various complicated interrelations between solubility and supersaturation phenomena, the formation and growth of crystal nuclei, the large growth in surface area of the hydration products, and the establishment of contacts, mechanical or chemical, in a framework of interlacing crystals.

Surface area in itself is a source of strength arising from surface forces and from friction between the particles restricting their mobility. Thus Czernin[1] has shown that a paste of finely ground quartz (20 000 cm^2/g specific surface) has a substantial compressive strength. The specific surface of set cement is much higher, some $2–3 \times 10^6$ cm^2/g, and the action of van der Waals' forces in drawing the particles together contributes to strength. These attractive forces may amount to thousands of pounds per sq. in. However, to account for the stability of set cement in water and its behaviour as a rigid gel we must look further to the formation of stronger bonds between the particles. It is the tobermorite 'gel' which is primarily responsible for strength and the factors responsible must reside in the nature and extent of its surface. The gel particles are essentially very thin sheets only two or three tobermorite molecular layers thick. The chemical attraction of the surface atoms could give rise to chemical bonding in various ways as for example by O–Ca–O bridges, by hydrogen bonding, or by condensation of $Si(OH)_4$ groups forming Si–O–Si (siloxane) linkages. The crystal chemistry of the tobermorite gel is too complicated to allow of any easy choice between the various possible chemical bonds or between the relative importance of these and the van der Waals' forces. A suggestion has been made[2] that the large difference, a factor of ten, between the compressive and tensile strengths of cement arises from the different forces involved. The relative strength of van der Waal bonds, hydrogen bonds and valency bonds, is of the order of 1 : 10 : 100. Failure in tension, it is suggested, involves working against van der Waals' forces between the particles, but in compression it is the valency forces within the tobermorite crystallites that have to be overcome. The large difference in strength is regarded as an indication that there is no extensive welding together of the crystallites. This perhaps involves a too clear-cut distinction between inter- and intra-particle forces, but it does suggest possible physico-mechanical lines of investigation.

[1] W. Czernin, *Zem. Bet.* **16,** 18 (1959).
[2] S. Brunauer, *Am. Scient.* **50** (1), 210 (1962); S. Chatterji and J. W. Jeffery, *Nature, Lond.* **214,** 559 (1967).

In an alternative theory[1] of the bonding mechanism it has been suggested that the particles are held together at solid-to-solid contacts having no regular atomic arrangement by bonds which cannot be simply classified as chemical or secondary (Van der Waal) bonds. The atoms at the surfaces in contact are envisaged as engaging a varying proportion of short- and long-range forces depending on the degree of disorder and the average spacing. This type of bond differs from a chemical bond which involves a more regular atomic arrangement in the coincident surfaces and from the van der Waal bond which is usually assumed to operate across an adsorbed water film between the particles. It is visualised that the type of bond suggested, like a van der Waal bond but unlike a chemical bond, can be remade after breaking, and evidence has been advanced to support this.

The strength of set cement can also be ascribed to the mechanical interlocking of crystals as proposed by Le Chatelier. We can picture the set cement as consisting of a matted felt or fine mesh-work of tobermorite crystals consisting individually of acicular particles or sheets which become folded into foils, ribbons and fibres and buckled. The reality of this network has been shown by photos taken in the scanning electron microscope.[2] As these crystals grow and become denser it seems logical to assume that they grow together and interlock mechanically with the structure, becoming strengthened by inter-particle surface or chemical bonds. The part played by calcium hydroxide crystals is still controversial. Some authors[3] consider that they contribute to strength while others[4] dismiss this on account of the large size of the crystals which is assumed to diminish interlocking and bonding. There is, it has been suggested,[5] a critical size for the optimum contribution of crystals to the strength of a network. Two opposing factors are involved. Precipitation from a highly supersaturated solution leads to a multiplicity of fine crystals favouring an increase in the number of 'concretion' contacts which leads to strengthening of the crystallisation network. At the same time, however, the crystallisation pressure increases and introduces stresses into the rigid network which may lead to partial breakdown and reduction in ultimate strength. There is, it is suggested, an optimum degree of supersaturation corresponding to growth of crystals of a certain size, which gives the best balance between these two opposing factors. Though many contributions to this subject of 'physico-chemical' mechanics have been made by Rebinder and his associates it is as yet uncertain how far these arguments on the effect of crystal size can be applied. Thus the hydration of magnesium sulphate at a suitable water content can produce a strong paste in which the crystals of $MgSO_4.7H_2O$ are large—from 25–200 μ. We cannot yet dismiss the possibility that calcium hydroxide crystals contribute to the strength of Portland cement.

[1] I. Soroka and P. J. Sereda, *Sym., Tokyo 1968*; R. F. Feldman and P. J. Sereda, *Mater. Structs.* **1** (6), 509 (1968).

[2] S. Chatterji and J. W. Jeffery, *Nature, Lond.* **209,** 1233 (1966); see also L. E. Copeland and E. G. Schultz, *J. Res. Dev. Labs Portld Cem. Ass.* **4** (1), 2 (1962).

[3] L. D. Ershov, *Sym., Moscow 1956*, 264; R. L. Berger, D. S. Cahn and J. D. McGregor, *J. Am. Ceram. Soc.*, **53,** 57 (1970).

[4] Ping-i Chou, E. E. Segalova and O. I. Lukyanova, *Kolloid. Zh.* **26,** 341, 373 (1964); A. Grudemo, *Sym., Washington 1960*, 615.

[5] E. E. Segalova and P. A. Rebinder, *Stroit. Mater.*, 1960 (1), 21; E. E. Segalova, E. A. Amelina and P. A. Rebinder, *Kolloid. Zh.* **25,** 229 (1963).

Apart from the theories as to the actual nature of the bond, an important contribution to strength is made by the filling of space in the set cement reducing voids and increasing its solidity, for as we shall see later strength is related to the gel : space ratio. The hydration products from 1 cc of anhydrous cement occupy rather over 2 cc of space so that on complete hydration rather over half of the products must migrate into the capillary spaces displacing water. For reasons which we shall discuss later there is, at least at w/c ratios below 0·35, a certain minimum porosity† of the cement gel, that is of space occupied by evaporable water, which cannot be eliminated. This value is about 0·28 of the total gel volume. The actual porosity of the gel depends on the original water : cement ratio and on the age and will normally be higher than this minimum value. Ageing, or drying, in fact increases the porosity because the shrinkage of the gel particles is greater than that of the external volume of the cement paste. Water storage decreases it because of the deposition of new hydration products in the pore spaces with only a small increase in the overall volume. Though, as we have seen, it is open to question how far calcium hydroxide crystals in set cement contribute appreciably to bonding, they must contribute to the density of the paste and thus indirectly to strength.

The cement-aggregate bond

Another factor to be considered in relation to the strength of concrete is the cement-aggregate bond.[1] This bond may arise from mechanical causes (surface rugosity), physical surface forces or in some cases chemical interaction which may be advantageous or the reverse. Studies on this bond have been made by measurements of shearing or tensile strength, by microhardness measurements across the interface, and by microscopic examination. With smooth inert hard aggregates the strength of the bond is less than that of the adjoining cement paste or of the aggregate itself. Chemical interaction can arise between lime and reactive silica or from the epitaxial growth of crystals at the interface. The latter can arise when the lattice of the aggregate mineral is similar to that of some component in the set cement so that the growing crystals can align themselves in an arrangement corresponding to that of the substrate. There is a two-degree orientation of the growing crystals such that two of their axis are parallel to two of the substrate. On calcium carbonate aggregates for example, Farran[2] found that calcium hydroxide crystals formed a contact layer between the cement paste and the aggregate in an epitaxial arrangement, and further suggested that a $Ca(OH)_2$–

[1] A review of the subject was given by K. M. Alexander, J. Wardlaw and D. J. Gilbert, *Proc. int. Conf. Structural Concrete, London 1965*, Cement and Concrete Association 1968.

[2] J. Farran, *C.r. hebd. Séanc. Acad. Sci., Paris* **237,** 73 (1953); *Revue Matér. Constr. Trav. publ.* **490–491,** 155, **492,** 191 (1956); J. Farran and J. C. Maso, ibid., 587–588 (1964); ibid., *C.r. hebd. Séanc. Acad. Sci.* **260,** 5195 (1965); J. Farran, *R.I.L.E.M. Bull.*, No. 24 (1958).

† Defined as the fraction of the volume of a saturated cement paste occupied by evaporable water. Evaporable water is defined as that having a vapour pressure greater than 0·5 μ Hg at 23° C (T. C. Powers).

$CaCO_3$ solid solution was formed. Failure at a cement-calcite interface occurred inside the calcite grains whereas at a cement-quartz interface it occurred at the interface. Blastfurnace slag aggregate is another example of a material that appears capable of surface chemical reaction with cement giving a bond of higher strength than with an inert aggregate. Certain dolomitic aggregates (see p. 576) on the other hand seem to undergo surface exchange reactions with cement which are expansive and detrimental to strength. In studies of the microhardness across an interface Lyubimova[1] found that the first 20–30 μ layer at the interface was harder than the outer layers before passing into the still harder main body of the set cement. This strengthening immediately adjacent to the aggregate face was attributed to a more intensive growth of crystal nuclei. Various other workers[2] have also found a strengthening of the bond at Portland cement-limestone interfaces and evidence of an interface reaction or orientation of the calcium hydroxide of the set cement. The bond strength as a proportion of the strength of the body of the set cement appears to decrease as the water-cement ratio increases. Amongst the silicious rocks Alexander[3] found that the bond to opal was substantially higher than to obsidian and in general was lower to large coarse aggregate pieces than to those of ½ in. or less in size. Bond was improved by vibration of a mix. Measurements of the permeability of an aggregate-cement interface by Valenta[4] have however yielded rather different results in that limestone behaved less favourably than some igneous rocks. There is scope for more work on the aggregate-cement bond since with many aggregates the bond is the weakest point of a concrete. The subject has been reviewed by Napper-Christensen.[5]

The setting and hardening process

A summary. The changes occurring in the setting of cement can be visualised as follows. The cement grains are acted upon by water to form a supersaturated solution from which the gel-like mass of crystals precipitates. Diffusion of water molecules to the surface, or even into the crystal lattice, to react *in-situ* must also play a part, at least in the later stages of hydration. Thus when an unhydrated ement core is surrounded by hydration products the water can only reach it by diffusion through this surface film. While still in a plastic condition the cement paste shrinks slightly because there is a contraction in volume of the system (cement+water) on hydration. Once the mass becomes rigid a small expansion sets in for the gel mass deposits around the cement grains and causes them to swell and to exert an outward pressure. The quantity of the gel mass progressively increases with time and it spreads into the inter-granular spaces.

[1] T. Yu. Lyubimova and E. R. Pinus, *Kolloid. Zh.* **24,** 578 (1962); T. Yu. Lyubimova and P. A. Rebinder, *Dokl. Akad. Nauk SSSR* **163,** 1439 (1965).
[2] A. D. Buck and W. L. Dolch, *Proc. Am. Concr. Inst.* **63,** 755 (1966); K. M. Alexander, *Tewksbury Sym. Fracture, Univ. Melbourne 1963*; T. T. C. Hsu and F. O. Slate, *Proc. Am. Concr. Inst.* **60,** 465 (1963).
[3] K. M. Alexander, *Proc. Am. Concr. Inst.* **56,** 377 (1959); *Nature, Lond.* **187,** 236 (1960)
[4] V. Valenta, *R.I.L.E.M. Sym. Durability of Concrete, Prague 1961*, Prelim. Report, p. 53.
[5] *Nord. Betong* **9,** 1 (1965).

The cement gel must be regarded as being formed initially in an unstable condition with the fibre-crystals of the gel particles, or the gel particles themselves, further apart and occupying a greater volume and enclosing more water than in their stable state. The gel thus has an inherent tendency to shrink and give off some of the water it contains. In water no measurable contraction occurs, but it would in any case be offset by the effects of continued hydration of previously unattacked cement. On drying the set cement it undergoes an irreversible contraction and reduction in water content as the gel changes into its more stable form. The expansion on subsequent wetting, and the amount of water taken up, then become reversible and are reproduced on successive cycles of wetting and drying. On very prolonged ageing under wet conditions a further slow change may occur by crystal growth. There is, however, a reduction in the surface area with ageing indicating that irreversible changes towards a more stable state are occurring.

We may set out the setting, hardening, and ageing processes diagrammatically as follows:

I. *Unhydrated cement + water*

↓ Transient expansion at end of induction period followed by contraction while still plastic.

II. *Plastic cement-water mass*

Expansion when rigid

IIIA. *Metastable gel of crystals of colloidal dimensions*

IIIB *Crystalline products above colloidal dimensions*: Calcium hydroxide, hydrated calcium aluminate and sulphoaluminate

Drying

Irreversible Shrinkage

Water storage slow change

IV. *Stable gel*

Water storage

Water storage

V. *Crystalline products of coarser dimensions*

The changes from I to IIIA and IIIB are part of the setting process. The change from stage IIIA to IV occurs when the set cement is allowed to dry in air, but under water storage the stage III persists though a slow change on ageing in water from stage IIIA towards IV occurs. The final change from stage IV to V seems to be exceedingly slow, and never to occur in mortars or concretes under normal conditions.

The ultimate fate of any cement product stored in air is to become converted to calcium carbonate and hydrated silica, alumina, and ferric oxide, since all cement compounds are decomposed by carbon dioxide. This action under ordinary conditions is limited, however, to the exposed surfaces of mortars or concretes and does not penetrate the inner mass.

The hardening of a set cement is to be attributed, as we have seen, to the production of increasing amounts of the crystals and gel in stage III. The additional hydrated material produced increases the impermeability of the mass, for it has a greater volume than the unhydrated material from which it was formed. A cement mortar bar stored in water undergoes a very slow increase in length due to the internal pressure produced by this further hydration. Gradual ageing of the gel with a change from stage IIIA to IV may also account for some increase in strength.

When stored in air, hydration seems to come practically to a stop when the relative humidity produced by the water in the material drops below about 80 per cent, though some slight reaction may occur at lower humidities.[1]

Bleeding. It is a common observation that, after a concrete has been placed in position, and while still plastic, water appears at the surface. This is essentially a sedimentation phenomenon, the solids settling in the plastic mass. The laws[2] governing it are essentially those of liquid flow in a capillary system rather than the normal Stokes law for rates of settling in a dilute suspension. Water often separates as a concrete is being worked into position. Bleeding is much influenced by the plasticity of a concrete mix and is increased by vibration. It is reduced by the addition of plasticising agents and air-entrainment.

Resetting of hardened cement. Set cement, even after very prolonged periods, still shows the presence of unhydrated cores. This is hardly surprising since it has been found that hydration of cement grains only proceeds to a depth of 6–9 μ in five months, and the subsequent increase must be very slow. Some 50 per cent or more by weight of cement consists of particles of diameter greater than 10 μ and ranging up to 100 μ. It is well known that cement which has once hardened, if reground and gauged again with water, will set again a second time and develop a degree of mechanical strength, which, although inferior to that attained at the first setting, is nevertheless considerable. For example, Grün[3] gauged a neat cement with water, allowed it to harden for three days, and then reground the set material. This was repeated a second time. The compressive strength (lb/in^2) developed by the cement in 1 : 3 sand mortar cubes was as follows:

	Fresh cement	After first hardening	After second hardening
3 days	3170	940	256
7 days	3940	1310	355
28 days	6000	1850	655

It is evident that even after a second gauging of cement with water some cores of unhydrated cement still remain.

[1] R. F. Feldman, P. J. Sereda and V. S. Ramachandran, *Highw. Res. Rec.*, No. 62, 106 (1964).
[2] T. C. Powers, *Portland Cement Association Bulletin*, No. 2 (1939); H. H. Steinour, ibid., No. 4 (1945).
[3] *Zement* **22,** 143 (1933).

Volume changes during setting

When Portland, or other hydraulic cements react with water, the system cement plus water undergoes a net volume diminution. Despite this decrease in volume of the system there appears[1] to be a small expansion if the plastic paste is kept saturated in the first few hours before contraction sets in, as shown below.

Initial volume expansion of Portland cement paste (w/c = 0·55)

Time (hours)	$1\frac{1}{2}$	$2\frac{1}{2}$	$4\frac{1}{2}$	$6\frac{1}{2}$†	$23\frac{1}{2}$
Volume expansion (%)	0	0·1	0·18	0·22	0·38

† Time of final set.

For five Portland cements the expansions ranged from 0·09 to 2·1 per cent at a w/c ratio of 0·55. Subsequent to this initial period the volume of the system as a whole contracts as would be expected from density calculations. In fact, in contradiction to the above results, most of the available data fail to show the small initial expansions. Some typical data found by Gessner[2] for the volume diminution calculated on the initial volume of a mixture of 100 parts cement and 33 parts water were:

	No.	Percentage volume diminution of system 1 day	7 days	28 days	100 days
Portland cement	1	2·8	4·8	6·0	6·9
,, ,,	2	1·7	4·4	—	6·3
Portland cement without gypsum	3	2·7	8·0	8·6	8·7
	4	2·6	6·3	7·5	7·6
High-alumina cement	5	11·1	13·8	15·2	16·3

Though the whole system shows a diminution in volume, the volume of the solid matter present increases. After hardening has commenced, the rigid mass can no longer accommodate itself to the localised growth of solid around the cement grains and an expansion occurs if a supply of water is maintained to continue the hydration. Thus cement pastes sealed in glass tubes with excess water on top eventually expand enough to crack the glass.

The final volume change can be estimated from the differences in density of Portland cement and fully hydrated cement. Thus Eiger[3] found a density of 3·20 for a sample of ground clinker and 2·26 for the completely hydrated material dried at 110°. The water content of the latter was 19·6 per cent of the anhydrous weight. It may thus be calculated that the increase in solid volume when 1 g of the anhydrous material is converted to 1·2 g hydrated product is about 70 per cent. This volume increase refers to the solid dried at 110° and as we shall see later there is additional water associated with the gel structure of the material which is lost at this temperature. In fact Brunauer[4] has given a value of 55 per

[1] H. Steinour quoted by T. C. Powers, *J. Res. Dev. Labs Portld Cem. Ass.* **3** (1), 50 (1961); R. F. Feldman, P. J. Sereda and V. S. Ramachandran, loc. cit.
[2] H. Gessner, *Kolloidzeitschrift* **46,** 207 (1928); **47,** 65 (1929). See also T. C. Powers, *Ind. Engng Chem.* **27,** 790 (1935); R. Nacken and A. Buhmann, *Zement* **30,** 385, 397 (1941); M. del Campo, *R.I.L.E.M. Bull.*, No. 4 New Series, 18 (1959).
[3] A. Eiger, *Revue Matér. Constr. Trav. publ.* **187,** 161 (1937).
[4] S. Brunauer, *Science of Engineering Materials*, p. 444, Wiley, New York 1957.

cent based on densities of 3·1 for cement and 2·5 for hydrated cement and Powers has calculated that the volume of the saturated cement gel is about 2·2 times that of the anhydrous cement.

The densities of some compounds of interest in connection with cements are as follows:

Anhydrous		Hydrated	
Portland cement	3·0–3·2	Completely hydrated	2·13
CaO	3·32	$Ca(OH)_2$	2·23
MgO	3·58	$Mg(OH)_2$	2·40
$3CaO.SiO_2$	3·12–3·15	Completely hydrated	2·15
Alite	3·14–3·25	$3CaO.2SiO_2.3H_2O$ (tobermorite)	2·44
α $2CaO.SiO_2$	3·035	$3CaO.2SiO_2.3H_2O$ (afwillite)	2·63
α' $2CaO.SiO_2$	3·40	$CaO.SiO_2.0{\cdot}35H_2O$	2·67
β $2CaO.SiO_2$	3·28	$3CaO.Al_2O_3.6H_2O$	2·52
γ $2CaO.SiO_2$	2·97	$4CaO.Al_2O_3.13H_2O$	2·02
$3CaO.Al_2O_3$	3·04	$4CaO.Al_2O_3.19H_2O$	1·80
$12CaO.7Al_2O_3$	2·69	$2CaO.Al_2O_3.8H_2O$	1·95
$CaO.Al_2O_3$	2·98	$CaO.Al_2O_3.10H_2O$	1·72
$CaO.2Al_2O_3$	2·91	$4CaO.3Al_2O_3.3H_2O$	2·71
$4CaO.Al_2O_3.Fe_2O_3$	3·97	$CaSO_4.2H_2O$	2·32
$CaCO_3$ (calcite)	2·71	$3CaO.Al_2O_3.3CaSO_4{\cdot}32H_2O$	1·73
		$3CaO.Al_2O_3.CaSO_4.12H_2O$	1·99

Water in set cement

The distribution of water in set cement has long attracted attention and many attempts have been made to divide it into free water, capillary water, gel water and water combined in hydrated cement compounds. Various divisions can be obtained by determining the water remaining after drying at reduced vapour pressures or at higher temperatures, or by studying the amount of water that can be frozen or removed with solvents.[1] In effect, all such methods divide the water content into groups corresponding to certain reduced vapour pressures, but the divisions must be somewhat arbitrary since the vapour pressure-water content curve for set Portland cement is a continuous one showing no breaks such as are obtained with crystalline hydrates. In view of the gel structure of the mass and its heterogeneous nature, it is hardly surprising that any such effects are masked. Further, there is no particular vapour pressure that divides gel water from that combined as water of hydration, or indeed gel water from capillary water. Similarly, on heating at, say, 105°, some water of hydration is lost for at this temperature the calcium sulphoaluminates and the hexagonal tetracalcium aluminate hydrate, for instance, lose part of their combined water. Despite these limitations, the study of the way in which water is lost from set cement on drying at progressively lower vapour pressures, or increasing temperatures, has thrown much light on the properties of the material.

[1] See F. M. Lea, *Cement* **5,** 395 (1932); S. Giertz-Hedstrom, *Sym., Stockholm 1938*, 505; T. C. Powers and T. L. Brownyard, *Proc. Am. Concr. Inst.* **43,** 101 (1947) et seq. for general surveys.

The content of water retained in a set cement on drying depends on the age of the specimen, the speed of hydration of the cement, the water : cement ratio, and the conditions of drying. The relative amounts retained under some different conditions of drying are shown below, taking the quantity retained over a magnesium perchlorate dihydrate-tetrahydrate mixture as unity.

Drying agent	Vapour pressure at 25°	Relative amount of water retained by set Portland cement
$Mg(ClO_4)2H_2O$–$4H_2O$	0·008 mm Hg	1·0
P_2O_5	0·00002 mm Hg	0·8
Conc. H_2SO_4	<0·003 mm Hg	1·0
Ice at −79°	0·0005 mm Hg	0·9
Heating at 50°	—	1·2
Heating at 105°	—	0·9

For a well-hydrated cement, the amount of water retained at 105° is of the order of 20 per cent by weight of the anhydrous material, and for a completely hydrated cement about 25 per cent.

Powers and Brownyard have classified the water in set cement into non-evaporable water—arbitrarily defined as that retained on drying to constant weight at 23° *in vacuo* over $Mg(ClO_4)_2.2H_2O$–$4H_2O$—and evaporable water which is the difference between the former value and the amount held in the saturated surface-dry condition. In a later revision[1] of the method, the samples were brought into equilibrium with the somewhat lower vapour pressure of water over ice at a temperature of −79° (dry-ice/alcohol). The non-evaporable water by the former method is about 1·08 times that obtained with the latter. The non-evaporable water is considered to be a measure of the chemically combined water, though it can only be an approximate one, since calcium sulphoaluminate, the hexagonal hydrated calcium aluminates and tobermorite lose some of their water of crystallisation at this vapour pressure. It reaches a maximum of about 28 g per 100 g anhydrous cement in a completely hydrated material. The evaporable water largely represents that which is present in capillaries in the mass or held by surface forces in the gel substance itself. The non-evaporable water content increases progressively as hydration proceeds, but the evaporable water content falls as the capillary voids in which it is contained decrease in volume because they become partly filled with hydration products. This is illustrated by some curves of Powers and Brownyard, shown in Fig. 67, in which the amount of water is plotted against the relative vapour pressure (p/p_s). The vapour pressure over a set cement remains at the saturation value as long as free water is present, but when this is removed and the water surface recedes into the capillaries, the vapour pressure drops. When a very low vapour pressure is reached the only water remaining is that which has been defined as non-evaporable. This is represented in Fig. 67 by the points at $p/p_s = 0$ and we see that it rises from 8 per cent at 7 days to 17 per cent at a year. Over the same period the total water content held at saturation ($p/p_s = 1$) increases only from 32·5 to 36·3 per cent and hence the evaporable water has fallen from 24·5 to 19·3 per cent.

[1] L. E. Copeland and J. C. Hayes, *Bull. Am. Soc. Test. Mater.*, No. 194 (1953).

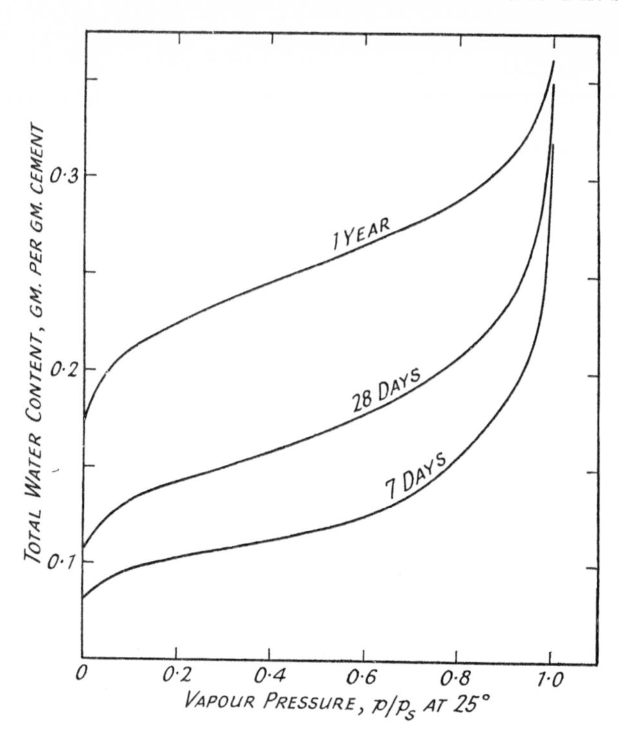

FIG. 67 Relation between water content and vapour pressure for set cement.

The surface area of set cement can be determined by the adsorption of water vapour.[1] The measurements depend on the use of the BET (Brunauer, Emmett and Teller) equation to determine the amount of adsorbed water required to form a monomolecular film over the surface of the solid. The specific surface of the paste is calculated from the equation

$$S = a_1 \frac{V_m N}{M}$$

where S is the specific surface in cm^2/g, a_1 the area covered by a single adsorbed gas molecule, N is Avogadro's number, M the molecular weight of the adsorbed gas, and V_m is the weight in grams of gas adsorbed per gram of the solid. The surface area of hydrated cement measured in this way is about 200 m^2/g dry weight. It does not vary much for different cements, as is illustrated by the data[2] in Table 33.

[1] T. C. Powers and T. L. Brownyard (loc. cit.); R. L. Blaine and H. J. Valis, *J. Res. natn. Bur. Stand.* **42,** 257 (1949).
[2] T. C. Powers, *Sym., Washington 1960*, 577.

TABLE 33 Calculated surface area of hydrated cement

Cement	Calculated compound composition				S_{cement}	$S_{gel\ only}$
	C_3S	C_2S	C_3A	C_4AF	m² per gram	
A	45	26	13	7	219	267
B	49	28	5	13	200	240
C	28	58	2	6	227	255
D	61	12	10	8	193	235

The surface area S_{cement} refers to the whole set paste. This contains usually some 20–30 per cent calcium hydroxide which has a negligible surface area and if this is corrected for we arrive at the values given under $S_{gel\ only}$. These large values for the surface area indicate that the individual particles making up the gel are of colloidal dimensions. Though, as we shall see later, there are uncertainties about the applicability of the BET equation, and therefore also of the precise values shown, this does not affect the conclusion regarding the order of size of the particles.

A comparison of the surface area, so measured, of the gel substance and the non-evaporable water content, has shown them to be more or less proportional. It is convenient for the present purpose to express the surface area in terms of the weight of water adsorbed in a mono-molecular layer and the following relationship then obtains:

$$V_m \approx kw_n$$

where V_m is the weight of water adsorbed per g of dry paste in forming the mono-molecular layer and w_n is the weight of non-evaporable water. The value of k seems to vary from about 0·2 to 0·3 for different cements, but Powers adopts an average value of 0·25. Thus a set cement containing, say, 20 per cent of non-evaporable water, can take up about 5 per cent of water in forming a mono-molecular film over the internal surface of the gel. This illustrates the very high surface area of the ultimate particles that make up the gel. The relationship between the surface area of the gel and the non-evaporable water content, and also the fact that the latter has been found to increase proportionately to the solid volume of the cement paste, indicates that the non-evaporable water is a measure of the quantity of cement gel present.

It was also found by Powers and Brownyard that the evaporable water could be divided into two parts, that held in the gel mass and that held in capillaries in the set cement structure. The division between these is somewhat arbitrary, but there is evidence to indicate that at saturation the maximum amount of evaporable water than can be held as gel water is of the order of $3V_m$.[1] This means that a set cement gel containing, say, 20 per cent non-evaporable water can adsorb in its internal voids some 15 per cent additional water at saturation, and at lower vapour pressure, of course, progressively less.

[1] T. C. Powers, *Sym., London 1952*, 426. In earlier papers a value of $4V_m$ had been used.

It also appears that below relative vapour pressures, variously estimated at 0·3 to 0·45, the capillaries are empty and all evaporable water present is in the gel. Up to this point, the amount of evaporable water held is found for any one cement to depend only on the content of non-evaporable water and thus on the quantity of cement gel. At higher relative vapour pressures, the water in the gel increases further, but water also starts to condense in capillaries for the increase in evaporable water no longer bears any relation to the amount of gel and varies with the water : cement ratio on which the volume of capillaries depends. Some evidence suggests that the capillary water does not become substantial until relative humidities approaching 0·8 are reached. A distinction is thus drawn between the ultra-fine pores in the hydrated cement gel and the capillary pores which form sub-microscopic channels in the set material. This distinction between gel and capillary pores is perhaps somewhat artificial and the real distinction is between small and large pores which overlap in their range and cannot be entirely assigned to the gel and the capillaries respectively.[1] The smallest pores in a cement paste of w/c 0·35 have a hydraulic radius (volume/internal surface) below 3·4 Å and the largest ones about 70 Å with no evidence of a bimodal distribution. The distinction between small and large pores is useful in that it is the large pores in the set cement that are mainly responsible for the permeability of set cement and its vulnerability to frost action.

In a set cement, one can thus picture the water as being held in three ways:

1. As water of hydration of set cement compounds;
2. As water adsorbed in the gel formed by the set cement compounds;
3. As water present in capillary voids in the set cement structure.

In a mortar or concrete there will also be capillary voids existing between the set cement and the aggregates. If we restrict our attention to a neat cement paste, we can represent its composition at different stages of hydration by some figures, derived from the data of Powers, given in Table 34.

TABLE 34 Composition of saturated neat cement pastes

	Original mix		Successive stages of hydration			
	Per cent volume	Per cent weight	Per cent volume	Per cent weight	Per cent volume	Per cent weight
Unreacted cement	38·5	66·6	29	50	12	20
Reacted cement with its non-evaporable water	—	—	16	20	44	56
Gel water	—	—	6	3	19	10
Capillary water	61·5	33·3	49	27	25	14

The data illustrate the rise in the content of gel water and the drop in the capillary water as hydration proceeds. In the limit the capillary water would drop to zero when the increase in solid volume produced by hydration became equal

[1] R. Sh. Mikhail, L. E. Copeland and S. Brunauer, *Can. J. Chem.* **42**, 426 (1964); see also E. Eipeltauer, W. Schilcher and W. Czernin, *Zement-Kalk-Gips* **17**, 543 (1964) for data on pores in C_3S pastes.

to the original water content, for all the capillaries would then be filled with solid hydration products. The composition of set cement at various stages of hydration is illustrated[1] diagrammatically in Fig. 68 which shows the volumes of the

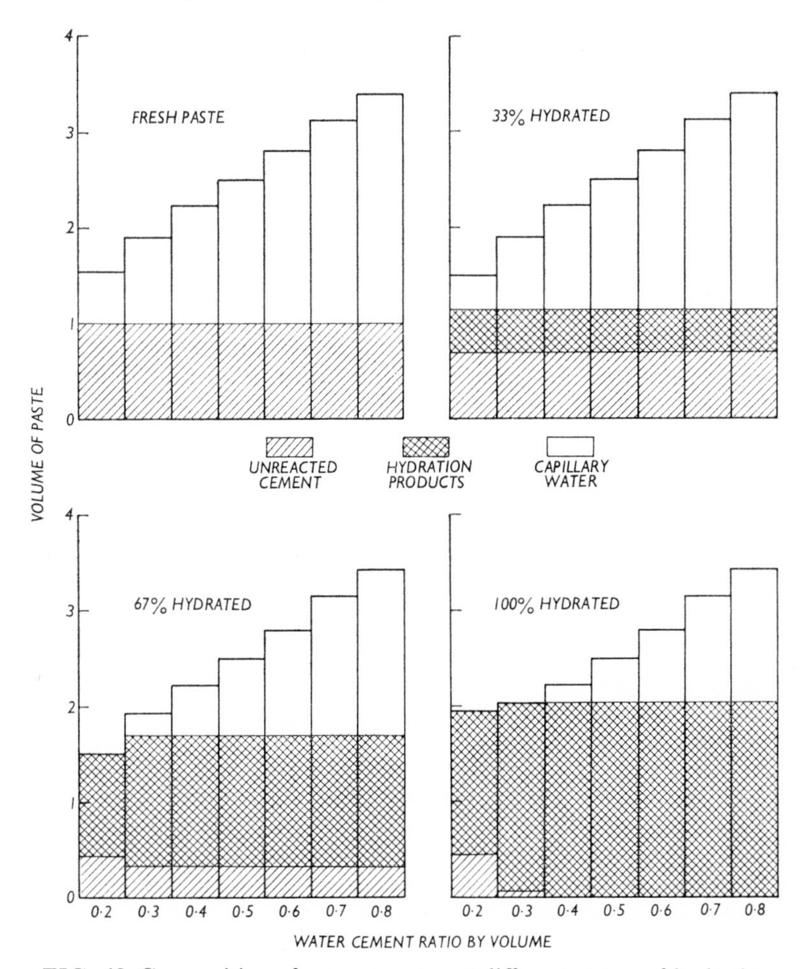

FIG. 68 Composition of cement pastes at different stages of hydration.

unreacted cement, hydration products and capillary water produced from one volume of cement and various water : cement ratios at different stages of hydration.

It would be expected that if the set cement gel could be transformed into a non-colloidal material, its capacity to absorb water would drop to a small value and that water would then only be taken up in the capillary channels. This in fact happens if the cement is cured in high-pressure steam, as may be illustrated by the data in Table 35.

[1] T. C. Powers, *Bull. Res. Dev. Labs Portld Cem. Ass.*, No. 90 (1958).

TABLE 35 Effect of high-pressure steam curing on water adsorption

Relative vapour pressure	Evaporable water content (g/g) Anhydrous cement	
	Cured 73° F 28 days	Autoclaved at 420° F for 6 hours
0·04	0·022	0·0015
0·53	0·069	0·0038
0·70	0·097	0·0052
0·81	0·128	0·0081
0·88	0·162	0·0131
0·96	0·223	0·0300
1·00	0·437	0·3010
Non-evaporable water content	0·154	0·162

Data by Powers and Brownyard on a cement-pulverised silica paste.

Though the amount of hydrated cement is about the same in both samples, the autoclaved material takes up most of its evaporable water at high vapour pressures, indicating that the set cement itself has lost its colloidal properties and that the water is condensing in the capillary pores of the set mass. This loss of colloidal material is also shown[1] by the fall in surface area of a set cement from about 100 m^2/g at ordinary temperatures to about one-tenth this value on autoclaving at 180°.

Since the extremely fine crystals that make up the cement gel at ordinary temperatures must be metastable with respect to larger crystals, there must be a tendency on long ageing for the gel to become less fine-structured. This would be shown by a gradual fall in the internal surface area as measured by V_m. The internal surface would thus increase to a maximum as hydration proceeded and then start to decrease as ageing effects come into play. There is some evidence that this does occur.[2]

We noted earlier that, on drying a set cement for the first time, some irreversible changes occur, and that these are to be regarded as one step in the transformation of the gel towards a more stable state, though it is only a small step and the material still remains a typical gel. Thus Powers and Brownyard, confirming earlier work by Jesser, found that the initial sorption curves, such as were shown in Fig. 67, are not entirely retraced on successive desorptions and adsorption. The initial curve is in fact unique and later curves show a somewhat lower water content at any given vapour pressure.

As we have seen, a considerable edifice has been built on the calculation of surface area of set cement from adsorption data and on the distribution of water

[1] W. C. Ludwig and S. A. Pence, *Proc. Am. Concr. Inst.* **52,** 673 (1956).
[2] C. M. Hunt, L. A. Tomes and R. L. Blaine, *J. Res. natn. Bur. Stand.* **64A,** 163 (1960).

based on an arbitrary distinction between evaporable and non-evaporable water. The question may be asked how valid are these conclusions.

The application of the BET equation to measure surface area assumes that the adsorption is a reversible process over the range of low relative vapour pressures, usually up to about a P/P_0 value of 0·35, involved in the formation of a monomolecular film.

There is evidence[1] however that the adsorption and desorption curves for water vapour on set cement are not identical even at these low relative pressures indicating that the process is not a reversible one. Part of the water taken up is considered not to be physically adsorbed on the surface but to enter into the layered structure of the cement gel as interlayer water. Evidence[2] has also been obtained by nuclear magnetic resonance studies on set cement indicating that much of the evaporable water has similar properties to interlayer water in clays. If adsorption and evaporation of interlayer water can occur then only an uncertain part of the water taken up will represent the surface adsorbed water which is the basis of the BET equation. The surface area can also be measured by the adsorption of nitrogen, for which the process is reversible, and the values so obtained are only about one-third to one-fifth of those calculated from adsorption of water vapour. Some workers[3] have attributed this to the pores being too narrow, or having too narrow an entrance, to admit nitrogen which has a larger molecular area, 16·2 $Å^2$, than water, 11·4 $Å^2$. If the nitrogen adsorption is preferred as a measure of surface area it still remains large and corresponds to particles of colloidal dimension, but some of the quantitative calculations, as for example of the size of the gel particles, made from the water adsorption data require appropriate adjustment.

The distinction between evaporable and non-evaporable water is not a distinction between physically and chemically bound water. Thus at a vapour pressure of 0·5 μ Hg there is a loss of water of hydration from some of the set cement compounds, e.g. calcium sulphoaluminate and tetracalcium aluminate hydrate. The water content of tobermorite decreases from 1·5H_2O per mole SiO_2 in the saturated state to 1·0 mole at the above vapour pressure and this may represent a displacement of interlayer water from the crystal lattice.[4] There are moreover real difficulties in drawing clear-cut distinctions between combined water and adsorbed water when dealing with crystals of only a few molecules thick with unbalanced atomic forces at the surface. The most firmly bound adsorbed water is more strongly held than the most loosely bound combined water. On the other hand the deductions from permeability measurements broadly confirm those arrived at from adsorption measurements as to the size of the gel particles. We can best conclude that the distinction between evaporable and non-evaporable water is in practice a useful one without being able to define too closely what it represents. We must also regard the calculated surface areas

[1] R. F. Feldman, *Sym., Tokyo 1968*; R. F. Feldman and P. J. Sereda, ibid.; *J. appl. Chem.* **14**, 87 (1964); S. J. Gregg and S. W. Sing, *Adsorption, Surface Area and Porosity*, Academic Press, London and New York, 1967.

[2] P. Seligmann, *J. Res. Dev. Labs Portld Cem. Ass.* **10** (1), 52 (1968).

[3] T. Sh. Mikhail, L. E. Copeland and S. Brunauer, *Can. J. Chem.* **42**, 426 (1964).

[4] G. L. Kalousek, *Proc. Am. Concr. Inst.* **51** (2), 233 (1955); S. Brunauer, *Am. Scient.* **50** (1), 210 (1962).

as an approximation rather than as a well-defined physical constant. There is nevertheless no doubt as to the decisive role played by the surface area in determining the mechanical properties of cement pastes. A more contentious deduction from the data is that the cement gel has a minimum porosity of about 28 per cent representing voids that cannot be filled by hydration products because these voids are of such small dimensions that crystal nuclei cannot form and grow in them and hydration ceases. This 28 per cent limit was arrived at by Powers[1] from consideration of the relation between the evaporable and total water content of mature pastes which indicated that below a w/c ratio of 0·35 the porosity did not continue to decrease with fall in w/c in the way expected. The limiting porosity corresponds to a specific volume of the cement paste of 0·567 cc/g of dry hydrated cement. However later work[2] has thrown doubt on this conclusion and it seems that hydration can proceed in the gel towards complete hydration to give a lower porosity. The point is perhaps of more academic than practical interest in that many years of hydration are required before even Powers' limit is reached.

From these various data, supplemented by the results of X-ray and electron-microscope studies, the cement gel is to be regarded as a porous solid made up by the aggregation of tobermorite gel particles each containing some 30 000–50 000 molecules and with a surface area of some 200 m^2/g. These particles are sheet-like in form and only some three or four molecules (30–40 Å) thick but extending in the two other dimensions to a hundred or more times the thickness with one dimension going up to one or several microns. The surface energy at 23·5° of tobermorite,[3] defined as the difference between the total energy of a molecule in the surface layer and inside the body of the substance, is 386 ± 20 ergs/cm^2, a value intermediate between that for calcium hydroxide (1180) and hydrous silica (129). The mean energy of binding of the evaporable water at the minimum porosity of 28 per cent is about 3600 cal per mole evaporable water.[4] This is the excess over the normal heat of condensation which is about 10 500 cal/g. The pores in the gel are smaller than the solid particles and are thin slits with an average width of some 15–30 Å. The solid volume of the hydrated cement is 2·2 times that of the anhydrous cement so that over half the hydration products have to be deposited in the spaces originally filled with water.

The gel is one component of the cement paste, a second is the residues of unhydrated cement and calcium hydroxide crystals, and the third is the residue of the space originally occupied by water which has not been filled by hydration products. These spaces are the capillary pores, as distinct from the pores within the gel itself, and they may range widely in size with diameters up to several hundred angstroms or even to 0·1 μ. In a mortar or concrete there are still larger pores formed by air entrapped under aggregate particles. Below a relative humidity of about 30–45 per cent all the adsorbed water is held in the gel while at higher humidities it is also held by surface adsorption or capillary condensation

[1] T. C. Powers and T. L. Brownyard, *Proc. Am. Concr. Inst.* **43,** 669 (1947); L. E. Copeland and J. C. Hayes, ibid. **52,** 633 (1956); T. C. Powers, *Sym., Washington 1960*, 577; R. H. Mills, *Spec. Rep. Highw. Res. Bd*, No. 90, 406 (1966); F. S. Fulton, *Rep. Lab. S. Afr. Portld Cem. Inst.*, No. SF2 (1962).

[2] R. Sh. Mikhail, L. E. Copeland and S. Brunauer (loc. cit.).

[3] S. Brunauer, D. L. Kantro and C. H. Weise, *Can. J. Chem.* **37,** 714 (1959).

[4] T. C. Powers and T. L. Brownyard, *Proc. Am. Concr. Inst.* **43,** 549 (1947).

in the capillary pores. Capillary condensation can hardly be operative until the capillary diameter is several water molecules wide. A relative humidity of 40 per cent corresponds by Kelvin's equation to a diameter of only about 12 Å, of 50 per cent to 15Å, and 60 per cent to 21 Å. A molecule of water has dimensions of about 2·7 Å. All the evaporable water is mobile[1] and thus takes part in permeation of water through the paste under hydrostatic gradient. The true density of hydrated cement is difficult to determine since the measurement of porosity gives differing values according to the displacing fluid used. Values[2] range from $d = 2{\cdot}43$ to 2·59.

The permeability of the paste is of the order of 1×10^{-12} cm/sec, an order of magnitude comparable with that of many dense natural rocks. The permeability of concrete is some hundreds of times higher than this owing to small fissures and cracks in the set cement around the aggregate particles.

Shrinkage

As well as changing in water content, a set cement also undergoes an irreversible diminution in volume on drying for the first time. This irreversible shrinkage takes place in two ways. The cement gel undergoes an inward shrinkage during which existing voids are increased in size, or new ones formed, and contemporaneously the whole structure suffers a small contraction in its outward dimensions. This contraction of the outward form of the mass is rather less than the inner shrinkage, thus accounting for a small increase in pore space that occurs. We can illustrate the changes in the water-content/vapour-pressure relation, and the irreversible shrinkage by the results of Jesser.[3] Plotting the total water content of a set cement against the vapour pressure as the material was dried Jesser found that on the first drying a curve *ABCD* was obtained (Fig. 69). If, after drying down to the point *D*, which corresponded to the condition when the material was stored over sulphuric acid, and only the non-evaporable water remained, the set cement was allowed to adsorb water again at increasing vapour pressures, the curve *DG* was obtained. On further wetting and drying this curve is retraced and it thus represents a reversible condition.[4] The initial curve *AD* is irreversible and can never be retraced. If drying is stopped at *B* and the set cement allowed to absorb water, the curve *BE* is obtained. This curve is approximately reversible as long as the water content is kept between *B* and *E*. Once, however, the material has been dried further, say to *C*, it can never be regained, and a new curve, *CF*, is followed on rewetting. This curve similarly is only reversible as long as the material is never dried beyond the point *C*. Both *BE* and *CF* belong, therefore, as does the curve *ABCD*, to unstable states which, once passed, cannot be regained.

If instead of plotting the water content as ordinate, the length of a specimen is plotted, a similar series of curves is obtained. The final curve *DG* then represents the reversible length change on wetting and drying, while *ABCD* represents

[1] T. C. Powers, H. M. Mann and L. E. Copeland, *Spec. Rep. Highw. Res. Bd*, No. 40 (1959).
[2] T. C. Powers, *Sym., Washington 1960*, 577.
[3] *Zement* **16,** 741 (1927); **18,** 158 (1929).
[4] This reversible curve, as Powers has shown, actually contains a hysteresis loop, the paths of desorption and adsorption not being identical as is common in colloidal materials.

the length change on the first drying. The irreversible shrinkage on first drying is given by an ordinate similar to *AG*. By drying only to intermediate points, such as *B* and *C*, and then rewetting, portions only of the irreversible and reversible movements are obtained.

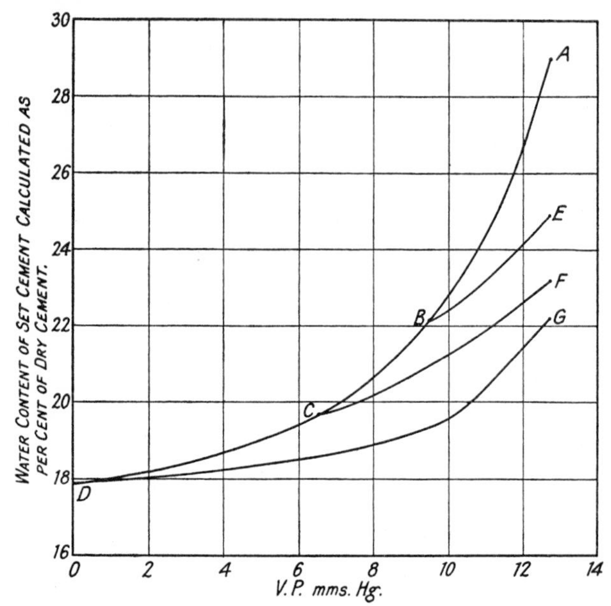

FIG. 69 Vapour-pressure/water-content curves for set Portland cement.

These reversible and irreversible shrinkages are illustrated in Fig. 70 by the results obtained[1] on slabs moist cured for eight months and only 1–2 mm thick,

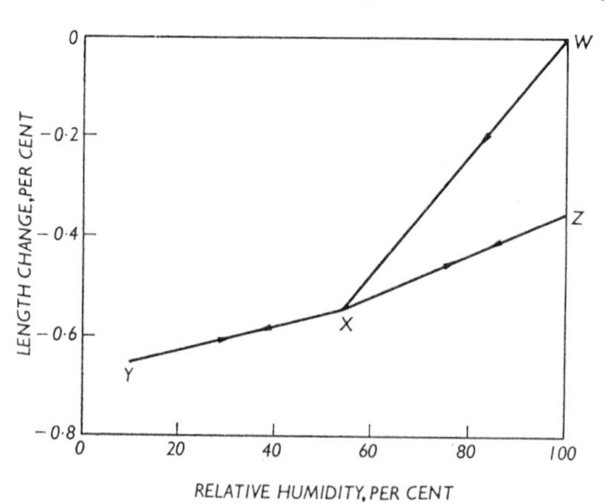

FIG. 70 Vapour-pressure/length-change curves for set Portland cement (w/c 0·6).

[1] R. A. Helmuth and D. Turk, *J. Res. Dev. Labs Portld Cem. Ass.* **9** (2), 8 (1967).

so that equilibrium could be reached quickly throughout the slab, and with careful protection against carbonation. On the first drying from 100 to 10 per cent relative humidity the curve WXY was obtained and on resorbing the moisture the curve YXZ. The length of the ordinate WZ represents the irreversible shrinkage and the curve WX can never be retraced. Successive drying and wetting cycles after the first follow fairly closely the path ZXY. Whilst the numerical values apply only to the particular specimen of neat cement measured, the general behaviour is typical also of mortars and concretes. The actual values are of course considerably lower for mortars and concretes than for neat cements. The irreversible shrinkage increases with the water-cement ratio and porosity of the cement paste but once the paste had been sufficiently dried the reversible shrinkage became independent of porosity. It is clear that the irreversible shrinkage is a sign of alteration in the structure of the set cement.

The length-change/water-loss relations at one particular relative humidity (47 per cent) are shown in Fig. 71 for the first drying of similar thin slabs for a

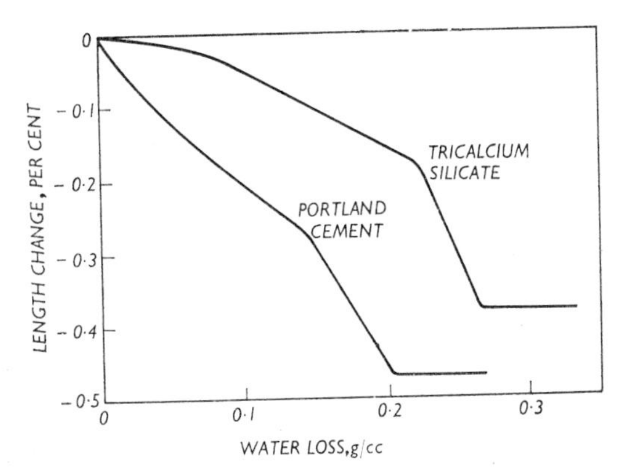

FIG. 71 Length-change/water-loss curves during drying of pastes (w/c 0·6) at 47 per cent relative humidity.

Portland cement and a tricalcium silicate paste. It will be clear from the change in slope of the curves that more than one shrinkage mechanism is operative. This is also evident from the initial vapour-pressure/length-change curve[1] for a 1 : 2 mortar shown in Fig. 72 over the range from 0 to 100 per cent relative humidity. Shrinkage does not commence until free water has disappeared from the material and the vapour pressure commences to drop. A typical curve for concrete is shown in Fig. 73.

Curves[2] showing the length change plotted against water loss over a wide range of relative humidities from 100 to 7 per cent are shown in Fig. 74 for the first drying of Portland cement pastes with water : cement ratios of 0·35 and 0·5.

[1] K. Kamimura, P. J. Sereda and E. G. Swenson, *Mag. Concr. Res.* **17** (50), 5 (1965).
[2] H. Roper, *Spec. Rep. Highw. Res. Bd*, No. 90, 74 (1966).

The curves show four sections of differing slope, again indicating that different mechanisms of shrinkage are operative.

The various theories of shrinkage fall into three broad groups:

1. Capillary tension
2. Surface adsorption
3. Interlayer water in the crystals.

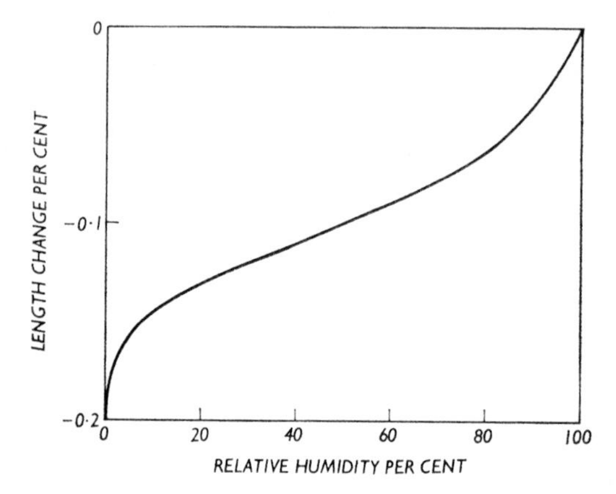

FIG. 72 Vapour-pressure/length-change curves for 1 : 2 mortar (w/c 0·4).

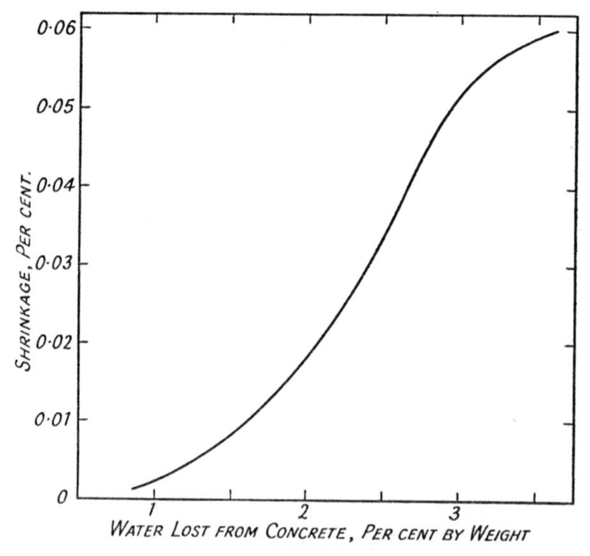

FIG. 73 Relation of drying shrinkage to water loss of concrete.

In capillary tension theory, shrinkage is associated with the increase of the tension at the water miniscus in the capillaries as drying proceeds. We know from

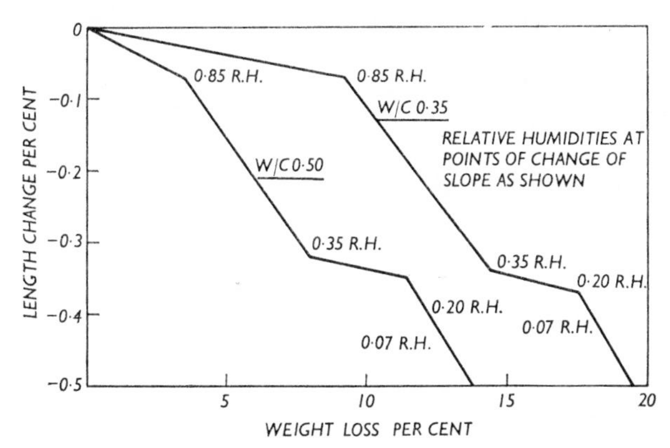

FIG. 74 Length-change/water-loss curves for cement pastes during drying to progressively lower humidities (*Roper*).

the Kelvin equation that a certain radius of curvature of a water miniscus corresponds to a given reduction in vapour pressure. As water recedes in the capillaries this curvature increases and the surface tension forces produce a tension in the water remaining in a capillary. This liquid tension must be balanced by an equivalent compressive stress in the solid, producing a shrinkage deformation. The force acting is the product of the fractional area of the water-filled capillaries in any cross section and the tension in the water. As the water progressively recedes in the capillaries the fractional area will decrease and at some point during drying the product will pass through a maximum and then start to decrease. On this theory the shrinkage should do likewise, but this does not occur with cement because there are other forces also operative.

The surface adsorption theory is derived from the effect of the adsorption of a gas or vapour at a solid surface in decreasing the surface tension of the solid and releasing the compressive stress which it creates in the solid. It is concerned with the adsorption of water molecules on the large surface area of the cement gel. This action can be pictured dynamically in another way. The adsorbed water molecules in a film at most a few molecules thick, though attached to the surface, have freedom of movement in directions parallel to it. The motion of these molecules exerts a wedge-like action from bombardment at sharp re-entrant angles in the surface and thereby brings about an expansion of the solid. As water is adsorbed therefore the solid expands, or conversely as it is removed the solid contracts. There is a simple relation,[1] analogous to Hook's law, between the expansion (x) as water is adsorbed and the two-dimensional pressure (F) exerted by the adsorbed water film.

$$x = \lambda F \text{ where } \lambda \text{ is a constant}$$

Yet another way[2] to look at the effect of adsorbed water is to regard the film as having a 'disjointing' effect forcing the solid surface of the gel particles apart.

[1] D. H. Bangham and N. Fakhoury, *J. chem. Soc.*, 1931, 1324.
[2] T. C. Powers, *Proc. int. Conf. Structure of Concrete, London 1965*, Concrete Association London 1969, 319.

Both these phenomena can be looked at in terms of the swelling pressure of a gel. If we consider a dry gel immersed in water it tends to take up water because its vapour pressure is very low. This causes the gel to expand, but if this tendency is resisted, as it is by the cohesion of a rigid gel, a swelling pressure develops which is related to the vapour pressures of the water inside the gel and that of the surrounding water by the Katz equation:

$$P = -\frac{RT}{MV_0} \log p/p_s$$

where P is the swelling pressure, V_0 the specific volume of water and p the vapour pressure of the gel water. This equation, being a purely thermodynamic relation, tells us nothing of the mechanism of the action and covers equally capillary and surface effects. Further, the pressure calculated is the potential pressure that could arise if no expansion occurred and not the pressures that actually exist when the material expands as it takes up water. In its actual application we should consider the reverse of this process for the cement sets in the fully saturated condition with its gel fully expanded. As it dries and water leaves the gel negative pressures, or tensions, will be set up inside it. The resulting contraction is resisted by the rigid framework of the mass, which is thus put into tension. The actual tensile forces developed are difficult to measure but as drying proceeds they mount[1] to tens of kilograms per square centimetre. If the test specimen is completely restrained from shrinking, then it breaks itself when the stress exceeds its tensile strength.

We have seen in a previous chapter that the lattice spacing of the hydrated calcium silicate crystals changes with their water content. A similar effect is found with calcium tetra-aluminate hydrate and with calcium sulphoaluminate. These crystals all have a layered structure and water can enter between the layers expanding the lattice. Such changes in lattice spacing would be accompanied by an overall movement of the solid. It is significant in this connection that the surface area of set cement as measured by the adsorption of water vapour by the BET method is several times as large as that obtained by the adsorption of nitrogen.[2] This can be attributed to the penetration of water molecules into the lattice layers, or into intercrystalline spaces that are not accessible to nitrogen.[3] There seems to be an analogy with the clay minerals. It has been found[4] that with kaolinite, which has a non-expanding lattice, the adsorption of nitrogen and water vapour takes place on the same surface, but with montmorillonite, which has an expanding lattice, the water enters the structure as well as being adsorbed on the outer surface, whereas nitrogen is adsorbed on the latter only. A variety of

[1] P. M. Pflier, *Zement* **26,** 659 (1937); K. E. C. Nielsen, *R.I.L.E.M. Bull.*, No. 1, 11 (1959); F. A. Blakey, *Proc. Am. Concr. Inst.* **30,** 591 (1958).

[2] R. L. Blaine and H. J. Valis, *J. Res. natn. Bur. Stand.* **42,** 257 (1949); L. A. Tomes, C. M. Hunt and R. L. Blaine, ibid. **59,** 357 (1957); C. M. Hunt, *Spec. Rep. Highw. Res. Bd*, No. 90, 112 (1966).

[3] S. Brunauer, L. E. Copeland and R. H. Bragg, *J. phys. Chem.* **60,** 116 (1956); S. Brunauer, D. L. Kantro and L. E. Copeland, *J. Am. chem. Soc.* **80,** 361 (1958).

[4] R. W. Mooney, A. G. Keenan and L. A. Wood, *J. Am. chem. Soc.* **74,** 1367, 1371 (1952).

evidence for the entry of water to form interlayer water in the lattice structure of the set cement crystallites has been advanced by Sereda[1] and his co-workers. Thus studies of the adsorption curves and of the influence of moisture on the elastic modulus of set cement have yielded data supporting this concept.

There is still some controversy about the relative part played by these various mechanisms in the shrinkage of cement. Capillary effects operate at the higher relative humidities down to 40 per cent relative humidity or even lower on drying.[2] Thus Feldman and Sereda[3] found a large water loss between about 40 and 30 per cent relative humidity with little change in length over this interval. This was attributed to a balance between the expansion that would be caused by the release of the compressive stress induced by the capillary menisci as the capillaries empty and the continuing contraction caused by removal of interlayer and adsorbed water. At lower vapour pressures the surface desorption effect becomes operative. Alongside the capillary and adsorption effects the evaporation of interlayer water occurs. Loss of interlayer water on first drying appears to occur primarily below about 30–35 per cent relative humidity. The continuity of the shrinkage process is due to the overlapping of these various mechanisms.

The first drying of a set cement is a special case since this leads to a large irreversible shrinkage. This can best be explained by the assumption that the closing of the gaps between adjacent surfaces leads to the formation of new additional chemical bonds that cannot be broken when water is taken up again. More generally we can say that the cement gel as formed initially is a metastable system of very small particles with high free surface energy. It tends to stabilise itself by reducing its surface area through the merging of small particles into larger aggregates or the growth of crystals. One sign of this is that the irreversible first-drying shrinkage decreases with prolonged wet curing.[4] Drying accelerates and extends this process and repeated drying and wetting cycles carry it rather further as shown by the tendency for the reversible movement to decrease slightly.

It must be noted that when set cement or concrete loses water the change in volume is only a small fraction of the volume of the water lost. This must be due to the restraints set up by the non-shrinking bodies present, e.g. the aggregate and unhydrated cement grains and be elastic forces in the gel itself.

There is relatively little information on the dimensional change produced by organic liquids. Benzene and carbon tetrachloride produce no significant movement but absorption of methyl alcohol[5] by dry set cement produces expansion, though less than that of water. Both the molecular diameter of the absorbent molecule and its type, polar or non-polar, are likely to be important factors.

[1] I. Soroka and P. J. Sereda, *Sym., Tokyo 1968*; R. F. Feldman, ibid.; P. J. Sereda, R. F. Feldman and E. G. Swenson, *Spec. Rep. Highw. Res. Bd*, No. 90, 58 (1966); R. F. Feldman and P. J. Sereda, *Mater. Structs.* **1** (6), 509 (1968).

[2] T. C. Powers (1965), loc. cit.; O. Ishai, *Appl. Mater. Res.* **5,** 154 (1966); R. F. Feldman, loc. cit.; R. G. L'Hermite, *Sym., Washington 1960*, 659; K. M. Alexander and J. M. Wardlaw, *Aust. J. appl. Sci.* **10,** 201 (1959).

[3] Loc. cit.; also *J. appl. Chem.* **14,** 87 (1964).

[4] R. G. L'Hermite, loc. cit.

[5] D. J. Hannant, *Mater. Structs.* **1** (5), 403 (1968).

Creep

Another property of set cement that is allied to shrinkage is creep,[1] that is the slow continued deformation under load. For engineering purposes creep has usually been regarded as independent of shrinkage, but it is clear that actually the two are interdependent. They are both manifestations of reversible and irreversible changes in the set cement paste. Since cement gel is formed in an unstable expanded condition we should expect the application of pressure, just as that of drying, to bring it closer towards a more stable condition. Such creep would be irreversible. Creep is in fact partly reversible and partly irreversible. On unloading a specimen there is an immediate elastic recovery followed by a delayed recovery leaving a part of creep not recoverable. Creep occurs under compression, tension, or torsion. The rate increases with temperature, it is greater in a wet than a dry specimen, and is greatest of all in a specimen that is simultaneously losing water by drying. Creep of concrete that has been autoclaved in high-pressure steam is much reduced.[2] Creep in concrete differs from that in metals in that it occurs down to the smallest loads at which measurements can be made.

There are numerous theories of creep and variants of them, but there is still no full explanation of all the creep phenomena.[3] The two main theories are the seepage and the viscous theories. In the seepage theory creep is ascribed to slow migration or expulsion under load of interlayer or adsorbed water or to its redistribution between its original sites and the capillary pores. When the water in these sites is first reduced by adequate drying, creep is much lower. There are numerous variants of this seepage theory according to the significance attached to the delay in re-establishing moisture equilibrium between the gel and its immediate environment or the external surroundings, or to the effect of load on the tensions in the adsorbed water films. The irreversible part of creep may involve some process in which a cement gel particle breaks its bond with another and then remakes it with a neighbouring particle thus causing a permanent deformation.

In the viscous theory the set cement is regarded as a highly viscous liquid whose viscosity increases with time. The viscous flow is itself irrecoverable but any elastic strain it imposes on the aggregate in a mortar or concrete is recoverable. The viscous flow itself we can perhaps regard as a gradual slip or sliding at surfaces bonded by weak van der Waal bonds though there is as yet no clear evidence of this. This theory and the delayed elastic effect theory mentioned below lend themselves to representation by models made up of viscous dashpots

[1] *Spec. Publ. Am. Concr. Inst.*, No. SP9 (1964); *R.I.L.E.M. Sym. Influences of Time on Strength and Deformation of Concrete, Munich 1958*; *R.I.L.E.M. Bull.*, Nos. 1, 3, 4, 5 (1959), No. 6 (1960); *R.I.L.E.M. Sym. Physical and Chemical Causes of Creep and Shrinkage of Concrete, Munich 1968*; *Mater. Structs.* **1** (6), 1968, **2** (7), 1969.

[2] F. E. Seaman, *Proc. Am. Coner Inst.* **53,** 803 (1957).

[3] I. Ali and C. E. Kesler, *Univ. Illinois Eng. Expt. Station Bull.* 476 (1965); W. Ruetz, *Materialprüfungsamt Tech. Hochschule Munich*, Report No. 62 (1965); O. Ishai, *Int. Conference on the Structure of Concrete, London 1965*, 345; T. C. Powers, ibid.; R. F. Feldman and P. J. Sereda, *Mater. Structs.* **1** (6), 509 (1968); R. G. L'Hermite, *Sym. Washington 1960*, 659; T. C. Hansen, *Swedish Cement and Concrete Res. Inst. Proc.* No. 31, *Stockholm 1960*; F. M. Lea and C. R. Lee, *Soc. Chem. Ind. 1947, Sym. on Shrinkage and Cracking of Cementive Materials, London*, p. 7.

and elastic springs in series and or parallel and from these models mathematical formulæ can be derived which express the creep and the recovery movements. Interesting as these models are they only represent the behaviour and do not explain it.

Another explanation related to the viscous flow theory is the delayed elastic effect which assumes that the load on a set cement is initially carried in part by the viscous fluid in the voids of the solid elastic skeleton delaying the full elastic response of the skeleton to the load. Gradual displacement of the viscous fluid slowly transfers more load on to the skeleton producing further elastic deformation. An older, and now largely discarded theory of Freysinnet[1] also invoked a delayed elastic effect based on an extension of the capillary condensation theory of shrinkage. In yet another theory, creep is ascribed to non-uniform shrinkage in a specimen which is simultaneously undergoing shrinkage by drying. This leads to non-uniform stresses, which change with time, between the exterior and the interior of a specimen.[2]

It seems evident that creep may well have more than a single cause and we cannot disregard the possibility that it could arise in part from slip at a cement-aggregate interface in a mortar or concrete. The formation of microcracks has also been held to be a factor particularly at loads above about half the failing load.

Thermal expansion

The coefficient of thermal expansion of set cement and concrete is influenced by the moisture content, and is higher at intermediate moisture contents than in the saturated or the dry condition. This can be related to the change in surface tension of water with temperature. A decrease in surface tension reduces the tension in the capillary water, or, what is equivalent, increases the swelling pressure of the gel, and the expansion this causes is added on to the purely thermal movement. In a saturated material no such change can occur because the capillary tension is zero, while in a dry specimen there is no evaporable water present, so there can be no capillary tension to change. If the length change for a water-soaked paste is plotted against temperature a hysteresis loop is found between the heating and cooling curves. This has been attributed[3] to diffusion of water from the cement gel to the capillaries on heating and the reverse on cooling.

Wet and dry strength

The strength of a mortar or concrete is another property that differs between the wet and the dry state. We are concerned here not with the effect of wet or dry conditions on the maturing of a concrete, but with the influence of its water content at the time it is tested. If a concrete that has been matured and dried is water-soaked there is a drop in compressive strength of some 20–40 per cent. In

[1] *Une Révolution dans le Techniques du Béton*, Paris 1936 (Librairie de l'Enseignement Technique); *J. Instr. struct. Engrs* **14,** 242 (1936); *Mag. Concr. Res.* **3** (8), 49 (1951).

[2] G. A. Maney, *Proc. Am. Soc. Test. Mater.* **41,** 1021 (1941); G. Pickett, *Proc. Am. Concr. Inst.* **38,** 333 (1942).

[3] P. A. Helmuth, *Proc. Highway Res. Board,* **40,** 315 (1961).

part this must be ascribed to the weakening of the cement gel as it adsorbs water and to the reduction of the surface forces between its particles, but it seems questionable if this is the whole story. Vogt[1] found that the bending strength decreased, instead of increasing like the compressive strength, as specimens were allowed to dry. He suggested that this could be ascribed to non-uniform shrinkage, resulting in tensile stresses at the surface over and above these applied in the bending test.

Degree of hydration of Portland cement and the strength developed

Since the development of strength in a set cement is due to its hydration, some relation may be expected to exist between the amount of hydration and the strength. The degree to which the different compounds in cement contribute to the strength is however not the same, for although the compounds C_3A, C_4AF and C_3S all hydrate rapidly, the last is responsible for the major part of the strength developed. Dicalcium silicate hydrates slowly but progressively, and its contribution to the strength follows a similar course. An exact relationship between degree of hydration and strength cannot therefore be anticipated.

The compressive strength has been considered variously to be related to the square[2] of the amount of cement hydrated and to be directly proportional[3] to it, but neither relation seems to fit the facts satisfactorily. The relation between strength, extent of hydration, and porosity, has also been expressed by Powers in terms of a function called the gel-space ratio.[4] This is the ratio of the volume of hydrated cement to the sum of this volume and that of the capillary pores. The former is related to the non-evaporable water content while the latter is the volume of water used to mix the material less the increase in space occupied by the cement that has hydrated. The Powers' equation is $S = Ax^n$ where S is the compressive strength of the cement paste, A is a constant, and n has a value between 2·5 and 3. The constant A may be taken to represent the strength of the cement gel and has a value of about 2000 to 3000 kg/cm^2. This relation of Powers is similar in principle to one used earlier by Werner and Giertz-Hedstrom,[2] Eiger[3] and others, and also subsequently by Dzulynski.[5] For a given cement a good relationship is obtained at different ages, but the strength at a given gel-space ratio, or at equal stages of hydration, is not the same for all cements and it also differs between neat cements and mortars. The former, of course, reflects the influence of cement composition on strength and the differing contributions made by the aluminate and silicate compounds. There is no data to show the effect of temperature on this gel-space relation, nor whether it applies equally to other types of cement than Portland cement.

A fair relation has been found to exist for Portland cements of widely differing composition between the compressive strength of concrete and the water com-

[1] F. Vogt, *The Flow and Extensibility of Concrete.* Norwegian Technical High School, Trondheim. 25th year Jubilee volume (1935).
[2] D. Werner and S. Giertz-Hedström, *Zement* **17,** 1002 (1928).
[3] A. J. Eiger, *Revue Matér. Constr. Trav. publ.* **335,** 161 (1937); *Cement* **7,** 231 (1935); *Tonind.-Ztg. keram. Rdsh.* **56,** 532, 558 (1932).
[4] T. C. Powers, *Bull. Am. Soc. Test. Mater.*, No. 158 (1949). A different definition of the gel-space ratio was used in the 1947 papers of Powers and Brownyard.
[5] M. Dzulynski, *Bull. Cent. Etud. Constrs. Génie civ., Liège* **6,** 99 (1953).

bined in the set cement.[1] The latter has for this purpose been arbitrarily defined as the water retained on heating to 110°, but lost at 550°, and is determined not on the concrete, but on the neat cement at the same age. In Fig. 75 is shown the

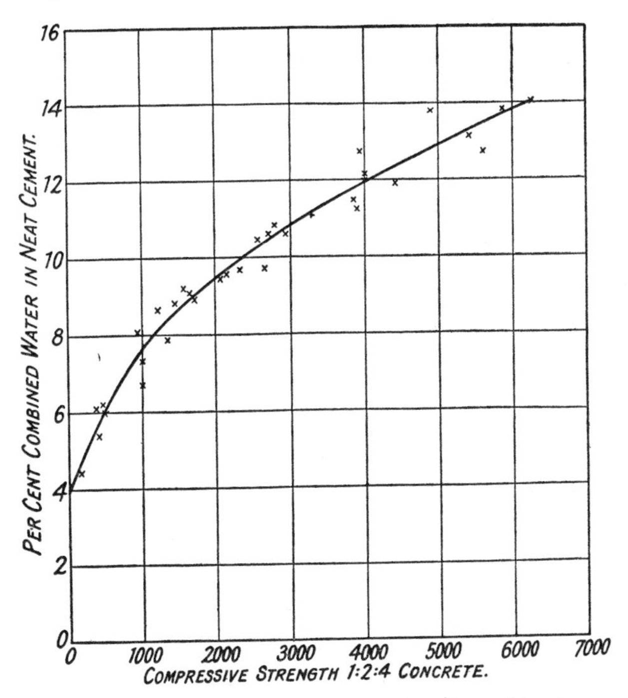

FIG. 75 Relation between compressive strength and combined water content.

compressive strength of concretes plotted against the combined water found at corresponding ages in neat cement cubes gauged with 25 per cent water. The points on the curve refer to ten different cements tested at ages from one day to six months and with calculated compound contents varying from 29 to 51 per cent C_3S, 15 to 42 per cent C_2S, 7 to 18 per cent C_3A, 2 to 10 per cent C_4AF, and the sum of the last two from 15 to 25 per cent. The curve does not start from the origin, but intersects the ordinate axis at a value of about 4 per cent combined water. A certain amount of hydration must be required before the products have sufficient cohesion to produce any strength. The curve is concave to the strength axis indicating that the strength increases at a rate more than directly proportional to the degree of hydration.

Since the tricalcium silicate which is considered mainly responsible for the early strength of cements liberates free calcium hydroxide on setting, some relation between the amount of this and the strength developed might be expected. Such a relation does exist to the extent that the calcium hydroxide content rises as the early strength rises, but the relationship is far from exact.

[1] F. M. Lea and F. E. Jones, *J. Soc. chem. Ind., Lond.* **54,** 63T (1935).

The amount of calcium hydroxide liberated at seven to twenty-eight days after gauging varies from about 5 to 10 per cent by weight of the cement.

Heat evolution in the setting of cement

The reaction of cement with water is exothermic and liberates a considerable quantity of heat. This is easily observed if a cement is gauged with water and placed in a vacuum flask or other insulated container and the temperature of the mass read at intervals. Much attention has been paid to the heat evolved during the hydration of cement on account of its considerable importance when large concrete masses are placed, as in the construction of dams.[1] Temperatures in the interior of such large masses up to 50° above the original temperature of the concrete mass at the time of placing have been recorded, and found to persist for a prolonged period; many years may elapse before the temperature in the centre of the mass sinks to its ultimate value. The principal cause of the serious cracking which has occurred in large concrete masses is the shrinkage which takes place as the mass cools from the high temperature attained during setting and early hardening. Although considerable attention has been paid to the temperature rise in dams, it seems that the highest temperatures may not always be reached in structures of this type. The mixes in these structures are usually leaner than those used in bridge abutments, retaining walls, piers, etc., and it is in such structures that some of the highest temperatures may be encountered.

Some examples of the temperatures recorded in the interior of masses of concrete are shown in Table 36.

TABLE 36 Temperature rise in concrete masses

Structure	Proportions cement : aggregate	Maximum temperature rise ° C	Maximum temperature reached ° C
Boonton Dam, U.S.A.	1 : 14	8	32
Lake Spaulding Dam Extension, U.S.A.	1 : 9	24	33
Stevenson Creek Dam, U.S.A.	1 : 5	19	47
Ariel Dam, U.S.A.	1 : 10	42	57
Laggan Dam, Scotland	1 : 8	25	45
Tongland Dam, Scotland	1 : 9	31	41
Retaining Wall, England	1 : 4	40	48
Caisson, U.S.A.	1 : 3·2	39	69

The maximum temperature rise observed is equivalent to over 80 calories per gram of cement. Artificial cooling has been resorted to in some modern dams to reduce the temperature rise and use has been made of cements with a reduced heat evolution.

The heat evolution of cements can be measured over a few days by vacuum flask methods, correcting for heat losses, or over a longer period in an adiabatic calorimeter. The only method suitable for long-period measurements, and the one commonly used for control purposes, is the heat of solution method described in BS 1370 : 1958 and ASTM C186–68, which are similar except for minor

[1] See A.C.1. Committee Report, Proc. Amer. Concr. Inst., **67,** 273 (1970). Mass Concrete for Dams and other Massive Structures.

details. This method depends on the measurement of the heat of solution of the original cement and of the set cement. By the application of Hess's law of constant heat summation to the difference between the heats of solution of fresh and set cements, the heat of hydration can be derived. The heat of solution of 3 g dry cement, or 4·2 g hydrated cement, is determined in a solvent consisting of 9·6 ml hydrofluoric acid (40 per cent w/w) and 388 ml 2·0 N nitric acid at a starting temperature of 18°. Both fresh and set cements dissolve in this solution in a few minutes. The heats of solution of the fresh and the set cements are calculated to an anhydrous basis by means of a determination of the loss on ignition at 900°, and the heat of hydration at any age obtained as the difference between the heat of solution of the set cement at that age and that of the fresh cement. The hydrated cement samples are prepared by mixing 60 g cement and 24 ml distilled water and sealing the mix in specimen tubes which are stored at 70° F (ASTM 73·4° F) until the time of test. The tube is then opened and the sample ground, as rapidly as possible to avoid carbonation, to pass a BS 18-mesh sieve. Portions of this ground material are used for the heat of solution and ignition loss determinations. The calorimeter used is an open-mouthed thermos flask with a cork lid, through which pass a Beckmann thermometer and a motor-driven stirrer. The inside of the flask and the thermometer, and the stirrer, are protected by a thin coating of paraffin wax. In order to reduce the corrections for external gain or loss of heat the thermos flask is placed in an insulated container. The actual heat of solution determination takes 20–25 minutes, including periods of about 5 minutes before adding the cement, and after solution is complete, to determine the initial and final heating or cooling corrections. The heat capacity of the system is determined with zinc oxide previously ignited at 900°. Its heat of solution in the solvent of the concentration specified is 256·1 cal g at 30° with a negative temperature coefficient of 0·1 cal per degC. The specific heat of ZnO is 0·1, unhydrated cement 0·2, and hydrated cement 0·3. Heat of solution values such as 597·9 cal/g for an anhydrous cement, and 528·6 for the hydrated cement are obtained, giving in this case a heat of hydration, to the nearest calorie, of 69 cal/g. The mean of three determinations on separately cured and prepared samples of the hydrated cement is required. The method requires considerable care and attention to the precise details laid down in the specification, since the result depends on the subtraction of two numbers which are considerably larger than their difference.[1] Carbonation of the hydrated cement during grinding is one serious source of error, for the adsorption of 1 per cent CO_2 causes an error of about 5·8 cal/g in the heat of hydration.[2] Drying of the sample during grinding results in an increased heat of solution and therefore in a decreased calculated heat of hydration. In an older form of the method, the hydrated cement was cured for 24 hours at 70° and thereafter at 100° F but this practice was later changed.

There has been some uncertainty in the application of the method to pozzolanic cements[3] since a proportion of the pozzolana usually remains insoluble

[1] See E. S. Newman, *J. Res. natn. Bur. Stand.* **45,** 411 (1950); *Proc. Am. Soc. Test. Mater.* **63,** 830, 852 (1963); T. W. Parker and R. W. Nurse, *Cem. Lime Mf.* **26,** 1 (1953).

[2] R. W. Carlson and L. R. Forbrick, *Ind. Engng Chem. analyt. Edn.* **10,** 382 (1938).

[3] E. S. Newman and L. S. Wells, *J. Res. natn. Bur. Stand.* **49,** 55 (1952); L. Santarelli, F. Piselli and C. Cavarelli, *Revue Matér. Constr. Trav. publ.* **533,** 31, **540,** 231 (1960); **548,** 262 (1961); A. Rio, A. Celani and A. Miele, *Industria ital. Cem.* **34,** 233 (1964).

and it is not quite the same for the fresh and the hydrated cements, though the difference does not usually amount to more than about 1 per cent of the cement. There is also at the end of the usual test period of 15 to 20 minutes a very slow continuing solution and slight heat evolution. It is not possible, therefore, to estimate the correction for heat loss or gain to the surroundings from the final temperature-time curve. This difficulty may be got over by means of a separate estimation of the final correction. The calorimeter is filled with the normal acid solution, warmed to a few degrees above room temperature, and the temperature-time cooling curve determined. The correction to be applied for the particular final temperature attained in the actual estimation of heat of solution can thus be derived.[1]

A similar difficulty arises with slag cements which leave an insoluble residue and continue to dissolve slowly and evolve heat at the end of the usual test period. This makes it impossible to determine the cooling correction in the usual way from the initial and final rates. Special procedures are therefore adopted when applying the heat-of-solution method to slag cements. These are described in BS 4246 : 1968 for low-heat Portland blastfurnace cement and BS 4248 : 1968 for supersulphated cement.

The adiabatic method[2] for determining heat evolution depends on placing a sample, usually of concrete, in an adiabatic calorimeter, i.e. a calorimeter from which no heat loss can occur. This condition is realised by surrounding the calorimeter with an outer bath, the temperature of which is caused to rise by automatic means at exactly the same rate as that of the concrete. The heat generated in the concrete serves, therefore, only to raise the internal temperature and no outward heat loss occurs. The temperature-time curve obtained also simulates very closely the actual thermal history of the interior of a large concrete mass where conditions remain close to adiabatic for at any rate the first few days after placing. The temperature rise can be converted into calories per gram of cement with the aid of values for the water equivalent of the concrete and any corrections necessary for the inner calorimeter. For a gravel concrete a specific heat of 0·25 may be assumed, but for other concretes, or where greater precision is required, the value needs to be determined over the range of temperatures involved.

Curves showing the temperature rise under adiabatic conditions up to 3 days are shown in Fig. 76. The method can be used satisfactorily up to 7 days, or 28 days, though the effect of slight cumulative errors becomes more appreciable, apart from the disadvantage of requiring the continuous use of the apparatus. Some typical curves for mass concretes with different types of Portland cement up to 28 days are shown in Fig. 77.

Another form of calorimeter, known as the conduction or vane calorimeter,[3] is also sometimes used. In this, the heat is conducted away through metal vanes or tubes and the rate of heat flow determined from the temperature gradient across

[1] R. W. Nurse and V. N. Pai, *Mag. Concr. Res.*, No. 22 (1956).
[2] N. Davey, *Concr. constr. Engng* **26,** 572 (1931); N. Davey and E. N. Fox, *Building Res. Tech. Pap.* No. 15 (1933); H.M.S.O. London. H. S. Meissner, *Proc. Am. Concr. Inst.* **30,** 21 (1934); D. Pirtz, *Mater. Res. Stand.* **2** (1), 22 (1962).
[3] R. W. Carlson, *Proc. Am. Soc. Test. Mater.* **34,** 322 (1934); G. E. Monfore and B. Ost, *J. Res. Dev. Labs Portld Cem. Ass.* **8,** 13 (1966).

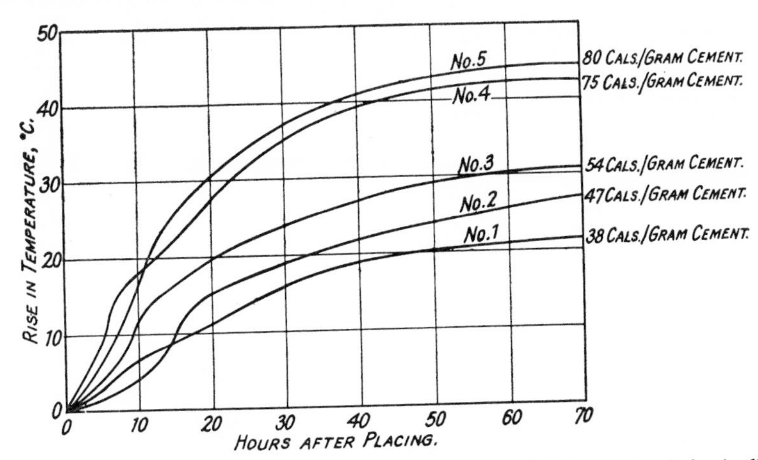

FIG. 76 Temperature rise in 1 : 2 : 4 concrete (w/c 0·60) cured adiabatically (*Davey and Fox*).

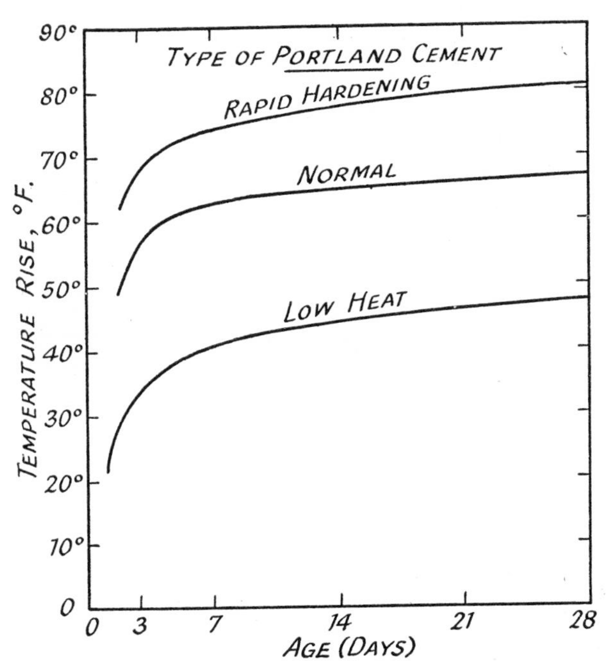

FIG. 77 Temperature rise in 1 : 9 (weight) concrete under adiabatic conditions.

them. Tests can be made in this calorimeter on neat cement specimens without any large rise in temperature occurring.

The rate of hydration of cements, and hence the heat evolution, increases with temperature, so that identical results are not to be expected from alternative methods in which the test specimens are subjected to different thermal histories. Heat evolution is also influenced by the water:cement ratio[1] as shown below for an ordinary Portland cement, and this differs as between a neat cement and a concrete.

Water : cement ratio	Heat of hydration at 70° F (cal/g)				
	3 days	7 days	28 days	90 days	1 year
0·4	61	79	96	104	109
0·6	66	88	107	115	120
0·8	66	89	112	120	122

Some comparative results by adiabatic and heat of solution methods are shown below:

Method	Heat of hydration at 7 days (cal/g) Cement		
	I	II	III
Adiabatic	78	49	89
Heat of solution:			
70° F curing	76	46	81
100° F curing	82	52	87

These results compare the adiabatic method on a concrete of water:cement ratio 0·6 with the heat of solution on neat cement pastes of w/c 0·4. If a higher water:cement ratio is used for the latter, the heat evolution is increased and may, even at 70° F, exceed the adiabatic value.[2]

Some average data for the heat evolution of different types of cement are given in Table 37.

TABLE 37 Heat of hydration at 70° F (cal/g)

Cement	3 days	7 days	28 days	90 days	1 year	6½ years	13½ years
Normal	61	80	96	104	109	117	118
Rapid hardening	75	92	101	107	113	121	121
Low heat	41	50	66	75	81	85	87
ASTM type II	47	61	80	88	95	98	101

The heat of hydration is influenced by fineness at the earlier ages, though little at very long ages. For an increase in 100 cm²/g (Wagner method) the heat evolution increases by some 4–5 cal/g at 1 day, 1–3 cal at 7 days and 28 days, and less

[1] G. J. Verbeck, *Sym., Washington 1960*, 453; V. Danielson, ibid., 519.
[2] See, for example, *Bur. Reclamation. Boulder Canyon Project., Final Reports*, Part VII, Bull No. 2, 164 (1949).

than 1 cal at one year.[1] Aeration or deliberate prehydration of a cement decreases the heat evolution by some 6 cal at 7 and 28 days and 4 cal at 1 year for each 1 per cent increase in ignition loss. These, however, are average figures and for individual cements the values may be halved or doubled.[2]

From determinations on a large number of cements, combined with the compound contents calculated from the composition, the heat evolved in the hydration of the constituent compounds can be derived (see p. 173). One such set of data obtained by Verbeck and Foster[3] are shown in Table 38.

TABLE 38 Heat evolved at 70° F (cal/g)

Compound	3 days	7 days	28 days	90 days	1 year	$6\frac{1}{2}$ years
$3CaO . SiO_2$	58 ± 8	53 ± 11	90 ± 7	104 ± 5	117 ± 7	117 ± 7
$2CaO . SiO_2$	12 ± 5	10 ± 7	25 ± 4	42 ± 3	54 ± 4	53 ± 5
$3CaO . Al_2O_3$	212 ± 28	372 ± 39	329 ± 23	311 ± 17	279 ± 23	328 ± 25
$4CaO . Al_2O_3 . Fe_2O_3$	69 ± 27	118 ± 37	118 ± 22	98 ± 16	90 ± 22	111 ± 24

The $\pm$ value indicates the calculated probable error on the deduced value.

The figures are only to be regarded as generally indicative of the relative contributions of the various compounds. Various similar sets of values have been derived by other workers, but the agreement even between the relative values for the four compounds is far from close.

The errors are particularly large in the case of the alumina and iron compounds. Thus the heats of formation of $3CaO . Al_2O_3 . 6H_2O$ and the hexagonal-plate hydrated aluminate from tricalcium aluminate were found by Thorvaldson to be 214 and 261 cal/g respectively.[4] The difference between the values obtained at the longest ages for tricalcium and dicalcium silicate is about equal to the heat of hydration of the lime which has to be removed from $3CaO . SiO_2$ to obtain $2CaO . SiO_2$; this amounts to 68 cal/g $3CaO . SiO_2$.

Lerch and Bogue[5] found from determinations of the heats of solution of unhydrated and completely hydrated pure compounds the following heats of hydration:

	cal/g
$3CaO . Al_2O_3$	207
$3CaO . SiO_2$	120
$4CaO . Al_2O_3 . Fe_2O_3$	100
$2CaO . SiO_2$	62

[1] R. E. Davis, *Proc. Am. Concr. Inst.* **30,** 485 (1934); M. J. M. Jaspers, *Revue Matér. Constr. Trav. publ.* **429, 430, 431** (1951).
[2] F. B. Hornibrook, G. L. Kalousek and C. H. Jumper, *J. Res. natn. Bur. Stand.* **16,** 487 (1936).
[3] G. J. Verbeck and C. W. Foster, *Proc. Am. Soc. Test. Mater.* **50,** 1235 (1950).
[4] T. Thorvaldson, W. G. Brown and C. R. Peaker, *J. Am. chem. Soc.* **52,** 3927 (1930).
[5] *J. Res. natn. Bur. Stand.* **12,** 645 (1934).

For the hydration of the solid solutions between $6CaO.2Al_2O_3.Fe_2O_3$ and $4CaO.Al_2O_3.Fe_2O_3$ ($Al_2O_3 : Fe_2O_3$ ratios of 1·277 and 0·64) Brisi[1] obtained values as follows:

$Al_2O_3 : Fe_2O_3$ ratio	1·36	1·12	0·88	0·64
Heat of hydration, cal/g	144	129	113	95

For the heat of hydration of $3CaO.SiO_2$ and $2CaO.SiO_2$ to form $3CaO.2SiO_2.aq$ and calcium hydroxide, Brisi obtained values of 125 and 63 cal/g, in fair agreement with Lerch and Bogue. It must, however, be noted that the products of the hydration are gels with surface areas that may vary with the conditions of hydration, or of preparation of the hydrated products. The measured heat of hydration is, therefore, a compositive quantity made up of the chemical heat of hydration and the heat of adsorption of the water contained in the gel. The actual chemical heats of hydration, excluding the heat of adsorption of water have been given[2] as $96{\cdot}5 \pm 1{\cdot}5$, and $24{\cdot}5 \pm 3{\cdot}5$, cal/g respectively for $3CaO.SiO_2$ and $2CaO.SiO_2$; the heat of adsorption of water by the gel is thus 20–40 cal/g of anhydrous compound. For practical purposes, it is the total heat that is required.

The heats of hydration of uncombined lime and magnesia are:

CaO	276 cal/g
MgO	203 ,,

These values vary with the physical nature of the hydration product and the physical condition of the oxide. Thus for CaO, values from about 265 to 279 cal/g have been reported. With MgO a sample prepared by dehydration of $Mg(OH)_2$ at 450° was found to give a heat of hydration as high as 249 cal/g.[3] The lower value of 203 cal/g refers to the hard burnt oxide.

The reaction:

$$3CaO.Al_2O_3.6H_2O + 3(CaSO_4.2H_2O) + 19H_2O = 3CaO.Al_2O_3.3CaSO_4.31H_2O$$

is accompanied[4] by a heat evolution of 195 cal/g SO_3. For the similar formation of the monosulphoaluminate the value is 187 cal/g SO_3. The heat of rehydration of any gypsum dehydrated to hemihydrate during the grinding of the cement amounts to 46 cal, and for soluble anhydrite 34 cal/g SO_3. At most this can only amount to about 1 cal/g cement.

The heat evolved in the first 48 hours from the pure cement compounds is, according to Lerch and Bogue, for materials ground to a fineness of 87·5 per cent through a 200 mesh, as follows:

	Heat evolved in 48 hours (cal/g)
$3CaO.Al_2O_3$	150
$3CaO.SiO_2$	100
$4CaO.Al_2O_3.Fe_2O_3$	40

[1] C. Brisi, *Ricerca scient.* **24,** 1436 (1954).
[2] S. Brunauer, J. C. Hayes and W. E. Haas, *J. phys. Chem.* **58,** 279 (1954).
[3] K. Taylor and L. S. Wells, *J. Res. natn. Bur. Stand.* **21,** 133 (1938).
[4] H. A. Berman and E. S. Newman, *J. Res. natn. Bur. Stand.* **67A,** 1 (1963).

The heat-of-solution method can be applied to any cement without assumption, but the calculation of the heat which will be evolved from the compound content by the use of data such as in Table 31 involves various errors. Thus, apart from differences in fineness, we have seen earlier that the calculated compound content is far from being quantitatively accurate. The calculated heat evolution may differ by up to 10 cal/g or more from that found by direct measurement. Such calculations can be useful as an initial guide to the formulation of a cement, provided their limitations are realised.

It is seen from Table 37 that heat continues to be evolved from a cement over a long period, but that a considerable proportion of the total is evolved in the first few days; further that different types of cements vary much in the amounts of heat they evolve. It is advantageous in order to limit the heat evolution to reduce the content of C_3A to a minimum. This can be done, without disadvantage to strength, by increasing the $Fe_2O_3 : Al_2O_3$ ratio so as to convert C_3A to C_4AF which has a low-heat evolution. A certain reduction in heat evolution can be obtained by this means, but to obtain any further lowering it is necessary to reduce the content of C_3S. This involves a corresponding increase in the content of C_2S and a reduction in early strength, though not in the ultimate strength.

Chemical factors affecting the setting of Portland cement

When Portland cement first became a commercial product, experience proved that the rapidity of setting varied in different specimens of cement. The setting time is a somewhat arbitrary magnitude, being the time which elapses after mixing the cement with the water before the paste will resist a certain arbitrary fixed pressure. It serves, however, for the comparison of the setting properties of different cements. In general, the greater the proportion of alumina and ferric oxide in a cement the more rapid is the initial set, on account of the rapid-setting properties of the alumina and iron compounds. The quick-setting properties of the old 'Roman' cements were due to their high alumina and ferric oxide content. The older varieties of Portland cement were slow in setting. This was due to the co-operative effect of several causes. They were relatively coarsely ground and contained appreciable amounts of free lime which tended to retard the set. All the sulphur contained in the fuel used also accumulated in the cement clinker as calcium sulphate which has a remarkable effect in retarding the set of cements. Clinker from rotary kilns was better burnt and contained less sulphate; with improvements in grinding methods it was also progressively more finely ground. The use of some agent to control the setting of cements then became necessary and the addition of gypsum during grinding of the clinker became general.[1]

The introduction of steam into the grinding mill was also at one time practised, but was eventually abandoned. About 1 per cent of water was absorbed by the cement and a smaller addition of gypsum, about 1 per cent, then sufficed to retard the setting time to the required extent.

The setting time of present-day Portland cements is controlled by the regulated addition of gypsum to the cement clinker as it is fed to the grinding mill.

[1] Some interesting details of the early history of the use of retarders are given by R. H. Bogue, *Rock Prod.*, Sept. 1928.

The amount added corresponds to about 1–3 per cent SO_3 and is limited by specification requirements, which in all countries permit only of the presence of a certain proportion in the finished cement. While the addition of a limited amount of gypsum is valuable in the control it affords of the setting time and in its favourable effect on strength, the presence of large quantities leads to slow expansion in the set cement; this is the reason for its stringent limitation.

The influence of gypsum on the setting of cement is not proportional to the amount added and its effect may appear quite abruptly, a small increase beyond a certain amount producing large changes in the setting time. The addition of lime, or its presence in the free condition in the clinker, assists in the retardation of set produced by the gypsum. In Fig. 78 are shown curves illustrating the effect of gypsum on the setting of Portland cement.

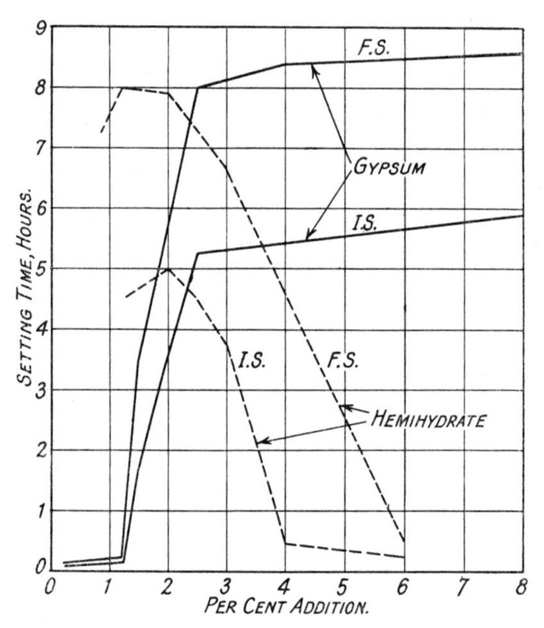

FIG. 78 Effect of calcium sulphate on setting time of Portland cement. I.S. = Initial set. F.S. = Final set.

Plaster of Paris and anhydrite retarders

In addition to gypsum there are other forms of calcium sulphate which have been tried as retarders.[1] Plaster of Paris, $CaSO_4 \cdot \frac{1}{2}H_2O$, retards the set of cements more vigorously than gypsum, and soluble anhydrite acts similarly. This seems to be due to the more rapid solution and higher concentration of sulphur trioxide which is obtained with these materials owing to the formation of supersaturated

[1] See reviews by H. H. Steinour, *Bull. Res. Dev. Labs Portld Cem. Ass.*, No. 98 (1958) W. C. Hansen and J. S. Offott, *Gypsum and Anhydrite in Portland Cement*, U.S. Gypsum Co. 1962.

solutions. It can be prevented, according to Schachtschabel,[1] by adding to the hemihydrate a small amount of gypsum, which, acting as nuclei for crystallisation prevents the formation of such highly supersaturated solutions. The addition of larger quantities of hemihydrate renders the cement quick-setting again, as shown in Fig. 78. This now generally attributed to the setting action of the plaster itself which leads to a stiffening of the cement paste. With natural anhydrite a higher addition of sulphur trioxide is required than with gypsum on account of the lesser reactivity of the material. Mixtures of gypsum and anhydrite in roughly equal proportion are sometimes used as a means of overcoming problems of false set. The mixture seems to give similar setting times and strength to gypsum alone.

'False' set

A phenomenon known as 'false' set in cements, or premature stiffening, sometimes occurs. On gauging the cement with water and mixing for a short time, the material stiffens and appears to set. Further working with the trowel breaks up this set and the material then exhibits a normal setting time. Little appreciable heat evolution occurs in false set and this differentiates it from the flash set shown by cements that have been insufficiently retarded. In a concrete mix[2] it may lead to stiffening of the material while being transported from the mixer to the point at which it is to be placed, or even during mixing. It can be overcome by adding more water to the mix, but at the expense of reducing the quality of the concrete. It is now generally agreed that the prime cause of false set is to be found in dehydration of the gypsum caused by the high temperatures attained in grinding the cement clinker and its recrystallisation. These on occasion may exceed 150°. When cements showing false set are mixed with water there is an immediate supersaturation of the solution with calcium sulphate to a much larger extent than with normal cements; this is indicative of the presence of a dehydrated form of gypsum, either the hemihydrate or soluble anhydrite. False set can usually be cured by cooling of the grinding mills by external water-spraying. Nevertheless, cements vary considerably in their susceptibility to develop false set, suggesting that some conditions of the cement clinker are more favourable to it than others. The presence of free calcium oxide, assisting the dehydration of the gypsum, may be one factor and it has often been suggested that underburning, or overburning, of the clinker can be a contributory factor. It cannot, however, be said that there is any real knowledge of how, or if, the physical or chemical characteristics of the clinker influence its sensitivity to false set. Carbonation of alkalies in the cement during storage has also been suggested as a cause, but, as we see below, this is more likely to produce a flash set than a false set.

A method of testing cements for false set is given in ASTM C451–68. It consists of a modification of the standard U.S. consistency test for neat cement pastes with a repetition of the test after a five-minute interval and a requirement that the penetration of the standard rod shall not have been reduced by more than a defined amount. There is also an ASTM test method (C359–67T) for determining false set of a mortar.

[1] P. Schachtabel, *Cement* **6,** 54 (1933).
[2] For a review, see R. F. Blanks and J. L. Gilliland, *Proc. Am. Concr. Inst.* **47,** 517 (1951); W. C. Hansen, *Mater. Res. Stand.* **1,** 791 (1961).

Aeration of cement

The properties of Portland cement undergo some change when it suffers aeration owing to access of moisture and carbon dioxide during storage. The setting time of cements stored in sacks changes in an erratic manner, sometimes being accelerated and sometimes retarded. Dry air, whether containing carbon dioxide or not, has no effect on cement and under such conditions no carbon dioxide is absorbed. In moist air both moisture and carbon dioxide are absorbed.[1] The absorption of moisture alone retards the set of Portland cement, but that of carbon dioxide accelerates it. The effect on setting time is thus dependent on the amount both of water and carbon dioxide absorbed, but the influence of the latter is the more important. In general, cements undergo alternate acceleration and retardation of set as aeration progresses and its effects are thus inevitably erratic.[2] The first effect of carbonation is to accelerate the set and this is generally attributed to the reduction in the immediate lime concentration of the solution when water is added, either as a result of direct carbonation or of formation of alkali carbonates. This delays the formation of ettringite.[3] An alternative view, suggested by Forsén,[4] is that decomposition of gypsum occurs leaving insufficient for retardation. In the mixed calcium hydroxide-calcium sulphate solution that arises when cement is mixed with water, there is no real difference between these alternative ways of expressing the action. Further carbonation seems to result in the complete covering of the surface of the most active particles with a film of calcium carbonate, reducing the rate of reaction and increasing the setting time. On further aeration the film flakes off again, probably because of slow hydration of the underlying cement, re-exposing the active cement surfaces. Thus, in the slow-setting carbonated state the setting time can be reduced by rubbing the cement gently in a mortar. Absorption of 1 or 2 per cent carbon dioxide only affects the strength at early ages, 1–3 days, and has little influence on the 28-day strength.

Cement stored in sealed containers at ordinary temperatures is usually little changed after long periods, but at high temperatures some effect is produced, for although the cement appears normal it may rapidly develop a flash set if subsequently exposed to air. Trouble with rapid setting is occasionally experienced with cements shipped from temperate to tropical climates. Development of a flash set also occasionally occurs in cement stored in bulk. The high temperatures attained during grinding probably persist for a considerable period in cement stored in silos, but the particular conditions which can lead to rapid setting and the corresponding changes occurring in the cement are far from clear.

Deliberate prehydration of cement by treatment with steam, without carbonation, reduces the heat of hydration (p. 295) and also the strength, particularly at early ages. The reduction in strength varies much for different cements and for each 1 per cent increase in ignition loss ranges from about 5 to 15 per cent at

[1] C. M. Hunt et al., *J. Res. natn. Bur. Stand.* **60,** 441 (1958).
[2] D. G. R. Bonnell, *Building Res. Tech. Pap.* No. 19 (1936) H.M.S.O. London; F. Matouschek, *Zement-Kalk-Gips* **16,** 483 (1963).
[3] T. Manabe and N. Kawada, *Proc. Am. Concr. Inst.* **56,** 639 (1960).
[4] L. Forsén, *Zement* **24,** 139 (1935); *Concrete* **45** (10), 250 (1937).

7 days and 3 to 6 per cent at 1 year.[1] The initial set is lengthened as moisture is absorbed and this was attributed by Roller[2] to the production of calcium hydroxide following his theory that this, rather than gypsum, is the effective retarding agent in cements. Other work, discussed later, indicates that calcium hydroxide is only an effective retarder for alkali-free cements and not for normal cements.

Effect of salts on the setting of cement

In addition to calcium sulphate there are many other salts which affect the setting of Portland cement. Some salts retard the set, others accelerate it; another group of salts retard the set when present in small amounts and accelerate it when larger amounts are added. The available data[3] are often conflicting and the effect produced often varies with the composition of the Portland cement used.

A clear distinction must also be drawn between the effect on setting and on subsequent hydration, for some substances which cause an immediate rapid stiffening can inhibit, or much reduce, the subsequent strength development. The effects given in the following paragraphs refer to the addition of salts to cements already retarded with gypsum; in certain cases they are quite different if the addition is made to a ground clinker containing no added gypsum.

The two salts which are of most importance, since they are sometimes deliberately added to Portland cement, are calcium chloride and sodium chloride. These salts are sometimes used to protect concrete laid in cold weather against frost. Though the addition of very small amounts of calcium chloride, less than 1 per cent, sometimes retards the setting of Portland cement, the addition of larger amounts produces an acceleration. The extent to which this occurs is, however, very variable. In some cases the addition of over 3 per cent calcium chloride causes a flash set. Sodium chloride produces less change in the setting of cement and the effect is erratic, an acceleration of set occurring with some Portland cements and a retardation with others.

The alkali carbonates,[4] like carbon dioxide, produce a very strong acceleration of the set and the addition of 1–2 per cent reduces the time of initial set to a few minutes. Ammonium carbonate is less active. The bicarbonates on the other hand often have a retarding effect. Of the chlorides, aluminium chloride and magnesium chloride have a considerable accelerating action, the alkali chlorides a small and variable effect, barium and strontium chlorides a slight accelerating action; ammonium, ferrous, ferric and cobaltous chlorides retard the set when present in amounts up to about 2 per cent, while larger amounts accelerate it.

The relative effect of some chlorides on the initial and final setting time of a Portland cement[5] are shown below. While these values afford an example of

[1] F. B. Hornibrook, G. L. Kalousek and C. H. Jumper, *J. Res. natn. Bur. Stand.* **16,** 487 (1936).
[2] P. S. Roller, *Ind. Engng Chem.* **26,** 669, 1077 (1934); **28,** 362 (1936).
[3] L. Forsén, *Zement* **19,** 1130 (1930); S. K. Biehl, *ibid.* **17,** 487 (1928); G. C. Edwards and R. L. Angstadt, *J. appl. Chem.* **16,** 166 (1966); H. H. Steinour, loc. cit.; J. Joisel, *R.I.L.E.M. Sym. Admixtures for Mortar and Concrete, Brussels 1967.*
[4] E. M. M. G. Niël, *Sym., Tokyo 1968.*
[5] R. Grün, *Z. Angew Chem.* **43,** 496 (1930).

the effects produced, the actual degree of acceleration or retardation varies with different Portland cements.

Percentage salts in mixing water	$CaCl_2$				$AlCl_3$				$FeCl_3$				$BaCl_2$			
	IS		FS		IS		FS		IS		FS		IS		FS	
	h	m	h	m	h	m	h	m	h	m	h	m	h	m	h	m
0	4	30	8	00	4	30	8	00	4	30	8	00	4	30	8	00
3	4	26	6	46	3	03	5	53	4	52	8	05	4	55	7	50
5	2	47	4	22	2	31	5	16	0	40	6	30	4	55	6	05
7	1	30	3	40	1	20	4	30	0	47	5	17	3	48	5	08
10	0	16	1	16	0	01	0	25	0	02	4	50	2	39	5	04
20	0	02	0	03	immediate				immediate				1	53	2	48

Sulphates and nitrates when added in dilute solutions to normally retarded cements have no very marked effect. Sodium and potassium hydroxides tend to have some accelerating action and also sodium silicate in higher concentrations. Sugar has a very marked effect and a 1 per cent solution almost completely inhibits real setting and hardening, though there may be an immediate rapid stiffening giving the appearance of a quick set. As little as 0·05 per cent by weight of the cement may result in no strength at 1 day and half the strength at 3 days, though in other cases it has been reported[1] that such a small amount can form a satisfactory water-reducing agent.

Theory of the action of retarders

When cement is mixed with water it becomes saturated with calcium hydroxide and sulphate within the first minute or two; indeed the solution is often super-saturated in some degree with respect to both components. The initial set is to be attributed to the alumina-containing compounds and also to tricalcium silicate, since this compound by itself shows an initial set within a few hours. The hydration of the various cement compounds must be expected to overlap, since hydration begins at the surface of a solid particle and progresses inwards. The availability as well as the relative reactivity of the compounds must thus be an important factor.

As we have seen in the previous chapter, both silica and alumina pass into solution to some extent, even if a very limited one, during the reaction of the anhydrous compounds with water. As new hydrated products are precipitated further anhydrous material passes into solution. It is the rate of reaction of the anhydrous solids with water and the interplay of the lime, alumina, silica and sulphate in solution that determines the nature of the products first produced and the speed of setting. It is this problem that we now have to examine.

It has long been known that retarders stop the rapid set shown by tricalcium aluminate and it has been variously suggested that this arises in cement from the formation of the high-sulphate of calcium sulphoaluminate, or of the low-sulphate form and its solid solutions with hydrated tetracalcium aluminate, or of the latter compound itself.

[1] R. Ashworth, *Proc. Instn Civ. Engrs* **31,** 129 (1964–5); R. F. Adams and J. M. Hemme, *Spec. tech. Publ. Am. Soc. Test. Mater.*, No. 266, 97 (1960).

Cements of high $Fe_2O_3:Al_2O_3$ ratio show a rapid, but not a flash set in the absence of gypsum, but the final set may be unduly slow. Gypsum retards the initial set, but it speeds up the final set. In the absence of gypsum, the iron compound $4CaO . Al_2O_3 . Fe_2O_3$ hydrates to form crystalline hydrated calcium aluminate and a hydrated iron oxide which by precipitation on the cement particles retards their further hydration. When gypsum is present a solid solution of sulphoaluminate with sulphoferrite is formed which does not have as prolonged an inhibiting effect. Gypsum also increases the rate of hydration of tricalcium silicate in a saturated lime solution. These effects exemplify the second action of gypsum, that of an accelerator of hydration.

Candlot,[1] who was amongst the first to investigate the action of retarders, considered that gypsum, or lime, in solution, retarded the hydration of tricalcium aluminate: Kühl[2] concluded that setting was due to the coagulation of a colloidal calcium silicate solution. A certain minimum concentration of aluonina in solution was necessary to effect this, but no such accumulation could arise as long as gypsum was available in the solution to precipitate it. An older view, due to Rohland,[3] was that alkalis catalysed the setting of cement and the addition of gypsum neutralised their action. A theory that lime was the primary retarding agent and that it acted by causing precipitation of hydrated tetracalcium aluminate was put forward by Roller.[4] Many of these conflicting views can now be seen to represent but one particular aspect of a many-sided phenomenon.

The work of Forsén[5] was the first to give a comprehensive picture of the action of addition agents in general on the setting of cement. From a study of the concentrations of the various ions set up in solution when cement reacts with water Forsén concluded that the effect of added salts could be closely related to the solubility of alumina. The passage of alumina into solution hindered the normal formation of the hydrated calcium silicates and led to rapid precipitation of an alumina-silica gel, comparatively poor in lime, leading both to quick set and to reduced strengths. Silica passing into solution from the tricalcium silicate was considered to coagulate with the alumina in a manner similar to that which occurs when an alkali silicate solution is mixed with an alkali aluminate solution. Thus, Forsén found that $3CaO . SiO_2$ set normally in a solution of 3 per cent KOH, whereas a quick set occurred if potassium aluminate was present in the solution. In normal setting the solution formed is one in which little alumina is present for the aluminates dissolved are at once precipitated as sulphoaluminate films around the cement grains, thus retarding their solution. Retarders, in Forsén's view, are compounds which precipitate the aluminates in this manner, while quick-setting occurs when no retarding agent is present in solution. The effect of different salts, or of differing concentrations of the same salt, could be related in this way. The typical retarders are those which form the double salts of the sulpho-aluminate type. The influence of alkalis was attributed by Forsén to their effect on the solubility of lime. A cement low in alkalis can be retarded with lime alone,

[1] *Cements and Hydraulic Limes*. Paris, 1906.
[2] *Zement* **13,** 362 et seq. (1924); **21,** 392, 405 (1932).
[3] *Z. angew Chem.* **16,** 622 (1903); *Kolloidzeitschrift* **4,** 233 (1909); **8,** 251 (1911).
[4] *Ind. Engng Chem.* **26,** 669, 1077 (1934).
[5] *Sym., Stockholm 1938*, 298; R. Hedin, *Proc. Swed. Cem. Concr. Res. Inst., Stockholm*, No. 3 (1945).

for in a saturated solution of lime in water the solubility of the hydrated tetracalcium aluminate formed is very low. Cements containing appreciable amounts of alkalis rapidly release alkali into solution, reducing the solubility of lime and increasing that of alumina, so that lime alone is no longer adequate as a retarder. Gypsum, on the other hand, still produces the insoluble sulphoaluminate. Forsén's theory in effect involves two mechanisms, the repression of the alumina solubility, resulting in the precipitation of the alumina as sulphoaluminates instead of an alumina-silica gel; and the retardation of the hydration of the tricalcium aluminate by the sulphoaluminate films formed round it. The relative importance of these two effects is still uncertain, for some workers[1] have found the alumina content of the solution rapidly extracted from a cement paste to be low whether gypsum is present or not. It is difficult, however, in such tests to be sure that precipitation of alumina is not so rapid as to invalidate the results.

We have seen in the previous chapter that the balance of evidence is in favour of the formation of ettringite as the immediate reaction product under normal conditions. Certain minimum concentrations of lime and calcium sulphate in solution are however necessary for this to occur. If the rate of reaction of the alite is reduced for any reason, such as surface carbonation, the calcium hydroxide concentration in solution may be too low in the first few minutes for ettringite to form and slow up the hydration of the aluminate compounds by producing surface coatings. Nor at such a low-lime concentration could a coating of tetracalcium aluminate hydrate be formed. The hydration of the aluminates then becomes so rapid as to lead to abnormally quick set. Again if the rate of solution of the gypsum is too slow the concentration of SO_4 ions in solution on mixing a cement with water may be too low for the immediate formation of the retardant coating of ettringite.[2] The compound formed may perhaps also vary from one cement to another with the alkali content and other factors.

The second action of gypsum, and of other calcium salts, in accelerating hydration, is attributed to the effect of the increased calcium, and consequent reduced hydroxyl, ion concentration on the speed of hydration of tricalcium silicate.

Forsén divided retarders into four groups according to the type of curve obtained when initial setting time was plotted against the quantity of retarder added, as shown in Fig. 79. The salts falling in these groups were as follows:

I $CaSO_4.2H_2O$, $Ca(ClO_3)_2$, CaI_2
II $CaCl_2$, $Ca(NO_3)_2$, $CaBr_2$, $CaSO_4.\frac{1}{2}H_2O$
III Na_2CO_3, Na_2SiO_3
IV Na_3PO_4, $Na_2B_4O_7$, Na_3AsO_4, $Ca(CH_3COO)_2$.

With type I retarders the alumina solubility remained low even with large amounts of additions. With type II, such as calcium chloride, alumina has a low solubility at low concentrations and the salt acts as a retarder, but at higher concentrations the solubility of alumina was found to increase in the cement solution and no retarding action occurs. The case of plaster is a special one, as

[1] G. L. Kalousek, C. H. Jumper and J. J. Tregonning, *J. Res. natn. Bur. Stand.* **30**, 215 (1943); P. P. Budnikov and M. I. Strelkov, *Spec. Rep. Highw. Res. Bd*, No. 90, 447 (1966).

[2] H. E. Schwiete and E. Niël, *Zement-Kalk-Gips* **18**, 157 (1965); **19**, 402 (1966); *J. Am. Ceram. Soc.* **48**, 12 (1965); Z. T. Jugovic and J. L. Gillam, *J. Materials* **3**, 517 (1968).

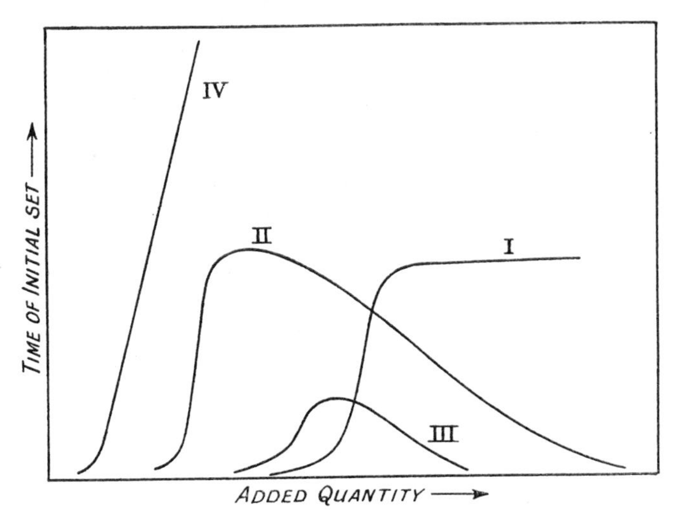

FIG. 79 Action of various retarders.

discussed earlier, and not related to dissolved alumina. Type III retarders are similar to type II. At low concentrations the alumina remains insoluble in the cement extract, but at higher concentrations dissolves in the caustic soda solution formed from interaction with the calcium hydroxide. Retarders of type IV hold up setting and hardening indefinitely and Forsén called them 'cement destroyers'. The various organic retarders used for oil-well cements are mainly of this type. Some retarding and water-reducing agents like calcium lignosulphonate also give a type IV curve though they are not 'cement destroyers'. The action of type IV agents seems to arise either from surface adsorption or the formation of surface precipitates which are too impermeable to permit hydration to proceed at any appreciable rate. Thus sugar, which falls in this group, combines with the lime solution to form calcium saccharate which is only slightly dissociated and the solubility of the alumina in the solution rises. On Forsén's theory a precipitation of a lime-poor alumina-silica gel occurs over the cement particles and severely retards hydration until it is gradually destroyed by lime released from the cement. If sugar is added in the form of calcium saccharate no such action occurs.[1] This may be an over-simplified explanation since while sucrose increases the amount of alumina in solution it also increases the alkalinity of the solution which will repress the hydrolysis of the silicate compounds and in this way also retard hydration.[2]

The alternative adsorption theory is favoured for most organic retarders.[3] Electrolytes, such as the sodium salts of complex hydroxylated carboxylic acids (e.g. tetrahydroxy adipic acid $(HO)_4C_4H_4(COOH)_2$) or of lignosulphonic acid, which are used as water-reducing and retarding agents contain CHOH groups

[1] U. E. Kornilovich and L. G. Gulinova, *Zh. prikl. Khim.* **26,** 876 (1953).
[2] M. H. Roberts, *R.I.L.E.M. Sym. Admixtures for Mortar and Concrete, Brussels* 1967.
[3] W. C. Hansen, *Sym., London 1952,* 598; *Spec. tech. Publ. Am. Soc. Test. Mater.,* No. 266, 3 (1960); H. N. Stein, *J. appl. Chem.,* 1961, **11,** 474, 482.

and this it is suggested is the effective group in these compounds leading to adsorption. Steinour[1] goes further and regards the effective group as the non-ionised (OH) group which is adsorbed by hydrogen bonding on to the surface of the cement particles. The greater the number of OH groups per molecule the greater he suggests is the possibility of adsorption and retarding action. The simple monohydric alcohols have slight retarding action, but the trihydric glycerol is a powerful retarder. The action of sugars according to Bruere[2] depends on their hydrolysing sufficiently to produce sugars which can be converted into saccharinic acids. The HO—C—C=O group is regarded as the active absorbing group.

Electrolytes with monomer or polymer organic cations act like the corresponding sodium salts but those with large complex organic anions retard or inhibit hydration. This is regarded by Stein as being in accordance with the concept of a positive surface charge on hydrating cement particles. Alternatively we may say that when an agent forms an insoluble calcium salt as a layer on the cement grains it acts as a retarder whereas if the calcium salt is soluble the hardening is not retarded. These organic retarders are also adsorbed on to the hydrated cement nuclei retarding crystal growth until all the retarder has been removed from solution. This as we have seen earlier (p. 258) happens with calcium lignosulphonate which, besides retarding hydration by adsorption, is also adsorbed on to the nuclei of the aluminate hydration products slowing up crystal growth until all the agent has disappeared from the solution. This results in the formation of much smaller crystals, as has been shown by electron-micrographs, and a higher degree of supersaturation of the solution with respect to CaO over the first few hours thus delaying the hydration of tricalcium silicate.

Sodium carbonate is an example of a salt with a different action on cement with, and without, gypsum. When it is added to a gypsum-retarded cement with a normal setting time it causes a quick set; with larger quantities a normal set occurs, and with still greater additions a quick set once more. The first action appears to be connected with decomposition of the gypsum and precipitation of calcium carbonate. On increasing the amount of sodium carbonate it then exerts its own normal action of type III. The addition of a small amount (0·25 per cent) of potassium carbonate to a cement causes some retardation of set but in larger amounts the setting time decreases and flash setting may occur. It seems that part of the available gypsum is converted to syngenite ($CaSO_4 . K_2SO_4 . H_2O$) delaying the formation of the ettringite film around the aluminate particles.[3] The presence of abnormally high content of soluble K_2O in a cement can similarly give rise to immediate formation of syngenite and an accelerated set.[4]

Particular attention has been paid by Lerch and other American workers[5] to the influence of gypsum on the initial rate of hydration of cement and to the most

[1] *Sym., London 1953*, 627.
[2] G. M. Bruere, *Nature, Lond.* **212,** 502 (1966); see also J. F. Young, *Sym., Tokyo 1968*; K. E. Daugherty and M. J. Kowadeski, ibid.
[3] E. M. M. G. Niël, *Sym., Tokyo 1968.*
[4] Z. T. Jugovic and J. L. Gillam, loc. cit.
[5] W. Lerch, *Proc. Am. Soc. Test. Mater.***46,** 1252 (1946); H. S. Meissner, *Bull. Am. Soc. Test. Mater.* No. 169, 39 (1950); A. G. Whittaker and V. E. Wessels, *Rock Prod.*, Aug. 1945, p. 95.

(i) Electron Micrograph of 2 $CaO.SiO_2.4H_2O$ (CSH(II)) (× 20 000). Grudemo

(ii) Electron Micrograph of Hydration Product from Tricalcium Silicate (× 10 000). Swerdlow

PLATE X

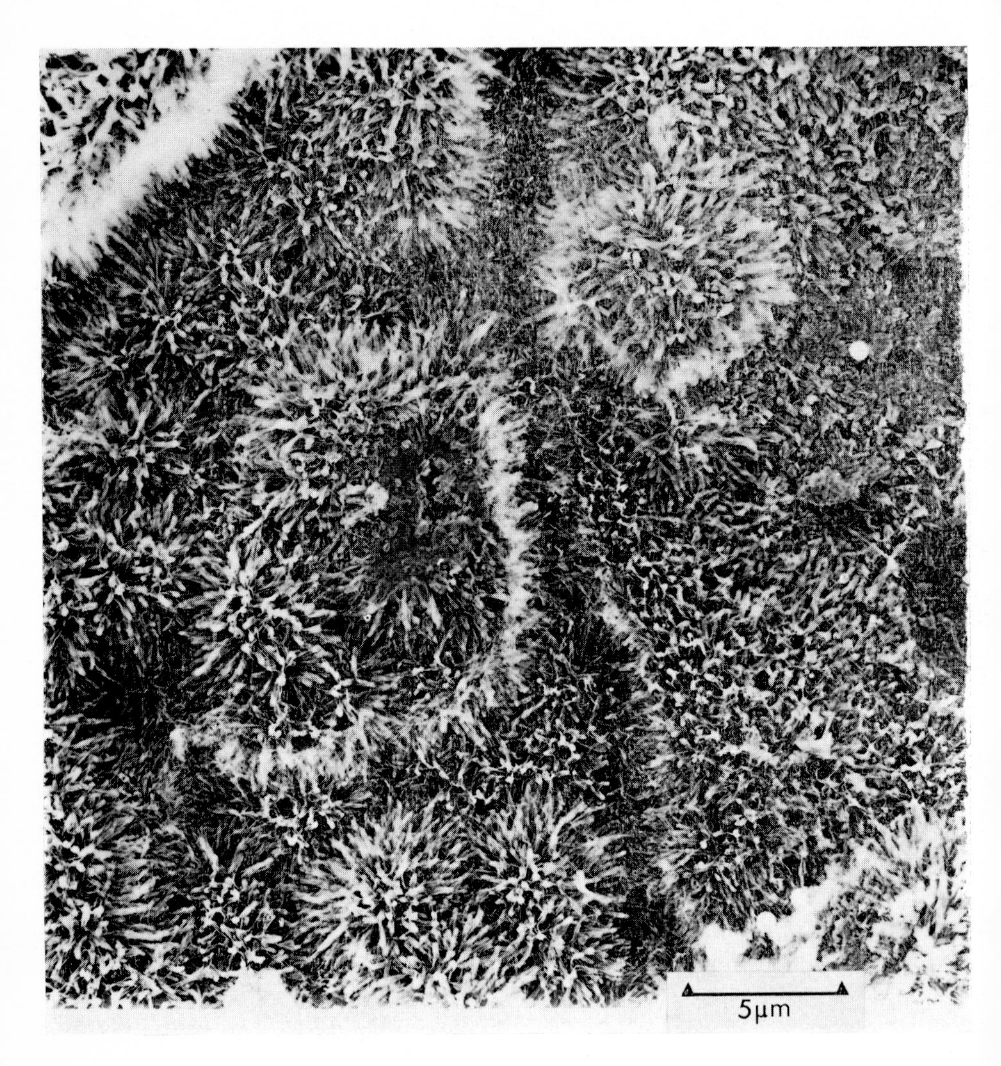

ELECTRON MICROGRAPH REPLICA OF FRACTURE SURFACE OF SET CEMENT (75 PER CENT C_3S, 25 PER CENT C_2S)

favourable proportion. By using a conduction calorimeter Lerch recorded the rate of heat evolution from cement pastes during the setting and early hardening period. A typical record is shown in Fig. 80. In general, on mixing the cement

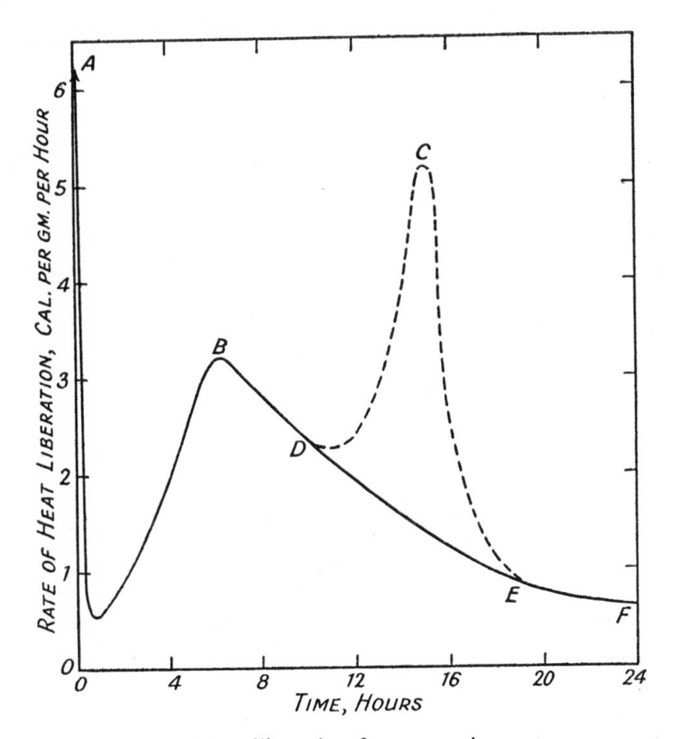

FIG. 80 Heat liberation from a setting cement.

with water there immediately occurs a rapid heat evolution, indicated by the peak *A*. This lasts only a few minutes but the rate of heat generation during this period may rise as high as a calorie per minute. This initial reaction soon ceases and the rate drops to a minimum to be followed by a gradually increasing heat evolution which reaches its peak (*B*) after some 4–8 hours for an average cement. The rate may then decline continuously (*DEF*), but in some cases a new 'delayed' rapid reaction was found to set in producing a third peak (*C*). The time of the first two peaks, *A* and *B*, is influenced by the aluminate, alkali, and gypsum content of the cements and, with sufficient gypsum, the third peak *C* can be eliminated and the reaction made to follow along *DEF*. The rate of the initial reaction at the peak *A* is reduced as more gypsum is added and Lerch attributed it to a very rapid hydration of the aluminate before a sufficient concentration of sulphate in solution had been established. If insufficient retarder is present then this reaction remains uncontrolled and a flash set results. During the following period of slow reaction a properly retarded cement retains much of its plasticity and initial hardening commences as the rate climbs again towards the second peak *B*.

The third peak *C* found with some cements occurs, according to Lerch, if all the gypsum has combined while some $3CaO.Al_2O_3$ is still available for reaction. The amount of gypsum required to prevent this increased with the content of $3CaO.Al_2O_3$, or of alkalis, in the cement. Lerch considered that just sufficient gypsum should be present to eliminate this peak, but its significance seems uncertain.

Systematic studies of the influence of gypsum content have shown that there is an optimum content which produces the highest strengths and the lowest drying shrinkage.[1] This amount varies with the contents of alkalis and tricalcium aluminate in the cement. For contents of $3CaO.Al_2O_3$ below about 6 per cent it is as low as 2 per cent SO_3 for low-alkali contents (below 0·5 per cent), increasing to 3–4 per cent SO_3 as the alkali content rises to 1 per cent or more. With higher $3CaO.Al_2O_3$ contents, e.g. 10 per cent or more, the optimum percentage of SO_3 is about 2·5–3 and 3·5–4 per cent with cements of low and high alkali contents respectively. With additions in excess of these optimum values a tendency to slow expansion in water appeared, indicating a continuing delayed reaction with the alumina compounds.

We noted earlier that there was still uncertainty about the relative importance in the action of gypsum of the parts played by the reduced solubility of alumina and the retardation of the hydration of tricalcium aluminate by surface films of sulphoaluminate formed round it. Forsén adopted both mechanisms, but Hedin later considered that it was the ionic concentrations in solution and the prevention of the alumina-silica gel formation that was of predominant importance. Lerch suggested that if the reduction of the initial peak of heat evolution were to be attributed to film formation it was difficult to explain the occurrence of the subsequent rapid reaction leading to the second peak (Fig. 80). He also, therefore, favoured the decrease in solubility of alumina in the presence of gypsum as the major factor. There are however various possible explanations. Thus the sulphoaluminate film may be disrupted by the osmotic pressure mechanism referred to earlier (p. 259). Alternatively disruption may be caused by the inception of the transformation of the high-sulphate form of sulphoaluminate into the low-sulphate form as the sulphur trioxide concentration in the solution falls below that required to stabilise the former. It has also been suggested[2] that the retarding action of gypsum does not arise primarily from the formation of the sulphoaluminate film which is said to be no more protective than a hydroaluminate film. On this view, retardation of setting is determined by the composition of the liquid phase and its influence on crystal structure formation in the cement paste. It is the change in this composition after the gypsum has been removed that leads to the acceleration of hydration and setting. While there is obviously still controversy as to the action of gypsum retardation, recent work by Ludwig[3] and others favours the formation of coatings of ettringite on the surface of the aluminate and ferrite compounds in the cement as the cause of retardation.

[1] See also G. Pickett, *Proc. Am. Concr. Inst.* **44,** 149 (1948); R. Kuhs, *Tonind.-Ztg. keram. Rdsh.* ,1959 (16), 388.
[2] Du Yu-Ju and E. E. Segalova, *Zh. prikl. Khim.* **34,** 50 (1961).
[3] U. Ludwig, *Zement-Kalk-Gips* **21,** 81, 109, 175 (1968).

TABLE 39 Effect of gypsum content on properties of cement: tests on 1 : 2·57 mortar (ASTM C109–47)

Cement No. 1 % composition				% SO_3 in Cement				Cement No. 2 % composition		% SO_3 in cement			
				1·06	2·23	3·21	3·97			0·98	1·98	2·49	3·46
C_3S	42	Compressive strength (lb/in²)	3 days	1370	2040	2360	2200	C_3S	44	1080	1400	480	1170
C_2S	32		7 days	2520	3140	3380	2320	C_2S	27	2150	2440	2540	2360
C_3A	9		28 days	4570	4880	4910	4970	C_3A	13	3580	4440	4390	4020
C_4AF	8							C_4AF	6				
Na_2O	0·13	% Shrinkage in 90 days in air at 50% relative humidity		0·142	0·132	0·099	0·101	Na_2O	0·15	0·149	0·126	0·114	0·112
K_2O	0·82							K_2O	0·12				
		% Expansion in water in 90 days		0·015	0·013	0·009	0·018			0·023	0·016	0·015	0·030
Cement No. 5				% SO_3 in cement				Cement No. 6		% SO_3 in cement			
				1·92	2·77	3·79	4·90			2·10	2·91	3·92	4·94
C_3S	49	Compressive strength (lb/in²)	1 day	1090	1630	1970	1660	C_3S	54	1200	1930	1430	1220
C_2S	19		3 days	2790	3130	3500	3710	C_2S	15	3100	3240	2830	1950
C_3A	14		7 days	4100	4230	4430	4680	C_3A	10	4860	4720	4170	2540
C_4AF	7							C_4AF	10				
Na_2O	0·18	% Shrinkage in 90 days in air at 50% relative humidity		0·140	0·135	0·116	0·104	Na_2O	0·14	0·136	0·119	0·105	0·140
K_2O	0·88							K_2O	0·11				
		% Expansion in water in 90 days		0·011	0·009	0·008	0·015			0·014	0·014	0·020	0·042
Cement No. 7				% SO_3 in cement				Cement No. 8		% SO_3 in cement			
				1·06	2·66	3·11	3·93			1·03	1·72	2·82	3·99
C_3S	28	Compressive strength (lb/in²)	3 days	900	1420	1580	1520	C_3S	22	640	820	590	540
C_2S	50		7 days	1460	2310	2300	2310	C_2S	59	970	1170	1070	920
C_3A	5		28 days	3030	3780	3610	3640	C_3A	4	2590	2570	2760	2470
C_4AF	9							C_4AF	8				
Na_2O	0·91	% Shrinkage in 90 days in air at 50% relative humidity		0·122	0·110	0·106	0·100	Na_2O	0·12	0·099	0·084	0·108	0·129
K_2O	0·39							K_2O	0·30				
		% Expansion in water in 90 days		0·017	0·010	0·008	0·012			0·015	0·012	0·021	0·036

Effect of gypsum on properties of cements

As we have seen from the foregoing discussion, the addition of gypsum to Portland cement not only controls the set, but also influences other properties; further there is an optimum content which varies with the composition of the cement. This is illustrated by the data[1] shown in Table 39 for pairs of cements of high and low alkali content for ordinary (Nos. 1 and 2), rapid-hardening (Nos. 5 and 6) and low-heat (Nos. 7 and 8) Portland cements. The results indicate that the amount of sulphur trioxide that can be added before any sign of unsoundness, as indicated by the expansion in water at 90 days, begins to appear, increases with the content of alkalis, or of the aluminate compounds, in the cement. There is a general tendency for the strength to be increased and the drying shrinkage reduced as the sulphur trioxide content is increased up to this point.

[1] H. S. Meissner, *Bull. Am. Soc. Test. Mater.*, No. 169, 39 (1950).

11 Structure of Cement Compounds

It has been shown earlier (Chapter 4) that some of the phases which go to make up unhydrated cement exist in different forms each of which has essentially the same chemical composition. The same is found with the hydrated compounds and, moreover, some of these compounds exhibit a wide range of composition while still possessing the same, or nearly the same, chemical and physical properties. In such circumstances the internal arrangement on the atomic scale becomes of great importance as a means of correlating and explaining chemical and physical behaviour. The structures of the initial and final products of a chemical reaction will often throw a great deal of light on the equilibria involved and the kinetics of the reaction itself. Many metastable conditions met with in cement chemistry probably owe their existence and continuity to a close resemblance between the structure of the metaphase and the phase from which it formed, the formation of the truly stable phase requiring a radical structural rearrangement with correspondingly high activation energy.

In the present stage of knowledge of the very complicated structures of cement compounds and their hydration products it is not yet possible to give any complete treatment of the part played by such structures in the process of setting and hardening. Much progress has recently been made, however, and this we shall review. It will first be necessary to deal with some points of the theory of X-ray analysis; the treatment must, of course, be very brief and the interested reader is referred to the various standard works for more detailed information.

The crystalline state

In many substances found in nature or produced synthetically the high degree of internal order is emphasised by their regular and beautiful crystalline form. However, if a piece of quartz, for instance, is reduced to a powder, the external symmetry will be lost, but the regularity of its internal structure remains. This is because each grain of powder is built up from the same structural unit composed of silicon and oxygen ions. If the subdivision is carried further, so that the size of the individual grains is of the same order as the fundamental units composing the quartz structure, the degree of order is reduced and the resultant powder may be described as amorphous. Amorphous structures can be formed, of course by processes other than subdivision to small enough fundamental units, as, for example, by precipitation as a gel. Many substances, including the principal component of set Portland cement, are normally produced in a fine state of subdivision; many such substances were at one time described as amorphous but have since been shown by X-ray methods to be crystalline. Another important tool for studying the structure of solids is electron diffraction. A beam of electrons behaves in many ways like a stream of electro-magnetic radiation similar in nature to light and to X-rays, but of shorter wavelength. When electrons are used to examine matter instead of X-rays, internal structure is revealed which is not resolved by X-rays. Some preparations of ferric hydroxide, for instance, give no X-ray powder pattern but a distinct electron diffraction pattern. There are, therefore, relative degrees of order in matter in the solid state and even glasses have a certain regularity of pattern in their internal structure which can, in fact, be resolved in part by X-ray or electron diffraction tests.

Bearing in mind that any real crystal will have imperfections which may affect its chemical and physical properties to some degree, the ideal crystal may be considered as an arrangement of points in three dimensions; these points are the mean positions occupied by the nuclei of the elements contained in the given substance. In a simple substance containing only one element, each point will be equivalent to any other, both in respect of nuclear mass and electric charge. The latter arises from the electron shell surrounding the nucleus; the existence of this shell implies that a given element in a given valency state requires a certain minimum space in the structure. It has, therefore, been possible to assign to each element one or more 'ionic radii' and these radii can be used to help in postulating structures by considering the most likely ways of packing a known number of ions into the given space.

The smallest element of space which, if repeated indefinitely in all directions, would reproduce the structure, is the unit cell. Many arbitrarily chosen volumes would satisfy this definition which has, therefore, to be qualified by the further statement that the simplest cell shall be chosen. In practice, this means the rhomb with shortest sides and the largest angles between the axes, which should correspond in direction with the normal crystallographic axes. The simplest case is the cube, where the sides are equal and the angles are right angles. However the unit cell is chosen, it will always contain an integral number of formula weights of the substance considered.

If the substance contains more than one element, then the points in space representing the crystal will not all be equivalent in respect of charge or mass. If the geometry of a system of such points (a 'lattice') is considered, it is found that there is a limit to the possible number of arrangements. Within this limit structures can be classified into 230 groups according to the symmetry operations by which the given arrangement can be extended. Crystals can, therefore, be assigned a 'space group' according to their structure; this space group is represented by a symbol such as $P2_1/c$ which shows in shorthand notation the type of lattice and the symmetry operations by which it is derived.

Although on an average the various ions contained in a predominantly ionic structure will be separated from their nearest neighbours by distances equal to the sum of the atomic radii, in practice there will be departures from these distances one way or the other according to the nature and strength of the bond between the ions considered. A really accurate determination of structures enables the bond lengths to be measured and therefore gives information on the chemical bonds in the compound concerned. In general each cation is surrounded by a fairly evenly spaced group of anions so that the angles between bonds are approximately equal. The number of ions in this group is the co-ordination number of the particular cation; usually the anion is oxygen.

If we consider for a moment the co-ordination of Si^{4+} in silicate structures, it is always found to be four. That is, each Si ion is placed in the centre of a regular tetrahedron at the corners of which are found oxygen ions. This group is so stable and is such a characteristic feature of silicate structures that it may be considered as a separate building unit on its own. In the orthosilicates the silicon-oxygen tetrahedron is isolated and acts, therefore, as an $(SiO_4)^{4-}$ ion; in other structures the SiO_4 tetrahedra are linked in various ways to form $(Si_2O_7)^{6-}$, $(Si_3O_9)^{6-}$ groups, etc. Similar arrangements are found in other series such as borates and phosphates.

Polytypism may effect the unit cell and the X-ray diffraction pattern. Polytypism is the stacking of modifications of the same basic structure; it may range from complete random arrangement to a simple doubling of the basic unit.

Diffraction of X-rays and electrons

A beam of X-rays and a stream of electrons may equally be considered, at any rate with respect to diffraction by matter, as beams of electromagnetic radiation. It is, therefore, unnecessary initially to distinguish between the two, and for convenience we shall speak only of X-rays.

In most methods of crystal analysis by X-rays, monochromatic radiation, that is, radiation of a single wavelength, is used. The X-rays are generated by the impact of an electron stream on a target and a characteristic spectrum of X-rays is produced, a single line of which is selected by means of filters. A frequently used radiation is the α_1 member of the doublet in the K series produced when copper is the target. This is known as Cu $K\alpha_1$ radiation of wavelength $\lambda =$ 1·537395 kX units. The kX unit is a length equal to 1·00203 Ångstrom units; the two units were intended to be equal, but a factor is necessary because of more exact determinations of Avogadro's number. In describing structures in this chapter, it has been assumed that the earlier determinations of structure were expressed in Å; modern determinations have been converted from kX to Å where the difference is significant.

X-rays are scattered by the atoms in a crystal, and since the atoms are in a regular arrangement the scattered rays will reinforce each other in certain directions and cancel each other out in others.

Consider a plane in a crystal defined by the Miller indices (*hkl*). It will be separated from similar planes by a distance *d*, the interplanar spacing, written as d_{hkl}. If a beam of X-rays falls on a single plane at an angle of incidence θ it can easily be shown that a diffracted beam will be formed (Fig. 81) in the plane of incidence emerging at angle θ, exactly as if the beam were reflected from the plane. The location and nature of the diffracting points which lie in the plane

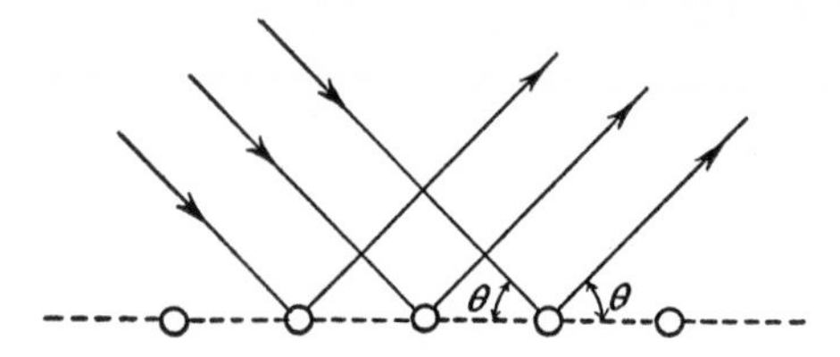

FIG. 81 Diffraction of X-rays by a single plane of atoms.

affect the intensity of this 'reflected' beam, but not its direction. When the beam is incident upon a three-dimensional array of points the diffracted rays from each layer must be in phase for an emergent beam to be formed and therefore the path difference AB–AC = $2d \sin \theta$ (Fig. 82) must be an integral number of wavelengths. We have, therefore:

$$n\lambda = 2d \sin \theta$$

In this case, therefore, a diffracted beam is formed only when the angle of incidence θ satisfies the Bragg equation written above. Note that a series of diffracted

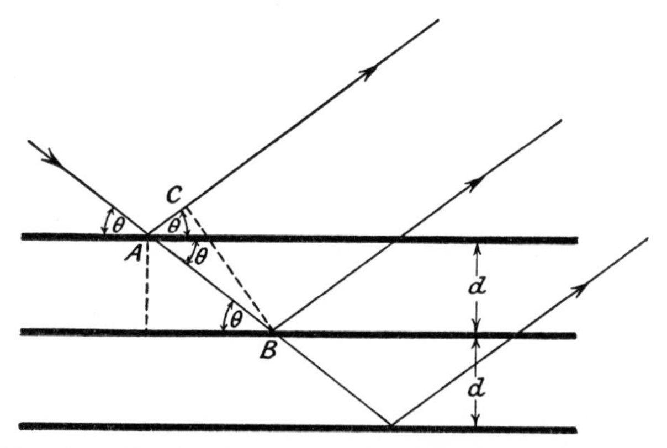

FIG. 82 Diffraction by a series of atomic planes of spacing *d*.

beams may be obtained by putting $n = 1, 2, 3$, etc. This is the same thing as saying that the spectrum for planes (*hkl*) will be formed at a different angle from that for (2*h*2*k*2*l*) and so on. If, for instance, our 'reflecting' planes are those parallel to (111) we shall get a series of spectra 111, 222, 333, etc. formed.

The powder X-ray diagram

Imagine a crystal placed at the centre of a cylindrical camera as in Fig. 83, and irradiated by a beam of X-rays. Each set of planes (*hkl*) in the crystal can produce

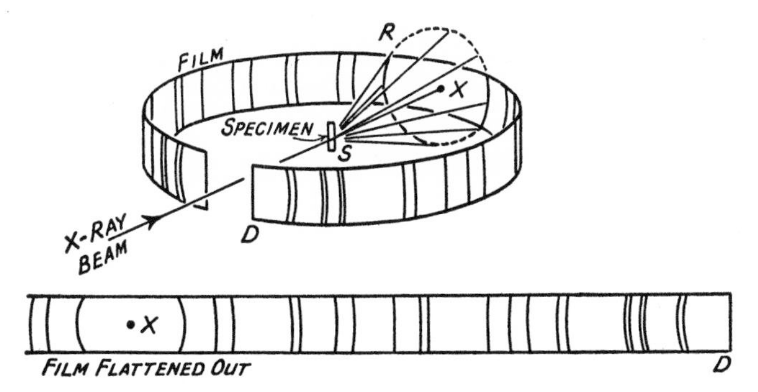

FIG. 83 Arrangement for taking powder photographs. The angle RSX is 2θ, where θ is the angle of incidence on a set of crystal planes.

a diffracted beam, but only if that set of planes is at an angle θ to the primary beam defined by the equation

$$d_{hkl} = \frac{n\lambda}{2 \sin \theta}$$

If, therefore, the crystal is turned about at random in the beam, reflected beams will flash out as the critical values of θ are passed. Instead of twisting and turning a crystal we may use a fine powder moulded into a rod or packed into a tube made of non-diffracting material. The crystals will be oriented in all possible directions, so there will be groups of them in a series of particular orientations corresponding to particular values of θ and so giving rise to reflected beams. All these reflected beams will form cones whose apices are the crystal and which cut across the photographic film to give arcs. From the measured position of each arc on such a powder photograph θ can be found and hence the spacings d/n. If the relative intensity of each arc is also recorded, the resulting table of spacings and intensities is characteristic of the substance under examination.

The American Society for Testing and Materials has compiled an Index of X-ray data in which substances are classified according to the spacings of the three strongest lines in the powder X-ray diagram. In Appendix III X-ray data are given for all the substances of interest in cement chemistry, the three strongest lines being marked.

For simple substances the powder X-ray diagram may be used to determine the unit cell. In a cubic structure of unit cell length 'a', for instance, $d_{hkl} = a/\sqrt{(h^2+k^2+l^2)}$. The process of assigning values of h, k and l to each line in the spectrum is called indexing. The indexing of diagrams for simple non-cubic structures is done graphically by comparing the given spectrum with sets of curves showing the variation of d_{hkl} with axial ratio and axial angle.

If a mixture of substances is examined by the powder X-ray method, the spectra are superimposed and, by careful comparison of the position and intensity of each line, the components of the mixture can not only be identified but a more or less accurate estimate of the relative proportions in which they are present can often be made. The minimum quantity which can be detected depends on the nature of the other substances present and is markedly affected by experimental factors contributing to the general background of scattered radiation.

Determination of structure

In a few simple cases, it has been possible to deduce the structure of a substance from the powder X-ray diagram. Usually, however, the much larger amount of information given by single crystal examination is required.

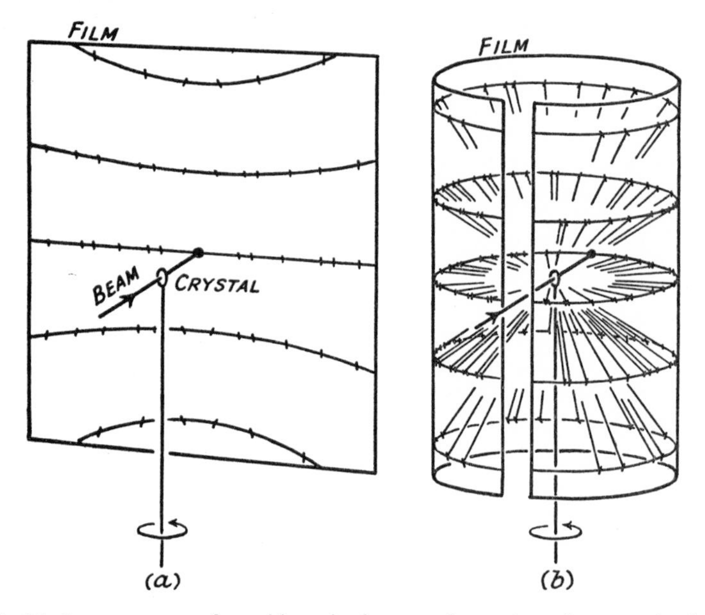

FIG. 84 Arrangements for taking single-crystal rotation photographs (*a*) on flat films and (*b*) on cylindrical films.

If the crystal is set up as in Fig. 84 with the X-ray beam passing through at right angles to one of its axes and is then rotated, the diffracted beams will appear as spots on the cylindrical film. When the film is opened out the spots will be found to be arranged in 'layer lines'. By measuring the distance between these layer lines and the equator, the axial length for the particular axis set up can be calculated. Thus from three rotation photographs the sides of the unit cell can be determined and, in the process of setting the crystal up, the axial angles are found. In practice more complicated cameras are frequently used for such work, but the basic principles are the same.

The next step is, graphically or by calculation, to index every reflection in the diagram, noting also its relative intensity. All the elements of symmetry present in the crystal are found also in the X-ray diagram and by an examination of the latter it is generally possible to assign the given structure a space group.

These two pieces of information, the unit cell dimensions and the space group, will often yield a great deal of information about the structure. From the volume of the unit cell, the density and the Avogadro number, the number of formula weights contained in the unit cell may be found. If M is the true molecular weight (which may be an integral number of times the ideal formula weight)

$$M = \frac{V\rho}{1{\cdot}66}$$

where ρ is the density and V is the volume of the unit cell in cubic Ångstrom units.

Suppose now that a unit cell of the given space group contains a certain number, say 2, of special positions. If in the molecular formula for the given substance only one element is represented twice, then clearly the positions of the two atoms of that element are known. Using such arguments, and trial arrangements for packing ions of known size in the unit cell, the structure can often be completely determined.

Sometimes more than one structure can be derived by such methods, each structure being of equal probability. In such a case, and in many cases where the simpler methods fail to give a guide, the final answer is obtained by trial and error. That is, a structure is postulated and the diffraction pattern calculated from that structure is compared in intensity and position with the pattern obtained experimentally. There are many methods of carrying out the work which cannot be detailed here, but the basic operation is the same. Eventually the position is reached where further refinement does not improve the agreement between calculation and experiment. At this stage, provided a certain agreed minimum of error has been reached, the structure is accepted until such time that a new postulation produces less error.

Electron microscopy and diffraction

It was stated earlier that the same principles governed the diffraction of both X-rays and electrons. The results obtained by the two methods do, however, differ in some respects. The scattering of electrons by an atom is a different process from the scattering of X-rays and the relative intensities of lines in the electron diffraction diagram are, therefore, different. The scattering unit is smaller with electrons and much finer structures can, therefore, be resolved; conversely, the electron beam is strongly absorbed by matter and in general only surface structure is revealed unless the specimen examined is very small.

Electron diffraction diagrams may be obtained by means of an arrangement of electron gun and flat specimen rather like some powder X-ray methods, or they may be obtained by transmission through thin specimens. The latter method is made use of in the technique by which diffraction diagrams are taken in the electron microscope; in favourable cases a single crystal diffraction pattern may be so obtained.

The 'optical' system in the electron microscope is the same as that in the visible light microscope except that electromagnetic or electrostatic focusing is employed. An electron beam is produced by acceleration of electrons emitted by a heated filament; the process requires a high vacuum and this creates experimental difficulties in preparing the specimens, some of which are decomposed by the effect of the vacuum and this is sometimes intensified by the effects of electron bombardment. Often the specimen is reinforced by the deposition on it of thin metallic films; if the film is deposited from one direction a shadow effect is obtained which reveals details not apparent in the untreated specimen. Sometimes a plastic 'replica' of the specimen is examined.

The specimen for electron microscopy is exceedingly small and must be in the form of a thin film or a well-dispersed finely divided powder. The chances of a sampling error or of contamination are, therefore, exceedingly high. When these possibilities are taken in conjunction with the risks already mentioned of modifying the specimen in the microscope, it will be seen that it is highly desirable to use the method in which the identity of the particles examined can be checked at once by means of the diffraction diagram. The results of electron microscopy of set cement have been discussed in Chapter 10.

Other techniques giving structural information

A great weakness of structure analysis by X-rays or electrons is the failure, except in the most favourable instances, to reveal the position of hydrogen atoms. Since the mode of bonding of hydrogen to oxygen in cement compounds is one of the most sought-after pieces of information concerning their structure, it is of interest to consider briefly what other types of investigation might give this information.

Neutrons are strongly scattered by hydrogen atoms, and neutron diffraction is therefore a means of locating hydrogen in a hydrated structure and deciding whether water molecules, hydroxyl ions or some more complicated type of ion is present. The method at present requires a very large crystal and is, therefore, unlikely to be of much assistance to the cement chemist for some time to come.

The method of studying ionic bonding by infra-red absorption spectra has has recently been extended to include the examination of powdered mineral specimens. By studying hydrates of known structure and classifying those with common absorption bands it should be possible to obtain valuable information concerning the hydrated compounds of cement. Very similar information might be obtained by electromagnetic microwave absorption measurements or by the method of nuclear magnetic resonance.

STRUCTURES OF THE ANHYDROUS COMPOUNDS

The oxides CaO and MgO

Both these compounds crystallise in the sodium chloride type structure shown in Fig. 85. In each case four molecules are contained in the unit cell, which is a cube of side $a = 4{\cdot}797$ for CaO and 4·203 for MgO. Each Ca or Mg is surrounded by six oxygens arranged in a regular octahedron; the co-ordination is, therefore, regular, and the co-ordination number 6. Note that, in drawing structures, it is

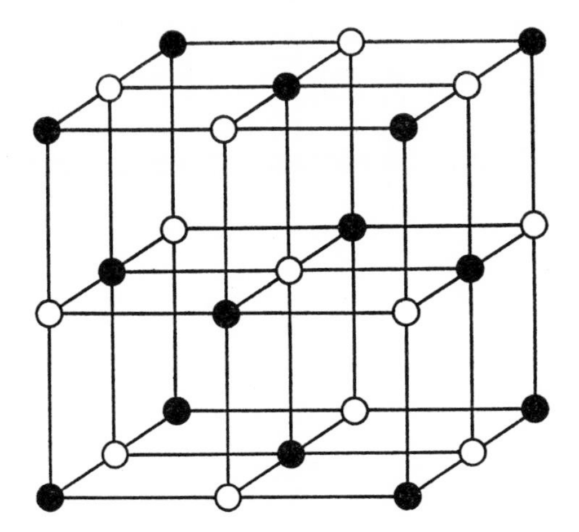

FIG. 85 Structure of CaO or MgO. The two types of circle are interchangeable and may be taken as oxygen or metal at will. See note in text on the size of atoms.

conventional to show the atoms smaller than they really are, so that bonds may be clearly seen. In MgO the oxygen atoms are actually in contact, with the small magnesiums packed in the interstices. The ionic radius of calcium is much larger than that of magnesium and in CaO the oxygens are forced apart. The CaO structure is, therefore, inherently less stable than that of MgO and this may be associated with the relative rates of hydration of the two oxides when dead burnt, and with the fact that MgO is stable enough to be found, if only rarely, as a natural mineral.

Tricalcium silicate

As discussed in Chapter 4, tricalcium silicate exists in more than one form. Pure tricalcium silicate is triclinic[1] and the exact structure is unknown; tricalcium silicate containing small amounts of iron, magnesium, potassium, sodium, aluminium and titanium[2] occurs in Portland cement clinker and is known as alite. In cements it occurs as the three main polymorphs, triclinic, monoclinic and trigonal. Five polymorphs have been proposed[3] but it has not been possible to identify them all in cement clinkers.

All these forms of C_3S are slight distortions of a much simpler pseudostructure, built up from SiO_4 tetrahedra and calcium and oxygen ions. An alternative structure containing 'free lime' groups was withdrawn after three-dimensional Fourier analysis had shown the superiority of the former one. This case provided a good example of the possibility of obtaining more than one credible structure by trial-and-error methods.

[1] J. W. Jeffery, *Acta Crystallogr.* **5**, 26 (1952); *Sym., London 1952*, 3.
[2] H. G. Midgley, *Mag. Concr. Res.* **20** (62), 41 (1968).
[3] M. Regourd, *Bull. Soc. fr. Minér. Cristallogr.* 2 (1964); *Sym., Tokyo 1968*.

The pseudo-hexagonal unit cell has $a = 7{\cdot}0$, $c = 25{\cdot}0$ Å and contains nine molecules of C_3S. Figure 86 shows a projection on to the base of the bottom one-third of the unit cell. The second and third layers are identical but are successively displaced by one-third in the direction of the long diagonal. Figure 87 is a vertical section through the long diagonal showing only those atoms in the plane of symmetry. Figures 86 and 87 are to the same scale, but in Fig. 87 the atoms are also shown full-scale in order to demonstrate the kind of packing.

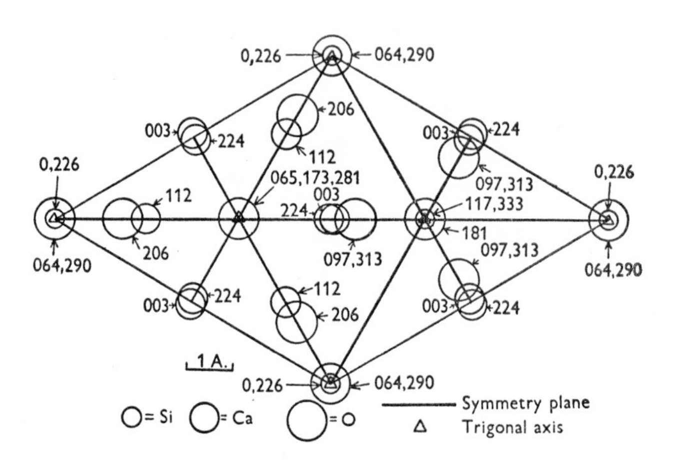

FIG. 86 Diagram of the bottom layer of the idealised or pseudo-structure of $3CaO.SiO_2$. The second and third layers are identical but displaced by one-third of the lattice spacing in the direction [1$\bar{1}$0]. The figures are thousands of the cell height.

Two of the main elements of the structure are shown in perspective in Fig. 88. Along each trigonal axis an element of type (*b*) alternates with two successive elements of type (*a*), the cross-linkage being both through direct O–O bonds and Ca-ions as shown in the figure. The O-ions in Fig. 88*b* are independent and bonded only to Ca, whereas those in Fig. 88*a* are built in to SiO_4 ions.

The co-ordination number of Ca in this structure is six and a notable feature is that the co-ordination is irregular, so that the oxygens are concentrated to one side of each Ca ion, leaving on the other side a 'hole' large enough to accommodate another Ca. The irregularity of the electrostatic field and open nature of the structure must imply a high lattice energy.

The alite-type structure is derived from the ideal structure by the substitution of one magnesium and two aluminium ions for two silicons. In the crystals so far examined this substitution occurred once in every 18 moles of C_3S, giving a formula $54CaO.16SiO_2.Al_2O_3.MgO$. The monoclinic cell dimensions are: $a = 33{\cdot}08$, $b = 7{\cdot}07$, $c = 18{\cdot}56$ Å, $\beta = 94° \ 10'$, two molecules to the unit cell, space group Cm.

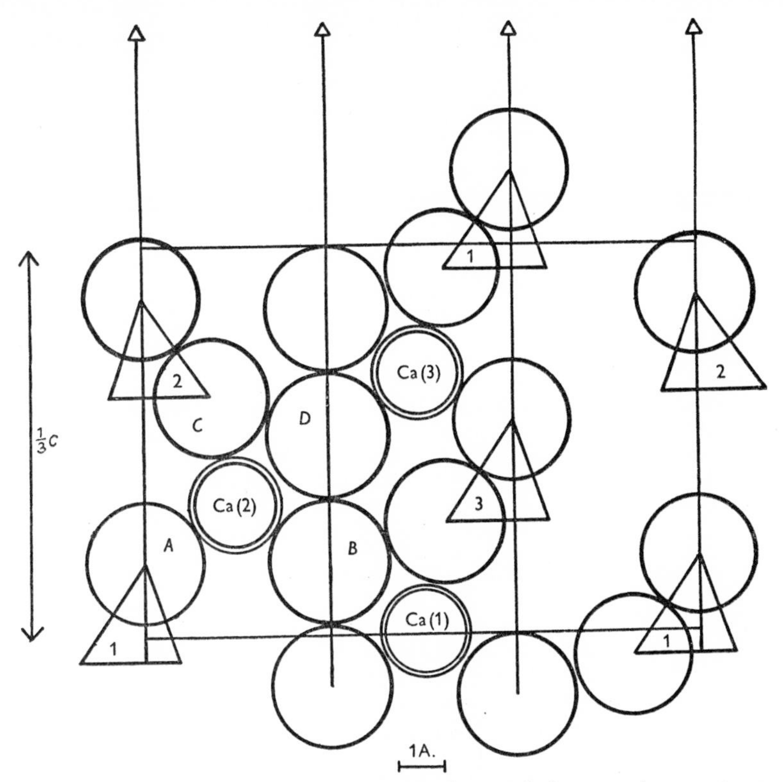

FIG. 87 Vertical section of the bottom layer of the pseudo-structure of $3CaO . SiO_2$ through the long diagonal of the cell. Only the oxygen atoms in the symmetry plane are shown as plain circles. 1, 2 and 3 are sections of SiO_4 tetrahedron. Calcium atoms are labelled.

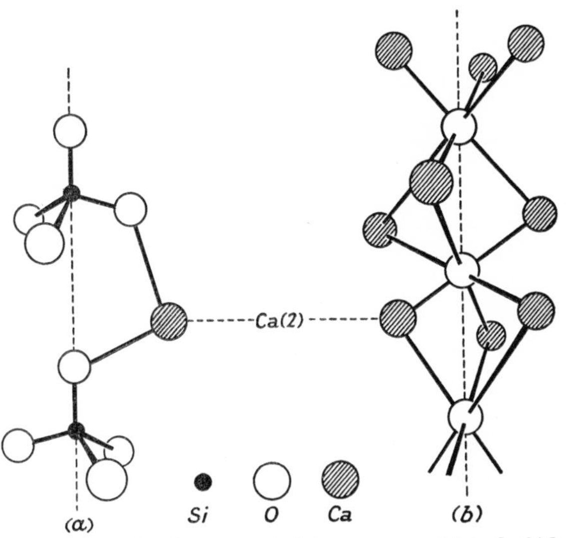

FIG. 88 Two of the main elements of the structure of $3CaO . SiO_2$.

TABLE 40 Structural data on modifications of $2CaO.SiO_2$ calculated from powder data

Modification	Unit cell dimensions Å				Composition of stabiliser
	a	*b*	*c*		
α	5·45	—	7·18		Pure at 1500° [1]
	5·38	—	7·10		Nagleschmidtite [2]
	5·419	—	7·022		2·5% Al_2O_3, 2·5% Fe_2O_3, 6·0% Na_2O [3]
α′	5·30	9·55	6·78		Pure at 700° [4]
	5·20	9·20	6·78		Merwinite [5]
	5·494	9·261	6·748		10% *CMS*, 4% K_2O [3]
	a	*c*	*b*	*β*	
β	5·507	9·317	6·754	94·62	Pure [6]
	5·513	9·326	6·760	94·60	0·5% B_2O_3 [7]
	5·511	9·321	6·752	94·56	0·25 mol% C_2B [7]
	5·547	9·207	6·690	94·50	2% C_2NP [7]
	5·514	9·330	6·772	94·60	0·25% Cr_2O_3 [7]
	a	*b*	*c*		
γ	5·06	11·28	6·78		Pure [5]
	5·083	11·232	6·773*		Pure [3]

* Axes interchanged to make compatible

[1] A. van Valkenburg and H. F. Macmurdie, *J. Res. natn. Bur. Stand.* **38,** 415 (1947).
[2] M. A. Bredig, *J. phys. Chem.* **49,** 537 (1945).
[3] G. Yamaguchi, Y. Ono, S. Kawamura and Y. Soda, *J. Ceram. Ass. Japan* **63,** 71, 2 (1963).
[4] G. Tromel, *Naturwissenschaften* **36,** 88 (1949).
[5] D. K. Smith, A. J. Majumdar and F. Ordway, *Acta Crystallogr.* **8,** 787 (1965).
[6] N. Yanaquis, *Revue Matér. Constr. Trav. publ.*, **480** (1955).
[7] H. G. Midgley, private communication (1968).

TABLE 41 Structural data on modifications of $2CaO.SiO_2$, single crystal measurements

Modification	Unit cell dimensions Å				Space group	Source of data
	a	*b*	*c*			
α	5·46	—	6·76			[1]
	21·80	—	21·54			Nagleschmidite [2]
α′	10·91	18·41	6·76		Pmnm	[1]
β	5·48	9·28	6·76	94·60	P21/c	Stabilised with B_2O_3 [3]
γ	5·06	11·28	6·78		Pbnm	[4]

[1] A. M. Douglas, *Mineralog. Mag.* **29,** 875 (1952).
[2] C. M. Midgley, *J. appl. Phys.* **3,** 277 (1952).
[3] C. M. Midgley, *Acta crystallogr.* **5,** 307 (1952).
[4] D. K. Smith, A. J. Majumdar and F. Ordway, *Acta crystallogr.* **18,** 487 (1965).

Dicalcium silicate

The structures of β and γ dicalcium silicate have been determined from studies on single crystals. Table 40 gives structural data from powder photographs, some of which were taken on pure material at high temperatures. Table 41 gives single crystal data, necessarily obtained with stabilisers present. The α' form is clearly very similar in structure to the β. Since β is the predominant form in cement clinker, it will only be necessary to discuss the structure of that form and of γ, which is of interest because it is non-hydraulic.

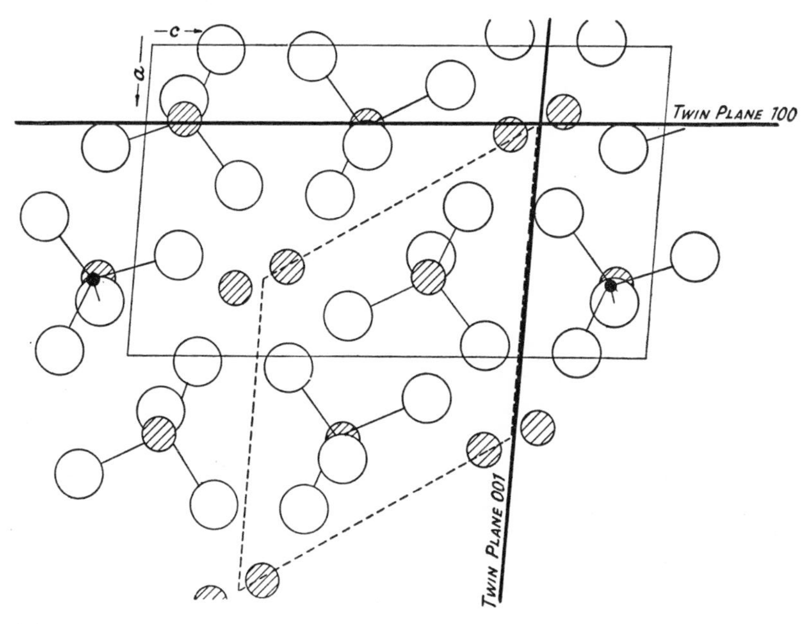

FIG. 89 Structure of β-$2CaO.SiO_2$. Projection along b axis. Twin planes in heavy lines. Si black, Ca hatched, O plain circles.

A photograph of the unit cell of β C_2S is shown in Plate XIII. Figure 89 is a projection of the structure along the b axis. The cell is built up of SiO_4 tetrahedra linked to two types of Ca ion. Four of the eight Ca ions (Ca I) are positioned alternately above and below SiO_4 tetrahedra in the y direction, and the remaining four (Ca II) are accommodated between tetrahedra. The co-ordination number of Ca I is six and of Ca II eight. The co-ordination is irregular, as can clearly be seen by examining the bonding of the two Ca atoms towards the front face of the unit cell in Plate XIII. The interstitial spaces thus formed in the structure are smaller than in C_3S.

By way of contrast the γ-C_2S structure shown in Fig. 90 is very regular. The oxygen ions are hexagonally arranged in two sheets; all are grouped tetrahedrally as independent SiO_4 ions. Every calcium atom is regularly co-ordinated by six oxygens. Although the structure is regular, it is to be regarded as open, the volume per oxygen atom being unusually large.

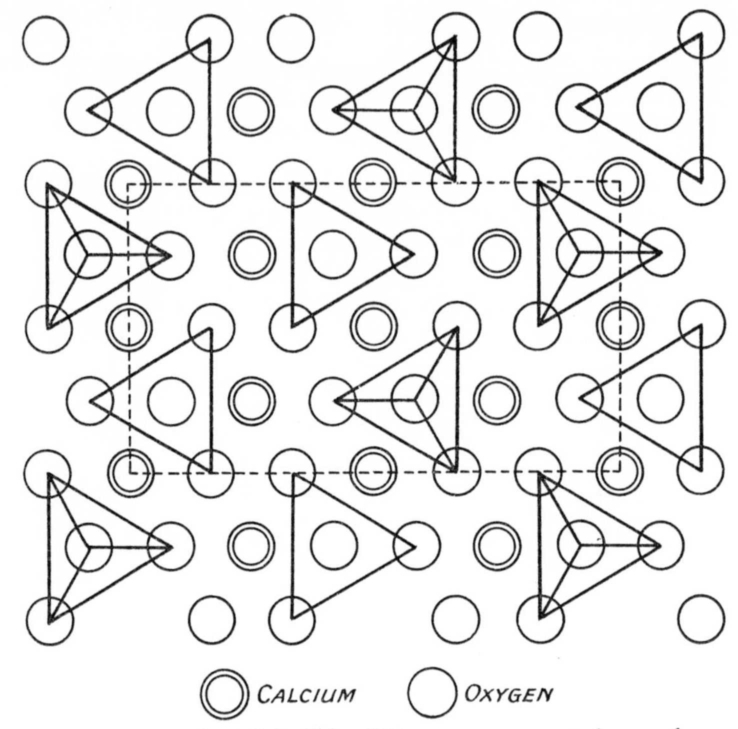

FIG. 90 Structure of γ-$2CaO.SiO_2$. Silicon atoms are not shown; they occur at the centre of the tetrahedra.

Other calcium silicates

Wollastonite $CaO.SiO_2$. There are two closely related forms of low-temperature *CS*, monoclinic and triclinic; and one for the high-temperature form, pseudo wollastonite. The relationships are given in Table 42.

The structure of the two low-temperature forms are similar and are based on infinitely kinked chains of SiO_3^{2-}. The chains are repeated at intervals of three tetrahedra and are known as 'Drierketter'[1]. A representation of the structure is

TABLE 42 Crystallographic data for wollastonite

	aÅ	bÅ	cÅ	α	β	γ
Monoclinic (para wollastonite) (CS)	15·42	7·32	7·07	90°	95° 24′	90°
Triclinic (wollastonite) (CS)	7·94	7·32	7·07	90° 02′	95° 22′	103° 20′
Triclinic (pseudowollastonite) (CS)	6·90	11·78	19·65	90°	90° 48′	90°

[1] F. Liebau, *Z. phys. Chem.* **206,** 73 (1956).

given in Fig. 91. The structure of the high-temperature form is not known but is thought to be made up of $Si_3O_9^{6-}$ rings. The mineral shows considerable polytypism. Rankinite and Kilchoanite are two minerals with a formula of 3CaO. $2SiO_2$. Their structures have not been determined, but the unit cells are:

Rankinite: *a* 10·55, *b* 8·88, *c* 7·85 Å; *β* 120·1°; $P2_1/a$.[1]
Kilchoanite: *a* 11·42, *b* 5·09, *c* 21·95 Å; Imam or Ima 2.[2]

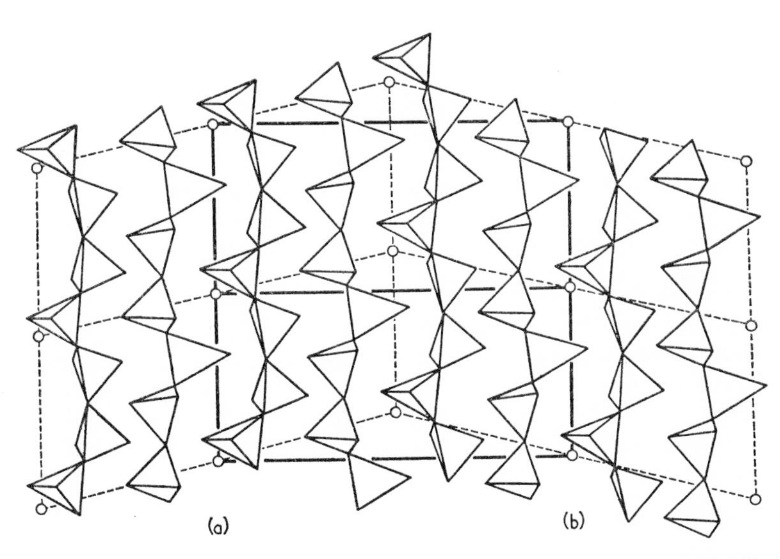

FIG. 91 The outline of the monoclinic (full lines) and triclinic (dashed lines) unit cells for β-$CaSiO_3$, showing the relationship between them and the Dreierketten.

Tricalcium aluminate

An approximate structure has been derived for tricalcium aluminate by means of single crystal studies on the isomorphous $3SrO.Al_2O_3$ and based on a structural resemblance of the two aluminates to $CaO.TiO_2$.[3] The unit cell and possible structure have been determined, giving a cubic cell of a 15·22 Å, space group Pa3, and the number of molecules per unit cell is 24.

The Al ions at the centres of the sides of the elementary cube are surrounded regularly by six oxygens. Other Al ions at the centres of the faces are surrounded by a plane ring of four oxygens. It is probable that this group is, in fact, a flattened tetrahedron and that the structure can be considered as made up of AlO_4 and AlO_6 groups connected into a network by shared oxygens. There are two kinds of Ca ion; those at the corners of the cube are regularly co-ordinated

[1] K. M. Moody, *Mineralog. Mag.* **30,** 79 (1952).
[2] S. O. Agrell and P. Gay, *Nature, Lond.* **189,** 743 (1961).
[3] F. Ordway, *Sym., London 1952*, 91.

by six oxygens, while those arranged in a small cube within the structure are very irregularly co-ordinated by nine oxygens.

A structure which is based on Al_6O_{18} rings of six tetrahedra has been proposed by Moore[1] with 64Ca at the corners of the a/4 lattice and a further 16 positions to be occupied by 8Ca or by 16Na. However the crystal structure reliability factor has not been reduced below 40 per cent. An orthorhombic modification of C_3A[2] has been identified in Portland cement clinker, and its relationship to the compound $Na_2O.8CaO.3Al_2O_3$ established.[3]

The compound $12CaO.7Al_2O_3$

This compound was originally thought to be $5CaO.3Al_2O_3$ but X-ray investigations first threw doubts on the accepted composition. The ratio $12CaO : 7Al_2O_3$ was suggested from X-ray evidence and later confirmed by a reexamination of the phase diagram.

The original published structure[4] consisted of two molecules of $C_{12}A_7$ placed in a unit cell of cube size 11·95 Å. More recent work[5] has shown that water may be an essential part of the structure.

The compound $5CaO.3Al_2O_3$

Although the cubic phase originally thought to be C_5A_3 has been shown to have a composition of $C_{12}A_7$ a phase with a composition of C_5A_3 does occur. It is probably equivalent to the 'unstable' C_5A_3.[6] It is orthorhombic, *a* 10·975, *b* 11·250, *c* 10·284 Å, C222; Z 4.[7] The structure has not been determined, but it has been postulated that it might consist of layers of Al_2O_7 ions interspersed with calcium ions.[8]

Compounds probably related to orthorhombic $5CaO.3Al_2O_3$

Three compounds related to C_5A_3 have been discovered in the system CaO–Al_2O_3–MgO.[9] A stable compound C_3A_2M, with a unit cell of *a* 16·77, *b* 10·72, *c* 5·13 Å, space group Pbma or $Pb2_1a$; it has been postulated that it may consist of layers of Al_2O_7, AlO_4, Al_2O_7 groups with alternate Ca ions between the layers and magnesium ions between the Al_2O_7 groups.[8]

An unstable compound of uncertain composition with a unit cell of *a* 44·34, *b* 10·76, *c* 5·13 Å; using the same criteria as for C_5A_3 and C_3A_2M the composition has been determined as $33CaO.5MgO.22Al_2O_3$[8] and it is suggested that it is made up of alternate layers of Al_2O_7 and AlO_4 groups.

The third compound is that thought originally to be the magnesium analogue of 'pleochroite'; recent work using an X-ray microprobe analyser gave the composition of the phase as $Ca_{35}Al_{53}Mg_4Si_5O_{128}$;[9] the phase has a unit cell of *a* 27·78, *b* 21·70, *c* 5·13 Å and is thought to consist of alternate layers of AlO_4 and Al_2O_7 groups.

[1] A. E. Moore, *Mag. Concr. Res.* **18** (55), 59 (1966).
[2] A. E. Moore, *Nature, Lond.* **199,** 480 (1963).
[3] K. E. Fletcher, H. G. Midgley, A. E. Moore, *Mag. Concr. Res.* **17** (53), 171 (1965).
[4] W. Büssem and A. Eitel, *Z. Kristallogr.* **95,** 175 (1936).
[5] R. W. Nurse, J. H. Welch and W. Gutt, private communication.
[6] E. S. Shepherd, G. A. Rankin and F. E. Wright, *Am. J. Sci.* **4,** 316 (1909).
[7] E. Aruja, *Acta Crystallogr.* **10,** 337 (1957).
[8] H. G. Midgley, *Trans. Br. Ceram. Soc.* **67,** 1 (1968).
[9] J. H. Welch, *Nature, Lond.* **191,** 559 (1961).

$CaO.Al_2O_3$

This compound is monoclinic with cell dimensions a = 8·837, b = 8·055, c = 15·250 Å, β = 90° 36′, 12 molecules to the unit cell, space group $P2_1/c$. The structure[1] closely resembles that of β tridymite with Al substituted for Si. The AlO_4 tetrahedra are linked by common oxygens into a continuous network, with calcium ions in the interstices. Co-ordination of the three kinds of Ca is six or seven and is irregular as shown in Fig. 92.

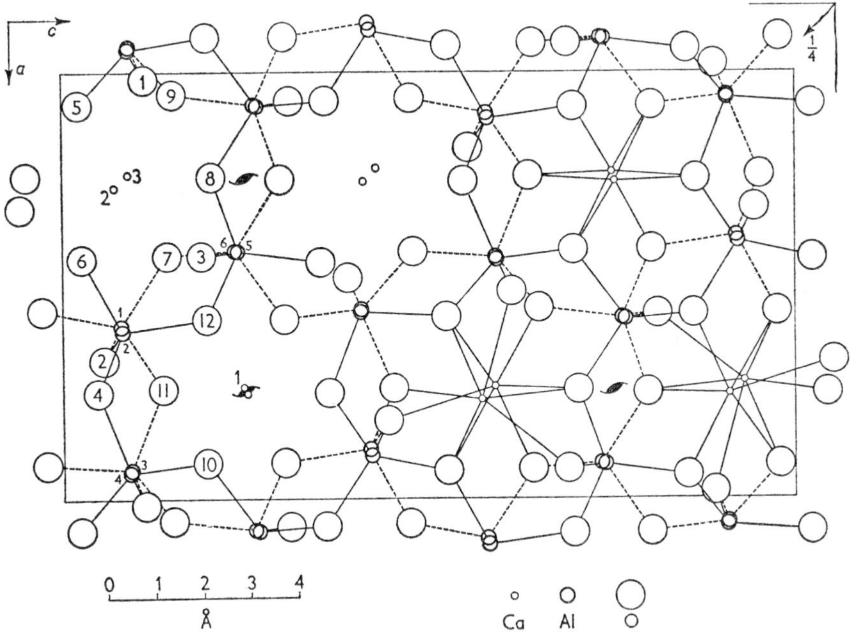

FIG. 92 Structure of $CaO.Al_2O_3$.

$CaO.2Al_2O_3$

Here again the determination of structural data by X-rays was of assistance in correcting the postulated formula.[2] Four molecules of $CaO.2Al_2O_3$ are arranged in a monoclinic unit cell of space group $C2/c$; a 12·89, b 8·88, c 5·45 Å, 107° 3′.[3] The structure consists of aluminium atoms tetrahedrally co-ordinated to oxygen with one oxygen at the common corner of three tetrahedra. The calcium atoms are very irregularly co-ordinated with five Ca–O bonds of about 2·6 Å, and four more of about 3·5 Å.

Gehlenite, $2CaO.Al_2O_3.SiO_2$

This compound is tetragonal a = 7·69, c = 5·10. Å Extensive solid solution occurs with other members of the melitite group, such as akermanite, with corresponding variations in the axial ratio.

[1] L. Heller, Ph.D. Thesis, Univ. London, 1951.
[2] K. Lagerqvist, S. Wallmark and A. Westgren, *Z. anorg. allg. Chem.* **234,** 1 (1937).
[3] E. R. Boyko and L. Wisyni, *Acta crystallogr.* **11,** 444 (1958).

All minerals of the melitite groups are built up from Si_2O_7 ions formed by the linking of two SiO_4 tetrahedra. In gehlenite the layers formed by these groups are connected by Ca ions which have an almost regular co-ordination of eight oxygens.

The ferrite series

A complete solid solution C_2F–'C_2A' exists up to a composition of $C_2F_{0\cdot31}A_{0\cdot69}$. The crystal structure of C_2F with *a* 5·42, *b* 14·77, *c* 5·60 Å, Pnma, has been described[1] as consisting of layers of FeO_6 octahedra perpendicular to *b*, joined to layers of FeO_4 tetrahedra with Ca atoms in holes between the layers having a very irregular ninefold co-ordination. All joins within and between layers are by corner only. The tertahedral layers consist of infinite chains of tetrahedra running parallel to *a*, but somewhat buckled in the *ac* plane by compression along the chain direction.

The solid solution series is not continuous,[2] the aluminium substitution takes place first in the tetrahedral layers allowing the tetrahedra to contract relative to the octahedral layers and also rotate into a more symmetrical position, making the space group Imma. Lattice constants[3] for some compositions are given in Table 43.

TABLE 43 Lattice constants of the ferrite series, Å

Composition	*a*	*b*	*c*
$2CaO.Fe_2O_3$	5·32	14·63	5·58
$4CaO.Al_2O_3.Fe_2O_3$	5·26	14·42	5·51
$6CaO.2Al_2O_3.Fe_2O_3$	5·22	14·35	5·48

Aluminium atoms substituting beyond $C_2F_{0\cdot67}A_{0\cdot33}$ are distributed equally between tetrahedral and octahedral sites. The composition C_4AF is not a distinct compound.

'Pleochroite'

This phase was originally thought to be $6CaO.4Al_2O_3.FeO.SiO_2$ by analogy with C_6A_4MS. Recent work on the composition of this phase which occurs occasionally in high-alumina cement, using the X-ray microprobe analyser, has shown the composition to be $Ca_{22}Fe_3Al_{14}(Al_2O_7)_8(AlO_4)_2(SiO_4)_2$.[4] A postulated structure indicates that the structure is made up of alternate Al_2O_7 and AlO_4 layers. The structure is related to orthorhombic C_5A_3, stable C_3A_2M and the unstable compound related to C_5A_3 referred to above.

[1] E. F. Bertaut, P. Blum and A. Sagnieres, *Acta crystallogr.* **12,** 149 (1959).
[2] D. K. Smith, *Acta crystallogr.* **15,** 1146 (1962).
[3] G. Malquori and V. Cirilli, *Sym., London 1952*, 120.
[4] H. G. Midgley, *Trans. Br. Ceram. Soc.* **67,** 1 (1968).

The glassy phase

X-ray structural studies on glasses have shown that oxygen polyhedra exist in these as in the more regular crystalline state. The glass-forming ion may be Si, Al, Be, B, or a number of others. The larger ions such as Na, K and Ca are embedded in the holes left in the network formed by the linking together of the polyhedra through the sharing of common oxygens. They may be removed by diffusion, leaving the glass apparently unaltered, as has been demonstrated by solution experiments on slag glasses. If too many modifying ions such as Ca are forced into the structure it crystallises spontaneously. This general regularity of glass structures gives rise to a series of diffuse lines in the X-ray diagram.

THEORIES OF REACTIVITY OF ANHYDROUS COMPOUNDS

Two main theories have been advanced to account for the cementing properties of the calcium silicates and aluminates. Both recognised the special importance of the Ca ions.

Brandenberger and Büssem,[1] suggested that calcium could exist in structures in an 'active' or 'inactive' form. At higher temperatures, such as are required for the formation of cement compounds, they assumed the co-ordination of calcium to be lower than normal; this low co-ordination is unstable at ordinary temperatures and the structure, therefore, tends either to form a new polymorphic arrangement or to hydrate, the new structure in either case containing calcium in its normal 'inactive' co-ordination of six.

Bredig[2] on the other hand pointed out that the normal effect of thermal expansion is to increase the co-ordination of calcium; he suggested (this was before the structure of β-C_2S was known) that the Ca co-ordination in β-C_2S was 8 and in α' or α that it would rise to 10. The tendency is therefore for compounds in which Ca has higher co-ordination to be unstable, inverting or hydrating readily to structures containing the normal sixfold arrangement. Büssem, although a supporter of Brandenberger's theory, laid great stress on the irregularity of the co-ordination of Ca in the unstable structures, with the consequent production of structure 'holes', and this view has been supported by Jeffery.[3]

From the review of structures given above, it will be seen that the facts tend to support Bredig rather than Brandenberger. In fact, however, the actual co-ordination number for Ca is not of very great significance. Thus C_3S, one of the prime cementing compounds, contains sixfold co-ordination of Ca. Büssem's point seems to have much more general validity; structural holes have been postulated in C_3S, C_2S, C_3A, $C_{12}A_7$, CA, and C_4AF. Of the calcium silicates or aluminates which do not hydrate and which are close in composition to these, structures have been suggested for γ-C_2S, β-CS and for C_2AS. The most interesting case is that of γ-C_2S where the composition is the same as that of the hydratable β-C_2S. In these three cases of 'inert' structures there are no 'holes'.

[1] W. Büssem, *Sym., Stockholm* 1938, 154.
[2] M. Bredig, *Am. Miner.* **28,** 594 (1943); *J. phys. Chem.* **49,** 537 (1945).
[3] J. W. Jeffery, *Sym., London* 1952, 30.

If cement hydration is to be regarded merely as a process of solution and precipitation, there seems to be no necessity to invoke a special mechanism. Sodium chloride, for instance, contains no 'holes' and the process of solution is a simple breaking of bonds between Na and Cl and the attaching of a hydration layer to the isolated ions so formed. The special facts in the case of cement hydration are that the hydration products are almost insoluble and that, therefore, a mechanism for the continuous production of a supersaturated solution has to be found.

The evidence from microscopic examination of set cement is that at any rate the later stages of hydration are to be considered as a solid state reaction. Each grain of cement mineral is seen to be surrounded by a layer of hydrated matter. If, therefore, hydration is to be a continuous process, it must be possible for water or OH ions to diffuse through the hydrated material and enter the anhydrous structure. The irregular types of structure containing 'holes' will lend themselves readily to diffusion processes. Such 'holes' are unnecessary in the hydrated structures (although they may be present) since they already contain OH ions or water molecules which can diffuse by an exchange process. There is some evidence that hydration of calcium oxide may proceed in part by a solid state reaction,[1] and that the hydration of mineral dicalcium silicate in nature has occurred by such a process. It has also been pointed out that the fracture by grinding of a structure containing irregular co-ordination will give rise to a surface containing local centres of unbalanced electrostatic charge. Such sensitive spots would form active nuclei for the beginnings of the hydration process.

Such structural work as has already been completed thus gives a general picture of what might be the special properties required in a cement mineral. Before it can be suggested why the products of hydration should knit together into a hard durable mass it is necessary to review the work on the structure of the hydrated compounds.

STRUCTURES OF THE HYDRATED COMPOUNDS

Because of the difficulty of obtaining sufficiently large crystals of the hydrates for complete structural analysis, only a few such analyses have been performed, mostly on natural minerals. However, before going on to a general discussion of the hydrates, such structures as have been determined will be summarised.

Calcium and magnesium hydroxides

These both crystallise in layer structures[2] of the type shown in Fig. 93. The Ca or Mg atoms lie in sheets perpendicular to the hexagonal *c*-axis and between other sheets of hexagonally close-packed hydroxyl ions. The sheets are strongly bonded together but layers so formed are attached to each other only by very weak forces.

[1] F. W. Birss and T. Thorvaldson, *Can. J. Chem.* **33,** 881 (1955); V. S. Ramachandran, P. J. Sereda and R. F. Feldman, *Nature, Lond.* **201,** 288 (1964); but see S. Chatterji and J. W. Jeffery, *Mag. Concr. Res.* **18** (55), 65 (1966).
[2] G. R. Levi, *G. Chim. ind. appl.* **6,** 333 (1924).

The $Mg(OH)_2$ layer (brucite) is one of the building units of the clay mineral structures and a more or less modified $Ca(OH)_2$ layer may well play the same part in the formation of the hydrated calcium silicates and aluminates.

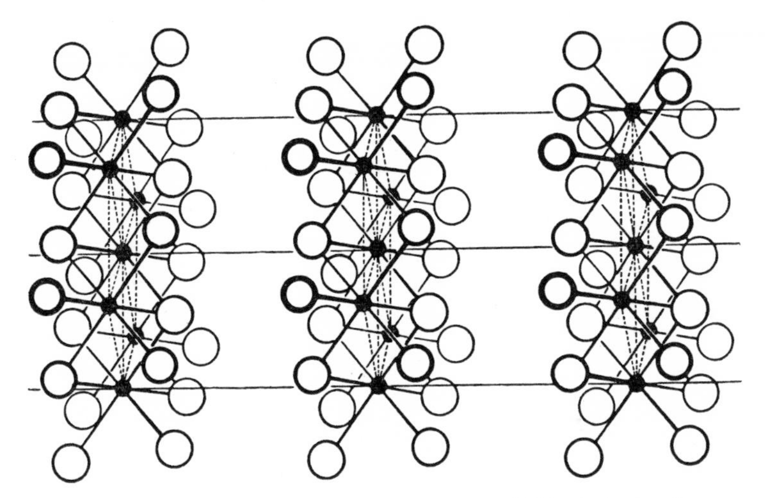

FIG. 93 $Ca(OH_2)$. Large circles OH, small black circles Ca.

Hexagonal cell dimensions are: $Ca(OH)_2$, $a = 3{\cdot}58$, $c = 4{\cdot}909$ Å; $Mg(OH)_2$, $a = 3{\cdot}11$, $c = 4{\cdot}74$ Å. Solid solutions do not form, the difference in ionic radius between Ca and Mg being too great.

Gibbsite $Al(OH)_3$

This compound[1] is monoclinic with $a = 8{\cdot}62$, $b = 5{\cdot}06$, $c = 9{\cdot}70$ Å, $\beta = 85°\ 26'$. The hydroxyl ions do not depart far from the hexagonal close packing found in $Ca(OH)_2$. Each Al is the centre of an octahedron of (OH) ions forming layers of $Al(OH)_6$ groups parallel to (001); each octahedron has a side in common with the neighbouring one so that every (OH) belongs to two octahedra.

The structure differs from that of brucite ($Mg(OH)_2$) in two important respects. Firstly there are only two Al ions in the gibbsite structure for every three Mg in brucite and one-third of the cation positions are therefore vacant. Secondly the stacking of the layers in gibbsite can be represented by the sequences ABBAAB as compared with the simple ABAB plan as in brucite.

Afwillite $3CaO.2SiO_2.3H_2O$

Afwillite occurs as a hydration product of Portland cement under special conditions and it is a possible product in materials treated hydrothermally. Moreover, the structure has been determined with very great accuracy[2] and therefore serves as a basis for work on the other hydrates.

[1] H. D. Megaw, *Z. Kristallogr.* **87,** 185 (1934).
[2] H. D. Megaw, *Acta crystallogr.* **2,** 419 (1949); **5,** 477 (1952).

Because of its low-scattering power the hydrogen atom cannot be easily located in a structure by X-rays. The contributions of O, OH and H_2O to the diffraction pattern will be almost identical; hydrogen atoms have therefore to be located as a result of very careful determination of bond lengths and confirmed by a consideration of the co-ordination of the surrounding atoms. This is the method used in determining the structure of afwillite, the bond lengths being determined to an accuracy of $\pm 0{\cdot}05$ Å units.

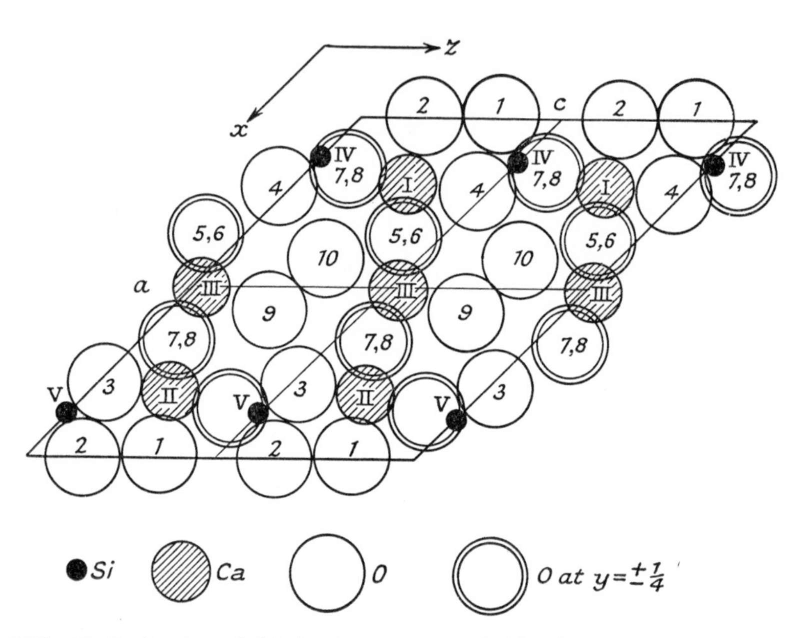

FIG. 94 Projection of idealised structure of Afwillite on (010), showing packing of atoms, with radii drawn to scale. All atoms are at $y = 0$ except those represented by double circles, which are at $y = \pm\frac{1}{4}$; atoms $y = \frac{1}{2}$ are omitted for clarity. Oxygen atoms include OH and H_2O.

A projection of the structure on 010 is shown in Fig. 94. It is built up from isolated $SiO_3(OH)$ ions, tetrahedral in shape, linked together by calcium polyhedra. The unit cell contains four molecules and has dimensions: $a = 16{\cdot}27$, $b = 5{\cdot}63$, $c = 13{\cdot}23$ Å, $\beta = 134° 48'$. The space group is Cc.

There are three kinds of Ca atom in the structure. The first (Ca I) is surrounded by six nearest neighbours, two (OH) and four O, arranged at the corners of a trigonal prism. A water molecule (O_9 in Fig. 94) is also part of the environment, but at a distance of 3·14 Å compared with the normal 2·4; this bond is, therefore, described as a Van der Waal contact. Ca II also has six nearest neighbours, four O and two (OH), somewhat irregularly arranged since in this case the water molecule (O_{10} in Fig. 94) is close enough to raise the co-ordination to seven. The third kind, Ca III, has six neighbours of which four are (OH) and two (H_2O).

It will be seen that the structural formula should be written $Ca_3(SiO_3OH).2H_2O$. Even this formula does not distinguish between the two water molecules which, as explained above, have different types of bonding. As might be expected from the structure, a differential thermal analysis shows that water is lost on heating in three stages.

Dicalcium silicate α—hydrate, $2CaO.SiO_2.H_2O$

This compound is C_2SH (A) in Bogue's nomenclature.

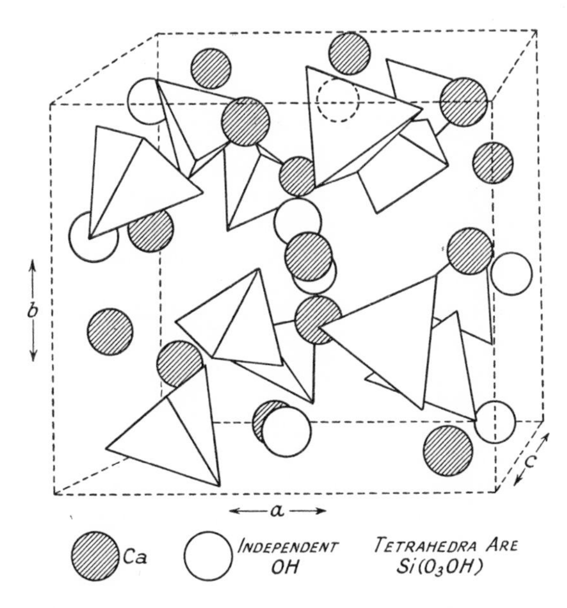

FIG. 95 Simplified representation of C_2S α-hydrate.

The unit cell dimensions are a = 9·34, b = 9·22, c = 10·61 Å, orthorhombic with eight molecules to the unit cell.[1] The structure obtained, shown in Fig. 95, is a simplified approximation to the true structure. It is made up of Ca polyhdera and $Si(O_3OH)$ tetrahedra. There are no water molecules; one hydroxyl in each formula unit is independent of the silicate tetrahedra. The structural formula is thus $Ca_2(SiO_3OH)OH$.

Tricalcium aluminate hydrate $3CaO\ Al_2O_3.6H_2O$

This compound is cubic with a = 12·56 Å, body-centred, space group Ia_3d; eight formula units are contained in the unit cell.[2] The structure resembles that

[1] L. Heller, *Acta crystallogr.* **5,** 724 (1952).
[2] E. Brandenberger, *Schweizer. Arch. angew. Wiss. Tech.* **2,** 45 (1936).

of garnet and an isomorphous series of garnet-hydrogarnets is formed according to the scheme:

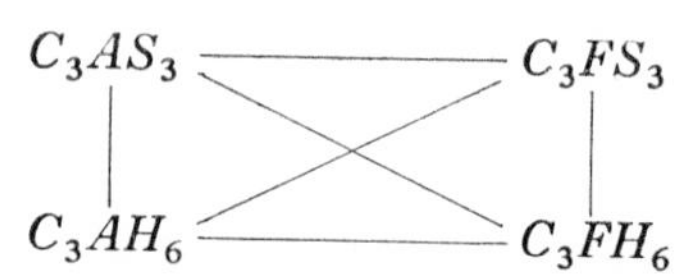

The structure[1] is derived from that of grossularite C_3AS_3 by ommission of the Si atoms from the centres of the SiO_4 tetrahedra and the substitution of $2H_2O$ for SiO_2. Each oxygen in the garnet structure thus becomes an OH in the hydrogarnet, thus balancing the charge of the missing silicons. The structure would be expected to be highly stable.

Lamellar calcium aluminate hydrates

The compound $Ca_2Al(OH)_7.3H_2O$ or C_4AH_{13} and several related basic salts form platy crystals which are hexagonal or pseudo hexagonal with basal cleavage. These hydrates are related to the natural mineral hydrocalumite, which is monoclinic *a* 9·6, *b* 11·4, *c* 16·84 Å, β 69°, the cell contains four formula units[2], space group $P2_1$.

C_4AH_{13} has been shown to be trigonal with a_H 5·73, c_H 47·16 Å, space group R $\bar{3}$c with six formula units per cell.[3] The unit cell is composed of six layers of 7·86 Å, each with a formula unit. The basic layers are composed of $Ca_2Al(OH)_6$ which lie in the *ab* plane and are of ordered structure which may be described as either distorted octahedral layers in which Ca^{2+} and Al^{3+} are regularly arranged or as agglomerates of Ca^{2+} and $Al(OH)_6{}^{3-}$ ions.

The interlayer regions are only partly ordered, with atoms statistically distributed among a number of sites. The remaining water molecules, hydroxide, etc., occupy cavities.

The structure of the related C_4AH_{19} polymorphs, α_1 and α_2, are not known, but the unit cells are closely related to C_4AH_{13}, the difference being in the *c* dimension, $\alpha_1\ c_0 = 64{\cdot}08$ Å, $\alpha_2\ c_0 = 21{\cdot}37$ Å.

The essential difference between the various lamellar calcium aluminate hydrates depends on the arrangement in the interlayer regions.

Ettringite (calcium sulphoaluminate)

The hexagonal unit cell of ettringite contains two molecules of $3CaO.Al_2O_3.3CaSO_4.31H_2O$ and has $a_0 =$, $b_0 =$, 11·10 and $c_0 = 21{\cdot}58$ Å. The space group[4] is $C\bar{3}1c$.

Four close-packed layers of O or OH ions make up the structure, with embedded Ca, S and Al. The latter form $Al(OH)_6$ groups in octahedral co-ordination and the sulphur is present as SO_4 tetrahedra. Four of the Ca ions have sevenfold co-ordination and eight sixfold.

[1] E. P. Flint, H. F. McMurdie and L. S. Wells, *Bur. Stand. J. Res.* **26,** 13 (1941).
[2] C. E. Tilley, H. D. Megaw and M. H. Hey, *Mineralog. Mag.* **23,** 607 (1934).
[3] S. J. Ahmed and H. F. W. Taylor, *Nature, Lond.* **215,** 622 (1967).
[4] F. A. Bannister, *Mineralog. Mag.* **14,** 324 (1936).

Forty-eight water molecules are located in channels parallel to the *c*-axis and two are more closely built into the structure. The structural formula thus becomes: $Ca_{12}Al(OH)_{24}(SO_4)_6.50H_2O$. On dehydration at about 100–110° the 48 interstitial water molecules are lost and the lattice shrinks so that a_0 = 8·4 and c_0 = 10·21 (= ½ 20·42) Å. In a differential thermal analysis[1] this gives rise to a large peak of 160°; a further small peak at 300° is attributed to decomposition of the hydroxyls bound to Al. One to two molecules of water are retained even at red heat.

Calcium silicate hydrates

A large number of artificial and naturally occurring hydrated calcium silicates have been examined by X-rays. The crystallographic data for many of these are given in Table 44. The compounds are divided into groups on the basis of what

TABLE 44 Crystalline calcium silicate hydrates

Name	Formula	*a*Å	*b*Å	*c*Å	*β*
Structures related to wollastonite					
Nekoite	$C_3S_6H_8$	7·60	7·32	9·86	Triclinic
Okenite	$C_3S_6H_6$	9·84	7·2	21·33	105°
Xonotlite	C_5S_5H	17·10	7·34	14·06	Triclinic
Hillebrandite	C_2SH	16·60	7·26	11·85	90°
Foshagite	C_4S_3H	10·32	7·36	14·07	106·4°
Tobermorite group					
14 Å Tobermorite	$C_5S_6H_9$				
11 Å Tobermorite	$C_5S_6H_5$	11·3	7·3	22·6	Triclinic?
9·3 Å Tobermorite	C_5S_6H				
12·6 Å Tobermorite	?				
10 Å Tobermorite	?				
All these cells are related to the 11 Å Tobermorite, with different *c* unit cell lengths. The description refers to the strong 002 reflection found in the X-ray diagram.					
Gyrolite group					
Z phase	CS_2H_2				
Truscottite	$C_6S_{10}H_3$	9·72	—	18·71	Trigonal
Gyrolite	$C_8S_{12}H_9$	9·72	—	132·8	Hexagonal
Structures related to γ C_2S					
Calciochondrodite	C_5S_2H	11·4	5·05	9·0	108·4°
Other structures					
Tacharanite	CSH				
Suolunite	CSH	11·15	19·67	6·08	90°
Rosenhahnite	C_3S_3H	6·946	9·474	6·809	108° 39′
Afwillite	$C_3S_2H_3$	16·27	5·63	13·23	134° 48′
Dicalcium silicate α hydrate	C_2SH	9·34	9·22	10·61	90°
Rustumite	C_4S_2H	7·62	18·55	15·51	104° 20′
Dellaite	C_6S_3H	6·80	6·91	12·85	91·58°
Tricalcium silicate hydrate	$C_6S_3H_3$	10·00	—	7·48	—

[1] G. L. Kalousek, G. L. Davis and W. E. Schmerz, *Proc. Am. Concr. Inst.* **45**, 693 (1949)

is known of their structures.[1] One of the most important minerals is tobermorite. The structure has been determined by Megaw and Kelsey;[2] Figure 96 is a representation of the structure. The structure is composed of layers parallel to (001) which are very similar to the layers in the layer clay minerals such as mica. Each layer is 11·3 Å thick. The central layer can be regarded as a distorted Ca(OH) sheet while on each side, but sharing some of the oxygen atoms, are silicon-oxygen 'dreierketten', that is SiO_3 chains containing kinks. This packing gives a

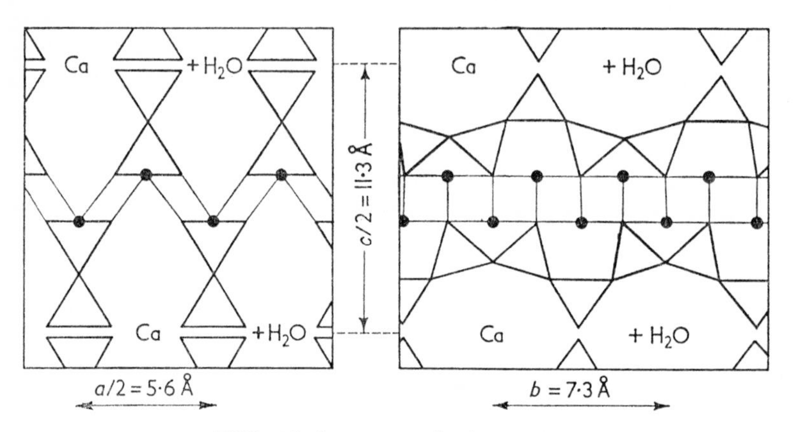

FIG. 96 Structure of tobermorite.

formula of $Ca_4Si_6O_{18}$; the remaining calcium in the formula unit and the four water molecules are packed into channels formed by the ribs produced by the kinking of the SiO_3 chains. The constitutional formula could thus be written $Ca_5(Si_6O_{18}H_2)4H_2O$. There are also a group of minerals which can be described as ill-crystallised *C-S-H*. There are two main types, *C-S-H* (I) and *C-S-H* (II), where the initials *C-S-H* stand for calcium silicate hydrate and do not represent a chemical formula. *C-S-H* (II) denotes material having a Ca : Si molar ratio of more than 1·5 and appears as fibres under the electron microscope. Within both categories considerable variations are possible in Ca : Si ratio, water content, basal spacing on the X-ray pattern, degree of crystallinity and other properties. The variation in the Ca : Si ratio of the *C-S-H* may be explained by the imperfections of the structure. Tobermorite has a structure which can be regarded as A-B-A-A-B-A where A is the SiO_3 chains and B the $Ca(OH)_2$ sheets. A structure that is made up of A–B–A–B, i.e. alternating silicon and calcium oxygen sheets is also possible and this would have a much higher Ca : Si ratio. As these are frequently ill-crystallised materials a mixed layer structure is possible, giving a considerable variation in the Ca : Si ratio of the product. Funk and Frydrych[3] have used the rate of solution of silicates, and Lentz[4] the complex-

[1] H. W. F. Taylor, *Sym., Tokyo 1968.*
[2] H. D. Megaw and C. H. Kelsey, *Nature, Lond.* **177,** 390, (1956).
[3] H. Funk and R. Frydrych, *Spec. Rep. Highw. Res. Bd,* No. 90, 284 (1966).
[4] C. W. Lentz, ibid., 269.

ing with organic silicones, to determine the degree of condensation of the silicate anion tetrahedra, that is, the degree of polymerisation. This has shown that single and double rings and chains may be detected in calcium silicates. So far the method has not been successfully applied to the *C-S-H* minerals in set cements. The variation in water content is similar to that found in the clay mineral vermiculite. The structure of *C-S-H* minerals produces sheets which are usually two unit cells thick, the two unit cells being bonded by a water layer. This layer may be varied and the amount of water present will depend on the water vapour pressure of the environment. The conditions are such that the layers may expand in steps which show up in the X-ray diffraction pattern, and with increasing water contents the 002 spacing will expand from 9·3 to 14 Å (see Table 44). It is usually assumed that ill-crystalline *C-S-H* is a tobermorite, but this may not be so and it may be more related to jennite[3] a sodium calcium silicate hydrate, $Na_2O.8CaO.5SiO_2.11H_2O$.

[3] H. W. F. Taylor, *Sym., Tokyo 1968.*

12 Action of Acid and Sulphate Waters on Portland Cement

The compounds present in set Portland cement are attacked by water and many salt and acid solutions, though fortunately in most cases the action on an impermeable set cement mass is so slow as to be unimportant. There are, however, certain conditions under which concretes may be exposed where these actions become very serious.

Pure water decomposes the set cement compounds, dissolving the lime from them, and to some extent the alumina; continued leaching eventually leaves only a residue of incoherent hydrated silica, iron oxide and alumina. This action on a mortar or concrete is so slow as to be negligible unless water is able to pass continuously through the mass. Waters which are acidic owing to the presence of uncombined carbon dioxide, of organic or inorganic acids, are more aggressive in their action, the degree and rate of attack increasing as the acidity increases and the pH value of the solution falls. The pH values of some solutions are shown in Table 45. In general, acid solutions which attack cement mortars or concretes by dissolving part of the set cement do not cause any expansion, but progresively weaken the material by removal of the cementing constituents. A soft and mushy mass is all that ultimately remains.

The main cause of acidity in natural waters is the presence of carbon dioxide and humic acid. The solubility of carbon dioxide in water under various partial pressures, and the pH values of the solutions, are shown in Table 46. In the presence of calcium carbonate the carbon dioxide becomes combined as calcium bicarbonate and the acidity of the solution is much reduced.[1]

In considering the carbon dioxide equilibria in water it is necessary to distinguish between the 'free' and the 'aggressive' carbon dioxide. The free carbon dioxide is that present over and above the amount required to form calcium

[1] The equilibrium constant $\dfrac{(H)(HCO_3)}{(H_2CO_3)} = 3\times10^{-7}$ and that for

$$\frac{(H)(CO_3)}{(HCO_3)} = 6\times10^{-11}.$$

TABLE 45

Solution	pH value
N NaOH	14
N/10 NaOH	13
Saturated limewater at 25°	12·42
Saturated magnesia solution at 25°	10·5
Neutral solution	7
Water saturated with CO_2 at the pressure (0·23 mm) present in normal air at 25°	5·72
Water saturated with CO_2 under 1 atmosphere pressure	3·95
Humic acid solution	3·5–4·0
1 per cent acetic acid solution	3·5
N/10 sulphuric acid	1
N sulphuric acid	0

TABLE 46

CO_2 content of atmosphere (% vol.)	Grammes CO_2 dissolved in 1000 g water at 18°	pH	Solutions saturated with $CaCO_3$	
			$CaCO_3$ dissolved in g/litre	pH
0·00	—	—	0·0131	10·23
0·03 normal air	0·00054	5·72	0·0627	8·48
0·30	0·0054	5·22	0·1380	7·81
1·0	0·0179	4·95	0·2106	7·47
10·0	0·1787	4·45	0·4689	6·80
100·0 (CO_2 at atmospheric pressure)	1·7870	3·95	1·0577	6·13

bicarbonate. The presence of some free carbon dioxide is necessary, however, to stabilise the calcium bicarbonate in accordance with the equation

$$CaCO_3 + H_2CO_3 \leftrightarrows Ca(HCO_3)_2$$

The free carbon dioxide which is required to maintain this equilibrium is incapable of effecting the solution of more calcium carbonate and is not therefore aggressive. Further the quantity so required increases with the amount of bicarbonate in solution. The aggressive carbon dioxide, capable of dissolving more calcium carbonate, is therefore not the free carbon dioxide present in excess of that required to maintain the existing bicarbonate equilibrium, but this quantity less that proportion of it which will be required to stabilise the additional calcium bicarbonate formed in the solution. In the following table are shown the amounts of free carbon dioxide required to maintain the bicarbonate equilibrium for various contents of calcium bicarbonate present in the water, and the pH value of this equilibrium solution.[1]

[1] Data recalculated by H. Gessner, *Schweiz. Verband für die Mat. Prüf. der Technik. Bericht*, No. 10, 61 (1928).

TABLE 47

CO_2 present as $Ca(HCO_3)_2$ (mg/litre)	Free CO_2 required to stabilise bicarbonate (mg/litre)	pH value of solution
44	0·32	8·64
88	2·45	8·05
132	8·00	7·70
176	18·3	7·46
220	35·0	7·28
264	58·8	7·13
308	92·0	7·00
352	133·5	6·89
396	187·0	6·79
440	253·0	6·70

The amount of free carbon dioxide in solution required to stabilise a given amount of calcium bicarbonate increases in the presence of other calcium salts such as calcium sulphate and decreases in the presence of salts of other bases such as sodium chloride. This is illustrated[1] by the curves in Fig. 97 which show the amount of free CO_2 required to stabilise increasing amounts of the bicarbonate. These curves are to be taken as an illustration of the effect of salts on the equilibrium rather than as a precise quantitative measure of it. The curve (3) refers to a

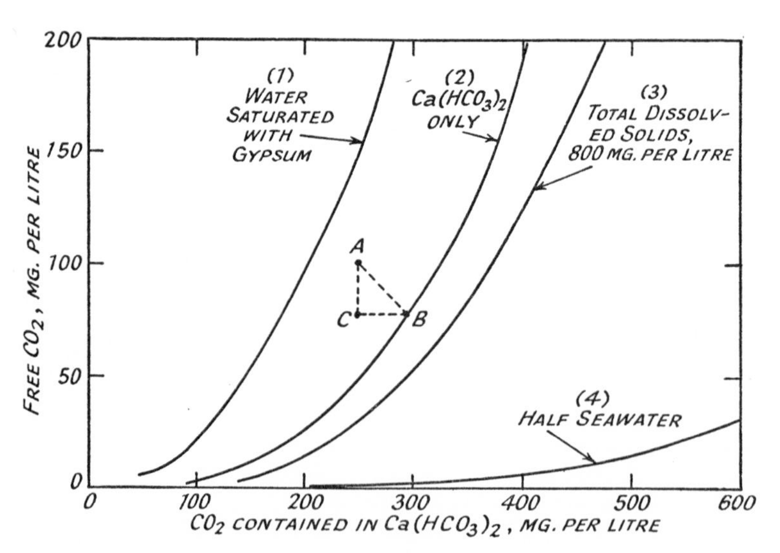

FIG. 97 Equilibrium between free carbon doxide and calcium bicarbonate in solution.

[1] R. D. Terzaghi, *Proc. Boston Soc. civ. Engrs* **36,** 136 (1949).

solution containing no other calcium salts than the bicarbonate, but with a total solid content (e.g. of sodium chloride), including the bicarbonate, of 800 mg per litre. It will be evident that for a given content of free CO_2 and calcium bicarbonate the amount of aggressive CO_2 is higher in the saline solutions, and less in the gypsum solution, than for water containing only the bicarbonate.

If the total amount of free carbon dioxide and calcium bicarbonate in water is estimated then the amount of aggressive CO_2 can be derived graphically from the CO_2–$Ca(HCO_3)_2$ equilibrium curve. We see from the equation:

$$CaCO_3+H_2CO_3 = Ca(HCO_3)_2$$

that for each molecule of free CO_2 consumed two molecules appear in the bicarbonate. Hence the solution composition must change in this reaction along a line such that the decrease in free CO_2 is half the increase in bicarbonate CO_2. If we plot, as in Fig. 97, the free CO_2 on twice the scale of the bicarbonate CO_2, then the path representing the change falls on a line at 45° to the axis. Thus if the original composition of a solution of CO_2 and $Ca(HCO_3)_2$ is represented by the point *A*, and it reacts with solid $CaCO_3$, it will change along the line *AB*. Reaction ceases at *B* since this point is on the equilibrium curve. The amount of free CO_2 consumed, i.e. the aggressive CO_2, is represented by *AC*. Thus of the 100 mg free CO_2 per litre at point *A* only 22 mg are aggressive. At lower, and more usual, free CO_2 concentrations, or at lower bicarbonate contents in the solution, the proportion of the free CO_2 which is aggressive is higher because of the shape of the equilibrium curve. For such conditions the values can be interpolated from the data in Table 48 which are calculated from the data of Tillmanns

TABLE 48

Free CO_2 (mg/litre)	CO_2 combined in $Ca(HCO_3)_2$ (mg/litre)	Aggressive CO_2 (mg/litre)	pH
1	0	1	5·6
1	4·4	1	7·2
5	0	5	5·2
5	4·4	5	6·5
5	8·8	4·9	6·8
5	22	4·7	7·2
10	0	9·9	5·1
10	4·4	9·9	6·2
10	8·8	9·8	6·5
10	22	9·5	6·9
30	0	29	4·8
30	8·8	29	6·0
30	22	28	6·4
30	44	27	6·7

and Gessner. It must be emphasised that these values are only applicable to $CaCO_3$–CO_2 equilibria; they will be altered as seen from Fig. 97, if other salts are present and also if part of the temporary hardness is caused by magnesium bicarbonate. In practice the pH value of natural waters of low hardness may differ by 0·5 units from that calculated from the carbon dioxide equilibria. If humic acid, or mineral acids, are present, these will determine the pH value which is then no index to the carbon dioxide content.

It must be remembered that if it is CaO, and not $CaCO_3$, that is being dissolved the reaction is:

$$CaO + 2CO_2 = Ca(HCO_3)_2$$

and that two molecules of free CO_2 are consumed for each two molecules that appear in the bicarbonate. However, in the case of a concrete exposed to a ground-water containing carbon dioxide the lime will first be converted to calcium carbonate and then, with a continuous supply of the water as will normally occur, the $CaCO_3$–CO_2 equilibrium relations will apply.

The content of calcium bicarbonate in a water can be determined by direct titration with N/10 HCl with methyl orange as indicator; 1 cc N/10 HCl = 0·0044 gm CO_2. The value obtained will include any other bicarbonates present. In waters in which the calcium bicarbonate content is fairly low, i.e. corresponding to less than 200 mg CO_2 per litre, and this includes most natural waters, the free CO_2 can be determined by direct titration with sodium hydroxide with phenolphthalein as indicator. In the presence of larger amounts of calcium bicarbonate the free CO_2 is determined by adding excess $BaCl_2$, and then excess $Ba(OH)_2$, and titrating back the $Ba(OH)_2$ with N/10 HCl with phenolphthalein as indicator. The difference between the titre of the $Ba(OH)_2$ solution added and that of the final solution represents the free CO_2 plus half the bicarbonate CO_2; 1 cc N/10 HCl = 0·0022 gm CO_2. The method is not accurate in presence of magnesium salts but a correction can be made if their content is known.

Free carbon dioxide is usually only present to an appreciable extent in rather pure natural waters, for in most cases where the water contains dissolved salts sufficient calcium carbonate is available to combine the carbon dioxide as calcium bicarbonate.

Considerable work has been done, first in Sweden and later elsewhere, on the leaching of lime from set cements by pure and acid waters. This work was stimulated by the serious troubles experienced with concrete dams in the Scandinavian countries. Various methods of test have been used in which specimens are subjected to the action of distilled water or rain or other natural waters. In the Swedish methods used by Virgin, Sundius and Assarsson, and Werner, crushed set cement was extracted either by shaking with successive quantities of water or by letting water percolate through a bed of the crushed material.[1] The most commonly used of these has been the Werner[2] method in which neat cement cubes, after curing for one to three months, are crushed and a fraction of grain size 0·21–0·09 mm (BS sieves Nos. 72 and 170) prepared for the test. A 1 gm sample is shaken for 5 minutes with 20 ml distilled water and this repeated fifteen times. The lime extracted is estimated in successive groups of three or five filtrates.

[1] See summary by J. O. Roos af Hjelmsäter, *Internat. Soc. Test. Mat. Zurich* **1,** 598 (1931).
[2] D. Werner, *Zement* **20,** 626 (1931).

In another method, used in various countries,[1] water is forced under pressure through concrete or mortar specimens and the lime extracted estimated. Specimens requiring high pressure, e.g. 100 lb/in^2 or more, to obtain any significant flow are preferable to lean mortar specimens needing only low pressures since the latter are more apt to contain fissures through which water flows rapidly without having time to dissolve much lime.

A third type of method, developed by Rengade,[2] depends on subjecting a 1 : 3 mortar specimen to the action of a jet of pure water impinging on its surface. The extent of wear or erosion as estimated by visual observation, or loss in weight, is taken as a measure of the solubility of the cement.

Percolation tests through concrete come closest to simulating practical conditions in a water-retaining structure, though a rather lean mix (e.g. 1 : 10) is necessary to obtain adequate permeability. The extraction method of Werner is rapid, simple in operation, and the reproducibility of results is good. Rengade's method is also simple, but takes some three months and involves a mechanical erosive as well as a solvent action.

Comparative tests on the three methods have shown[3] that the results from the extraction method generally place different cements in a similar relative order to those obtained from percolation tests on concrete. This is illustrated by Fig. 98 in which the amounts of lime extracted are plotted against the volume of

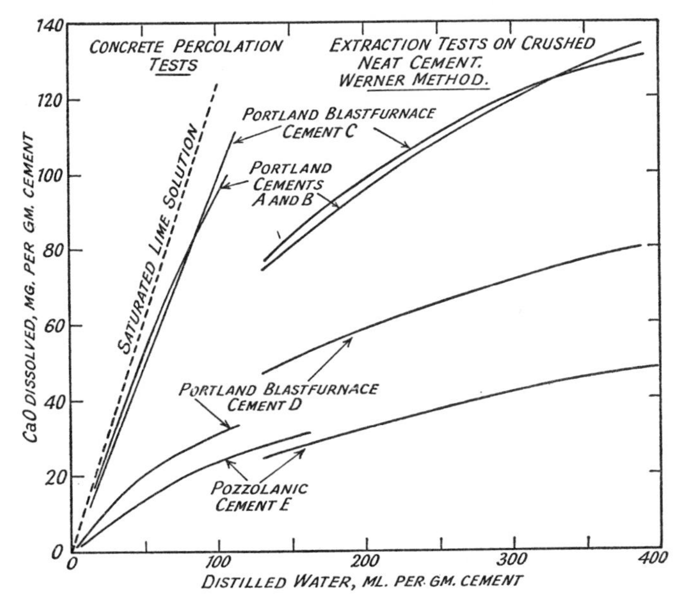

FIG. 98 Dissolution curves for cements.

[1] M. Mary, *Annls Ponts Chauss.* **103,** 467 (1933); **104,** 421 (1934); A. Ruettgers, E. N. Vidal, S. P. Wing, *Proc. Am. Concr. Inst.* **31,** 382 (1935); **32,** 378 (1936); F. M. Lea, *Building Res. Tech. Pap.* No. 26 (1939). H.M.S.O. London.

[2] E. Rengade, *Rev. Mat. Constr.* **275,** 309; **276,** 363 (1932); *Proc. 14th int. Congr. appl. Chem., Paris 1934*; 15th ibid., Brussels 1935.

[3] F. M. Lea, loc. cit. (1939).

water percolating through the concrete or shaken with the crushed set cement. The maximum volume of water used in the former test is only about the minimum used in the latter. The extraction curves do not follow on the percolation curves and the extraction method cannot therefore reproduce quantitatively the differences between cements found in the concrete tests. No significance is to be attached to small differences in the results it yields, but it shows whether different cements have a markedly different resistance to leaching by pure water.

The Rengade method is not so useful as a measure of the leaching produced where water percolates through cement, but it is perhaps the best indicator of the liability of cements to suffer attack when concrete is exposed to running water. It shows for instance high-alumina cement to have a high resistance to such surface solution, in agreement with practical experience, whereas the other methods do not. This type of cement is vulnerable when water acts on the crushed material or percolates through concrete but it is highly resistant to surface action.

In both the extraction and percolation methods with Portland cements and pure water the concentration tends to keep near to saturation with lime until some 10–15 per cent of the original weight of cement has dissolved. This probably represents the free calcium hydroxide present as a result of the hydration of the cement. The remaining lime, representing that present in hydrated calcium silicates and aluminates, is only extracted much more slowly. With hard waters there is an initial solution of lime but it eventually ceases and lime may even be deposited again in the specimen. There is not any appreciable difference between the behaviour of Portland cements of different types, but Portland blastfurnace cements show a more variable behaviour and some lose lime less readily. The presence of pozzolanas, as shown in Fig. 98, and also from the results of other percolation tests,[1] reduces the rate at which lime can be extracted and the proportion dissolved of the total lime present in the cement, owing to the combination of the free calcium hydroxide in the lime-pozzolana compound. Assarsson[2] also found that the addition of trass and moler reduced the rate of solution of the lime somewhat, but that arsenious oxide was particularly effective. The latter owing to its poisonous nature seems unlikely, however, to come into practical use. Virgin[3] as the result of tests on the solubility of set cements in waters of varying carbon dioxide contents has concluded that the attack of water on set Portland cement depends partly on the content of bicarbonate and free carbon dioxide in the water, and partly on the carbonation of the cement surface. For a constant content of bicarbonate in the water the lime-removing power increases with the content of free carbon dioxide. When the content of free carbon dioxide remains unchanged the dissolving power increases as the content of bicarbonates in the water falls. Carbonation of a cement surface by exposure to air reduces the action of waters which have a high solvent power.

The leaching of lime from the set cement progressively reduces the strength of a mortar or concrete but the loss in lime has to become substantial before the

[1] L. Santarelli and C. Cesarini, *Industria ital. Cem.*, May 1949; G. Goggi, ibid., **30** (8) 1960.

[2] *Zement* **21,** 64 (1932).

[3] See summary by J. O. Roos of Hjelmsäter, *Internat. Soc. Test. Mat.*, Zurich, vol. 1, 598 (1931).

strength is seriously affected. Thus in 1 : 3 Portland cement mortars a loss of 18 per cent of the lime was found to reduce[1] the compressive strength by about 15 per cent. For Portland blastfurnace cement the loss in strength was higher. In some tests on 4 in. concrete cubes (w/c = 0·6) immersed in running moorland water with zero temporary hardness Halstead[2] found that nearly half the compressive strength was lost after four years with little difference between Portland, Portland blastfurnace and supersulphated cements. With high-alumina cement the loss in strength was about one-third less. The 50 per cent loss in strength Halstead suggests is equivalent to a complete loss of strength in the outer $\frac{5}{8}$ in. of the 4 in. concrete cubes. From tests on concrete specimens stored in flowing water with a temporary hardness of 55 parts $CaCO_3$ per million into which carbon dioxide was injected Tremper[3] found that over the range pH 6 to 7 the loss of lime by solution was inversely proportional to the pH value. With a loss of 16 per cent of the lime from the cement the loss in compressive strength was about 20 per cent, with 27 per cent lime loss about 40 per cent. When the lime loss was over 50 per cent the strength was almost entirely lost. A 25 year-old concrete lining to a reservoir which had deterioated to a soft mass through percolation of water was found to have lost some 57 per cent of its original lime content thus supporting the above conclusion. In another case[4] reported a concrete which had lost about a quarter of its original lime content had lost half its strength. As a rough approximation it seems that for Portland cement the strength falls by 1–2 per cent for 1 per cent loss in lime.

Sulphate solutions

Sulphates of various bases attack set cements very markedly. Sodium, potassium, ammonium and various other sulphates react both with the free calcium hydroxide in a set cement to form calcium sulphate, and with the hydrated calcium aluminates to form the more insoluble calcium sulphoaluminate. The reactions with sodium sulphate, for example, can be formulated as:

1. $Ca(OH)_2 + Na_2SO_4.10H_2O = CaSO_4.2H_2O + 2NaOH + 8H_2O$
2. $4CaO.Al_2O_3.19H_2O + 3(CaSO_4.2H_2O) + 16H_2O = 3CaO.Al_2O_3.3CaSO_4.31H_2O + Ca(OH)_2$

The extent to which reaction (1) proceeds depends on the conditions. In flowing waters, with a constant supply of sodium sulphate and removal of the sodium hydroxide, it will eventually proceed to completion. If, however, the sodium hydroxide accumulates an equilibrium is reached depending on the sodium sulphate concentration.[5] Thus with 5 per cent Na_2SO_4 solution only about one-third of the sulphur trioxide is deposited as calcium sulphate when equilibrium is reached, and with 2 per cent Na_2SO_4 only about one-fifth. With calcium sulphate only reaction (2) can occur. The alkali sulphates do not attack

[1] K. Takimoto, H. Takahashi and S. Takaqi, *Japan Cement Engineering Association: 13th General Meeting, 1959*, p. 30.
[2] P. E. Halstead, *Mag. Concr. Res.* **6** (17), 93 (1954).
[3] B. Tremper, *Proc. Am. Concr. Inst.* **28,** 1 (1932).
[4] R. D. Terzaghi, *Proc. Am. Concr. Inst.* **44,** 977 (1948).
[5] W. C. Hansen and E. E. Presler, *Ind. Engng Chem.* **39,** 1280 (1947).

the hydrated calcium silicates to any appreciable extent, for they are more insoluble than the calcium sulphate and alkali silicates which would result. The calcium hydroxide liberated in the setting of tri- and dicalcium silicates reacts according to equation (1). Thus in sodium sulphate solution crystals of gypsum are soon formed from tricalcium silicate, but with dicalcium silicate the reaction proceeds much more slowly, corresponding with the very slow rate at which this compound splits off calcium hydroxide in water.

Magnesium sulphate has a more far-reaching action than other sulphates and decomposes the hydrated calcium silicates in addition to reacting with the aluminates and calcium hydroxide. If tri- or dicalcium silicate is placed in a magnesium sulphate solution, formation of gypsum crystals occurs rapidly. The hydrated calcium silicates react in the following general manner:

$$3CaO.2SiO_2.aq+3MgSO_4.7H_2O \rightarrow CaSO_4.2H_2O+3Mg(OH)_2+2SiO_2.aq$$

The reason why this action proceeds completely, while with sodium sulphate it does not occur, is to be found in the low solubility of magnesium hydroxide and the resulting low pH value of its saturated solution. It is soluble only to the extent of about 0·01 g per litre and its saturated solution has a pH of about 10·5. This is lower than the pH required to stabilise the hydrated calcium silicate. The silicate liberates lime to the solution to establish its equilibrium pH, but when magnesium sulphate is present the lime reacts with it, forming magnesium hydroxide and calcium sulphate. The former at once separates from solution, reducing the pH value to 10·5 again; more lime passes into solution by decomposition of a further part of the hydrated calcium silicate to re-establish the pH and so the action proceeds. The calcium sulphate accumulates in solution until it becomes saturated and crystals of gypsum then separate out.

The above equation is not the end of the reaction for Steopoe[1] found that the magnesium hydroxide and the silica gel can react very slowly to form a hydrated magnesium silicate. This was confirmed[2] by the finding in a deteriorated concrete seawall of a soft white material which gave an X-ray pattern not unlike $4CaO.Al_2O_3.12H_2O$ but which proved to have a composition approximating to $4MgO.SiO_2.8{\cdot}5H_2O$. This hydrated magnesium silicate appears to have no binding power, in contrast to silica gel, and its formation represents therefore a final stage in the deterioration of concrete attacked by magnesium sulphate solution. It may in practice, however, only be reached after long periods, and, indeed, Steopoe attributes the favourable influence of pozzolanas on resistance to seawater or magnesium sulphate primarily to the increased amount of silica gel that is formed when the sulphate acts not only on the hydrated silicates formed from the cement, but also on those arising from the reaction of lime with the pozzolana.

Magnesium sulphate has initially a similar action to that of other sulphates on the hydrated calcium aluminates, and calcium sulphoaluminate is first formed together with magnesium hydroxide. Calcium sulphoaluminate is itself, however, unstable in the presence of a magnesium sulphate solution, and by the continued action of this salt is ultimately decomposed again to form gypsum, hydrated

[1] A. Steopoe, *Tonind.-Ztg. keram. Rdsg.* **60,** 487, 503, 944 (1936).
[2] W. F. Cole, *Nature, Lond.* **171,** 354 (1953); W. F. Cole and H. V. Heuber, *Silic. ind.* **22,** 77 (1957).

alumina and magnesium hydroxide. This action must also be attributed to the pH value of the saturated magnesium hydroxide solution being below that required to stabilise calcium sulphoaluminate. It can be seen to occur if calcium sulphoaluminate is placed in a magnesium sulphate solution, when crystals of gypsum gradually form. The external skin of cement mortars which have been exposed for a fairly long period to the action of magnesium sulphate solution sometimes appears quite free from calcium sulphoaluminate crystals, while gypsum is present in large amount. In the interior of the mortar, where access of the solution has been much slower, both calcium sulphoaluminate and gypsum are found.

In concretes exposed to seawater there is a similar effect. Thus analyses of samples of Portland cement concretes from a structure exposed to seawater for 10 years showed the following results:

Sample	Per cent CaO	Per cent MgO	Per cent SO_3	Ratio $\frac{MgO}{CaO}$	Ratio $\frac{SO_3}{CaO}$
A	20·3	0·36	0·60	0·018	0·03
B	16·2	2·9	1·7	0·18	0·10
C	16·4	5·9	2·17	0·36	0·13
D	12·6	5·5	0·93	0·44	0·07

Sample A showed no signs of deterioration while B, C and D had been progressively more attacked. Part of the aggregate had been separated before these analyses were made.

The conversion of calcium hydroxide to gypsum more than doubles the solid volume, the respective molecular volumes of $Ca(OH)_2$ and $CaSO_4.2H_2O$ being 33·2 and 74·2 cc. The combination of hydrated calcium aluminate and gypsum in solution to form calcium sulphoaluminate also rather more than doubles the solid volume. These reactions account for the expansion and disruption of mortars and concretes attacked by sulphate solutions. Magnesium sulphate forms a hard dense skin on mortars and concrete owing to the deposition of magnesium hydroxide in the pores and this tends to hinder penetration of the solution. A mortar attacked by sodium or calcium sulphate ultimately becomes soft and incoherent but with magnesium sulphate the disrupted mass more often consists of hard granular particles.

The changes in solid volume[1] for various other reactions with sulphates can be calculated from the densities and molecular volume given in Table 49.

Though it is clear that sulphate attack leads to an increase in solid volume there is still some uncertainty as to the precise mechanism. It has been suggested[2] for example that it is the solid-state conversion of $4CaO.Al_2O_3.13H_2O$ to $3CaO.Al_2O_3.CaSO_4.12H_2O$ that is responsible for sulphate expansion, but the evidence[3] points to the presence of the $19H_2O$, and not the $13H_2O$ hydrate of

[1] For various other calculations of volume change, see W. C. Hansen, *Proc. Am. Soc. Test. Mater.* **61,** 1038 (1961); *Highw. Res. Rec. No. 113,* 1 (1966); *Canadian Building Series No. 2*; *Performance of Concrete,* p. 18 (1968).

[2] S. Chatterji and J. W. Jeffery, *Mag. Concr. Res.* **15** (44), 83 (1963); **19** (6), 185 (1967).

[3] M. N. Roberts, ibid. **16** (49), 236 (1964).

tetracalcium aluminate in set cement. As can be seen from Table 49 this conversion for the $19H_2O$ hydrate would lead to a decrease, not an increase, in solid volume. Other evidence[1] aginst this hypothesis has also been advanced.

TABLE 49

Compound	Molecular weight	Density	Molecular volume (cc)
$Ca(OH)_2$	74·1	2·23	33·2
$Mg(OH)_2$	58·34	2·38	24·5
$CaSO_4.2H_2O$	172·2	2·32	74·2
$4CaO.Al_2O_3.19H_2O$	668	1·81	369
$3CaO.Al_2O_3.6H_2O$	378	2·52	150
$3CaO.Al_2O_3.3CaSO_4.31H_2O$	1237	1·73	715
$3CaO.Al_2O_3.CaSO_4.12H_2O$	622	1·99	313

There are, however, difficulties in ascribing expansion in sulphate solutions directly to the increased volume of solids produced for there is little correlation between the amount of calcium sulphoaluminate or gypsum formed and the degree of expansion observed. The distinction between solid and liquid phase reactions is also important. Lafuma[2] suggested that combination of sulphate as a solid state reaction caused expansion, but not when the reaction occurred in solution and the products could be redistributed throughout the mass. Thorvaldson[3] from a review of the influence of electrolytes and other factors on sulphate expansion has suggested that the volume changes are controlled by osmotic forces concerned with the shrinkage and swelling of the gel system. The chemical reactions condition the gel system and destroy the cementing substances. The formation of new crystalline substances as a result of these reactions is, on this view, incidental to them and not the prime cause of expansion. Hansen has carried this theory further and calculated that, provided water can diffuse freely out of cavities or capillaries in the cement paste, it is not possible by a solid-liquid reaction for the volume of growing new crystals to exceed the space available to accommodate them. The pressure is conceived to arise from the diffusion of the sulphate salts into the gel pores. This disturbs the equilibrium between the gel and its surrounding liquid phase resulting in the movement of more water from the outside into the gel pores causing them to expand. It still remains to be seen if such a theory can account for the gross expansion that can occur in sulphate solutions and whether direct crystal thrust by anisotropic growth can be ignored.

The presence of other electrolytes influences the expansion of cement mortars in sulphate solutions. As far as chlorides are concerned the evidence is conflicting, some authors finding the expansion increased and others decreased. The more complex solution present in seawater definitely retards the expansion of concrete

[1] P. K. Mehta, *J. Am. Ceram. Soc.* **50** (4), 204 (1967).
[2] H. Lafuma, *Revue Matér. Constr. Trav. publ.* **243,** 441 (1929).
[3] T. Thorvaldson, *Sym., London 1952*, 436.

by sulphate reaction. This was attributed by Batta[1] to the greater solubility of gypsum and calcium sulphoaluminate in chloride solutions. Failure arising from chemical attack by seawater is not preceded by swelling to the same degree as in solutions of sodium or magnesium sulphate. The rate of expansion, for example, in a solution of 3 gm $MgSO_4$ per litre, about the same amount as in seawater, is much higher than in seawater where the effect of leaching actions tend to predominate. The presence of sodium bicarbonate, but not of chloride, was found by Locher[2] somewhat to decrease attack by sulphate solutions suggesting that the carbon dioxide content of seawater plays some part in decreasing its aggressive effect. Carbonation of mortars does increase their resistance to sulphate attack,[3] but the presence of sodium hydroxide also represses sulphate expansion.[4] There does not seem as yet to be any sufficient explanation of the reduced aggressiveness of seawater but it is evident that test in pure sulphate solutions do not reproduce the effects of seawater. Perhaps unexpectedly in view of the usual increase of the rate of chemical reactions with temperature, expansion in sulphate solutions seems to decrease[5] as the temperature rises from 20° to 40°. There may be some change in the incongruent solubility of calcium sulphoaluminate with temperature that could account for this but no data are available.

The effect of sulphate solutions on mortars or concretes can be followed by observing the change in strength or, more conveniently, by measuring the gradual expansion. This can easily be measured on mortar rods, 5 inches long and 1 inch diameter, made in cylindrical moulds which are filled from one end. Rectangular bars made in moulds such that one surface is finished with a trowel tend to bend during expansion. This is due to the formation of a rich cement skin on the trowelled surface which is attacked less rapidly than the other surfaces by the sulphate solution. As end-points for the measurement, steel balls $\frac{1}{4}$ inch diameter, cemented with a high-alumina cement grout into small cavities at the ends of the specimen, are satisfactory. The changes in length of the specimen can be measured by means of a screw-gauge micrometer reading to 0·0001 inch mounted on a stand of fixed dimensions as illustrated in Fig. 99. The micrometer reading is checked by means of a steel rod used as a standard of length and corrections made for any changes found in the reading given by this.

The rate of attack on mortars and concretes by sulphate solutions depends on the strength of the solution. It increases rapidly up to about an 0·5 per cent concentration for magnesium sulphate and 1 per cent for sodium sulphate, but beyond these limits only at a diminishing rate. Thus the rate of loss of strength in a 5 per cent solution may only be about two or three times that in a 0·5 per cent solution. At a concentration of 5 per cent, magnesium and sodium sulphates have roughly the same effect but at 0·5 per cent the sodium sulphate has the less rapid action. This is illustrated by Fig. 100 where data are reproduced for the strength of 1 : 2 : 4 concrete cubes after storage for a year in water and in 0·5 per cent and 5 per cent sodium and magnesium sulphate solutions. Results are shown for

[1] G. Batta, *Annals Trav. publ. Belg.*, August 1948.
[2] F. W. Locher, *Sym., Tokyo 1968.*
[3] M. G. Arber and H. E. Vivian, *Aust. J. appl. Sci.* **12,** (3) 330 (1961).
[4] T. Thorvaldson, *Sym. London 1952*, 436.
[5] J. D. Richards, *Mag. Concr. Res.*, **17,** (51) 69 (1965).

two Portland cements of differing sulphate resistance. The cement composition also affects the relative action of magnesium and sodium sulphate; the former tends to have the more severe action on cements of lower C_3A content, and sodium sulphate on those of higher content. This is to be expected from the chemistry of their actions. Calcium sulphate being soluble only to the extent of 0·2 per cent has a much slower action on dense mortars and concretes than the stronger solutions that can be obtained with the more soluble sulphates, but its eventual effect is comparable with that of a sodium sulphate solution of rather higher sulphate trioxide content. On porous mortars and concretes its action can be rapid.

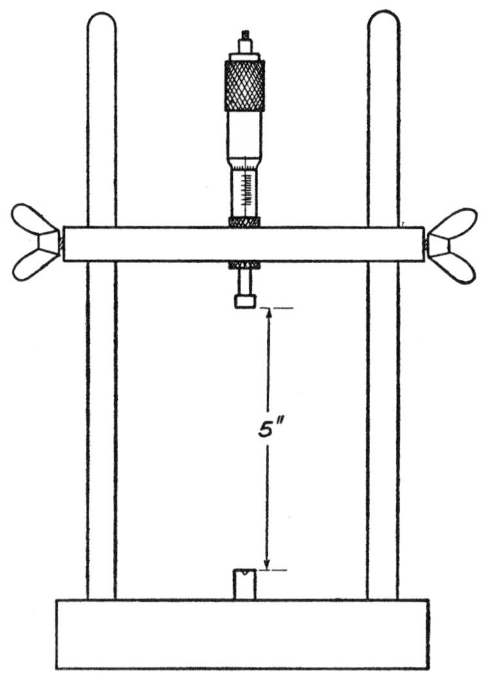

FIG. 99 Apparatus for measurement of expansion.

Some expansion curves[1] for 1 : 10 cement-sand (20–30 mesh) mortars in sulphate solutions are shown in Fig. 101 and 102. The relative resistance of cements and cement compounds to magnesium, sodium and calcium sulphate solutions as measured by the time required to expand 0·5 per cent is also shown in Table 50.

Tricalcium silicate expands rapidly in magnesium sulphate solutions, and very slowly in sodium or calcium sulphate solutions. Dicalcium silicate behaves similarly in magnesium sulphate solution but is very resistant to the other sul-

[1] T. Thorvaldson, D. Wolochow and V. A. Vigfusson *Can. J. Res.* **6**, 485 (1932); T. Thorvaldson, *Sym., London 1952*, 436.

phates. The substitution of $3CaO.Al_2O_3$ for either of these compounds to the extent of 20 per cent, rather above the maximum in a Portland cement, renders the expansion very rapid in all three sulphate solutions. The substitution of $4CaO.Al_2O_3.Fe_2O_3$ also increases the expansion rate of $3CaO.SiO_2$, but to a lesser extent. Though the expansion of mortars as lean as 1 : 10 is very rapid, the conclusions drawn as to the relative resistance of the different compounds are

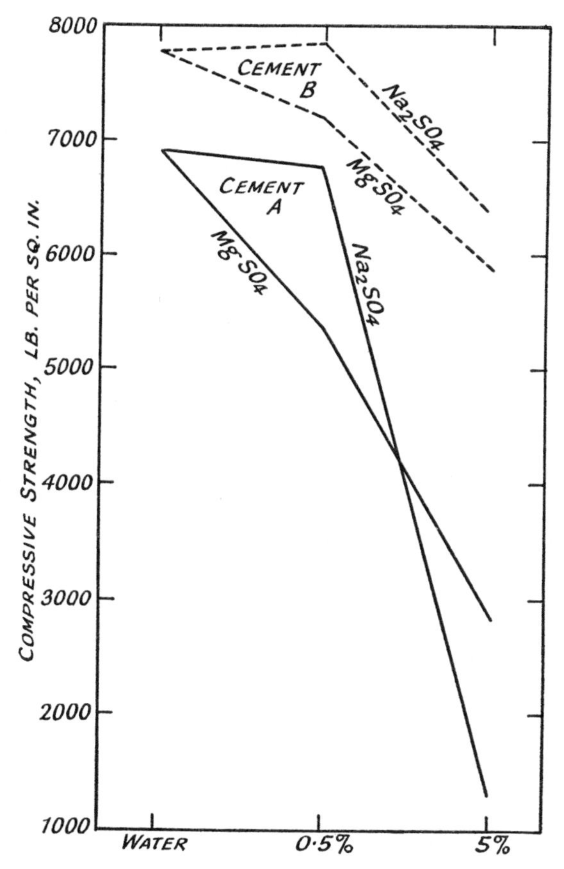

FIG. 100 Strength of concrete after one year's storage in water, 0·5 per cent and 5 per cent sulphate solutions.

borne out by tests on richer mortars such as 1 : 2 or 1 : 4 except that, in cements containing all the four main compounds, there appears to be little change in resistance with alterations in the relative proportions of tri- and dicalcium silicate.[1].

The relation between the reduction in tensile strength of mortars and the expansion[2] is illustrated by Fig. 103. While the general form of this curve pro-

[1] R. H. Bogue, *Portland Cement Association Fellowship Paper*, No. 55 (1949).
[2] T. Thorvaldson, *Can. J. Res.* **1,** 273 (1929).

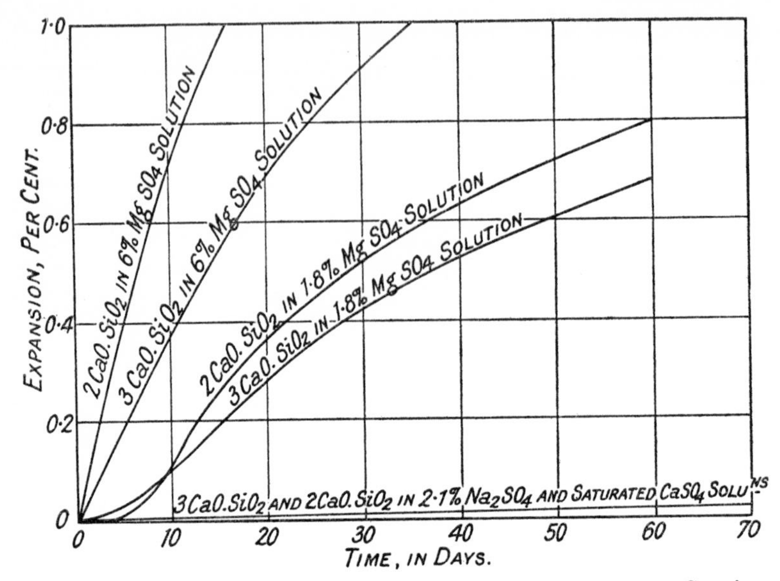

FIG. 101 Expansion of 1 : 10 cement/standard-sand mortars. Specimens aged 8 weeks in water before immersion.

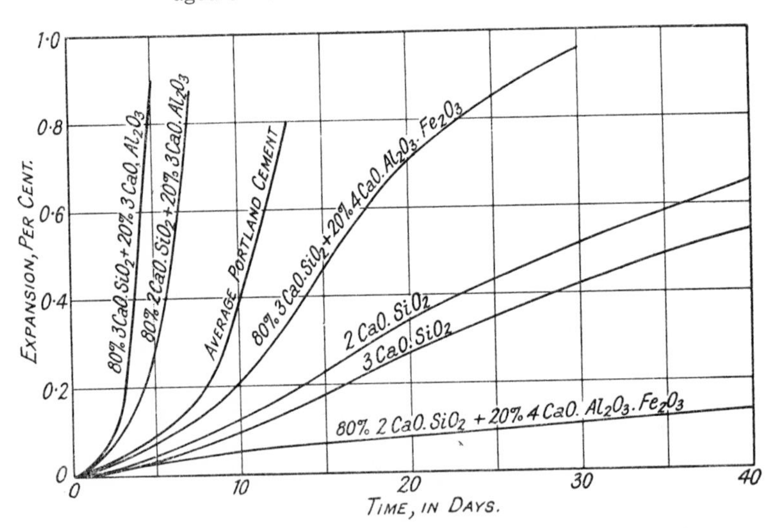

FIG. 102 Expansion of 1 : 10 cement/standard-sand mortars in 1·8 per cent $MgSO_4$ solution. Specimen aged 8 weeks in water before immersion.

bably holds in all cases, considerable variation in the ratio of diminution of strength to expansion occurs with different cements and under different conditions. Thus Miller and Manson[1] found that when Portland cement concrete

[1] D. G. Miller and P. W. Manson, *Tech. Bull. Minn. agric. Exp. Stn*, 194 (1951); D. G. Miller, *Proc. Am. Concr. Inst.* **49**, 217 (1953).

TABLE 50 Relative rates of expansion of 1 : 10 mortars at 21°

Cement	Time in days to expand 0·5 per cent		
	1·8% $MgSO_4$	Saturated $CaSO_4$	2·1% Na_2SO_4
$2CaO.SiO_2$	28	Negligible expansion in 18 years	
$3CaO.SiO_2$	35	0·22% in 9 years	12 years
80% $2CaO.SiO_2$ 20% $3CaO.Al_2O_3$	6	10	4
80% $3CaO.SiO_2$ 20% $3CaO.Al_2O_3$	4	11	7
80% $3CaO.SiO_2$ 20% $4CaO.Al_2O_3.Fe_2O_3$	16	0·15% in 3 years	400
40% $3CaO.SiO_2$ 40% $2CaO.SiO_2$ 20% $4CaO.Al_2O_3.Fe_2O_3$	43	About 0·06–0·07% in 3 years	
50% $3CaO.SiO_2$ 50% $2CaO.SiO_2$	65	0·19% in 18 years	0·04% in 12 years; then more rapid expansion
Mean of 8 Portland cements	11	—	13

Rectangular mortar bars 0·625 inch thick aged 8 weeks in water before immersion in the solution.

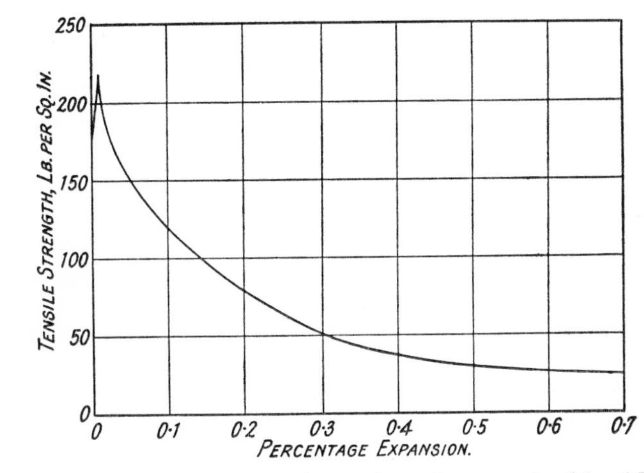

FIG. 103 Relation between expansion and tensile strength of 1 : 5 Portland-cement/standard-sand mortar in 6 per cent $MgSO_4$ solution.

cylinders had expanded 0·25 per cent in length the loss in compressive strength was about 65–70 per cent for specimens stored in 1 per cent sodium or magnesium sulphate solution, but this relation was not independent of the C_3A content of the cement. For pozzolana cements large falls in strength were found, particularly in magnesium sulphate solutions, with little or no expansion.

The rate of expansion of mortars in sulphate solutions is somewhat diminished by increasing periods of water curing before exposure to the sulphate solution. This does not always apply to very short periods of water storage, and specimens exposed at one day old may sometimes suffer damage less rapidly than specimens exposed after a few days in water. An intermediate period of drying in air also increases the resistance owing to the formation of an impermeable calcium carbonate skin. Resistance to sulphate attack of mortars or concretes increases broadly in line with the cement content of the mix. This is to be attributed to the reduction in permeability and a relation has been found between the gas permeability and the rate of sulphate attack.[1]

One method of increasing the sulphate resistance of Portland cement lies in the substitution of the compound C_4AF for C_3A. This substitution of ferric oxide for alumina was suggested by Le Chatelier before the end of the nineteenth century and later taken up in Germany where it led to the manufacture of Erz cement, a material with so low an alumina : iron-oxide ratio as to contain the C_4AF–C_2F solid solution. By 1925 the A : F ratio had been changed to about 0·7. About 1920, Ferrari in Italy produced a cement with an Al_2O_3 : Fe_2O_3 ratio below unity and this material is known in some European countries as Ferrari cement. The troubles experienced in Canada with concrete in sulphate soils led in 1930 to the introduction of a cement low in tricalcium aluminate and this was followed later by the production of similar sulphate-resisting Portland cements in Great Britain, the U.S.A. and elsewhere.[2] There is ample evidence that a low C_3A content increases the resistance of a cement to attack by sulphate solutions, and that there is a general parallelism between C_3A content and sulphate resistance, but there are many anomalies. As a result of the long series of tests carried out both in the laboratory and Medicine Lake (see p. 640) on over 100 cements, Miller recommended that the C_3A content of sulphate resisting cements should not exceed 5·5 per cent; a slightly higher value has been suggested from the Sacramento tests (p. 641). Miller divided his cements into twelve groups of which the most resistant averaged 4·4 per cent C_3A and the least resistant 11·9 per cent. Such averages indicate the broad trend but they conceal the exceptions. Thus some cements with 6 per cent C_3A fell into the three groups of lowest resistance while some cements with up to 10 per cent C_3A were in the first four groups. The maximum content in the most resistant group of all was, however, 5·4 per cent. Broadly similar results have been found by many other investigators. Amongst the reasons for the anomalies that arise are the errors inherent in the calculation of the content of C_3A and the effect of the heat treatment of the clinker (p. 150).

The connection between sulphate resistance and the content of the other compounds in cement is less clear but Bogue (loc. cit) has shown that too high a content of C_4AF decreases the resistance of low-C_3A cements and that it should not exceed about 15 per cent. The ASTM specification (C150–68) limits the C_4AF content by placing a maximum of 20 per cent on the sum of the C_4AF plus twice the C_3A content. There is no clear evidence as to the effect of the relative

[1] U. Ludwig and H. E. Schwiete, *Zement-Kalk-Gips* **20,** 555 (1967).

[2] References to much of the literature are to be found in *Bibliography on sulphate resistance of Portland cements, concrete and mortars*, by D. G. Miller, P. W. Manson and R. T. H. Chan, Univ. Minnesota (1952).

proportions of C_3S and C_2S on sulphate resistance. The results of Thorvaldson showed that C_2S has intrinsically the higher resistance, but his later studies,[1] while confirming the higher rate of expansion of C_3S in lean mortars, indicated that in rich mortars the expansion of a cement with a $C_3S : C_2S$ ratio of 0·5 eventually exceeded that of one with a ratio of 2·5. This would seem to suggest that the more rapid development of a dense structure in a high C_3S cement offsets its greater chemical vulnerability, but tracer studies,[2] using radio-active ions, have not shown significant differences in the rates of diffusion of sulphate ions into 1 : 2 mortars of ordinary and sulphate-resisting Portland cements.

We have still to explain why the compound C_4AF is more resistant to sulphate attack than C_3A, since it should be capable of forming calcium sulphoaluminate and the corresponding sulphoferrite, producing expansion. A clue to this is to be found in some Italian work. When cements containing C_3A and C_4AF are hydrated in the presence of lime and calcium sulphate, Cirilli[3] found that the amount of calcium sulphate taken up corresponded to the complete formation of $3CaO.Al_2O_3.3CaSO_4.aq$ and $3CaO.Fe_2O_3.3CaSO_4.aq$ whatever the A/F ratio of the cement. If, however, the cement was hydrated before being exposed to the calcium sulphate solution the amount of sulphate taken up steadily decreased with decreasing A/F ratio. In this case a solid solution can first be formed between tetracalcium aluminate and ferrite, the rate of attack of which by calcium sulphate decreases with decreasing A/F ratio. The tetracalcium ferrite is more resistant even than the solid solutions and if any free calcium aluminate is formed it may be protected by films of the ferrite over it. Another contributory cause to the increased resistance is suggested by the observation[4] that whereas the high-sulphate calcium sulphoaluminate formed in the setting of ordinary Portland cements later converts into the monosulphate sulphoaluminate this appears not to occur with low C_3A cements. If this monosulphate sulphoaluminate comes into contact later with additional sulphates it is converted back into the high-sulphate compound with an accompanying increase in volume. Some other evidence[5] bearing on this suggestion comes from an observation that the severity of sulphate attack on a cement mortar was greater at a temperature of 5° C than at higher temperatures. This was attributed to the formation of the monosulphate sulphoaluminate, instead of the trisulphate compound, during the hydration of cement at low temperatures and its subsequent destructive conversion to the trisulphate sulphoaluminate under sulphate attack.

Effect of steam curing on sulphate resistance

The resistance of cement mortars and concretes to sulphate waters is much increased by curing in steam under suitable conditions. Such treatment makes concretes almost completely resistant to the action of sodium sulphate and calcium sulphate, and increases very materially the resistance to magnesium sulphate. Curing in saturated steam at temperatures below 100° does not increase

[1] *Sym., London 1952*, 436.
[2] J. W. T. Spinks, H. W. Baldwin and T. Thorvaldson, *Can. J. Technol.* **30,** 20 (1952).
[3] V. Cirilli, *Ricerca scient.* **11,** 973 (1940); G. Malquori and V. Cirilli, ibid. **14,** 78 (1943).
[4] P. K. Mehta and A. Klein, *Highw. Res. Rec.* **192,** 36 (1967).
[5] J. H. P. van Aardt, *Sym., Tokyo 1968*.

the resistance, and may even decrease it, but treatment at 100° or, better, at higher temperatures under pressure is more effective.

The relative effect of various temperatures of steam curing in delaying the expansion of very lean mortars in sulphate solutions is illustrated by the data[1] in Table 51, where the times required for mortar bars to expand a given amount are shown.

TABLE 51 Times required for expansion of 1 : 10 cement-sand mortars in sulphate solutions

Solution	Linear expansion per cent	Temperature of saturated steam curing (24 hours)				
		21°	100°	125°	150°	175°
2·1% Na_2SO_4	0·02	3 days	10 days	1 year	1 year	> 1 year
Saturated $CaSO_4$	0·02	7 days	18 days	1 year	1 year	> 1 year
1·8% $MgSO_4$	0·10	5 days	13 days	25 days	25 days	65 days
	0·20	6 days	32 days	75 days	130 days	> 1 year

It will be seen that the resistance to magnesium sulphate is not increased by autoclaving in the same degree as it is to sodium sulphate. This has been confirmed by tests on denser 1 : 3 mortars.[2]

The resistance is raised by an increase in time as well as temperature of steam curing. In tests on concrete exposed for 17 years in Medicine Lake, Miller found that curing in water vapour at temperatures up to 88° had little influence on sulhpate resistance, but that at 100° and above the resistance increased, the best results being obtained by curing for 6 hours and upwards at 175° corresponding to a steam pressure of 120 lb/in^2. Under these conditions concretes remained sound after 17 years even though made from cements which, under normal curing, gave concretes that lost all their strength within a few years.

From tests on the pure compounds, Thorvaldson[3] found that steam curing at 150° for 24 hours substantially reduced the rate of expansion of C_3S in sulphate solutions, but had little effect on that of C_2S. In mixtures of 80 per cent C_3S or C_2S with 20 per cent C_3A or C_4AF the steam curing also reduced the rate of expansion, the C_2S mix being the more stable.

When cured in high pressure steam various changes occur in the nature of the set cement compounds, while the permeability of a mortar or concrete is also reduced. With siliceous aggregates the free calcium hydroxide reacts to form a hydrated mono-calcium silicate (see p. 199). The removal of free calcium hydroxide increases the resistance to sodium sulphate by suppressing the reaction $Ca(OH)_2 \rightarrow CaSO_4.2H_2O$, but it gives less protection against magnesium sulphate which can attack the hydrated calcium silicates. With a limestone aggregate this hydrothermal reaction cannot take place and only slight improvement in

[1] T. Thorvaldson, V. A. Vigfusson and D. Wolochow, *Can. J. Res.* **1,** 359 (1929).
[2] J. H. P. van Aardt, *Bull. natn. Bldg Res. Inst. Un S. Afr.*, No. 17 (1959).
[3] T. Thorvaldson and D. Wolochow, *Proc. Am. Concr. Inst.* **34,** 241 (1938). *Sym. London 1952*, 436.

resistance to sodium sulphate occurs on steam curing.[1] It should however be noted that the formation of a compound approximating to $CaSiO_3.CaCO_3.Ca(OH)_2.3H_2O$ on autoclaving a mixture of CaO, $CaCO_3$ and SiO_2 has been reported.[2] If such a reaction occurred it would take up free $Ca(OH)_2$.

Both tri- and dicalcium silicate form crystalline hydrates in the high pressure steam. The former yields either $C_3S.1{\cdot}5H_2O$, or one of the dicalcium silicate hydrates (C_2SH) (A) or (B) and the latter either C_2SH (A) or (B). These crystalline compounds are more unreactive than the hydrated silicates formed at ordinary temperatures. There is more uncertainty about the alumina and ferric oxide compounds. On its own tricalcium aluminate hydrates to the stable cubic $C_3A.6H_2O$ instead of the more reactive $C_4A.19H_2O$ and the still less reactive $C_4A_3.3H_2O$ may appear to some extent. Calcium sulphoaluminate on its own decomposes to $C_3A.6H_2O$ and $CaSO_4$. However in autoclaved cement pastes, or cement-quartz mixtures, most investigators have failed to identify $C_3A.6H_2O$. It is also uncertain how far a hydrogarnet type solid solution of the general formula

$$3CaO.(Al, Fe)_2O_3[(H_2O)_2, SiO_2]$$

may be formed. These solid solutions are very resistant to the action of sodium and calcium sulphates.[3]

These various transformations add to the inherent chemical resistance of set cement to attack by sulphates and they are favoured by increasing temperature, and time, of steam curing.

Tests for sulphate-resistance

Though a general indication of the relative sulphate resistance of Portland cements can be obtained from the calculated content of $3CaO.Al_2O_3$, a more direct, and at the same time rapid, test has long been sought. This has been desired not only for differentiating between Portland cements, but also for comparing cements of different types, e.g. Portland, pozzolanic, slag-containing, and air-entrained cements. Tests on the strength, expansion, or modulus of elasticity of concrete, or dense mortar, specimens immersed in sulphate solutions are too slow for use as specification tests, though suitable for research purposes. Lean mortars are affected much more rapidly by sulphate solutions and give a measure of the relative chemical resistance of cements, but they take less account of any differences in physical structure produced between different cements in concretes. Even with richer mortars the expansion test does not entirely parallel the behaviour of concretes. Thus, for example, the results of expansion tests on 1 : 3 cement : standard-sand mortar bars of plastic consistence immersed in 5 per cent magnesium sulphate solution tend to show a greater increase in resistance from the substitution of pozzolanas than is found in strength tests on concrete specimens in a similar solution; the results in fact more closely parallel the relative behaviour of the corresponding concretes in 0·5 per cent magnesium sulphate

[1] F. M. Lea, *Can. J. Res.* **B27,** 297 (1949).
[2] M. S. Shvartszaid *et al., Chem. Abstr.* **57** (3), 3038 (1962).
[3] E. P. Flint and L. S. Wells, *Bur. Stand. J. Res.* **27,** 171 (1941); E. T. Carlson, *Bur. Stand. Bldg Sci. Ser.*, No. 6 (1966); B. Marchese and R. Sersale, *Sym., Tokyo 1968.*

solution. In 5 per cent sodium sulphate solution, on the other hand, the results of expansion tests on the mortars and strength tests on concretes are more closely similar. For specification purposes it is also necessary that any test should be closely reproducible in different laboratories.

Considerable attention has been paid to the refinement of the expansion test on mortar bars, with the object of so standardising the details of making, curing, and storing the specimens in the sulphate solution that similar rates of expansion can be obtained in different laboratories. In trials on 1 : 4 by weight Portland-cement graded-sand mortars of a standard consistence, immersed in 5 per cent sodium-sulphate solution at 7 days old, it was found in a series of ASTM tests[1] that the expansion after 28 days was sufficient to detect Portland cements of low resistance when the average results obtained in seventeen laboratories were considered; the reproducibility between different laboratories was not, however, sufficiently good for some definite expansion value to be fixed that could apply to all laboratories. Such a form of test is useful within one laboratory, particularly if some 'standard' cements are kept as yard-sticks and use in comparative tests with other cements. In one form of this test[2] developed in Germany $1 \times 1 \times 6$ cm prisms of 1 : 2 : 2 cement : fine-standard-sand : coarse-standard-sand are immersed at an age of 21 days in a 10 per cent $Na_2SO_4 . 10H_2O$ solution. The prisms are tested in bending after various periods of immersion up to 77 days and the sulphate taken up by the prisms is also determined. This test is intended to be applicable to all types of cements though cements of different types vary in their relative susceptibility to attack by sodium sulphate and other sulphates such as magnesium. Further work[3] has also suggested that a longer test period than 77 days is needed.

Various forms of test in which thin neat cement slabs are immersed in a sulphate solution and the deterioration assessed from appearance, softening, warping, etc. have also been tried, but none have proved sufficiently reproducible or reliable. This also applies to a chemical test, intended to be a measure of the reactivity of the alumina compounds in a cement, in which a cement-sand mix is shaken in saturated lime-water for six hours and the amount of sulphur trioxide combined as calcium sulphoaluminate determined. Various of these methods have been reviewed by Wittekindt.[4]

A somewhat different form of sulphate-resistance test for Portland cements in which a mortar specimen is not immersed in a sulphate solution but, instead, additional gypsum is added to the cement mortar which is then immersed in water has been developed in the U.S.A. This method was found to be sufficiently reproducible[5] between different laboratories to be standardised in ASTM C452–68. The method is an option in the ASTM specification for sulphate-resisting Portland and is intended primarily for use in research. Mortar bars, $1 \times 1 \times 11\frac{1}{4}$ in., are made from a 1 : 2·5 graded sand mortar and the gypsum

[1] D. Wolochow, *Proc. Am. Soc. Test. Mater.* **52,** 250 (1952).
[2] A. Koch and H. Steinegger, *Zement-Kalk-Gips* **13,** 317 (1960).
[3] O. Fernandez-Pena, *Inst. E. Torrocha de la Construction y del Cemento, Madrid;* Report No. 6 (1967); P. G. Parades Gaibrois, ibid., Report No. 13 (1968).
[4] W. Wittekindt, *Zement-Kalk-Gips* **13,** 565 (1960); see also F. W. Locher, ibid. **9,** 204 (1956).
[5] Committee Report, *Bull. Am. Soc. Test. Mater.,* No. 212, 37 (1956).

content of the cement is brought up to 7 per cent SO_3 by adding to it extra gypsum. After curing in moist air for 22–23 hours the bars are immersed in water at 23° C and the expansion at 14 days (or longer if desired) measured.

A comparison of the relative behaviour of cements in this test with the performance of the corresponding concrete specimens after 17 years exposure in the Sacramento sulphate soil trials (see p. 641) showed a fair relationship[1] but one which was not significantly freer from anomalies than the relation between the trial results and the calculated C_3A contents of the cements. In reviewing this test Tuthill[2] has suggested that it should be regarded as a supplement to, and not a substitute for, compound content requirements. The test has not been found suitable for Portland blastfurnace cements.

A form of test intended for application to cements of all types that has enjoyed a certain vogue in some European countries is known as the Anstett test.[3] The neat cement is gauged to a plastic paste with 50 per cent water by weight, allowed to harden for some weeks, and the specimen then crushed to 5 mm size, and dried at not above 40°. Gypsum to the extent of 50 per cent by weight of the dry set cement is then added and the mixture ground to zero residue on the 900-mesh metric sieve. The ground material is gauged with 6 per cent distilled water, filled into a cylindrical mould (not oiled), 8 cm diameter by 3 cm high, and compressed under a pressure of 20 kg/cm^2 for one minute. The cylinder is placed on filter paper, kept damp by letting the ends dip in water, and covered with a bell jar to form an airtight joint. The diameter is measured at 24 hours and then again after 28 and 90 days. The expansion should not exceed 1·25 per cent. A rather more elaborate procedure designed to ensure a standardised degree of hydration of the cement before mixing with the gypsum has been described.[4] This test is a rather severe one and exposes the cement to conditions rather different from practice. High-alumina cement and supersulphated cement behave well under this test while other slag-containing cements and Portland cements, even of the sulphate-resisting type, do not. Pozzolanic cements prove resistant in the test if made with Portland cements of low tricalcium aluminate content but not with those of high content,[5] except when the pozzolana has an exceptionally high content of reactive silica.[6] Blondiau found by repeatedly extracting a pozzolanic and a blastfurnace cement with water before mixing with the gypsum that expansion did not occur when the lime concentration in solution set up by the cement had fallen below 1 gm CaO per litre. This was considered to support Lafuma's theory that when the concentration of lime in solution is less than 1·08 g CaO per litre, the alumina becomes more soluble and reacts with gypsum in the liquid, and not the solid, phase. Under this condition expansion is considered not to occur, but as discussed later (p. 524) this theory is not free from objection.

[1] W. Lerch, *Proc. Am. Soc. Test. Mater.* **61,** 1043 (1961).
[2] L. H. Tuthill, *Spec. Publ. Am. Soc. Test. Mater.*, No. 169A, 275 (9166).
[3] F. Anstett, *Revue Matér. Constr. Trav. publ.* **162,** 51 (1923).
[4] L. Blondiau, *Revue Matér. Constr. Trav. publ.* **314,** 261 (1935); **546,** 189 (1961).
[5] L. Santarelli, *Chimica Ind., Milano* **24,** 323 (1942).
[6] R. Turriziani and A. Rio, *Sym., Washington 1960*, 1067.

13 Physical and Mechanical Properties of Portland Cement

The necessity of distinguishing by means of tests which may be rapidly applied in the laboratory between different qualities of cement has been the cause of much of the attention devoted to the physical properties of cement. The object has been to fix values of certain readily determined properties for cements which are found to be satisfactory in practice, in order that inferior materials may be detected by their deviation from such standards. Properties of this type form the basis of all specifications for cement. They include fineness, setting time, soundness and strength. There are other physical properties of cements which, though of importance in practice, find no place in specifications; these we shall also have to consider. They include various properties which may be classed under volume change, also permeability, absorption, etc.

Standard specifications

The test methods used in the national specifications in different countries vary so that often it is not possible to compare directly the results obtained under different specifications. Correlations can however be established between them as has been done for instance for strength tests by Calleja.[1] Amongst these specifications are BS 12. *Ordinary and Rapid-Hardening Portland Cement*, BS 1370 *Low-Heat Portland Cement* and BS 4027 *Sulphate-Resisting Portland Cement;* the ASTM C150; the German DIN 1164; the French NFP 15–302 and the Swiss SIA 115. International methods have been developed by R.I.L.E.M.-CEMBUREAU[2] to provide a common basis for testing.

It is obviously important that the methods used in a standard specification should give as closely similar results as possible when tests are made in different laboratories and much attention has been paid to the reproducibility of the test methods by comparative tests carried out in a number of laboratories. The causes of variation have been examined in the U.S.A. by Bean and Dise[3] and the statis-

[1] J. Calleja, *Inst. tech. de la Construccion y del Cemento*, No. 211, Madrid (1961).
[2] *R.I.L.E.M. Bull.*, No. 34, 1 (1957); *The Testing of Cement*, CEMBUREAU, Paris 1967; ISO draft recommendations.
[3] B. L. Bean and J. R. Dise, *Monogr. natn. Bur. Stand.*, No. 28 (1961).

tical aspects of comparative testing in other papers.[1] The R.I.L.E.M.–CEM-BUREAU test method for the strength of mortars[2] and the relation between the mortar and concrete tests used in the British Standard have been similarly examined.[3] An interesting account of the history of cement testing, particularly in the U.S.A., has been given by Gonnerman.[4]

Density

The density of a cement is not a property normally determined for its own sake, but it is required in the measurement of specific surface. It is determined in the usual manner by displacement of liquid in a density bottle using kerosene (paraffin oil) purified by re-distillation, collecting the fraction condensing at 200–240°. The bottle containing a weighed quantity of cement is half-filled with kerosene and evacuated for at least half-an-hour to remove entrapped air bubbles before completing the filling. The density of Portland cements ranges from about 3·0–3·20.

Apparent density

The weight of a given volume of cement is not a property that is of much importance since, except for the crudest concrete, the cement is proportioned by weight. Precise values for the weight of a given volume of a fine powder like cement cannot be obtained, for the value varies considerably with the method by which the container is filled and the degree of shaking employed. Thus Portland cement loosely filled into a box weighs only 950–1300 kg/m^3 (60–80 lb/ft^3), while on consolidation the value rises to 1600–1800. The values shown in Table 52 can be taken for the weight per cubic foot of materials filled into the container with gentle shaking.

TABLE 52

Material	lb/ft^3	kg/m^3
Portland cement	80– 90	1280–1440
High-alumina cement	90–100	1440–1600
Portland blastfurnace cement	75– 85	1200–1360
Pozzolanic cement	70– 80	1120–1280
Hydrated lime	30– 40	480– 640
Sand (dry)	95–105	1520–1680
Gravel	80– 90	1280–1440

As filled into a gauge box in practical use, the weight of a cubic foot of cement averages rather less than these values and ranges in the case of Portland cement

[1] W. J. Jouden, *Proc. Am. Soc. Test. Mater.* **59,** 1129 (1959).
[2] *On the Testing of Cement,* CEMBUREAU, Malmo 1957; R. Dutton, *Zement-Kalk-Gips* **13,** 64 (1960); J. Nielson, *R.I.L.E.M. Bull.,* New Series No. 31, 225 (1966).
[3] A. J. Newman, *Proc. Sym. Concrete Quality, London 1964,* Cement and Concrete Association, London 1964.
[4] H. F. Gonnerman, *Bull. Res. Dev. Labs Portld Cem. Ass.,* No. 93 (1958).

from 75 to 85 lb/ft^3. In proportioning concrete by volume it is usually required that 90 lb of cement shall be taken for each cubic foot, independently of the actual apparent density of the cement used, and a 1 cwt bag as representing $1\frac{1}{4}$ ft^3. After metrication the standard bag will be 50 kg (110 lb). In the U.S.A. proportioning is based on the 94 lb sack of cement.

Determination of setting time

The chemical changes involved in the setting of cements have been discussed in previous chapters and it now remains to describe the methods of determining the time occupied in the process of setting. Two periods are distinguished, the 'initial set', or interval between the gauging and partial loss of plasticity, and the 'final set', or time required for the gauged cement to acquire sufficient firmness to resist a certain definite pressure. The Vicat needle in its original form, or in one of its modifications, is now used almost universally. This instrument (Fig. 104), as used in Great Britain and the U.S.A., consists of a slender rod,

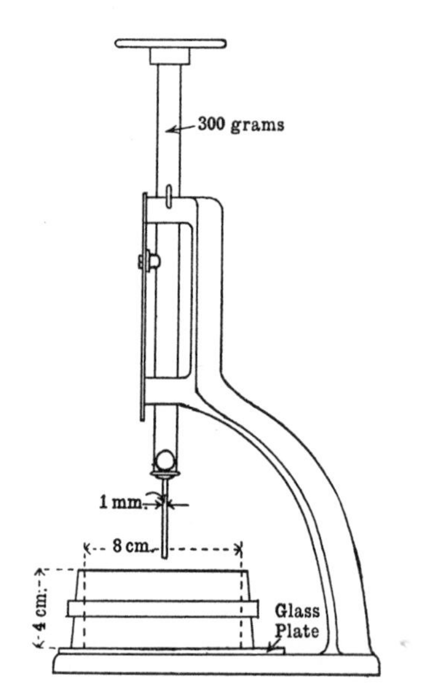

FIG. 104 Standard Vicat needle.

square or round, of 1 millimetre square cross-section, with a flat end, the weight of the cap, rod and needle being 300 g. After the cement has been mixed with water to a defined consistence and filled into a cylindrical mould, the 'initial set' is said to be reached when the needle is no longer able to pierce the block to within about 5 mm from the bottom. For the 'final set' a needle similar in dimensions, but fitted with a metal attachment hollowed out so as to leave a circular cutting edge 5 mm in diameter, the end of the needle projecting 0·5 mm beyond this edge, is used. 'Final set' is said to be attained when the needle, gently lowered to the surface of the test block, makes an impression on it, but the circular edge of the attachment fails to do so.

The determination of setting time is affected to a marked degree by the quantity of water used in gauging, the temperature, the hydrometric state of the atmosphere, and other factors; comparable results can only be obtained by working under standard conditions. The temperature is fixed at $18{\cdot}9 \pm 1{\cdot}1°$ C ($66 \pm 2°$ F) and the relative humidity at not less than 90 per cent. The water used in gauging is required to be that necessary to bring the cement paste to a defined consistence. The method by which this is measured is given in the British Standard.[1] It is prescribed therein that for normal cements the initial setting time shall be not less than 45 minutes, nor the final more than 10 hours. Quick-setting cements with an initial set of not less than 5 minutes and a final set of not more than 30 minutes are sometimes specified for special purposes.

The setting time of Portland cement usually decreases as the temperature is raised, but the rate of this change varies with different cements. Thus two cements which have the same setting time at, say, 16° may show quite different setting times at 3°. At temperatures of 30° and above, the setting time does not appear to decrease and the reverse effect may be found. This is occasionally observed at much lower temperatures. Table 53 shows some data on the effect of temperature:

TABLE 53

Cement		A	B	C	D	E	F	G
Initial set	3° C	4·50	0·75	1·17	3·50	1·08	2·92	0·50
(hours)	16° C	1·42	1·33	1·17	2·67	1·33	2·67	1·83
Final set	3° C	7·50	3·75	4·83	5·00	3·25	4·92	1·58
(hours)	16° C	2·83	1·83	2·83	3·33	2·25	4·42	2·25

Various investigations have been made of other physical properties such as heat evolution, electrical conductivity, and viscosity, as a measure of the time of set.

On mixing a cement with water there is an immediate heat evolution during the first few minutes, amounting to a few calories per gram of cement.[2] The heat evolution then drops to a low value after which there is a gradual increase reaching a maximum after some hours about, or soon after, the time of final set.

Measurements of the electrical conductivity of cement pastes show a certain characteristic form in the conductivity-time curve, an example[3] of which is shown in Fig. 105. The conductivity of a cement paste gauged with 25–30 per cent water rises to a maximum which is reached after 5–20 minutes and is

[1] The quantity of water is defined as that required to give a paste in which a flat-ended Vicat plunger 1 cm in diameter, weighing 300 g, when released at the surface of the paste, penetrates to a point 5–7 mm from the bottom of a Vicat mould filled to a depth of 40 mm.

[2] R. W. Carlson and L. R. Forbrick, *Ind. Engng Chem. analyt.* **10,** 382 (1938); W. Lerch, *Proc. Am. Soc. Test. Mater.* **46,** 1252 (1946).

[3] K. E. Dorsch, *Cement* **6,** 131 (1933); G. Baire, *Revue Matér. Constr. Trav. publ.* **272,** 182 (1932); J. Calleja, *Proc. Am. Concr. Inst.* **50,** 249 (1954).

followed by a rapid fall which ceases after about an hour. It then remains roughly constant for a period, after which a further rapid fall sets in. The initial maximum probably corresponds to the formation of a saturated, or supersaturated, solution of calcium hydroxide, together with any soluble alkalis present in the cement. The subsequent inflexion points must be attributed to stages in the setting process which reduce the conductivity until it reaches a low value in the hardened mass.

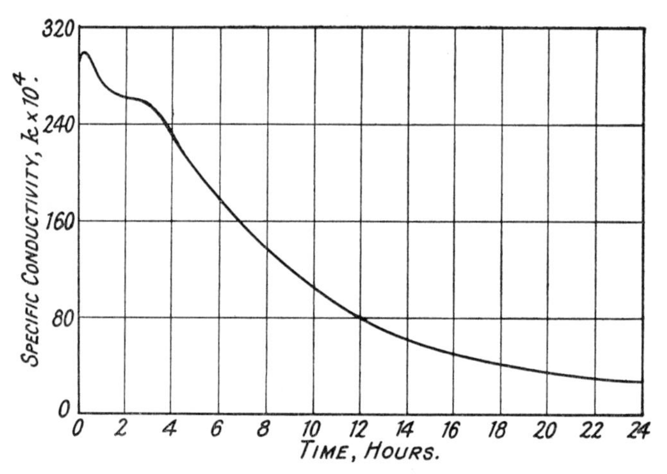

FIG. 105 Conductivity of Portland cement paste at 16°.

Rheology of cement paste

The flow behaviour of cement pastes is complicated by the development with time of a structure in the paste as hydration proceeds. Since there is a rapid reaction of cement with water in the first minute or two after mixing, hydration products are present in the system at the earliest times at which measurements can be made. At low w/c ratios, 0·25–0·35, cement pastes have a consistence of a thick cream while at high w/c ratios the mix becomes increasingly fluid in its characteristics.

Cement pastes show a yield value, i.e. a certain minimum shearing stress is required to produce flow, followed at higher stresses by a behaviour which may be reversible thixotropic or anti-thixotropic (rheopectic).[1] When tested in a rotational concentric cylinder viscometer the relation between the rate of shear (the angular velocity of rotation of one of the cylinders) and the torque produced on the other cylinder can take one of the three forms shown in Fig. 106. The intercept on the torque axis is a measure of the yield value and the slope of the curve of the plastic viscosity.

[1] M. Ish-Shalom and S. A. Greenberg, *Sym., Washington 1960*, 731; S. A. Greenberg and L. M. Meyer, *Highw. Res. Rec.*, 1963, No. 3, 9; J. P. Blombled and O. Kalvenes, *Revue Matér. Constr. Trav. publ.* **617,** 39 (1967); G. H. Tattersall, *Br. J. appl. Phys.* **6,** 165 (1955); A. G. B. Ritchie, *Cem. Lime Mf.* **38,** 9 (1965); S. A. Nessim and R. L. Waida, *Mag. Concr. Res.* **17** (51), 59 (1965).

Thixotropic behaviour implies a breakdown of structure under stress resulting in a hysteresis loop with the return curve displaced towards a lower stress and with successive repeated cycles resulting in a shift of the curves in a similar direction. This means that on 'working' the paste it becomes more fluid. Anti-thixotropic behaviour is the reverse of this and the paste stiffens on working. The normal definition of thixotropy also implies a recovery of structure in a paste on standing, but this recovery seems at most to be slow with Portland cement pastes; it would be hard to prove if it occurred too slowly owing to the stiffening effect of further hydration.

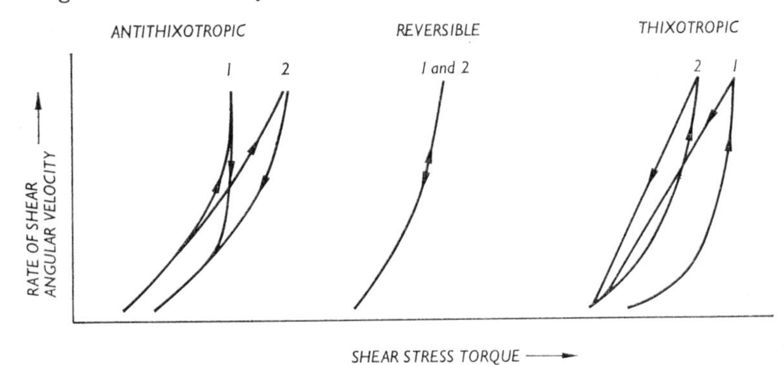

FIG. 106 Three types of flow behaviour of Portland cement pastes. Successive cycles 1 and 2.

At w/c ratios up to 0·5 the plastic viscosity increases with time after mixing and with decreasing w/c ratio. At higher w/c ratios (e.g. 0·7) the value changes little over the first hour or so after mixing. Portland cements usually show thixotropic curves though at an early time after mixing (15 min) Ish-Shalon and Greenberg found an anti-thixotropic behaviour with a high C_3A (14 per cent) cement; this behaviour changed with time through a reversible to a thixotropic behaviour. The yield stress, and also the extent of the thixotropy, was found to rise with the C_3A content of the cement for contents above a few per cent, but the results depended also on the duration of mixing and the extent to which any initial structure was broken up in the process. Pastes of alite are thixotropic at all stages so the more complicated behaviour with Portland cement can be attributed to the aluminate component, or the gypsum. As work in the U.S.S.R. has shown (see p. 256), the hydration of C_3A leads to the development of thixotropic properties and this is influenced both by the presence of gypsum and of surface-active agents.

Thixotropy implies that the bonds between the particles in a paste are sufficiently strong to hold the system together but not strong enough to withstand small shearing forces. The inter-particle bonds are to be ascribed to the balance of Van der Waal's and electrostatic forces and these will be affected by other ions in the solution. It is hardly surprising therefore that the rheological behaviour of cement paste is variable and as yet not fully explained. The anti-thixotropic behaviour sometimes observed may be due to the presence of dehydrated gypsum and the formation of a plaster set, or, it has been suggested,[1] to the dilatancy of a granular suspension before much colloidal material is present.

Constancy of volume and soundness

It is highly important that a cement, after gauging with water and setting in place, shall not undergo any appreciable change of volume. From the discussion in a previous chapter of the colloidal properties of set cement and from a later section in this chapter, it is evident that all set cements undergo some shrinkage when allowed to dry out, and some expansion when continuously stored under water. These movements are quite small, and of a much lower order of magnitude than those with which we are concerned in considering the soundness of cements. Certain defective cements have been found to undergo a large expansion after setting, leading to the disruption of the hardened mass. This is a very serious defect in a cement, involving as it does the risk of destruction of any structure in which it is used. The testing of the soundness of cements, so as to ensure that no material showing such a subsequent expansion shall be used, has always therefore been considered of prime importance.

A defective cement expands after setting, sometimes after a period of months or years have elapsed. The expansion, which produces cracks, occurs in mortars or concretes as well as in the neat cement, though the effects in the latter are more magnified.

The expansion of defective cement is due to the slow hydration of certain of its constituents. The phenomenon very closely resembles the slaking of lime, differing from it only in the slowness with which it proceeds. The expansion of the calcium oxide in this reaction is only apparent, as the volume of the slaked lime is actually less than the sum of the volumes of the quicklime and the water from which it is formed, but, as in the hydration of plaster (p. 251), the particles of hydrate formed grow in one direction preferentially, producing an outward thrust and an expansion of the mass. It follows that the mass must expand as the particles are thrust apart. Unsoundness cannot be attributed to the hydration of quicklime in the physical condition in which it is best known to us as a loose, friable solid, for such lime would undergo hydration rapidly in contact with water and would be rendered harmless before the setting process began. Crystallised lime, however, as mentioned on page 29, only becomes hydrated after long exposure to water and then exhibits expansion in a very marked degree. Cements prepared from mixes of too high lime content, or of correct lime contents but insufficiently well burnt, contain free calcium oxide in an unreactive, dead-burnt condition; when the amount of such material present exceeds certain limits, which vary with the fineness to which the cement is ground, the cement shows unsoundness. Actually three errors in composition are known which may give rise to unsoundness:

1. An excess of lime above that which, under the conditions of manufacture, can become combined with the acidic oxides of the cement mix.
2. An excessive proportion of magnesia.
3. An excessive proportion of sulphates.

[1] A. J. Gaskin, *Sym., Washington 1960*, 744.

Since unsoundness in cements often does not exhibit itself until after a considerable period of time, accelerated tests are required to detect it. There are a number of such tests in common use. One of these accelerated tests used in the British Standard is that due to Le Chatelier. The gauged cement is filled into a small brass cylinder, split longitudinally and provided with two needles, the ends of which diverge if the cylinder is forced open by the expansion of the material contained in it. The form and dimensions of the apparatus are shown in Fig. 107.

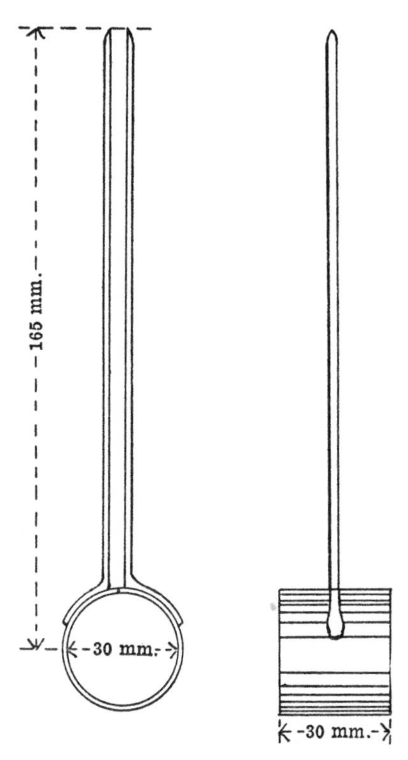

FIG. 107 Le Chatelier's apparatus for measurement of expansion.

The cement is gauged[1] and filled into the mould on a small piece of glass, the edges of the mould being pressed gently together. When the mould has been filled, it is covered with a plate of glass held down by a small weight, and the whole is immersed in water at $18{\cdot}9 \pm 1{\cdot}1°$ C ($66 \pm 2°$ F) for 24 hours. The distance between the indicator needles is then measured and the mould replaced in cold water which is raised to 100° in 25 to 30 minutes, and is kept boiling for one hour. After cooling, the distance between the indicator needles is again measured. The difference between the two measurements represents the expansion of the cement. This must not exceed 10 mm when first tested; if this is exceeded

[1] The quantity of water to be used in gauging the cement is defined in a similar way to that for setting time (p. 363).

the cement may be spread out and aerated for 7 days, after which the expansion must not exceed 5 mm. A large copper water-bath, with constant level supply, is used for immersing the apparatus, and is made of sufficient capacity to accommodate a number of moulds.

Another test in common use, as in Germany, is the hot pat test. The cement is gauged to a plastic paste, placed on a glass plate, and formed into pats, 8–10 cm in diameter and 1–1·5 cm thick at the centre, tapering to a thin edge. The pats are kept in moist air for 24 hours and then boiled in water for 2 hours. Unsoundness in these pat tests is shown by the appearance of distortion or cracking. The appearance of shrinkage cracks during the first 24 hours, or the failure of the pats to remain attached to the glass, are not indications of unsoundness. Such defects may appear in pats improperly made, or exposed to drying during the first 24 hours owing to the moist air in which they are stored being insufficiently humid. A cold water pat test in which the pat is stored in cold water for 28 days also appears in some European specifications. A later form of test, the autoclave expansion tests, has been introduced into the specifications in the U.S.A. and other countries as a safeguard against expansion of concrete arising from very slow hydration of free magnesia. None of the other forms of expansion test detects unsoundness due to this cause and the test is, in fact, a revival of an old one proposed by Erdmenger towards the end of the nineteenth century. In this test (ASTM C151–68) mortar bars, 1 × 1 inch cross-section, of neat cement, are cured for 24 hours in moist air and their length measured on a suitable comparator with a dial gauge or micrometer. They are then placed in an autoclave which is raised to a steam pressure of 295 lb/in^2 (216°C, 420°F) in 45–75 minutes and maintained at this pressure for three hours. After cooling in the autoclave and finally in water the length is again measured. The expansion must not exceed 0·8 per cent. Under these conditions of high steam pressure, crystals of MgO which hydrate very slowly at ordinary temperatures, producing a risk of long-term unsoundness, and which do not hydrate fast enough for their effect to be detected in tests at 100°, become hydrated and expand. There is also some evidence[1] that crystals of free CaO above about 20 μ in size do not hydrate completely in boiling tests, but do so in the autoclave test. Since the autoclave test detects expansion arising both from free CaO and MgO it has entirely replaced the other forms of soundness test in the ASTM specification.

Unsoundness due to the presence of free lime may arise from an over-limed mix, inadequate burning, or insufficiently fine grinding and mixing of the raw materials fed to the kiln. Since the more finely the overburnt free lime is ground the more rapidly it hydrates, it is probable that higher contents can be carried in the modern finely ground cement than was possible in the coarser ground cements of the early twentieth century without the appearance of unsoundness. In fact modern cements rarely fail in the Le Chatelier or hot pat soundness tests even when the free lime content is relatively high. There is, however, another reason. We have shown in a previous chapter that a large proportion of the free lime in a Portland cement becomes hydrated to calcium hydroxide during grinding. Free lime present as $Ca(OH)_2$ is clearly harmless and the total free lime

[1] L. S. Brown and M. A. Swayze, *Rock Prod.* **41** (6), 65 (1938).

(oxide and hydroxide) in a cement does not therefore afford a measure of potential unsoundness. It has been found[1] that unsoundness commences to appear in the boiling tests when the free CaO is appreciably in excess of 2 per cent, but that cements with a total free lime content as high as 5 per cent are quite sound when it is mainly present as hydroxide and the actual free CaO content is below 2 per cent.

Unsoundness in cements due to too high a magnesia content first attracted attention as far back as 1884 when a number of bridges and viaducts in France failed from expansion of the concrete two years after erection and about the same time the town hall of Cassel in Germany had to be rebuilt because of expansion cracking.[2] Both cements contained much magnesia, the French from 16 to 30 per cent, and the German 27 per cent, as they had been made from a dolomitic limestone. It was the investigations which followed that led to the introduction in cement specifications of limits for the total magnesia content. Lightly calcined magnesia hydrates rapidly and does not cause expansion in the hardened mass, as is evidenced by the good record of some of the lightly burnt natural cements of high magnesia content of the U.S.A. At the temperature of burning of Portland cement the magnesia is dead-burnt and crystals of periclase (MgO) only react with water over a period of years at ordinary temperatures.

It was for long accepted that contents of magnesia up to 5 per cent could be safely carried, but some experiences of expansion in concrete structures in the U.S.A., and long-term expansion tests on cements, indicated that some additional safeguard beyond that provided by the Le Chatelier and hot-pat soundness tests was desirable. This led to the introduction of the autoclave test in the U.S.A. and, subsequently, in some other countries where a proportion of the cements had magnesia contents in the upper part of the permitted range. The test has not been introduced into the British Standard, for the raw materials available for making Portland cement in Great Britain are low in magnesia and it is rare to find a content above 2 per cent in the cement, the average indeed being around 1 per cent.

The results of numerous studies on the influence of magnesia and free lime on the behaviour of cements in soundness tests has been very fully detailed.[3] We have seen earlier that some MgO is taken up in solid solution in the other clinker compounds. It is only that which has crystallised as periclase (MgO) which may cause unsoundness. The degree of unsoundness depends also on the crystal size since the smaller crystals tend to hydrate the more rapidly or without setting up excessive internal pressures. Thus Keil[4] found that a content of 4 per cent periclase crystals below 5 μ in size produced only about the same autoclave expansion as 1 per cent of crystals of 30–60 μ size. About 1·5–2 per cent MgO, calculated on the total cement composition, is present in the liquid phase at the clinkering temperature. The quicker the clinker is cooled the smaller will be the periclase crystals formed by the crystallisation of this liquid (see p. 150). Hence clinkers

[1] G. E. Bessey, *J. Soc. chem. Ind. Lond.* **52,** 219T (1933); W. Lerch, *Concrete* **35** (1), 109 (1929).
[2] See H. Le Chatelier, *Mortiers hydrauliques*, pl. 42.
[3] H. F. Gonnerman, W. Lerch and T. W. Whiteside, *Bull. Res. Dev. Labs Portld Cem. Ass.*, No. 45 (1953); G. Goggi, *Zement-Kalk-Gips* **11,** 383 (1958).
[4] F. Keil, *Revue Matér. Constr. Trav. publ.* **503/504,** 262 (1957).

that are cooled rapidly can carry more magnesia safely than slowly cooled clinkers. Cements with as much as 5 per cent magnesia will pass the autoclave test if quickly cooled, and the free CaO is also low, even though only about half of this MgO content can be made up of magnesia held in solid solution in other clinker compounds or as small crystals formed from the clinker liquid. The extent to which the periclase crystals may themselves have impurities in solid solution also appears to influence their speed of hydration.[1] In slowly cooled clinkers, failure to pass the autoclave test may occur with magnesia contents of the order of 3 per cent. Cements with free lime contents below about 2 per cent will pass the test when the total MgO content is low (1–2 per cent), but with high contents the free CaO may need to be below 1 per cent to be sure of conformity. A further factor that influences the expansion in the autoclave test, though it does not lead to long-term expansion in practice, is the content of tricalcium aluminate; even when the magnesia and free lime contents are low, the autoclave expansion increases as the calculated content rises above about 8 per cent in a well crystallised clinker and may exceed the permitted limit at about 14–16 per cent. The iron compound, $4CaO.Al_2O_3.Fe_2O_3$, has little effect.

It will be apparent that expansion in the test is the integrated effect of a number of separate factors and that the limits given above cannot therefore be taken as more than broad indications. Though behaviour in the autoclave test is related in a general way to the risk of long-term expansion in practice, it is not an exact guide and various anomalies are apparent in the available data. The addition of pozzolanas, or granulated slag, reduces expansion in the autoclave test.

The third compound liable to cause expansion is calcium sulphate. The increase of volume is not due in this case to the hydration of the sulphate, as the hydrated salt, gypsum, produces the same effect if added to the cement in more than a small quantity. The expansion is attributed to the formation of calcium sulphoaluminate. This is harmless when formed in small amounts during the setting of a cement, but if larger amounts of gypsum are present, such that formation of the sulphoaluminate salt continues after setting and hardening, expansion occurs. The maximum amount of gypsum that can safely be added is thus related to the ability of the cement to combine with it during the setting, or very early hardening period. The maximum limits allowed in the specifications of most countries vary from 2·5 per cent SO_3 for cements of low C_3A content to as high as 4·0 per cent for cements of high C_3A content. Such limits have been shown[2] to be on the safe side. Unsoundness due to an excessive gypsum content is not detected with any certainty in the usual soundness tests. An excessive SO_3 content is, of course, easily detected by analysis.

Fineness

The reduction of the clinker to a very fine state of division by grinding is usually considered to increase its value as a cementing material in two ways. In the first place, a fine powder is able to coat the surfaces of grains of sand or other inert

[1] F. Gille, *Zement-Kalk-Gips* **5**, 142 (1952).
[2] W. Lerch, *Proc. Am. Soc. Test. Mater.* **46**, 1252 (1946); H. S. Meissner, *Bull. Am. Soc. Test. Mater.*, No. 169, 39 (1950).

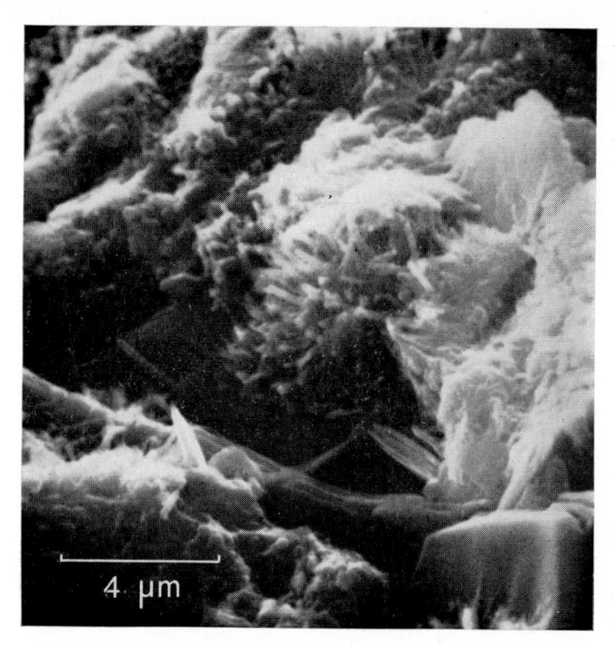

(i) Scanning Electron Micrograph of CSH Crystals in Set Cement

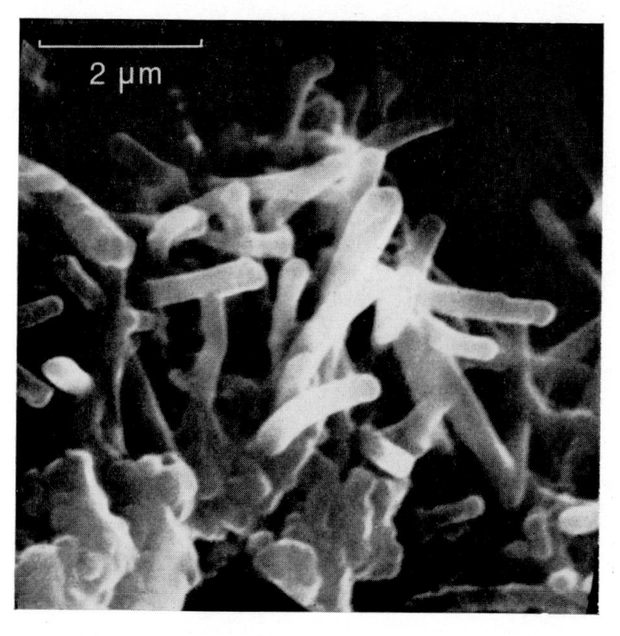

(ii) As XI (i) but at Higher Magnification

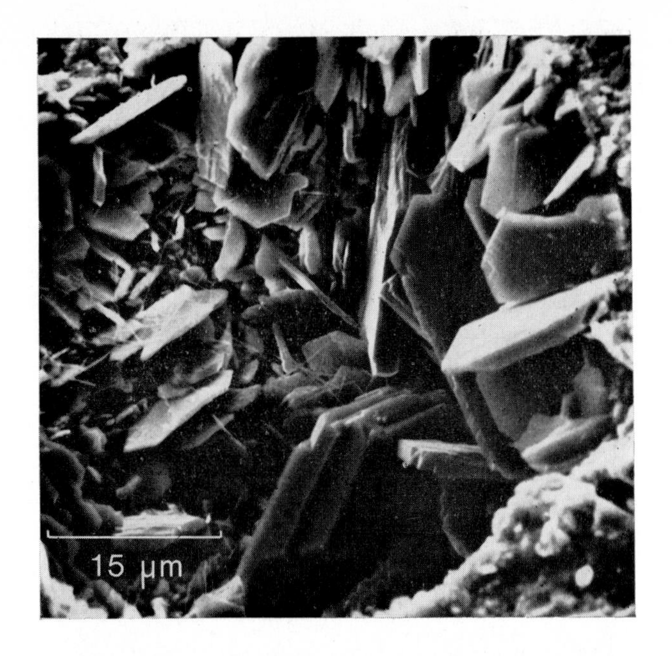

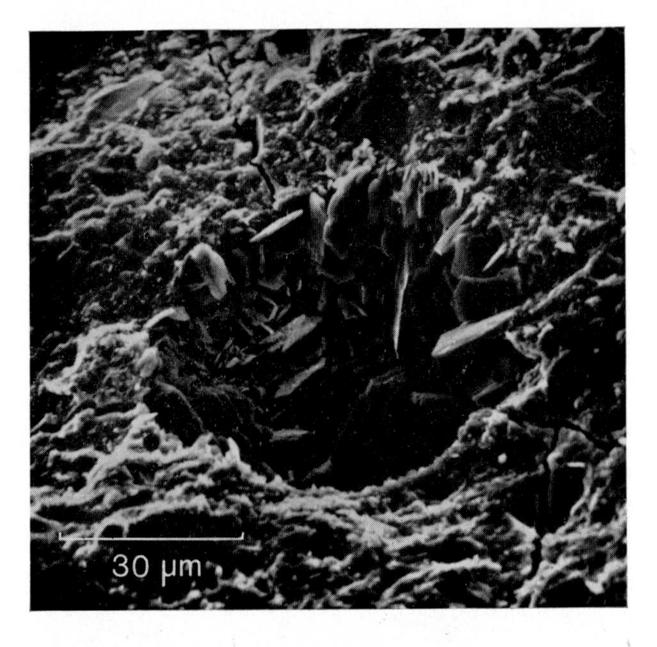

SCANNING ELECTRON MICROGRAPH OF $Ca(OH)_2$ PLATES IN SET CEMENT

material more completely than a coarse one, so that a more intimate contact of the components of the mortar is assured. In the second place, the reaction between cement and water takes place only at the surface of the solid particles, further action being hindered by the accumulation of the products of the reaction coating the unattacked material. Hence the more finely ground a cement, and the greater the surface exposed in proportion to its mass, the more rapid is the rate of hydration and the greater is the proportion of the cement which reacts.

The cost of grinding cement clinker increases with the fineness to which it is ground, and this in itself would set some limit to fineness, but there are also limits imposed by the type of grinding machinery used and indeed by what is desirable. After a long period in which fineness of grinding steadily increased it became stabilised during the years immediately preceding the second world war and has since changed relatively little. It varies for different types of cement, rapid-hardening Portland cement being the most finely ground.

Fineness was formerly determined for specification purposes solely by sifting through sieves having meshes of certain dimensions and weighing the residue which fails to pass through each sieve after shaking. The limit of fineness which can be determined in this way is, however, too restricted for modern cements and in many specifications (including the British and the ASTM) it is now replaced, or supplemented, by a determination of specific surface, i.e. the surface area per gramme of cement, by either the air permeability or Wagner turbidimeter methods described later. Air elutriators and sedimentation methods are also used for laboratory purposes to classify materials down to much smaller particle sizes than can be obtained by sieve tests. The air permeability method is now increasingly becoming the standard method for determining the specific surface of cement, either in its original form as developed by Lea and Nurse or in a variation of it by Blaine. The older Wagner turbidimeter method is still retained in the ASTM specification as an alternative to the air-permeability method. For reasons discussed later the results from these two forms of test differ by a factor that has an average value of about 1·8. The minimum specific surfaces by the air-permeability method required in the British Standards are 2250, 3250, 3200 and 2500 cm^2/g respectively for ordinary, rapid-hardening, low-heat and sulphate-resisting Portland cements. The ASTM minimum average values for all types of cement, except Type III for which no value is specified, are 2800 cm^2/g by the air-permeability method and 1600 by the turbidimeter method. The value in any one sample is permitted to fall 200 and 100 cm^2/g respectively below these limits. In countries in which sieve tests are still retained the maximum residue permitted on a sieve with an opening of about 90 μ varies from 10–18 per cent for ordinary Portland cement and 5–10 per cent for rapid-hardening Portland cement. The sieves used are the BS 170 (90 μ), ASTM 170 (88 μ) or 200 (74 μ), or the European 4900 (90 μ) or their counterparts under the International Standards Organisation proposals.

There has been a steady increase in the fineness of cements over the years as measured by the decrease in the sieve residues that were permitted. The average results recorded by C. R. Redgrave and C. Spackman for 1879 showed a residue of 10 per cent on a sieve having 30 meshes per linear inch (140 per cm^2) and 45 per cent on one with 100 meshes per inch (1550 per cm^2). According to Blount, a residue of 10 per cent on a sieve having 50 meshes per linear inch (385 per cm^2) would not have been considered unreasonable in 1886. By 1910 the residue on the 180-mesh for many British and Continental cements had dropped well below 10 per cent. Most present-day ordinary Portland cements average around 5 per cent on the 170-mesh sieve, and the rapid-hardening cements from 1 to 3 per cent. Similar general trends in the U.S.A. have been reported by Gonnerman and Lerch[1] in a review of the changes in Portland cement over the first half of the present century.

In order to determine the proportions of the finer particles in a cement, finer sieves have sometimes been used. Thus 240- and 300-mesh-per-linear-inch sieves can be obtained. Such sieves are very troublesome and tedious to use owing to the liability of the meshes to become clogged with the cement.

The particle-size distribution of powders finer than the 170-mesh sieve is normally determined by sedimentation or elutriation.[2] Both methods depend on the difference in the rate of fall of particles of different diameter and make use of Stokes' law which, assuming spherical particles, gives the terminal velocity of fall under gravity of a fine particle through a gaseous or liquid medium.

It is:

$$v = \frac{10^{-8}g}{18}\frac{(\sigma-\rho)D^2}{\eta}$$

where v is the velocity of fall in centimetres per second, g the gravitational constant (981 dynes per gram), σ and ρ the density of the solid and the medium respectively, D the diameter of the solid particle in μ (10^{-4} cm), and η the viscosity of the medium in absolute c.g.s. units.

The first requirement in sedimentation analysis is to find suitable conditions for dispersing the powder in the liquid medium, and maintaining dispersion throughout the sedimentation process. There is no simple test of the degree of dispersion, and although the onset of sudden flocculation may readily be detected by the fact that the suspension settles out as a whole with a well-defined layer of clear liquid above, it is difficult to establish that dispersion has been quite complete in the first place, and that partial flocculation is not occurring during the experiment. Generally, the test is repeated with variation of the concentration of solid to liquid medium, variation of the amount of mechanical dispersion, and the addition of varying amounts of surface active agent. A set of conditions is then selected which gives the maximum yield of the finer fractions and consistent results in replicate experiments.

In the case of cement, a further requirement is that the liquid medium shall not react chemically with the cement. Many methods of sedimentation analysis have been described differing in the way in which the rate of settling is measured. The commonest is the Andreasen pipette method in which a sample of the sus-

[1] ASTM Special Publication No. 127 (1951).

[2] For general surveys see *Symposium on New Methods for Particle Size Determination*, ASTM 1941; *Symposium on Particle Size Analysis*, ASTM Special Technical Publication No. 234 (1958); *Symposium on Particle Size Analysis*, supplement to *Trans. Instn chem. Engrs.* **25** (1947); *Physics of Particle Size Analysis*, supplement No. 3 to *Br. J. appl. Phys.* (1954); *Proceedings of Conference on Particle Size Analysis*, Society for Analytical Chemistry, London 1967.

pension is withdrawn from a given depth at various time intervals and the concentration of solids in the sample determined. Alternative methods make use of the fact that the density and pressure at a given depth will decrease as the particles fall out. A hydrometer[1] or series of divers can be used to detect the changes in density, or a sensitive manometer to detect the change in pressure.[2] In a more direct method the particles falling out are caught on a balance pan at the bottom of the sedimenting column and a continuous record of the weight fallen out is obtained.

Another method, the turbidimetric[3] method of sedimentation analysis, has the advantage that there is no disturbance of the fluid, or interference with the free fall of the particles. A beam of light is passed through the suspension at a known depth below the surface and the percentage of light transmitted is measured photo-electrically. If the intensities of the light transmitted through the fluid alone and through the suspension are I_0 and I_s respectively (in arbitrary units), c is the concentration of solid in g per cc, D the diameter of the particles and l is the length of suspension in centimetres through which the light passes then

$$I_s = I_0 e^{\frac{-kcl}{D}}$$

where k is a constant, depending in part on the opacity of the particles.

Undoubtedly the most accurate method is that in which the tracks of individual particles are photographed,[4] and their velocities calculated from the known time of exposure.

Estimates of particle-size distribution can also be made by microscopic examination,[5] by direct measurement, by a microprojection method throwing an image of the particles on a screen, or by a photomicrographic method. The process is lengthy and tedious and is particularly difficult for particles of irregular shape. Since the amount of powder examined is very small, sampling errors can be high.

Elutriation is the reverse of sedimentation and air is the fluid commonly used. The cement is introduced into a long vertical column up which air flows at a suitable rate. Those particles having a rate of fall greater than this velocity fall to the bottom while the finer ones are carried out at the top. Taking the density of cement as 3·1, the viscosity of air at 20° as $\eta = 1{\cdot}82 \times 10^{-4}$ c.g.s. units, and neglecting its density as negligible, the rates of fall of cement particles of various sizes in air are as shown below.

Particle diameter in microns	5	10	15	20	30	40	50	60
Rate of fall (cm/s) for cement, σ 3·1	0·229	0·917	2·06	3·66	8·24	14·65	22·9	33·0

[1] A. Klein, *Proc. Am. Soc. Test. Mater.* **41,** 953 (1941).
[2] R. T. Knapp, *Ind. Engng Chem. analyt.* **6,** 66 (1934).
[3] L. A. Wagner, *Proc. Am. Soc. Test. Mater.* **33** (II), 553 (1933); H. E. Rose and H. B. Lloyd, *J. Soc. chem. Ind., Lond.* **52,** 65 (1946).
[4] W. F. Carey and C. J. Stairmand, *Trans. Instn. chem. Engrs* **16,** 57 (1938).
[5] BS 3046: Part 4: 1963; ASTM E20–62T.

There are two main types of air elutriations in use, the Gonell[1] pattern and the Roller[2] pattern. Details of these are to be found in BS 3406 : Part 3 : 1963 and ASTM B293. They are both applicable to the separation of particles in the 5–75 μ range and involve the successive use of three or four vertical elutriation columns of decreasing diameter and with increasing air velocities. The sample is introduced into a special container connected at one end by a cone to the bottom of the elutriation column and at the other end to the air supply, the flow rate being measured by a flowmeter. Elutriation is carried out at a succession of air velocities and the residue in the container weighed in the Gonell method or the elutriated particles collected in a thimble filter and weighed in the Roller method. The cement sample is dried at 100–110° and the air supply is also dried. In another more recent form[3] a blower serves a set of jets of different size giving varying air velocities up a single column.

The form of apparatus employed in determining the size-weight distribution of cement by the Andreasen sedimentation method[4] is shown in Fig. 108. Details are given in BS 3046 : Part 2 : 1963. The tip of the pipette is at a standard distance of 20 cm from the topmost mark of the measuring cylinder. The medium most commonly employed has been water-free ethyl alcohol containing a little anhydrous calcium chloride (from 16 to mg 2–3 g/litre), or methyl alcohol saturated with sodium pyrophosphate. Quinoline[5] and *iso*propanol[6] have been recommended as alternatives requiring no dispersing agent. Enough cement, previously dried at 110°, is weighed out to give a concentration of 2 g per 100 ml when the Andreasen bottle is filled to the upper mark. A little of the sedimentation fluid is added and the cement is mixed, mechanical stirring being used if desired. The paste so obtained is washed into the sedimentation cylinder with the medium, and the volume is made up. After closing with the ground glass stopper the cylinder is inverted several times and then placed to settle out in a constant-temperature room. Samples are withdrawn at times corresponding to a 20-cm fall of particles of diameters greater than a series of values, such as 112, 25, 12, 6 and 2·5 μ respectively. The samples may either be evaporated to dryness or filtered on sintered glass. Commencing with a uniform dispersion, it is evident that at the end of a given time, t, all particles above a certain diameter, D, will have fallen more than 20 cm and that the sample obtained at this time is an average sample for all particles of diameter less than D. The weight of the fraction, divided by the calculated amount of cement present in the same volume in the initial dispersion and multiplied by 100 gives the percentage content in the cement of particles of diameter less than D. A curve of the type shown in Fig. 109 can thus be obtained.

The results are often quoted in terms of the Andreasen cubical dimension rather than the Stokes' law sphere. This is the side of the cube having the same

[1] H. W. Gonell, *Zeit. Ver. dt. Ing.* **72,** 945 (1928).
[2] P. S. Roller, *Tech. Pap. Bur. Mines, Wash.*, No. 430 (1931); *Proc. Am. Soc. Test. Mater.* **32,** II, 607 (1932).
[3] *SchrReihe ZemInd.*, Vol. 33 (1967); *Cement Fineness*, Beton-*Verlag* GmbH, Dusseldorf.
[4] H. M. Andreasen, *Zement* **19,** 698, 725 (1930); *Trans. ceram. Soc.*, 1930 (Wedgwood Bicentenary volume), p. 239; *Ingvidensk Skr.*, 1939, No. 3, Copenhagen (in English).
[5] F. Gille, *Zement* **31,** 316 (1942).
[6] A. Klein, loc. cit.

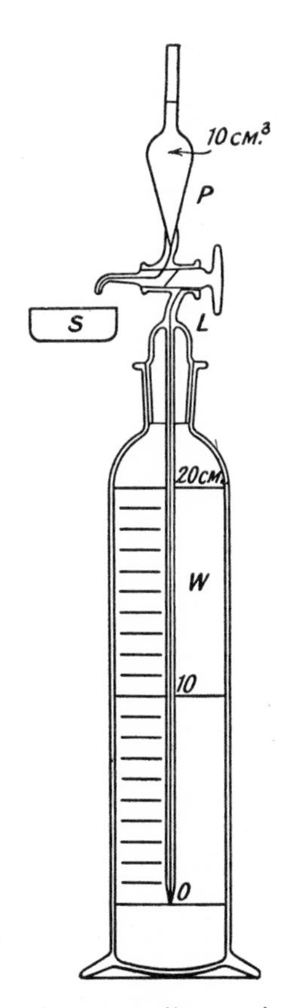

FIG. 108 Andreason sedimentation apparatus.

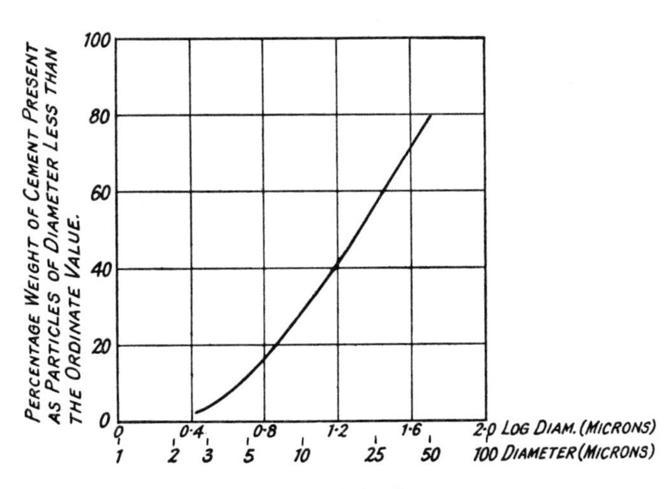

FIG. 109 Particle-size curve.

volume as the sphere generated by the Stokes' law radius. If A is the Andreasen diameter and S the Stokes' law diameter, A = 0·806S.

In the Wagner method, the turbidity of a suspension of cement in kerosene is measured. Since the settling of the suspension is not disturbed in any way, numerous measurements can be made, and, by moving the light beam upwards so that measurements are made at small heights in the suspension, results for the finer fractions can be obtained in a short time. The sedimentation fluid used in the ASTM specification (C115–67) is kerosene. A sample of 0·3–0·5 g cement is stirred in a test tube with 10–15 ml of kerosene, to which five drops of oleic acid or aged linseed oil has been added, and the suspension then transferred to the glass settling tank and the volume made up to 335 ml with kerosene. A parallel beam of light of constant intensity is passed through the cell and the emergent beam allowed to fall on a photocell. Readings are taken at known times and known heights below the surface of the fluid. For each reading:

$$S_d = c(\log I_0 - \log I_d)$$

where S_d = surface in cm^2 of all particles in the sample smaller than *d* microns in diameter, I_0 = the intensity of light transmitted through the clear sedimentation fluid, I_d = the intensity of light transmitted through the suspension at the time required for particles of diameter *d* to settle from the surface of the suspension to the centre of the light beam, and *c* is a constant. The value of *c* is derived by a summation process over the whole particle size range in which it is assumed that the particles are spherical and that *c* does not vary with particle size. The turbidimeter is calibrated with a standard sample of cement supplied by the Bureau of Standards. The specification should be consulted for details.

While the results can be calculated to give a size-weight distribution curve,[1] it is used in the ASTM specification to give a measure of the total specific surface of a cement, that is, the total surface area divided by the weight and expressed as cm^2/g. Instead of taking readings at a series of times corresponding to different diameters of particles in suspension, the turbidimeter can also be used to estimate the surface area from a single initial reading of the turbidity. This is only recommended for samples from the same plant which do not vary greatly in fineness, colour or density, for this rapid method in effect assumes that the value of *c* remains unchanged. There are, in fact, difficulties about the assumptions made in connection with the transmittancy constant *c* even in the full method, but they are not too serious as applied to cement.

The specific surface of a cement can always be derived from an analysis of a particle-size distribution curve however obtained. Considering Fig. 109, we may obtain from the curve given a factor termed the 'weight frequency per micron', or $W/\Delta D$, where W is the percentage weight of cement falling between any two diameters differing by a small amount ΔD. If this factor is plotted against the mean diameter of the range over which ΔD is taken, and more conveniently we may use the logarithm of this, we obtain a curve as that shown in Fig. 110. This curve shows the weight frequency particle-size distribution in the cement. By integrating the area under this curve by counting squares the surface area per unit weight of cement is obtained.

[1] W. G. Hime and E. G. La Borde, *J. Res. Dev. Labs Portld Cem. Ass.* **7** (2), 66 (1965).

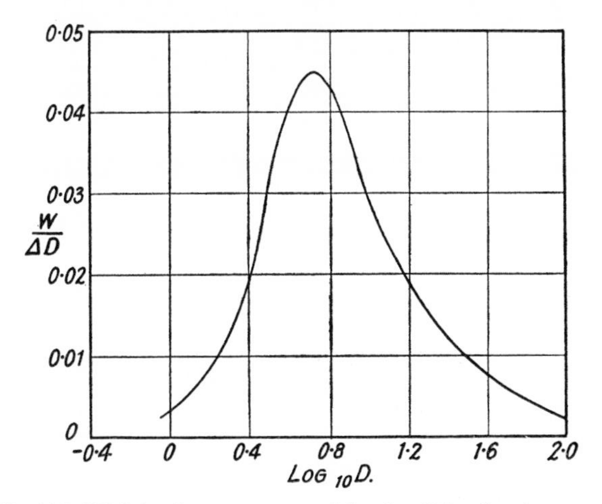

FIG. 110 Weight frequency particle-size Distribution curve.

Surface Area $\propto$ Weight/Diameter

$$= K\frac{W}{D} = K\Sigma\frac{\text{weight frequency per micron}}{D}$$

$$= K\Sigma\frac{W}{\Delta D}\cdot\frac{1}{D}$$

$$= K\int\frac{W}{\Delta D}\cdot\frac{1}{D}\,dD = K\int\frac{W}{\Delta D}\,d\log D$$

where K is a constant. K is evaluated as follows:

For a particle of diameter D we have

$$\text{Surface area} = \pi D^2$$

$$\text{Weight} = \tfrac{1}{6}\pi D^3\sigma, \text{ where } \sigma \text{ is the density}$$

$$\frac{\text{Area}}{\text{Weight}} = \frac{A}{W} = \frac{\pi D^2}{\frac{1}{6}\pi D^3\sigma} = \frac{6}{D\sigma}$$

$$\therefore \text{Area} = A = \frac{6}{D\sigma}W$$

Now D is in microns or centimetres $\times 10^{-4}$.

$$\text{Area (cm}^2) = \frac{6\times 10^4}{\sigma}\left(\frac{W}{D}\right)$$

$$\text{Whence } K = \frac{6\times 10^4}{\sigma} \text{ and}$$

$$\text{Surface area} = K\int \frac{W}{\Delta D}\, d \log D$$

$$= \frac{6 \times 10^4}{\sigma} \int \frac{W}{\Delta D}\, d \log D$$

or using logarithms to base 10

$$= \frac{6 \times 2{\cdot}3 \times 10^4}{\sigma} \int \frac{W}{\Delta D}\, d \log_{10} D$$

The integral represents the area under the curve in Fig. 110. If a total weight of 1 g of cement is considered, the surface area per gram is then obtained.

The method by which the curve in Fig. 110 is obtained from the curve in Fig. 109 is shown in Table 54.

TABLE 54 Calculation of weight frequency particle-size distribution

Diameter range (microns)	ΔD (microns)	W (grams)	$\frac{W}{\Delta D}$	Average diameter (microns)	Log_{10} D
50–100	50	0·214	0·00428	75	1·87
40–50	10	0·066	0·0066	45	1·65
30–40	10	0·088	0·0088	35	1·54
20–30	10	0·116	0·0116	25	1·40
15–20	5	0·089	0·0178	17·5	1·24
13–15	2	0·043	0·0216	14	1·15
11–13	2	0·053	0·0265	12	1·08
9–11	2	0·056	0·028	10	1·0
7–9	2	0·072	0·036	8	0·90
5–7	2	0·088	0·044	6	0·78
4–5	1	0·044	0·044	4·5	0·65
3–4	1	0·037	0·037	3·5	0·54
0–3	3	0·039	0·0077	1·5	0·18

The air-permeability method depends on the fact that the resistance offered to gas flow by a plug of powder compacted to a known volume and porosity depends on the specific surface of the powder. This method was first developed by Lea and Nurse[1] following the work of Carman[2] on the flow of liquids through powders coarser than cement. In the apparatus developed by Lea and Nurse, and adopted in the British Standards for cements, where full details are to be found, the cement is formed into a bed in a cylindrical container A (Fig. 112) and a stream of dry air passed through it. The rate of air flow is measured by a flowmeter which consists of a capillary placed in the circuit with a manometer connected across its ends (Fig. 111). Once the capillary has been calibrated, the rate of air flow can be obtained from the drop in pressure across the capillary

[1] F. M. Lea and R. W. Nurse, *J. Soc. chem. Ind., Lond.* **58,** 227 (1939); *Symposium on Particle Size Analysis,* supplement to *Trans. Instn. chem. Engrs* **25** (1947), p. 47.
[2] P. C. Carman, *J. Soc. chem. Ind., Lond.* **57,** 225 (1938); **58,** 2 (1939).

measured on the manometer. The resistance of the bed of cement to the air flow causes a fall in pressure across it, and this is measured by a manometer connected to the top and lower sides of the bed. This resistance of the bed varies with the closeness of packing of the bed, i.e. with its porosity, which is the proportion of the bed volume which consists of pores or air spaces. The porosity is fixed by the weight and density of the cement used to make the bed which, in the apparatus described, has a fixed volume. The specific surface of the powder can then be

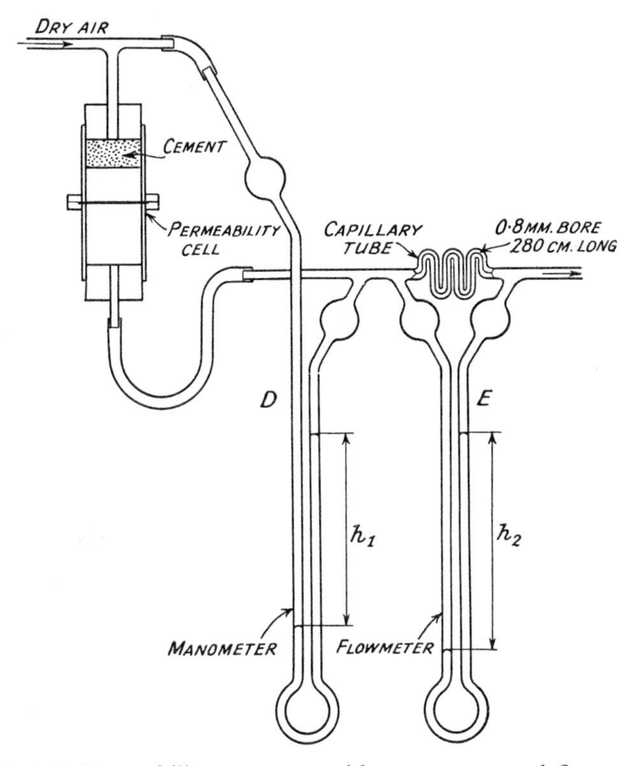

FIG. 111 Permeability apparatus with manometer and flowmeter.

obtained from the rate of air flow and the fall in pressure across the bed. The measurement of specific surface thus consists of the following operations:

1. Weighing out the required amount of cement to form a bed of porosity 0·475. The weight required varies with the density of the cement and, if the density is not known, it has first to be determined.
2. Forming the bed, compacting it with the plunger C, and connecting up the apparatus.
3. Passing a steady stream of air through the bed for five minutes and then measuring the difference in level of the two columns of the manometer across the capillary which forms the flowmeter and also the difference in level of the two columns of the manometer across the bed.

4. Calculating the ratio of the difference in level of the manometer across the bed to the difference in level of the manometer across the capillary. The square root of this ratio is then taken and multiplied by a constant. The result is the specific surface of the cement in sq. cm. per g.

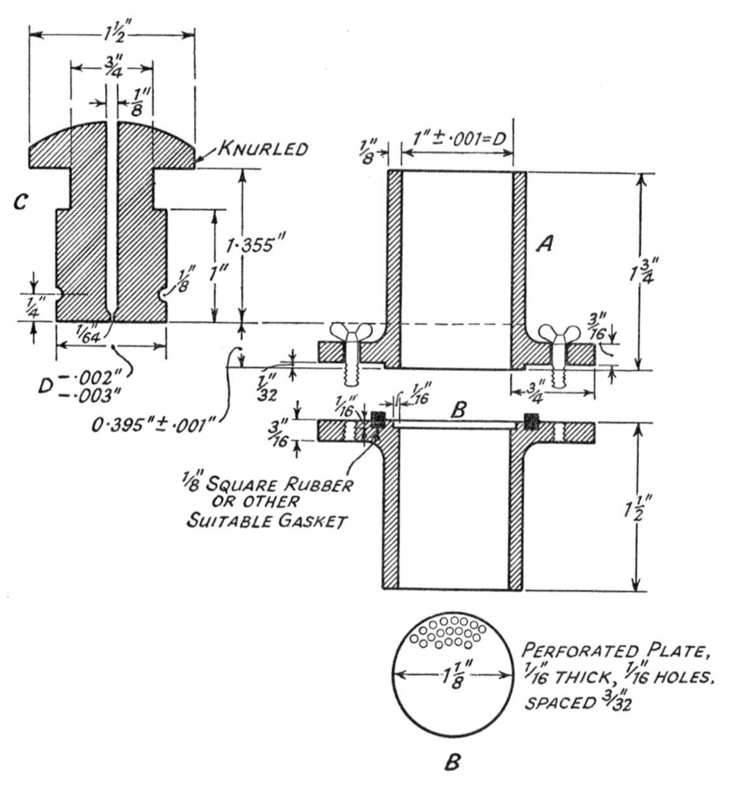

FIG. 112 Details of permeability cell.

The method depends on the validity of an equation developed by Carman

$$S = \frac{14}{\sigma(1-\varepsilon)}\sqrt{\frac{\varepsilon^3 A i}{v Q}}$$

where S is the specific surface (cm^2/g)
σ is the powder density
ε is the porosity of the powder bed
A is the cross-sectional area (cm^2)
i is the hydraulic gradient
v is the kinematic viscosity of the permeating fluid (Stokes)
Q is the rate of flow (cc per second).

For the given apparatus this equation reduces to:

$$S = \frac{14}{\sigma(1-\varepsilon)} \sqrt{\frac{\varepsilon^3 A h_1}{C L h_2}}$$

where C is the flowmeter constant
L is the height of the powder bed
h_1 is the manometer reading
h_2 is the flowmeter reading.

For any one apparatus, this further simplifies to:

$$S = K \sqrt{\frac{h_1}{h_2}}$$

where K is a constant varying only with the cement density. An automatic form of the apparatus, designed for checking the output of grinding mills, has been described.[1]

The U.S. Standard method (ASTM C204–68) is based on a variation of the method of Lea and Nurse. It was first described, and its theory discussed, by Rigden,[2] but it was also developed independently in the United States by Blaine[3] who devised the apparatus shown in Fig. 113. In this method air does not pass through the bed at a constant rate, but at a steadily diminishing rate. Oil in the manometer U tube serves the double purpose of producing the flow of air and measuring the rate of flow. Air is withdrawn through the side tube until the oil in the manometer has risen above the uppermost timing mark; the valve is closed and the oil level at once begins to fall, drawing air through the bed. The time T (seconds) taken by the oil meniscus to travel between the two innermost timing marks is noted by means of a stop-watch. For an apparatus of the dimensions given and a bed of standard porosity,

$$S_w = K_A \sqrt{T}$$

The constant K_A is not so easily specified from the dimensions of the apparatus as the corresponding constant K in the British Standard method, as it is affected by the diameter of the U tube. In U.S.A. practice, K is therefore determined by using a standard sample of cement issued by the U.S. National Bureau of Standards. The standard porosity is 0·500 ($\pm$0·005) and a mercury-displacement method is employed to measure the bulk volume of the bed of cement when calibrating the apparatus.

Comparative tests show that there is close agreement between the two variations of the method.[4]

The air-permeability method gives a higher value for the specific surface of cements than the Wagner method because it measures directly the surface area of the finest particles whereas in the Wagner method the average diameter of all the particles below 7·5 μ in diameter is assumed to be 3·8 μ, a value which is too

[1] M. Papadakis, *Revue Matér. Constr. Trav. publ.* **569,** 48; **570,** 79 (1963).
[2] P. J. Rigden, *J. Soc. chem. Ind., Lond.* **62,** 1 (1943).
[3] R. L. Blaine, *Bull. Am. Soc. Test. Mater.* **123,** 51 (1943).
[4] R. W. Nurse, *Cem. Lime Mf.* **24** (2), 17–21 (1951).

high. The ratio of the value by the air permeability method to that given by the Wagner method varies with the fineness of the cement and its gypsum content and can range from about 1·6 to 2·2. For conversion purposes a value of 1·8 can be used without very serious error for most cements.

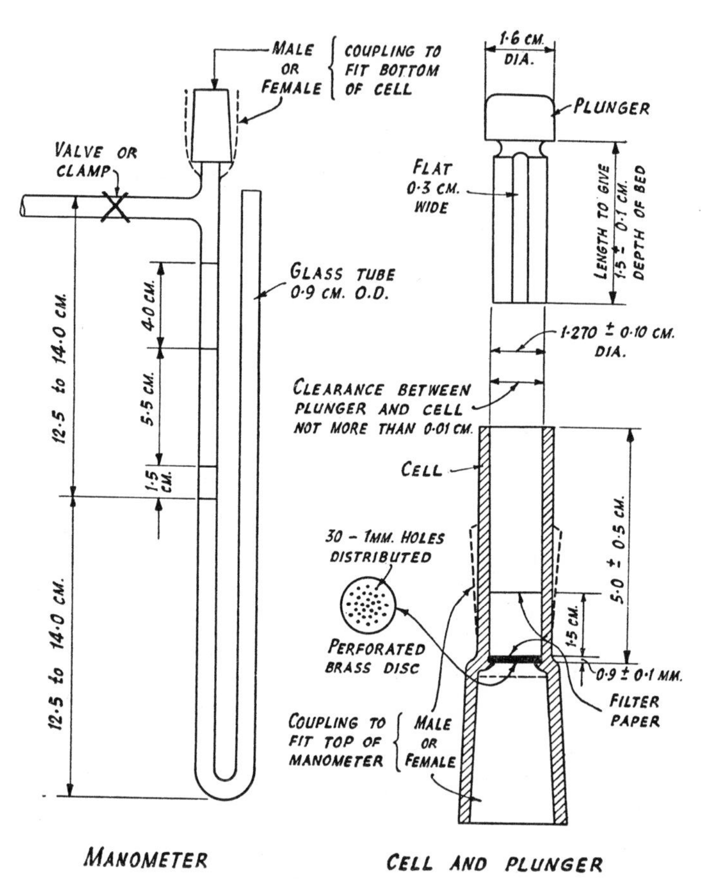

FIG. 113 Blaine permeability apparatus.

A further direct method which has sometimes been used for cement is that in which the isotherm for the adsorption of a gas or vapour on the powder is determined. A certain inflexion in the curve connecting relative vapour pressure with amount adsorbed is assumed to represent the completion of a monolayer of the adsorbate on the adsorbent. If the area occupied by a single molecule of the adsorbate is known, the specific surface can then be calculated. The theory used is due to Brunauer, Emmett and Teller and the adsorbate is usually nitrogen, at the temperature of liquid oxygen. Water vapour has been used at room temperature to determine the specific surface of set cement and the difference between the results obtained with water vapour and with nitrogen has been considered of significance in connection with the physical structure of cement pastes (see p. 277).

TABLE 55 Specific surface (cm^2/g)

Cement	Andreasen method		Wagner method	BS air permeability	Nitrogen adsorption
	Stokes' spherical dimension	Andreasen cubical dimension			
I	2310	2860	1790	2600	7900
II	3110	3860	2270	4150	10000

Andreasen sedimentation carried down to 0·6 μ and particles below that assumed to have an average diameter of 0·3 μ.

The results obtained for the specific surface of the same cement by the various methods differ as may be seen from Table 55. This difference is not due to any lack of physical meaning in the results, but arises from the assumptions regarding the size of the finest particles made in the sedimentation methods and differences in the degree to which minor surface irregularities, or internal surfaces, are measured by the different methods. For instance, the nitrogen-adsorption method, which probably represents the closest approach to an absolute measurement, includes surface arising from the inner porosity of the grains. Certain corrections[1] are strictly speaking to be applied to air-permeability results which somewhat increase the specific surface, but by common consent these are omitted in the calculation of the results of specification tests. The Andreasen and Wagner methods in which surface area is calculated from sedimentation data have been shown[2] to agree well provided the same assumption is made regarding the contribution of the finest particles. This is illustrated in Table 56 where the mean diameter of all particles below 7·5 μ has been assumed to be 3·8 μ in both methods.

TABLE 56 Specific surface of cements calculated from size-weight distribution for particles below 25 μ

Laboratory	Method	Specific surface (cm^2/g)		
		Cement A	Cement B	Cement C
1	Wagner	2005	1735	1620
2	Andreasen	2140	1930	1705
3	Wagner	2195	1850	1775
4	Andreasen	2240	2005	1695
5	Wagner	2180	1770	1650

Mean diameter of particles below 7·5 μ assumed to be 3·8 μ in all cases.

[1] F. M. Lea and R. W. Nurse, loc. cit. (1947); P. C. Carman and J. C. Arnell, *Can. J. Res.* **A26,** 128 (1948); S. B. Ober and K. J. Frederick, *Spec. tech. Publ. Am. Soc. Test. Mater.*, No. 234, 279 (1959).

[2] F. M. Lea, *J. Soc. chem. Ind., Lond.* **58,** 146 (1939).

There is no consistent relationship in general between the residue on the 170-mesh sieve and the specific surface. This is not surprising, as the sieve residue is small and there is no reason why the amount of this small coarse fraction should bear any relation to the overall fineness. While the specific surface is not so informative as a full-size distribution curve it is rapidly and conveniently obtained and has proved to be a satisfactory means of controlling the output of cement mills. There is a general tendency for the ratio of the air permeability to Wagner values to be higher for open circuit grinding mills than in closed circuit mills in which the ground particles are swept away by an air stream and less of the very finest particles are produced.

It has been estimated from electron microscope photographs that the finest particles present in cement lie within the range 0·03–0·12 μ.[1] This lower limit must, however, depend to a great extent on the method of transporting and packing the cement on the plant and the air velocities used in dust extraction.

Selective grinding of clinker

The various size fractions into which a cement can be divided by air separators are not identical in composition, showing that there is some separation of the compounds present in the cement. It is found that the tricalcium silicate content tends to be higher in the finer fractions, and the dicalcium silicate content higher in the coarser fractions. This is probably due to the tricalcium silicate being easier to grind than the dicalcium silicate. This may be attributable to internal strain in the tricalcium silicate crystals for it has been noticed that they can break up spontaneously, or before hydrating on immersion in water. The tricalcium aluminate and tetracalcium aluminoferrite are present to about the same extent in all fractions. Some analyses of various fractions of ground cement clinkers are shown in Table 57.

Fineness of cement and strength

It has long been known that the more finely a cement is ground the greater is the strength, and particularly the early strength, it yields. The question has arisen, however, whether there is not some limit to this and whether there is not some particular type of particle-size distribution which is better than others. The influence of the specific surface of a cement can be determined by grinding samples of clinker to varying fineness and determining the strength, but to ascertain the relative contribution of different size fractions, it is necessary to separate them by air elutriation. As we have seen above, these different size fractions also vary in composition and this makes the interpretation of the data more difficult.

Tensile tests on neat cements show that, beyond a certain point, which lies between 1 and 10 per cent residue on a 170-mesh, the strength often decreases with finer grinding, and a similar effect has occasionally been reported for the tensile and compressive strength of mortars of dry consistence, i.e. with a water: cement ratio below 0·35. This led Kühl[2] and some others to suggest that a definite

[1] G. Haegermann, *Zement* **31,** 441 (1942).
[2] *Cement Chemistry in Theory and Practice*, p. 55, London, 1931.

TABLE 57 Analysis of various size fractions of ground cement clinker[1]

Cement	Fraction size in microns	Ignition loss	Per cent composition				Per cent compound content			
			SiO_2	CaO	Fe_2O_3	Al_2O_3	C_3S	C_2S	C_4AF	C_3A
A	Whole	2·4	21·0	65·5	3·7	6·9	56	19	11	12
	0– 7	6·4	20·3	65·3	3·6	7·3	59	14	11	13
	7–22	2·5	20·4	65·8	3·5	7·1	62	11	11	13
	22–35	1·5	21·2	65·5	3·6	7·2	52	22	11	13
	35–55	1·1	21·1	64·8	3·7	7·4	49	24	11	13
	Above 55	0·9	21·1	64·4	3·7	7·5	47	25	11	14
B	Whole	1·3	22·0	66·2	2·5	6·6	54	23	8	13
	0– 7	2·8	22·0	66·2	2·1	6·4	57	20	6	13
	7–22	1·2	22·3	66·5	2·2	6·5	54	23	7	14
	22–35	0·7	22·2	66·0	2·6	6·0	56	22	8	12
	35–55	0·6	21·6	65·6	3·0	6·7	54	22	9	13
	Above 55	0·7	21·2	65·2	3·5	6·9	53	22	11	12
C	Whole	0·7	23·8	65·4	5·6	3·5	55	27	17	1
	0– 7	2·6	21·0	65·7	7·1	4·2	64	13	20	0
	7–22	0·8	23·0	66·3	5·4	3·5	64	18	16	0
	22–35	0·5	24·7	65·6	5·0	3·4	51	33	15	1
	35–55	0·4	24·8	65·2	5·4	3·3	46	37	16	0
	Above 55	0·3	25·0	65·4	5·8	3·7	43	39	18	0
X	Whole	—	23·3	66·0	3·5	4·5	51	28	11	6
	0– 3	—	21·3	62·5	4·2	5·1	41	31	13	7
	3–10	—	21·9	64·9	4·1	5·0	48	26	12	6
	7–15	—	22·5	66·3	3·7	4·6	57	21	11	6
	10–25	—	23·8	66·9	3·2	4·2	55	27	10	6
	25–50	—	24·0	66·4	3·4	4·3	50	31	11	6
	Above 45	—	23·6	65·5	4·4	4·5	46	34	13	5

optimum limit of fineness exists. Such tests do not, however, provide any adequate measure of the influence of cement fineness on the strength of plastic mortars and concretes and for these there is, in general, an increase in strength as a given cement is more finely ground, at least as far as present commercial limits of fineness are concerned. This is illustrated[2] in Fig. 114, from which it will be seen that the effect of increased fineness is proportionately much greater at early than later ages. The influence of fineness on strength varies with the cement and the concrete mix. An increase in specific surface from 1800 to 2500 cm^2/g (Wagner, equivalent to 3200 to 4500 by air-permeability method) raises the compressive strength at 1 day by 50–100 per cent, at 3 days by 30–60 per cent, and at 7 days by 15–40 per cent. Increases in fineness beyond about 5000 cm/g

[1] J. A. Swenson, L. A. Wagner and G. L. Pigman, *J. Res. natn. Bur. Stand.* **14,** 419 (1935) for cements A, B, C; A. Rio and F. von Baldass, *Industria ital. Cem.* **24,** 87 (1954) for cement X.

[2] W. H. Price, *Proc. Am. Concr. Inst.* **47,** 417 (1951); see also H. C. Entroy, *Trans. 8th int. Congr. large Dams,* **3,** 193 (1964); M. Venuat, *Revue Matér. Constr. Trav. publ.,* Nos. 550–553 (1961); 595–596 (1965).

(air-permeability) only produce a relatively small increase in strength except at less than 1 day. Ordinary Portland cements usually have a specific surface (air permeability) of about 3000 to 4000 and rapid hardening cement up to about 5000 cm^2/g. The particle-size fraction below 3 μ has been found to have the predominant effect on the strength at 1 day while the 3–25 μ fraction has a major influence on the 28 day strengths.[1]

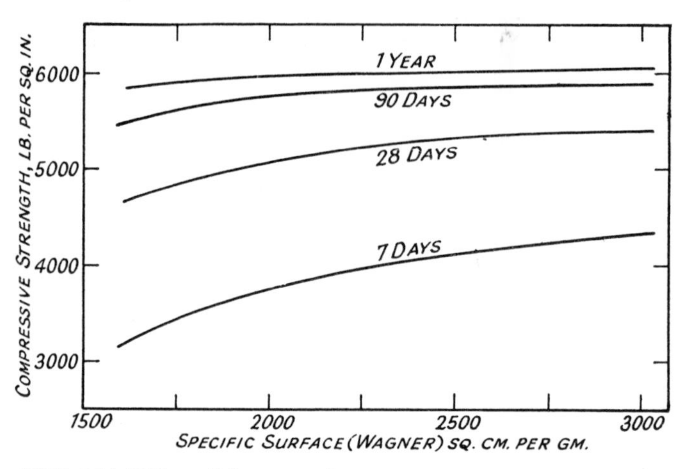

FIG. 114 Effect of fineness of cement on strength of concrete.

An increase in fineness also affects other properties of a cement or concrete. It tends to raise the amount of gypsum required to control setting or to give optimum strength. It may slightly increase the amount of water required to give a neat cement paste of standard consistence but, in contrast, reduce slightly the amount needed to obtain a given workability in a concrete mix. This latter effect does not extend indefinitely as fineness is increased, for if the fraction below, say, 7 μ is separated from a cement by air elutriation it is found to require considerably more water to produce a workable concrete than the parent cement. When gauged neat it gives a sticky mass with a high power for retaining water and it is this fine fraction that must be of primary importance in restricting the tendency of water to separate from a concrete mix during placing. This fraction also produces mortars with a higher drying shrinkage than the parent cement.[2] It will be evident that, while an adequate proportion of grains below 7 μ are desirable in a cement, there is a limit beyond which the adverse effects will start to outweigh the beneficial ones. The specific surface of the fraction below 7 μ is two to three times that of a cement as a whole, so that it represents a much more finely ground material than present commercial cements in which some 15–40 per cent by weight is below 7 μ. It has been suggested[3] that 25–30 per cent is the maximum desirable below 7 μ, but it appears that this can be increased somewhat.

[1] F. W. Locher, J. Wohrer and K. Schweden, *Tonind.-Ztg. keram. Rdsh.* **90,** 547 (1966); B. Beke, *Zement-Kalk-Gips* **13,** 419 (1960).
[2] J. Wuhrer, *Zement-Kalk-Gips* **3,** 148 (1950); A. Rio and F. von Baldass (loc. cit.).
[3] A. Eiger, *Tonind.-Ztg. keram. Rdsh.* **55,** 1389 (1931); **56,** 552, 558 (1932); J. A. Swenson, L. A. Wagner and G. L. Pigman (loc. cit.).

A given specific surface in a cement could be produced with a variety of particle-size distributions, but no final answer can yet be given to the question whether there is some optimum grading. Basically, strength and other properties of cements are functions of the amount hydrated and initially this will be related to the specific surface rather than to particle grading. As hydration proceeds, however, the finest particles become completely hydrated and the surface area of the remaining anhydrous material will become progressively less related to the original value, and more dependent on the original particle-size distribution. Thus, there is evidence[1] that at 1-day strength is related more or less linearly to the specific surface, but that at longer ages it is influenced also by the kind of particle-size distribution. The maximum strength at relatively short ages with a given specific surface would be produced by particles of roughly uniform size, but this is not necessarily true at longer ages nor, in fact, could such a gradation be produced with existing types of grinding machinery.

THE STRENGTH OF CEMENT

The value of cement when employed as a structural material depends primarily on its mechanical strength in the set and hardened condition, a strength due to the cohesion of the particles of the cement and to their adhesion to the grains of sand or other aggregate with which they are mixed; when used as a mortar, the adhesion of the material to the surface of brick or stone is also of prime importance. Mechanical tests therefore play a most important part in determining the quality of cement, and every specification requires a certain minimum strength that must be attained under given conditions. The strength is in a high degree dependent on the conditions of mixing, preparation of the specimens, and testing, and it is necessary to prescribe the exact manner of testing in order to obtain comparable results.

Strength tests take three different forms, the specimen being subjected to tension, compression or bending. The last, commonly known as a transverse test, is not much used in cement testing in Great Britain though it has found favour abroad. The subject of strength testing is one which requires a volume on its own for adequate treatment and only the bare outlines of specification methods, with some notes on the significance of the results, can be included here.

The strength developed by a cement, either when tested neat, in a mortar, or a concrete, is dependent on many factors, such as the grading of the sand or aggregate, the proportion of water used, the degree of mixing, and the temperature and humidity of the atmosphere in which it is conducted, the method by which the material is placed in the moulds and the specimen made, the curing conditions, the method of testing and the age at which tests are carried out. For testing purposes it is necessary to define all these conditions, and in all standard specifications this is done as closely as possible.

[1] J. A. Swenson, L. A. Wagner and G. L. Pigman (loc. cit.); see also M. J. M. Jaspers, *Revue Matér. Constr. Trav. publ.* **429,** 169; **430,** 207; **431,** 254 (1951); W. Czernin, *Zement-Kalk-Gips* **7,** 160 (1954).

Strength tests on neat cement were used for many years but they were abandoned because they afforded no reliable indication of the quality of modern finely-ground cements. All countries in their specifications for cement use tests on cement-sand mortars in compression, bending or tension. The tensile tests have now been abandoned in the British specification, and in some others such as the German and the Swiss, but they are still retained in the French specification. In the British specification a compression test on concrete was introduced in 1958 as an alternative to a compression test on mortar.

The sand to be used in mortars for standard tests has to be carefully specified. In some countries a sand of closely uniform size is used but an increasing number of countries have adopted a graded sand. In Great Britain, the standard sand is obtained from Leighton Buzzard and consists of round quartz grains all passing a BS No. 18 sieve (0·85 mm) with not more than 10 per cent passing a No. 25 sieve (0·6 mm). In the ASTM specification a uniform Ottawa silica sand passing an ASTM No. 20 sieve (0·84 mm) and retained on a No. 30 sieve (0·595 mm) is used for the tensile test, but for the mortar for the compressive test a sand graded between a No. 16 sieve (1·19 mm) and a No. 100 sieve (0·149 mm) is used. In order to promote uniformity between different countries in testing methods a R.I.L.E.M.–CEMBUREAU method has been developed in which tests for compressive and bending tests are carried out on a 1:3 graded sand (2·0–0·08 mm) mortar with a w/c ratio of 0·5. Most standard specifications[1] in European countries and the R.I.L.E.M.–CEMBUREAU method[2] contain a compressive and a bending test in which the former is made on the broken halves of the mortar prisms used for the latter. While some countries still retain a tensile and a compressive test on separate specimens, the use of the R.I.L.E.M.–CEMBUREAU method is spreading. It has been adopted for instance in the 1967 revision of the German specification DIN 1164 and in the International Standards Organisation (ISO) draft recommendations[2] for the testing of cement. In this method the mortar is mixed mechanically and the specimens compacted on a jolting machine. In the ASTM method for tensile tests (C190–63), now in suspence, a semi-plastic 1:3 mortar of about 10 per cent water content is used and pressed into the mould with the thumbs. The ASTM compressive tests (C109–64) on 2 in. cubes make use of a 1:2·75 graded sand mortar of about 13–14 per cent water content which is tamped into the moulds. In those European countries which still retain the tensile test on briquettes (Fig. 115) it is commonly made on 1:3 mortars of dry consistence with a water content of about 8 per cent, the test specimens being compacted by mechanical ramming.

The British Standard (BS 12/1958) permits the use of a compressive test on either mortar or concrete. In the former the test cubes (side length 7·06 cm (2·78 in.) giving an area of each face of 50 cm^2) made from a 1:3 mortar with 10 per cent water content are compacted on a vibrating machine. For the tests on concrete cubes (4 in.–10·16 cm–side length) the coarse aggregate, which may be flint, granite, limestone, porphyry or quartzite, must pass a $\frac{3}{4}$ in. BS mesh and be substantially retained on a $\frac{3}{16}$ in. mesh and the amount passing a BS 200 mesh

[1] A review of Portland cement specifications of most countries has been published by CEMBUREAU, Paris, France (1968).
[2] Loc. cit. (p. 360).

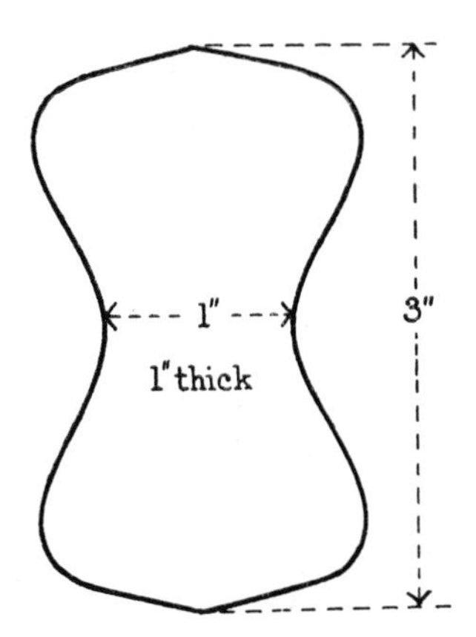

FIG. 115 Standard form of briquette for tensile tests.

must not exceed 0·5 per cent. The fine aggregate is a siliceous sand passing a $\frac{3}{16}$ in. mesh with not more than 10 per cent passing the BS 100 mesh and not more than 2 per cent passing the BS 200 mesh. For each 4 in. test cube 325 g of water and 195 g cement are mixed with such a quantity of dry aggregate, usually around 1950 g, that the finished surface of the fully-compacted concrete is level with the upper plane of the mould. The relative proportions of coarse and fine aggregate, which are determined by trial, must be such as to give a concrete with a slump of between $\frac{1}{2}$ in. and 2 in. and be reasonably workable. These requirements in effect replace the closer control of aggregate grading that is found in mortar tests. The specimens are compacted in the moulds with a tamping bar. In both the mortar and the concrete test the specimens are stored in their moulds for 24 hours at a temperature of $18{\cdot}9 \pm 1{\cdot}1°$ C ($66 \pm 2°$ F) in an atmosphere of at least 90 per cent relative humidity. The specimens are then removed from their moulds and stored in water at the same temperature until the time of test arrives. The specimens are kept in water because the rate of strength development in water and air differs. Further they must be tested immediately they are removed from the water without being allowed to dry at all, for the strength of otherwise similar specimens differs considerably between the wet and the dry states. The control of the temperature is also very important because of its influence on the rate of strength development. The minimum compressive strengths required are set out in Table 58.

The strength in the concrete test is lower than in the mortar test because the water:cement ratio is 0·6 as against 0·4.

A specification test for strength has to fulfil two requirements. Firstly, it must give reproducible results when tests are done on the same material in different laboratories, and, secondly, the results of the test should bear some reasonable relation to the strengths that will be developed in actual use. A test that placed the relative strengths of different cements in a quite different order from that obtained when they are used in concrete can obviously be misleading. Much attention has been given over many years to the problem of cement testing, and it is now widely recognised that the tensile tests are unsatisfactory on both counts

TABLE 58 British Standard: Compressive strength requirements

Type of Portland cement	Units	Mortar test			Concrete test		
		3 days	7 days	28 days	3 days	7 days	28 days
Ordinary	lb/in²	2200	3400	—	1200	2000	—
	kg/cm²	154	239	—	84	140	—
Rapid-hardening	lb/in²	3000	4000	—	1700	2500	—
	kg/cm²	210	281	—	119	175	—
Low-heat	lb/in²	1100	2000	4000	500	1000	2000
	kg/cm²	77	140	281	35	70	140
Sulphate-resisting	lb/in²	2200	3400	—	1200	2000	—
	kg/cm²	154	239	—	84	140	—

The strength in the concrete test is lower than in the mortar test because the water: cement ratio is 0·6 as against 0·4.

mentioned above. Any form of test on mortars which relies on manual compaction of the mix in the moulds also tends to be less reproducible than those in which mechanical means of compaction are used. The degree of correlation between the strength of mortars and that of concrete improves as the water: cement ratio of the test mortar is increased and for this reason contents of 10 per cent and above in 1:3 mortars, i.e. water:cement = 0·40 upwards, are better than the lower value of 8 per cent used in the earth-dry mortars. It is the recognition of these facts that has led to the changes in cement testing.

The strength results obtained under the specifications of different countries cannot be compared directly because of the differences in methods of compaction, water content and general technique.

Tests on concrete are widely used for design and control purposes in constructional work. Compression tests on concrete are carried out on cubes of, for instance, 4 or 6 in. side or on cylinders with a height equal to twice the diameter such as 6 × 3, 8 × 4 or 12 × 6 in. The strength varies with the size and shape of the test specimen, that of an 8 × 4-in. cylinder, for example, being about three-quarters of that of a 6-in. cube. This ratio of cylinder:cube strength is not constant but increases with the strength of the concrete and ranges over values from 0·7 to 0·9. The size of test specimens also influences the strength which decreases somewhat as the size increases, the difference being of the order of 10 per cent between a cube of 10 cm edge length and one of 30 cm, or between a 20 × 10 cm and a 50 × 25 cm cylinder. The rate of application of the load in the compression test also affects the strength recorded, this being higher for rapid than slow rates. Transverse tests on concrete beams are also used. Another form of test on concrete is a tensile splitting test (BS 1881:1970, ASTM 496–66) in which cylinders are loaded on their side and split longitudinally. The tensile strength measured in this way is about 60–70 per cent of the modulus of rupture measured in a bending test. The ratio of the compressive strength of concrete cubes to the modulus of rupture of beams varies with the aggregate and the age of the concrete over a range of about 6 to 10. At short ages and low strengths it is lower than at longer ages.

Accelerated strength tests

It is usual in the construction of concrete work of importance to require that test cubes be made from samples of the concrete that is being placed. Since these will not be tested until they are 7 or 28 days old there has long been a desire for some rapid testing procedure which could give within a day or so an estimate of the strength the concrete will attain at 28 days. Numerous forms of rapid test have been examined[1] and they all involve accelerating the strength development by heating, or by curing in water or steam at 50–100° C. Curing at these temperatures accelerates the development of strength of different cements in rather different degrees so that no precise relation between the accelerated strength and that obtained after 7 or 28 days normal curing can be expected. Nevertheless some such methods have been found to show a sufficient correlation to be useful even though inevitably there are some cements for which the margin of error is large.

A systematic examination of a variety of accelerated curing procedures at temperatures from 35° to 100° was made by an 'Accelerated Testing Committee' of the Institution of Civil Engineers.[2] The method recommended is to place the test cubes in their moulds, covered by a top plate, in a water bath at $55 \pm 2°$ at $\frac{1}{2}$ hr$\pm\frac{1}{4}$ after gauging and leave them there for 24 hr$\pm\frac{1}{4}$. They are then tested at $\frac{1}{2}$ hr$\pm\frac{1}{4}$ after removing from the bath. The strength obtained was found to predict the 28 day strength with a coefficient of variation of about 8·5 per cent. In other methods the accelerated test is made on the standard mortar specimen used for normal specification testing in order to speed these up rather than to estimate the strength of a concrete being placed in constructional work. Thus, in a method used in Switzerland,[3] the standard mortar prisms are cured 24 hours at 18° in air of 95 per cent relative humidity, then heated in an autoclave in one hour to 12 atmospheric pressure, maintained at that pressure for 3 hours, and then allowed to cool to about 90° C in three hours. The autoclave is then opened, and the specimens cooled to room temperature in about one hour before testing. The compressive strength is claimed to correspond generally to within ± 10 per cent of the strength at 28 days normal curing and almost certainly to within ± 20 per cent and to be satisfactorily reproducible between different laboratories.

Effect of water content

The strength of mortar or concrete with a given cement, and at a given age and temperature of curing, is dependent on the water:cement ratio (w/c) and the

[1] For reviews, see *R.I.L.E.M. Bull.*, New Series No. 31 (1966), Symposium by Correspondence; P. Smith and H. Tiede, *Highw. Res. Rec.*, No. 210, 29 (1967).
[2] *Proc. Instn civ. Engrs* **40,** 125, May 1968.
[3] E. Brandenberger, *Zement-Kalk-Gips* **12** (9), 385 (1959); *Eidgenossische Mat. prüfüngsanstalt*, Report No. 188 (1957).

degree of compaction. For a workable concrete that can be fully compacted so as to contain only a small proportion (e.g. 1 per cent) of air voids, the strength is an inverse function of the water:cement ratio and can be expressed by the relation $S = A/B^x$ where S is the compressive strengths, x the w/c ratio, and A and B are constants depending on the materials and conditions of test. This is the well-known Abrams water:cement ratio law which is in fact only another form and a limiting case of Feret's law which had been formulated much earlier. Feret's law was expressed in the form:

$$S = K\left(\frac{c}{c+e+a}\right)^2$$

where K is a constant and c, e and a are the absolute volumes of cement, water and air in the mix. It thus takes account of air voids as well as the water content of the mix. In essence, these laws state that strength is dependent on the concentration of cement in the cement paste. With air-entrained concretes (see p. 598) the volumetric ratio of (water and air)/cement is found to be roughly equivalent to the w/c ratio of ordinary concrete in its effect on strength. In practice, it is more convenient to express the w/c ratio on a weight, rather than a volume, basis, and this method is adopted in all references to water:cement ratio in this book. If the concrete strength at a given age and for given test conditions is plotted against the w/c ratio, a curve is obtained. There is an approximately linear relation between strength and the inverse ratio, c/w, which holds for values of this ratio between about 1·2, and 2·5, i.e. for w/c ratios from 0·4 to 0·8. It has been expressed by Inge Lyse in the form S = X+Y (c/w) where X and Y are constants. These various relations are essentially approximations and assume that the aggregate is sound and of adequate strength and that the concrete is matured at ordinary temperatures.

Typical data showing the influence of water content on the strength of a 1:2:4 concrete[1] are shown in Table 59.

TABLE 59

Water:cement ratio	Compressive strength (lb/in²)			
	3 days	7 days	28 days	90 days
0·5	2400	3600	5200	7000
0·6	1700	2700	4200	5500
0·7	1100	2000	3100	4300
0·8	700	1500	2500	3500

The water:cement ratio that has to be used in practice is dependent on the cement content, the aggregate size, the method of compaction of the concrete, and various other factors.

[1] The expression 1:2:4 concrete means 1 part cement, 2 parts fine aggregate, and 4 parts coarse aggregate. In this book, proportions are by weight, and the aggregates are sand and gravel, unless otherwise stated.

Effect of curing conditions

The relative rates of development of strength of concrete stored in water and in air are illustrated by the data in Table 60.

TABLE 60 Compressive strength, lb/in², 1:2:4 concrete cubes, w/c = 0·60

Cement	Days in water				Days in air			
	3	7	28	90	3	7	28	90
A	1420	2200	3880	5050	1415	2250	3490	4450
B	1640	2570	4130	5530	1700	2540	4140	4470
C	1970	2810	4130	5360	2000	2840	4020	4670
D	2490	3920	5710	6630	2560	3560	5140	6080
E	3010	4300	5900	6990	3100	4120	5520	6610

Better development of strength occurs in water storage than in air. In the case of the above data the various specimens were tested in the same condition as stored. These data include therefore not only the effect of the storage condition but also the condition of testing. If two mortar (or concrete) specimens are stored under the same conditions and tested respectively wet and dry, the strengths are found to differ. The ratio of the strength of two otherwise similar specimens when tested in the wet and dry condition is sometimes known as the wet/dry ratio. Its value varies considerably with the material and may range from 0·6 to 0·9, or even more widely. In general, the more dense and strong the concrete, the higher is the value of this ratio.

Effect of temperature

The strength developed in mortars is appreciably affected by relatively small variations in temperature, particularly at early ages, and the following data illustrate the need for strict temperature control in cement testing.

Cement	Temperature of curing °F	Compressive strength 1:3 mortar (lb/in²)	
		1 day	3 days
A	58	1220	3940
	64	1540	4230
B	58	2110	4480
	64	2360	4710

The rate of development of strength in concretes is similarly affected (Table 61) and account has to be taken of this in constructional work in cold weather.
While higher temperatures lead to an increase in strength at the earliest ages the strength at later ages starts to decrease beyond a certain temperature which

TABLE 61

	Temperature of maturing	Compressive strength (lb/in²)			
		1 day	3 days	7 days	28 days
Ordinary Portland cement	2°	—	199	800	1916
	11°	—	564	1158	2248
	17°	373	925	1472	2625
	25°	493	1096	1586	2946
	35°	620	1349	1956	3036
Rapid-hardening Portland cement	2°	50	459	1961	3846
	11°	225	1443	2469	3604
	17°	575	1951	2831	4076
	25°	1106	2221	2865	4263
	35°	1686	2698	3004	4043

Tests on 9 × 3-in. cylinders. 1:2:4 concrete, w/c 0·6.

TABLE 62

	Temperature of maturing		Compressive strength (lb/in²)			
	°F	°C	1 day	3 days	7 days	28 days
I Ordinary Portland cement	73	23	1150	2600	4060	5440
	90	32	1490	3090	4160	5240
	105	41	1850	3200	3850	4720
	120	49	2040	2840	3440	4110
II Rapid-hardening Portland cement	73	23	2210	3460	4610	5460
	90	32	2760	4290	5010	5710
	105	41	3010	3900	4580	5120
	120	49	3280	3850	4160	4720

Concrete cubes. 517 lb cement per yd³; w/c I 0·45, II 0·49.

is lower for rapid-hardening than for ordinary Portland cement. This is illustrated by the data[1] in Table 62.

The effect of curing temperature, over the range 0–40°, and age on strength, can be expressed approximately in terms of a single parameter for if the strength is plotted against a product of age and temperature the points fall roughly in a single curve.[2] To obtain this relation the base temperature is taken as −10° which is assumed to be the lowest temperature at which any strength development can

[1] P. Klieger, *Proc. Am. Concr. Inst.* **54,** 1063 (1958).
[2] S. G. Bergstrom, *Bull. Swed. Cem. Concr. Res. Inst.*, No. 27 (1953); J. M. Plowman, *Mag. Concr. Res.* **8** (22), 13; **8** (24) 169 (1956); W. Brand, *Zement-Kalk-Gips* **9,** 328 (1956) has examined the applicability of the parameter to other types of cement.

continue in a concrete already partially hardened. The strength is, therefore, plotted against (age in days) multiplied by $(t+10)$ where t is the temperature of curing in degrees centigrade, i.e. for 7 days curing at 18° the product is $7 \times 28 = 196$. Different concrete mixes and cements give different curves as is illustrated in Fig. 116. This relation can only hold over a limited range of ages and temperatures, which will vary with different cements. It does not hold if the concrete is allowed to dry, nor for a regime of varying temperatures.

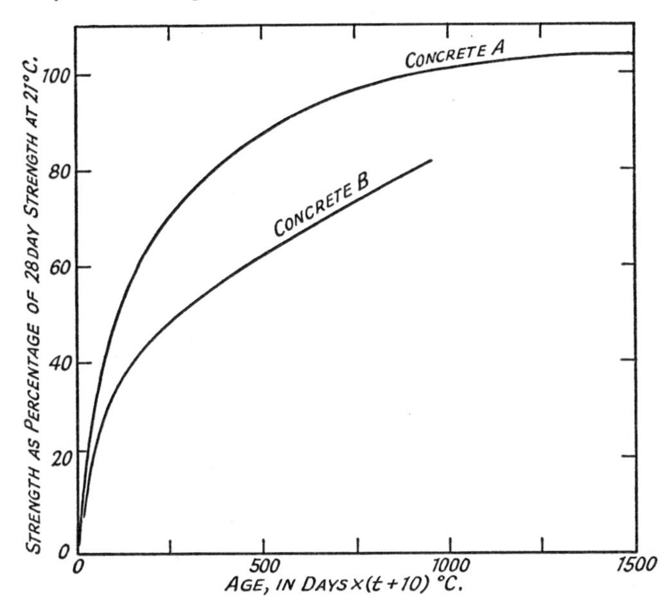

FIG. 116 Compressive strength and time-temperature of curing.

Electrical heating of concrete is sometimes used to accelerate hardening, both for concreting at low winter temperature and for precast concrete products. An alternating current is passed through the concrete, using either sheet iron plates on the surface, or any reinforcement, as electrodes.[1]

Effect of steam curing on strength of concrete

The initial rate of development of strength of mortars and concretes may be considerably accelerated by curing in steam at temperatures up to 100° and still more in high-pressure steam at temperatures above 100°. The degree to which this takes place varies with different Portland cements and the aggregate also has some effect. Thus in high-pressure steam a quartz sand will react with lime, adding to the strength, while if any pozzolanic materials, such as pulverised fuel ash are present, their reaction is accelerated in steam curing at 100° or below. With calcareous aggregates there is no additional strength gain from reaction

[1] See papers in *R.I.L.E.M. Sym. on Problems of Accelerated Hardening of Concrete, Moscow 1964.*

with the aggregate and the strength gain arises only from the increased rate of hydration of the cement and follows the maturity law discussed later. It has been claimed[1] that under autoclave conditions calcium carbonate reacts with lime to form $CaCO_3.Ca(OH)_2$ but, if so, it would not seem that it contributes to strength. There have been reports that with some marble aggregates containing coarse calcite crystals steam curing may have adverse effects because the large difference between their thermal expansion in two directions causes an abnormal expansion. A study[2] of the influence of the mineralogical composition of aggregates on the tensile strength of mortars cured in high-pressure steam at 150° has shown that minerals containing combined silica give substantially lower strength than aggregates containing free silica such as flint and quartz. This must be attributed to a reduction in the lime-silica reaction. There have been some indications that prolonged autoclaving (e.g. 96 hours) may cause some retrogression in strength attributed to various chemical changes.[3] Some investigators have also found that over a range of temperatures of steam curing up to 100° there is an optimum temperature at variously ascribed values between 70° and 90°. However this may well be due to the influence of the pre-steaming period and the rate at which the specimens are heated. Many aspects of steam curing below and above 100° are discussed in papers presented to the R.I.L.E.M. Symposium on 'Problems of Accelerated Hardening of Concrete', held in Moscow in 1964.

The general effect of steam curing at various temperatures and for various times on a 1:2:4 gravel concrete is illustrated by the data[4] in Table 63.

TABLE 63 Effect of steam curing

Initial moist curing period	Steam curing period	Compressive strength (lb/in²) for steam curing temperatures (°C)			
hours	hours	50°	65°	80°	95°
½	4	760	1020	1460	2200
	8	1710	1985	2360	2630
	16	2560	3125	3270	3485
2	4	755	1470	1745	2295
	8	1790	2550	2590	2760
	16	2480	3030	3380	3495
4	4	925	1265	1545	2200
	8	1770	2280	2480	2715
	16	2645	2985	3330	3240
8	4	1000	1720	1745	2560
	16	2695	2970	3280	3630

1:2:4 concrete, w/c 0.60. Strength moist cured 28 days at 18°C 5700 lb/in²

[1] M. D. Shvartszaid et al., *Chem. Abstr.* **57** (3), 3083 (1962).
[2] T. Thorvaldson, *Proc. Am. Concr. Inst.* **52,** 771 (1956).
[3] T. Thorvaldson, loc. cit.; *Sym., Stockholm 1938*, 261.
[4] R. W. Nurse and T. Whittaker, *R.I.L.E.M. Sym. Accelerated Hardening of Concrete, Moscow 1964.*

Similar data for air-entrained gravel concrete and lightweight aggregate concrete have been given by Hanson.[1]

The results of some tests[2] on 1 : 3 graded sand mortars subjected to steam at 100° immediately after making indicate how much its effect can vary with different cements (Table 64).

TABLE 64 Effect of steam curing at 100° on cement mortars

	In steam 100° for 16 hours; then in water at 22° until tested						In moist air at 22°					
	Compressive strength (lb/in^2)			Tensile strength (lb/in^2)			Compressive strength (lb/in^2)			Tensile strength (lb/in^2)		
Days	1	7	28	1	7	28	1	7	28	1	7	28
Cement A	2460	2330	2360	195	200	200	460	3095	4870	75	345	470
B	3215	3630	3690	300	305	355	290	1915	4435	75	290	465
C	2715	2625	2850	215	210	220	1170	3450	6880	200	500	520
D	1295	1580	1890	160	170	185	210	1560	3550	55	255	405

Steam curing results in a high immediate strength, but the subsequent gain at later ages is reduced and the ultimate strength is lower than that obtained by curing at ordinary temperatures. More favourable results tend to be obtained with lightweight than with ordinary aggregates.

While the influence of steam curing is influenced by many variables the strength can be roughly expressed as a function of the product of the temperature and duration of steam curing.[3] This can conveniently be done by expressing the strength as a percentage of the strength at 3 days at normal temperature in order to obtain a curve that is common for different mixes. The curve derived by Nurse from data from a variety of sources is shown in Fig. 117. While it applies broadly to concretes made with aggregates that are not reactive at temperatures from about 50° to 100°, individual results for different cements and aggregates may fall appreciably above or below the curve. With many lightweight aggregates which react with lime to some extent during steam curing, the strength obtained may be much above the curve. The addition of ground silica sand was similarly found by Menzel[4] to have a favourable influence.

Steam curing is much used in the manufacture of precast concrete blocks in order to speed up the manufacturing process.[5] The usual procedure is to allow the blocks to harden for a few hours before starting the steam curing and then to steam cure for some 10–15 hours at 65–80°. The rate of temperature rise of the

[1] J. A. Hansen, *Proc. Am. Concr. Inst.* **60,** 75 (1963); **62,** 661 (1965).
[2] P. H. Bates and R. L. Blaine, *Proc. Am. Concr. Inst.* **28,** 531 (1932).
[3] R. W. Nurse, *Mag. Concr. Res.*, No. 2, 79 (1949); R. W. Nurse and T. Whittaker, loc. cit.; A. G. A. Saul, *Mag. Concr. Res.*, No. 6, 127 (1951).
[4] C. A. Menzel, *Proc. Am. Concr. Inst.* **31,** 125 (1935); **32,** 51, 621 (1936).
[5] See Recommended practice for atmospheric pressure steam curing of cements, *Proc. Am. Concr. Inst.* **66,** 629 (1969) which includes a review of the literature on steam curing.

blocks must be restricted to not more than 20–30° per hour to avoid any damage from thermal shock. By this means a strength equal to about 60 per cent of the ordinary moist-cured 28-day strength is obtained in 24 hours after making. The initial drying shrinkage may be somewhat reduced by steam curing perhaps because the hot blocks lose moisture and undergo part of the irreversible shrinkage when removed from the steam chambers. The reversible moisture movement is not changed.

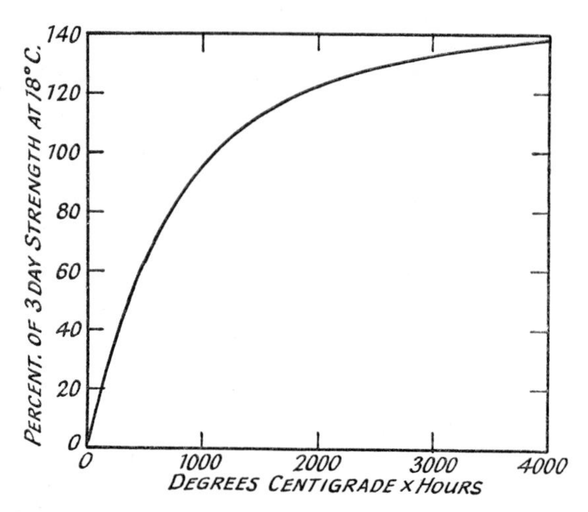

FIG. 117 Strength and time-temperature of steam curing.

Curing under high-pressure steam results in the development of higher strengths which may in some cases exceed those obtained at 28 days by normal curing. The effects vary with the aggregate, with the cement and the addition of ground silica, with the period and temperature of autoclaving and other factors. Though data have been published in various countries there is still, on account of the number of variables involved and general differences in commercial technology, a lack of systematic comparative data. It is common to use the steam gauge pressure and not the absolute pressure in defining the autoclaving conditions. The latter is 1 atm or 14·7 lb/m^2 (1·03 kg/cm^2) higher. The relation between gauge pressure and temperature for saturated steam is:

Temperature °C	120	140	160	170	180	190	200
°F	248	284	320	338	356	374	392
Steam gauge pressure							
lb/in^2	14	38	75	100	130	167	210
kg/cm^2	1	2·7	5·3	7·0	9·1	11·7	14·8

The actual temperature-pressure relation in an autoclave may differ slightly from these data owing to the presence of residual air.

Some results[1] on 1:1·9:2·8 concrete mixes are shown in Table 65. The specimens were stored for 24 hours in moist air before steam curing.

[1] J. C. Pearson and E. M. Brickett, *Proc. Am. Concr. Inst.* **28,** 537 (1932).

TABLE 65 Effect of steam curing on concrete: Compressive strength (lb/in^2)

Steam pressure (lb/in^2)	68		103		197		Cured moist air (days)	
Period of steam curing (hours)	18	42	18	42	18	42	7	28
Cement A	2660	5480	5350	7140	6970	7600	5280	6340
B	3330	5040	5120	5470	6000	7030	5480	6200
C	3550	5470	4760	6670	5730	6790	2740	5080
D	3460	5760	5580	7160	6110	6760	2560	5080

TABLE 66 Effect of steam curing for six hours

Steam pressure (lb/in^2)	21	43	85	170
Temperature (°F)	252	282	316	360
Compressive strength as percentage of 28-day moist-cured strength	67	71	89	106

1:5:6 Weight Concrete. w/c 0.45. Gravel aggregate.
Rapid-hardening Portland cement.

Some other data[1] given in Table 66 illustrate the effect of high-pressure curing for a period of six hours at maximum pressure, which corresponds more closely to the conditions in commercial practice.

The influence of the aggregate is illustrated by some data[2] in Table 67 for concretes autoclaved for nine hours at a steam pressure of 120 lb/in^2. The cement was composed of 60 per cent Portland cement, 40 per cent ground silica (75–25 with the lightweight aggregate) and the content was about 620 lb/yd^3 (about 1:5 cement:aggregate ratio).

It is not only additions such as ground silica, which can contribute to strength by their reaction with calcium hydroxide liberated from the cement, but also in some cases the finest fraction of the aggregate. This is illustrated by the data[3] in Table 68 on the strength of lime mixes autoclaved for 6 or 16 hours.

The contribution to strength made by the fine particles depends essentially on the readiness with which the silica contained in them can react with lime. In the case of slag which already contains some 40 per cent CaO the reaction is a direct hydration which is stimulated by a small addition of lime.

In general not much extra gain in strength is obtained by curing at gauge pressures above 150 lb/in^2 for periods of autoclaving, at the maximum pressure, of more than about 8 to 12 hours; at higher pressures, strengths sufficiently close

[1] L. Blondiau, *Revue Matér. Constr. Trav. publ.* **571,** 125; **572,** 163; **573**, 204 (1963).
[2] C. A. Menzel, *Proc. Am. Concr. Inst.* **32,** 621 (1935).
[3] H. G. Midgley and S. K. Chopra, *Mag. Concr. Res.* **12** (35), 73 (1960).

TABLE 67 Effect of steam curing on concrete with different aggregates

Aggregate	Compressive strength (lb/in^2)	
	Steam-cured Tested at 7 days	Moist-cured* 10 months
Gravel-sand	6250	6730
Granite	6230	6630
Marble	4870	6920
Light weight expanded clay	4270	3750

* Cement 100 per cent Portland.

TABLE 68 Influence of additions and aggregate fines in steam-cured lime mixes

Aggregate (through 300 mesh)	Mix proportions (weight)		Compressive strength (lb/in^2) on autoclaving at 160 lb/in^2	
	Lime	Fines	6 hours	16 hours
Pulverised fuel ash	10	90	2760	2430
	20	80	4640	4370
Quartz sand	10	90	6500	4640
(only 47 per cent through 200 mesh)	20	80	12180	9310
Expanded shale	10	90	6330	6210
	20	80	8290	9590
Foamed slag	10	90	5465	3040
	20	80	2630	1870

to the maximum are obtained in 6 to 8 hours. Autoclaving at pressures below 100 lb/in^2 does not develop such high strengths except for extended steaming periods. These figures can only be rough guides for they are influenced by the cement, the aggregate, the concrete mix and by additions such as ground silica. An initial hardening period of a few hours is necessary before commencement of autoclaving and the rate of heating has to be limited to avoid any disruption of the concrete.

High-pressure steam curing not only gives an immediate high strength, but it also reduces the drying shrinkage of concrete to about a half of the value for normally cured material and it substantially increases the resistance to sulphate attack. The strength of the bond to reinforcement is however reduced. Autoclave treatment is used to a considerable extent in North America[1] in the manufacture of precast concrete blocks, though rarely in Western European countries. Much

[1] *Proc. Am. Concr. Inst.* **62,** 869 (1965), Report of Committee on 'High Pressure Steam Curing'.

STRUCTURE OF β $2CaO.SiO_2$ SHOWING OUTLINE OF THE UNIT CELL. Si ATOMS BLACK, Ca ATOMS LIGHT GREY, O ATOMS DARK GREY

(i) Portland blastfurnace cement concrete.
Specimens immersed in 5 per cent $MgSO_4$ solution at 28 days old
Age when photographed, 2 years.

(ii) Normal Portland cement concrete.
Specimens immersed in 5 per cent $MgSO_4$ solution at 28 days old
Age when photographed, 2 years.

use of it is made in the U.S.S.R.[1] for plain and reinforced concrete units. In commercial practice in the U.S.A. it is common to hold the blocks for about two hours after moulding and then to raise the steam pressure gradually over a period of about 3 hours. Steam gauge pressures of 120 lb/in^2 (350° F) with 8 hours at the maximum pressure, or 150 lb (365° F) and 7 hours, are recommended.[2] Pressures as low as 80 lb (325° F) with longer autoclaving periods are sometimes used. Concretes both with ordinary and lightweight aggregates are cured in this way and an addition of ground silica, pulverised fuel ash or natural pozzolana may be made to the mix. The addition may vary from 20 to 40 per cent by weight of the mix of Portland cement and addition according to the cement and aggregate. Though such additions can increase strength they also tend to increase shrinkage. Curing in high-pressure steam also forms part of the process of manufacture of aerated (cellular) concrete and sand-lime bricks, and of some asbestos cement products.

Non-destructive test methods

The normal methods of testing the strength of concrete involve the destruction of the specimen. Methods which permit of repeated tests on the same specimen, at different ages, for example, are known as non-destructive test methods.[3]

The modulus of elasticity bears a fair relation to the strength of concrete and can be measured in a number of ways. In the static methods a load is applied to a concrete prism and the resulting strain measured, or, more conveniently, the deflection of a beam under a load is measured and the value of E calculated from the appropriate bending equation. For many purposes the dynamic method which depends on the determination of the natural resonant frequency of vibration, either longitudinal or transverse, of a beam, is more convenient and rapid. The specimen is usually set into vibration by electrical means and the method is often known as the 'sonic' method. Under certain conditions the modulus of elasticity can be calculated directly from the natural frequency of longitudinal vibration (BS 1881:1970). For transverse vibrations the value of Poisson's ratio is also required in order to calculate E (ASTM C215–60).

Another dynamic method (ASTM 597–68T) that can be applied to specimens of more varied shape, and to concrete *in situ*, is the so-called supersonic method, in which short-period pulses of high-frequency vibrations are sent into the concrete and the time taken for a pulse to traverse a known length measured. The velocity of the ultrasonic wave can thus be calculated and from this, combined with the value of the density and Poisson's ratio, the value of E can be derived. Where prisms can conveniently be used as test specimens, the 'sonic' method is to be preferred to this, but for cubes or observations on large specimens, or on concrete *in situ*, the supersonic method is the only practicable one. It has proved

[1] S. A. Mirinov and L. A. Malinina, *Autoclaved Cured Concrete*, State Publishing Office of Literature of Structural Engineering, Moscow 1958. (U.S. Portland Cement Association translation); G. L. Kalousek, *Proc. Am. Concr. Inst.* **63,** 817 (1966).

[2] C. A. Menzel, *Concr. Prod.* **63** (5), 24 (1966).

[3] See R. Jones, *Non-destructive Testing of Concrete*, Cambridge University Press (1962); E. A. Whitehurst, *Spec. tech. Publ. Am. Soc. Test. Mater.*, No. 169A (1966); *Monogr. Am. Concr. Inst.*, No. 2 (1966).

useful for the detection of deteriorated portions of concrete structures, the poorer the quality of the concrete the lower being the wave velocity. Since the wave velocity is influenced not only by the strength of the concrete, but also by the cement and moisture contents and other factors, care is necessary in the interpretation of the results.

With the ready availability of radio-active sources, γ rays are now used for the study of concrete and in particular of variations in density from point to point and for the detection of faults. The attenuation of the rays after passing through a known thickness of material is influenced by both these factors. The moisture content of concrete can be measured by neutron scattering.

Another very simple, but rough, method of assessing the quality of concrete depends on measuring the surface hardness. One form[1] of apparatus is based on the same principle as the Brinell hardness test for metals in which the diameter of the indentation produced when a hard steel ball is pressed into the surface under standardised conditions is measured. On concrete this form of test is apt to give very variable results, particularly when larger pieces of aggregate are close to the surface at the point at which the test is done. Another form of this test, using a simple instrument known as a sclerometer, depends not on measuring an indentation, but on measuring the rebound when a metal rod with a rounded head is projected on to the concrete by a calibrated spring under standardised conditions. There is a general relation between the rebound and the strength of the concrete, but the dispersion of the results is rather large.[2] The test is, nevertheless, very easy to carry out and a considerable number of measurements can be made in a short time.

PERMEABILITY AND ABSORPTION

The permeability of a mortar or concrete is a measure of the rate at which a liquid will pass through it, while the absorption is a measure of the pore space in the material. The two are not necessarily related and the values obtained for them vary considerably with the conditions of measurement.

The absorption is usually determined by drying a specimen to constant weight, immersing in water and noting the increase in weight.[3] The penetration of water into the specimen may be assisted by boiling, or by evacuating it before flooding with water. Drying is often carried out at 100–110°, but this temperature is sometimes considered too high. At a lower temperature such as 50°, effective drying can be obtained, but the time required is much longer. Drying *in vacuo* over sulphuric acid at ordinary temperatures results in a smaller loss of water during drying and a consequent smaller absorption on subsequent immersion in water. It may variously be argued that drying at ordinary temperatures is in-

[1] K. Gaede, Deutscher Auschuss für Stahlbeton, Report No. 107 (1952); German Standard DIN 4240.
[2] E. Schmidt, *Schweizer Arch. angew. Wiss. Tech.* **17,** 139 (1951); E. A. Whitehurst, *Spec. tech. Publ. Am. Soc. Test. Mater.*, No. 169A, 176 (1966); J. Kolek, *Mag. Concr. Res.* **10** (28), 27 (1958); *Concrete* **2,** 402 (1968).
[3] One form of this test is described in BS 1881:1970.

effective because of the density of the material, or that the larger losses on drying at higher temperatures represent water combined in the set cement. This undoubtedly occurs to some extent. A comparison of absorption values obtained in some different ways is given below.

	Concrete	A	B	C	D
Dried 100°	1. Immersed in water for ½ hour	4·7	3·2	8·9	12·3
	2. ,, ,, ,, ,, 24 hours	7·4	6·9	9·1	12·9
	3. ,, ,, ,, ,, 48 hours	7·5	7·0	9·2	13·1
	4. As 3, plus 5 hours' boiling	8·1	7·3	14·1	18·2
Dried 65°	Boiled 5 hours	6·4	6·4	13·2	17·2

Concrete	E			F		
Period of immersion in water of previously evacuated specimen	1 hour	24 hours	7 days	1 hour	24 hours	7 days
Dried 105° to constant weight	3·0	3·4	3·5	7·4	7·7	7·8
Dried at 20° *in vacuo* over lime for 30 days	1·9	2·2	2·3	5·9	6·3	6·4

The absorption of concretes dried at 100° and immersed in water for 48 hours may fall as low as 2 per cent by weight in exceptional cases and rise to 15–20 per cent. Most good concretes will fall below 10 per cent. In certain types of concrete which are purposely made with aggregates so graded as to leave a large proportion of voids, with the object of obtaining a light material, the absorption may be much higher. For comparable types of concrete with similar aggregates the quality tends to decrease as the absorption increases. This is probably particularly the case where the concrete has to resist the action of aggressive waters.

The permeability of a mortar or concrete to the flow of water may be determined under a small or large head.[1] In many cases a fairly high water pressure, up to 200 lb/in^2 or more, is used, and the rate at which water passes through the specimen determined. The permeability is affected by many factors, but in general increases as the proportion of cement is decreased or the water:cement ratio increased, except for too dry mixes which may show a high permeability. It is influenced somewhat by the grading of the aggregate and very considerably by age and curing conditions. Concrete cured in water becomes more impermeable than concrete cured in air.

Tests of permeability at high pressures are most commonly made on concrete cylinders or circular slabs in an apparatus which permits of water under the pressure required being fed to the upper side of the specimen and the water percolating through being collected from the underside and measured. Alternatively, the rate of inflow of water may be measured. The specimens are usually sealed into metal containers with suitable watertight caps. The rate of flow is

[1] Cf. W. H. Glanville, *Tech. Pap. Bldg Res. D.S.I.R.*, No. 3, 1931 (revised edition); A. Ruettgers, E. N. Vidal and S. P. Wing, *Proc. Am. Concr. Inst.* **31,** 382 (1935); H. K. Cook, *Proc. Am. Soc. Test. Mater.* **51,** 1156 (1951); *U.S. Bureau of Reclamation*, Concrete Manual, p. 39 (1966); I. L. Tyler and B. Erlin, *J. Res. Dev. Labs Portld Cem. Ass.* **3** (3), 2 (1961).

more rapid at the start, but decreases to a steady value after one or several days depending on the dimensions of the specimen and other factors. The form of the flow rate-time curve is very similar to that shown in Fig. 119 for low-pressure tests.

Tests may also be carried out under small pressures such as a 20-cm head of water or less. This form of test is often used in testing surface films of waterproofers which may be intended for use under conditions where a mortar or concrete will only be subject to small water pressures. It is also useful for testing thin concrete products; the results in general run parallel with those obtained by a high-pressure test. A test of the permeability under a 20-cm head of water is included in the British Standard for concrete roofing tiles (BS 473, 550:1967; also BS 1881:1970). An apparatus for carrying out low-pressure permeability tests is shown in Fig. 118. The specimen, of suitable size, is dried to constant

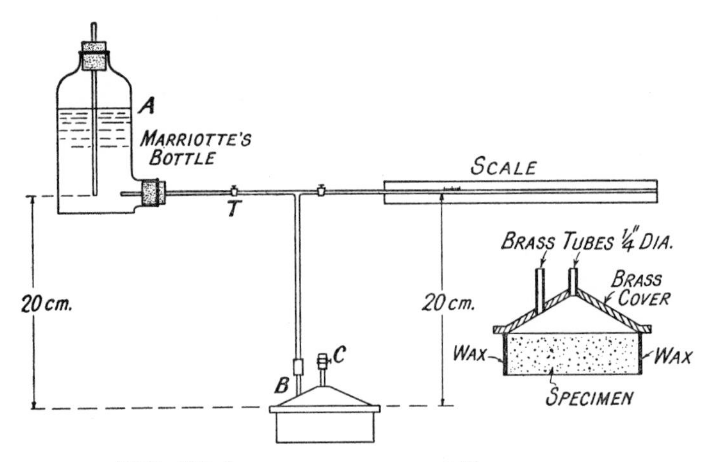

FIG. 118 Low-pressure permeability apparatus.

weight at 50–100° and waxed to a brass cover with Faraday wax; the sides of the specimen are also covered with wax. The cover is fitted with two tubes, *B* and *C*. The space between the specimen and the cover is filled with water by allowing the water to flow in through *B* and out through *C* which is closed when the space is full. Water is supplied from the Marriotte's bottle, *A*, which maintains a constant head of water. Measurements of the rate of penetration of water are made at intervals by cutting off the main supply at *T* and measuring the rate of movement of the meniscus in the calibrated capillary tube. The head of water is shown as 20 cm, but this can easily be varied if desired. Blocks $4 \times 4 \times 2$ in. are convenient as specimens, but either the size or thickness can be varied.

The rate of flow of water into the specimen is at first determined not only by the applied head, but also by the capillary forces in the material. These capillary forces cease to operate once water has passed through the specimen and wetted the undersurface. The rate of flow of water into the specimen therefore decreases fairly rapidly until a steady value is reached. This represents the flow-through rate when water has entirely penetrated the specimen to the undersurface. This steady state may be reached in minutes or hours, depending on the thickness and permeability of the specimen. A typical curve showing the initial rapid absorption of water into a concrete specimen 2 in. thick, and the final steady state, is shown in Fig. 119. The final flow-through rate, or permeability, is about 1×10^{-6}

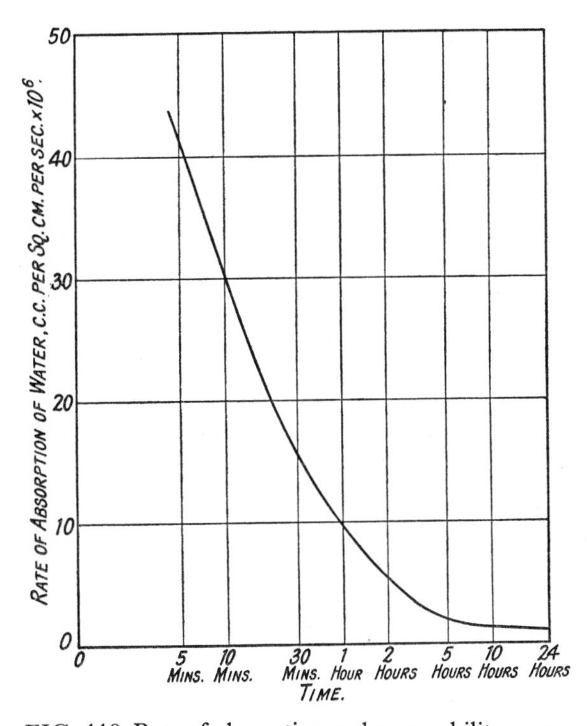

FIG. 119 Rate of absorption and permeability curve.

cc/cm^2/s. If the test is continued over a long period some further slow change in the flow rate may occur owing to silting up of pores, further hydration of the cement, and other factors.

SMALL VOLUME CHANGES IN SET CEMENT PRODUCTS

All cement products, neat cement, mortar, or concrete, are subject to small changes in their volume with changes in moisture conditions. Though small, these changes are very important, since they are considerably greater than the maximum elastic deformation a concrete suffers when subjected to the normal constructional stresses. A concrete under load also undergoes a slow deformation beyond the initial elastic movement which occurs on application of the load. This further movement takes place at a decreasing rate for a long period and does not reach its limit for some years. It is usually termed 'creep' or 'flow under load'. It is analogous to the creep which occurs in metals stressed beyond their elastic limit, but differs from it in that it occurs under any sustained loads, however small. Cement products also undergo a shrinkage when subjected to atmospheric carbonation. This is of little importance in determining the movement of a concrete where carbonation can only very slowly penetrate beyond the surface, but is important in connection with the more porous concrete blocks and with surface crazing. The subject of carbonation shrinkage and crazing falls more conveniently into a later chapter; the remaining factors leading to small volume changes, including the ordinary thermal expansion, are very briefly discussed here.

Effect of moisture conditions

We have seen in Chapter 10 that a set cement undergoes a shrinkage when first dried, and that on subsequent wetting and drying an approximately reversible expansion and contraction occurs.[1] On long storage in water a set cement undergoes a further expansion due to a slow hydration of unhydrated cement contained in it. When short periods only of water storage are considered the initial shrinkage of the set cement exceeds any subsequent expansion in water, but on more prolonged storage the further hydration of the cement may cause the initial volume to be exceeded. The various changes are shown in an exaggerated diagrammatic form in Fig. 120.

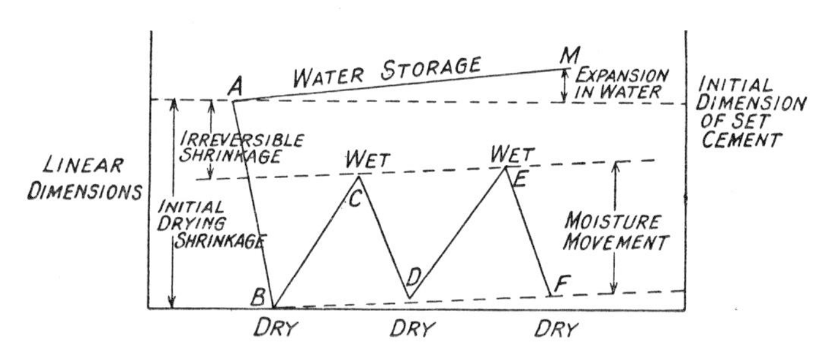

FIG. 120 Diagrammatic representation of changes in dimensions of concrete with moisture conditions.

A concrete continuously stored in water increases in dimensions as represented by the line *AM*, *A* representing the initial state. A similar concrete which is dried shrinks to *B*, and if immersed in water again expands to *C*. On subsequent drying and wetting it follows the course represented by *CDEF*. The movement between *B*, *D*, *F*, and *C*, *E*, we shall term the moisture movement. On successive repetition of the dry-wet cycle the moisture movement may tend to diminish slightly in good concretes, though for poor ones it may increase. At the same time the length may tend to increase slightly as further hydration occurs during the periods in water. Whether this takes place or not depends very much on the

[1] For review of shrinkage see *R.I.L.E.M. Sym. Shrinkage of Hydraulic Concretes, Madrid 1968.*

relative duration of the wetting and drying periods, and the reverse effect may occur. The expansion from B to C, or D to E, usually tends towards a limit in 3–7 days, the denser the concrete the longer being the time required; the slow further expansion which will take place if the water storage is prolonged for weeks and months may legitimately be assigned to the effect of further hydration.

The simplest method of measurement of the small changes in length occurring in mortar and concrete specimens is a screw-gauge micrometer or dial gauge reading to 0·0001 in., erected in a stand of fixed dimensions. An Invar steel rod is used as a standard of length for checking the readings of the gauges and enabling corrections to be made for any changes in the dimensions of the apparatus.[1] Alternatively any of the various forms of extensometers and other strain gauges can be used.

The initial drying shrinkage varies with the method of drying. Materials stored in air shrink progressively further as the humidity of the air is decreased; a still greater shrinkage is obtained by drying in an oven at 50–100°. Some typical curves[2] for the contraction in air and expansion in water of a dense concrete, in specimens of 3 x 3-in. cross-section, are shown in Fig. 121.

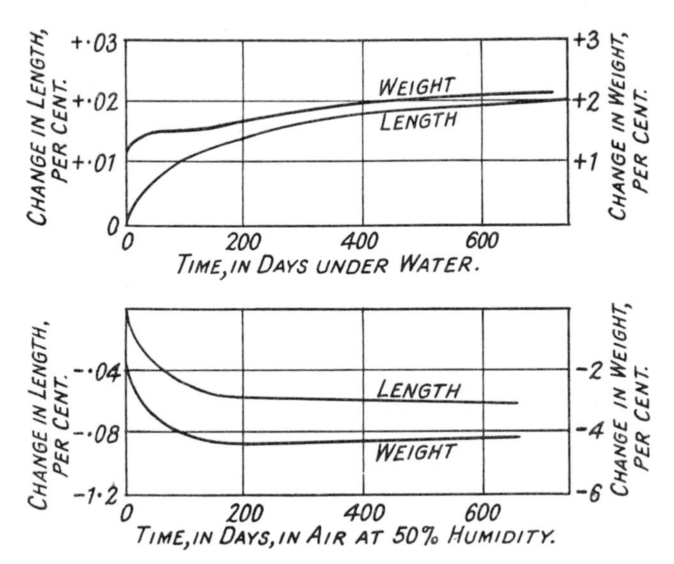

FIG. 121 Changes in length and weight of 1:5 crushed-granite concrete.

Some typical values for the drying shrinkage of mortar and concretes of relatively thin section (5 in. square) exposed for six months to air of 50 per cent relative humidity at 70° F, are shown in Table 69.

For a relative humidity of 65–70 per cent the values would be about three-quarters of those shown, while on prolonged exposure at 50 per cent relative humidity they would be increased by about one-third. With specimens of smaller

[1] Such an apparatus is described in BS 1881: 1970.

[2] R. E. Davis and G. E. Troxell, *Proc. Am. Concr. Inst.* **25,** 201 (1929).

TABLE 69

Concrete mix weight proportions	Drying shrinkage, per cent			
	Six months in air at 70° F and 50 per cent R.H.			
	Water: cement ratio			
	0·4	0·5	0·6	0·7
1:3	0·08	0·12	—	—
1:4	0·055	0·085	0·105	—
1:5	0·04	0·06	0·075	0·085
1:6	0·03	0·04	0·055	0·065
1:7	0·02	0·03	0·04	0·05

cross-sectional area, e.g. 3 × 3 in., this maximum shrinkage is obtained in about six months. These values must only be taken as indicative of the order of the movements, for the drying shrinkage also varies with the cement and the aggregate. Since shrinkage is caused by loss of moisture it is dependent on the rate of diffusion of water or water vapour from the interior to the surface. This rate decreases rapidly with distance from the surface and it has been calculated,[1] for example, that under conditions where a point 1 in. below the surface would lose 50 per cent of its evaporable water in one month, a point 7 in. below the surface would take 10 years to reach the same state. Shrinkage is, therefore, very much slower for units of large cross-sectional dimensions,[2] and its total amount somewhat reduced owing to internal restraints arising from the differential movements caused by the moisture gradient. This is illustrated by the curves in Fig. 122 for a 1:2:4 concrete exposed to air at 64° F and 65 per cent relative humidity. The shrinkage of concrete increases with, and is determined primarily by, the water content per unit volume of the mix and only by the

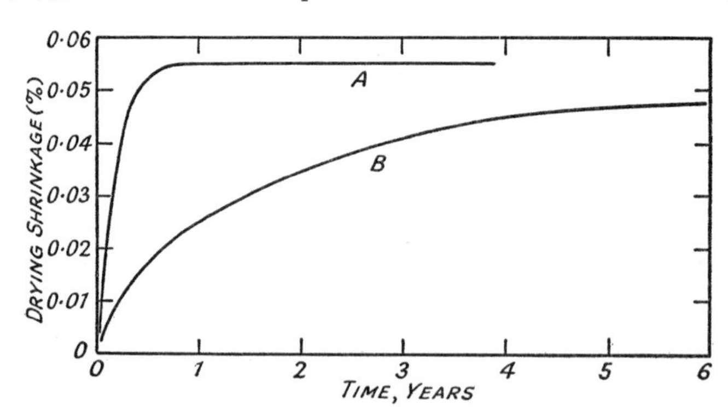

FIG. 122 Effect of size on drying shrinkage. A: specimen 36 × 4 × 4 in. B: specimen 9 ft × 30 in. × 20 in.

[1] R. W. Carlson, *Proc. Am. Concr. Inst.* **33,** 327 (1937); G. Pickett, ibid. **42,** 361 (1946); F. S. Rostasy, *Zement-Kalk-Gips* **13,** 93 (1960).
[2] See T. C. Hansen and A. H. Mattock, *Proc. Am. Concr. Inst.* **63,** 267 (1966).

cement content in so far as it affects the water content required in mixing. The comparative drying shrinkage of concrete, mortar and neat cement is shown[1] in Fig. 123.

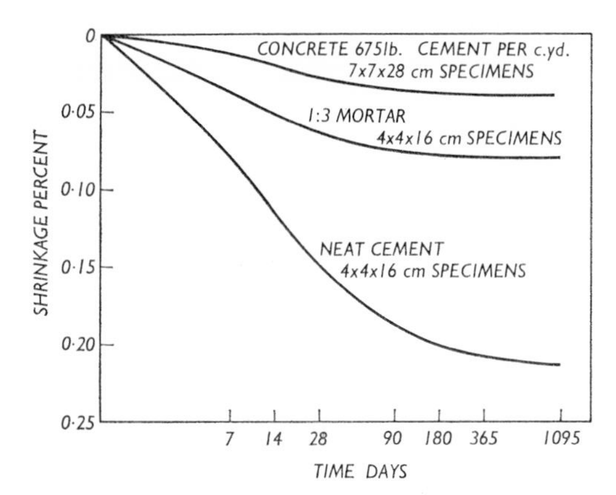

FIG. 123 Comparative drying shrinkage of concrete, mortar and neat cement at 50 per cent relative humidity.

In the shrinkage tests required in specifications for precast concrete blocks (BS 2028, 1364:1968), which must be fairly quick to carry out, the specimens, after 4–7 days' immersion in water are dried in an oven at 55–60° C and 17–20 per cent relative humidity controlled by circulating through the oven air saturated with water vapour at 27° C. In previous editions of the specification the temperature was 50° and the humidity controlled at 17 per cent relative humidity by placing saturated calcium chloride solution in trays in the oven. The new form of test leads to slightly higher shrinkage values so in the 1968 edition the permissible limits were slightly raised (see p. 581). A similar test method, varying in some details, is given in ASTM C426–66T. The accelerated shrinkage at 50° and 17 per cent relative humidity has been found to correlate[2] well with the longer period shrinkage in air at 73° F and 50 per cent relative humidity for lightweight aggregate concretes, but not for aerated cellular concrete.[3]

The expansion of concrete on prolonged storage in water is usually of the order of 0·01–0·02 per cent.

The total drying shrinkage of neat cement is rather variable and values up to 0·4 per cent on oven drying have been reported. Neat cement specimens stored in water may expand up to 0·1 per cent in three months. The rate at which these movements occur again depends on the thickness of the specimen as is illustrated by the following data:

Thickness of specimen (in.)	$\frac{1}{8}$	$\frac{1}{2}$	2
Expansion in 3 months in water (per cent)	0·11	0·065	0·017
Shrinkage in 3 months in air (per cent)	0·16	0·13	0·07

[1] M. Venuat, *Revue Matér. Constr. Trav. publ.* **538/539,** 169; **540,** 215; **541,** 251 (1960).
[2] J. O. Bryson and D. Watstein, *Proc. Am. Concr. Inst.* **58,** 103 (1961); R. C. Valore and W. H. Kuenning, ibid. **59,** 1391 (1962).
[3] W. Kinniburgh, *R.I.L.E.M., Sym. Lightweight Concrete, Gothenburg 1960.*

The shrinkage of different Portland cements varies by as much as 50 per cent, and extreme cases can be found of cements differing by a factor approaching two. Shrinkage is increased somewhat by air-entrainment. The type of aggregate also has an influence, but no general rules can be given except that the more porous sandstones tend to give the higher values. The cement shrinkage is influenced, as we have seen elsewhere, by the composition and gypsum content.[1] Fineness of the cement has relatively little effect on the shrinkage of concretes, but that of mortars tends to increase somewhat with fineness. It has, however, an indirect influence on the extensibility of concrete and its susceptibility to shrinkage cracking because as the rate of strength development of a concrete increases the more prone it is to be subject to high stresses arising from drying shrinkage in its early life. Increasing the fineness of a cement raises its rate of strength development and decreases the ability of a concrete to deform by creep and adjust itself to shrinkage stresses without cracking. This may be illustrated by the curves in Fig. 124 which show the tensile stress developed in concrete specimens held at

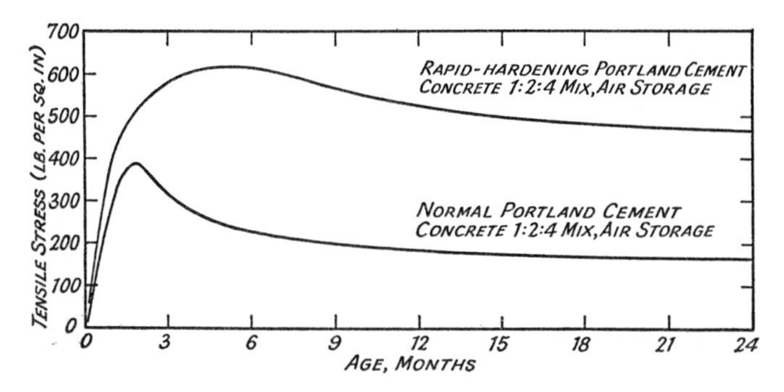

FIG. 124 Stress development in completely restrained concrete members due to shrinkage.

constant length while exposed to drying in air. The stress developed is greater in the specimen made with the more finely ground and more rapid-hardening cement. It is fairly common experience in practice that concretes made with coarse ground cements usually show less shrinkage cracking than those made with more finely ground cements *of the same type*. Any wider application of this generalisation can be very misleading, since rate of strength development is also a function of the cement composition, while shrinkage cracking is also dependent on a number of other factors.

The moisture movement, as measured by the expansion of specimens on immersion in water for 7 days, after previous drying at 50°, is usually about two-thirds of the initial drying shrinkage. Typical values for materials of good quality are shown below. Mortars or concretes of poorer quality may show considerably larger values.

[1] G. Pickett, *Proc. Am. Concr. Inst.* **44,** 149 (1948).

Moisture movement (per cent)	
Neat cement	0·10
1:1 mortar	0·04
1:2 ,,	0·03
1:3 ,,	0·02
1:2:4 concrete	0·03

Creep of concrete

The creep of concrete[1] has been intensively studied (see p. 286), but any adequate consideration of it lies outside the scope of this book. Creep is the continuous deformation which occurs in concrete under a sustained load. A typical curve is shown in Fig. 125 for a concrete in compression under a load of 600 lb/in^2. The creep is seen to occur rapidly in the first few weeks after loading and then to proceed at a steadily diminishing rate. The creep is directly proportional to the applied stress up to about half the ultimate strength and this holds down to the smallest loads for which measurements are practicable. The earlier the age at which the concrete is loaded, the greater is the creep obtained. It is influenced by the various factors of cement content, water-cement ratio, aggregate, etc., which influence the strength of concrete, and in general diminishes as the strength increases. Creep is also influenced by the type of aggregate, tending to be lower for limestone and higher with sandstone than with gravel aggregates.

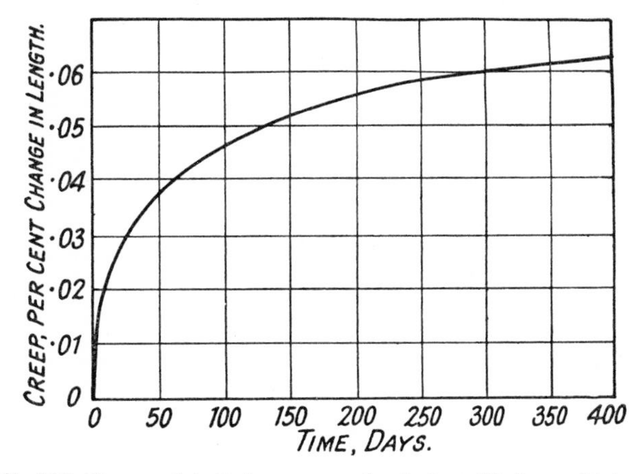

FIG. 125 Creep of 1:2:4 concrete loaded at 28 days old in air.

The effect of creep is to cause a reduction in stress in a concrete under a constraint. This action can be seen from Fig. 124 in which the stress in the restrained specimen undergoing drying builds up to a maximum and then decreases again as the effects of creep start to outweigh those of drying. For structural design purposes the effect of creep is conveniently regarded as equivalent to a reduction in the modulus of elasticity.

[1] See R. E. Philleo, *Spec. tech. Publ. Am. Soc. Test. Mater.*, No. 169A, 160 (1966).

The creep of Portland cement concretes is greater in air than in water, and is increased when a specimen is shrinking by drying at the same time as it is undergoing creep under load. With high-alumina cement the creep is less in air than in water, but no adequate explanation of this is known.

Thermal expansion of concretes

The thermal expansion of concretes varies only a little with the proportion of aggregate present, increasing slightly with the richness of the mix, but it varies much with the type of aggregate.[1] Dry and wet specimens show about the same value, but at an intermediate condition of dryness the value is some 20 per cent higher. Typical values for materials in the wet state are:

Neat cement	8×10^{-6} per °F
Gravel concrete	7×10^{-6} per °F
Granite concrete	5×10^{-6} per °F
Limestone concrete	3×10^{-6} per °F

There is a tendency for the coefficient to decrease slightly at long ages. For large structures where considerable restraint in movement exists a value about one-half the above may be used as, for example, in estimating the opening of contraction joints caused by the cooling of a large concrete mass subsequent to setting and hardening.

Thermal conductivity. The thermal conductivity of concrete varies with the density, the type of aggregate, and the moisture content. For wet, dense concretes, weighing from 140 to 160 lb per cubic foot, the values range from about 10 to 25 British thermal units per hour per square foot per inch of thickness per degree Fahrenheit of temperature change[2] (Btu/h/ft^2/in./°F). The corresponding values in c.g.s. units (cal/s/cm^2/cm/°C) are obtained by dividing by 2903.Typical values for a 1:2:4 gravel concrete are 15–16 wet and 12 air-dry. Glassy aggregates give lower values than crystalline ones, e.g. whinstone and vitreous air-cooled slag tend to give concretes having conductivities at the lower end, and quartzite at the top end, of the range given above. The specific heat varies from about 0·20 to 0·25 Btu/lb/°F or cal/g/°C. The diffusivity, which is the thermal conductivity divided by the heat capacity per unit volume, is important in calculations of heat flow. It is designated as h^2 and has the dimensions ft^2/h, or cm^2/s in c.g.s. units, values in the British units being converted to the c.g.s. units by dividing by 3·87. Values of diffusivity for dense concretes range from about 0·02 to 0·06 ft^2/h.

Modulus of elasticity. The value of Young's modulus is roughly proportional to the compressive strength of a concrete and for normal concretes falls within a range of $3–8 \times 10^6$ lb/in^2. The value of Poisson's ratio—the ratio of lateral to

[1] D. G. R. Bonnell and F. C. Harper, *Natn. Bldg. Stud. tech. Pap.*, No. 7 (1951); S. L. Meyers, *Proc. Highw. Res. Bd*, **30**, 193 (1950); G. E. Monfore and A. E. Lentz, *J. Res. Dev. Labs Portld Cem. Ass.* **4** (2), 33 (1962); H. Dettling, Deutscher Auschuss fur Stahlbeton, Report No. 164, Part 2, 1 (1964).

[2] *6th int. Congr. large Dams, New York 1958*, Report C24; A. E. Lentz and G. E. Monfore, *J. Res. Dev. Labs Portld Cem. Ass.* **7** (2), 39 (1965); **8** (3), 27 (1966); H. W. Brewer, ibid. **9** (1), 48 (1967).

longitudinal deformation—is commonly assumed to be one-sixth, but actually it varies from about 0·16 to 0·25 or even higher.

Small-scale testing. Apparatus for measuring the setting time, soundness and strength of cements by small-scale test methods has been devised by Parker, Kühl and others.[1] Compressive strength tests have been carried out on mortar cubes down to $\frac{1}{2}$-inch size and bending tests on $1 \times 1 \times 3$ cm bars. For the former the sand was graded between the BS 52 and 72 mesh. A small Le Chatelier cylindrical mould, 15 mm diameter by 15 mm depth, serves for the soundness test, and one of similar dimensions for setting time. Parker has correlated the results with those obtained on specimens of normal size. A volumetric method for small-scale shrinkage tests using 1-in. cubes has been described by Nurse.[2]

[1] T. W. Parker, *J. Soc. chem. Ind., Lond.* **58,** 203 (1939); H. Kuhl, *Zement* **15,** 840 (1926); *Cem. & Cem. Mf.* **3,** 1301 (1930); J. A. Forrester and R. A. Keen, *J. appl. Chem.* **10,** 358 (1960); F. Keil and H. Mathieu, *Zement-Kalk-Gips* **17,** 279 (1964).

[2] R. W. Nurse, *J. Soc. chem. Ind., Lond.* **58,** 37 (1939).

14 Pozzolanas and Pozzolanic Cements

Pozzolanas are usually defined as materials which, though not cementitious in themselves, contain constituents which will combine with lime at ordinary temperatures in the presence of water to form stable insoluble compounds possessing cementing properties. The pozzolanas can be divided into two groups, natural and artificial.

The natural pozzolanas are for the most part materials of volcanic origin, but include certain diatomaceous earths. The artificial pozzolanas are mainly products obtained by the heat treatment of natural materials such as clays and shales and certain silicious rocks, and pulverised fuel ash (fly-ash). The early history of pozzolanas has already been discussed in Chapter 1 and much of the later literature can be traced through a survey of the chemistry of pozzolanas made in 1938 by Lea,[1] the ASTM symposium[2] in 1949 on pozzolanic materials, the symposium[3] on pozzolanas and their use held by the Italian Chemical Society in 1953, and the review by Malquori[4] in 1960.

Natural pozzolanas

Volcanic deposits

The pozzolanas of volcanic origin consist of glassy incoherent materials or compacted tuffs arising from the deposition of volcanic dust and ash. They may occur as does for example the Rhenish trass, in a consolidated rock-like form underlying material deposited subsequently, or in a more fragmentary and unconsolidated state like some of the Italian pozzolanas. In Europe the materials which have been the most exploited are the Italian pozzolanas, Santorin Earth obtained from the Grecian Isle of Santorin, and the German trass which is found in the neighbourhood of Coblenz on the Rhine and in Bavaria. Trass deposits are also worked in Rumania and the U.S.S.R. There are no volcanic ash deposits suitable for use as pozzolanas in Great Britain. A volcanic ash pozzolana known as tosca is obtained from Teneriffe in the Canary Islands and a material known as tetin

[1] *Sym., Stockholm 1938*, 460.
[2] *ASTM Special Technical Publication*, No. 99 (1950).
[3] *Annali Chim.* **44**, 569–768 (1954).
[4] G. Malquori, *Sym., Washington 1960*, 983.

from the Azores. Deposits are numerous in the United States and Canada and some of the former have been used in pozzolanic cements. As might be anticipated, there are numerous deposits in Japan and New Zealand.[1]

The Rhenish trass is obtained from quarries in which, after removing the overburden, there come layers of pumice and sand, followed by transition layers of tuffstein containing pumice, and then tuffstein, brown in the upper part and blue in the lower. The material as quarried is called tuffstein and only the ground material is called trass. It was, at one time, thought that the blue trass was better than the brown, but later it was found that there was little difference between them provided the brown material was taken from below the transition layer. Leucitic tuffs of identical composition, but different mineralogical nature and microscopic appearance, which have only poor pozzolanic value, are also found. Rhenish trass has been known for some 2000 years, for mortars containing trass have been found in old Roman buildings along the Rhine. Since about the end of the seventeenth century it has been increasingly used, first in lime mortars and subsequently in Portland cement concretes.

The Italian pozzolanas occur in the neighbourhood of Vesuvius and around Naples as incoherent trachytic tuffs, and around Rome, where they are present in a more compact form, though still soft and easily worked. They are obtained from open pits and quarries, some of which have been in use for centuries.

Some analyses of volcanic ash pozzolanas are shown in Table 70.

TABLE 70 Percentage composition of volcanic ash pozzolanas

Pozzolanas	Ignition loss	SiO_2	Al_2O_3	Fe_2O_3	TiO_2	CaO	MgO	Na_2O	K_2O	SO_3
Rhenish Trass	10·1	54·6	16·4	3·8	0·6	3·8	1·9	5·1	3·9	0·4
Rhenish Trass	8·5	54·8	17·2	4·4	0·6	2·3	0·9	7·0	3·8	0·1
Bavarian Trass	14·5*	57·0	10·9	5·6	0·5	6·0	2·2	1·8	1·5	0·2
Santorin Earth	4·9	63·2	13·2	4·9	1·0	4·0	2·1	3·9	2·6	0·7
Santorin Earth	3·1	65·2	12·9	6·3	—	3·2	1·9	2·6	4·2	—
Rome: Segni	9·6	44·1	17·3	10·7	—	12·0	2·0	1·4	3·1	—
Segni	5·3	48·2	21·9	9·6	—	7·5	3·2	4·1		0·3
S. Paolo	4·1	45·2	20·0	10·7	—	9·8	3·8	6·2		0·3
Naples: Bacoli	4·8	55·7	19·0	4·6	—	5·0	1·3	3·4	3·9	—
Baia	4·4	59·5	19·3	3·3	—	2·1	0·2	11·3		0·2
Rumanian Trass	13·9	62·5	11·6	1·8	—	6·6	0·7	2·9		—
Crimean Tuff	11·7	70·1	10·7	1·0	—	2·5	0·3	3·7		—
U.S.A. Rhyolltic	3·4	65·7	15·9	2·5	—	3·4	1·3	5·0	1·9	—
Pumicite	4·2	72·3	13·3	1·4	—	0·7	0·4	1·6	5·4	Trace

* Includes a considerable amount of carbon dioxide.

The volcanic pozzolanas are composed of a mixture of silicates and contain both glass and crystalline particles. The volcanic ash and dust from which they were formed has undergone rapid cooling, and in some cases has subsequently suffered considerable chemical alteration leading to the formation of zeolitic

[1] J. Henderson, *N.Z. Jl Sci. Technol.* **18**, 781 (1937).

compounds. The alteration is generally attributed to the action of superheated steam and carbon dioxide below the earth's surface. The effect of this action has been to convert much of the original material into a more chemically reactive modification, whilst the basic constituents have been partially removed under the combined influence of the carbon dioxide and water. This conversion has proceeded further in trass and some of the older Italian pozzolanas than in Santorin Earth. Pozzolanic activity is shown in some degree even by massive lavas and to a feeble extent by some of the products of the decay of igneous rocks such as arènes and psammites. These are of no practical importance.

The zeolitization of a vitreous volcanic ash has been reproduced in the laboratory[1] by hydrothermal treatment at 200–400° in an alkaline medium (e.g. 1 per cent or more alkali hydroxide). Volcanic ashes, such as those of Vesuvius, which contain a high content of crystalline minerals are much more stable and show only low pozzolanic activity. The volcanic materials owe their pozzolanic properties both to the volcanic glass and to the altered zeolitic compounds. The zeolites found include herchellite, phillipsite, chabazite and analcite. These are compounds of the type $(R_2Ca)O.Al_2O_3.4SiO_2.xH_2O$ differing in their contents of calcium and alkali cations. Analcite approximates to $Na_2.Al_2O_3.4SiO_2.2H_2O$, chabazite to $(R_2Ca)O.Al_2O_3.4SiO_2.6H_2O$ rich in lime, herschellite to a corresponding compound richer in alkalies and phillipsite to a potash-rich compound.

The Rhenish trass is a trachytic (alkali feldspar) tuff which has been subjected to the action of carbon dioxide-bearing waters for such a long period that a large part of the minerals originally present has become hydrated and decomposed. It consists of an isotropic ground mass containing various crystalline mineral constituents such as feldspar, leucite and quartz with small amounts of augite, hornblend, mica, etc. The glassy matrix, amounting to about half of the trass, is the material that has undergone alteration and consists of zeolitic compounds amongst which are analcite and chabazite or herschellite.[2]

The two main Italian deposits are the Laziale group near Rome and the Flegrean group around Naples. The Roman pozzolanas belong to the leucitic type and the Naples to the alkali-trachitic group. Crystals of leucite ($K_2O.Al_2O_3.4SiO_2$), present both as clear unaltered crystals and in a turbid altered form, and of pyroxenes occur in the amorphous ground mass of the Roman pozzolanas while the Napolitan ones contain inclusions of sanidine (a potassium feldspar) with small amounts of other minerals.[3] The glass in the Roman pozzolanas is considerably altered, but not in the more fragmentary Napolitan ones.[4] The glasses also differ in composition, that of the Roman materials having a composition near to that of the feldspar labradorite while that of the Napolitan lies close to oligoclase or between that of orthoclase and anorthite. The zeolitic compounds in the Roman pozzolanas are herschellite, chabazite and phillipsite. The yellow compacted Napolitan tuffs contain herschellite and the green ones analcite. The latter owe their colour to the presence of glauconite ($K_2O.Fe_2O_3.4SiO_2.aq$).

[1] R. Sersale, *Rc. Soc. miner. ital.* **17,** 499 (1961) and a series of papers in *Rc. Accad. Sci. fis. mat., Napoli*, Ser. 4, **25** (1958), **26** (1959), **27** (1960).
[2] V. Ludwig and H. E. Schwiete, *Zement-Kalk-Gips* **15,** 160 (1962); R. Sersale and R. Aiello, *Silic. ind.* **30,** 13 (1965).
[3] B. Tavasci, *Il Cemento* **43,** 4 (1946); **44,** 1 (1947); **45,** 114 (1948).
[4] N. Parrovano and V. Cagliotti, *Ricerca scient.* **8,** 271 (1937).

In practice only the incoherent glassy materials seem to be used in the Naples region though the altered compact tuffs when ground give the more active pozzolana. The Flegrean pozzolanas tend to be more acidic than the Roman with lower contents of R_2O_3 and higher SiO_2 and to show a slower initial action, but a greater increase in strength at long ages.

Santorin Earth consists mainly of a granular isotropic material mixed with pumice, obsidian and fragments of crystalline feldspars, pyroxenes and quartz, etc. Of the three main size fractions, pumice, fine sand, and dust, into which it can be separated, only the latter appears to be an active pozzolana. The combined water content of Santorin Earth is lower than that of trass or the Italian pozzolanas and it gives lower strengths at early ages in lime mortars, though the ultimate strength is not inferior. The activity has been attributed by Vournazos[1] and others to the presence of hydrated silica, though this does not exclude the presence also of the kind of active material found in other volcanic pozzolanas.

The natural volcanic pozzolanas used in the U.S.A. have been mostly tuffs,[2] containing a rhyolitic glass with an index of refraction ranging from 1·480 to 1·507 corresponding to a silica content of 76–70 per cent. The glass content varies from about 50 to nearly 100 per cent; the remaining constituents include quartz, feldspars, biotite, horneblende, hypersthene, sanidine, calcite and small amounts of opal, together with varying amounts of montmorillonite-type clays. The rhyolitic glass is probably to be regarded as a continuous random network of silica with atoms of other elements occurring at random. The activity must be expected to alter with the composition and variations in the atomic arrangement. Loss of volatile components on heating will lead first to the development of a porous structure and at higher temperatures to a collapse of the structure.

The source of the pozzolanic properties of trass, the Italian pozzolanas, and other similar materials has long been a subject of controversy. Though the term 'pozzolana' is sometimes applied locally, as in the Naples region, to any fragmentary deposit of volcanic action, not all volcanic ashes possess any pozzolanic properties and some are suitable only for use as an inert sand. Thus in the Naples region it is the materials of Bacoli and Baia which are examples of true pozzolanas and not numerous other volcanic deposits. It is further not possible to establish from microscopic examination to what extent a material is active. It is generally agreed that the active part is the amorphous or vitreous portion and the zeolitic compounds, and that little activity can be attributed to the crystalline minerals for these are well-defined stable compounds inert to lime. The activity of the amorphous and vitreous portion does not arise directly from its composition for, apart from the combined water content, the analyses of the different fractions of one pozzolana are very similar and also approach closely to that of the lavas.

Three theories have been advanced to explain the origin of the activity of volcanic pozzolanas:

1. The zeolitic material is regarded as an alteration product of the original minerals, produced by prolonged exposure, after deposition, to waters containing carbon dioxide and to superheated steam.

[1] A. Vournazos, *Z. anorg. allg. Chem.* **200**, 237 (1931).
[2] R. C. Mielenz, L. P. Witte and O. J. Glantz, *Spec. tech. Publ. Am. Soc. Test. Mater.*, No. 99, p. 43 (1949).

2. Pozzolanas originate from materials carried by volcanic eruption from geological strata of essentially clay-like composition; materials that during eruption have undergone heating sufficient to produce complete dehydration and chemical alteration, but not fusion.
3. The vitreous portion of pozzolanas is essentially a very porous aerogel of high internal surface area.

As we have seen, there is much evidence to support the first theory that the zeolitic materials are alteration products. Santarelli[1] has cited the gradual diminution of pozzolanic activity from the base to the top of the Italian quarries and the greater abundance of crystalline constituents in the materials of more recent origin. Materials of similar composition, but of more recent eruptions, such as the ashes and lapilli of Vesuvius, show but weak pozzolanic properties. There is, however, no relation between the content of combined water in trass and its hydraulic properties[2] and the same applies to other volcanic pozzolanas.

In the second theory, propounded by Gallo, emphasis is laid more on the nature of the erupted material than on any subsequent alteration though this may cause some activation.

The third theory, due to Parraveno and Caglioti,[3] attributes pozzolanic activity primarily to the physical state of the material, an 'aerogel' being formed by the pulverisation of the fused magma by gases during eruption. The pozzolanic character of the material depends then not only on its volcanic origin but also on the eruptive conditions of its formation. Subsequent alteration is regarded as only a secondary phenomenon and the activity is attributed to the high internal surface area and the gel-like structure, for synthetic aerogels were found to resemble pozzolanas quite closely. This theory has been elaborated further by Penta[4] who considers that some degree of devitrification of the glass, and pneumolytic alteration produced by the action of gases and subterranean waters, accompanied by some loss of more soluble constituents, contribute to the activity of the micro-cellular vitreous material.

The effect of heating on natural volcanic pozzolanas seems variable. At temperatures up to 500–600° the materials may be dehydrated without any collapse of the atomic or physical structure on which their activity depends, but at a red heat the activity is much decreased or lost. Thus trass is not adversely affected at 500° and tests on American rhyolitic materials showed a tendency for the activity to increase after treatment at 500–800°. Santorin Earth loses most of its activity at 750°, while the Italian pozzolanas benefit by short heat treatment (a few hours) at 300–700°, but prolonged heating at 600° or above decreases their activity. Though the elimination of combined water, provided no collapse of the structure occurs, may improve reactivity by increasing the porosity, it seems that the main effect of heat treatment may often be to calcine any clay impurities present, changing them from deleterious constituents to inert or pozzolanic ones. A heat treatment at 500° is used in processing some Roman pozzolanas to be used in pozzolanic cements.

[1] L. Santarelli, *Annal. Chim. appl.* **27,** 3 (1937).
[2] H. Burchartz, *Tonind.-Ztg. keram. Rdsh.* **48,** 1221 (1924).
[3] N. Parravano and V. Caglioti, *Ricerca scient.* **8,** 271 (1937).
[4] F. Penta, *Annal. Chim.* **44,** 572 (1954).

High silica materials

This class of naturally occurring materials showing pozzolanic properties include the diatomaceous earths, which are known by many different names and trade descriptions, and some silicious rocks which have lost much of their basic oxides. Some materials of this latter type used in Italy contain over 85 per cent silica and only a few per cent alumina plus ferric oxide. The diatomaceous earths are composed of the silicious skeletons of diatoms deposited from either fresh- or sea-water. In many cases the deposits are mixed with sand or clay. The largest known deposits are those occurring in California, but large deposits are also found in Canada, Algeria, Denmark and Germany. Within the British Isles there are small deposits in Scotland, the Western Isles and Northern Ireland. In their raw condition many diatomaceous earths have little value as pozzolanas, for, although they combine very actively with lime, their physical state is such that a large proportion of water is needed to render mixes of them with lime or cement plastic and the cementing value then becomes poor. A deposit of Tertiary age of diatomaceous earth containing a considerable proportion of clay which occurs in Denmark is used as a pozzolana under the name of Moler. Though this material was formerly used in its raw condition, a superior product is obtained on burning.

The main constituent of diatomaceous earths is opal. This is an amorphous form of hydrous silica containing up to 10 per cent water, but sub-microscopic crystals of beta-cristobalite may also be present in it. The water is progressively lost at temperatures up to 1000°, but the more important effect of heat treatment is probably to dehydrate the clay impurities when these are present in appreciable amounts.

Artificial pozzolanas

The chief artificial pozzolanas are burnt clays and shales, spent oil shales, burnt gaize, burnt moler, and pulverised fuel ash (fly-ash).

Burnt clays and shales

The pozzolanic properties of burnt clays were well-known to the Romans, who utilised ground clay bricks and tiles as a substitute for the natural volcanic ash pozzolanas. The value of burnt clay as an addition to fat lime mortars to obtain hydraulic properties has also long been known in India and Egypt, where it passes under the names Surkhi and Homra. Experiments on burnt clay-lime mortars were carried out a century ago in connection with the rebuilding of London bridge,[1] while still earlier Smeaton had tried them, without success, as a mortar ingredient for the Eddystone lighthouse. A Swedish engineer, Baggé of Gothenburg, and Count Chaptal in France, both experimented in the eighteenth century with burnt-clay pozzolanas. Vicat mentions burnt-clay pozzolanas in papers published in 1843 and 1857, and they are discussed at some length in Captain Smith's translation of Vicat's *Mortiers et Ciment Calcaires*. Frémy[2] seems to have been the first to link the development of pozzolanic properties in clays on burning with the temperature at which loss of water of hydration occurs. The

[1] J. White, *Phil. Mag.* **11,** 333 (1832).
[2] *C. r. hebd. Séanc. Acad. Sci., Paris* **67,** 1205 (1868).

use of burnt clay in lime-mortars was also common at one time in Great Britain. A pozzolanic cement consisting of ground burnt clay and Portland cement was described in 1909 by Potter, who stated that over 5000 tons of it, known as Potter's Red Cement, were used in freshwater and seawater construction about that time.[1]

For a period of many years up to about 1915 the Lafarge Company in France manufactured an 'undecomposable' cement composed of a mixture of burnt clay and the French grappiers cement. This pozzolanic cement was produced for use in the gypsiferous territories of France. Its manufacture was discontinued following on the production of high-alumina cement by the same firm.

The use of a lime-surkhi mortar, prepared by grinding lime and crushed bricks or burnt clay together in a mortar mill, has long been common practice in India, where it has formed the cementing material in dams and structures under water. In Egypt a lime-burnt clay mortar was used initially for the hearting of the Assuan Dam and found to yield a more watertight mortar than a 1 : 4 Portland cement and sand mix.[2] Its use was, however, given up owing to difficulties in producing the material at the required rate. The slow rate of setting of the material also hindered the progress of construction. In the building of the Sennar Dam[3] on the Blue Nile in 1919–1925 the masonry was set in a cement composed of 70 per cent Portland cement and 30 per cent burnt clay, both materials being produced in a factory on the site of the construction.

Burnt clays have in the past been used in mortar and concrete mainly when Portland cements were not available, or when a saving in cost has resulted. The materials used were, however, produced under crude and uncontrolled conditions of burning and, as it has since been shown that a degree of care in burning similar to that required in the manufacture of Portland cement is required to produce the best products, it seems probable that only relatively inferior materials were obtained. The investigations of Feret[4] in France, of the Building Research Station in Great Britain,[5] and of various American workers,[6] have shown that excellent burnt-clay pozzolanas can be obtained. Such a material was, for instance, used in the Bonneville dam in the U.S.A. and in oil-well cements manufactured in Great Britain. Modern work in India[7] has also been directed to the production of more controlled materials than the traditional surkhi.

Burnt-clay pozzolanas are produced by burning suitable clays or shales at a temperature which varies from 600° to over 900°, depending in the nature of the clay and the conditions of burning; the product is ground to cement fineness. Certain spent oil shales which have undertone burning subsequent to distillation also yield, when selected with due care, a pozzolana of similar type and quality.

Raw clays consist essentially of a group of hydrated aluminium silicates, though alumina may be replaced to varying extents by ferric oxide, and to a lesser extent by bases such as MgO, Na_2O and CaO. The water contents can

[1] C. J. Potter, *J. Soc. chem. Ind., Lond.* **28,** 6 (1909).
[2] M. Fitzmaurice, *Proc. Instn civ. Engrs* **152,** 71 (1902–3).
[3] O. L. Prowde, *Proc. Instn civ. Engrs* **222,** 81 (1925–6).
[4] *Annls Ponts Chauss.* **92** (iv), 5 (1922); *Additions de Matières Pulvérulentes aux liants hydrauliques*, Paris 1925.
[5] F. M. Lea, *Bldg Res. tech. Pap.*, No. 27 (1940), H.M.S.O. London; *Proc. Third Congress on Large Dams, Stockholm 1948*, Vol. III.
[6] *ASTM Special Technical Publication*, No. 99 (1950).
[7] R. C. Hoon, *Indian Concr. J.* **26,** 225, 252 (1952).

also vary considerably. Kaolinite, $Al_2O_3.2SiO_2.H_2O$, is a well-defined mineral and the kaolinite group also includes other minerals of the same composition such as nacrite, dickite and metahalloysite. The montmorillonite group covers a variety of minerals with $Al_2O_3 : SiO_2$ ratios varying from 1 : 3 to 1 : 6 and their isomorphous solid solutions. They can be typified by 'ideal' formulae such as montmorillonite $Al_2O_3.4SiO_2$.aq, nontronite $Fe_2O_3.3SiO_2$.aq, beidellite $Al_2O_3.3SiO_2$.aq and saponite $3MgO.4SiO_2$.aq. The hydromicas and illites form yet another group.

All clays have a considerable content, some 10–15 per cent, of combined water. In the kaolinite type clays this is not lost on heating until a temperature of 500° or above is reached, but the montmorillonite and illite types commence to lose water at rather lower temperatures. As we have seen earlier (p. 120) there is still some controversy about the changes that clays undergo when the combined water is lost.

Burnt gaize

Gaize is a soft, porous, highly silicious, sedimentary rock containing a proportion of clay, which is found distributed over a considerable area in the Ardennes and Meuse valley in France. It contains silica in a gelatinous condition and has a density of only about 1·4 after drying. The rock has been used in the raw state as a pozzolana, but usually it is burnt at a temperature of about 900°.

A pozzolanic cement composed of burnt gaize and Portland cement has been manufactured in France and was used at Boulogne, Havre, St. Malo, Dieppe and other French ports for concrete work in sea-water.

Moler

A moler cement composed of 75 per cent Portland cement and 25 per cent moler has been manufactured in Denmark and used for concrete work in seawater. Initially the raw moler was used, but the burnt moler was found to be superior and used instead of the raw product.

Pulverised fuel ash (fly-ash)

When pulverised coal is burnt in boilers the ash is carried forward in the gases as fused particles which solidify into a roughly spherical shape. The properties of this material as a pozzolana was first reported on in 1937 by Davis[1] and his co-workers and its use in concrete started to develop in the U.S.A. in the following decade particularly for mass concrete for dams. Later its use became wider and spread to many other countries.[2] The ash is very finely divided with a specific surface (air-permeability method) commonly falling between about 2000 to 5000 cm/g though higher values are sometimes found. Its specific surface is thus of the same order as that of cement. The composition and properties of pulverised fuel ash depend both on the coal burnt and the efficiency of the combustion process so that the value of the material from different power stations can vary widely. The product from a modern power station when working on base-load will normally be more consistent. In composition the ash approximates

[1] R. E. Davis, R. W. Carlson, J. W. Kelly and H. E. Davis, *Proc. Am. Concr. Inst.* **33**, 577 (1937).
[2] E. A. Abdun-Nur, *Bull. Highw. Res. Bd*, No. 284 (1961); M. Kokubu, *Sym., Tokyo 1968*.

to that of a burnt clay high in alumina and iron oxide but can vary widely as illustrated by the data in Table 71. The major constituent, some 60–90 per cent, is glass with quartz, mullite, hematite and magnetite as the more important crystalline components.[1] The glass consists chiefly of silica and alumina with some iron oxide, lime, alkalies and magnesia. Some sulphur trioxide is present as anhydrite, or gypsum, or in the glass. Combustible matter is always present, but in well-burnt materials it is below 10 per cent, and often below 3 per cent, though it can rise to 20 per cent. The ash contains constituents with a range of specific gravities and the mean value can vary from about 2·0 to 2·7, rising with the iron oxide content and tending to be lower as the carbon content increases.

TABLE 71 Percentage composition of some artificial pozzolanas[2]

Pozzolana	SiO_2	Al_2O_3	Fe_2O_3	CaO	MgO	Na_2O and K_2O	SO_3	Ignition loss*
Burnt clay	58·2	18·4	9·3	3·3	3·9	3·9	1·1	1·6
Burnt clay	60·2	17·7	7·6	2·7	2·5	4·2	2·5	1·3
Spent oil shale	51·7	22·4	11·2	4·3	1·1	3·6	2·1	3·2
Raw gaize	79·6	7·1	3·2	2·4	1·0	—	0·9	5·9
Burnt gaize	88·0	6·4	3·3	1·2	0·8	—	Trace	—
Raw moler	66·7	11·4	7·8	2·2	2·1	—	1·4	5·6
Burnt moler	70·7	12·1	8·2	2·3	2·2	—	1·5	—
Raw diatomite (U.S.A.)	86·0	2·3	1·8	Trace	0·6	0·4	—	8·3
Burnt diatomite (U.S.A.)	69·7	14·7	8·1	1·5	2·2	3·2	—	0·4
Fly-ash (U.S.A.)	47·1	18·2	19·2	7·0	1·1	3·95	2·8	1·2
Fly-ash (U.S.A.)	44·8	18·4	11·2	11·6	1·1	3·14	2·0	7·5
Fly-ash (British)	47·4	27·5	10·3	2·1	2·0	5·7	1·8	0·9
Fly-ash (British)	45·9	24·4	12·3	3·6	2·5	4·2	0·9	4·1

* Includes carbon in case of spent oil shale and fly-ash.

The glass is the active material in pulverised fuel ash, the mineral constituents being inert. The value of the ash as a pozzolana depends therefore on its glass content, but also on its fineness and composition, though there are no close correlations between these parameters and the contribution made to strength development. Increasing SiO_2, or $SiO_2+Al_2O_3$, content seems to have a favourable influence on the pozzolanic value. There is a British Standard (BS 3892 : 1965) for pulverised fuel ash for use in concrete which classifies the material in three fineness zones, 1250–2750, 2750–4250, and above 4250 cm^2/g. The MgO content is limited to 4 per cent, SO_3 content to 2·5 per cent and the loss on ignition to 7 per cent. The ASTM specification (C618–68T) requires a minimum specific surface of 6500 cm^2/cm^3 and limits the SO_3 content to 5 per cent and the loss on ignition to 12 per cent. The sum of $SiO_2+Al_2O_3+$

[1] H. S. Simons and J. W. Jeffery, *J. appl. Chem.* **10,** 328 (1960); J. D. Watt and D. J. Thorne, ibid. **15,** 585, (1965); L. J. Minnick, *Proc. Am. Soc. Test. Mater.* **59,** 1155 (1959); M. Mateus and D. T. Davidson, ibid. **62,** 1008 (1962).

[2] D. Watt and D. J. Thorne, *J. appl. Chem.* **15,** 585 (1965) give analyses for many British fly-ashes and R. H. Brink and W. J. Halstead, *Proc. Am. Soc. Test. Mater.* **56,** 1161 956) for many U.S.A. fly-ashes.

Fe_2O_3 must be at least 70 per cent. Tests are also required on the contribution of the ash to strength development as noted later.

Though one of the major uses of pulverised fuel ash has been in dam construction, partly on economic grounds and partly on account of its influence on workability and heat evolution, it is also used in other concrete work. Thus it is used in some countries in ready-mixed concrete, while in France up to 20 per cent is permitted in one class of Portland cements. Pulverised fuel ash can have a depressing effect on air-entrainment, the amount of air-entraining agent required to obtain a given air content increasing with the carbon content of the ash. There has been considerable controversy as to the effect of grinding the ash either with, or separately from, the Portland cement. The effect must depend on the original fineness of the ash but it seems that grinding has some favourable effect on early strength development. The decisive factor in practice is probably the additional cost of grinding.

Bauxite

The use of bauxite as a pozzolana in some ancient mortars is reported by Le Chatelier,[1] who has found mortars made from pulverised bauxite and lime in the ruins at Baux in the Rhone valley in France. Ferrari states that burnt bauxite forms an excellent pozzolana.

Granulated blastfurnace slag

Granulated blastfurnace slag is sometimes described as a pozzolana but, for reasons discussed in the next chapter, this material should be classified as a latent hydraulic cement. The importance of the types of cement containing granulated slag is such as to warrant the devotion of a separate chapter to them.

The use of pozzolanas

Pozzolanas are used in lime-pozzolana mortars, in blended pozzolanic cements, and as direct additions to a concrete mix. In the traditional use of pozzolanas in lime mortars the material was used in a relatively coarse state, as indeed it still is for example, in Italy, where the specification for natural pozzolanas used in building mortars only requires that the maximum size of grains shall not exceed 5 mm. The coarser particles act as more-or-less inert aggregate and only the finer material, which grades down to a fine dust, as an active pozzolana. With the development of the use of pozzolanas with Portland cement the practice has become common of grinding them to cement fineness. Thus the Italian pozzolanic cements, the German trass cements, and the pozzolanic cements used in the U.S.A. are interground mixtures of pozzolana and Portland cement clinker. In the case of pulverised fuel ash, which may be added as a separate ingredient at the concrete mixer, the material is already finer than Portland cement. The activity of pozzolanas is increased by fine-grinding and indeed it is essential if the maximum value is to be secured from their use.

Lime-pozzolana mortars are used as building mortar in many parts of Europe and enjoy a good reputation. They have been shown by the experience of 2000

[1] *7th int. Congr. appl. Chem., London 1909.*

years to possess the durability claimed for them by Vitruvius in his Treatise on Architecture, where, speaking of pozzolana, he says that when mixed with rich lime and stone it sets solidly, not only in all kinds of structures, but particularly in those built in the sea. Roman monuments such as the bridges of Fabricus, Aemilius, Elius and Milvius, the arches of Claudius and Trajan at Ostia and Nero at Antium, together with many maritime works built in the time of the Roman Emperors, some of which, such as those erected by Trajan at Ancona and Civitavecchia, are in use today, all stand as a tribute to the permanence of lime-pozzolana mortars.

Lime-pozzolana mortars and concretes, or pozzolanic cement mixes, are used in Italy for seawater work in place of Portland cement. The traditional mortar going back to Roman times was 1 volume of slaked lime putty to 2 volumes unground pozzolana. The slaked lime putty contains about 560–640 kg/m^3 (35–40 lb/ft^3) $Ca(OH)_2$ while the pozzolanas vary from 960–1200 kg/m^3 (60–75 lb/ft^3). In modern practice a 1 : 3·5 volume ratio is preferred, corresponding to about 150 kg hydrated lime per cubic metre of the mortar mix. The traditional lime-pozzolana concrete was composed of 1 volume lime putty, 2 volumes unground pozzolana, 3–4 volumes aggregate (about 1 : $3\frac{1}{2}$: 8–12 by weight) but again in more recent practice a 1 : 3·5 volume ratio of lime to pozzolana was adopted, with the ratio of the lime-pozzolana mortar to aggregate maintained at about 1 : 2 by volume. The rate of hardening of lime-pozzolana concretes is slow but it can be increased by grinding some 15–20 per cent of the pozzolana to cement fineness. Alternatively an addition of 100–150 kg cement per cubic metre (170–250 lb/yd^3) of concrete is made. Tribute has been paid by numerous Italian engineers to the excellent performance of these lime-pozzolana concretes, and it is stated that those with the accelerated rate of hardening can be exposed to the wash of waves within a day of placing.

The desire with modern constructional methods for a more rapid hardening of the concrete led to the introduction of pozzolanic cements. These have been widely used for seawater construction in Italy over the last half century. The Portland cement used as base is of the sulphate-resisting type and in one recent form[1] of these cements the pozzolana used has a high content (about 90 per cent) of reactive silica. This latter type of cement combines a very high chemical resistance with a rate of strength development similar to that of Portland cement.

The use of lime-pozzolana concretes in seawater is not favoured in northern Europe on account of the still slower rate of hardening caused by the lower prevailing temperatures. Mixtures of lime, trass and Portland cement have been used in concretes to some extent in Germany, as in the locks of the Rhine-Herne canal in which a 1 : $1\frac{1}{4}$: 2 : $6\frac{3}{4}$: $13\frac{1}{2}$ volume mix of Portland cement, lime, trass, sand and ballast respectively was employed.[2]

The original reason for the use of pozzolanic Portland cements was the improved durability, combined with some economy, obtained in concrete in marine, hydraulic, and underground structures. The reduction in heat evolution later provided another reason for their use in large mass concrete works and then,

[1] A. Rio and A. Celani, *RILEM. Sym. on the Behaviour of Concretes exposed to Seawater, Palermo 1965*; R. Turrizianni and A. Rio, *Sym., Washington 1960*, 1067.
[2] K. Ostendorf, *Bautechnik* **5** (39), 566 (1927).

with the discovery of the alkali-aggregate reaction (p. 569), it was found that the addition of pozzolanas was often a useful means of preventing expansion arising from this cause. Finally pozzolanic cements have been developed as general constructional cements. Thus, in Italy, cements with a 30–40 per cent pozzolana content are widely used in the same mix proportions as Portland cement for concrete in general building construction, for dams, and for reinforced concrete in seawater. A fairly high-strength Portland cement is used to help offset the loss in strength at early ages. Use is made in the U.S.A. and other countries of pulverised fuel ash in ready-mixed concretes for general building work.

The addition of trass to Portland cement concrete mixes has been common practice for many years in Germany, Holland, and Belgium, where it has been used for docks and harbours, inland waterways, drainage systems, railway bridges and other structures. It has similarly been much used in conjunction with the Portland blastfurnace types of cement. Formerly the common practice was to add the trass as a separate constituent to the concrete mix, but it has become the practice to premix the trass with the cement. Trass-Portland cements containing 20–40 per cent trass, are marketed in Germany. In general, the proportions of trass recommended in Germany are:

High-lime Portland cement	0·66 cement : 0·34 trass
Low-lime Portland cement	0·75 cement : 0·25 trass
Portland blastfurnace cement	0·80 cement : 0·20 trass

Considerable use has been made in the western regions of the U.S.A. of cements containing up to 30 per cent pozzolana. Thus the cement used in the large bridges at San Francisco Bay contained 15 per cent of a calcined shale pozzolana, and that in the Bonneville Dam on the Columbia river 25 per cent. Other examples are the 20 per cent pumicite in the cement for the Friant Dam and from 24 to 32 per cent of pulverised fuel ash in the Hungry Horse and many later dams. Use has also been made in Great Britain of pulverised fuel ash in dams in Scotland and in Devon.

The addition of ground sand or crushed rock to Portland cement to form what were known as 'sand cements' attained a certain degree of prominence many years ago and numerous tests on them were made by Feret and other investigators. For most purposes no benefit was found to accrue from their use and their manufacture eventually ceased, to be revived again as an economy measure during the 1939–45 war (see p. 537). Though ground sand has no pozzolanic properties, it was used successfully in a few U.S.A. dams built in the early part of the present century.

For many purposes a pozzolana can be regarded as a substitute for a proportion of the Portland cement in a concrete, but in some cases it is preferable to use it, in part at least, as an addition. Thus for seawater work in northern Europe, where frost action is a major cause of deterioration, it is now generally considered that trass should be used as an addition.

It should be noted that, since the specific gravity of most pozzolanas is about 2·3–2·6 compared with 3·1 for Portland cement, the absolute solid volume of a pozzolana is about 15–25 per cent greater than that of an equal weight of Portland cement. Thus a substitution of 40 per cent by weight of cement by pozzolana increases the solid volume of the cement plus pozzolana by about 6–10 per cent, and *pro rata* for other degrees of substitutions. The substitution is therefore sometimes made on the basis of equal solid volumes.

Mechanism of the lime-pozzolana reaction

The fundamental property of a pozzolana is its ability to combine with lime and, in explanation of this, two main theories have been advanced, base exchange and direct combination.

The idea that natural pozzolanas are zeolitic compounds and owe their properties to base exchange runs through much of the older literature. As we have seen, modern work has shown that zeolitic compounds are present in various volcanic deposits that have undergone alteration, but not in the fragmentary unconsolidated glassy deposits such as those of Naples. The zeolites are a group of insoluble hydrated alumino-silicates of the alkalis and alkaline earths which have the property of exchanging some of their base constituents for others when immersed in salt solutions. This property is utilised for water-softening in the permutite process. The permutites are artificial zeolites in which the bases are very easily replaceable. When a hard water containing, say, calcium sulphate, percolates through sodium permutite the soda is replaced by lime and sodium sulphate formed. The permitute can be regenerated again by passing a strong solution of sodium chloride through the mass, when sodium permutite is reformed together with calcium chloride.

Various earlier workers[1] found that small amounts of alkalies were released in mixes of lime with volcanic ash pozzolanas or when the pozzolana was shaken with solutions of calcium nitrate. More recent studies by Rio[2] on mixtures of Italian pozzolanas with Portland cement, kept in contact with water at 40°, showed a release of up to one-third of the total alkalies in the pozzolana after 8 days and as much as 85 per cent at 1 year for a Segni pozzolana. In mixtures with lime, Rio[3] found that about half the alkalies were released in three months at 40°, but that the rate was much slower at 20°. Rhenish trass has a high alkali content and in mixes of this with lime about 0·04 g K_2O+Na_2O per g trass has been found to be released in 28 days. This is far from equivalent to the lime combined which amounted 0·2 g CaO per g trass.[4] A similar conclusion[5] has been reached from other studies on the reaction of Napolitan pozzolanas with saturated lime solution at room temperature. The results are shown in Table 72. The zeolite, herschellite, is a constituent of the yellow Napolitan tuff.

Both these data, and X-ray evidence, showed that the major reaction was not base exchange but the formation of new hydrated compounds. It is evident that base exchange makes only a minor contribution to the combination of lime by natural pozzolanas and it seems doubtful if it can make any contribution to strength development. In base exchange the lattice of the zeolitic compound remains

[1] See F. M. Lea, *Sym., Stockholm 1938*, 460 for review.
[2] A. Rio, *Annali Chim.* **44**, 684 (1954).
[3] A. Rio, *Rass. chim.* per *Chimica Ind.* **9** (4), 3 (1957).
[4] U. Ludwig and H. E. Schwiete, *Sym., Washington 1960*, 1093.
[5] R. Sersale and V. Sabatelli, *Rc. Accad. Sci. fis. mat., Napoli*, Ser. 4 **27**, 263 (1960); **28**, 45 (1961).

unchanged and one base ion is exchanged for another fitting into a similar position in the crystal lattice. It is unlikely that this action will be cementitious even though it may be valuable in removing free calcium hydroxide from a set cement, and indeed normal zeolites show no cementing properties. Base exchange does not occur to any appreciable extent with burnt clay pozzolanas.

TABLE 72 Reaction of Napolitan pozzolanas with lime solution

Material	Time (days)	CaO combined g/100 g material	Na+K released to solution- g/100 g material	CaO (g) equivalent to alkalis- released
Herschellite	28	35·5	6·36	7·30
Yellow napolitan Tuff	29	34·3	2·80	2·14
Flegrean pozzolana	29	—	0·88	—
	91	30·6	2·07	2·01

The course of the combination of lime with pozzolanas can be followed by the Florentin method which is based on the insolubility of pozzolanas in cold (below 5°) hydrochloric acid ($d = 1{\cdot}12$) and the solubility in the same reagent of the lime-pozzolana reaction product. Much use of this method has been made by Steopoe,[1] and later authors. As used by Feret[2] 1 g of the pozzolana, ground to pass a 900-mesh/cm^2 sieve, is stirred for 10 minutes with 100 ml of the cold hydrochloric acid and the amount of SiO_2, Al_2O_3 and Fe_2O_3 dissolved determined and calculated as a percentage of the ignited weight of the pozzolana. A mix of equal parts by weight of pozzolana and high-calcium lime is gauged to a plastic consistence and moulded into prisms. These are stored 3 days in moist air and then in water at 15°. After the required period a prism is dried (avoiding carbonation), ground to pass a 900 mesh as before, and the solubility determined on a 2-g sample. The loss on ignition is also determined and the amounts of soluble SiO_2, Al_2O_3 and Fe_2O_3 calculated back to the original ignited pozzolana weight. Data obtained by Feret are shown in Table 73.

There is a progressive increase with age in the amount of soluble silica and alumina in the lime-pozzolana mix indicating that these constituents of the pozzolana have reacted with the lime, but little increase in the soluble Fe_2O_3. Very similar results obtained by Italian workers[3] also show the slower rate of reaction of the Naples pozzolanas compared with the Roman. Pulverised fuel ash also shows a similar progressive increase in the soluble silica and alumina, and little in the soluble ferric oxide; there is also a general parallelism between the increase in soluble matter and strength up to six months, though not at longer ages.[4]

[1] A. Steopoe, *Tonind.-Ztg. keram. Rdsh.* **52,** 1609 (1928).
[2] R. Feret, *Revue Matér. Constr. Trav. publ.* **281,** 41; **282,** 85; **288,** 293 (1933).
[3] G. Malquori and F. Sasso, *Ricerca scient.* **6,** 3 (1935); N. Fratini and A. Rio, *Annali Chim.* **41,** 274 (1951).
[4] J. D. Watt and D. J. Thorne, *J. appl. Chem.* **16,** 33 (1966).

TABLE 73 Acid soluble constituents of pozzolanas and set pozzolana-lime mixes

		Trass	Roman pozzolana	Bacoli pozzolana	Burnt clay
Total per cent in pozzolana	SiO_2	60·0	47·8	57·1	60·1
	Al_2O_3	18·2	17·3	18·3	24·3
	Fe_2O_3	4·1	9·8	4·6	7·2
Per cent soluble in cold HCl ($d=1{\cdot}12$)	SiO_2	0·4	1·0	0·3	1·2
	Al_2O_3 }	10·0	8·3	1·1	0·4
	Fe_2O_3 }		0·7	0·4	5·0
Lime-pozzolana mix stored in water at 15°. Per cent soluble in cold HCl calculated as percentage of pozzolana	1 week SiO_2	2·4	5·0	2·3	7·9
	Al_2O_3 }	11·3	11·9	3·1	7·0
	Fe_2O_3 }		0·7	0·4	6·0
	4 weeks SiO_2	6·7	8·3	4·5	10·7
	Al_2O_3 }	11·3	12·9	3·5	12·0
	Fe_2O_3 }		0·9	0·4	6·1
	26 weeks SiO_2	13·7	14·1	10·6	14·9
	Al_2O_3 }	12·5	16·4	5·3	14·4
	Fe_2O_3 }		1·0	1·0	6·4

The identification of the nature of the products formed was difficult. As long ago as 1908 Gallo observed the formation of hexagonal plate crystals from Italian pozzolanas and in 1930 Malquori obtained evidence for the presence of $4CaO.Al_2O_3.aq$. Later the cubic $3CaO.Al_2O_3.6H_2O$ was identified in a lime-burnt clay reaction product.[1] Subsequently Tavasci,[2] examining several-year-old mortars of lime and Segni and Bacoli pozzolanas, identified crystals with optical properties apparently corresponding to the hexagonal hydrated calcium aluminates. Though not identified microscopically, evidence was produced at the Stockholm Cement Symposium[3] that a hydrated silicate corresponding to $3CaO.2SiO_2.aq$ and its solid solution with $CaO.SiO_2.aq$ was present in lime-pozzolana pastes.

A major advance came with the work of Strätling[4] who showed that a hydrated calcium alumino-silicate, $2CaO.Al_2O_3.SiO_2.aq$, previously unknown, was formed from burnt kaolin and lime-water at 20°. The compound, which had a distinctive X-ray pattern, formed weakly birefringent plates with a refractive index of 1·500–1·505. In addition, evidence was obtained for the presence of $3CaO.2SiO_2.aq$ and Strätling concluded that the reaction was:

$$2(Al_2O_3.2SiO_2)+7Ca(OH)_2 \rightarrow 3CaO.2SiO_2.aq+2(2CaO.Al_2O_3.SiO_2.aq)$$

[1] G. Malquori and F. Sasso, *Ricerca scient.* **8,** 144 (1937).
[2] B. Tavasci, *Il Cemento* **44,** 106 (1947); **45,** 114 (1948).
[3] F. M. Lea, *Sym., Stockholm 1938*, 471.
[4] W. Strätling, *Zement* **29,** 311 et seq. (1940); W. Strätling and H. zur Strassen, *Z. anorg. allg. Chem.* **245,** 257 (1940).

The hydrated gehlenite compound dissolved incongruently in water to give a solution containing 0·001 gm SiO_2, 0·0013 gm Al_2O_3, 0·008 gm CaO per 100 ml.

Later work[1] has shown that $4CaO.Al_2O_3.aq$ is also a product of the reaction with burnt kaolin and that at a temperature of 42° the cubic aluminate $3CaO.Al_2O_3.6H_2O$ is formed and at 50° a hydrogarnet of approximate composition $3CaO.Al_2O_3.0{\cdot}33SiO_2.5{\cdot}3H_2O$.

Both Rhenish trass[2] and the Italian pozzolanas[3] in mixes with lime form a hydrated calcium silicate similar to *CSH* (I) ($0{\cdot}8$–$1{\cdot}5CaO.SiO_2.aq$) and tetracalcium aluminate hydrate. When calcium sulphate is also present ettringite appears initially and later transforms slowly into the monosulphate, $3CaO.Al_2O_3.CaSO_4$ aq, as gypsum is removed from the solution. Similar products are also formed from pulverised fuel ash.[4] In mixtures of pozzolana with Portland cement the reaction products obtained from the pozzolana are similar to those in mixes with lime and gypsum. The extent to which hydrated gehlenite is formed in lime-pozzolana mixes or pozzolanic cements is still open to question. Apart from burnt kaolinite, many workers have failed to find it in the hydration products. However its formation has been reported[5] from the Italian pozzolanas and trass and it is suggested that it is favoured by a high silica content in the pozzolana and a low lime concentration in the solution, caused by a high concentration of alkali released from the pozzolana. The presence of hydrogarnets has also occasionally been reported and said to be favoured by a high alumina content in the pozzolana.

The combination of lime in pozzolana mixes

The estimation of the amount of uncombined lime in set lime-pozzolana mixes or pozzolanic cements is subject to similar difficulties to those encountered with set Portland cement with the addition of any complications arising from the presence of the pozzolana. Of the various solvent extraction methods, the Franke method using acetoacetic ester-isobutyl alcohol as a solvent, or the later refinements of it by extrapolation to zero solvent volume or zero extraction time (see p. 247), seems the most reliable. The *CSH* (I) compound formed by pozzolanas is more resistant to attack by the solvent than *CSH* (II) arising from Portland cement, but the tetracalcium aluminate hydrate formed from pozzolanas seems to lose some lime to this solvent.[6] Other methods[7] that have been used are the calorimetric method, differential thermal analyses and extraction with partially saturated calcium hydroxide solutions.

[1] R. Turriziani and G. Schippa, *Ricerca scient.* **24,** 366, 2645 (1954).
[2] U. Ludwig and H. E. Schwiete, *Sym., Washington 1960*, 1093; *Zement-Kalk-Gips* **16,** 421 (1963); H. E. Schwiete *et al., Sym., Tokyo 1968*.
[3] R. Turriziani, *Ricerca scient.* **24,** 1709 (1954); **26,** 3387 (1956); *Silic. ind.* **23,** 181, 265 (1958); G. Malquori, *Sym., Washington 1960*, 983.
[4] J. Jambor, *Mag. Concr. Res.* **15** (45), 131 (1963); *Zement-Kalk-Gips* **16,** 177 (1963); E. V. Benton, *Bull. Highw. Res. Bd* **239,** 56 (1959).
[5] R. Sersale and P. G. Orsini, *Sym., Tokyo 1968*; R. Turriziani, *Industria ital. Cem.* **32,** 67 (1962).
[6] H. E. Schwiete *et al., Sym., Tokyo 1968*.
[7] G. E. Bessey, *Sym., Stockholm 1938*, 484.

A comparison[1] of the values for free calcium hydroxide in pozzolanic cements by the Franke method and by differential thermal analysis (DTA) has shown agreement to within about 0·5 per cent free CaO. The DTA method depends on the measurement of the area under the 'peak' caused by dehydration of calcium hydroxide at 500–650° on a differential heating curve done under standardised conditions. This DTA method is essentially an alternative to the calorimetric method.[2] Calcium hydroxide does not dissociate at 350° but is almost completely decomposed to give calcium oxide at 550°. In the calorimetric method separate samples of the hydrated cement are heated for 30 minutes at 350° and 550° and the heat of hydration on subsequent immersion in water determined. The difference between these two values, with certain small corrections which are determined for the apparatus and test conditions used, gives a measure of the amount of calcium hydroxide present. The heat of hydration of CaO is 276 cal/g.

On lime-pozzolana mortars the results obtained by the calorimetric and the lime solution extraction methods do not differ in any consistent way though the difference between them was found by Lea[3] to vary from 0 to 1 per cent free CaO.

The lime-extraction method was developed for use in lime mortars.[4] The method consists in shaking a known weight of a ground sample containing not more than 0·08 g of free CaO with 250 ml half-saturated lime water for 24 hours. The suspension is then allowed to settle for 4–6 hours and the CaO in the clear solution estimated with N/30 HCl with phenolphthalein as indicator. The final solution lime concentration should lie between 0·8 and 1·0 g CaO per litre to avoid hydrolysis of hydrated calcuim silicates. The principal source of error is probably the further reaction of pozzolana with free lime during the extraction and the liberation of lime by hydrolysis of any hydrated tetracalcium aluminate present. The method has also been used[5] for mixes of Portland cement and pozzolana, though there is the risk that some further hydration of the Portland cement in the ground sample, with liberation of additional calcium hydroxide, might occur during the extraction.

The free lime contents of some pozzolana-lime mortars after storing for various periods in water at 18° are shown in Table 74.

The maximum amount of lime that is taken up at an age of a year is about 50 g per 100 g pozzolana in 1 : 1 by weight hydrated lime-pozzolana mixes. With higher proportions of pozzolana, or in mixes with Portland cement, the proportion is lower. In pozzolanic cement concretes, where the ratio of free calcium hydroxide produced on hydration to pozzolana is usually relatively low, it seems that a pozzolana will not at most combine with more than about 20 per cent of its weight of lime in a year. Thus, in a 1 : 2 : 4 concrete (w/c 0·6) stored in water, the free calcium hydroxide content at an age of one year was 2·1 per cent CaO with Portland cement, corresponding to some 15 per cent by weight of the anhydrous cement, and 0·3 per cent when 40 per cent of the Portland cement was replaced by a pozzolana. The reduction in the amount of Portland cement present

[1] R. Turriziani and A. Rio, *Annali Chim.* **44**, 787 (1954).
[2] G. E. Bessey, *Bldg Res. tech. Pap.*, No. 9 (1931). H.M.S.O. London.
[3] F. M. Lea, *Bldg Res. tech. Pap.*, No. 27 (1940). H.M.S.O. London.
[4] B. Bakewell and G. E. Bessey, *Bldg Res. spec. Rep.*, No. 17 (1931). H.M.S.O. London.
[5] A. Celani, P. A. Moggi and A. Rio, *Sym., Tokyo 1968.*

would, in itself, reduce the free CaO to about 1·3 per cent so that the pozzolana, amounting to rather over 5 per cent of the concrete, had taken up about 1 per cent CaO. Tests[1] on suspensions of pozzolana and lime in water (water : solid ratio = 5) show broadly similar results with amounts of lime combined by natural pozzolanas after 90 days ranging from 20–40 g CaO per 100 g pozzolana.

TABLE 74 Free lime content* (per cent CaO) of hydrated Lime : pozzolana: standard sand mortars

Mortar weight proportions	1 : 1 : 6				1 : 2 : 9				1 : 4 : 15			
Pozzolana	Days				Days				Days			
	0	7	28	365	0	7	28	365	0	7	28	365
Burnt shale	8	6·1	5·5	3·0	5·3	3·2	2·3	1·4	3·2	1·0	0·5	0·1
Burnt clay	8	4·9	4·5	3·2	—	—	—	—	3·2	2·0	1·2	0·5
Trass	8	5·2	4·6	2·7	5·3	2·7	2·3	1·7	3·2	1·2	0·4	0·2
Santorin earth	8	4·8	4·3	2·4	5·3	—	2·3	1·3	3·2	2·1	1·1	0·2

* Values determined by calorimetric method.

It must be noted that there is little relation between the strength developed and the amount of lime combined by different pozzolanas even in the same mix.

Estimation of pozzolana content in cements

The proportion of pozzolana present in pozzolanic Portland cement can be approximately estimated by selective extraction of the Portland cement with a suitable reagent leaving the pozzolana as undissolved residue. One method[2] involves the use of an alkaline solution of dimethylamine citrate. This is prepared by adding 5 g pure citric acid to 150 ml of an aqueous solution of 33 per cent pure dimethylamine and cooling. The solution should have a pH value of 12·6. One gram of Portland cement stirred for 8 hours with this amount of reagent generally leaves a residue of less than 5 per cent whereas natural pozzolanas dissolve only to the extent of about 5 per cent and a burnt kaolin pozzolana about 15 per cent. An alternative method[3] is to use a solution of picric acid in methyl alcohol. A sample (0·5 g) of cement is stirred with 5 g picric acid and 30 ml methyl alcohol for ten minutes and 20 ml aq dist. added and the stirring continued for 30 minutes. It is then filtered and the residue washed with methyl alcohol, followed by warm water (40°), dried and ignited at 800–1000°. Portland cement dissolves to the extent of about 99 per cent while the pozzolana is insoluble. In the case of a natural pozzolana containing a substantial amount of combined water which is largely lost in the ignition the result will be correspondingly low.

[1] H. E. Schwiete *et al.*, loc. cit.
[2] N. Fratini and R. Turriziani, *Ricerca scient.* **25**, 2834 (1955); **27**, 77 (1957).
[3] S. Takashima, *Semento Gijutsu Nenpo* **111**, 188 (1957); V. Trkulja and V. Ducic, *Cement Jugoslavia* **10** (2), 64 (1966).

The chemical evaluation of pozzolanas

The evaluation of pozzolanas by chemical tests that can be carried out rapidly has long attracted attention. The analytical composition of a pozzolana affords no criterion of its activity. It is true that the combined water content has sometimes been considered of significance and a minimum content is to be found in the German specification for trass, but its value even for distinguishing between materials of this single origin has been disputed and for pozzolanas as a whole it has no practical value.

Many attempts have been made to assess the value of pozzolanas by their content of material soluble in acid or alkali solutions and much of the older work has been summarised by Feret[1] and Sestini.[2] Extraction with hydrochloric, nitric or sulphuric acid, or with alkali hydroxide or carbonate solution, have all failed to give results that showed any relation between the content of soluble constituents and the strength developed by the pozzolana in lime or cement mortars. Later attempts, based on an initial acid attack followed by an alkali treatment, have been more successful in effecting in at least a partial separation of the more active part of natural pozzolanas from the inactive crystalline constituents. Thus the work of Sestini and Santarelli,[3] and of Malquori and Sasso,[4] of which a short review is to be found in the proceedings of the Stockholm Symposium (p. 481), has shown, for Italian pozzolanas, some general relations between the silica soluble on successive treatment with hydrochloric acid and potassium hydroxide solutions and the eventual capacity of the pozzolana to combine with lime, but no relation to the strength developed. Similarly acid-alkali treatments of pulverised fuel ash have not yielded any satisfactory measure of the potential strength properties.[5] A method of test involving the determination of the silica extracted by treatment of pulverised fuel ash with 1N NaOH was required in a former ASTM Standard for fly-ash for use with lime, but abandoned in a later revision (C593–66T). Though the determination of the soluble SiO_2 and R_2O_3 in pozzolanas may have some value for comparing materials of closely similar origin, or type, or for eliminating inert materials, it must be concluded that for any more general use such methods afford no reliable index of quality.

The oldest test for pozzolanas is that of Vicat in which the rate of absorption of lime from a calcium hydroxide solution is measured. This test, both in its original form and in various later modified forms, has some limited value for distinguishing rapidly between active and inert materials, but again it affords no adequate guide to the value of a pozzolana in use. The increase in solid volume that occurs when a pozzolana is placed in a lime solution, usually known as the flocculation test, has also been shown by many investigators to have little value.

[1] R. Feret, *Additions de Matières pulvérulentes aux liants hydrauliques* (Paris 1925). An English translation was issued by the U.S. Bureau of Reclamation, Denver, as *Technical Memorandum*, No. 418 (1934).
[2] Q. Sestini, *Annali Chim. appl.* **27,** 105 (1937).
[3] Q. Sestini and L. Santarelli, *Annali Chim. appl.* **26,** 533 (1936).
[4] G. Malquori and F. Sasso, *Ricerca scient.* **6** (ii), 3, 237 (1935).
[5] J. Forest and E. Demoulin, *Silic. ind.* **29,** 265 (1964); J. D. Watt and D. J. Thorne, *J. appl. Chem.* **16,** 33 (1966).

We have seen earlier that the content of alumina and silica in a lime pozzolana mortar soluble in HCl ($d = 1{\cdot}12$) increases progressively as reaction takes place. When the total content of soluble material determined in this way is plotted against the strengths developed in lime mortars the results for a limited group of pozzolanas may fall approximately in a common curve, as was found by Feret and by Watt and Thorne for pulverised fuel ash, but it is clear that no broad relation exists for other tests on a group of Italian pozzolanas[1] did no more than pick out the materials that made the least contribution to strength.

Properties of lime-pozzolana mortars

Pozzolanas are normally used in conjunction with high-calcium limes, though there is no objection to their use with hydraulic limes, except for test purposes when it is desired to ascertain the properties of the pozzolana itself.

The setting time of lime-pozzolana mixes may be determined by the usual Vicat needle method. The amount of water to be used for the paste is ascertained by the Vicat plunger as for Portland cement. The setting time is variable and though the initial set may occur in 1–3 hours the final set does not usually occur in much less than 10–12 hours.

The strength developed in lime-pozzolana mortars varies with the ratio of lime to pozzolana in the mix. At early ages the maximum strength is obtained with a lime-pozzolana ratio of about 1 : 4 with finely ground materials, but at longer ages the optimum ratio moves towards mixes of higher lime content and approaches about 1 : 3 to 1 : 2 at one year. Feret found for plastic mortars containing 1 part (weight) of lime and pozzolana to 3 parts graded sand the results shown below on specimens tested at the age of 1 year.

Pozzolana	Maximum strength (lb/in^2)		Percentage* of lime giving maximum strength	
	Bending	Compression	Bending	Compression
Trass	550	1640	35	26
Roman	504	1850	32	28

* Percentage in the lime-pozzolana mix.

TABLE 75 Effect of pozzolana : lime ratio on strength of mortars

Pozzolana	Mix proportions (weight)			Tensile strength (lb/in^2)			
	Hydrated lime	Pozzolana	Standard sand	7 days	28 days	90 days	1 year
Burnt shale	1	1	6	107	207	341	521
	1	2	9	133	322	459	560
	1	4	15	203	371	514	533
Trass	1	1	6	213	361	447	477
	1	2	9	225	390	425	495
	1	4	15	234	363	412	433

[1] N. Fratini and A. Rio, *Annali Chim.* **41,** 274 (1951).

The figures in Table 75 illustrate some results[1] obtained on mortars of dry consistence with finely ground trass and burnt-shale pozzolanas.

Plastic mortars show similar trends with variation in the lime : pozzolana ratio, as also do concretes. Some data, on concretes of 2-inch slump, stored in water at 18° after an initial 14 days in moist air, are shown below:

Concrete mix (weight) $1 : 1\frac{1}{2} : 2\frac{2}{3}$, lime + pozzolana : sand : gravel

Cementing agent (weight)	1 Hydrated lime 2 Pozzolana	1 Hydrated lime 4 Pozzolana
Compressive strength (lb/in^2)		
28 days	243	378
3 months	786	1158
6 months	1240	1670
1 year	1725	2020
2 years	2160	2365
Ratio W/(C+P)	0·68	0·65

Tests on 8 × 4-inch cylinders.

The strengths[2] developed by some lime-pozzolana mortars tested according to the German trass specification are shown in Table 76.

TABLE 76 Tests on 0·8 : 1 : 1·5 hydrated lime : pozzolana : standard sand mortars

Pozzolana	Percentage residue on 170-mesh sieve	Tensile strength (lb/in^2)		Compressive strength (lb/in^2)	
		7 days	28 days	7 days	28 days
Italian	33·8	214	356	1405	2710
Trass	35·5	120	255	1120	1700
Moler	49·2	178	270	411	990
Pumice	29·2	72	270	495	1480

In tests carried out at the Berlin Materialprüfungsamt, Burchartz[3] found for 30 samples of trass tensile strengths varying from 204 to 396 lb/in^2 at 28 days with an average value of 283, and compressive strengths varying from 1490 to 3080 with an average value of 2220.

The effect of the fineness of grinding of pozzolanas is illustrated by Table 77.

The rate of development of strength of lime-pozzolana mortars is much increased by rise in temperature. The effect is indeed so marked that a pozzolana which is comparatively inert for weeks at the average temperatures prevailing in this country, and therefore of relatively little value, may prove satisfactory under the temperatures attained in the Mediterranean summer and in more tropical climates.

[1] F. M. Lea, *Building Res. tech. Pap.* No. 27 (1940). H.M.S.O. London.
[2] R. Grün, *Internat. Soc. Test. Mat.*, Zurich **1**, 781 (1931).
[3] *Tonind.-Ztg. keram. Rdsh.* **48**, 1221 (1924).

TABLE 77 Effect of fineness of grinding of pozzolanas 1 : 4 : 15 Lime : Trass : graded sand mortars*

Percentage residue on 170-mesh sieve	Strength (lb/in^2)			
	Bending		Compression	
	28 days	1 year	28 days	1 year
43	183	372	341	1190
14·5	269	475	500	1280
3·0	306	480	690	1460

* 14·8 per cent water. Data after Feret.

TABLE 78 Effect of temperature on strength of lime-pozzolana mortars*

Pozzolana	Tensile strength (lb/in^2)							
	7 days				28 days			
	0°	12°	25°	35°	0°	12°	25°	35°
Santorin earth	0	8	44	103	0	86	317	408
Burnt shale	48	86	151	256	92	241	348	455

* 1 : 1 : 6 Lime : Pozzolana : standard sand mortars of dry consistence.

Small additions of gypsum, 1–3 per cent, to the lime-pozzolana mix sometimes have a favourable effect on the strength of mortars at short ages, but the effect is erratic and cannot be predicted. Somewhat larger additions,[1] 5 per cent, are used in Italy and are stated to accelerate setting and hardening of the volcanic ash pozzolanas and to give a high resistance to attack by seawater.

Pozzolana-lime mortars attain a much higher ultimate strength when cured in water than in air, though the initial effect is normally the reverse. With pozzolana-lime concretes in particular a long period of moist curing is essential to the development of high strengths, and rapid drying has most injurious effects.

Properties of pozzolanic cement concretes

Pozzolanas were originally used in Portland cement mortars and concretes on account of their property of combining with lime and so removing the calcium hydroxide liberated during the setting of the Portland cement. The calcium hydroxide, which is readily subject to chemical attack, is thus removed and in its place a lime-pozzolana compound is formed.

[1] C. Vittori, *Internat. Soc. Test. Mat.*, Zurich **1,** 854 (1931); R. Turriziani and G. Schippa, *Ricerca scient.* **24,** 1895 (1954).

The substitution of pozzolana for Portland cement reduces the strength obtained at the earlier ages though the ultimate strength attained may be increased. This is illustrated in Fig. 126 where the relative strength at different ages of a concrete made with pozzolanic cements containing various proportions of a burnt shale pozzolana are shown.

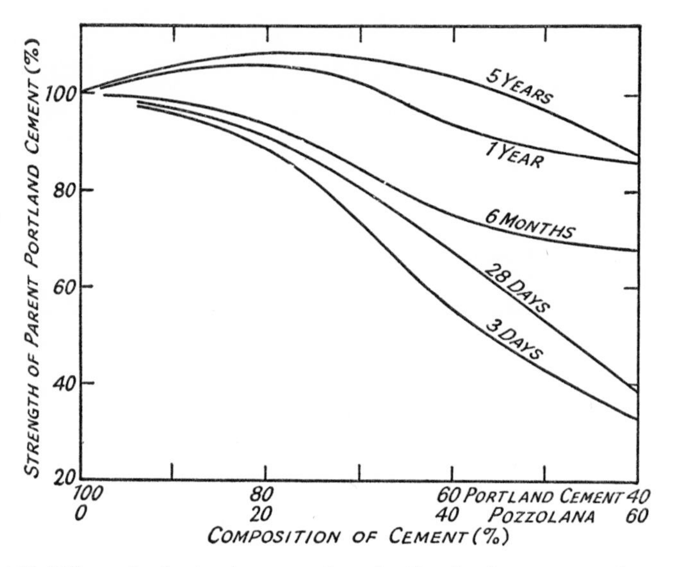

FIG. 126 Effect of substituting pozzolana for Portland cement on the strength of 1 : 2 : 4 : 0·6 concrete stored in water at 18°.

The difference between good and indifferent pozzolanas is much more marked at long than at short ages when in any event most of the strength comes from the Portland cement constituent. Some comparative data on a good and poor burnt shale and clay pozzolana are given in Table 79.

TABLE 79 Compressive strength of pozzolanic cement concretes
1 : 2 : 4 concrete, W/(C+P) = 0·60, 8 × 4-inch cylinders stored in water at 18°. lb/in²

Cement	3 days	28 days	6 months	1 year	5 years
100 per cent Portland cement A	1210	3345	4400	4560	5540
60 per cent Portland cement A 40 per cent good pozzolana	660	2230	3535	4290	5790
60 per cent Portland cement A 40 per cent poor pozzolana	670	2000	3090	3345	4230

Some comparative data on a variety of pozzolanas are shown in Table 80 from which it can also be observed that the proportionate loss in strength at ages up to a year on substituting pozzolana is greater with a rapid than with a slower-hardening Portland cement.

TABLE 80 Compressive strength of pozzolanic cement concretes
1 : 2 : 4 concrete, W/(C+P) = 0·60, 4-inch cubes stored in water at 18°

Cement	Per cent compressive strength*					
	Portland cement A			Portland cement B		
	28 days	180 days	1 year	28 days	180 days	1 year
100 per cent Portland Cement	100	100	100	100	100	100
60 per cent Portland cement; 40 per cent pozzolana						
Pozzolana:						
Burnt clay	68	82	93	61	69	77
Burnt shale	74	88	95	64	74	80
Rhenish trass	69	75	80	58	66	71
Naples	60	71	79	52	63	70
Ground sand	45	52	57	45	49	50
Actual strength of 100 per cent Portland cement (lb/in²)	3710	5950	6350	5800	7990	8350

* Expressed as percentage of the strength of the unsubstituted Portland cement at the same age.

For Italian pozzolanas, tested in concretes containing 300 kg cement per cubic metre, and cured at 20°, Ceresto and Rio[1] have given strengths, for a 40 per cent substitution, of 62, 75, 85 and 86 per cent of that of the unsubstituted Portland cement at 7, 28, 180 and 365 days for Segni pozzolana and 63, 86, 95 and 97 per cent for Bacoli pozzolana. With a substitution of only 20 per cent the strength at ages of 28 days and upwards was about the same as that of the parent Portland cement.

Typical data on pulverised fuel ashes with carbon contents below about 10 per cent show that their substitution for 20 per cent Portland cement reduces the strength of concrete cured at 18–20° by 10–30 per cent at 7 days, 0–25 per cent at 28 days, and leaves it substantially unchanged or slightly increased at 1 year. For a 30 per cent substitution the corresponding reductions range from 25 to 50 per cent at 7 days, 5 to 45 per cent at 28 days and from a 15 per cent loss to a strength increase at 1 year. The fineness of the ash has more influence on the strength developed than the combustible content when this is below about 10 per cent. The results are also influenced by the amount of water required for a

[1] A. Ceresto and A. Rio, *Il Cemento* **48** (10–12), 3 (1951).

given workability of the concrete. Because of the rounded shape of the particles this may be reduced by up to 10 per cent with some ashes but in other cases it can be increased by a similar amount.

The tensile strength of pozzolanic cements is always a higher proportion of the strength of the unsubstituted parent Portland cement than is found for the compressive strength, as may be seen from Table 81. The data here refer to tests at a standard consistence and not a constant ratio of water to pozzolana plus cement. With only a 20 per cent substitution of pozzolana the tensile strength at 18° often equals that of the Portland cement at 90 days and exceeds it at longer ages.

TABLE 81 Tensile and compressive strength of pozzolanic cement mortars percentage strength*

	Per cent water in mortar	Tensile			Compressive		
		7 days	28 days	180 days	7 days	28 days	180 days
100 per cent Portland cement	11·0	100	100	100	100	100	100
60 per cent Portland cement; 40 per cent pozzolana							
Pozzolana:							
Burnt clay	10·5	82	92	102	56	79	84
Burnt shale	11·5	71	86	89	50	70	81
Trass	11·2	67	86	94	52	73	—
Naples	10·3	74	82	87	55	71	81
Ground sand	10·5	66	66	69	49	54	—

* Expressed as percentage of strength of the unsubstituted Portland cement at the same age. Tests on 1 : 3 standard sand mortars of standardised plastic consistence, stored in water at 18°.

We have seen earlier that the rate of reaction and strength development of pozzolana-lime mixes increases much with rise in temperature and this is reflected in the behaviour of pozzolanic cements. Thus in some tests[1] made in the U.S.A. a pozzolanic cement concrete that developed about 50 and 66 per cent at 7 and 28 days of the strength obtained with the unsubstituted Portland cement when both were cured at 70° F, gave 79 and 84 per cent of the strength when both were cured under adiabatic conditions with rising temperature corresponding to the conditions in the interior of a large concrete mass. Conversely at lower temperatures the rate of strength development of pozzolana cements is affected more adversely than that of Portland cements.

Pozzolana concretes require a long period of wet curing if the best results are to be obtained and if cured in air they only develop lower strengths. This is, of course, also a characteristic of Portland cements, but it is more pronounced, to a degree which seems to vary with the pozzolana, with pozzolanic cements.

[1] R. Blanks, *Proc. Am. Concr. Inst.* **46,** 89 (1950).

Concretes containing pozzolanic cements usually have a rather lower slump than those made with unsubstituted Portland cement at the same water content though there are exceptions, such as with some pulverised fuel ashes of low carbon content and high fineness which actually reduce the amount of water required. The plasticity of the concrete and its freedom from segregation and bleeding, at a given slump, is usually much improved and this is particularly notable in lean mixes. The workability, as measured by compaction factor or remoulding tests, differs little from that of Portland cement at the same water content.

The heat evolved during the hardening of Portland cement concretes is reduced by the substitution of pozzolanas though not proportionately to the degree of substitution since the reaction between lime and pozzolana itself evolves some heat. As a rough approximation the percentage reduction in heat evolution at 7 to 28 days may be taken at about one-half the percentage substitution, but it varies very appreciably for different pozzolanic cements.

The expansion of Portland cement mortars and concretes in water and the reversible moisture movement are little influenced by the substitution of pozzolanas for cement, but the initial drying shrinkage usually tends to be slightly increased. It has been found in practice in mass concrete construction that pozzolanic cement concretes show rather better behaviour in respect of cracking than ordinary Portland cements. They share this property with low heat Portland cements and it arises not merely from a reduced thermal contraction in the mass, but also from an increased ability to creep under load and to extend to a greater extent before rupture occurs.

The permeability of concretes at early ages is not much influenced by the substitution of pozzolana for Portland cement, but at longer ages under conditions of wet curing it becomes progressively lower.

The frost resistance of pozzolanic cement concretes tends to be lower than that of Portland cement when tests are commenced at an early age such as 28 days. This doubtless follows from the reduction in strength at such ages, for there is evidence that if the tests are not made until an age of six months or more their performance equals that of Portland cement. It has been found that the substitution of pulverised fuel ash, and this may apply to other pozzolanas, reduces the air content of an air-entrained concrete unless additional air-entraining agent is added. The resistance to abrasion of pozzolanic cements seems to be rather lower than that of Portland cements though again the difference may decrease at long ages.

The addition of suitable pozzolanas has been found to be a very useful method of inhibiting expansion of concrete arising from alkali-aggregate reaction but the discussion of this is more conveniently postponed to Chapter 18.

Resistance of pozzolanic cements to chemical attack

One of the important reasons for using pozzolanic cements has been the increased resistance they offer to attack by chemical agencies and particularly seawater.

Feret conducted tests at the Laboratoire des Ponts et Chaussées at Boulogne extending over very many years. Mortars composed of 1 part Portland cement and pozzolana and 3 parts graded sand immersed for 24 years in seawater showed the beneficial effects of the substitution of trass, Roman pozzolana, and particularly burnt gaize, for part of the cement. Another long series of tests included mortar cubes of 40 cm edge immersed in the open sea when one month old. Three mixes with cement-sand ratios (weight) of 1 : 2·7, 1 : 3·6 and 1 : 5·3 respectively were used and a large number of Portland cement and pozzolanic cements tested. The pozzolanic cements were composed of 2 parts Portland cement to 1 part (by weight) of Roman pozzolana, trass, burnt gaize or burnt clay pozzolana. Feret reported that, after 10 years' immersion, the substitution of burnt gaize and burnt clay pozzolanas had rendered even the leanest mortars practically immune from attack and that they were more effective than either the trass or Roman pozzolana.

Some data showing the relative rates of expansion in sulphate solutions of 1 : 3 plastic cement mortars with an ordinary (Type I) Portland cement as base is shown in Table 82.

TABLE 82 Expansion of cement mortars in sulphate solutions

Cement		Pozzolana	Time in weeks to expand 0·1 per cent	
Per cent Portland cement	Per cent pozzolana		5 per cent sodium sulphate	5 per cent magnesium sulphate
100	0	None	18	10
80	20	Burnt clay	52	20
60	40	,, ,,	> 200	220
60	40	Burnt shale	> 200	> 200
80	20	Trass	50	14
60	40	,,	> 200	170
60	40	Ground sand	22	9
60	40	Ground brick	44	21

1 : 3 Cement : standard sand mortars. w/c about 0·5. Immersed in sulphate solution at 18° at 7 days old.

Data such as these are only relative and in fact tests on mortars tend to overestimate the benefit gained in concrete, but they nevertheless illustrate the increased chemical resistance obtained with pozzolanas. In concrete a substitution of 30–40 per cent pozzolana for ordinary Portland cement will considerably increase the resistance to 5 per cent sodium sulphate and to 0·5 per cent magnesium sulphate solutions, but is less effective against 5 per cent magnesium sulphate as is shown by the data in Table 83.

The partial replacement of sulphate-resisting Portland cement by pozzolana does not necessarily give a comparable improvement in sulphate resistance and the result obtained is related to the $SiO_2 : R_2O_3$ ratio in the pozzolana. Thus pulverised fuel ash which usually has a high R_2O_3 content has in various tests not proved an effective substituent for enhancing the sulphate resistance of sulphate-resisting Portland cement. On the other hand a highly silicious Italian

TABLE 83 Effect of sulphate solutions on pozzolanic cement concrete

Cement	Water storage compressive strength (lb/in^2)		Compressive strength as percentage of strength at same age of concrete stood in water			
			5 per cent sodium sulphate		5 per cent magnesium sulphate	
	6 months	12 months	6 months	12 months	6 months	12 months
100 per cent ordinary Portland cement	6575	6930	45	19	73	42
60 per cent Portland cement; 40 per cent pozzolana						
Pozzolana:						
Burnt clay	5165	6270	115	94	84	63
Trass	4730	5260	86	74	78	51
100 per cent sulphate-resisting Portland cement	7050	7460	92	80	92	82

Tests on 4-in. cubes 1 : 2 : 4 concrete, w/c 0·60.

pozzolana containing nearly 90 per cent SiO_2 and only a few per cent R_2O_3 has been found to increase the resistance of sulphate-resisting Portland cement.[1]

The increased resistance to sulphate and seawater attack obtained by the addition of pozzolanas has long been the subject of discussion. In part it is attributable to the removal of the free calcium hydroxide, formed in the hydration of Portland cements, by combination with the pozzolana. This removal is illustrated in Fig. 127 in which the free calcium hydroxide content of mortars made with Portland cement and with a pozzolanic cement containing 60 per cent of the same Portland cement and 40 per cent burnt clay are shown. This, however, cannot be the sole explanation, since the expansion accompanying the reaction of the alumina compounds in set Portland cement with sulphate solutions to form ettringite, $3CaO.Al_2O_3.3CaSO_4.31H_2O$, is also a cause of their disintegration.

Lafuma[2] has advanced a theory to explain the resistance of high-alumina cement to sulphate attack (see p. 524) which has also been applied to pozzolanic cements. He suggests that the combination between an insoluble set cement compound in the solid state and a substance in solution always causes expansion, but that if the cement compound passes into solution, reacts, and then precipitates as a solid, no expansion occurs. It is known that the solubility of the hydrated

[1] R. Turriziani and A. Rio, *Sym., Washington 1960*, 1067; *Industria ital. Cem.* **27,** 145 (1957); **32,** 313 (1962); A. Rio and A. Celani, *R.I.L.E.M. Sym. Behaviour of Concrete Exposed to Sea-water, Palermo 1965.*

[2] H. Lafuma, *Revue Matér. Constr. Trav. publ.* **243,** 441 (1929); **244,** 4 (1930).

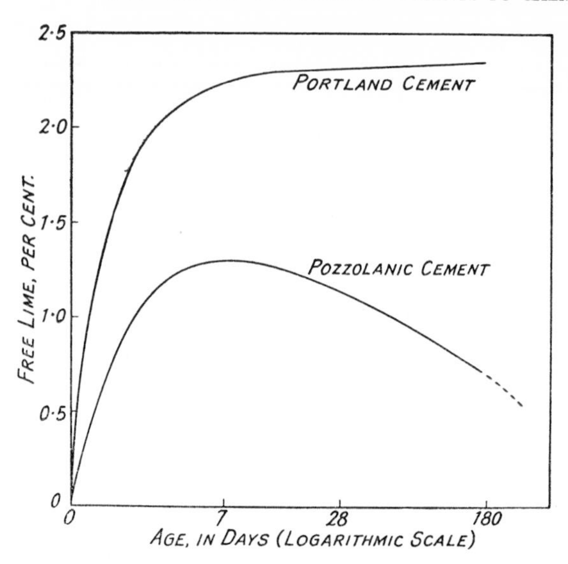

FIG 127 Free lime content of 1 : 3 cement-sand mortars.

calcium aluminates is very low in a saturated lime solution, but increases as the lime in solution falls, so that on this theory the removal of free calcium hydroxide establishes conditions under which sulphate reaction can occur without expansion. A somewhat different explanation has been advanced by Turriziani and Rio[1] in the light of later knowledge of the equilibria in the system $CaO–Al_2O_3–SO_3–H_2O$. We have seen earlier (p. 236) that in set Portland cement the aluminate compounds and the gypsum eventually form a solid solution between $3CaO.Al_2O_3.CaSO_4.12H_2O$ (the 'low' sulphate form) and $4CaO.Al_2O_3.12H_2O$. The action of sulphates is to convert this to ettringite, but the sulphate concentration required increases as the concentration of calcium hydroxide in solution decreases. Further, since the reaction between the pozzolana and lime itself produces hydrated calcium aluminates, more reaction with a sulphate solution is required before the formation of ettringite, which causes the swelling, commences, for the first action will be to form the 'low' sulphate solid solution.

These explanations, though partial reasons, are still not adequate for the resistance of pozzolanic cements seem to depend more on the inhibition of sulphate attack than on its mode of action. An increased impermeability of the set cement mass may be another factor but considerable importance must, it seems, be attributed to the formation of protective calcium silicate hydrate films over the vulnerable aluminate compounds. The alumina in a pozzolana, at least in part, forms hydrated tetracalcium aluminate though there is evidence that the pozzolana aluminate products are not as susceptible to sulphate attack as those

[1] R. Turriziani and A. Rio, *Annali Chim.* **44,** 787 (1954); see also Chassevent and Stiglitz, *C. r. hebd Séanc. Acad. Sci., Paris* **222,** 1499 (1946); W. Eitel, *Proc. Am. Concr. Inst.* **53,** 679 (1957).

from Portland cement. The proportion of *C-S-H* gel in a hydrated pozzolanic cement is higher than in hydrated Portland cement and its protective action therefore increased. The mode and time of formation of the hydrated silicate in relation to that of the hydrated aluminates from the Portland cement may also play a part. Turriziani and Rio have noted that, microscopically, the aluminate reaction products in pozzolanic cements appear more protected by the *C-S-H* gel than are those in Portland cements. The difference between the effect of substituting high-alumina pozzolanas in ordinary and in sulphate-resisting Portland cement suggest that the increase in total hydrated aluminates can with the latter cement outweigh the increase in protective effect of the *C-S-H* gel. With a pozzolana of high-silica and low-alumina content on the other hand the protective effect of additional *C-S-H* gel adds to the resistance of a sulphate-resisting Portland cement.

Another theory which has been advanced by Steopoe[1] is particularly applicable to the action of seawater, though its author does not restrict it to this. Steopoe considers that the lime-pozzolana reaction products are decomposed by seawater, with removal of lime, leaving silica and alumina gel, and that the resistance to deterioration is attributable to a change in the cement-binding material from an unstable complex of lime compounds to an inert gel.[2] This gel is considered to form a protective coating over the remaining lime compounds, as had been suggested earlier by Ferrari and other workers. The presence of pozzolanas increases the amount of silica gel ultimately remaining as the bonding constituent and so preserves the concrete. It is well known that silica gel can form an excellent bonding material, and also that lime is progressively leached from concrete in seawater. Thus, lime-pozzolana mortars immersed in seawater for long periods show a much reduced lime content, though remaining hard and sound, and Portland cement concretes also slowly lose lime. Steopoe[3] has also found that of the silica and alumina and ferric oxide gels which finally result from the action of seawater, or magnesium sulphate, on set cement, the silica gel has the lowest permeability to sulphate ions. This supports his view that the beneficial action of pozzolanas in seawater work arises ultimately from the extra silica gel which results from the decomposition of the hydration products. As far as the action of sodium sulphate or calcium sulphate, which do not decompose the calcium silicate hydrates, is concerned the protective effect can only arise from the *C-S-H* gel itself. This probably applies generally to mixed sulphate ground waters, where leaching action is not usually a major factor, and to laboratory tests with any sulphate salt.

[1] A. Steopoe, *Tonind.-Ztg. keram. Rdsh.* **60,** 487, 503, 944 (1936); *Zement* **26,** 169, 643 (1937); and numerous other papers. For a general review, see F. M. Lea, *Sym., Stockholm 1938*, 460.

[2] Much of Steopoe's conclusions are based on the Florentin method for separating the silica in Portland cement and hydrated silicates from insoluble silica gel and unreacted pozzolana. In this method the sample is treated with cold HCl ($d = 1{\cdot}12$) for 10–15 minutes. The crucial point is the rate of solution of the 'soluble' silicates, since any not dissolved is classed as insoluble. It is doubtful if the method does completely separate the soluble calcium silicates from silica gel and unreacted pozzolana, but it is a useful method for comparative purposes.

[3] A. Steopoe and L. Vaicum, *Zement-Kalk-Gips* **14,** 348 (1961).

Testing of pozzolanas and pozzolanic cements

The testing of pozzolanas is complicated by the fact that they have no cementitious properties in themselves and only develop them when mixed with lime or cement. Further, in pozzolanic cements, it is the properties contributed by the pozzolana at long ages that are important, for at short ages the influence of the Portland cement fraction of the mix is dominant in determining the strength. We have, in effect, two separate, though related, problems to consider, the testing of pozzolanas and the testing of pozzolanic cements.

We have seen earlier that chemical tests have no general value for assessing pozzolanas, though they may find a place as an initial sorting test when considering only a limited class. In the German specification for trass[1] the combined water content, defined as that lost on ignition after first drying to constant weight at 98°, is required to be not less than 6 per cent. A few countries impose some general limitation on the chemical composition of pozzolanas. Thus the U.S.A. standard for natural pozzolanas (C618–68T), and for pulverised fuel ash requires that the content of $SiO_2+Al_2O_3+Fe_2O_3$ shall be at least 70 per cent, while that of Japan sets a maximum limit for SiO_2 of 60 per cent. The latter unusual limitation is presumably intended for pulverised fuel ash and not natural pozzolanas which are not used in Japan. These overall composition requirements do not appear in European specifications and indeed the U.S.A. minimum requirement comes close to rejecting some of the Roman pozzolanas. It cannot be said that such overall limitations on composition have much practical value and they can be unnecessarily restrictive. The SO_3 content of pulverised fuel ash is limited to 2·5 and 5·0 per cent respectively in the British and ASTM specifications, and the MgO content to 4 per cent in the British. Most other countries make no such requirement for the pozzolana but only for the mixed pozzolanic cement. There is no consistency between the requirements in different countries as to the proportion of pozzolana permissible in a pozzolanic cement. Some countries, such as Italy, impose no direct limitation while in others it varies from about 15 to 50 per cent. Thus the U.S.A. requirement is 15–40 per cent and the German 20–40 per cent.

Pozzolana-lime mixes

Strength tests on pozzolana-lime mixes are carried out in a variety of ways in different countries. The Italian standard[2] specifies tests on a mortar composed of 1 part by weight hydrated lime and 3 parts pozzolana passing a 3-mm sieve, sufficient water being added to give a consistence such that the mix will agglomerate under hand pressure. The mortar is filled into the moulds with a metal spatula and levelled off after about 6 hours. For the more active pozzolanas the tensile and compressive strengths at 28 days of specimens cured for 7 days in moist air, and subsequently in water, at 15–20° must not be less than 5 and 25 kg per sq. cm respectively. The specimens are dried before testing, an unusual procedure. In Germany, trass is tested, and used, in a more finely-ground condition, with a maximum residue of 20 per cent by weight on the 900-mesh/cm^2

[1] Deutsche Normen. DIN 51043 (1931). Now under revision.
[2] Consiglio Nazionale delle Richerche, Norme, Rome 1952.

sieve. The test mortar is composed of a 0·8 : 1 : 1·5 weight mix of hydrated lime-trass-standard sand gauged with 11–12½ per cent water. This gives a mix of rather dry consistence and the tensile briquettes and compression cubes are compacted with the Boehme hammer machine. Specimens are cured for 3 days in moist air at 17–20° and then in water at that temperature. The minimum tensile strengths required are 5 and 16 kg per sq. cm at 7 and 28 days, and for compressive strength 45 and 140 kg.

In some other countries pozzolanas are tested in mortars composed of 1 part lime and pozzolana to 3 parts standard sand by weight. Feret, at the Brussels meeting of the International Association for Testing Materials in 1906, recommended these prportions and also that the mix should be plastic. The consistence test suggested made use of the Vicat plunger (see p. 363) but loaded so as to have a total weight of 2 kg. The test was carried out on the mortar, not on the neat lime-pozzolana mix, and the standard consistence was taken as that at which the rod, released at the surface of the mortar, penetrated 34 mm into a mass 40 mm deep. The amount of water required varied from 12 to 15 per cent by weight of the mortar. This form of consistence test, which is a very useful one for plastic mortars, has been used by the author in a slightly modified form.[1] The addition load in this case was 2 kg, making, with the plunger weighing 300 g, a total load of 2300 g, the standard consistence being that at which the plunger sank 30 mm into the 40-mm deep mould. Mortars of this consistence can be pressed into the moulds with the thumbs. Feret recommended a ratio of pozzolana to lime of 4 : 1 by weight, but a 2 : 1 ratio was used in the work in England. It is necessary that the strength shall not be influenced appreciably by the hydrated lime used. This difficulty is got over in Germany by specifying the source of the lime and the manner of its hydration to a dry powder, while in Italy it is required that the lime shall contain at least 94 per cent CaO and that the dry hydrate shall pass a 0·18-mm sieve. The ordinary specification for hydrated lime is not sufficiently stringent to ensure the high degree of uniformity required in a lime used for testing pozzolanas, but it is possible to exclude any serious influence of the hydrated lime if this is required to conform to the British Standard for high-calcium hydrated lime (BS 890 : 1966) with the following *additional* requirements:

(*a*) The calcium compounds present in the lime, and calculated as CaO, shall not be less than 93 per cent by weight of the ignited sample.

(*b*) The loss on ignition, after correction for the carbon dioxide content, shall be between 21 and 25 per cent by weight.

(*c*) The expansion in the Le Chatelier soundness test laid down in the standard shall not exceed 2 mm.

(*d*) The residue on the BS 85 mesh sieve (180 μ) shall not exceed 1 per cent and after passing this sieve not more than 2 per cent shall remain on the BS 170 sieve (90 μ).

For use in practice in mortars for building purposes, the requirements of BS 890 are, of course, quite adequate. The ASTM specification for pozzolanas (C593–66T) gives rather similar requirements.

[1] F. M. Lea, *Building Res. tech. Pap.* No. 27, H.M.S.O. London (1940).

The tensile strengths of plastic 1 : 2 : 9 hydrated lime : pozzolana : standard sand mortar briquettes cured seven days in moist air (90 per cent relative humidity), at 18° and then in water, may be quite low at 14 or even 28 days and the differentiation between materials that show good and indifferent performance at longer ages is not adequate. An accelerated test may be used in which the briquettes are cured for 7 days in moist air at 18°, then immersed in water in a container which is placed in a thermostat at 50°. The temperature of the water in the container should reach 50° within 1½ hours and the briquettes are cured under this condition for 46 or 94 hours from the time of placing in the thermostat. The briquettes are then placed in water at 18° for 2 hours before testing. Both these curing treatments give strengths that show a fair correlation with the strength developed at 90 days at normal temperatures, but the 94-hour curing at 50° gives the better correlation with the strength at 180 days. Longer periods of curing at 50°, or curing at 100°, are not suitable, since relatively inert materials then develop strength. A compressive strength test on cubes can alternatively be used.

Some typical data are shown in Table 84 for pozzolanas ground to cement fineness.

TABLE 84 Tensile strength of 1 : 2 : 9 hydrated lime : pozzolana : standard sand plastic mortars

Pozzolana	Per cent water in mortars	Tensile strength (lb/in²)					
		Water storage* at 18°				Curing at 50°	
		14 days	28 days	90 days	180 days	46 hours	94 hours
Good burnt clay	13·5	173	294	414	445	404	405
Good burnt clay	15·1	57	164	300	312	313	355
Poor burnt clay	15·5	34	67	119	143	66	91
Good burnt shale	14·1	41	105	303	351	204	395
Trass	13·7	92	184	257	275	285	317
Bacoli (Naples)	12·8	56	122	300	300	269	315
Santorin earth	13·0	0	63	284	361	181	283
Lightly burnt brick	12·8	0	0	98	233	37	147

* First 7 days in moist air.

Pozzolanic cements

The testing of pozzolanic cements may be based on tests on the blended cement or separate tests may be required on the pozzolana and the Portland cement before blending. The nature of the tests required must depend to some extent on the purpose for which the cement is to be used. Thus, if a pozzolanic cement is to be used in mass concrete on account of its reduced heat evolution or better workability, or for economy, it suffices to know that its strength is adequate, both at short and long ages, and that it shows the other properties mentioned. If inhibition of alkali-aggregate reaction is required then an additional test for this purpose is necessary. In all these cases the proportion of pozzolana present will often be only 20–30 per cent. If, however, the cement is required for resistance to seawater or sulphate waters, it is not only necessary that the pozzolana should be an active material, but also that it should be present in a sufficient proportion, usually 30–40 per cent.

Whatever the purpose for which the cement is to be used, the methods of testing must be such as not only to check that it reaches the minimum strength values required at the normal test ages up to 28 days, but also to ensure that the pozzolana present is an active material that will contribute substantially to strength at long ages.

Tests on the strength of lime-pozzolana mortars at 28 days can distinguish between inert and active materials, but as has been seen from Table 84, the 28-day strength bears a very variable relation to that at 180 days for different pozzolanas. The accelerated test with curing for 94 hours at 50° does, however, give results which correlate fairly well with the relative performance of different pozzolanas in pozzolanic cement concretes at the age of 1 year. The temperature and duration of the curing at the higher temperature which gives the best relation can only be determined empirically, but the conditions mentioned seem satisfactory.

The ASTM specifications for pozzolanas make use of such an accelerated test on a lime-pozzolana mortar to measure the activity of a pozzolana. Compressive tests are made on 2×4 inch cylinders of a mix of 1 : 9 hydrated lime-graded sand plus an amount of pozzolana equal to twice the weight of lime multiplied by the factor (sp. gr. of pozzolana)/(sp. gr. of lime); that is the solid volume of pozzolana is twice that of the lime. The mortar is cured in the mould first for 24 hours at 23° C and then for 6 days (less 4 hours) at 55° and a final 4 hours cooling to 23°. This test is called the 'pozzolanic activity index with lime'. Comparative compressive strength tests are also required in the ASTM specifications on 2 in. cubes of a 1 : 2·75 Portland cement : graded sand mortar (as specified in ASTM C109–64) of a standard consistence (w/c about 0·5) and of a 0·65 : x : 2·75 Portland cement : pozzolana : graded sand mortar with the same Portland cement. The amount x of the pozzolana is 0·35 multiplied by the factor (sp. gr. of pozzolana)/(sp. gr. of Portland cement). In other words 35 per cent by weight of the cement is substituted by an equal solid volume of pozzolana. The specimens are cured in moist air first for 24 hours at 23° C (73° F) and for 6 or 27 days at 38° C (100° F). The ratio of the strength of the second to the first mortar at 7 or 28 days is taken as another index of the activity of the pozzolana. It is called the 'pozzolanic activity index with Portland cement'. (ASTM C618–68T).

The strength shown by pozzolanic cements in plastic mortars or concretes at ages up to 7 days is determined almost entirely by the Portland cement present, and largely so at 28 days, for it is only at longer ages that the pozzolana-lime reaction contributes substantially to the strength. The early strengths can be controlled by the normal specification tests for cement, provided these are not carried out on mortars of dry consistence. Tensile tests on mortars with a water : cement ratio of about 0·32, as formerly used for the tensile test in Great Britain and many other countries, show even at the earliest ages an unchanged or increased tensile strength when pozzolanas are substituted in limited proportions, up to about 30 per cent, for cement. The compressive strength of similar mortars does show a reduction in strength when pozzolana is substituted in more than small amounts, but even so it does not reflect the more substantial drop found in plastic mortars and concretes.[1] Strength tests on mortars with a water : cement ratio of 0·40, as used in the British Standard compression test, or at a standard mortar consistence determined with a Vicat plunger with a total weight of 2300 g as described on page 445, are free from this defect and can be applied to pozzolanic cements. Some typical data for the compressive strength of pozzolanic cement mortars of this standard plastic consistence are shown in Table 85.

TABLE 85 Compressive strength of plastic pozzolanic cement mortars*

Cement				Compressive strength (lb/in²)		
Per cent Portland cement	Per cent pozzolana	Per cent water	Pozzolana	7 days	28 days	180 days
100	0	11	None	3640	5380	8380
60	40	11·5	Burnt shale	1810	3860	7190
		10·5	Burnt clay	2180	4600	7650
		11·0	Trass	2040	4230	6400
		10·3	Bacoli (Naples)	2080	4140	7470
		10·5	Ground sand	1710	2800	4650
100	0	11·0	None	6330	8200	11380
60	40	11·5	Burnt shale	3010	5080	8610
		10·5	Burnt clay	3240	5940	8820
		11·0	Trass	3020	5470	7870
		10·3	Bacoli (Naples)	3320	5220	8240
		10·5	Ground sand	3150	4490	6270

* Mortars of standard plastic consistence, vibrated according to BS 12—1947.

The ASTM specification (C618–68T) requires for pulverised fuel ash a strength of 800 lb/in² at 7 days in the 'pozzolanic activity index with lime' test and one of 85 per cent of the control at 28 days in the 'pozzolanic activity index with Portland cement' test. There is also a requirement that the compressive strength of a 1 : 0·25 : 2·5 Portland-cement : pozzolana : graded-sand mortar shall not be less than that of a 1 : 2·75 Portland-cement : graded-sand mortar using the same cement at 7 or 28 days. In this test the pulverised fuel ash is treated as a substitute for sand and not for cement. The specification requires for natural pozzolanas a strength of 800 lb/in² at 7 days in the lime index test and 75 per cent of the control at 28 days in the cement index test. For premixed pozzolanic cements (ASTM C595–68) compressive strength tests are carried out as for Portland cement on a 1 : 2·75 graded-sand mortar. The minimum

[1] For a discussion of the water : cement ratio effect, see K. M. Alexander, *Aust. J. appl. Chem.* **6,** 61 (1955).

strengths required are the same as for Type V Portland cement (see p. 168), 1500 and 3000 lb/in^2 at 7 and 28 days. The proportion of pozzolana is required to be between 15 and 40 per cent by weight of the cement and the pozzolana, when ground to a residue not above 12 per cent on a 325 mesh (44 μ), must give a minimum compressive strength of 800 lb/in^2 the lime pozzolanic activity index test at 7 days.

In the Italian specification for pozzolanic cements the maximum content of pozzolana is indirectly controlled by the requirement that the amount insoluble after treatment with 1 : 1HCl and 5 per cent sodium carbonate solution shall not exceed 16 per cent.

Much attention has been directed to the development of chemical methods by which some composite measure of the quantity and activity of the pozzolanic constituent of a cement can be assessed. The amount of free lime combined is one obvious measure of this, but tests are needed at too long ages for use in routine testing. It is also argued that in set pozzolanic cements some of the calcium hydroxide is surrounded and protected by the *C-S-H* gel and takes no part in maintaining the calcium hydroxide concentration in the liquid phase. In determinations of free lime on a powdered sample of set cement such calcium hydroxide would be included in the amount estimated and the result therefore would not be a true guide to the effect of the pozzolanic addition. A method for overcoming these difficulties by using accelerated curing and an estimate of whether the solution in contact with the set cement is, or is not, saturated with calcium hydroxide has been developed by Fratini.[1] This method is now included in the Italian specification (1968) for pozzolanic cements and also in the International Standards Organisation (ISO) recommendation (R 863—1968). In this test 20 g of the cement are placed in 100 ml distilled water at 40° in an Erlenmeyer flask of alkali-resistant or wax coated glass or of a plastics material, which is sealed and shaken vigorously for 20 seconds to prevent lumps of cement remaining on the bottom. The flask is placed in a thermostat at 40°, with the bottom horizontal to ensure that the cement settling from suspension is of uniform thickness. After 8 days the solution is filtered through a sintered glass filter and the filtrate allowed to cool to room temperature. The total alkalinity is estimated on a 50 ml sample by titration with 0·1N hydrochloric acid with methyl-orange indicator, and the CaO as oxalate by titration with potassium permenganate or by other rapid methods. The results, expressed in mmol per litre, are plotted on Fig. 128 which shows the solubility of lime at 40° in solutions of varying total alkalinity. For the cement to be accepted the 8-day result must fall below the solubility curve showing that the solution is not saturated with lime. If the result is close to the solubility curve a further test may be made after 15 days storage and this must show the solution to be unsaturated. The test shows whether the pozzolana in the cement is present in sufficient quantity and of is adequate activity to combine with the available lime liberated from the Portland cement.

An alternative approach to the assessment of the pozzolanic constituent in a cement that has been used by the author takes advantage of the difference between the acceleration of hardening of Portland cement and pozzolanic cements

[1] N. Fratini, *Annali Chim.* **44,** 709 (1954); **40,** 461 (1950); **39,** 41 (1949); and A. Rio *Annali Chim.* **41,** 274 (1951); **42,** 526 (1952).

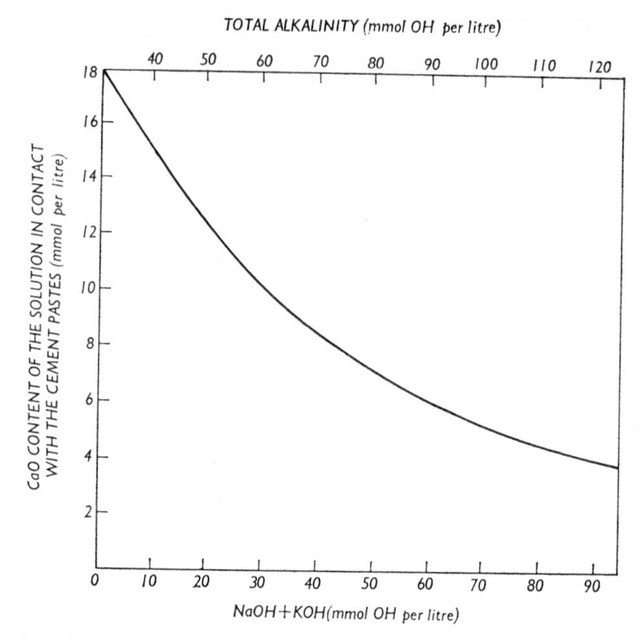

FIG. 128 Solubility of $Ca(OH)_2$ in solutions of alkali hydroxides at 40°.

when cured at 50°. At ages up to 7 days the pozzolana contributes very little to strength at ordinary temperatures, but at 50° it makes a substantial contribution depending in magnitude on the proportion and activity of the pozzolana present. This increase in strength from the pozzolana is considerably greater than any change in the Portland cement strength between the two conditions so that the difference between the strengths developed at 18° and 50° affords a measure of the value of the pozzolana present. Either tensile or compressive tests can be made, using a 1 : 3 cement : standard-sand mortar of a standard plastic consistence defined as previously by the Vicat plunger method, with a total load of 2300 g, or at a fixed water : cement ratio of 0·40. One set of specimens (6 for tensile briquettes, 3 for vibrated mortar cubes) is cured at 18° for 7 days (1 day moist air, 6 days water) and another set for 1 day in moist air and 4 days in water at 18°, then 46 hours in water at 50°, followed by a final 2 hours in water at 18°.

Some typical data obtained in such tests are shown in Table 86 from which it will be seen that the strength difference increases with the proportion of pozzolana and is considerably greater for good pozzolanas than for a poor one or an inert material. The data in Table 87 show, for the same pozzolanas, that the difference values do not vary too seriously with the Portland cement used as base, nor are they appreciably affected by using a fixed water : cement ratio of 0·40 instead of the value determined by the consistence test.

For cements with a fixed pozzolana content, the strength difference shows a correlation with the contribution the pozzolana will make to the strength of concrete at an age of 6 months or a year, as is illustrated in Fig. 129 for cements

TABLE 86 Difference in compressive strength of mortars* cured at 50° and 18°

Cement		Pozzolana	50°—18° Strength difference (lb/in²)	
			Portland cement	
Per cent Portland cement	Per cent pozzolana		X	Y
100	0	None	510	905
90	10	Burnt clay A	1270	1380
		Burnt shale	1295	1025
		Trass	950	1165
		Bacoli (Naples)	940	880
80	20	Burnt clay A	2290	2470
		Burnt shale	2240	—
		Trass	1720	—
		Bacoli (Naples)	2020	1890
		Poor burnt clay B	950	1050
		Ground sand	350	—
60	40	Burnt clay A	3090	3250
		Burnt shale	2930	2955
		Trass	2160	2470
		Bacoli (Naples)	2900	2795
		Poor burnt clay B	1250	1580
		Ground sand	460	560

* 3-inch cubes 1 : 3 standard sand vibrated mortars to B.S. 12.

TABLE 87 Difference in compressive strength of vibrated mortars cured at 50° and 18°

Pozzolanic cements: 60 per cent Portland cement, 40 per cent pozzolana, by weight

Portland cement	50–18° strength difference (lb/in²)						
	Pozzolana						
	None	Burnt clays A	Burnt clays B	Burnt shale	Trass	Bacoli (Naples)	Ground sand —
1	510	3090	1250	2930	2160	2900	460
2	905	3520	1580	2955	2470	2795	560
3	670	3580	1335	3060	2420	2710	335
4	1495	3620	1790	2610	2750	2995	700
5	970	3665	1385	3230	2405	3070	380

made by blending 40 per cent pozzolana with one Portland cement. The strength difference is not, of course, a measure of the absolute strength for this depends also on the Portland cement. For a 40 per cent pozzolana content, the strength difference for this particular form of strength test should not be less than about 2000 lb/in^2 (140 kg/cm^2) and for a 20 per cent content about 1500 lb/in^2 (110 kg/cm^2).

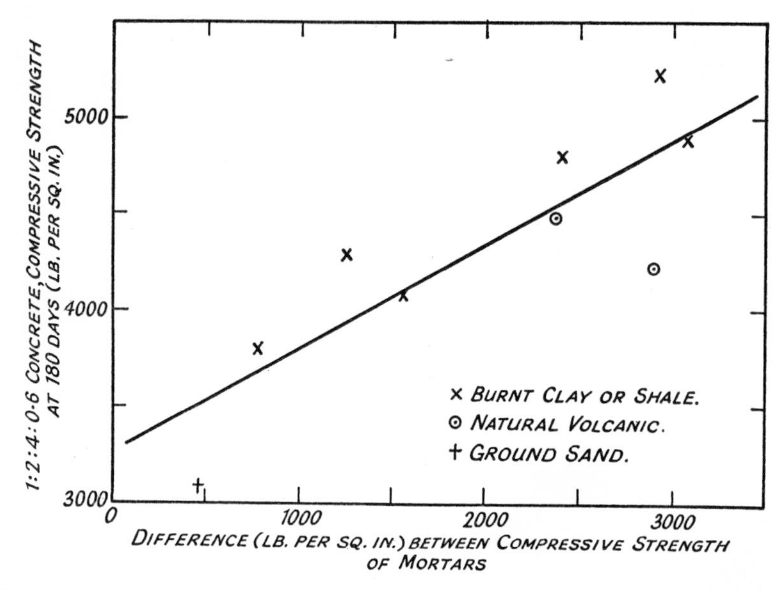

FIG. 129 Relation between strength of pozzolanic cement concrete at 180 days and the difference in strength of mortars cured at 50° and 18°.

Comparison of the durability of mortars in magnesium and sodium sulphate solutions shows that the strength difference also affords an index to the contribution a pozzolana will make to increasing resistance to attack by sulphate solutions. For this purpose it is necessary that the pozzolana content should not be less than about 30 per cent for, as we have seen earlier, contents of say 20 per cent are much less effective, and also that the strength difference should exceed 2000 lb/in^2 (140 kg/cm^2). The test is solely a measure of the efficiency of the pozzolana and does not distinguish between base Portland cements of the sulphate-resisting and ordinary types.

It will be evident from the foregoing discussion that finality in the routine testing of pozzolanic cements has not yet been reached. The position can be summarised as follows:

1. Strength tests on plastic mortars at 7 or 28 days will give an index to the strength of concrete at the same ages and one that can be directly compared with Portland cements.

2. The strength, even at 28 days, gives no indication in general of the quality of the pozzolana present in a pozzolanic cement, since it depends primarily on the Portland cement present, but when comparing different pozzolanas blended in a constant proportion with the same base Portland cement, it affords some limited differentiation between good and poor pozzolanas.

3. The proportion of pozzolana present in a pozzolanic cement can be estimated approximately by selective extraction methods or very roughly controlled by limitations on the composition of the cement.

4. The content of material soluble after acid and alkali treatment can be used to distinguish between inert and active materials only as long as the pozzolana is limited to a particular type and the content present in the cement is roughly constant. Even so, it affords only the broadest of distinctions.

5. Tests which afford a measure of the rate of removal of free calcium hydroxide from the set cement give a composite assessment of the quality and quantity of the pozzolana present. Their results may afford an index to chemical resistance, but not to the strength the pozzolana will contribute at long ages.

6. Strength tests with accelerated curing for a limited period at temperatures up to about 50° can, when suitably conducted, afford a measure of the contribution the pozzolana will make to strength at long ages, and, when combined with a requirement for a certain minimum pozzolana content, of resistance to sulphate attack.

15 Cements made from Blastfurnace Slag

Blastfurnace slag is a by-product obtained in the manufacture of pig-iron in the blastfurnace, and is formed by the combination of the earthy constituents of the iron ore with the limestone flux. The proportion of slag to iron produced by the blastfurnace varies with the richness in iron of the ore and the size and efficiency of the furnace, but usually ranges from about 0·3–1·0 tons of slag per ton of iron. The essential components of slag are the same oxides as are present in Portland cement, namely lime, silica and alumina, but their proportions differ.

Slag has found a considerable use in the road and building industries in the production of cementing materials, as an aggregate in concrete and tarmacadam, in the production of lightweight aggregate, and in the manufacture of slag wool for thermal insulation. In Great Britain its major use has been as a road-stone, the other uses representing only a relatively small proportion of total slag production. The use of slag as a concrete aggregate is discussed in a later chapter.

The extent to which slag is used in cements varies much in different countries. It is particularly well developed in several western European countries and several different types of cement are produced.

1. Ground slag, mixed with a suitable proportion of limestone, is used as a raw material for the manufacture of Portland cement.
2. Granulated blastfurnace slag is ground with Portland cement clinker in various proportions. Such cements form the British and U.S.A. Portland blastfurnace cements, the British low-heat slag cement, the German Eisenportland and Hochofen cements, the French Ciment Portland de Fer, Ciment métallurgique mixte, Ciment de Haut Fourneau and Ciment de Latier au clinker, and analogous cements in other countries. In a Belgian development, known as the Trief Process, the granulated slag is ground wet and added either as a slurry, or after drying, to the cement and aggregate in the concrete mixer.
3. Granulated slag is ground with a small proportion of dead-burnt gypsum or anhydrite together with a smaller addition of cement or lime. Such products are known as supersulphated cements. This type of cement is made mainly in Belgium, but also in France, Germany and England.

4. Ground granulated slag is mixed with hydrated lime. The French and Belgian Ciments de Laitier á la Chaux and the American slag cement are of this type; it was also formerly made in Germany and in Great Britain, when it was sometimes known as cold-process slag cement.

Composition of blastfurnace slag

The composition of slag[1] can vary over a wide range depending on the nature of the ore, the composition of the limestone flux, the coke consumption, and the kind of iron being made. It can also change over the years with alterations in the sources and kinds of ore being smelted. These variations affect the relative contents of the four major constituents, lime, silica, alumina and magnesia, and also the amounts of the minor components, sulphur in the form of sulphide, and ferrous and manganese oxides. Some analyses of blastfurnace slags are given in Table 88. In general the lime content may range from 30 to 50 per cent, silica

TABLE 88 Composition (per cent) of blast furnace slags

Source	Type	CaO	SiO_2	Al_2O_3	MgO	Fe_2O_3	MnO	S
Britain	Basic	37–42	30–36	12–22	3–11	0·3–2·1	0·4–2·2	0·9–1·9
,,	Foundry	39–44	30–38	15–23	3– 8	0·1–1·0	0·2–0·7	1·4–2·5
,,	Hematite	38–42	32–37	10–22	3– 9	0·4–1·3	0·3–1·0	1·1–2·4
Germany	Thomas	38–46	29–35	10–16	4–12	0·2–1·0	0·1–1·3	0·6–1·9
,,	Foundry	35–43	30–40	11–16	5–11	0·2–1·3	0·4–2·0	0·9–1·8
,,	Hematite	40–44	34–35	11–13	6– 8	0·3–0·4	0·5–1·1	1·4–1·8
France	,,	40–48	29–36	13–19	2– 8	0·5–3·8	0·1–1·0	0·4–1·5
U.S.A.	,,	36–45	33–42	10–16	3–12	0·3–2·0	0·2–1·5	—
S. Africa	,,	28–39	28–38	10–22	7–21	0·4–3·0	0·2–0·9	0·7–1·4
U.S.S.R.	,,	29–48	34–35	5–23	0–18	0·3–2·4	0·1–2·1	1·1

28 to 38 per cent, alumina 8 to 24 per cent, magnesia 1 to 18 per cent, sulphur 1 to 2·5 per cent, and ferrous and manganese oxides 1 to 3 per cent, except in the special case of ferro-manganese production when the manganese oxide content of the slag may be considerably higher.

In the operation of a blastfurnace the iron oxide ore is reduced by means of coke to metallic iron, while the silica and alumina constituents combine with the lime and magnesia to form a molten slag which collects on top of the molten iron

[1] See F. Schröder, *Sym., Tokyo 1968* and W. Kramer, *Sym., Washington 1960*, 957 for a discussion of the factors determining slag composition.

at the bottom of the furnace. The iron itself dissolves some of the constituents in a reduced form, notably silicon, manganese and sulphur. The partition of sulphur between iron and slag is important. In order to favour solution of sulphur in the slag either the temperature or the lime content of the slag is increased; the latter in any event necessitates an increase in temperature in order to retain fluidity. An increase within certain limits in magnesia content also lowers the slag viscosity and tends to be favoured in modern practice with low slag volumes. Dolomite or magnesian limestones are also used in some countries for economic reasons. A variation of the normal process is to separate the desulphurising function from the reduction operation in the blastfurnace and to carry it out as a separate process after the molten iron has been tapped from the furnace. In such cases it is possible to use a fluid slag with a much lower lime content.

Processing of blastfurnace slag

Blastfurnace slag issues from the blastfurnace as a molten stream at a temperature of 1400–1500°. Its conversion into products suitable for various uses depends on the subsequent processing; widely different products are obtained according to the kind of process used in cooling the molten slag.

When the slag is allowed to cool slowly, it solidifies into a grey, crystalline, stony material, known as air-cooled, lump, or dense slag. This forms the material used in a road-stone and as a concrete aggregate. The various cooling procedures and the properties of the slag as an aggregate are discussed in Chapter 18. More rapid chilling with a limited amount of water, applied in such a way as to trap steam in the mass, produces a porous, honeycombed material which resembles pumice. This lightweight material is called foamed slag and after crushing and grading is used as a lightweight aggregate.

Slag wool is normally made by reheating old slag from a slag bank sometimes with added silicious or other materials. The process is carried out in a cupola from which the remelted slag is tapped in a thin stream. As it issues from the tap hole it is chilled quickly by means of a jet of air, or air and steam, causing the slag to form vitreous threads which felt together into a lightweight mass. The product is an excellent thermal insulator.

Slag which is to be used in the manufacture of the various slag cements is chilled very rapidly either by pouring into a large excess of water or by subjecting the slag stream to jets of water, or of air and water. The purpose is to cool the slag so quickly that crystallisation is prevented and it solidifies as a glass. At the same time the quenching breaks up the material into small particles varying from glassy beads to a lightweight froth. The product is called granulated slag and is principally used as cement of one type or another. Minor quantities are sold for glass-making, as abrasive, or as a sand for concrete.

The object of granulation is to obtain as glassy, non-crystalline a material as possible. The ease with which this can be done depends on the chemical composition of the slag and on the temperature at which it issues from the furnace. Silicious slags cool to glasses most easily while those of high lime content are more difficult to prevent crystallising. The slags of higher lime content are, however, chosen for manufacture of granulated slag because of their greater activity as cements. After quenching, the granulated product contains water and, since it needs to be dried for all but one of the subsequent processes for conversion to cement, there is an advantage in using a quenching process which leaves the slag with the lowest practicable water content. Simple quenching by pouring the slag into a tank of water tends to produce a light, frothy product which retains a high water content even after draining. Slags of high-alumina content also tend to form this kind of product more readily than those of low, or medium, alumina content. Generally, the production of a well-quenched glass as dense beads seems to be most easily achieved by quenching with high-pressure water jets. Such a product drains to a relatively low water content with resultant economies in drying.

The chemical composition of the slag is one factor determining the use which can be made of it. High-lime slags may exhibit a phenomenon known as 'falling' when cooled slowly and such slags are unsuitable for use as a dense aggregate. A 'falling' slag is one which disintegrates into a powder after solidifying. This is caused (see p. 44) by the presence of crystalline dicalcium silicate in the slag, the conversion from the high to the low temperature form of this compound causing a breakdown identical with that of the 'dusting' of Portland cement clinker. 'Falling' does not occur with the more rapid methods of chilling used for manufacture of either foamed or granulated slag because crystallisation of the dicalcium silicate is prevented. The production of foamed slag formerly required a fairly high lime, hot slag, but processes are now used which produce foamed slag successfully from a much wider range of compositions. Slag wool is best made from silicious slags.

Mineral composition of air-cooled slag

The nature of the minerals formed when slag cools slowly is of concern in the use of the material as a dense aggregate. It is of less direct interest for slag in the glassy, or granulated form, but some knowledge of it is essential to a discussion of the hydraulic activity of such materials.

The probable assemblages of minerals formed by the four major constituents CaO, SiO_2, Al_2O_3 and MgO, in the range of proportions characteristic of slag, cannot be predicted with certainty from phase equilibrium diagrams because the quaternary system has not yet been completely worked out. Estimates which have been made[1] are, however, in reasonably close agreement with the results of microscopic studies on thin sections, or polished and etched surfaces, of slags.

Slags of relatively low MgO content, up to 5 per cent, usually contain melilite as the main constituent. Melilite is the name given to an isomorphous series of solid solutions of which the two principal end members are gehlenite (C_2AS) and akermanite (C_2MS_2). In some slags the solid solution may be replaced by akermanite itself. Other minerals that may occur in slags are bredigite (α' C_2S), larnite (β C_2S), γ C_2S, pseudowollastonite and wollastonite (CS), rankinite (C_3S_2), merwinite (C_3MS_2), spinel (MA), diopside (CMS_2), monticellite (CMS), anorthite (CAS_2), and forsterite (M_2S). The only one of these minerals

[1] R. W. Nurse and H. G. Midgley, *Silic. ind.* **16** (7), 211–217 (1951); *J. Iron Steel Inst.* **174,** 121 (1953).

that is active and will hydrate is dicalcium silicate in the α' or β form; the γ form is inert. Periclase (MgO) has only been observed in slags of high MgO content (> 16 per cent) when remelted and heat-treated. Theoretically it can occur in the assemblages shown in Table 89.

TABLE 89 Some possible assemblages of minerals from the four components CaO, Al_2O_3, SiO_2, MgO in slags

No. of slags of 21 examined	C_2AS/C_2MS_2 solid solution	C_2MS_2	C_2S	CS	C_3S_2	C_3MS_2	MA	CMS_2	CMS	CAS_2	M_2S	MgO
5	X		X		X							
5	X			X						X		
4	X			X	X							
3	X					X	X					
2	X						X			X		
1		X					X			X	X	
1		X				X	X		X			
Other possible assemblages not observed in the 21 slags	X		X			X						
		X		X				X		X		
		X						X		X	X	
		X					X		X		X	
			X			X	X					X
						X	X		X			X

There are a considerable number of possible combinations of these minerals that might theoretically occur in any one slag. Some of these possible assemblages are set out in Table 89. The upper part of the table shows the number of slags out of a group of 21 from different ironworks, 19 British and 2 overseas, which had a particular assemblage. The lower part of the table shows some other assemblages that seem possible, but which did not occur in any of these 21 slags. Some of the minor constituents may not be detected under the microscope, either because of the small amount present or of incomplete crystallisation leaving a glassy phase. More phases may also sometimes be observed than would be possible under conditions of complete equilibrium. The 'zoned' appearance of the melilite solid solution also indicates incomplete equilibrium. It is not, therefore, possible to rely entirely on calculation from the oxide composition for an estimation of the minerals present in a slag and supplementary microscopic examination is necessary. Small changes in oxide composition may sometimes markedly affect the mineral assemblage deduced by calculation.

Some of the minerals can appear in more than one form. Thus both pseudowollastonite and wollastonite (the high and low temperature forms of *CS*) may be observed, but the former is the more common. C_2S may also appear in more than one form. The high-temperature forms observed are α' or β into which α' may transform on cooling; 'falling' slags contain the γ form often mixed with β.

In addition to the above minerals, all blastfurnace slags contain calcium sulphide, occasionally in the form of well-shaped but small crystals, but more often having the appearance of a non-crystalline or dendritic mass. Other minor constituents such as alkalies or iron and manganese oxides, when present in small amounts, of the order of 1 per cent, probably occur in solid solution in one of the other minerals. At higher contents, of the order of 3 per cent, they may, however, occur as sulphides.[1]

Granulated slag: Constitution and properties of slag glasses

Glasses, such as granulated slag, are to be regarded as supercooled liquids. It is a characteristic feature of silicate melts that on rapid cooling from the liquid state they tend to form a glass. The passage from the liquid to a crystalline solid condition is accompanied by a rearrangement of the ions which take up a definite orientation in the crystals, and the viscosity of molten silicates near the freezing-point is so large that this rearrangement only takes place slowly. If the cooling is rapid the ionic groups largely retain their irregular arrangement, and, the viscosity increasing rapidly as the temperature falls, the slag passes from the liquid state to one in which the rigidity approaches that of a solid, but without the development of a crystalline structure. Glasses are thus greatly undercooled liquids having a very high internal viscosity. Glasses have no definite melting-point, but soften and gradually pass into a liquid state on heating. Below the freezing-point of the crystallised mix the stable condition is that of a completely crystallised solid, and the glasses are unstable and tend to pass into the crystalline state; this tendency is restrained by the high viscosity which reduces the mobility of the constituent ions. At ordinary temperatures a substance may persist indefinitely in the glassy condition, but, if it be annealed at a temperature not too far below its melting-point, devitrification (the formation of crystals) may begin and continue more or less rapidly after cooling.

When a powdered granulated slag is examined on a microscope slide by transmitted light, the glass is seen to consist of clear isotropic transparent grains. Less perfectly granulated slags may show brown or black zones where incipient crystallisation has begun or even birefringent areas under crossed Nicols.

Ordinary air-cooled slag has no, or very little, cementing properties. Granulated slag alone has similarly a negligible cementing action, but, if some suitable activator is present, all except the more silicious granulated slags show marked cementitious properties. The activator may be lime, Portland cement, alkalies such as caustic soda or sodium carbonate, or the sulphates of the alkalis, lime, or magnesia. It is, of course, this property which has led to the use of the material in various forms of cement.

In 1862 Emil Langens discovered that the process of water granulation of slag yielded a material which when mixed with lime had good cementing properties. This observation was the foundation of two industries, slag cement manufacture and the use of granulated slag as a lightweight aggregate. The properties of granulated slag were later investigated by Michaëlis, Prüssing, Tetmajer, Prost,

[1] W. Fischer and S. Wolf, *Schwefel und Schlackenwolle*, E. Schweizerbart, Stuttgart 1951; F. Schröder, *Sym., Tokyo 1968.*

Feret and Grün amongst others. Passow, who introduced the process of air granulation, played a large part in the development of the utilisation of blastfurnace slags. A slag-lime cement was first used commercially in Germany in 1865, while in 1883 slag was used as one of the raw materials for Portland cement manufacture. The first production of a Portland blastfurnace type of cement by grinding together Portland cement clinker and granulated slag occurred in Germany in 1892.

In general, the more basic the slag, the greater its hydraulic activity in the presence of alkaline activators. When sulphate activators are used, the basicity of the slag is not the only criterion; the slag needs also to contain at least an average or moderately high, alumina content. The cause of this hydraulic activity and its relation to the chemical composition and physical state of a slag has been studied extensively, without reaching finality as yet. The glass content is a prime factor and an increase in the temperature of the molten slag when granulated promotes the hydraulicity of the product. Thus for a series of hematite slags quenched to give varying glass contents, Schwiete and Dolbor[1] obtained a roughly linear relation between strength and glass content. Increasing contents of crystalline components reduce the cementing properties though some of them may make some contribution to strength. This would be expected for α' or β dicalcium silicate but there is also evidence that gehlenite and akermanite may not be inert. The influence of composition[2] on the hydraulicity of the granulate is complex since it affects the slag temperature and granulation condition as well as the intrinsic properties of the glass. The practicable alterations in slag composition are determined by the requirements of the blastfurnace operation. The hydraulic value increases with the CaO/SiO_2 ratio up to a limiting point when increasing CaO content makes granulation to a glass difficult. At constant basicity the strength increases with the Al_2O_3 content and a deficiency in CaO can be compensated by a larger amount of alumina. Both iron and manganese oxides have an adverse effect but a relatively high calcium sulphide content is favourable, possibly because it is associated with hot slags. The influence of MgO as a replacement for CaO seems to depend both on the basicity and the alumina content of the slag. Variations in the MgO content up to about 8–10 per cent may have little influence on strength development, but high contents have an adverse effect.

There is no direct relationship between the contents of C_2S, the melilite solid solution and C_2MS_2 that glassy slags would yield on crystallisation and their hydraulic activity but the more markedly active slags are those which contain these components as potential phases. Slags which on complete crystallisation give phases of low lime content such as CS and CMS_2, have poor hydraulic properties.[3]

Numerous investigations have been made on pure synthetic glasses in an endeavour to determine the optimum composition. The results of tests on ground glasses of pure compound composition are not entirely concordant. Thus Keil[4]

[1] See F. Shröder, *Sym., Tokyo 1968*.
[2] For surveys of the extensive literature see F. Shröder, *Sym., Tokyo 1968*; W. Kramer, *Sym., Washington 1960*, 957; F. Keil, *Sym., London 1952*, 530.
[3] J. E. Kruger, K. H. L. Sehlke and J. H. F. Van Aardt, *Cem. Lime Mf.* **37**, 63, 89 (1964).
[4] F. Keil, *Sym., London 1952*, 530; F. Keil and F. Gille, *Zement-Kalk-Gips* **2**, 229 (1949).

has reported that a glass of akermanite composition develops strength when mixed with water alone, a conclusion supported by Butt,[1] but Budnikov[2] found no strength development. It may well be that the temperature of preparation of the glass has some influence. A glass of gehlenite composition does not develop strength with water. In the presence of lime on the other hand, a gehlenite glass develops a higher strength than an akermanite glass. The compressive strength of glasses[3] within the $CaO–Al_2O_3–SiO_2$ system with 5 per cent MgO added are shown in Fig. 130. The glasses were mixed with Portland cement and gypsum

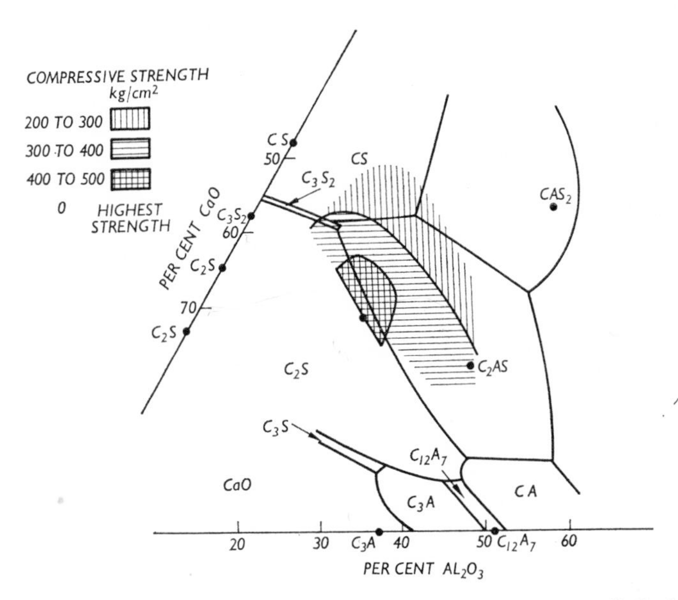

FIG. 130 Compressive strength at 28 days of glasses in the system $CaO–Al_2O_3–SiO_2$ plus 5 per cent MgO, activated by Portland cement clinker.

in the proportions 77 : 18 : 5 by weight and the tests made on cement mortars according to the German standard (DIN 1164). The most favourable composition of the glass was 50 per cent CaO, 31 per cent SiO_2 and 19 per cent Al_2O_3. A similar composition also proved the most favourable for activation with 15 per cent anhydrite and 2 per cent Portland cement clinker, corresponding to a super-sulphated cement. Broadly similar results were obtained by Tanaka[4] and by Solacolu[5]. Thus Tanaka found the optimum glass composition for Portland cement clinker activation was 51·5 per cent CaO, 33 per cent SiO_2 and 15·5 per

[1] Yu. M. Butt, U. M. Astreeva and Z. S. Krasnoslobodskaya, *Tsement* **26,** (3) 8 (1960).
[2] P. P. Budnikov and V. Pankratov, *Dokl. Akad. Nauk SSSR* **146** (1), 156 (1962).
[3] F. Keil and F. W. Locher, *Zement-Kalk-Gips* **11,** 245 (1958); F. W. Locher, *Sym., Washington 1960*, 267.
[4] T. Tanaka, *Rock Prod.* **59** (7), 106 (1956); **60** (3) 100; **60** (4) 107 (1957); T. Tanaka, T. Sakai and J. Jamani, *Zement-Kalk-Gips* **11,** 50 (1958).
[5] S. Solacolu, *Zement-Kalk-Gips* **11,** 125 (1958); S. Solacolu and P. Balta, *Revue Matér. Constr. Trav. publ.* **583,** 95 (1964).

cent Al_2O_3. For anhydrite activation the optimum composition was 49–50 per cent CaO, 31–33 per cent SiO_2 and 18–19 per cent Al_2O_3.

While these results give a general picture of the influence of slag composition on reactivity, it is not a complete one since the amount and kind of activation that gives maximum strength varies with the composition of the slag. Thus Tanaka found that in supersulphated cements the content of Portland cement clinker that gave maximum strengths varied from 2 to 6 per cent with different slags. The action of lime as an activator also seems to differ from that of Portland cement or calcium sulphate. This has been attributed by Locher to differences in the hydration products and the strengths they develop. It is evident that there are many factors influencing strength development which have yet to be fully worked out.

The reactivity of granulated slag must be related to the structure of the glass and the ease with which hydration can occur. A number of workers[1] have used Zachariasen's[2] network theory of glass formation to explain the constitution of slag glasses. This theory postulates the formation of networks of oxygen tetrahedra such that, although there is no periodicity of structure which would yield a sharp X-ray pattern, the statistical arrangement is similar to the ordered arrangement of the crystalline state. In slags the network-forming cations are silicon and aluminium which link together by sharing the oxygen atoms of the tetrahedra. Calcium and magnesium ions occupy the relatively large voids in the network structure and are known as network modifiers. Aluminium ions can act as a network former or modifier, but they can only replace Si ions to a limited extent and in so doing they increase the interionic distance between the cation and the oxygen ions. This increases the specific volume of the glass and produces a more open network that can accommodate more of the larger Ca ions in the voids in the structure. For tetrahedral-type stable glasses the ratio of the network forming cations to oxygen should lie between 0·33 and 0·5. This theory accounts for the instability of glasses from very high lime slags and enables an upper limit of slag compositions for granulation to be estimated. Parker and Nurse, for example, found that a granulated slag with a ratio of 0·28 was devitrified while another with a ratio of 0·31 was highly vitrified when originally made but devitrified almost completely in a period of three years. Tanaka also found a boundary between glassy and crystallised granulates at about $n = 0{\cdot}33$. The hydraulic behaviour of the slag may be expected to depend on the extent to which aluminium ions are present as network formers, giving an aluminium silicate network, and the extent to which they are present as network modifiers occupying voids in the structure. The nature of the hydration products will be determined by the lime concentration in the solution and this will depend on the activator and on the rate of supply of Ca ions from the voids in the glass network.

Another theory[3] of glass structure, known as the crystallite theory, postulates a 'micro heterogeneity' in glasses. A glass is considered to contain short-range domains in which there is a certain degree of order, or, as they have been termed,

[1] T. W. Parker and R. W. Nurse, *Bldg Res. Tech. Pap.*, No. 3 (1949); F. Keil and F. W. Locher, loc. cit.; T. Tanaka, loc. cit.; F. Shröder, loc. cit.; S. K. Chopra and C. A. Taneja, *J. appl. Chem.* **15,** 157 (1965); *Sym., Tokyo 1968.*
[2] W. H. Zachariasen, *J. Am. chem. Soc.* **54,** 3841 (1932).
[3] E. A. Porai-Koshits, *Glastech. Ber.* **32,** 450 (1959).

'embryos of crystallisation'. These regions contain most of the cations and are linked together by amorphous regions formed by the remaining anions. De Langavant[1] has suggested that, in the liquid slag, there is a tendency for polymerised Si_xO_y groups to exist as a framework of chains or bands leaving a higher ratio of base to acid ions amongst the other constituents of the melt. Very hot slags cannot sometimes be chilled to a glass and this is attributed to fracture of the silica network with the loss of its inhibiting effect on crystallisation. The theory of liquid slag structure and its possible relations to the properties of the granulated slag were reviewed by Shröder at the Tokyo Symposium in 1968.

The difference between the network and the crystallite theories of glass structure may well be reduced as more information becomes available but it is clear that the mineral compounds do not exist as such when a slag is in the vitreous state and that there is some form of two-region structure.

The function of the activator has been conceived by Locher as providing a chemical composition in the solution such that the ions dissolved from slag can rapidly form strength-producing hydrate phases. But the activator must also open up the silicate structure of the glass so as to accelerate the rate of hydration. Thus a lime activator prevents the formation of an impermeable alumino-silicate film and progressively combines with the slag. In the case of sulphate activators the initial attack will be on the alumina to form calcium sulphoaluminate and this must be presumed to open up the silicate structure (see p. 485).

Composition requirements

Various formulae have been evolved, from experience and from tests on the strength of granulated slag mixes, for the compositions desirable in the slag. In the absence of any quantitative theory of the relation between the activity and composition of granulated slag, such formulae cannot be more than useful working rules, but such rules are obviously necessary as a guide in production.[2]

In Germany the modulus first adopted for granulated slags to be ground with Portland cement in the manufacture of Eisenportland and Hochofen cements required that the ratio of the percentage contents of the oxides should be:

$$\text{(I)} \qquad \frac{CaO+MgO+\frac{1}{3}Al_2O_3}{SiO_2+\frac{2}{3}Al_2O_3} \geqq 1$$

This was superseded in 1942 by the formula which still appears in the specification DIN 1164 : 1967:

$$\text{(II)} \qquad \frac{CaO+MgO+Al_2O_3}{SiO_2} \geqq 1$$

A third formula proposed subsequently though not used in specifications is:

$$\text{(III)} \qquad \frac{CaO+CaS+\frac{1}{2}MgO+Al_2O_3}{SiO_2+MnO} \geqq 1{\cdot}5$$

[1] J. C. de Langavant, *Revue Matér. Constr. Trav. publ.* **401,** 38 to **411,** 425 (1949); **448,** 1 to **453,** 175 (1953).
[2] F. Schröder, *Sym., Tokyo 1968*, has reviewed the various formulae.

This last formula, which refers only to the constituents soluble in hydrochloric acid, gives, according to Keil,[1] an improved distinction between very good (1·9 or more), good (1·5–1·9) and merely usuable slags.

From statistical correlations of various moduli with the strength of a wide range of British granulated slags mixed with Portland cement, Parker and Nurse concluded that the German Modulus I gave the best correlation when combined with a measure of the percentage of glass in the slag. The product GM, where G is the percentage of glass and M the Modulus I, was accordingly proposed as a guide to the activity of a slag. The range of slags tested had values of M varying from 0·8 to 1·7 and glass contents from 14 to 100 per cent. The higher the value of GM the more hydraulic is the slag provided that M has a value between 0·72 and 1·5. At lower values of M the slag has no hydraulic activity and above 1·5 the slag cannot be granulated without devitrification occurring. The adoption of Modulus I implies that MgO is as effective as CaO in producing hydraulic properties in a slag, but the slags on which the product GM was tested only covered samples with a MgO content up to 7·5 per cent. Other workers have found that up to contents of about 10 per cent MgO the hydraulic value of granulated slags is not affected adversely, though its effect cannot be considered independently of the associated changes in the CaO : SiO_2 ratio. Studies on slags of high-magnesia content have indicated that with contents of 15–20 per cent the MgO contributes little to the activity[2] though useful Portland blastfurnace cements can still be made from them, as is done for example in South Africa.[3] There is evidence that an Al_2O_3 content of about 12 per cent or more is then advantageous. It seems probable that for high-magnesia slags the Moduli I and II need modification by counting the MgO content as worth less than its percentage content as is done in Modulus III.

Such a partial allowance for magnesia is found in the 'index of quality' (i) of de Langavant[4] in France.

$$i = 20 + CaO + Al_2O_3 + 0{\cdot}5MgO - 2SiO_2$$

Inserting percentage weight contents into this formula, which is very similar to the German Modulus III, de Langavant classifies the hydraulic properties of slags as inferior if $i < 12$ and very good > 16; but Blondiau[5] has put the minimum desirable value at 19 for slags of low (3·5 per cent) MgO content.

Blondiau[6] from tests on Belgian slags, normally of low magnesia content, in both slag-Portland and supersulphate type cements, found that the most reliable results were obtained with slags with a CaO : SiO_2 ratio between 1·45 and 1·54, a SiO_2 : Al_2O_3 ratio between 1·8 and 1·9 and a 'hydraulic potential' between 70 and 80 cal/g. The latter is a measure of the degree of vitrification. It is determined as the difference between the heats of solution, in a nitric-hydrofluoric acid solvent (see p. 291), of the granulated slag and of the same slag after heating for

[1] F. Keil, *Sym., London 1952*, 530.
[2] N. Stutterheim and R. W. Nurse, *Mag. Concr. Res.* **9,** 101 (1952).
[3] N. Stutterheim, *Proc. Am. Concr. Inst.* **56,** 1027 (1960); R. J. Davies, *S. Afr. ind. Chem.* **11,** 232 (1957).
[4] Loc. cit.
[5] L. Blondiau, *Silic. ind.* **25,** 545 (1960).
[6] L. Blondiau, *Rev. Mat. Constr.* **424,** 6; **425,** 42 (1951).

four hours at 1000°, both samples being ground to a specific surface of 2700 cm^2/g (Wagner).

Though high-magnesia contents have an adverse effect on the hydraulic properties of granulated slags, they do not cause unsoundness as is found in Portland cements of high-magnesia content. The composition of slags is such that, even when fully crystalised, no free MgO appears until the magnesia content exceeds 16 per cent at least.[1] Portland blastfurnace cements containing granulated slags with up to 20 per cent magnesia content have shown no unsoundness in the ASTM autoclave test while concretes made with them gained strength in a normal manner up to three years and showed no indication of dimensional instability.[2] Granulated slag which is to be used in the manufacture of Portland cement clinker must, of course, have a magnesia content such that the requirements for the latter are not exceeded.

Various rapid qualitative tests have also been proposed. The refractive index of slag glasses varies from about 1·635 to 1·67, but lime-rich slags usually have an index above 1·65. There is no close correlation between activity and the heat of crystallisation of vitreous slags of differing composition. For one composition activity is related to heat of crystallisation in so far as this is a measure of the efficiency of the initial granulation. Studies by DTA methods on granulated slags have also failed to show any relation between the exothermic peaks observed and the hydraulic properties.[3]

Considerable use is made in Germany of a fluorescence test[4] as a rapid method for assessing granulated slags, though experience with British slags indicates that it is only useful for slags produced by one particular process or even by a single furnace. In this test the sample is examined under ultra-violet light. Vitreous slag particles are claimed to emit red and pink colours whereas highly-crystalline particles emit blue and violet colours. The product of the percentage of red fluorescing particles in a granulate multiplied by the CaO : SiO_2 ratio of the slag is considered to be a reliable index of hydraulic value. The fluorescence test is claimed to distinguish between glasses of different hydraulic value and therefore to be preferable to the estimation of glass microscopically. It is evident from this test, and that used by Parker and Nurse, that neither a composition modulus nor a glass content alone are sufficient to characterise the hydraulic properties of a slag and that some product of the two is required.

A chemical method for assessing the activity of granulated slags has been described by Lieber.[5] Five grammes of the slag ground to a specific surface of 3200 cm^2/g are shaken for 3 hours at 80° with 50 ml 5 per cent NaOH solution, filtered, washed four times with 40 ml methanol and twice with 25 ml diethylether. The residue is then dried for 30 minutes at 60° and the loss or ignition at 600° determined as a measure of the bound water. A fair relation was found

[1] B. G. Baldwin, *J. Iron Steel Inst.* **186,** 388 (1957); J. G. H. Steyn and M. D. Watson, ibid. **203,** 445 (1965).
[2] N. Stutterheim and R. W. Nurse, loc. cit.; N. Stutterheim, *Sym., London 1952*, 573; *Sym., Washington 1960*, 1035.
[3] Von W. Schrämli, *Zement-Kalk-Gips* **16,** 140 (1963).
[4] W. Kramer, *Sym., Washington 1960*, 957; F. Shroder, *Sym., Tokyo 1968.*
[5] W. Lieber, *Zement-Kalk-Gips* **19,** 124, (1966).

between the amount of bound water and the strength developed in a 50/50 Portland cement-slag cement tested in German standard mortars (DIN 1164).

Composition moduli are convenient for the rapid control of slag quality, since regular chemical analysis of the slag is necessary in any event in the control of pig-iron production, but the only final guide is the strength developed in cements. For the purpose of assessing the contribution to strength made by the slag a 'hydraulic index' is recommended by Keil.[1] This is based on a comparison of the compressive strength at 28 days of mortars containing (I) the Portland-slag cement, (II) the same Portland cement alone, and (III) a Portland-ground quartz sand cement (sand ground to a specific surface of 4000 cm^2/g air-permeability method). If the strengths developed in mortars I, II and III are a, b and c respectively, and the amount by weight of slag, or ground quartz, in cements I and III is 30 per cent, then

$$\text{Hydraulic index } 70/30 = \frac{a-c}{b-c} \times 100$$

If the slag is non-hydraulic its effect should be similar to the inert ground sand and the hydraulic index be zero. If it is as cementitious as the Portland cement itself the index should be 100. In practice the index is found to vary with the proportion of slag present in the cement and it is not necessarily the same in a 50/50 or 30/70 Portland cement-slag mix as in the 70/30 mix. Mortars of low water : cement ratio (e.g. w/c = 0·32) must not be used for the test since, in such mortars, appreciable substitutions of fine inert material for cement cause little or no reduction in strength. Keil has used the German standard mortar (DIN 1164) in which the water : cement ratio is now 0·5 (formerly 0·6), but ratios down to 0·4 should be suitable (see p. 390). The reduction in strength on substituting an inert addition for cement tends to be somewhat greater than the percentage substitution and to bear a roughly linear relation to it. With granulated slag the

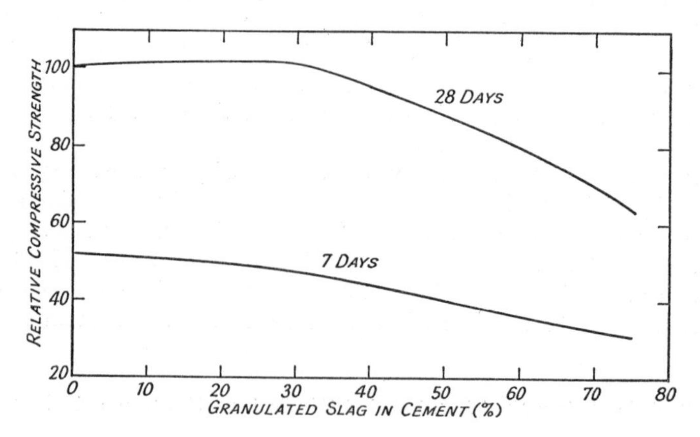

FIG. 131 Effect of granulated slag on compressive strength of plastic mortars. Strength of Portland cement at 28 days taken as 100.

[1] F. Keil, *Sym., London 1952*, 530.

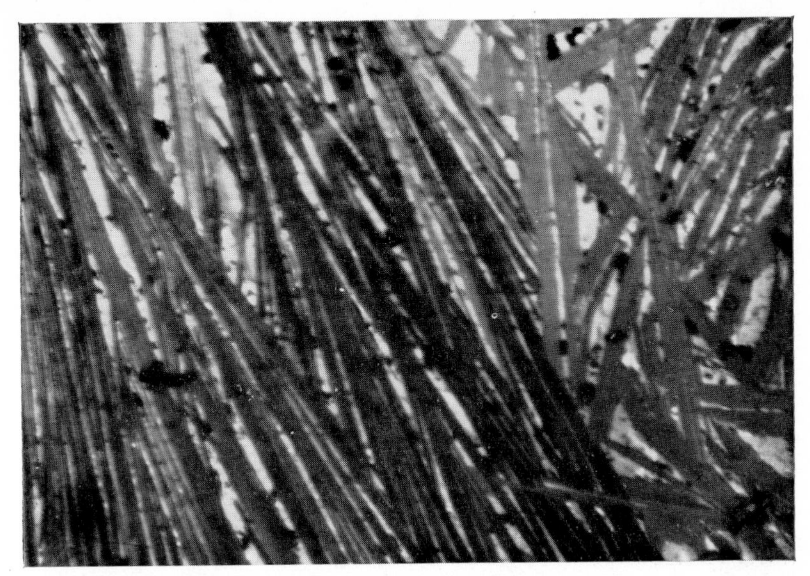

(i) Polished Surface of High Alumina Cement Clinker, Showing Pleochroite. Etched with 1 per cent Borax Solution for 30 Seconds at 100°C (×200)

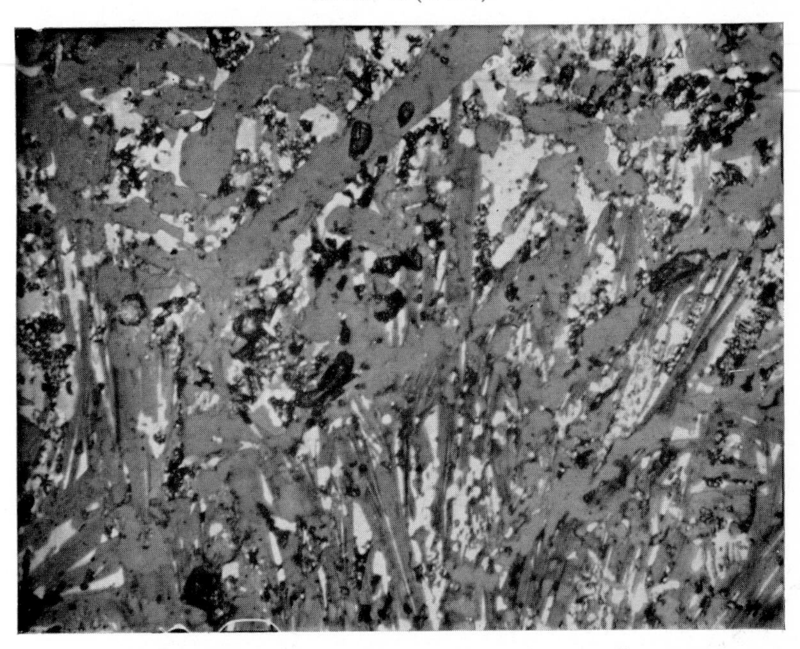

(ii) Polished Surface of High Alumina Cement Clinker, Showing Grey Prismatic $CaO.Al_2O_3$, a little Pleochroite and White, Highly Reflecting, Iron Constituents. Borax Etch (×560)

PLATE XVI

(i) HIGH ALUMINA CEMENT CONCRETE.
SPECIMENS IMMERSED IN 5 PER CENT. $MgSO_4$ SOLUTION
AT 28 DAYS' OLD
Age when photographed, 2 years.

(ii) CARBONATED SKIN ON PORTLAND CEMENT CONCRETE

effect varies with the age of testing and departs progressively more from a linear relation as the age increases. This is illustrated in Fig. 131 where some data for one granulated slag, tested by Keil in the German standard mortar, are plotted. The activity shown by a granulated slag may vary somewhat with different Portland cements, but the effect is not a major one.

Lime-slag cements

Cements made of mixtures of hydrated lime and ground granulated slag were the earliest of the cementitious products made from slag. They were first produced in Germany but their use spread to many other countries. In Great Britain production never attained large dimensions and ceased years ago. The use of the cement has, in fact, now practically died out in most countries although some, notably France and Belgium, still retain standard specifications for the material which is used as a masonry cement. In Great Britain the cement was known as slag-cement, or 'cold-process' cement; in the United States in the early years of the present century it was sometimes known by the confusing name of pozzolan cement. It is now known there as 'slag cement' and is covered by ASTM specification C595–68 which defines it as a mix consisting essentially of granulated slag and hydrated lime containing at least 60 per cent slag. It is intended for use as a blend with Portland cement in concrete or with hydrated lime in masonry mortars. In Germany the term 'zement' may only be used for materials that attain the lowest grade of Portland cement; otherwise they are known as 'mischbinder' or 'schlackenbinder'. The French nomenclature is 'ciments de laitier à la chaux', while in Belgium the material is known as 'ciment de laitier'; it was also sometimes called 'tilleur' cement.

With the normal granulated slag of high-lime content, about 10 per cent of high-calcium lime is required to give optimum results[1] although this figure varies somewhat according to the test criteria adopted, e.g. bending or of compressive strength, or suitability for use in sulphate-bearing waters. Hydraulic lime may be used instead of high-calcium lime without affecting the strength, but a higher content of lime is needed. In practice, higher lime contents, up to 30 per cent, are used to offset the deleterious effects of carbonation during storage which can render the cement inactive. As the slag-lime mix sets rather slowly, an addition of sodium sulphate to the extent of 1 per cent or less was sometimes made, but it had the disadvantage of producing efflorescence. A gypsum addition can be made with advantage instead of sodium sulphate. Slag cement specimens after storing in water show a strong greenish tint when fractured and the odour of hydrogen sulphide can often be detected.

The Belgian specification (NBN 49) recognises three types of Ciment de laitier, called, in terms of increasing strength requirements, normal (LN), special (LS), and special 400 (LS 400). The cements must contain not less than 75 per cent of granulated slag. The French specification (P 15–306) requires a mix containing not less than 70 per cent granulated slag and a maximum SO_3 content of 5 per cent; the cement is designated as CLX 160.

[1] R. Feret, *Revue Matér. Constr. Trav. publ.* **355,** 61 (1939); P. Jolibois and A. Nicol, ibid. **437,** 33 (1952).

Lime-slag cements have been specially used for seawater and underground foundation works because of their resistance to attack by sulphates and because of their good plasticity. Feret cited as an example a breakwater at Calais which was removed after 16 years when the concrete was found to be in good condition. De Langavant[1] quotes their satisfactory use in the construction of the underground Métropolitain railway in Paris in 1900 because of their resistance to sulphate groundwaters. A lime-slag cement was also probably used in the seawater jetty at Skinningrove which was built about 70 years ago and is still in good condition.

The abandonment in most countries of production of this form of cement must be ascribed to its sensitivity to deterioration in storage and its low strength in comparison with modern Portland cements.

Slag as a raw material for the manufacture of Portland cement clinker

The use of slag in place of clay or shale as a raw material for Portland cement manufacture falls strictly into the chapters dealing with Portland cement, since the product obtained is a true Portland cement; it is convenient, however, to discuss it briefly here.

A slag containing 35–50 per cent lime, 30–40 per cent silica, 10–18 per cent alumina and small amounts of magnesia, and of manganese and iron oxides, is, except for its lower lime content, similar in composition to Portland cement. If mixed with the necessary quantity of limestone its composition can be brought to that required in a Portland cement raw mix. Blastfurnace slag forms therefore a very useful raw material for the manufacture of Portland cement.

The method of manufacture is similar to that used when clay or shale are the raw materials. The slag and limestone are finely ground and mixed in the requisite proportions and burnt in a rotary kiln in the usual manner. The resultant clinker often has a somewhat brownish colour, owing to the manganese oxide present, but does not differ in any other respect from the clinker obtained from a clay or shale raw mix. High lime slags are usually preferred, since a smaller addition of lime is required in preparing the raw mix. The fuel consumption is somewhat reduced by the use of a slag-limestone mix, since a smaller amount of calcium carbonate has to be dissociated during burning. In ordinary Portland cement manufacture the heat necessary to dissociate the calcium carbonate corresponds to an expenditure of about 7 per cent standard coal[2] on the weight of clinker produced.

Air-cooled lump slag is sometimes used, but it is more usual to granulate the slag. The water-granulated slag contains from 10 to 40 per cent water and is dried in rotary driers. It is then ground and mixed with the ground limestone. Both the drying and grinding of granulated slag are relatively expensive operations. The dry process of cement manufacture is most often used, but some plants utilise the wet process, which avoids the necessity of drying the slag. In Germany the slag is used both in the manufacture of the Portland cement clinker by the dry process and for grinding subsequently with the clinker to form the Portland blastfurnace cement.

[1] C. de Langavant, *Ciments et Betons*, Collection Armand Colin, Paris, 1953, p. 128.
[2] Defined as 12,600 Btu/lb, or 7000 ca/g.

One difficulty which is peculiar to this raw material is encountered in the manufacture of Portland cement by the wet process from blastfurnace slag. A slurry containing a ground high-lime blastfurnace slag has a definite tendency to thicken and gradually set into a cake. This is due to the latent cementitious properties of the slag which thus exhibit themselves in slight degree without the presence of any accelerator. Cases have occurred where a whole silo of a ground slag cement slurry has set into a solid mass which had to be dug out. The tendency to setting of the slurry is more marked with granulated slag than with the air-cooled form. The setting action increases rapidly with temperature and above about 36° may become very rapid.[1] This trouble has been surmounted by the addition of a small amount of sugar which inhibits the setting of the slurry and by storing the slurry for as short a time as possible before use.

Portland blastfurnace cement

Portland blastfurnace cement is a mixture of Portland cement and granulated slag containing not more than 65 per cent of granulated slag according to BS 146 : 1958. In the U.S.A., where the cement is called Portland blastfurnace slag cement, the granulated slag content is from 25 to 65 per cent (ASTM C595–68). In Germany two varieties are specified (DIN 1164) namely, Eisenportland cement, containing not more than 40 per cent, and Hochofen cement, from 41 to 85 per cent granulated slag. In France, Ciment de Fer contains 25–35 per cent granulated slag, Ciment Métallurgique Mixte, 45–55 per cent, Ciment de Haut Fourneau, 65–75 per cent, and Ciment de Laitier au Clinker at least 80 per cent. Various grades, according to the rate of strength development, are specified in France, Germany and other European countries. In Great Britain and the U.S.A. there is one grade corresponding to ordinary Portland cement together with a low heat cement (BS 4246 : 1968) in Great Britain and moderate heat and moderate sulphate-resistant cements in the U.S.A.

The Portland cement clinker is manufactured as usual or may be made from slag and limestone and burnt in the usual manner in a rotary kiln. The resulting clinker is then fed, together with the required proportion of dry granulated slag and added gypsum to control the set, to the grinding mill. The ground cement is then stored in silos and bagged in the usual way.

Though it has been usual to grind the Portland cement clinker and the slag together they may be separately ground and subsequently mixed.[2] When ground together it is the softer material which will be preferentially ground and this is usually the clinker. With slag contents up to about 50–60 per cent the early strength is mainly determined by the fineness of the clinker fraction and later strengths by that of the slag fraction. With cements of higher slag content the fineness of the slag is of major importance at all ages as is shown in Fig. 132 for tests on German standard mortars (DIN 1164). In cements of lower slag content the influence of slag fineness is to be expected to be rather less, and that of the clinker fineness more pronounced at early ages. For the same fineness of the components there seems to be little difference in strength development between

[1] L. N. Bryant, *Rock Prod.* **34** (14), 52 (1931).
[2] F. Schröder, *Sym., Tokyo 1968*; N. Stutterheim, ibid.

intergound or separately ground materials or, in the latter case, between slag ground dry or wet. Separate grinding is used as an easy means of varying the slag-clinker proportion in the finished cement to meet market demands. Separate grinding of the slag is also beneficial in improving the strength developed by slags of below optimum activity. This is illustrated by the data in Table 100.

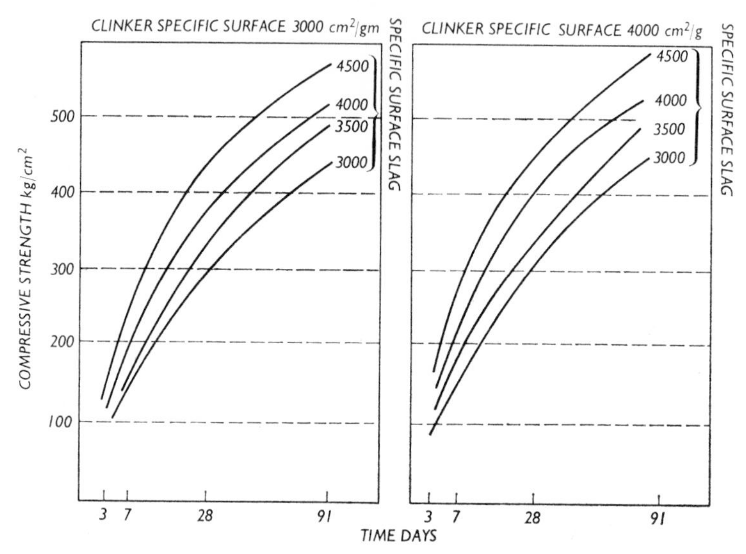

FIG. 132 Compressive strength of mortars: cements with 75 per cent slag of differing fineness (after *Schroeder*).

Portland blastfurnace cement has been manufactured in the Glasgow area for over 50 years. Originally a high-limed slag was used[1] but changes in the iron production have led to lower lime slags being granulated. Such changes, which would normally result in a reduced strength can be compensated by increasing the ratio of Portland cement clinker to granulated slag. Some analyses of Scottish and U.S.A. Portland blastfurnace cement are given in Table 90.

The range of compositions[2] of German Eisenportland and Hochofenzement is shown in Table 91, with the corresponding range for German Portland cements.

Portland blastfurnace cement is similar in physical properties to ordinary Portland cement and the British Standards for the two are identical in respect of fineness, setting time and soundness. The strengths required are somewhat lower than for Portland cement and are shown in Table 92.

The cement clinker used in the manufacture of Portland blastfurnace cement is required to conform to the same composition limits as are laid down for Port-

[1] E. H. Lewis, *Jl W. Scotl. Iron Steel Inst.* **26,** 8 (1918).
[2] F. Schröder, private communication.

TABLE 90 Composition (per cent) of Portland blastfurnace cements

	Scotland				U.S.A.[1]		
CaO	60·90	56·83	57·34	58·15	54·97	56·32	59·36
SiO_2	23·90	24·70	24·50	24·12	26·29	25·94	24·12
Al_2O_3	8·16	9·48	8·40	8·23	8·25	7·64	6·55
Fe_2O_3	2·34	1·89	2·48	2·65	2·22	2·00	2·52
Mn_2O_3	—	—	0·44	0·25	0·33	0·40	0·10
TiO_2	0·25	0·28	0·44	0·40	—	—	—
MgO	1·50	3·22	2·55	1·85	3·76	3·69	2·31
Na_2O	0·28	0·32	0·26	0·31	0·07	0·21	0·06
K_2O	0·56	0·54	0·71	0·55	0·19	0·76	0·63
SO_3	1·32	1·36	1·41	1·97	2·39	1·67	2·07
Loss on ignition	0·73	0·82	1·62	0·15	0·48	0·94	1·97

[1] P. Klieger and A. W. Isberner, *J. Res. Dev. Portld Cem. Ass.* **9** (3), 2 (1967).

TABLE 91 Composition range (per cent) of German Portland cement, Eisenportlandzement and Hochofenzement

Type	CaO	SiO_2	Al_2O_3	MgO	FeO* Fe_2O_3	MnO* Mn_2O_3	SO_3	S
Portland	61–68	19–24	3– 7	0·7–3	1·5–7·2	0 –0·2	1·3–3·0	0·0–0·1
Eisenportland	56–61	22–26	7– 8	1·7–4	1·3–3·3	0·1–0·5	2·0–3·0	0·3–0·7
Hochofen	45–58	24–30	7–12	1 –7	0·9–2·9	0·4–1·1	1·5–3·9	0·4–1·6

* FeO and MnO—Hochofenzement.
Fe_2O_3 and Mn_2O_3—Portland cement and Eisenportlandzement.

TABLE 92 Compressive strength requirement for Portland blastfurnace cement

			3 days	7 days	28 days
Method 1 (vibrated mortar cubes, w/c 0·4)	Ordinary	lb/in²	1600	3000	5000*
		kg/cm²	112	210	351
	Low heat	lb/in²	1100	2000	4000
		kg/cm²	77	141	281
Method 2 (concrete cubes, w/c 0·60)	Ordinary	lb/in²	800	1600	3200*
		kg/cm²	56	112	224
	Low heat	lb/in²	500	1000	2000
		kg/cm²	35	70	141

* Test optional.

land cement (see p. 167). The finished cement, which may not contain more than 65 per cent of slag, is required to conform to the following maximum values:

	Per cent
Insoluble residue	1·5
Magnesia	7
Sulphur trioxide	3
Sulphur present as sulphide	1·5
Ignition loss	3

The physical properties of some Scottish Portland blastfurnace cements[1] are shown in Table 93.

TABLE 93 Physical properties of Portland blastfurnace cements

Residue (per cent) on BS 170 sieve		5·06	5·10	3·70	1·92
Specific surface (cm^2/gm)		—	—	—	3150
Le Chatelier expansion (mm)		1·0	2·0	nil	1·0
Compressive strength (lb/in^2)					
Vibrated mortar cubes	1 day	—	—	530	990
	3 days	—	—	1680	2700
	7 days	—	—	3410	3970
	28 days	—	—	5830	6890
Concrete 1 : 2 : 4 w/c 0·60 water stored	1 day	323	430	390	600
	3 days	1340	1390	1220	1510
	7 days	2170	2380	2270	2230
	28 days	3620	4350	3920	3990
	3 months	4620	5990	5600	5850

The ASTM specification (C595–68) places no limit on the MgO content and limits the SO_3 content to 3·0 per cent and the sulphide sulphur to 2·0 per cent. For the moderate sulphate resistant cement the C_3A content of the Portland cement used must not exceed 8·0 per cent. The strength requirements for the ordinary cement are the same as for Type I Portland cement.

The German standards require that the slag shall conform to the formula

$$\frac{CaO+MgO+Al_2O_3}{SiO_2} \geqq 1$$

The SO_3 content of the cement is limited to 3·5 per cent for Eisenportland cement with a specific surface not above 4000 cm^2/g and 4·5 per cent for higher specific surfaces. For Hochofen cement the corresponding limits are 4 and 4·5 per cent with the additional provision that when the slag content is above 70 per

[1] For data on U.S.A. cements see P. Kleiger and A. W. Isberner, *J. Res. Dev. Labs Portld Cem. Ass.* **9** (3), 2 (1967).

cent the limit is 4·5 per cent for any specific surface. There is no limit to the MgO content of the cements but the Portland cement clinker used must not contain above 5 per cent.

It is generally recognised that the rate of hardening of Portland blastfurnace cement in mortar or concrete is somewhat slower than that of Portland cement during the first 28 days but thereafter increases so that at 12 months the strengths become close to, or even exceed, those of Portland cement. Table 94 is derived from data given by Keil[1] on the relation of compressive strength to age for German Portland, Eisenportland and Hochofen cements. These are average values for a large number of cements and have been calculated in terms of the strength of Portland cement at 28 days expressed as 100.

TABLE 94 Relative compressive strengths of Portland cement, Eisenportlandzement and Hochofenzement

Age	7 days	28 days	3 months	1 year
Portland cement	66	100	119	135
Eisenportlandzement	61	101	—	—
Hochofenzement	50	90	114	144

Portland blastfurnace cement is rather lighter in colour than Portland cement and has a slightly lower average specific gravity, about 3·0 to 3·05 as compared with 3·1 to 3·15. The setting time is affected by the addition of salts in a manner similar to that of Portland cement. Thus, gypsum acts as a retarder while chlorides act as accelerating agents and may also increase the initial rate of hardening. Slag which has been granulated with salty water, as from tidal river water, may show this effect. The moisture movement, drying shrinkage and creep of blastfurnace cement are very similar to those of Portland cements, as is also the bond to steel reinforcement.

The heat evolution during the setting and hardening of Portland blastfurnace cement varies with the composition of the Portland cement clinker and of the granulated slag, as well as with their relative proportions. Some comparative values for Portland and Portland blastfurnace cements are shown in Table 95.

TABLE 95 Comparative values of heat of hydration of cements

Cement	Number of cements tested	Heat evolved Cal/g at end of		
		1 day	2 days	3 days
Ordinary Portland	13	23–42	42–65	47–75
Rapid-hardening Portland	13	35–71	45–89	51–94
Portland blastfurnace	6	18–28	30–51	33–67

[1] *Sym., London 1952*, 530.

The data in Table 96 illustrate the relative heat evolutions of a Portland cement and of its mixes with three granulated slags in a 70 : 30 proportion.[1] The heat of hydration of granulated slag is lower than that of Portland cement and in mixtures of the two Allègre found it to be related linearly up to 7 days to the slag content.[2] However, this linear relation seems not always to hold.

TABLE 96

Cement		Slag	Heat of hydration (cal/g)		
Portland cement	Granulated slag		7 days	28 days	90 days
100	0	—	81	93	103
70	30	A	68	83	86
70	30	B	59	73	78
70	30	C	62	—	79

The British standard for low-heat Portland blastfurnace cement (BS 4246 : 1968) requires that the heat of hydration at 7 and 28 days shall not exceed 60 and 70 cal/g respectively. It is determined by a modified heat of solution method (see p. 292). The strength requirements are shown in Table 92. The proportion of granulated slag must lie between 50 and 90 per cent by weight of the cement. The maximum permitted content of MgO is 9 per cent and of sulphide sulphur 2 per cent. These limits are both higher than in BS 146 : 1958 for ordinary Portland blastfurnace cement reflecting the increase in knowledge. The SO_3 content is limited to 3 per cent and the insoluble residue to 1·5 per cent.

The lower heat of hydration is used to advantage in mass concrete structures to reduce the temperature rise, but it is a disadvantage in cold weather when the rate of hardening may become slow.

TABLE 97 Compressive strengths of 1 : 1¾ : 3½ by vol. concrete, w/c 0·515, expressed as percentages of the values at the same age for samples stored at 17° C

Age	Portland				Portland blastfurnace				Supersulphate			
	8° C	6° C	4° C	2° C	8° C	6° C	4° C	2° C	8° C	6° C	4° C	2° C
3 days	86·7	66·7	40·3	18·6	37·4	21·5	12·1	10·3	35·9	25·7	24·3	23·3
7 days	83·2	74·5	58·5	49·7	54·7	52·3	39·0	30·8	67·5	48·2	36·6	34·2
28 days	85·5	79·0	76·1	68·2	62·5	61·8	60·0	53·2	75·9	76·0	63·3	60·5
1 year	95·5	92·5	85·8	82·4	76·5	74·5	73·6	69·4	100·0	90·0	84·5	78·9
° F	46·4	42·8	39·2	35·6	46·4	42·8	39·2	35·6	46·4	42·8	39·2	35·6

[1] F. M. Lea, *3rd Congress, Commission Int. Grands Barrages*. R11. Stockholm 1948.
[2] R. Allègre, *Revue Matér. Constr. Trav. publ.* **548,** 247 (1961).

The effect of low temperatures in retarding strength development of Portland blastfurnace cement seems to be comparable to that of Portland cement when the water : cement ratio of the concrete is low[1] (0·4). At w/c ratios of 0·5 and upwards the blastfurnace cement is the more affected as is shown by the data[2] in Table 97. The loss in strength at low temperatures is also more pronounced in cements with slag contents approaching or above 70 per cent.

Since Portland blastfurnace cement often shows a somewhat lower strength at early ages at normal temperatures than Portland cement, the reduction in rate of strength development at low temperatures can be correspondingly more troublesome.

Steam curing can be used to accelerate the development of early strength of Portland blastfurnace cements, as with Portland cements. Some data[3] obtained on Belgian rapid-hardening blastfurnace and Portland cements and on supersulphated cement concretes are given in Table 98. The immediate gain in strength of Portland blastfurnace cement is somewhat greater than with Portland cement and the loss in strength at 28 days is lower.

TABLE 98 Effect of steam curing on strength

Pre-steaming period hours	Period of steam curing at 80° C	Portland blastfurnace cement		Portland cement		Supersulphated cement	
		w/c 0·40	0·45	0·40	0·45	0·40	0·45
		Compressive strength as percentage of strength after 48 hour normal curing					
0	4	65	48	49	67	41	38
	6	95	76	72	80	50	41
	12	145	131	75	92	54	46
8	4	113	87	81	87	51	39
		Compressive strength at 28 days as percentage of strength after 28 days normal curing					
0	4	78	86	70	64	60	52
	6	82	92	70	75	53	49
	12	98	94	80	77	48	43
8	4	104	101	84	84	56	43
Normal curing		Compressive strength (kg/cm²)					
	24 hours	117	145	246	181	102	105
	48 hours	343	348	419	361	229	282
	28 days	560	577	643	610	393	460

Concrete 1 : 1¾ : 3½.

[1] P. Klieger and A. W. Isberner, loc. cit.
[2] L. Blondiau, *Revue Matér. Constr. Trav. publ.* **500,** 141 (1957).
[3] L. Blondiau, *Ibid,* **569** 52 (1963).

Supersulphated cement is adversely affected by curing in steam at atmospheric pressure above 50°C. Tests on U.S. Portland blastfurnace cements shown in Table 99, give a broadly similar pattern.

TABLE 99 Effect of steam curing on strength

Curing	Cement	Compressive strength (lb/in^2)			
		1 day	3 days	7 days	28 days
4 hrs. 73° F 16 hrs. in steam at 160° F Then at 73° F	Portland blastfurnace cement	4000	4450	4700	6200
	Portland cement	3700	4050	4450	5700
73° F	Portland blastfurnace cement	1300	2700	4150	6500
	Portland cement	1450	3350	4700	6500

Concrete w/c 0·42.

Portland blastfurnace cements also behave similarly to Portland cements on autoclaving at high steam pressures according to Blondiau.

Trief process

A modification of the usual process for making blastfurnace cement was introduced in Belgium by V. Trief[1] and has become known as the Trief process. In this process the slag, after granulating, is ground wet and stored as a wet slurry. It is kept as a separate constituent until the concrete is being mixed in the concrete mixer, where the Portland cement, slag slurry and aggregate are added together. The advantages claimed are a saving in fuel for drying the slag and a greater efficiency of grinding in the wet state, so that a ground granulated slag of much higher specific surface than would be got by dry grinding may be obtained for the same power consumption in the mills. With the finer slag good strengths can still be obtained with low proportions of Portland cement and in the application of the Trief process only 30 per cent or less of Portland cement is commonly used. In a further modification the slag is ground wet to a high fineness and then dried and bagged so as to adapt the process to conditions which do not permit of the grinding of the slag at, or near to, the place of use.

It is clearly essential that the slurry should remain inactive during the period of grinding and storage. Experience has shown that this is so, providing the temperature of the slurry does not rise too high. Slurry temperatures of 40° do not cause the slurry to begin setting and hardening, but at 70° the slurry is unstable.[2] Badly granulated high-lime slags may also act as activators and cause

[1] *Brit. Pat.* 673,866 (1949); 674,913 (1950).
[2] C. de Langavant, *Revue Matér. Constr. Trav. publ.* **440,** 124 (1952).

the slurry to set. Difficulties can also arise from sedimentation of the slurry, which contains about 30 per cent water, leading to the formation of a stiff paste.

The Trief process has been used for the manufacture of concrete products in Belgium and also for the construction of dams, as for example, Bort-les-Orgues[1] in France and the Cluanie[2] and Avon dams[3] in Great Britain.

An increase in the fineness of a granulated slag increases its activity as is indicated by the data[4] in Table 100. This shows the compressive strengths of 1 : 2 : 4 concrete made with cements composed of 30 per cent Portland cement and 70 per cent of a granulated slag, somewhat low in lime (hydraulic modulus I = 1·0), the latter being ground to various finenesses.

TABLE 100 Effect of slag fineness on strength of concrete

Specific surface (air permeability) of slag (cm^2/g)	Compressive strength (lb/in^2)				
	1 day	3 days	7 days	28 days	90 days
3095	104	360	855	2460	3950
3930	127	417	1220	3160	4690
4850	152	522	1535	3600	5270
6140	173	707	1810	3925	5670
100 per cent Portland cement	915	2415	3510	4950	6370

30 : 70 Portland cement : slag in 1 : 2 : 4 (vol.) gravel concrete, w/c 0·55, water stored.

The strength at 1 or 3 days of Portland cement-granulated slag mixes with Portland cement contents of 30 per cent or less is often very low and very fine grinding does not compensate for this. Accelerators such as NaCl or $CaCl_2$ give some improvement. The former was used at the Bort dam and the latter at the Cluanie and Avon dams. Parker and Ryder have observed that a much greater increase in strength can be obtained by using air-cooled slag aggregate in the concrete with this type of cement, as is shown in Table 101.

Ternary cements. Cements composed of Portland cement, granulated slag and pulverised fuel ash are manufactured in France[5] and are stated to meet the strength requirements for ordinary Portland cement and to offer particular advantages in workability and chemical resistance.

Estimation of granulated slag in cement

The presence of slag in a cement can be detected by moistening a sample with HCl and noting the evolution of H_2S, recognisable by its smell. Under the microscope, granulated slag grains can be differentiated from Portland cement by

[1] M. Mary, *Ann. Inst. Bâtiment*, New Series No. 8 (1951); C. de Langavant, *Revue Matér. Constr. Trav. publ.* **438,** 81; **439,** 105; **440,** 123 (1952).
[2] C. M. Roberts, J. H. Thornton and H. Headland, *Proc. Instn civ. Engrs* **11,** 41 (1958).
[3] J. M. N. Bogle, R. M. Ross and T. McMillan, ibid. **12,** 83 (1959).
[4] T. W. Parker and J. F. Ryder, private communication.
[5] P. Fouilloux, *Revue Matér. Constr. Trav. publ.* **502,** 191 (1957); *Sixth International Congress of International Commission on Large Dams, New York 1958*, Vol. III, 955.

differences in the refractive index, that of the slag being about 1·65–1·66 while the Portland cement particles have indices of about 1·70. If a small amount of a Portland blastfurnace cement on a microscope slide is moistened with a lead acetate solution acidified with acetic acid (the reagent is made up by mixing 2 parts of a 5 per cent lead acetate solution with 1 part of a 50 per cent acetic acid solution) and examined under the microscope, the slag grains are found to exhibit a rapid progressive brown to black coloration. The Portland cement grains do not show this change in coloration, but, as some of them are brownish originally, the differentiation is not absolutely clear.

TABLE 101 Early strength of Trief concrete

Aggregate	Slag fineness (cm^2/g)	Compressive strength (lb/in^2)		
		1 day	3 days	7 days
Gravel	3930*	127	417	1220
Slag	3930	510	1400	2450
Gravel	4930*	147	1100	2610
Slag	4930	550	1820	3450
Gravel	Portland cement†	915	2415	3510

30 : 70 Portland cement : slag in 1 : 2 : 4 (vol.) concrete, w/c 0·55, water stored.

* The granulated slags ground to 3930 and 4930 cm^2/g fineness are different slags.
† Same Portland cement as used for the Trief concrete.

A rapid control method in manufacture when the compositions of the separate constituents are known is based on a determination of the reducing power by means of titration with $KMnO_4$. Rapid polarographic methods are also available making use of a suitable reference component.[1]

Quantitative estimation when the composition of the constituents is unknown is less easy. Simple separation by sedimentation in heavy liquids, making use of the density difference between granulated slag and Portland cement, is not practicable because the finest particles do not separate owing to coagulation effects. A preliminary separation on a very fine sieve is necessary to prepare a coarse fraction. The slag content of the latter is then determined either by flotation in heavy liquids (usually assisted by centrifuging) or by planimetric measurements under the microscope.[2] Since the coarse fraction may differ in Portland cement : slag ratio from that of the whole sample, a check is advisable by analysis of a reference component in the original and in the coarse fraction. For this purpose sulphide sulphur[3] is the most obvious, being present in all slags

[1] P. Janssens, *Silic. Ind.* **16,** 129 (1951); *Sym., London 1952,* 571.
[2] F. Keil, *Sym., London 1952,* 565; F. Shröder, *Sym., Tokyo 1968*; see also M. Duriez and M. C. Houlnick, *Rev. Matér. Constr. Trav. publ.* **422,** 323; **423,** 351 (1950); J. Brocard, ibid. **424,** 10 (1951).
[3] F. Keil and F. Gille, *Zement* **27,** 541 (1938); R. E. Cromarty, *Cem. Lime Mf.* **38,** 33 (1965).

but absent, or only present in negligible quantities in clinker. Alternatively MnO may be estimated.[1] When the separation of slag and Portland cement is not sufficiently complete the proportion of cement in the slag fraction and of slag in the cement fraction is measured microscopically. If the Portland cement : slag ratio in the coarse fraction examined is not the same as in the whole cement a correction can be applied by estimating the content of the reference component in the slag and Portland cement fractions and in the original cement. In any event the estimation is not a precise one but it is not usually necessary to know the slag content of the cement to within more than a few per cent to identify its type.

The coarse fraction of grain size 0·06–0·09 mm is prepared by washing with alcohol between two appropriate sieves. The prepared fraction is dried and heated in a covered crucible for 5 minutes at 700–800°. For sedimentation analysis mixtures of methylene iodide and benzene (density range 3·3–0·9), methylene iodide and acetylene tetrabromide (3·3–3·0) or acetylene tetrabromide and benzene (3·0–0·9) may be used. The density of clinker is taken as 3·2; that of slag as 2·9. In the planimetric work, the same grain size fraction is used, setting the sample for examination in Canada balsam on a microscope slide. Three slides are prepared and the measurements carried out with the aid of an integrating table or planimeter ocular on the microscope. The values obtained correspond to volume ratios and need correction to weight ratios using the approximate densities given above.

The relative accuracy of these methods has been examined by Grade[2] who recommends that for arbitration purposes both microscopic counting of particles in a 30–40 μ sample and the heavy liquid separation combined with analysis of a reference component should be used.

Other physical methods have also been recommended. Thus when granulated slag is subjected to differential thermal analysis an exotherm is found corresponding to the heat of crystallisation of the slag. Kruger[3] has used this effect to estimate the slag content of cements. An X-ray method for determining the proportion of crystalline material in the cement has also been described.[4] It is based on the assumption that the slag is all vitreous and the Portland cement fraction all crystalline.

Hydration of Portland blastfurnace cement

The hydration of Portland cement-granulated slag mixes is more complex than that of Portland cement, since both constituents react with the water. There seems little doubt that the Portland cement grains hydrate in the manner described in previous chapters. We have here, therefore, to consider the hydration of the slag as influenced by the calcium hydroxide liberated from the Portland cement. No hydration products can be observed when ground granulated slag is placed in water. This is ascribed to the formation of acidic surface films as a

[1] P. Catharin, *Zem. Bet.*, 1957 (8) 14.
[2] K. Grade, *SchrReihe Zemind* **29,** 113 (1962); Verein Deutscher Zementwerke E. V. Dusseldorf.
[3] J. E. Kruger, *Cem. Lime Mf.* **35,** 105 (1962).
[4] K. H. L. Sehkle, *Cem. Lime Mf.* **40,** 57 (1967).

small amount of Ca^{++} ions is released into the solution. In a calcium hydroxide solution reaction occurs removing this film and continued hydration, as the lime breaks into the silica framework of the slag, takes place. It was formerly considered that little combination occurred between lime in solution and the slag, but later work has shown that there is a progressive take-up of lime and that the hydration products are richer in lime than the slag. Provided a continuous supply of lime is available the slag can combine with up to about a tenth of its weight of CaO in three months.[1] Granulated slag contains some calcium sulphide[2] which reacts with water to give $Ca(OH)_2$ and $Ca(SH)_2$, but the amount of lime so produced from the sulphide on the surface of the slag particles can only provide a very minor amount of lime to the solution.

If a blastfurnace cement is placed with a drop of water on a microscope slide the production of amorphous matter from both slag and cement grains can be observed within a few hours. After a day large numbers of plates and needles of hydrated calcium aluminate, and large columnar crystals of calcium hydroxide can be seen.

The hydration products clearly identified[3] in set blastfurnace cements are:

1. A tobermorite-like calcium silicate hydrate phase which probably contains alumina in solid solution for not all the alumina can be accounted for in the other hydration products. This phase may also contain magnesia in solid solution because free $Mg(OH)_2$ cannot be detected in the set cement.[4]
2. Hexagonal tetracalcium aluminate hydrate or its solid solution, or intergrowth, with $3CaO(Al_2O_3, Fe_2O_3).CaSO_4.12H_2O$.
3. An ettringite-type phase $3CaO(Al_2O_3, Fe_2O_3).3CaSO_4.31H_2O$.
4. Calcium hydroxide.

The relative extent to which the tri- and mono-forms of calcium sulphoaluminates are formed in the early stages of hydration is considered by Ludwig to depend on the reaction velocity of the alumina compounds and the rate of solution of the calcium sulphate retarder. An increase in the former, it is suggested, favours the formation of the mono-compound.

The extent to which calcium hydroxide is present will vary with the Portland cement : slag ratio in the cement. In mixes of low Portland cement content it may be absent in the later stages of hydration. The curves[5] given in Fig. 133 show the calcium hydroxide content up to 28 days for cements of various Portland cement : slag ratios. These data were determined by X-ray analysis but Franke's method should also be applicable. Other compounds that might sometimes be formed are hydrated gehlenite(C_2ASH_8) and a hydrogarnet. Locher[6] found both these compounds in the hydration products of $CaO–Al_2O_3–SiO_2$ glasses and

[1] R. Sersale and G. Orsini, *Ricerca scient.* **30,** 1230 (1960); C. H. Schmitt, *Zement-Kalk-Gips* **16,** 321 (1962); N. Fratini, *Silic. ind.* **29,** 285 (1964).
[2] F. Kämpfe, *Zement* **24,** 257 et seq. (1935).
[3] F. Shröder, *Sym., Tokyo 1968*; V. I. Satarin and Y. M. Syrkin, ibid.
[4] U. Ludwig, *Zement-Kalk-Gips* **21,** 81 (1968).
[5] W. Kramer and H. G. Smolczyk, *Atti del Convegno sulla Produzione e le Applicazioni dei Cementi Siderurgici, Naples 1960*, p. 249, Gennaro d'Agostini, Naples 1961.
[6] F. W. Locher, *Sym., Washington 1960*, 267. See also H. E. Schwiete, V. Ludwig and K. E. Würth, *Zement-Kalk-Gips*, **22,** 154 (1969).

that the C_2ASH_8 subsequently took up more lime to transform into a hydrogarnet with a composition about C_3ASH_4. However, Smolczyk[1] in the examination of a large number of set blastfurnace cements was only able to identify C_2ASH_8 in one case, and a hydrogarnet in another case after a year's storage in water. Similarly Sersale[2] did not find any hydrated gehlenite in set cements stored for a year or more in water. It seems that in cements these two compounds can at most be only minor hydration products. Another product[3] that may occur rarely is C_2AH_8.

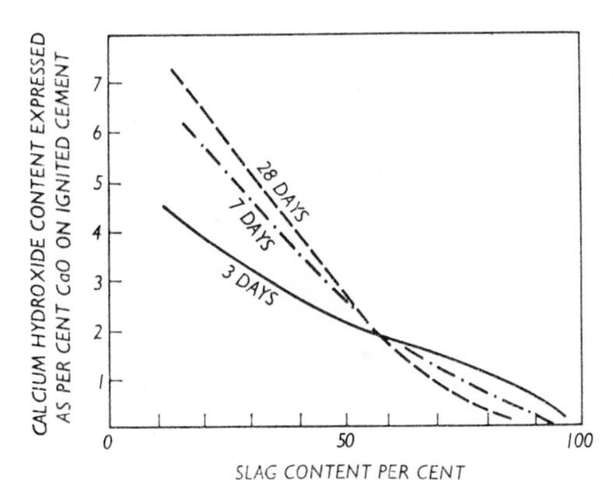

FIG. 133 Calcium hydroxide content of set blastfurnace cements.

Supersulphated cement

Cement made from granulated blastfurnace slag activated by means of calcium sulphate is known in Great Britain as supersulphated cement (BS 4248 : 1968). In Belgium, which has been one of the main sources of commercial production, the name given is 'Ciment métallurgique sursulfaté' (Belgium specification NBN 132). In France the material is known as 'Ciment sursulfaté' (Norme P.15–313) and in Germany 'sulfathüttenzement' (DIN 4210).

The cement is made by grinding a mixture of 80–85 per cent granulated slag, 10–15 per cent anhydrite or hard burnt gypsum, and about 5 per cent Portland cement; the product is ground more finely than is characteristic of Portland cement. The British Standard requires the content of granulated blastfurnace slag to be not less than 75 per cent, MgO not greater than 9 per cent, SO_3 not less than 4·5 per cent, sulphur other than present as SO_3 not more than 1·5 per cent and insoluble residue not greater than 3 per cent. The specific surface must be not less than 4000 sq cm per g. The German standard requires a minimum content of 75 per cent slag and 3 per cent SO_3, and a maximum content of 5 per

[1] H. G. Smolczyk, *Zement-Kalk-Gips* **18,** 238 (1965).
[2] R. Sersale, R. Arello and V. Amicarelli, *Ricerca scient.* **36,** 453 (1966).
[3] W. Richartz, *Tonind.-Ztg. keram. Rdsh.* **90,** 449 (1966).

cent Portland cement or other alkaline activator. The slag must contain at least 13 per cent Al_2O_3 and conform in composition to the formula

$$\frac{CaO+MgO+Al_2O_3}{SiO_2} \geqq 1{\cdot}6$$

The French standard requires a minimum content of 5 per cent SO_3 but does not specifically control the proportions of slag and activator.

A moderately high-alumina content is usually preferred in the slag used. Keil gives values of 15–20 per cent but cements are made satisfactorily with slags of rather lower alumina content. Blondiau[1] has suggested that the CaO : SiO_2 ratio should be between about 1·45 and 1·54 and the SiO_2 : Al_2O_3 ratio between 1·8 and 1·9. Keil and Gille[2] recommend a modulus of at least 1·8 and preferably 1·9 or higher as calculated by the Modulus III (p. 463). Some typical cement compositions are shown in Table 102.

TABLE 102 Compositions (%) of supersulphated cements

	Belgian			German	British				
No.	305	558	586	550	335	585	613	628	629
CaO	45·41	44·71	44·45	41·40	45·23	41·52	41·01	41·67	42·26
SiO_2	27·00	25·88	28·56	25·47	24·40	30·52	28·89	30·39	29·74
Al_2O_3	11·93	11·86	12·38	13·30	13·04	13·79	13·85	14·04	13·74
Fe_2O_3	1·07	0·80	0·95	0·54	0·64	1·18	0·93	0·81	0·83
Mn_2O_3	1·18	0·76	0·81	0·73	0·75	1·18	0·95	1·01	1·08
TiO_2	0·50	0·48	0·60	0·83	0·57	0·56	0·52	0·48	0·51
MgO	2·61	3·63	3·69	6·75	3·34	3·75	3·64	3·56	3·38
Na_2O	0·34	0·29	0·42	0·63	0·47	0·27	0·28	0·33	0·34
K_2O	0·75	0·54	0·52	0·65	0·79	1·03	1·14	1·06	1·08
SO_3	6·72	7·80	5·65	8·53	7·57	5·29	5·76	5·04	5·37
S (sulphides)	0·84	0·96	0·98	1·41	1·07	0·98	0·98	1·01	1·00
P_2O_5	—	0·25	0·32	0·10	—	0·09	—	—	—
Loss on ignition	1·00	0·22	0·14	0·49	2·35	0·17	0·85 (gain)	1·36 (gain)	1·23 (gain)
Total[1]	98·93	97·70	98·98	100·13	99·69	99·84	98·95	100·40	99·33

[1] Totals less O for S. Ignition *gains* excluded from totals.

The strength requirements for supersulphated cement required by the British Standard are:

		3 days	7 days	28 days
Method 1 (vibrated mortar cubes) w/c 0·4	lb/in^2	2000	3400	5000
	kg/cm^2	141	239	352
Method 2 (concrete cubes) w/c 0·55	lb/in^2	1000	2400	3700
	kg/cm^2	70	169	260

[1] L. Blondiau, *Revue Matér. Constr. Trav. publ.* **424,** 6; **425,** 42 (1951).
[2] F. Keil and F. Gille, *Zement-Kalk-Gips* **2,** 81 (1949).

The expansion in a modified Le Chatelier test for soundness in which the test specimen is immersed in *cold* water (66° F, 19° C) for 24 hours must not exceed 5 mm. When a low-heat cement is specified the heat of hydration determined by a modified heat of solution method must not exceed 60 and 70 cal per g at 7 and 28 days respectively. The initial setting time must not be less than 45 minutes and the final setting time must not exceed 10 hours, the same as for Portland cement. Some typical physical properties of supersulphated cement are shown in Table 103.

TABLE 103 Physical properties of supersulphated cements

		British			Belgian		German
Specific surface cm²/g		5055	4930	—	4980	4150	3340
Compressive strength lb/in² vibrated mortar cubes	1 day	350	290	560	970	930	1770
	3 days	2380	2440	2510	4420	4770	2280
	7 days	4100	4000	5240	7030	6500	6370
	28 days	6640	6580	8040	9530	9240	8550
Compressive strength concrete 1 : 2 : 4 w/c 0·6 water stored	1 day	580	430	400	710	90	350
	3 days	1380	1450	1010	2480	680	1490
	7 days	2630	2440	2260	4190	2400	3440
	28 days	4490	4380	5140	5300	5530	6000

Supersulphated cement, like high-alumina cement, combines with more water on hydration than does Portland cement. The curve for strength against water : cement ratio has been reported to be unusually flat. The strength falls off more rapidly than with other cements as the aggregate content of the mix is increased, so that it is not desirable to use mixes much leaner than 1 : 2: 4. The cement is rather more sensitive to deterioration during storage than Portland cement, owing to the effects of carbonation; the setting time is lengthened and the early strength very seriously affected. Cement that has deteriorated in this way can be restored by the addition of 1 per cent hydrated lime.[1] Supersulphated cement has a low heat of hydration, about 40–45 cal/g at 7 days and 45–50 at 28 days. The rate of hardening increases with temperature up to about 40° and the cement has been used in tropical climates. At higher temperatures its strength drops seriously and a similar effect is found on curing in high-pressure steam or in steam above 50° (see p. 475). This fall in strength is ascribed to the dehydration, or decomposition, of calcium sulphoaluminate. Because of its high sulphate-resistance, the cement has been particularly used below ground, although it is also used for structures above ground. In the latter case, care needs to be taken in the initial curing stages to maintain the surface damp, otherwise there is a tendency for it to become friable or dusty. Alternatively the shuttering should be left in place as long as possible or a curing compound applied. The depth of this friable layer does not increase progressively with time. According to Keil the chances of it

[1] F. Keil, *Sym., London 1952,* 530.

occurring can be reduced by increasing the Portland cement content of the cement.

Supersulphated cement is resistant to a variety of aggressive agents though there may be some variation with the slag composition. Blondiau[1] found a satisfactory behaviour during test periods of up to 3 years in contact with saturated gypsum solution; natural ground waters, strongly sulphated, of the Liége area; $MgSO_4$ solution (20 g/litre); $Al_2(SO_4)_3$ (20 g/litre); $(NH_4)_2SO_4$ (5 g/litre); humic acid (2 g/litre) in the presence of $MgSO_4$ (20 g/litre) and NaCl (30 g/litre); and artificial and natural seawater. Good resistance was also shown against HCl (5 g/litre) and also to linseed oil, an agent which damages both Portland cement and Portland blastfurnace cement.

Other tests[2] have shown the cement to be resistant to calcium and sodium sulphates, and to be slightly affected by magnesium sulphate, at a concentration of 350 parts SO_3 per 100 000, but to be more severely attacked by ammonium sulphate of this concentration. The resistance to sodium and calcium sulphates can be attributed to the absence of calcium hydroxide and the fact that part of the alumina is already combined in ettringite in the set cement. Magnesium sulphate has a more severe action because it also attacks the calcium-silicate hydrate and decomposes ettringite (see p. 346). There is evidence that supersulphated cement concrete of good quality will resist acid conditions with or without the presence of sulphate ions down to a pH of about 3·5. It has also proved serviceable in some conditions of use, e.g. floors, when exposed to mineral acids of lower pH values. Both in practice and in laboratory tests the results obtained under these more acid conditions will depend on the extent to which the strength of the acid in contact with the concrete is maintained and on how long coatings of reaction products can give some protection. Thus in laboratory tests with stagnant solution the concrete showed a good resistance to 0·25 per cent sulphuric acid up to 2 years and then lost strength.

The cement is also resistant to weak concentrations of organic acids (e.g. below 0·5 per cent) such as lactic, acetic, citric and tartaric acids and phenols and cresols. Stronger solutions such as 1 per cent cause damage.[3] Chlorides and alkali hydroxides and carbonates seem to have no action.

Supersulphated cement has been used in harbour and breakwater construction on the Belgian coast, and it was also included in the Belgian seawater trials showing a generally good performance (see p. 631).

Hydration. The initial setting and hardening of supersulphated cement is associated with the formation of calcium sulphoaluminate from the slag constituents and the added calcium sulphate.

The formation of the high-sulphate form of calcium sulphoaluminate (ettringite) does not take place readily in saturated lime solution and the monosulphate tends to be formed as an unstable phase; neither can ettringite be formed in water with no calcium hydroxide present. The Portland cement addition to the cement is required to give the correct alkalinity to enable the ettringite to be formed.

[1] L. Blondiau, *Revue Matér. Constr. Trav. publ.* **350** (1938) to **362** (1939).
[2] D. J. Baxter and J. M. Boardman, *Civ. Engng publ. Wks Rev.* **57,** 472, 627, 778 (1962); G. H. Thomas, *Sym., Tokyo 1968.*
[3] D. N. Evans, R. L. Blaine and P. Worksman, *Sym., Washington 1960*, 871.

Blondiau found the most favourable concentration to be about 0·2 g CaO per litre in the liquid and later workers[1] indicate a range from 0·15 to 0·5 g CaO per litre. Combination of lime by carbonation leads to a too low CaO concentration in the solution and this accounts for its deleterious effect. An excessive addition of Portland cement, giving a saturated lime solution, results in low strengths. In manufacture hydrated lime can be used, in a smaller proportion, in place of Portland cement, but the latter is rather less sensitive to carbonation and is, therefore, preferred.

Examination of the set cement by X-ray and DTA methods has shown that the main hydration products are ettringite and a tobermorite-like phase.[2] The ettringite has been detected after a few hours hydration and the tobermorite-like phase after a day. The ettringite phase develops rapidly and seems to reach a maximum within three to seven days. The early strength seems, therefore, to be largely attributable to ettringite formation, but the development of strength from about three days onwards must come from the increasing formation of the calcium-silicate hydrate. There is less certainty about the extent to which, and at what stage, the monosulphate form of calcium sulphoaluminate is produced. Blondiau[3] had earlier suggested that it was formed during the first few hours when the CaO concentration in the solution was high, and that as the lime in solution became depleted by this reaction, the reserves being small, the formation of ettringite set in. It has not been detected at this stage, but Smolczyk was able to identify it in the set cement at seven days. Whether it is formed directly, or by decomposition of ettringite as D'Ans and Eick[4] suggested, is still not clear, though it may be questioned if a change between two compounds of such different molecular volume could occur without some marked effect on strength. The amount of sulphur trioxide in a supersulphated cement is sufficient only for conversion of about a quarter of the alumina to ettringite, so the excess must be present in other forms, hydrated or unhydrated. Even formation of the monosulphate still leaves substantial excess alumina. From consideration of the phase equilibrium in the system $CaO–Al_2O_3–CaSO_4–H_2O$, D'Ans and Eick deduced that hydrated alumina must be a product of the hydration and considered that the final product is a mixture of the two forms of calcium sulphoaluminate and hydrated alumina together with a hydrated calcium silicate. The hydrated alumina has not been detected by X-rays but it would be expected to be amorphous. Some of this missing alumina is probably present in the calcium-silicate hydrate phase. The ettringite formed is probably not the pure compound. Firstly ferric oxide may substitute for alumina and, secondly, Nurse[5] has advanced evidence that the phase in supersulphated cement contains some $Ca(OH)_2$ substituting for $CaSO_4$.

[1] P. P. Budnikov, *Chem. Abstr.* **42,** 6505 (1948); F. Köberich, *Zement-Kalk-Gips* **2,** 109 (1949).
[2] H. G. Smolczyk, *Zement-Kalk-Gips* **18,** 238 (1965); U. Ludwig and H. E. Schwiete, *Tonind.-Ztg. keram. Rdsh.* **89,** 174 (1965); J. Chi-Sun Yang, *Sym., Tokyo 1968.*
[3] L. Blondiau and V. Blondiau, *Revue Matér. Constr. Trav. publ.* **453,** 165 (1953).
[4] D'Ans and H. Eick, *Zement-Kalk-Gips* **7,** 449 (1954).
[5] R. W. Nurse, *Atti del Convegno sulla Produzione e le applicazione dei Cemente Siderurgici, Naples 1960,* p. 187.

Utilisation of slag cements

When cements of the Portland blastfurnace type were first introduced they met with considerable suspicion and opposition and the literature of 70 years ago contains many articles condemning them. The suspicion was not without foundation, since some manufacturers had ground air-cooled or even dusted slag with Portland cement and such additions were correctly regarded as a valueless adulteration. The same opinion was advanced in regard to granulated slags and it was only gradually as experience was gained, and it became understood that only glassy granulated material could be used, that their cementing qualities became generally accepted. The fear was also expressed that the sulphide sulphur present in granulated slag would oxidise, causing expansion of a concrete and corrosion of any reinforcement. It was only after years of use and the results of many tests that these suspicions were shown to be groundless.

The addition of granulated slag to Portland cement was first introduced in Germany in 1892, but despite the advocacy of Michaelis and Passow, its industrial development was delayed. By 1901, however, the manufacture had sufficiently developed for Eisenportland cement to become an accepted name and for the German Association of Eisenportland cement manufacturers to be formed. Hochofen cement was first made in 1907. By 1930 Eisenportland cement production had reached 550 000 tons per annum and Hochofen cement 460 000 tons. By 1961 these figures had risen to 4·2 and 3·2 million tons respectively, together with 8000 tons of supersulphated cement.

The history of cements of these types in other countries followed much the same course as in Germany. The extent of their manufacture now depends primarily on economic factors and the availability of suitable slags in sufficient proximity to sources of Portland cement clinker. In Great Britain manufacture is limited to the Glasgow and Lincolnshire areas, but it is much more widespread in Belgium, France and other European countries including the U.S.S.R. There is a large production in Japan and a significant one in the U.S.A.

For many years after the first introduction of the Portland blastfurnace type of cement in Germany their use was restricted to seawater work, foundations, and other structures in which the concrete was not in contact with air. It was considered that while blastfurnace cement hardened excellently under water or in damp conditions, it was uncertain that in air the strength development would be so satisfactory. In 1915–1916, however, an official decree was issued in Germany placing blastfurnace cement on the same footing as Portland cement. This decree followed as a result of a series of tests which showed that the blastfurnace cement developed strengths either in air or water, and in rich or lean mixes, which were equal to those given by Portland cement, and after it had also been established by extensive investigations by the German Reinforced Concrete Committee that reinforcement in concrete made with blastfurnace cement did not behave any differently from that in Portland cement concrete.[1] Tests carried out at the Berlin Material-prüfungsamt showed that the protection against corrosion afforded to embedded steel by blastfurnace cement was as good as that given by Portland cement and that there was no objection to its use for reinforced

[1] Cf. *Eisenportlandzement Taschenbuch*, Düsseldorf, 1931, p. 54.

concrete structures. This conclusion has been confirmed by many subsequent investigations and observations on the condition of concrete structures made from slag-containing cements. It has also been found[1] that no difference in the susceptibility to corrosion of pre-stressing wires in Portland or Portland-slag concretes could be detected. Much of the earlier opposition to slag cements was based on the supposition that instability might be caused by the presence of sulphides in the slag. Gutt and Hinkins[2] showed that most of the sulphide in slag cement concrete disappeared after a year without causing any expansion or instability. Cements containing more sulphide than is permitted in the British Standard were stable in the ASTM autoclave soundness test.

Blastfurnace type cements have now long been used in many countries for general building work, including reinforced concrete, for water-retaining structures, and for precast products such as concrete pipes. They develop strengths comparable to Portland cement but require more careful handling at low temperatures when the rate of strength development is depressed rather more than that of Portland cements unless the water : cement ratio is low. The frost resistance of the concrete is similar to that of Portland cements and the expansion with alkali-reactive aggregates is considerably lower than that with Portland cement of similar alkali content.[3] Cements of the Portland blastfurnace, but not of the Hochofen type have been included in later extensions of the 'Long Time Study of Cement Performance in Concrete' being carried out by the Portland Cement Association in the U.S.A.[4]

The claim has always been made for Portland blastfurnace cement that it is more resistant to attack by seawater and other chemical agencies than Portland cement, but this applies to the comparison with ordinary Portland cement rather than with the modern sulphate-resisting Portland cement. However, the sulphate resistance of Portland blastfurnace cements depends on the cement clinker as well as the slag and can be raised further by the use of a sulphate-resisting cement clinker. Grün found that the lower lime slags, and also those of the higher alumina contents (above 15 per cent) gave better results in sulphate resistance tests than those of higher lime content (above 43 per cent) or higher magnesia content (above 7 per cent). Blastfurnace cements have been widely used in Germany and Holland and elsewhere for seawater work, but the discussion of this must be left to a later chapter.

The increase in resistance to sulphate solutions caused by the substitution of granulated slag in cements is indicated in Fig. 134 which is based on the average results from 26 slags tested by Grün.[5] The strength of the various mixes after 2 years in a 10 per cent $MgSO_4$ solution is expressed as a percentage of the strength attained by the same Portland cement-granulated slag mix after a similar period in water.

Later work has shown that the sulphate resistance of slag type cements is dependent on the C_3A content of the Portland cement fraction and the alumina

[1] B. Ost and G. E. Monfore, *J. Res. Dev. Labs Portld Cem. Ass.* **5,** 23 (1963).
[2] W. Gutt and B. Hinkins, *J. Iron Steel Inst.* **203,** 580 (1965).
[3] P. Kleiger and A. W. Ishberner, loc. cit.; W. E. Grieb and G. Werner, *Proc. Highw. Res. Bd* **40,** 409 (1961); B. Mather, *Proc. Am. Concr. Inst.* **54,** 205 (1957).
[4] P. Klieger, *J. Res. Dev. Labs Portld Cem. Ass.* **5** (1), 2 (1963).
[5] R. Grün, *Internat. Assoc. Test. Mat.*, Zürich **1,** 781 (1931).

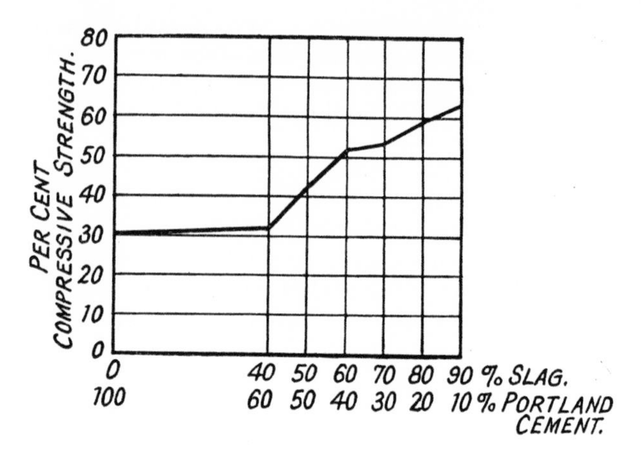

FIG. 134 Strength of 1 : 3 mortars after 2 years in 10 per cent $MgSO_4$, as percentage of strength in water.

content of the slag. Increasing alumina contents in the slag can have an unfavourable influence. From tests on mortar in 4·4 per cent Na_2SO_4 solution Locher[1] found that for cements with at least 65 per cent slag content the sulphate resistance was always greater than that of the parent Portland cement, though the improvement was much more marked with Portland cements of 8 and 11 per cent C_3A content than with one of zero content. When, however, the slag content was between 20–50 per cent the use of a high-alumina (17·7 per cent) slag decreased the sulphate resistance whereas that of a low-alumina (11 per cent) slag increased it independently of the C_3A content of the cement in both cases. Other tests[2] have confirmed these influences of the Al_2O_3 content of the slag and the C_3A content of the Portland cement and shown that they also apply to exposure in 5 per cent $MgSO_4$ solution. There is however still a need for more tests using larger concrete specimens rather than small mortar specimens since the results from the latter are the more dependent on the specific test conditions.

In Plate XIV there are shown 1 : 2 : 4 concrete prisms of a Portland and a Portland blastfurnace cement after storage for 2 years in 5 per cent magnesium sulphate solution.

The increased resistance of blastfurnace cements to attack by sulphates is generally associated with the lesser content of free calcium hydroxide present in the set cement and the less basic nature of the hydrated calcium silicates. Obviously the amount of alumina from C_3A available for conversion to ettringite is reduced because of the smaller Portland cement content. No balance sheet for the hydration products of the alumina in the slag can be drawn up, but part of it is taken up in the calcium silicate hydrate where it is less readily open to sulphate attack.

[1] F. W. Locher, *Zement-Kalk-Gips*, **19**, 395 (1966).
[2] J. H. P. van Aardt and S. Visser, Research *Bull National Bldg. Res. Inst. Un. S. Africa.* 47 (1967).

Supersulphated cement is still more resistant than Portland blastfurnace cement to attack by seawater, sulphates, and other aggressive agencies. It has been used successfully in a variety of aggressive conditions, for example in Belgium for seawater work and for reinforced concrete pipes in groundwaters that had proved very destructive to Portland and Portland blastfurnace cements. It has also found some use in general building construction, a notable example being the Palais de Chaillot in Paris built in 1937. As noted earlier, care is required in moist curing to avoid the formation of a friable or dusty surface layer. In Great Britain, it has been used in chemical works under conditions involving exposure to high concentrations of sulphates and for the underside of bridges over railways to resist the action of locomotive flue gases.

16 High-alumina Cement

High-alumina cement is obtained by fusing or sintering a mixture, in suitable proportions, of aluminous and calcareous materials and grinding the resultant product to a fine powder.

History of high-alumina cement

Though it was realised by several earlier investigators that the less basic aluminates of calcium had excellent cementing qualities, the commercial development of high alumina cement is associated with the work of Bied in the laboratories of the Société J. and A. Pavin de Lafarge at La Teil, France. Frémy[1] in 1865 prepared various melts of lime and alumina and found them to possess good hydraulic properties and this was confirmed by Michaëlis[2] some years later. Schott[3] also, in 1906, had published a memoir showing the high strengths which were given by fused calcium aluminates. Even as long ago as 1888[4] a patent was taken out for a limestone-bauxite cement and in the early years of the present century a number of patents appeared.

The work of Bied[5] arose out of the serious troubles which were experienced in France owing to the decomposition of mortars and concretes in grounds containing large contents of sulphates, notably gypsum. The deterioration of concretes in sea water had long been known and as early as 1853 the French Society for the Encouragement of National Industry had offered a reward for the best study on the causes of this action. The problem of sulphate attack had become of increasing importance by 1890 owing to the failure of concrete work on the P.L.M. Railway in the South of France. Bied had investigated some of these failures in 1898 and come to the conclusion that they were due to the attack of the concrete by gypsum in the soil and ground waters, the disintegration being attributed to the formation of calcium sulphoaluminate and to the crystallisation

[1] *C. r. hebd. Séanc. Acad. Sci., Paris* **60,** 993 (1865).
[2] *Hydraul. Mörtel,* Leipzig, 1869, 35.
[3] Thesis for Doctorate, Heidelberg, 1906.
[4] *Br. Pat.* 10312 (1888).
[5] Cf. J. Bied, *Recherches Industrielles sur les Chaux, Ciments et Mortiers,* Dunod. Paris, 1926.

of gypsum in the pores of the concrete. In 1900 and 1902 further troubles due to gypsum were experienced in tunnels at Alicum, Rimont and La Plagnotte and also, due to magnesium sulphate and sodium sulphate, in structures on the Bou-Saada line in Southern Algeria. Following a further inquiry into these troubles, Bied commenced a prolonged investigation into the production of cements resistant to sulphate action.

Vicat had suggested in 1846 that a cement in which the ratio of (silica+alumina) to (lime+magnesia) was greater than unity would be resistant to sulphate action, and it was from this suggestion that Bied's work commenced. Apart from work on the use of pozzolanas, Bied concentrated on the study of the cements of high-alumina content, knowing that the less basic silicates had little cementing value. This work resulted in 1908 in the patenting[1] of a cement manufactured by fusing together a bauxite, or other aluminous and ferruginous material of low silica content, with lime in proportions approximating to the formula $CaO.Al_2O_3+2CaO.SiO_2$. The cement which was thus obtained proved not only to have the desired properties of sulphate resistance, but to show in addition a rapidity of hardening far surpassing that of the Portland cements then manufactured. Considerable trouble was experienced in production on a commercial scale and it was found that while some batches of cement were excellent, others of similar chemical composition would prove useless. Further investigation into the causes of these variations was necessary, but by 1913 a satisfactory method of manufacture had been achieved. Because, however, of the previous difficulties, a series of trials extending over five years was carried out on the product, both by the Lafarge Company itself and by the French Government, before the cement was finally put on the market in 1918. During the war years the cement was supplied to the French army for the construction of gun emplacements and other special work, and also in 1916 to the P.L.M. Railway for reconstruction work in the Brauss Tunnel[2] on the Nice–Coni line. This tunnel passes through a solid mass of anhydrite which had had a most destructive effect on the mortars used in the original construction.

Contemporaneously with the work of Bied, patents were taken out by H. S. Spackman[3] in the United States. Spackman had worked along different lines from those of Bied, his object being the production of a material rather of the nature of a pozzolana to be added to lime or Portland cement to enhance its cementing qualities. Spackman's aluminate compounds were prepared by adding bauxite to high-alumina slags. A number of natural cements containing additions of Spackman's calcium aluminates were marketed in the United States about 1910 under the name of Alca natural cements, but their manufacture was later abandoned. A further study of high-alumina cements was carried out by P. H. Bates[4] at the U.S.A. Bureau of Standards, the first results of which were published in 1921. Bates prepared high-alumina cements by clinkering in a small (20-foot) rotary kiln and tested their properties in concrete.

[1] *Br. Pat.* 8193/1909.
[2] Touche, *Le Ciment* **31,** 240 (1926).
[3] *Cement Age* **11** (2), 76 (1910); *Proc. Am. Soc. Test. Mater.* 1910, 315; *Br. Pats.* 10110/1908 and 18345/1908.
[4] *Bur. Stand. tech. Paper,* No. 197 (1921).

Manufacture

The raw materials used for the manufacture of high-alumina cement are limestone and bauxite. Although alumina is very widely distributed in nature, bauxite is the only suitable material available commercially on a scale adequate for cement production.

Bauxite belongs to the group of naturally occurring materials known as laterites which are formed when an igneous rock has weathered in such a way that the silica, lime, magnesia and alkalis have been gradually removed in solution, leaving behind hydrated oxides of aluminium, iron, and titanium contaminated with residual silica. Vast deposits of laterites occur, particularly in tropical regions, but it is only those which contain a high-alumina content and are known as bauxites which have any commercial value. The utilisation of laterites in general awaits the development of methods of manufacture suited to them.

Bauxite also forms the raw material for the manufacture of aluminium metal, of alum and aluminium sulphate, and of certain types of refractories. For cement manufacture a considerable content of iron oxide is permissible and in practice ferruginous bauxites are normally used. The silica content should be fairly low; the presence of several per cent of titanium oxide is not objectionable.

Typical analyses, on a moisture-free basis, of bauxites used in high-alumina cement manufacture are as follows:

	French	Greek	Yugoslavian
Combined H_2O	12	11	19
SiO_2	4·5	3	2
Al_2O_3	53	55	53
Fe_2O_3	24	27	23
TiO_2	3	3	3
CaO	2	2	2

The material used in Great Britain comes from France and Greece. During the last war, when bauxite was short, use was made of aluminium dross and the red mud from the Beyer process for production of alumina.[1]

The method of manufacture of high-alumina cement is as yet far from standardised and a variety of furnaces are in use.[2] High-alumina cement was originally manufactured in a water-cooled vertical furnace, lined with refractory material, rather similar to, but much smaller than, a blastfurnace. The bauxite and limestone, mixed with the required proportion of coke, were charged into the top of the furnace. A hot-air blast, preheated in recuperating chambers, was introduced through tuyeres near the bottom of the furnace; this part of the furnace jacket was water-cooled. As the mix passed down the furnace, combination between the limestone and bauxite occurred and a part of the iron oxide was reduced to metallic iron. At the bottom of the furnace, complete fusion occurred and the molten cement poured continuously from a tap hole.

This type of furnace has been superseded by a reverberatory open-hearth furnace,[3] which is the most common method of manufacture at the present time,

[1] A. V. Hussey, *Chemy Ind.*, 1947 (42), 635.
[2] See T. D. Robson, *High Alumina Cements and Concretes*, Contractors Record Ltd., London 1962.
[3] Cf. *Br. Pats.* 222426 and 222427.

and is used in England and France. This open-hearth furnace is arranged with a long vertical stack into which the bauxite and limestone, or chalk, are charged. It is fired with pulverised coal or oil with a hot-air blast. The furnace gases, passing through the charge of raw materials, drive off the moisture and carbon dioxide. Fusion takes place at the point where the charge drops from the vertical stack into the furnace proper. The cement is maintained in a liquid state in the hearth by heat radiated from the arch and pours out continuously from a tap hole. It is run into moulds and cooled. Furnaces of this type about 9 feet wide and 16 feet long, excluding the vertical stack, produce about 70 tons of cement a day. The fuel consumption is about 25 per cent of the weight of cement produced. The temperature reached is about 1550–1600°.

Electric furnaces[1] have also been used at a number of plants in Europe. These are of the arc type and employ two vertical carbon electrodes. It is not possible to utilise a resistance furnace owing to the poor conductivity of the fused aluminates. The electric furnace appears to offer no particular advantages except where electric power is very cheap, for the wear on the electrodes is very severe and the life of the refractory lining short. The furnaces are tapped intermittently, every few hours, and yield an output of some 15–40 tons per day.

High-alumina cement is also produced by fusion in a rotary kiln of a type similar to that used in Portland cement manufacture. This method is used in the United States. Although Bates succeeded in clinkering high-alumina cements with a low iron content (< 4 per cent Fe_2O_3), it has not been found possible to clinker the raw materials containing a high iron content that are used commercially. The temperature range between incipient melting and fusion in these mixes is too small to permit of clinkering successfully. Despite this difficulty, several patents have been taken out for the production of high-alumina cement by clinkering, but they have not yet been applied commercially. A process has, however, been developed for manufacture by sintering the very finely ground raw mix in a Hoffman-type kiln at a temperature not above 1250°. At least one European plant at the present time is working this sintering process. In a patent taken out by the Lafarge Company[2] it is stated that heating for six hours at 1000° is sufficient to produce complete combination.

After cooling, the cement mass resembles a dark, fine-grained compact rock approaching in structure and hardness a basalt. In the former method of manufacture with vertical furnaces a product resembling scoria, black, porous, and vitrified in appearance, was occasionally obtained. This usually disintegrated on cooling, owing probably to a change to the γ form of the dicalcium silicate it contained. Suggestions have been made that such disintegration should be aimed at in manufacture in order to reduce the cost of grinding the fused cement.[3] It seems, however, that to obtain cements which disintegrate it is necessary to raise the silica content, which reacts unfavourably on the properties of the cement. This method has not therefore been adopted in practice.

The pigs of fused cement, after cooling, are crushed and then ground in tube mills to a fineness of about 2–6 per cent on a 170-mesh sieve and a specific surface

[1] A. Brissaud, *Chim. Ind.* **9,** 1187 (1923).
[2] *Br. Pat.* 250246.
[3] Voisin, *Br. Pat.* 259203.

of about 3000 cm^2/g. The material is very hard to grind and the power consumption high. The wear on the mills and screens is also heavy. No additions are made during grinding, the setting time being controlled primarily by the composition. The compound $12CaO.7Al_2O_3$ sets in a few minutes whereas $CaO.Al_2O_3$ has a setting time of an hour or more and the proportion of the former present in the cement is kept as low as possible. The so-called 'unstable' form of $5CaO.3Al_2O_3$ which is now known to be a more complex aluminate, apparently has less effect in quickening setting. Very fast cooling markedly reduces the rates of strength development though not the ultimate strength, but there is a wide range of rates over which the properties of the cements are little affected.

Some data due to Blanchet[1] illustrate the effect of very fast cooling on the pure calcium aluminates.

Compound	Method of cooling	Setting time	Compressive strength 1 : 3 mortar (kg/cm^2) 1 day	2 days	3 days	7 days	28 days
"$5CaO.3Al_2O_3$"	Air	5–15 min	260	130	—	270	—
	Water-quenched	1–7 min	—	25	—	76	—
$CaO.Al_2O_3$	Air	5 hr	305	303	330	430	445
	Water-quenched	10–15 hr	0	—	370	500	—

Experiments by Berl and Löblein[2] showed that the strength tended to increase the more slowly the mix was cooled and the more completely crystalline the product. The methods of cooling vary with the type of furnace used for production. Thus, with a rotary kiln, the melt may be allowed to drop into water from which it is removed while still red-hot and subsequently cooled in air to allow crystallisation of the glass. Very slow cooling is apt to give rapid-setting cements.

High-alumina cements are manufactured in England (Ciment Fondu), France (Ciment Fondu), Germany (Rolandshütte, Lübeck, made in a blastfurnace), U.S.A. (Lumnite), Yugo-Slavia (Istra) and Czecho-Slovakia (Citadur, made by sintering). A specially pure type of high-alumina cement, white in colour, is made in England, France and the U.S.A. for use as a bonding agent for castable refractories for use at high temperatures. Alumina instead of bauxite is used as the raw material and the cement which contains 70–80 per cent Al_2O_3 is practically free from silica and iron oxide. This pure cement contains $CaO.Al_2O_3$, $CaO.2Al_2O_3$ and Al_2O_3 and has a much higher melting temperature than ordinary high-alumina cement. It is made by a sintering or clinkering process.

Methods have been proposed[3] for the simultaneous production of metallic iron and high-alumina cement from iron ore and bauxite, or aluminous iron ore, and limestone. This method is used for the cement made at Lübeck in Germany, in a blastfurnace, from scrap iron, bauxite, and limestone.[4] Aluminous cement com-

[1] *Revue Matér. Constr. Trav. publ.* **219,** 397 (1927).
[2] *Zement* **15,** 642 et seq. (1926).
[3] E. C. Eckell, *Concrete* **26** (5), 117 (1925); *Br. Pat.* 237779; J. C. Séailles, *Les Fabrications Liés en Cimenterie, Conférence Maison de la Chimie,* Paris 1943.
[4] E. Reitler, *Cem. & Cem. Mf.* **9,** 190 (1936).

positions have also been made in the U.S.S.R. by adding bauxite, or alumina-titania slags from the production of ferro-titanium to the molten lime-alumina slag in steel refining.[1] The use of blastfurnace slag and bauxite for the manufacture of high-alumina cement, and the replacement of limestone by calcium sulphate in order to manufacture sulphuric acid and high-alumina cement, have also been suggested. The simultaneous production of phosphorus and high-alumina cement is another suggestion.[2] A further proposal[3] is to treat a mix of bauxite and calcined lime with steam at high pressure, e.g. 200 lb/in^2, to yield a mixture of hydrated calcium aluminate and excess alumina. The product is then heated to 1000° to decompose the hydrated products and form the less basic aluminates.

A cement known as sulphoaluminate cement made by grinding together high-alumina cement and gypsum or anhydrite was patented[4] by the Lafarge Company and worked for a time in French Indo-China. It apparently proved difficult to control its properties, and its manufacture was discontinued. A cement based on a mixture of high-alumina cement, gypsum and hydrated tetracalcium aluminate is used in the U.S.S.R. as a shrinkage-compensating cement (see p. 535). Studies[5] on mixtures of high-alumina cement and gypsum, dehydrated at 600–700°, suggested that a cement of this type has a resistance to chemical attack similar to high-alumina cement, but a lower heat of hydration, and that it does not suffer from loss in strength when cured at temperatures up to 50°. This type of cement is, in fact, very similar to the supersulphated slag cements, for, in both, a major product of hydration is calcium sulphoaluminate. This compound is stable up to about 50° but at higher temperatures it loses water and at 75° there is a considerable loss in strength with supersulphated cement.[6] It also does not appear possible to get the very high strengths at one day, which are characteristic of high-alumina cement at ordinary temperatures, from its mixes with calcium sulphate.

The high-alumina cements made commercially have been classified by Robson[7] into four types as shown in Table 104.
By far the largest proportion of world production is of Type 1, red bauxite being the most widely available and cheapest source of alumina. White bauxite, as used for type 3, is much less freely available.

Composition and constitution of high-alumina cement

High-alumina cement is composed essentially of roughly equal proportions of alumina and lime, each usually within the range 36–42 per cent, a fairly large proportion of iron oxides, running up to 20 per cent, and a small percentage of silica of between 4 and 7 per cent. The iron oxides are present in both the ferrous and ferric states, the relative proportions depending on the extent to which an

[1] V. F. Krylov, *Tsement* **28** (1), 8 (1962); V. F. Krylov and V. K. Pomyan, ibid. **26** (2), 1 (1960).
[2] Albright and Wilson Ltd. (G. King), *Br. Pat.* 747016 (1956).
[3] N. V. S. Knibbs, *Br. Pats.* 303639, 385032.
[4] *Br. Pat.* 317783.
[5] P. P. Budnikov and I. G. Goldenberg, *Zh. prikl. Khim.* **20** (11), 1155 (1947).
[6] M. J. Brocard, *Ann. Inst. Bâtiment.*, New Series No. 54 (1948).
[7] T. D. Robson, *Sym., Tokyo 1968.*

oxidising atmosphere is maintained in the furnace during burning. Of the minor constituents titania occurs to the extent of about 2 per cent, magnesia usually 1 per cent or less, and sulphate or sulphide less than 0·5 per cent. The content of alkalis is usually less than 0·5 per cent; too high an alkali content may, in fact, be troublesome in some cements by producing undesirably quick setting. Insoluble matter may range around 2 per cent. The cements usually show a gain on ignition owing to the oxidation of the ferrous oxide.

TABLE 104 Types of high-alumina cement

Type	Colour	Al_2O_3 per cent	Iron oxides as per cent Fe_2O_3	SiO_2 per cent	CaO per cent	Source of alumina	Process of manufacture
1	Grey to black	37–40	11–17	3–8	36–40	Red bauxite	Fusion
2	Light grey	48–51	1–1·5	5–8	39–42	Red bauxite	Reductive fusion with removal of Fe metal
3	Cream or light grey	51–60	1–2·5	3–6	30–40	White bauxite	Sintering Clinkering Fusion
4	White	72–80	0–0·5	0–0·5	17–27	Alumina	Sintering Clinkering

The present British Standard for high-alumina cement (BS 915:1947) requires a minimum content of Al_2O_3 of 32 per cent and a ratio of Al_2O_3 to CaO, by weight, of not less than 0·85 nor more than 1·3.

The analysis of some high-alumina cements of British and foreign origin are shown in Tables 105 and 106.

In the early period of the development of high-alumina cement, a considerable number of investigations were carried out to determine the most favourable compositions. Bied[1] believed that the composition should lie along the line of C_2S–CA and that, although the strength decreased as the ratio of C_2S to CA increased, the ratio could vary from 0·25 to 0·90 before the products ceased to be appreciably hydraulic. The cementing value of lime-alumina-silica mixes, with and without ferric oxide, of high-alumina cement composition was studied by Bates,[2] Endell,[3] Berl and Löblein,[4] Solacolu,[5] Kühl and Ideta[6] and Richter.[7]

[1] *Le Ciment* **25,** 295 (1920); **26** 135 (1921); **27,** 101 (1922).
[2] *Bur. Stand. Tech. Paper,* No. 197 (1921).
[3] *Zement* **8,** 319 (1919).
[4] Ibid. **15,** 642 et seq. (1926).
[5] Ibid. **22,** 13, 114, 250, 311 (1933).
[6] Ibid. **20,** 261 (1931).
[7] Ibid. **21,** 445 et seq. (1932).

TABLE 105 Composition (per cent) of high-alumina cements*

Country	Type of manufacture	SiO_2	Al_2O_3	CaO	Fe_2O_3	FeO	TiO_2	MgO	S″	SO_3
England	Reverberatory furnace fusion	4–5	38–40	36–39	8–10	5–7	< 2	1	Trace	0·1
France	,,	3·5–4·5	38–40	36–39	9–11	4–6	< 2	1	Trace	0·1
Spain	,,	4–5	36–38	39–42	10–12	4–5	< 2	1	Trace	0·1
Yugoslavia	,,	6–8	38–40	36–39	8–10	4–7	< 2	1	Trace	0·1
U.S.A.	Rotary kiln fusion	8–9	40–41	36–37	5–6	5–6	< 2	1	0·2	0·2
Germany†	Blastfurnace Reductive fusion	5–8	48–51	39–42	0·1	< 1	1·5	1	1	0·5
Czechoslovakia	Brick-kiln-type furnace sintering	6–8	40–45	37–42	12–14	Trace	< 2	1	Trace	0·5

* T. D. Robson, *High Alumina Cements and Concretes*, Contractors Record Ltd., London 1962. All cements made from red bauxite.
† Also 1 per cent metallic Fe.

TABLE 106 Detail composition (per cent) of some high-alumina cements

	British			German
SiO_2	4·37	4·44	4·58	5·69
Al_2O_3	38·18	37·78	39·02	50·45
CaO	37·92	37·90	37·10	39·21
Fe_2O_3	10·56	9·74	9·72	—
FeO	5·91	6·55	5·98	0·70*
TiO_2	1·80	1·94	2·02	0·94
MgO	0·52	0·98	0·70	1·01
Na_2O	0·04	0·06	0·06	0·10
K_2O	0·14	0·14	0·08	0·05
Mn_2O_3	0·07	0·08	0·07	0·05
S″	0·03	—	—	0·97
SO_3	0·01	0·11	0·05	0·27
CO_2	} 0·37	0·17	0·43	0·37
H_2O		0·27	0·35	0·43

* Also 0·5 per cent metallic Fe.

In general, this early work concentrated mainly on consideration of the compositions in terms of the three components $CaO–Al_2O_3–SiO_2$. The high-alumina cement zone then falls in the shaded part of the diagram shown in Fig. 135, the compounds to be expected in the cement thus being $C_{12}A_7$, CA and C_2S; CA, C_2S and C_2AS; or CA, C_2AS and CA_2. This and subsequent work has shown that the strength of the cement is primarily determined by its content of CA. A relatively low silica content is important in order to retain it in the form of C_2S rather than C_2AS. Pure C_2AS has little or no hydraulic activity though the impure compound appearing in high-alumina cement may not be so unreactive.[1] Tricalcium aluminate is not a normal constituent of high-alumina

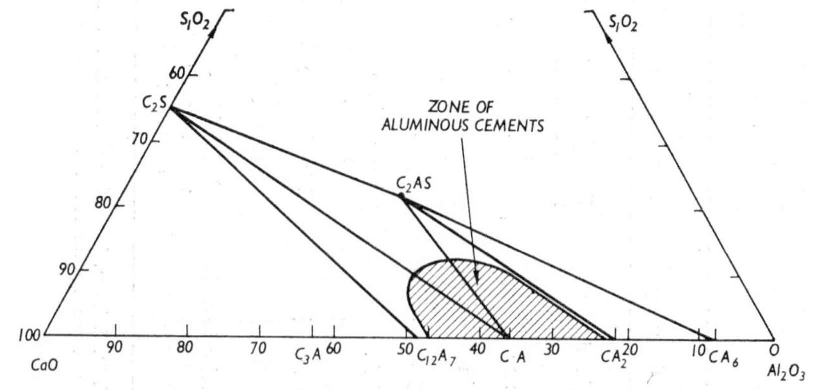

FIG. 135 Zone of aluminous cements in system $CaO–Al_2O_3–SiO_2$.

[1] P. P. Budnikov and A. F. Chirkasova, *Dokl. Akad. Nauk SSSR* **102** (4), 793 (1955).

PLATE XVII

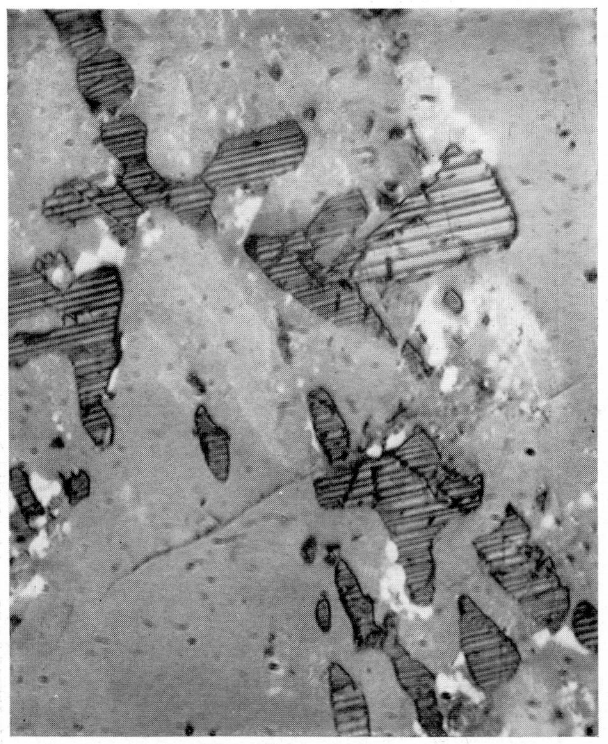

(i) Etched Striated Crystals of Dicalcium Silicate in Unsound Blastfurnace Slag. Polished Surface Etched for One Minute in 10 per cent $MgSO_4$ Solution at 50° (×100)

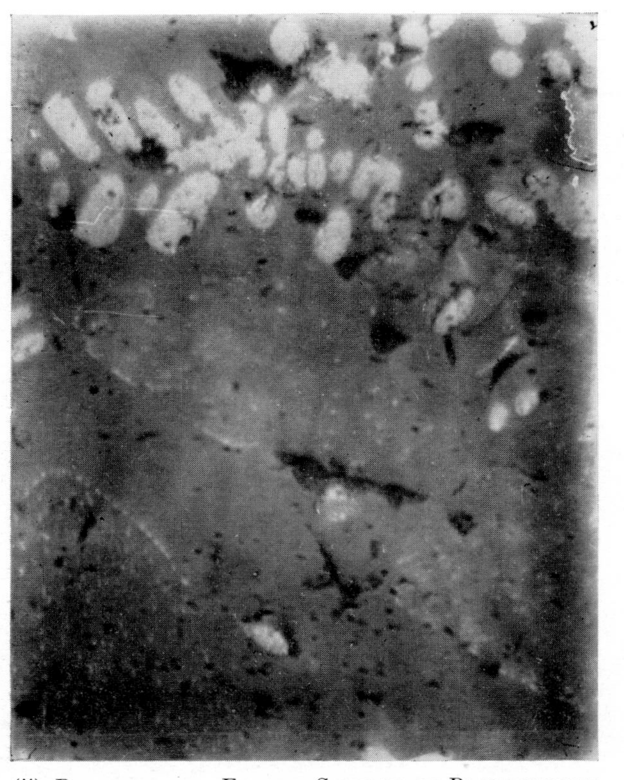

(ii) Polished and Etched Surface of Blastfurnace Slag. White Crystals are Calcium Sulphide (×360)

(i) Cracking of Concrete caused by high moisture movement of a sandstone aggregate (South Africa)

(ii) Cracking of Concrete caused by Alkali-Aggregate Reaction (U.S.A.)

cement though it could possibly form in sintered cements owing to local unhomogenieties in the raw mix.

The petrographic examination of thin sections of high-alumina cement has confirmed that these compounds do occur although the structure of the clinker is such that identification of minerals is often not easy by this method. In addition, the appearance under the microscope shows much more variability from one clinker to another, or even between different parts of the same clinker, than is common with Portland cement clinker. It is accordingly more difficult to describe a characteristic appearance. The edges of clinker pieces which have been in contact with, or near, the mould in which the material solidifies from the molten state frequently have a glassy appearance with dendritic crystals set in the glass. In the interior, where cooling has been slower, crystallisation is more complete. The crystal size may vary considerably from fine grained to coarse, almost macrograins. In fine-grained samples the general apperarance is of transparent prismatic, lath, and fibre-shaped crystals set in a brown to opaque ground mass, there being usually much more of the latter than of the matrix in Portland cement. The larger transparent crystals are CA, gehlenite, and the so-called 'unstable C_5A_3' although not all of them may occur together in the same section. The distinction between CA and gehlenite is often difficult. Sometimes the CA is tinged brown, while the gehlenite occasionally exhibits a slight blue pleochroism. 'Unstable C_5A_3' which is actually a quaternary compound now called pleochroite, is easily recognised by its characteristic fibrous form and strong violet pleochroism. Very occasionally samples can be obtained in which this compound predominates among the aluminates present and the fibre crystals may then appear in very large size. The ground mass cannot be identified with certainty in thin sections although ferrites are usually assumed to be one of the constituents.

Knowledge of the constitution of the cement was carried a very considerable step forward by Sundius and by Tavasci, contributing to the Symposium on the Chemistry of Cement at Stockholm in 1938. Sundius reviewed the data on the possible compounds present and applied this to a systematic examination of thin sections of four cements. One of these contained a high proportion of pleochroite and it was possible to separate this from the remainder by centrifuging the powdered clinker in a methylene iodide-benzene solution and, after making approximate corrections for impurities, to derive an approximate analysis for the compound of SiO_2, 5·1 per cent; TiO_2, 2·0 per cent; FeO, 6·0 per cent; Al_2O_3, 45·6 per cent; MgO, 0·6 per cent; CaO, 40·8 per cent. These figures give a molecular ratio:

$$\frac{CaO+MgO+FeO}{Al_2O_3+SiO_2+TiO_2}=\frac{3}{2{\cdot}01}$$

Sundius also observed in the clinkers a glassy residue which could be separated by solution of the remainder in 0·5 N HCl solution. Analyses of these residues, which amounted to 5–10 per cent of the clinkers, showed them to contain perovskite ($CaO.TiO_2$) and a wüstite (FeO)-bearing glass, or a glass containing iron oxide of composition varying between FeO and Fe_3O_4. The groups of compounds in the clinkers were not found to agree at all well with predictions from phase equilibrium data in the $CaO–Al_2O_3–SiO_2$ system and Sundius concluded that the effects of constituents other than these three, and particularly the effects of the iron oxides and titania, were the chief cause of the lack of agreement.

Tavasci introduced a new technique in the examination of polished and etched sections by means of the metallographic microscope, using a series of etching reagents having specific effects on the various minerals. The normal petrographic methods using transmitted light fail with the dark coloured, and frequently almost opaque, constituents, whereas the reflected light technique is well adapted to their examination. From this work, Tavasci identified the solid solution of C_2F and C_4AF, magnetite, and a further ferrite, said to be a solid solution of CF_2 with CAF_2. His method also enabled identifications to be made of the aluminates and he noted the presence of CA, C_2AS, pleochroite and also perovskite.

This work was carried a stage further by Parker, who reported at the Symposium on the Chemistry of Cements in London in 1952, on studies carried out with co-workers. By analytical methods, the conclusion of Sundius that pleochroite contained both FeO and SiO_2 and TiO_2 was confirmed. It was also observed that a compound resembling the FeO compound could be made in which MgO replaced the FeO and a study was, therefore, made of the appropriate part of the quaternary system $CaO–Al_2O_3–SiO_2–MgO$. A field of primary crystallisation of the MgO analogue of pleochroite was found and the most probable composition was concluded to be $6CaO.4Al_2O_3.MgO.SiO_2$. The phase relations in a plane of constant 5 per cent MgO content in which the primary phase field of this compound occurs are shown in Fig. 30 (p. 69). By analogy it appeared probable, therefore, that pleochroite in high-alumina cement might have the formula $6CaO.4Al_2O_3.FeO.SiO_2$ in which FeO replaces MgO. Titania may replace part of the SiO_2. Photomicrographs of the pleochroite appearing in some high-alumina cements and of the more typical calcium monoaluminate are shown in Plate 15.

From a consideration of the MgO system, and taking into account possible modifications due to the change from MgO to FeO, Parker concluded that assemblages of minerals in high-alumina cement could include:

$$CA–C_6A_4F''S–C_{12}A_7–C_2S$$
$$CA–C_6A_4F''S–C_2S–C_2AS$$
$$CA–C_6A_4F''S–C_{12}A_7–\mathrm{FeO}$$
$$C_6A_4F''S–C_{12}A_7–C_2S–\mathrm{FeO}$$
$$CA–C_6A_4F''S–C_2AS–\mathrm{FeO}$$

These suggestions are in agreement with the earlier findings of Sundius, which were also confirmed experimentally by Parker and colleagues, and account for the presence of a wüstite-bearing glass in some high-alumina cements.

The nature of the pleochroite ('unstable C_5A_3') phase has been further investigated by Majumdar[1] and Midgley.[2] From electron microprobe analyses of the mineral present in high-alumina cement the latter has proposed the formula $22CaO.17Al_2O_3.3FeO.2SiO_2$ which agrees better with X-ray structural work than Parker's formula. However the qualitative deduction of Parker and Ryder

[1] A. J. Majumdar, *Trans. Br. Ceram. Soc.* **63,** 347 (1964).
[2] H. G. Midgley, ibid. **67,** 1 (1968) and private communication.

as to the compatible phase assemblies are not affected by this change in composition. The sample of pleochroite examined by Midgley was strongly pleochroic, pale green to blue; it occurred in needles with straight extinction, negative elongation, refractive indices $\alpha = 1{\cdot}676$, γ 1·680, 2 V 40° positive.

In addition to these compounds there are also the ferrites, based on Fe_2O_3, to consider and for these we must turn back to the relevant part of the CaO–Al_2O_3–Fe_2O_3 system. The probable primary phase relations and solidus relations have been shown in Fig. 17 (p. 57). From this diagram we can deduce that the phase assemblies derived from lime, alumina and ferric oxide that may occur are:

$C_{12}A_7$(s.s.) and CA(s.s.)
CA(s.s.), $C_{12}A_7$(s.s.) and C_6A_2F
CA(s.s.) and a solid solution on the line C_6A_2F–C_2F
CA(s.s.), CF(s.s.) and a solid solution on the line C_6A_2F–C_2F

The composition of the ferrite in high-alumina cement seems to be variable. From measurements of micro-reflectivity Parker and Ryder found higher A/F ratios than those later deduced from the measurement of X-ray parameters. Majumdar, from X-ray measurements on several cements, found the composition of the ferrite phase to correspond to compositions on the C_2F–'C_2A' join having 52–75 mole per cent C_2F. Recent analyses using the electron beam microprobe technique suggest that both the optical and, to a lesser extent, the X-ray, results are affected, but in different directions, by the considerable solid solution of TiO_2 and SiO_2 in the ferrite. Mme Jeanne[1] found three interstitial components by microscopy. The first which was white in reflected light appeared to be wüstite (FeO) with possibly some CaO and Al_2O_3 in solid solution. The second which was pale grey gave an analysis showing more TiO_2 than Fe_2O_3 and approximated to the formula $C_{10}A\,(0{\cdot}4F\;1{\cdot}2T)_4$. The third which was dark grey gave a composition $C_{10}A\,(0{\cdot}9F\;0{\cdot}2T)_4$. Both the pale and dark grey materials also contained SiO_2, about 4·5 and 6·4 per cent respectively. The general composition of these two ferrites was 20 mole per cent C_2A, 80 mole per cent C_2F, in agreement with X-ray measurements, with isomorphoric substitution of TiO_2 and SiO_2.

In the system CaO–Al_2O_3–Fe_2O_3 the aluminates take up ferric oxide in solid solution. A similar behaviour is found in high-alumina cement for the refractive indices of the aluminate crystals are higher than those of the pure compounds. Mme Jeanne found up to 4·5 per cent Fe^{+++} in CA, 3·5 per cent in CA_2 and 3·5 per cent in C_2AS in one cement examined and suggested that the contents might vary from 0–5 per cent.

It will be evident that the mineralogical composition of high-alumina cement is variable and influenced by the ratio of ferrous to ferric iron. The compounds that can appear are $C_{12}A_7$, CA, CA_2, C_2AS, C_2S, CT, the pleochroic phase, FeO and glass. The predominant constituent is usually CA with a small amount of C_2S, and the iron-bearing interstitial material consisting of a 'C_2A'–C_2F ferrite with other oxides in solid solution. A small amount of C_2AS, with other oxides in solid solution, is also commonly found and the content of this increases rapidly as the silica content of the cement rises above the preferred amount of

[1] Mme Jeanne, *Revue Matér. Constr. Trav. publ.* **629**, 53 (1968).

about 4–5 per cent. Ferrous iron may be present as wüstite or in a glass phase and titania as perovskite or in solid solution in the ferrite. In cements containing the quaternary pleochroic compound the content of CA is much reduced and the strength is usually lowered. In the white high-aluminous cements the major compound is CA along with CA_2 and Al_2O_3. These cements are not equilibrium products. When cements containing the pleochroic compounds are reheated to over 1000° C, so that the ferrous iron is oxidised, this compound disappears and is replaced by CA. There is insufficient evidence to determine the contribution to strength made by the pleochroic compound but its replacement by CA leads to an increase in strength.

Present knowledge does not permit of ready methods of calculation of the compound content of high-alumina cement, but Parker and Ryder[1] have suggested some approximate methods based on the analytical data from the original cement and on the insoluble residue after solution in 0·5 N HCl. Sundius[2] has also derived the compound content of three clinkers he investigated.

Hydration of high-alumina cement

The hydration products of the calcium aluminates have, as seen in an earlier chapter, been the subject of many investigations and the results throw much light on the hydration of the cement.

On shaking together high-alumina cement and excess water at room temperature, an unstable supersaturated solution with a $CaO : Al_2O_3$ molecular ratio a little above unity is formed, just as when the compounds CA and C_5A_3 are similarly treated. The concentration of this solution rises fairly rapidly to a maximum, the $CaO : Al_2O_3$ molecular ratio remaining around 1·1–1·2, and then falls abruptly[3] (Fig. 136). The time at which this maximum is reached increases as the ratio of water to cement is increased, while the maximum concentration attained decreases slightly. The ratio of lime to alumina in the solution at the maximum point remains, however, roughly the same. After passing the point of maximum concentration the solution becomes opalescent and deposits a solid. The solution which remains has a much increased lime : alumina ratio and a slowly increasing pH value. On prolonged standing it attains a final pH of about 11·7 and a final concentration of about 0·3–0·4 g CaO and 0·15–0·20 g Al_2O_3 per litre. The precipitated solid, the $CaO : Al_2O_3$ ratio in which may be as low as 0·5, consists of a gelatinous product together with hexagonal plate crystals occurring in a spherulitic formation as white globular masses. The unstable equilibrium that is obtained can be seen from the phase diagram for the $CaO–Al_2O_3–H_2O$ system (Fig. 56). The solids precipitated correspond to those to be expected at the unstable invariant point Q for $2CaO.Al_2O_3.8H_2O$ and alumina gel, or for an analogous unstable invariant point for $CaO.Al_2O_3.10H_2O$, and the change in solution concentrations with time to the change in this invariant point as the

[1] *Sym., London 1952*, 493.
[2] *Sym., Stockholm 1938*, 395.
[3] L. S. Wells, *J. Res. natn. Bur. Stand.* **1,** 951 (1928); Koyanagi, *Concrete* **40** (8), 40 (1932); J. D. Ans and H. Eick, *Zement-Kalk-Gips* **6,** 197 (1953); M. J. Brocard, *Ann. Inst. Bâtiment*, New Series No. 12 (1948).

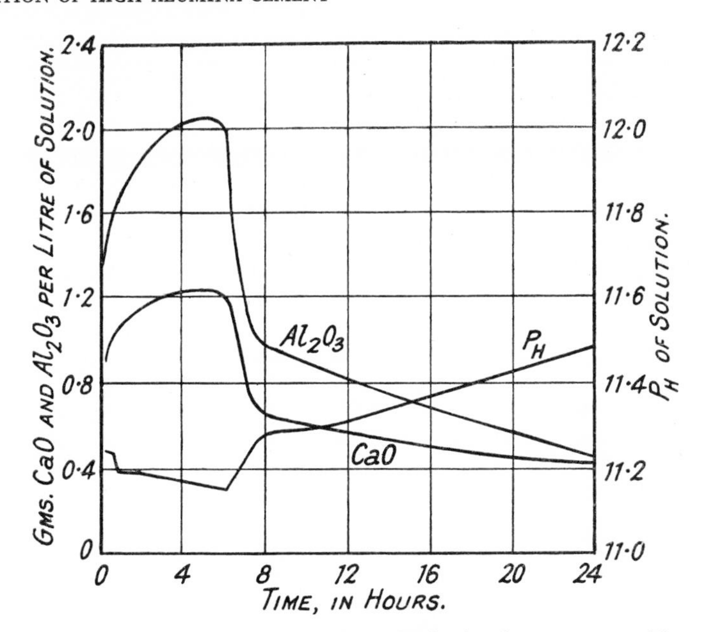

FIG. 136 Solution obtained on shaking high-alumina cement with excess water (50 g/litre).

alumina gel ages. The globular masses can be separated from the residual cement and the gel products by sifting on a 240- or 300-mesh sieve, using absolute alcohol as a washing agent. Usually the crystalline product so obtained is contaminated with gelatinous products, but Koyanagi, from a suitable cement, obtained a totally crystalline preparation which proved to be the hydrated dicalcium aluminate $2CaO.Al_2O_3.7{\cdot}5H_2O$. The formation of these coarse spherulitic crystal masses permits a separation to be made by decantation of the products obtained on hydrating high-alumina cement in excess water. Thus Kühl and Berchem[1] on shaking coarsely ground (88–120 μ) high-alumina cement with ten times its own weight of water for different periods at 12° were able to separate the hydration products into gel, crystals and solution fractions. The composition of the crystal masses approximated to that of hydrated dicalcium aluminate. Alumina was the predominant constituent in the gel mass but this also contained lime, pointing to the presence of either hydrated mono- or dicalcium aluminate.

Assarsson[2] obtained evidence that calcium aluminate gels can be formed and play a considerable part in the setting and hardening of high-alumina cement. The first hydration product found to appear was a gel of refractive index about 1·52–1·53 which Assarsson considers is a hydrated calcium monoaluminate gel of unknown water content. At a later stage, some six hours after gauging the

[1] *Zement* **21,** 547, 561 (1932).
[2] *Zement* **23,** 15 (1934); **26,** 293 (1937); *Swedish Geological Survey. Årsbok, Series C,* 27 (4) No. 379 (1933); 30 (6) No. 399 (1936).

cement, a different gel of refractive index 1·50 and composition $CaO.Al_2O_3.10H_2O$ was found. A small amount of hydrated dicalcium aluminate crystals was also observed. Assarsson has also found that when high-alumina cement is treated with limited amounts of water the concentrations of the supersaturated solutions produced alter with time rather differently from those in Fig. 136 obtained with a large excess of water. A supersaturated solution with a $CaO : Al_2O_3$ ratio (molecular) of about 1·1 : 1 is still formed initially, but subsequently the lime content decreases much more than the alumina; their ratio continuously decreases up to 24 hours and may fall below 0·1, in marked contrast to the behaviour in a large excess of water. Figure 137 illustrates this influence of the proportion of

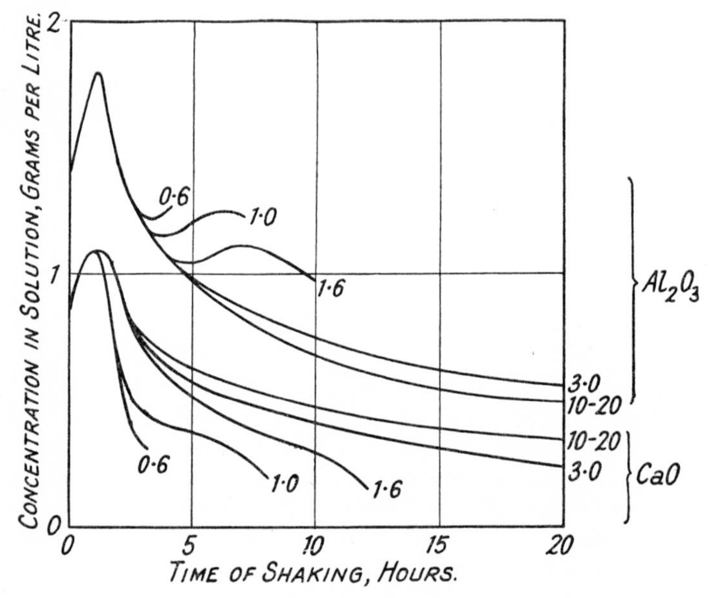

FIG. 137 CaO and Al_2O_3 contents of high-alumina cement solutions. Weight ratio water : cement shown. (After *Assarsson*.)

water used and may be compared with Fig. 136. The pure calcium aluminates show no such behaviour when treated with very limited amounts of water, and the solution curves remain similar to those obtained in the presence of a large excess of water. The different results with high-alumina cement seem to be connected with the presence of small amounts of alkalis in the material. These are dissolved rapidly, as is seen from Fig. 138, and their concentration in the solution increases as the water : cement ratio is decreased. The great excess of alumina over lime in the cement solutions is thus probably attributable to the alkalis in solution substituting for lime as a base. It is evident that the hydration of high-alumina cement in excess water may differ from that in pastes with the lower water : cement ratios such as are used in practice.

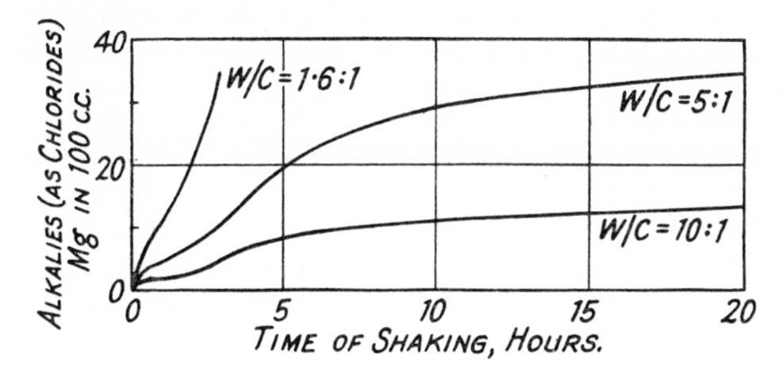

FIG. 138 Alkali content of solutions obtained from high-alumina cement.

Assarsson[1] finally concluded that at low temperatures the hydrated monocalcium aluminate was formed in high-alumina cement pastes and that at about 20° the hydrated dicalcium aluminate and alumina gel appeared, while at higher temperatures transformation to the cubic $3CaO.Al_2O_3.6H_2O$ occurred. The relative extent to which the mono- and the dicalcium compounds appear is likely to be influenced by the alkali content of the cement, since the presence of alkali considerably modifies the equilibria in the system $CaO–Al_2O_3–H_2O$.

More recent investigations using X-ray analysis, electron microscopy and differential thermal analysis confirm that the main hydration reactions in high-alumina cement pastes can be formulated as follows:

$$CaO.Al_2O_3 \longrightarrow CaO.Al_2O_3.10H_2O$$

$$\searrow \qquad \downarrow$$

$$2CaO.Al_2O_3.8H_2O + Al_2O_3.aq$$

$$\downarrow$$

$$3CaO.Al_2O_3.6H_2O + Al_2O_3.aq$$

Some gel appears to be formed in the first few hours and this seems from DTA to be alumina gel[2] rather than a CAH_{10} gel as suggested by Assarsson. This alumina gel on ageing gradually crystallises as gibbsite ($Al_2O_3.3H_2O$). The extent to which CAH_{10} or C_2AH_8 predominate in the hydration products depends on the temperature and also the cement.[3] If $C_{12}A_7$ is present in the cement the C_2AH_8 appears more freely, as hexagonal plates. If there is any exposure to carbon dioxide, hexagonal plates of $3CaO.Al_2O_3.CaCO_3.11H_2O$ are also formed. The predominance of CAH_{10} increases as the temperature is decreased and at 15° or below the content of C_2AH_8 reported by various investigators has usually been small. When pure CA is hydrated the critical temperature above which C_2AH_8 is formed is about 23°. However the amount of C_2AH_8 reported by different investigators from high-alumina cement at temperatures

[1] *Sym., Stockholm 1938*, 441.
[2] H. G. Midgley, *Trans. Br. Ceram. Soc.* **66,** 161 (1967).
[3] H. Lehmann and K. J. Leers, *Tonind.-Ztg. keram. Rdsh.* **87,** 29 (1963); K. S. Leers, ibid. **88,** 426 (1964); H. E. Schwiete, V. Ludwig and P. Müller, *Betonsteinzeitung* **32,** 141, 238 (1966).

between about 20° and 23° varies considerably. It may perhaps be influenced by the alkali content of the cement as well as by the $C_{12}A_7$ content. The dicalcium aluminate hydrate, with ferric oxide replacing part of the alumina, $C_2(AF)H_8$, is also formed from the ferrite compound in the cement.[1] There is also a tendency for more of the dicalcium aluminate hydrate to appear on ageing, either by conversion of CAH_{10} or from further hydration of the ferrite compound in the cement. At 25° and above the initial main hydration products are C_2AH_8 and hydrated alumina, and as the temperature is raised further the isometric compound C_3AH_6 increasingly becomes the dominant product within days or weeks.

The presence of hydrated gehlenite, $2CaO.Al_2O_3.SiO_2.8H_2O$, in the set cement within a few weeks has been detected by Midgley and it has also been found in a thirty-year-old concrete.[2] Pure gehlenite does not react with water but the compound present in high-alumina cement contains other oxides in solid solution and there is evidence that it is more reactive. This hydrated gehlenite might also arise from hydration of C_2S in the alumina-rich solution formed by the cement.

The hydration reactions of the ferrites in high-alumina cement are still not clear. As we have seen, ferric oxide is present in $C_4AF.C_2F$ solid solution and ferrous oxide as wüstite or in pleochroite. Hydration of ferrites[3] with an A/F molecular ratio below unity leads predominantly to the formation of the tetracalcium hydrate, $C_4(AF)H_x$ together with an amorphous form of Fe_2O_3 or its hydrate. With A/F ratios above unity the ferrite hydrates to give predominantly the dicalcium hydrate $C_2(AF)H_x$. Since the A/F ratio of the ferrite in high-alumina cement is normally below unity the $C_4(AF)H_x$ hydrate would be expected. It has been detected by Schwiete in cement suspensions (w/c = 10), but not in cement pastes. At temperatures of 25° and upwards the isometric C_3AH_6 is formed from the ferrite in amounts increasing with time and temperature. The extent to which Al_2O_3 is replaced by Fe_2O_3 in the hydration product from high-alumina cement is still uncertain.

The hydration of high-alumina cement can be followed on a microscope slide provided the cement is protected from the air by waxing the cover-glass round the edges. A gelatinous product commences to form round the cement grains after an hour or two, producing a furred appearance. This gel increases rapidly in amount, and crystals also commence to form. After one day the original anisotropic grains of the cement have largely disappeared and have been replaced by the gel together with the typical spherulitic growths of hexagonal plate groups and needles. The production of gelatinous material is so heavy that the slide becomes fogged. A clearer picture has been obtained by examining fractured surfaces of set cement in the stereoscopic electron microscope. These show needles of CAH_{10} which grow into a hexagonal prismatic form, plates of $C_2(AF)H_8$ and carbo-aluminate, and amorphous alumina gel which within three months had changed to microcrystalline needles.

[1] S. Chatterji and J. W. Jeffrey, *Trans. Br. Ceram. Soc.* **67,** 171 (1968); S. Chatterji and A. J. Majumdar, *Indian Concr. J.* **40,** 245 (1966).
[2] F. M. Lea, Canadian Building Series, *Performance of Concrete,* edited by E. G. Swenson, 1968, p. 56.
[3] E. T. Carlson, *J. Res. natn. Bur. Stand.* **68A,** 453 (1964); A. J. Majumdar and S. Chatterji, *Indian Concr. J.* **40,** 153 (1966).

Effect of temperature on hydrated high-alumina cement

We shall see later (p. 516) that high-alumina cement cured at high temperatures, or subsequently exposed to them in a moist state, has a reduced strength. This is closely connected with the change in the nature of the hydration products. The compounds $CaO.Al_2O_3.10H_2O$ and $2CaO.Al_2O_3.8H_2O$ produced at ordinary temperatures are metastable, and on prolonged ageing tend to change into the cubic compound $3CaO.Al_2O_3.6H_2O$. At ordinary temperatures this change is very slow and may never occur in concretes kept dry, but it must be expected to occur, even though very slowly, in wet concretes. At high temperatures it occurs rapidly. Midgley[1] has defined 'half conversion' as the state in which the quantity of C_3AH_6 is equal to that of CAH_{10} as determined by DTA. For neat cements (w/c 0·26) the time for half conversion was about a week at 50°, 100 days at 40° and an estimated 20 years or so at 25°. A concrete cube stored in water for 27 years at 18° was found to be about half converted but samples from a 30-year-old concrete pile extracted from the seabed showed little signs of conversion,[2] the predominant compound still being CAH_{10}. The maximum temperature of the sea in the region concerned was about 18°. It is evident that at temperatures of 18° or below the rate of conversion is very slow.

When high-alumina cement is hydrated at relatively high temperatures, e.g. 35–45°, the cubic compound is rapidly formed and can readily be observed under the microscope when hydration takes place in excess water, and detected by X-rays in pastes of the consistence used in practice. Its presence is also indicated by the manner in which the hardened cement loses water on heating. Thus the compound $3CaO.Al_2O_3.6H_2O$ loses most of its water between 225° and 275° whereas the other hydrated aluminates lose water more steadily over a much wider temperature range. It is found that the loss occurring between these temperatures increases markedly in cements cured at higher temperatures, or subsequently exposed to such temperatures in a wet condition. This is illustrated by the following data on neat cement cubes.

	275–225° loss per cent
High-alumina cement cured at 15°	1·7
High-alumina cement cured at 45°	9·2
High-alumina cement cured at 15° and then boiled	7·9

There is also a corresponding change in the density of the hydrated cement since the pure compounds have densities as follows:

$CaO.Al_2O_3.10H_2O$	1·72
$2CaO.Al_2O_3.8H_2O$	1·95
$3CaO.Al_2O_3.6H_2O$	2·52
$Al_2O_3.3H_2O$	c2·4

The volume change for the reactions

$$(1)\ 3\ CAH_{10} \rightarrow C_3AH_6+2AH_3+18H$$

and

$$(2)\ 3\ C_2AH_8 \rightarrow 2C_3AH_6+AH_3+12H$$

[1] Loc. cit.
[2] F. M. Lea, loc. cit.

can be calculated from the densities assuming a density of about 2·4 for hydrated alumina. In reaction (1) the volume of the solids is reduced to just below half, and in reaction (2) to about two-thirds, of the original volume.

Samples of cement hydrated to the extent of about 90 per cent by successive regrinding, and then dried over calcium chloride, gave densities of 2·11 for hydration at 18° and 2·64 for hydration at 45°.

Calcium monoaluminate shows a fall in strength, parallel to that of high-alumina cement,[1] on curing at high temperatures, as indicated by the data in Table 107.

TABLE 107 Tests on 1 : 2 calcium monoaluminate : sand mortars

Age	Compressive strength (lb/in^2)		275–225° Weight loss* per cent for storage at	
	Stored in water at 18°	Stored in water at 45°	18°	45°
1 day	8715	6425	3·3	16·8
7 days	10140	3415	3·9	21·3
28 days	10770	2450	5·7	18·3

* Calculated on $CaO.Al_2O_3$ content.

The change from the less basic forms of the hydrated calcium aluminates to the cubic $3CaO.Al_2O_3.6H_2O$ is thus closely associated with the marked loss in strength. There is much evidence, however, that it is the volume changes on conversion, rather than the mineralogical nature or morphology of the hydrates formed, that is responsible for the fall in strength. Unless there are compensating factors these volume changes make the converted cement much more porous than the original.[2] As will be shown later this development of porosity is partly compensated in mixes of low water : cement ratio by the continuing hydration of anhydrous cement. It has also been found[3] that C_3AH_6 and AH_3 can give as high strength as CAH_{10} if the porosity is low, and that the increase in porosity on conversion decreases with the w/c ratio.

The resistance of high-alumina cement to attack by sulphate salts is also decreased by the inversion of the hydrated aluminates, even more seriously in magnesium sulphate than sodium sulphate solutions. This may also be ascribed to the increase in porosity for it is less apparent in mixes of low w/c ratio where the porosity increase is reduced. High-alumina cement mortars have been found resistant to attack under the acidic conditions arising in gas-washing plants, even though, because of the temperatures and wet conditions involved, the inversion and the concomitant fall in strength must have occurred to the full. For non-load-bearing uses the latter may be of little importance provided the resistance to attack by acid gases is retained.

[1] F. M. Lea, *J. Soc. chem. Ind., Lond.* **59,** 18 (1940).
[2] R. Tsukayama, *Sym., Tokyo 1968.*
[3] R. Alègre, *Revue Matér. Constr. Trav. publ.* **630,** 101 (1968).

Water combined on setting. High-alumina cement combines with considerably more water during hydration than does Portland cement. If the initial product of hydration is a mixture of the mono- and dicalcium hydrates, up to 70 per cent of water would be required; if $2CaO.Al_2O_3.8H_2O$ and $Al_2O_3.3H_2O$ about 50 per cent; and if $3CaO.Al_2O_3.6H_2O$ and $Al_2O_3.3H_2O$ about 35 per cent. Such calculations are very rough, since they assume complete hydration of the alumina and silica compounds and make no allowance for the iron compounds. In set cement which has been nearly completely hydrated, and the aluminate converted to the cubic tricalcium aluminate hydrate by treatment at 100°, the water content approximates to the 35 per cent indicated above. At ordinary temperatures, and with the incomplete hydration that occurs in practice, up to 50 per cent water is combined. If lesser amounts of water are used to gauge the mix more of the cement must remain unhydrated and, once hardening has occurred, storage in water for some weeks does not appreciably increase the amount of water combined in the cement. This is one reason why high-alumina cement concrete must be kept wet, and loss of water prevented, during the first day after placing.

Volume change in setting. High-alumina cement appears to swell considerably more than Portland cement during hydration, but there is actually a greater diminution in the combined volume of solid and water. This decrease amounts to about 10–12 cc per 100 g cement hydrated, compared with about 5 cc for Portland cement.[1] The contraction in the volume of the system (cement plus water) is practically complete within 1 day with high-alumina cement, compared with 7 days for Portland cement (see page 269).

The swelling of high-alumina cement begins at about the time of final set. The extent to which it occurs is affected by the addition of salts which change at the same time the rate of evolution of heat and also the pH of the high-alumina cement solution.[2]

There are no direct experimental data showing the change in the volume of the solid during the hydration of the cement. The change can, however, be calculated from the density of hydrated and unhydrated cement. For these we have the following values:

$$\text{High-alumina cement } d_{25^\circ} = 3{\cdot}27$$
$$\text{Cement completely hydrated* at } 15^\circ\ d_{25^\circ} = 2{\cdot}11$$
$$\text{Cement completely hydrated* at } 45^\circ\ d_{25^\circ} = 2{\cdot}64$$

* Dried over calcium chloride. ($CaCl_2.4H_2O$.)

The increase in the volume of the solid on hydration calculated from these values is about 50 per cent for cement hydrated at 15° and 23 per cent for cement hydrated at 45°.

Properties of high-alumina cement

The cement is characterised by its very rapid-hardening properties and the very high strength it develops within twenty-four hours. The specific gravity ranges

[1] Le Chatelier, *Le Ciment* **32,** 82 (1927); H. Gessner, *Kolloidzeitschrift* **46,** 207 (1928); **47,** 65, 160 (1929).

[2] R. Salmoni and H. E. Schwiete, *Zement* **22,** 523 (1933).

from about 3·20 to 3·25. The weight per cubic foot varies like that of Portland cement according to the method of filling the container, but for proportioning concrete mixes by volume a value of 90 lb/ft^3 is taken. High-alumina cement concrete tends to give a harsher concrete mix than Portland cement and to offset this a somewhat higher proportion of sand or fine aggregate is required. On the other hand with an appropriate aggregate grading a lower w/c ratio can be used for equivalent workability. Relatively high w/c ratios were formerly recommended because of the larger consumption of water in the hydration process, but the discovery that a low w/c ratio overcomes the disadvantageous effect of conversion on strength has changed this view. Mixes should not be richer than is required to obtain the desirable w/c ratio, which is 0·5 for reinforced concrete and 0·4 for prestressed concrete. Lean concrete mixes such as 1 : 9 do not show the characteristic resistance to chemical attack.

The methods of testing high-alumina cement follow those in use for Portland cement.

Fineness

The British Standard for high-alumina cement (BS 915 : 1947) prescribes a maximum residue of 8 per cent on the 170-mesh or a minimum specific surface of 2250 cm^2/g. The average product has a residue of about 5 per cent and a specific surface ranging from about 2500–4000 cm^2/g and commonly about 3000.

Specific gravity

The ferrites present in the cement have a higher specific gravity than the aluminates so that the specific gravity of the cement rises with the iron oxide content. For cements made by the reverberatory furnace method the value is about 3·20 to 3·25 while for the blastfurnace method it is about 3·0. The loose bulk density is about 70–85 lb/ft^3 rising to 115–120 lb on consolidation; these values are slightly higher than for Portland cement.

Soundness

High-alumina cement contains no free lime and very little sulphur trioxide, and no risk of unsoundness due to these causes arises. As a precaution the British Standard requires that the expansion on boiling in the Le Chatelier test shall not exceed 1 mm. It is doubtful if either this test, or the boiled pat test, has any value for this type of cement.

Setting time

The setting time of high-alumina cement is similar to that of Portland cement. It is tested by the Vicat needle method on a paste made by gauging the cement with 22 per cent water. The British Standard requires that the initial setting time shall be not less than two, nor more than six, hours and the final setting time not more than two hours after the initial set. Some typical comparative data on the setting time of high alumina and Portland cements are shown in Table 108.

The available data on the effect of temperature on setting time are conflicting; in some cases little change with temperature has been found. There is a tendency for the setting time to decrease as the temperature is raised from 0° to 10° and

TABLE 108 Setting times of high-alumina and Portland cements

	Initial		Final	
	h	m	h	m
High-alumina cement	3	00	3	40
	4	00	4	36
	4	32	5	16
	5	08	5	35
Rapid-hardening Portland cement	0	42	1	14
	1	30	2	00
	2	26	2	51
	4	00	5	12
Ordinary Portland cement	1	30	3	00
	2	12	3	00
	3	30	4	14
	4	33	5	25

then to increase up to about 30°. Above this temperature the setting time progressively decreases. These effects are not however found with all cements.

The effects of the addition of salts and other substances on the setting time may be summarised as follows, though the effects in some cases vary with the particular cement tested. This may account for the conflicting results reported from different tests.

TABLE 109 Effect of various substances on setting time of high-alumina cement

Accelerate	Retard when present in small amounts and accelerate when present in large amounts	Retard
Calcium hydroxide	Magnesium chloride	Sodium chloride
Sodium hydroxide	Calcium chloride	Potassium chloride
Sodium carbonate	Barium nitrate	Barium chloride
Sodium sulphate	Acetic acid	Sodium nitrate
Ferrous sulphate	Calcium sulphate	Hydrochloric acid
Sulphuric acid		Glycerine
		Sugar

In general, substances which increase the pH value of the solution tend to accelerate the set. Small additions of calcium sulphate, e.g. 0·25 per cent retard the set, but 1 per cent may reduce the time of final set to below 30 minutes. Sodium sulphate in an 0·5 per cent concentration has some accelerating effect, and magnesium sulphate a retarding one; it seems probable that the behaviour may vary with concentration as with calcium sulphate. Sugar retards the set very markedly and an addition of 1 per cent may delay the set for a day or more; even with considerably smaller amounts the retardation may be marked. The presence of sugar may even inhibit setting and hardening entirely, as occurs with Portland cement. Parker has reported the effect of a variety of other substances.[1]

The pH value[2] is normally about 11·6 during the first few hours of setting and then rises to about 12·2–12·4 after 12 hours. Where conversion of the hydrates occurs the pH drops to about 11·5.

Strength

The British Standard for high-alumina cement requires a compressive strength test on a 1 : 3 vibrated mortar cube similar to that used for Portland cement. The mortar is mixed with 10 per cent of water. A minimum strength of 6000 lb/in² (422 kg/cm²) at 1 day, and 7000 lb/in² (492 kg/cm²) at 3 days, is required on curing at 61 ± 1°F in moist air, of at least 90 per cent relative humidity for the first day, and in water at 58–64°F subsequently.

TABLE 110 Compressive strength (lb/in²)

	1 : 3 Vibrated mortar			1 : 2 : 4 Concrete cubes w/c 0·60		
	1 day	3 days	7 days	1 day	3 days	7 days
High-alumina cement	8500	9720	10550	5870	7610	7960
	9550	10780	11750	7110	7990	8780
	10400	11200	11750	6640	7590	8380
	11500	11530	13120	8110	8310	8950
Rapid-hardening Portland cement	1620	3940	5460	1100	2730	4000
	1750	4920	6370	1130	2950	4030
	1990	4490	6200	1260	2690	3770

High-alumina cement attains a strength close to its maximum in 24 hours at normal temperatures; some comparative data are given in Table 110 and a curve showing the strength development over the first 24 hours in Fig. 139.

The relative strengths developed by average high-alumina and Portland cements at low, normal and high temperatures are shown in Table 111 where compressive strength values for 4-inch cubes of 1 : 2 : 4 concrete w/c 0·60 are given. While high-alumina cement shows much higher strengths than Portland cement at low and normal temperatures, it is considerably inferior at high temperatures when the w/c ratio is 0·6. However at lower w/c ratios, as discussed later, the loss in strength on conversion is reduced and at a w/c ratio of 0·4, or less, the strength still remains equal or superior to that of Portland cement.

[1] *Sym., London 1952*, 512; see also M. de Tournade and P. Bourrelly, *J. chem. Phys.* **63**, 257 (1960) for the effect of salts on the rate of hydration.

[2] H. E. Schwiete, V. Ludwig and P. Müller, *Betonsteinzeitung* **32,** 141, 238 (1966); K. J. Leers and E. Rauschenfels, *Tonind.-Ztg. keram. Rdsh.* **90,** 155 (1966).

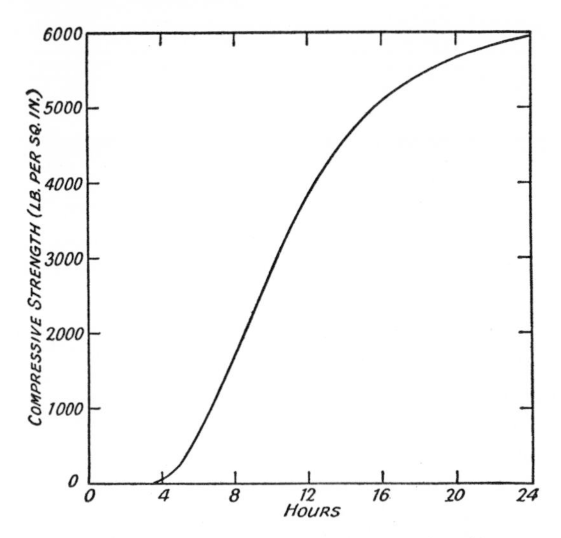

FIG. 139 Development of strength of 1 : 2 : 4 high-alumina cement concrete.

The strength of high-alumina cement concretes at 18° shows a change with w/c ratio similar to that found with Portland cement. Some data on concrete cubes are shown in Table 112.

TABLE 111 Effect of temperature on strength of concretes, w/c 0·6

Type of cement	Age (days)	Compressive strength (lb/in²) Low temperature 2°	Normal temperature 18°	High temperature 35°
Normal Portland	1	—	460	1450
	3	420	1640	2450
	7	1040	2680	3200
	28	3820	4150	4190
	56	4480	4570	4650
Rapid-hardening Portland	1	—	790	1970
	3	490	2260	3200
	7	1230	3300	3940
	28	4620	4920	5020
	56	5310	5410	5310
High-alumina	1	2130	5710	4260
	3	6300	7330	5110
	7	7330	7670	2560
	28	8090	8520	2130
	56	8690	8950	2130

TABLE 112 Effect of water : cement ratio on strength at 18°

W/c ratio	Compressive strength lb/in² (kg/cm²)			
	12 hours	1 day	3 days	7 days
0·4	7000(492)	9000(633)	10000(703)	10750(756)
0·5	6250(439)	8000(562)	9250(650)	10000(703)
0·6	5500(386)	7000(492)	8250(580)	9000(633)

The strength of high-alumina cement is affected very adversely by the presence of calcium chloride in the mixing water and also by gauging with seawater.[1] Though seawater has often been used without ill effect in mixing Portland cement concrete, it should never be used with high-alumina cement concretes. The cement is also rather more sensitive than Portland cement to impurities in the mixing water and to excess fine dust in aggregates. Air-entraining and water-reducing agents are not normally used in high-alumina cement concrete. Though there is no specific objection to their use, initial trials should be made to check that there are no adverse effects. No addition of lime, calcium chloride or integral waterproofers should be made as these materials may interfere with the setting or seriously affect the strength. Sodium silicate is not effective for surface treatments of high-alumina cement concrete, but magnesium silicofluoride or drying oils may be used.

Other properties. Relatively little has been published on the moisture movements of high-alumina cement concretes and mortars, but they appear to be very similar in magnitude to those for Portland cement. The shrinkage in air is also very similar in total magnitude to that of Portland cement, but under comparable conditions it occurs considerably more rapidly. The permeability of high-alumina cement concrete is of a similar order to that of a rapid-hardening Portland cement concrete. The coefficient of thermal expansion is about the same as that of Portland cement concrete. The modulus of elasticity increases with the strength of the concrete and it is found in general that the variation of E with strength can be represented by a common curve for both high-alumina and Portland cement concretes.

A phenomenon not infrequently observed in test pats of neat high-alumina cement and in concretes is a powdering or dusting of the surface. This results in the formation of a thin surface layer which can easily be removed by rubbing with the finger. It has been attributed both to too rapid drying of the surface and to the effect of atmospheric carbonation. The latter seems to be the correct explanation,[2] but the carbonation is facilitated by rapid drying of the surface. Such surface dusting does not occur if the surface of the material is kept moist during the hardening period. Apart from its effects on the appearance of the concrete it is of no special importance.

[1] H. Burchartz, *Zement* **16,** 77 (1927); R. Feret, *Revue Matér. Constr. Trav. publ.* **223,** 135 (1928).
[2] H. W. Gonell, *Zement* **15,** 714 (1926).

When high-alumina cement mortars or concretes are cured in water a loose, somewhat gelatinous, white deposit can often be seen on the surface of the concrete or on the bottom of any container in which it is immersed. This deposit seems to arise from some unhydrated cement, on the surface of the concrete, dissolving in the water to form a super-saturated solution which subsequently deposits alumina gel and hydrated calcium aluminate. The occurrence of this rather curious phenomenon is not unusual and need occasion no alarm. The solution of cement does not proceed beyond a thin surface film.

Mixtures of high-alumina and Portland cements. The addition of Portland cement to high-alumina cement, or vice-versa, reduces both the setting time and the strength of the single cements. For certain ranges of mixes the setting time becomes very short and a flash set may occur. Curves are shown in Fig. 140

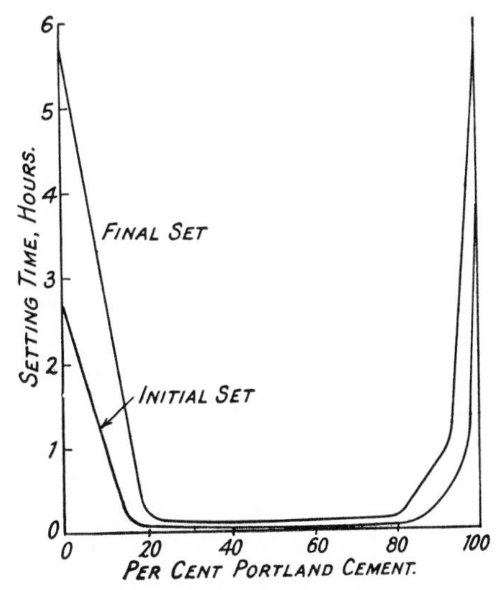

FIG. 140 Setting time of mixtures of high-alumina and Portland cement.

of the setting times of the mixed cements. While these illustrate the type of behaviour which arises, it should be emphasised that the range of mixes over which very rapid setting occurs varies with different Portland and high-alumina cements and cannot be predicted without a trial. Thus Robson[1] found a flash set for mixes containing between 15 and 60 per cent Portland cement. It seems that a number of different factors may combine to cause the rapid setting observed. Both gypsum and calcium hydroxide accelerate the setting of high-alumina cement. The former is present in Portland cement while the latter is usually present to some extent, and is also produced by hydration; these together are probably sufficient to explain the acceleration of the set of high-alumina cement by Portland cement additions. The removal of the gypsum from Portland cement, by combination with the hydrated calcium aluminates formed from high-alumina cement, is probably responsible for the acceleration of the set of Portland cement produced on adding high-alumina cement.

[1] T. D. Robson, *Chemy Ind.*, 1952 (1), 2; see also J. L. Miners, *Concrete* **44** (8), 38 (1936).

Mixtures of the two cements are sometimes used in mortars when a rapid set is required, as for example for sealing leaks, and very occasionally in concrete.

The ultimate strength of mixes of the cements decreases progressively as Portland cement is substituted for high-alumina cement until a minimum is reached which is much lower than that of Portland cement alone. The strength developed over the first few hours is, however, greater for certain mixes than that of high alumina cement alone, even though the ultimate strength is low. These trends are illustrated by the data of Robson given in Table 113, but it must be emphasised that in other batches of the cements the proportions giving the quick early strength, and the minimum ultimate strength, may differ somewhat.

TABLE 113 Mixes of high-alumina and Portland cements
Compressive strength (lb/in^2), 1 : 2 : 3 concrete, w/c 0·55

High-alumina cement (per cent)	Portland cement (per cent)	3 hours	6 hours	24 hours	7 days	28 days
100	0	0	1200	6800	9000	10200
90	10	0	1500	5500	7500	8200
75	25	250	1600	4000	5800	6200
60	40	300	1000	2500	3100	3200
40	60	800	900	1200	1500	1600
25	75	500	600	800	1400	1800
10	90	50	100	300	1100	1700
0	100	0	0	700	3000	5000

The resistance to sulphate attack of mixes of 5–20 per cent high-alumina cement with 95–80 per cent Portland cement has been found[1] to be lower than that of Portland cement alone, but Robson reports that when less than 20 per cent Portland cement is present in the mix the resistance is comparable with that of high-alumina cement.

Evolution of heat from high-alumina cement on hydration: effect of temperature on strength

Much attention has been directed to the study of the temperatures reached in high-alumina cement concrete masses and to the resultant effects on the strength developed.

While the total heat evolved during hydration is of the same order as for Portland cements, the much more rapid hydration leads to its evolution over a shorter period of time. Comparative temperature rise curves[2] for adiabatically cured concretes are shown in Fig. 141, from which it is seen that the greater part of the heat evolution from alumina cement occurs over a period of some 10 hours, and that after 24 hours the further evolution is slight. The ultimate value of the heat of hydration seems to be similar to that of rapid-hardening Portland cement.

[1] D. G. Miller and P. W. Manson, *Tech. Bull. U.S Dep. Agric.*, No. 358 (1933).
[2] N. Davey and E. N. Fox, *Bldg Res. tech. Pap.*, No. 15 (1933). H.M.S.O. London.

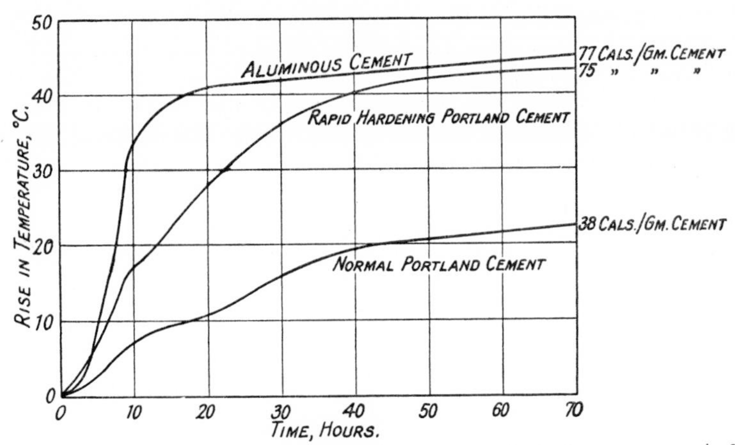

FIG. 141 Temperature rise in adiabatically cured 1 : 2 : 4 concrete, w/c 0·6 (*Davey*).

The very rapid evolution of heat leads to the development of a high internal temperature in high-alumina cement concrete unless it is placed in quite thin sections. This has an important bearing on the methods adopted in using high-alumina cement, since high temperatures are very deleterious to the strength of the concrete, in marked contrast with Portland cement. It offers, however, a definite advantage when concrete has to be placed in freezing weather.

The comparative strengths of 1 : 2 : 4 concrete cylinders made and cured for the first day at 21° (70°F) and 38° (100°F) respectively, and thereafter at 21° (70°F) are shown in Table 114.[1]

TABLE 114 Effect of temperature of curing over first 24 hours on strength of high-alumina cement concrete cylinders, w/c 0·53

Cement	Curing temperature first 24 hours	Compressive strength (lb/in²)			
		1 day	3 days	7 days	28 days
1	21°	4080	4290	4200	3660
	38°	2730	2800	2800	1820
2	21°	3510	4070	4280	4560
	38°	2280	2550	2750	1730
3	21°	2950	4110	4190	4040
	38°	2350	2500	2760	2260
4	21°	3710	4050	4440	4030
	38°	2750	3290	2680	1620
5	21°	3520	4390	4700	4990
	38°	2870	2740	2890	1920
6	21°	4480	5680	5850	6200
	38°	3180	3180	3190	2680

[1] P. H. Bates, *First Com. Internat. Assoc. Test. Mat.*, Zürich 1930, 211.

A high temperature during the setting and hardening period over the first 24 hours thus causes a considerable loss in strength and a regression at later ages.

The influence of continuous storage at 16° (61°F) and 35° (95°F) on the strength is illustrated by Fig. 142, and that of similar storage after an initial curing period of 24 hours at 16° by Fig. 143.[1] It is observed from both these figures that storage at a temperature of 35° under moist conditions very seriously

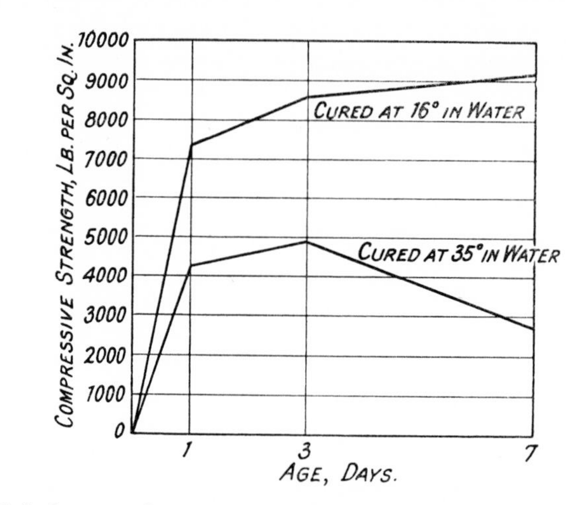

FIG. 142 Influence of temperature of storage on strength of 1 : 2 : 4 high-alumina cement concrete, w/c 0·6.

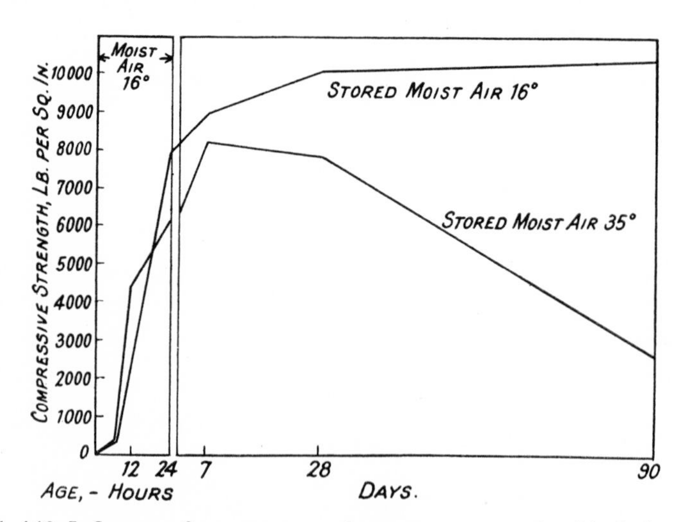

FIG. 143 Influence of temperature of storage on strength of 1 : 2 : 4 cement concrete, w/c 0·6, after curing for 24 hours at 16°.

[1] N. Davey, *Bldg Res. tech. Pap.*, No. 14 (1933). H.M.S.O. London.

reduces the ultimate strength of a concrete with a water : cement ratio of 0·6. This drop in strength occurs with increasing rapidity as the temperature is raised but it is negligible, except over long periods, below about 25°. Once a concrete is dry the conversion phenomenon which leads to this loss in strength ceases to occur. The loss in strength on conversion varies with the w/c ratio and at low water contents the strength of the converted concrete still remains high. The general form of the relation between the strength of fully converted concrete and the 1-day strength of unconverted concrete at various w/c ratios is shown in Fig. 144. The loss in strength in conversion is substantially the same whether the temperature of the concrete rises above 25° during the initial hardening period or after a period at lower temperatures. There is however some evidence that the loss in strength is less when conversion proceeds slowly over a period of years.

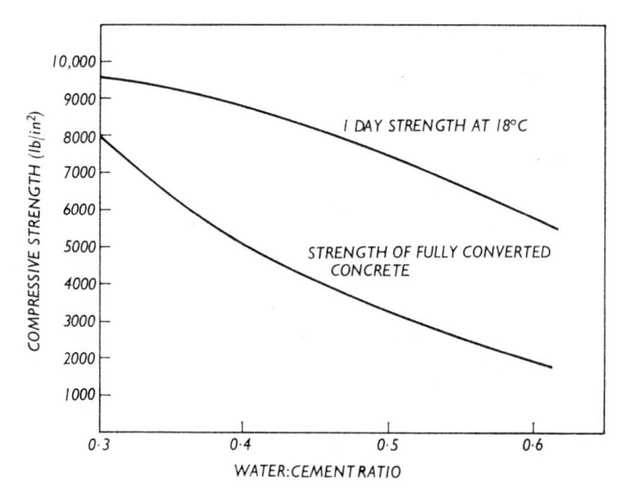

FIG. 144 Comparative strengths of converted and unconverted high-alumina cement concretes.

Under adiabatic conditions, such as are approached in the interior of a concrete mass of appreciable size, the temperature rise in high-alumina cement concretes is large and in a 1 : 2 : 4 mix exceeds 40°. This maximum is attained in less, and sometimes considerably less, than 24 hours after placing. The difference in strength between a concrete with a w/c ratio of 0·6 cured adiabatically, as in the centre of a mass, and one cured at ordinary temperatures is large, as is to be anticipated from the data previously given.

1 : 2 : 4 High-alumina cement concrete w/c 0·6

	Compressive strength (lb/in²)	
	Cured at 17°	Cured adiabatically
1 day	4300	3005
2 days	4479	1568
3 days	4623	1680

As we have seen, high-alumina cement concrete should not have a w/c ratio above 0·5 and this will usually mean a minimum cement content of about 550 lb cement per cubic yard concrete (330 kg/cm^3). In masses it should not be placed in lifts of more than 12 in. at a time nor at intervals of less than 24 hours. In thin sections, e.g. up to 2 ft thick, the lift can be any depth desired.[1] The surface of the concrete should be sprayed with water as soon as it has set and the formwork stripped as early as possible in order to reduce the rise in temperature and so obtain the strength of which the material is capable. Air-dry high-alumina cement concrete seems to be much less affected by rise in temperature and hence, once a concrete has been matured and dried, its strength is probably not affected to the extent suggested by the foregoing data by subsequent exposure to high temperatures. In tropical waters, however, it seems that only a very moderate ultimate strength can be expected from high-alumina cement and its suitability for such purposes is open to doubt.

High-alumina cement concretes which have suffered from too high temperatures during hardening show a characteristic chocolate to grey colour; this may often be best seen by breaking a piece of the concrete and examining the interior. Several instances have been reported of defective concretes which were easily distinguished in this way. Thus during the construction of the foundations of a bridge at Plougastel, near Brest, in 1926, several faulty patches of concrete which failed to harden properly were distinguished by this characteristic coloration. This failure seems to have been due to the abnormal temperature of the aggregate and raw materials used, owing to exposure to the sun during a spell of hot weather while this concrete was being laid. The brown colour is only incidental to the deterioration occurring in high-alumina cement concrete exposed to high temperatures, for Rengade states that it can be observed in any porous high-alumina cement concrete and is caused by the oxidation of the ferrous iron compounds existing in the cement. The reason for the effect of high temperatures on high-alumina cement has been discussed earlier.

Behaviour of high-alumina cement at long ages

Since high temperatures have been found to decrease considerably the strength of high-alumina cement concretes, the question may be asked whether a similar fall in strength may not occur over years at ordinary temperatures. Miller and Manson[2] found that concretes stored in water in the laboratory at 25–30° for many years showed a definite decrease in strength, but that this was only found in lesser degree, or not at all, in similar specimens stored in the cool water of a lake. Some data illustrating these effects are shown in Table 115.

Other specimens stored in water in the laboratory showed losses in strength at 5 years, varying from 0 to 50 per cent, but the strength after 20 years of specimens buried in the ground at a depth of 4½ feet exceeded the initial value. The tendency of the specimens stored in the laboratory to lose strength is almost certainly due to the higher average temperature of storage.

[1] *The Use of High-alumina Cement in Structural Engineering.* Institution of Structural Engineers, London 1964.

[2] *Tech. Bull. Minn. agric. Exp. Stn*, No. 194 (1951).

TABLE 115 Strength at long ages of high-alumina cement concrete w/c 0·52

Cement	Compressive strength (lb/in²)						
		1 year		5 years		20 years	
	Initial (7 days)	Laboratory in water	In lake*	Laboratory in water	In lake*	Laboratory in water	In lake*
X	7730	7480	7010	4030	7340	5680	5230
Y	6700	6260	7420	7770	8950	6970	8570
Z	7080	8160	8210	6880	7830	5960	6890

* A water with a high sulphate content.

There has been considerable controversy[1] on the long-term effect of storage of concrete in water temperatures around 18° and the data are somewhat conflicting. The data shown below do not indicate any loss in strength up to 5 years but there is some indication of retrogression at 10 years.

Cement	Compressive strength (lb/in²)			
	7 days	1 year	5 years	10 years
A	6100	7720	7960	7720
B	6470	6890	8030	7520
C	4800	6220	6390	6030

In some other tests Thomas[2] found strengths at 20 years varying from 57 to 97 per cent of the 1-year strength for specimens stored in water at 18° and 67 to 87 per cent for specimens stored in air at the same temperature. The w/c ratio was 0·6. Tests by Talabér[3] in Hungary on concretes with w/c ratios of 0·5 to 0·6 stored in the open air with summer temperatures above 18° showed strengths at 20 years around a half of the 1-year strength. Some tests[4] on the strength of specimens cut from high-alumina cement concrete piles, the feet of which had been immersed in seawater, showed that after 30 years the strength of the concrete at the bottom of the piles was above the 7-day strength whereas that from the tops of the piles which had been exposed to the prevailing atmospheric conditions in south-east England showed about 80–90 per cent of the 7-day strength for concrete of a w/c ratio of 0·3 and 50–60 per cent for a w/c ratio of 0·48.

After experience in France of the use of high-alumina cement in massive structures had revealed various instances of severe deterioration, its use was discontinued in 1943 for permanent public works. It must be remembered, however, that much of the concrete concerned is likely to have been placed at relatively high w/c ratios and better results might have resulted from modern practice. Experience has shown that some seawater structures, for example, are

[1] A. M. Neville, *Proc. Instn civ. Engrs* **25**, 287 (1963); O. J. Masterman, *Civ. Engng publ. Wks Rev.* **56**, 483 (1961).
[2] F. G. Thomas, *R.I.L.E.M. Bull.*, No. 9, 17 (1960).
[3] J. Talabér, *R.I.L.E.M. Sym. Durability of Concrete, Prague 1961*, Final Report, p. 109 (1962).
[4] F. M. Lea, Canadian Building Science Series, *Performance of Concrete*, edited by E. G. Swenson, p. 56 (1968).

still in excellent condition after 40 years in cool climates, as are also many concretes used in industrial buildings. The strength of high-alumina cement concrete after 20 years at 18° is likely, as the foregoing data show, to have decreased, but for w/c ratios of 0·5 or less it should still be adequate for structural purposes even when conversion is complete. At lower temperatures the time periods involved before the maximum loss in strength occurs are likely to be more prolonged. Reviewing the situation in 1964 the Institution of Structural Engineers concluded that high alumina cement concrete could be safely used for load-bearing structures provided that the various precautions as to maximum water : cement ratio and conditions of curing outlined earlier are taken. It should not be used in load-bearing structures in tropical conditions or as structural members in industrial structures where it will be subject to warm and moist conditions. It may however continue to serve well as a protective material even under warm moist conditions where conversion is inevitable.

High-alumina cement concrete at low temperatures

The rapid heat evolution of high-alumina cement concrete during hardening is beneficial when concrete is placed at low temperatures as may be seen from the data[1] in Table 116 on concrete prisms 7 × 7 × 28 cm with a cement content of 590 lb/yd^3 (350 kg/m^3) and a water : cement ratio of 0·50 stored at various temperatures.

TABLE 116 Effect of low temperature on strength of concrete
w/c 0·50

Temp.	Cement	Units	6 hours	16 hours	1 day	2 days
18°	RHPC	lb/in^2	0	510	1720	3250
		kg/cm^2	0	36	121	229
	HAC	lb/in^2	3410	5520	5670	6400
		kg/cm^2	240	388	399	450
12°	RHPC	lb/in^2	0	199	483	1950
		kg/cm^2	0	14	34	137
	HAC	lb/in^2	3000	5110	5570	6070
		kg/cm^2	211	358	392	427
6°	RHPC	lb/in^2	0	85	185	938
		kg/cm^2	0	6	13	66
	HAC	lb/in^2	2630	5060	5270	5470
		kg/cm^2	185	356	371	385
0°	RHPC	lb/in^2	0	0	43	342
		kg/cm^2	0	0	3	24
	HAC	lb/in^2	768	4630	4950	5560
		kg/cm^2	54	326	348	391

RHPC Rapid-hardening Portland cement.
HAC High-alumina cement.

[1] E. Rengade, Internat. Assoc. Test. Mat., London 1937, Congress Book, p. 245; see also N. Davey, ibid., p. 342.

In practice it is found that provided the temperature at which the concrete mix is placed is sufficient to keep it above freezing until hydration starts, that is for about three hours, the heat evolution will suffice to raise the temperature enough to ensure hardening. Calcium chloride must not be added to high-alumina cement.

Resistance of high-alumina cement to sulphate waters

The investigations which led to the production of high-alumina cement were, as we have seen, stimulated by a desire to obtain a cement which was resistant to the action of sulphate-bearing waters. These investigations were successful in their aim and in high-alumina cement we have a material which, under ordinary temperature conditions, has a resistance to attack by sulphates and seawater unequalled by any other construction cement.

Many laboratory investigations and tests under practical conditions have since been made on the resistance of high alumina cement mortars and concretes to sulphate action. High alumina cement was used in 1916 for constructional work on the P.L.M. Railway in France in certain regions where the subsoil strata consisted of anhydrite or gypsum, and concrete test pieces were at the same time immersed in running water saturated with calcium sulphate coming from these strata. After years these test specimens were unattacked, and even specimens in which anhydrite had been used as an aggregate were unchanged.[1] In the years to 1923 the P.L.M. Railway used over 7000 tons of high-alumina cement for constructional work in gypsiferous soils with satisfactory results.

Miller and Manson found the behaviour of high-alumina cement concretes immersed in the highly sulphate waters of Medicine Lake, S. Dakota, for 20 years, to be excellent and many of the test specimens retained a strength above the initial value. Under the same conditions ordinary Portland cements had mostly failed within 5 years and even sulphate-resisting Portland cements with a content of $3CaO.Al_2O_3$ below 5 per cent had lost 20 per cent of their strength in 5 years and nearly half in 10 years. Laboratory tests have similarly shown the great resistance of well-made high-alumina cement concrete to sulphate waters at normal, or low, temperatures. In Plate 16 (i) are shown some 1:2:4 high-alumina cement reinforced concrete prisms after immersion for 2 years in 5 per cent magnesium sulphate solution. A comparison of these, which are quite unattacked and still show sharp edges, with similar specimens of Portland cement shown in Plate 14 (ii), indicates the markedly greater resistance of high-alumina cement.

The relative resistance of Portland and high-alumina cement to sulphates is illustrated by the data in Table 117, where the rates of expansion of mortars immersed in the various solutions at a week old are shown.

The cause of the great resistance shown by high-alumina cement to the action of sulphates is still in dispute. It may be attributed in part to the absence of any free calcium hydroxide in the set cement, in contrast to Portland cement, but it is evident that this is by no means a full explanation. The aluminate compounds in Portland cement are one of the main points of attack by sulphate solutions with

[1] Touche, *Ciment* **31,** 240 (1926).

TABLE 117 Expansion of 1 : 3 cement mortars in sulphate solutions (linear expansion per cent)

Solution	Portland cement				High-alumina cement
	4 weeks	12 weeks	24 weeks	1 year	
5% Na_2SO_4	0·018	0·070	0·144	0·32	No expansion in 1 year
5% $MgSO_4$	0·018	0·054	0·25	0·91	,, ,, ,, ,,
5% $(NH_4)SO_4$	0·100	3·800	—	—	,, ,, ,, ,,

formation of calcium sulphoaluminate, and in general the resistance of Portland cement decreases as the alumina content increases. Yet high-alumina cements with a content of approximately 40 per cent alumina are resistant. Lafuma[1] has suggested that combination of a cement compound in the solid state with a substance in solution always leads to expansion, but that if the cement compound passes into solution, reacts, and then precipitates as solid, no expansion occurs. It is considered by Lafuma that the formation of calcium sulphoaluminate during the action of sulphate solutions on set Portland cement is an action of the first type, because the free calcium hydroxide present renders the hydrated calcium aluminates entirely insoluble, but that in set high-alumina cement the action is of the second type, as the hydrated calcium aluminates are slightly soluble in water when additional lime is not present. The resistance of high-alumina cement to attack by sulphates seems, however, to be due to the immunity of the aluminate compounds from attack rather than to any difference in the manner in which such a reaction occurs. Thus high-alumina cement concretes, and even lean mortars, immersed in sulphate solutions for long periods, show only small sulphate contents on subsequent analysis, indicating that little reaction had occurred. Specimens which have suffered attack, either because of poor quality of the concrete or from the effects of high temperatures, show on the other hand considerable sulphate contents. This can be illustrated by the analysis of some samples taken from a concrete tank which showed disintegration of the surface to a depth of an inch after holding a 5 per cent magnesium sulphate solution for 8 years. In the outer inch of concrete all the lime in the cement had been converted to calcium sulphate and calcium sulphoaluminate, and magnesium hydroxide had been deposited. In the concrete 2 in. below the surface about a quarter of the lime had been so converted. As a contrast, concrete from immediately below the surface of a cylinder which showed no sign of deterioration after 12 years in a similar solution contained only enough sulphur trioxide to be equivalent to the conversion to calcium sulphate of less than 1 per cent the lime in the cement. It seems, therefore, that the primary cause must be sought in the formation of protective surface films.

[1] *Revue Matér. Constr. Trav. publ.* **243,** 441 (1929); **244,** 4 (1930); *Zement* **16,** 1179 (1927); *Revue gén. Sci. appl.* **1** (3), 66 (1952).

Resistance of high-alumina cement to acid waters

The relative degree of resistance shown by high-alumina cement and other concretes to chemical attack is discussed fully in a later chapter, but some mention may be made here of the resistance of high-alumina cement to acid waters. Waters which are acidic owing to the presence of free carbonic acid, as occurs in some pure mountain waters containing very little dissolved solids, slowly dissolve the cement from the surface of Portland cement concretes exposing the aggregate. High-alumina cement concretes are much more resistant to this action and also to attack by very weak solutions of mineral acids.

There is no conclusive evidence as to the cause of the increased resistance of high-alumina cements to such acid waters, but the presence of alumina gel enveloping the more susceptible lime compounds is an important factor. Alumina gel is precipitated from aluminium salt solutions when the pH becomes less acid than about 4, and hence will not redissolve in acid waters until the pH of the solution falls below 4, whereas the hydrated calcium aluminates form solutions with pH of 11 or higher and tend to lose lime to the solution if placed in a medium of lower pH. All the constituents of set Portland cement tend to lose lime to solutions of pH below 10–11 and there is no hydrated compound present which is entirely resistant to solutions of lower pH value. The suggestion that the alumina gel in the set cement is responsible for the increased resistance shown seems, therefore, to be not unlikely and is supported by the fact that, if set high-alumina cement is crushed and distilled water allowed to percolate through it, alumina and lime are leached out. Porous high-alumina cement concrete will not resist the action of pure waters permeating through it. It is also significant that the resistance to acid solutions is limited to a minimum pH of about 4. The presence in set Portland cement of a considerable quantity of free calcium hydroxide which is readily dissolved, and its absence from set high-alumina cement, is almost certainly a contributory factor. High-alumina cement is more resistant to the action of weak solutions of organic acids than Portland cement, but against strong solutions, in which in any case the degree of attack is heavy, it is less resistant. This seems to be the case for acetic acid where the critical value is around 0·5–1 per cent. The reason for the existence of such a critical point is not evident unless it is connected with the solubility points of the gel constituents of the set cement.

Effect of alkaline solutions on high-alumina cement

Alkali carbonates react with the hydrated calcium aluminates in high-alumina cement according to the following equations:

$$K_2CO_3 + CaO.Al_2O_3.aq \rightarrow CaCO_3 + K_2O.Al_2O_3$$
$$2K_2CO_3 + 2CaO.Al_2O_3.aq \rightarrow 2CaCO_3 + K_2O.Al_2O_3 + 2KOH$$

The alkali hydroxide can in turn dissolve alumina to form more alkali aluminate and atmospheric carbonation can regenerate the alkali carbonate so that it can react again. The alkali carbonate thus in effect acts as the vehicle for promoting attack by atmospheric carbon dioxide.

An action of this type caused the deterioration in France of the bases of rather porous high-alumina cement concrete electricity transmission poles set in Portland cement blocks.[1] The alkali carbonate, derived from the Portland cement, permeated into the high-alumina cement concrete and was drawn by capillarity to the surface higher up where both concentration by evaporation and recarbonation could occur. This cause of deterioration from an external agent is unlikely to affect dense concrete, but a similar action can arise from within the concrete if the aggregates used contain soluble alkalis. For this reason it is important that crushed igneous rock aggregates for use in high-alumina cement concrete should be free from crusher dust that may liberate alkali. The presence of alkalis is also said by Lafuma to encourage the inversion of the hydrated calcium aluminate in the set cement.

Refractory concrete

An important use of high-alumina cement is in refractory concrete[2] to withstand medium or high temperatures. The aggregate must have heat-resisting properties, for it plays an important part in determining the maximum temperature the concrete will withstand. Crushed firebrick is the aggregate most commonly used and that from bricks containing over 40 per cent alumina yields concretes stable up to 1300–1350°. Ordinary clay building brick aggregate gives concrete suitable for use up to 900–1000°. Refractory concretes for use at 1400–1600° can also be obtained with special aggregates such as sillimanite, carborundum, chrome, chrome-magnesia, dead burnt magnesite, and fused alumina. From lightweight aggregates, such as calcined diatomite, expanded vermiculite, expanded clay, etc., there are obtained insulating concretes suitable for use up to 900–950° and with a thermal conductivity down to 1·5–2·5 Btu/ft^2/h/°F/in. at the operating temperature and a weight of 30–60 lb/ft^3. A white calcium aluminate cement is also available which, with a fused alumina aggregate, gives concretes suitable for service at temperatures up to 1800°.

For concrete, mixes such as 1 cement, 2 crushed firebrick ($\frac{1}{8}$ inch down), 3 crushed firebrick ($\frac{3}{4}$–$\frac{1}{8}$ inch), or 1 cement, 2 fine aggregate ($\frac{1}{8}$ inch down), 2 coarse aggregate ($\frac{3}{8}$–$\frac{1}{8}$ inch), by volume, are used, and for mortar a mix of about 1 : $2\frac{1}{2}$ cement–crushed firebrick ($\frac{1}{8}$ inch down). The aggregates should be soaked with water and allowed to drain before mixing with the cement. Little further addition of water may then be needed to give a suitable working consistence which should not be too wet. The concrete is wet-cured for 24 hours and allowed to dry before exposure to heat.

Refractory concretes set in the ordinary way and develop strength by the hydraulic bond. When exposed to heat this bond is gradually reduced and at a temperature of 900° a minimum strength of some 1000 lb/in^2 remains. From about 700° upwards, depending on the type of aggregate, a new form of ceramic bond starts to develop by solid reactions between the cement and the fine aggregate, and increases with the temperature. After losing all their water in service,

[1] E. Rengade, P. L'Hopitallier and P. Durand de Fontmagne, *Revue Matér. Constr. Trav. publ.* **318,** 52; **319,** 78 (1936).
[2] See T. D. Robson (loc. cit.) for details of refractory concretes.

refractory concretes with firebrick aggregates weigh about 110–120 lb/ft^3 (1760–1290 kg/m^3) and have a thermal conductivity of about 6–8 Btu/ft^2/h/°F/in. thickness. With the more heat-resistant aggregates the weights rise to 140–170 lb/ft^3 (2240–2720 kg/m^3) and the thermal conductivity to 20 units. The value in SI units for the thermal conductivity (J/m^2/s/°C/m thickness) is obtained by multiplying by 0·144 and in c.g.s. units (cal/cm^2/s/°C/cm thickness) by multiplying by 2903.

The heat-resistant concretes, i.e. those stable up to 1000°, are used for foundations for furnaces, coke ovens, boiler settings, and similar purposes, while the refractory concretes proper find a wide variety of uses for furnaces and kilns. They are used, for example, in the construction of brick kilns and tunnel kilns and for the car tops and decks, for kiln doors and linings, and in electric furnaces. Since the concrete is moulded in a plastic condition, special shapes can readily be made true to size. Refractory mortars are used as a jointing medium for firebricks and for patching and repairs.

17 Some Special Cements and Cement Properties

White cements

The manufacture of white Portland cement requires suitable raw materials and special care against contamination during production. The grey colour of ordinary Portland cement is due to the iron oxide present, and for a white cement the content of this needs to be kept low. The raw materials are a suitable limestone, or chalk, and china clay, both of which must be low in iron. Cryolite (sodium aluminium fluoride) has at times been added as a flux to aid in burning. Contamination with iron, during the preparation of the raw mix and burning, has to be avoided. The cement is burnt in a rotary kiln, usually with an oil or gas fuel to avoid the contamination with coal ash which is inevitable when pulverised coal is used for firing. The presence of small quantities of manganese oxide is said to have a marked influence on the colour of white cements.

A white cement is manufactured in France from the well-known Tiel deposits of iron-free limestone. These deposits, known as Chaux du Tiel, contain sufficient clayey matter to yield a hydraulic cement. This material owing to its physical nature, requires only light burning and the product is akin to an eminently hydraulic lime. As is common with the French hydraulic limes, it is partially hydrated in such a way as to hydrate the free lime present, but not the calcium silicates. Its composition is such that it falls outside the limits of Portland cement specifications and its physical properties are more closely akin to those of a hydraulic lime than a Portland cement.

Analyses of white cements are shown in Table 118 (p. 529).

The strengths of white Portland cements are adequate to pass the British Standard for Portland cement, but are rather lower than those of normal Portland cement.

Coloured cements

Coloured cements fall into two groups with white and grey Portland cement respectively as their basis. The former class is naturally much the more costly, but a wide range of colours is obtainable by the use of suitable pigments. The cements of the latter class are usually either red or brown, as with other colours it is not possible to cover up satisfactorily the grey colour of the ordinary Portland cement. Coloured Portland cements consist simply of Portland cement with 5–10 per cent of pigment. The pigment cannot be satisfactorily distributed throughout the mass by mixing, and it is usual in production to grind the cement and pigment together.

TABLE 118 Analyses of white cements

	British cements		Tiel white cement
CaO	67·70	67·44	67·43
SiO_2	22·13	22·16	21·56
Al_2O_3	4·07	4·62	0·77
Fe_2O_3	0·46	0·41	0·53
MgO	0·46	0·20	1·72
Na_2O	0·11	0·34 (Na₂O + K₂O)	—
K_2O	0·12		—
SO_3	2·39	2·34	0·89
Loss on ignition	2·64	2·17	6·48
Free CaO	3·4	4·8	—

Not all the pigments which can successfully be used in paints are suitable for cement colour. The properties required in a cement pigment are durability of colour under exposure to light and weather, a fine state of subdivision, a chemical composition such that the pigment is neither affected by the cement nor detrimental to it, and the absence of soluble salts. These requirements are met by the following pigments which are in common use:

Iron oxides	Red, yellow, brown, black
Manganese dioxide	Black, brown
Chromium oxide	Green
Cobalt blue	Blue
Ultramarine blue	Blue
Carbon pigments	Black

Many pigments are heavily adulterated with fillers of various types. In general, a pure pigment, at a correspondingly higher cost, is preferable; the amount of pigment required, provided it is sufficiently finely ground, is proportionately less. Common adulterants in pigments are chalk (calcium carbonate), barium sulphate and gypsum (calcium sulphate). The two former are inert and harmless, but the last-named, if present in large proportions, is definitely dangerous, owing to the slow expansion which occurs as it reacts with the set cement. Pigments containing more than a few per cent of gypsum should not be used. The content of sulphur trioxide in a Portland cement is limited by specification to 2·75 per cent; the addition of, say, 10 per cent of a pigment containing an appreciable amount of gypsum obviously nullifies the safeguard afforded by this limitation. In the British Standard (BS 1014) for pigments for use with cement a limit of 2·5 per cent water soluble matter is imposed for most pigments. This in effect limits the permissible gypsum content to 5·4 per cent or less. Soluble salts other than calcium sulphate in pigments should be, and usually are, low, 1–2 per cent or less. For carbon black the BS 1014 restricts the water soluble matter to 1 per cent and includes requirements as to the content of volatile matter and ash and of constituents soluble in ether. Restrictions on volatile matter are also imposed on the other pigments.

Certain types of pigments, such as those containing Prussian blue, zinc and lead chromates, and cadmium lithophone, cannot be used. Organic colours are also, in general, unsuitable, though there are certain exceptions. Chrome green, which is a mixture of chrome yellow (a basic lead chromate) and Prussian blue, needs to be carefully distinguished from chromic oxide green. Lead oxide pigments must not be used, as lead oxide has a very serious effect on the setting and hardening of cement.

Black concretes are the most difficult of all colours to attain. Neither the iron oxide nor manganese dioxide pigments give a jet black. Carbon black is better as a pigment, but it has a rather more serious effect on the strength developed. High contents of inert pigments tend to lower to some extent the strength of the cement to which they are added, but provided the addition does not exceed 10 per cent the loss in strength should not exceed about one-third of the normal concrete strength. Indeed it has been claimed[1] that with some of the mineral pigments additions of up to 15 per cent by weight of the cement can be made to 1 : 4 mortar mixes with no adverse effect on strength. Carbon black has a more adverse effect on strength, but only a smaller amount of this than of other black pigments is required to produce the desired colour.

Ultramarine blue is not an entirely stable pigment since it combines with free calcium hydroxide, and its colour very slowly fades.[2] Under normal conditions it seems, however, to be used successfully as atmospheric carbonation of the exposed surface of a concrete occurs sufficiently rapidly to stop the action. Cobalt blue has been recommended in preference to ultramarine blue; it is considerably more expensive.

The fading of coloured concretes and renderings made with pigments of approved types is often a serious difficulty in their use. It is not, apart from the case of ultramarine blue, due to any loss of colour of the pigment, but to the formation of a white calcium carbonate film over the surface. The free calcium hydroxide liberated during the setting of the cement is carried towards the surface and there reacts with atmospheric carbon dioxide. When this reaction occurs below the surface the calcium carbonate formed does not show externally, though it can usually be seen if a piece of the mortar or concrete is broken away and examined. The conditions which determine whether carbonation occurs at, or just below, the surface are very varied. The surface texture and the curing conditions are important factors. A calcium carbonate film can be removed from the surface by washing over with dilute hydrochloric acid (1 : 10)—followed by a wash-down with water. The acid should not be allowed to remain more than a short time on the surface of the concrete which should be first wetted to reduce the suction if it is at all porous.

[1] B. Kroone and F. A. Blakey, *Constr. Rec.* **41,** 25 (1968).
[2] B. Kroone, *Chemy Ind.*, March 2nd, 1968, 287.

Attempts are sometimes made to colour the surface of concrete by absorption of dyes or salts. Various aniline colours, ferric sulphate ($\frac{1}{2}$–1 lb per gallon of water), and copper sulphate ($\frac{1}{3}$–$\frac{2}{3}$ lb per gallon) are the materials commonly used. The solution of the colouring matter is applied to the concrete when it is a few days' old and it has the effect of reducing the strength development.

Waterproofed cements

Certain special cements are marketed in Great Britain and elsewhere as waterproofed cements. The claim is made for them that they yield more impermeable mortars and concretes than ordinary Portland cement. Most of these cements are formed from a standard Portland cement to which has been added in grinding a small percentage of calcium, aluminium, or other metal stearate. A non-saponifiable oil has also been used for the same purpose. Another proprietary brand is stated to be composed of a normal Portland cement ground with gypsum which has been treated with tannic acid. Considerable controversy exists as to the value of waterproofed cements, and the use of integral waterproofers in general. In the laboratory it is difficult to show that any definite advantage in the direction of reduced permeability is gained by their use, but it is fairly widely held that in practice their use enables an impermeable concrete to be attained more easily (see p. 604).

Masonry cements

The traditional mortar material for building work was lime, but later to an increasing extent it was ousted by cement. Adequate strength in fully-hardened mortar, as well as rapid development of strength in the early stages, is most conveniently obtained by using Portland cement. But it is not practicable to adjust the strength simply by varying the ratio of cement to sand, because lean mixes of cement and sand are harsh and unworkable. As with lime mortars, about one volume of cement is needed for three volumes of sand to give a workable mix, and a mortar of this kind is too strong for general use, either as a jointing mortar or as a rendering.

These difficulties with cement mortars led to the introduction of mortars made with appropriate proportions of lime and cement, which combine the advantages of each.

Using a 1 : 3 cement mortar as a basis, lime-cement mortars are designed on the principle that part of the cement is replaced by an equal volume of lime so that the binder paste still fills the voids in the sand. In this way, good working qualities, water retention, bonding properties, and early strength can be secured without the mature strength being too high. Experience has shown that a choice from a range of cement-lime-sand mixes in the proportions of 1 : $\frac{1}{2}$: 4$\frac{1}{2}$, 1 : 1 : 6, 1 : 3 : 12 by volume will meet most requirements. Subsequently, air-entrained cement-sand mortars were also introduced. These provide good working properties even with lean mixes, since the air bubbles increase the volume of the binder paste and so help to fill the voids in the sand. Masonry cements, which were first introduced in the U.S.A.,[1] have been developed in various countries and are now widely used. Although earlier developments included mixtures of cement with lime, granulated slag, pulverised fuel ash or other finely ground materials, masonry cements now generally consist of a mixture of Portland cement with a finely ground inert material usually limestone, and an air-entraining agent. Swedish masonry cements are indeed restricted to such mixtures by a composition clause in the Swedish standard, and this also requires a minimum of 40 per cent of Portland cement clinker in the mix. American and Canadian masonry cements also generally contain about 50 per cent of cement, so that when mixed with three volumes of sand they provide a mortar roughly equivalent in strength to a 1 : 1 : 6 cement/lime/sand mix. Stronger mortars are obtained by adding more Portland cement during site mixing. British masonry cements, however, contain about 75 per cent of Portland cement clinker, and the mortar strength is adjusted simply by altering the proportion of sand in the mix, from three volumes of sand for each volume of masonry cement in the strongest mixes, to seven volumes in the weakest mixes.

The aim of the manufacturer of masonry cements is to improve plasticity and water-retentivity and to reduce shrinkage, but there is still much to be learnt about the best forms of specification tests to evaluate their performance in these and other respects. The choice of a sand suitable for determining mortar properties is also a problem. The ASTM standard specification for masonry cements, C91–66, includes requirements for fineness, soundness and setting time of the cement, and for compressive strength, minimum air content and water retention of a 1 : 3 mortar made with Ottawa blended sand. The Swedish standard includes tests on a 1 : 5 mortar made with the RILEM–Cembureau graded sand, and substitutes a water separation or bleeding test for the water retention test. The British Standard is still in preparation but it will be based on the mortar testing methods of BS 4551 : 1970. Recent studies of the air content of masonry cement mortars in relation to the strength of bond have indicated that there may be a need for standards to include a maximum value for air entrainment, in addition to a minimum value.

Oil-well cements

Cements are used in the petroleum industry for cementing the steel casing of oil wells to the rock formation and to seal off porous water-, or gas-bearing, formations through which the drilling passes. The cement in the form of a slurry has to be pumped into position, often at considerable depth, and may be subject to temperatures up to 350° F and pressures up to 18 000 lb per sq in. The slurry must remain sufficiently fluid to be pumped under these conditions for periods up to several hours and then harden fairly rapidly. It may also have to resist corrosive conditions from sulphur gases or waters containing dissolved salts.

The type of cement required varies according to the conditions, but its setting time and the rate of thickening must be adequately retarded. The types of cement used include slow-setting Portland cements with an $Al_2O_3:Fe_2O_3$ ratio rather

[1] See C. E. Wuerpel, *Sym., London 1952*, 633 for an account of the development of masonry cements.

below 0·64, pozzolanic cements, and coarsely ground ordinary Portland cements with special retarders. Many different agents have been patented as retarders, but according to Hansen[1] the commonest ones consist of starches or cellulose products, sugars, or acids or salts of acids containing one or more hydroxy groups. These three classes of compounds all have the group HO–C–H in common, though that alone is not sufficient since other compounds, such as alcohols, containing this group, do not act as retarders. The manner in which these retarders function is not known, but it seems probable that the HO–C–H group is adsorbed on to the grains of $3CaO.SiO_2$ and $3CaO.Al_2O_3$ protecting them for a time from attack by the water. Calcium or sodium lignosulphonate are also used and their retarding action is ascribed to the adsorption of the anion.

Special methods of testing oil well cements under conditions of high temperature and pressure have been developed.

Expanding and non-shrinking cements

The value of a cement which would suffer no overall change in volume on drying has long been realised, while for some purposes, such as underpinning and other types of repair work, a small controlled overall expansion which would tighten the joint between new and old work would have advantages. Another use of an expanding cement which was appreciated later is as a 'self-stressing' cement to induce tensile strength in steel reinforcement, that is to make prestressed concrete not by the normal method of direct tensioning of the steel but by inducing this tension through the expansion of the concrete bonded to the steel. The various cements that have been successfully used to date depend on the formation of calcium sulphoaluminate in the concrete during hardening and vary only in the materials used to secure a controlled rate of formation of the sulphoaluminate and the accompanying expansion. The use of hard-burnt magnesia or lime as an expanding agent has also been explored particularly in the U.S.S.R.[2] but has not so far proved sufficiently reliable for practical use. The development of expanding cements stems from the work of Lossier[3] in France and the firm of Poliet Å Chausson which undertook the first manufacture,[4] and continued to produce the material over a period of years though manufacture now seems to have ceased in France. The later developments which have led to a wider use of expanding cements are due mainly to work in the U.S.A. and the U.S.S.R. Two main types of material have been produced, a 'shrinkage-compensating' cement and a 'self-stressing' cement.

As we have seen elsewhere, the addition of excess gypsum to Portland cement leads to unsoundness owing to the continued gradual reaction of the gypsum with the tricalcium aluminate in the cement after it has hardened. It is necessary in expanding cements that the formation of calcium sulphoaluminate should occur at a time when the concrete has developed some strength but is still suffi-

[1] W. C. Hansen, *Sym., London 1952*, 598, has reviewed the properties required in oil well cements and methods of testing.
[2] W. Statanoff and N. Djabaroff, *Silikattechnik* **11,** 473 (1960).
[3] H. Lossier, *Mém. C. r. Trav. Soc. Ing. civ. Fr.* **101,** 189 (1948); *Génie Civ.* **131,** 341 (1954); *Struct. Engr* **24,** 505 (1946).
[4] H. Lafuma, *Sym., London 1952*, 581.

ciently extensible to accommodate some limited expansion without significant cracking. The formation of calcium sulphoaluminate requires the presence of calcium, aluminium and sulphate ions in an alkaline environment and this can be achieved in several ways. It is also necessary that the materials used should be such that the time and amount of expansion are under control. The French cement consisted of three elements: (1) a Portland cement (2) an expanding agent and (3) a stabiliser. The expansive agent was prepared by burning a mix of 50 per cent gypsum, 25 per cent red bauxite and 25 per cent chalk which it was thought might produce an anhydrous calcium sulphoaluminate which would expand on reaction with water. This composition in fact gives a considerable excess of calcium sulphate and ground blastfurnace slag was added as a stabiliser to take up excess calcium sulphate slowly and bring the expansion to an end.

Later studies[1] showed that a stable compound $4CaO.3Al_2O_3.SO_3$ is formed on burning suitable mixes of lime, alumina and calcium sulphate and that its hydration in the presence of additional lime and calcium sulphate leads to the formation of calcium sulphoaluminate by a reaction which can be written as $4CaO.3Al_2O_3.SO_3+8CaSO_4+6CaO+aq\rightarrow3(3CaO.Al_2O_3.3CaSO_4.31H_2O)$. In mixes of higher $Al_2O_3 : SO_3$ and $CaO : SO_3$ ratios than represented by this equation the monosulphate calcium sulphoaluminate appears also to be formed and to coexist with ettringite.[2] Formation of the monocompound seems to lead to little expansion. In mixes of low SO_3 content hydrated alumina is formed along with ettringite, as, for example

$$4CaO.3Al_2O_3.SO_3+2CaSO_4+aq\rightarrow3CaO.Al_2O_3.3CaSO_4.13H_2O+4Al(OH)_3$$

The expanding cement developed in the U.S.A.[3] is based on a mixture of Portland cement with an expansive agent containing $4CaO.3Al_2O_3.SO_3$, CaO and $CaSO_4$ prepared[4] by clinkering a mix of chalk, bauxite and gypsum at 1300–1400°. The content of calcium sulphate is lower than in the Lossier cement and no stabilising agent is used. The expansion increases with the proportion of the expansive component and also as the fineness of the cement increases and as the C_3A content of the Portland cement decreases. The shrinkage-compensating cement contains about 10–15 per cent of the expansive component and the self-stressing cement about 25–40 per cent. The aim in the former is to obtain a relatively low unrestrained expansion potential of about 0·2 per cent and in the latter a much higher one, about 3–6 per cent. The strength of the self-stressing cement, if allowed to expand freely, is low owing to the disruptive effects of the large expansion. The strength is a maximum if the expansion is completely restrained, but then there is no self-stressing. With normal amounts of reinforcement the expansion is restricted to about 0·2 to 0·3 per cent.

The U.S.A. shrinkage-compensated cement is available commercially and has been used for various types of concrete products and structures. The use of

[1] P. E. Halstead and A. E. Moore, *J. appl. Chem.* **12,** 413 (1962); T. A. Ragozina, *Zh. prikl. Khim.* **30,** 1682 (1957).
[2] A. Klein and P. Mehta, *Sym., Tokyo 1968*; N. Fukuda, ibid.
[3] A. Klein and G. E. Troxell, *Proc. Am. Soc. Test. Mater.* **58,** 986 (1958); A. Klein *et al.*, *Proc. Am. Concr. Inst.* **58,** 59 (1961); **60,** 1187 (1963); *Proc. int. Conf. Structure of Concrete, London 1965*, Cement and Concrete Association (1968), p. 467.
[4] P. K. Mehta and A. Klein, *Spec. Rep. Highw. Res. Bd*, No. 90, 328 (1966).

self-stressing cement for prestressed concrete is not yet out of the development stage for there are many variables influencing the product on which more information is still needed.[1] Nevertheless trials have been made on its use in concrete pipes, beams, slabs and shells. The unrestrained expansion of the cement is a measure of the potential of the cement for expansion, but to be useful the expansion must be restrained by inducing tensile stress in reinforcement. Restraint is also essential to the development of strength in the concrete. Thus in one set of tests the strength of a concrete at 28 days was only 330 lb per sq in. when the longitudinal expansion was 2·75 per cent and 2340 lb per sq in. when it was restrained to 0·16 per cent. There is also a transverse expansion which may have to be restrained for satisfactory strengths to be obtained.

The expansive cements developed in the U.S.S.R.[2] are of two kinds. The shrinkage-compensating variety is based on a mixture of high-alumina cement, gypsum and the hydrated aluminate $4CaO.Al_2O_3.13H_2O$ and the self-stressing cement on a mixture of Type I Portland cement, high-alumina cement and gypsum in a ratio such as 66 : 20 : 14, The former cement is quick-setting with a final set within 10 minutes and was developed in the first instance for the repair of damaged structures and later used for caulking and waterproofing. The formation of calcium sulphoaluminate is completed and the expansion ends shortly after setting. The second cement was developed for the self-stressing of prestressed concrete steam-cured at 60–80° C but is applied more generally for precast reinforced concrete units similarly cured because of its speed of hardening compared with Portland cement. A further type of low expansion cement is made in the U.S.S.R. from high-alumina cement ground with hard-burnt gypsum or anhydrite. This seems similar to the sulphoaluminate cement patented earlier by the Lafarge Company (see p. 495).

The expansion of the U.S.S.R. type self-stressing cements is controlled by the relative proportions of the components and the temperature of curing. At normal temperatures the sulphoaluminate formed initially in mixes of low w/c ratio is $3CaO.Al_2O_3.CaSO_4.12H_2O$ and on immersion of the concrete in water this is converted to $3CaO.Al_2O_3.3CaSO_4.31H_2O$ with a destructive expansion.[3] At higher w/c ratios the high sulphate compound is formed rapidly during the setting period and there is little expansion. The cement has been used in rich 1 : 1 mortars w/c 0·2–0·3 applied by guniting, centrifuging and vibro-squeezing for the production of prestressed concrete pipes but it appears that technical problems still remain in full scale manufacture. To obtain a satisfactory product it is necessary at an age of 24 hours to cure the mortar at 70–100° C for some 6 hours followed by water curing. Some tests by Monfore[4] illustrate influence of temperature. For mortars (1 : 2 w/c 0·4) cured initially for 18 hours at 23° C and then in water for 24 hours at various temperatures the ultimate free expansions recorded were 9 per cent at 23°, 5·5 per cent at 38°C and 2·3 per cent at 88° C. The restrained expansion will of course be much lower, about one-tenth or less.

[1] M. Polivka and V. V. Bertero, *Proc. int. Conf. Structure of Concrete, London 1965* Cement and Concrete Association (1968), p. 479; V. V. Bertero, *Proc. Am. Concr. Inst.*, **64**, 84 (1967).
[2] V. V. Mikhailov, *Sym., Washington 1960*, 927; P. P. Budnikov and I. V. Kravchenko, *Sym., Tokyo 1968*.
[3] V. V. Mikhailov, loc. cit.
[4] G. E. Monfore, *J. Res. Dev. Labs Portld Cem. Ass.* **6** (2), 2 (1964).

The expansion of expanding cements only continues as long as the concrete is kept moist and on subsequent drying the usual shrinkage occurs though the extent of this may be diminished if expansion is still continuing in the inner part of the concrete. The volume still remains greater than the initial volume as cast with the self-stressing cements and about the same volume with the shrinkage-compensating cements. The frost resistance of shrinkage-compensating cement concretes seems to be parallel to that of ordinary Portland cement concretes, but that of the self-stressing cement concretes to be lower.[1] The setting time of expansion cements is less than that of ordinary Portland cements.

Hydrophobic cement

A type of cement that has been used particularly in the U.S.S.R., called hydrophobic cement, is obtained by grinding ordinary Portland cement clinker with a film-forming substance such as oleic acid. The water-repellent film formed around each grain of cement reduces the rate of deterioration of the cement during storage, or transport, under unfavourable conditions. The film is broken down when the cement and aggregate are mixed together and normal hydration can then proceed. Such additions, as is well-known, also act as a grinding aid in manufacture. They entrain the air and act as water-reducing agents, particularly in lean concrete mixes. The U.S.S.R. standards specify the use of 0·08–0·1 per cent oleic acid, 0·1–0·12 per cent napthenic acids ($Cn\ H_{2\ n-1}$ COOH where $n = 8$ to 13) derived from certain crude oils, or 0·2–0·25 per cent soap naphtha derived from the alkali treatment of the oils and containing 50 per cent water. These hydrophobic cements are used in the U.S.S.R. primarily as a protection against deterioration during storage, but they also find use as a masonry cement on account of their plasticity, water-retentivity, and ease of pumping.[2] Concretes made from them are stated to have strengths similar to ordinary cement, any retarding influence of the additive being offset by the reduced water : cement ratio. A beneficial influence on frost resistance is also claimed.

Tests by Nurse[3] have confirmed that oleic, lauric and stearic acids, and also pentachlorophenol,[4] act as hydrophobic agents and that cement ground with them will float on the surface of water indefinitely and withstand exposure to air for a period without deterioration. Oleic acid was the most effective and clinker ground with 0·35 per cent produced a cement that withstood six months' exposure to air without loss of strength, whereas the untreated cement lost over three-quarters of its strength at 7 days after aeration for a month. Calcium stearate was not effective, in agreement with the U.S.S.R. conclusion that the agent should be such as to form an oriented adsorbed film at the surface of the particles.

[1] A. H. Gustaferro, H. Greening and P. Klieger, *J. Res. Dev. Labs Portld Cem. Ass.* **8** (1), 10 (1966); P. Klieger and N. R. Greening, *Sym., Tokyo 1968.*

[2] A detailed account is given by M. I. Khigerovich, 'Hydrophobic cement and hydrophoboplasticizing additives', 'Promstroyizdat', Moscow 1957, English translation issued by U.S. Naval Civil Engineering Laboratory, Port Hueneme, California.

[3] R. W. Nurse, *Cem. Lime Mf.* **26,** 47 (1953); see also U. W. Stall, *Spec. Publ. Am. Soc. Test. Mater.*, No. 205, 7 (1958); H. Wiegler and J. Nicolay, *Betonsteinzeitung* **34** (1) 16 (1968).

[4] *Br. Pat.* 22714 (1951).

Cement with inert additions

Cement made from Portland cement clinker ground with sand has occasionally been used in the past for mass concrete construction in dams, but during the war of 1939–1945 various countries became interested in the use of inert additions as a means of economy in the use of cement. The materials commonly used as additions were limestone, sand, and air-cooled slag.

At the same fineness the dilution of cement by grinding sand with it leads to a reduction in strength in concrete of about 10, 25–30, and 45–50 per cent for substitutions of 10, 20 and 35 per cent respectively of sand for cement. If tests are made on mortars of dry consistence (e.g. $w/c = 0{\cdot}32$) the results can be quite misleading for little or no reduction in strength is found for substitutions up to 20 per cent. The effect of grinding limestone or marl with the cement clinker is more variable. With certain hard limestones and marls, combined with certain clinkers, at constant times of grinding, a substitution of 10 per cent does not reduce the strength of concrete at constant slump. A small substitution, 2 per cent, in certain cases may increase the strength probably owing to its acting as a grinding aid. With substitution of 20 per cent the strength is reduced by about the same percentage. These effects appear to be very specific and the same hard marl has been found when substituted to the extent of 10 per cent with one clinker to reduce the strength by 10–15 per cent and with another to leave it more or less unchanged. This may be due in part to differential grinding actions.

Cement-emulsion mixes

Cements with special properties are obtained by mixing Portland cement, and in certain cases high-alumina cement, with bituminous rubber, and synthetic resin emulsions.

The oldest of these is the bituminous emulsion product. Portland cement mortar gauged with a bituminous emulsion instead of water is used as an industrial flooring, and for patching worn concrete, and to a limited extent for non-industrial floors. The mix can be pigmented to give dark colours and the hardness adjusted to suit different conditions of wear. The product, laid in thicknesses of $\frac{1}{2}$–$\frac{5}{8}$ inch, is softer than concrete flooring.

Mixtures of cement and rubber latex provide the basis of some jointless floor coverings and of chemically resistant jointings and linings. Originally only high-alumina cement was used, but later it was found possible to stabilise the rubber latex so that it did not coagulate prematurely on mixing with Portland cement. For jointless flooring, aggregates or fillers such as sawdust, cork or wood chips are incorporated to give resilient surfaces and materials such as crushed stone, marble chips or sand for harder surfaces. The mixes, which can be pigmented, are laid in thicknesses of $\frac{1}{4}$ inch upward. Cement-latex mixes adhere very strongly to most other surfaces. Following the development of various synthetic resin emulsions it was found possible to use these as alternatives to rubber latex.[1] Amongst the resin emulsions so used are polyvinyl acetate and chloride, and

[1] *R.I.L.E.M. Bull.* **37** (1967); *Materials and Structures*, Nos. 1 and 3 (1968); *R.I.L.E.M. Sym. on Experimental Research on New Developments brought by Synthetic Resins to Building Techniques, Paris 1967.*

neoprene latex the products from which are much more resistant to oils and greases than the natural rubber latex-cement mixes. Styrene-butadiene, epoxy and furfural emulsions and various others are also used.

Rubber latex-sand high-alumina cement mixes are used as jointing materials to withstand attack by chemically corrosive agents. They will withstand mineral acid solutions up to 5 per cent concentration, and dilute alkalies, at temperatures up to about 50°, for although the cement is attacked at the surface by acids to a depth of $\frac{1}{8}$–$\frac{1}{4}$ inch the rubber-sand mixture left swells and protects the underlying material. When used as jointing in floors they also have a good resistance to grease, again because the rubber swells and protects the material underneath. They are also resistant to sulphate solutions and dilute organic acids, but not to organic solvents which attack rubber, or to strong mineral acids. These cements have the valuable property that they are resilient and not brittle and can yield to small movements without breaking. As floor surfacings they withstand many mildly corrosive conditions, those based on high-alumina cement being more acid-resistant than those containing Portland cement. The same general rule applies to the synthetic resin emulsion mixes, those based on high-alumina cement being used in preference to the Portland cement mixes where resistance to corrosive agencies is important. The polyvinyl chloride and acetate mixes do not possess the same valuable flexibility as the rubber ones.

Cement injection processes[1]

Injection of cement grout under pressure is sometimes used for sealing contraction joints in mass concrete structures such as dams, for repairing cracks in concrete, and for tightening water-retaining structures. The grouts are commonly of neat cement with or without the addition of a fluidising agent, but sometimes finely ground aggregates such as sand, foamed slag or pulverised fuel ash are added. It is important that the grout should have good flow characteristics and be relatively free from bleeding. For fine cracks, 0·01 in. or less, all the cement should pass a 100 or even 200 mesh sieve. The water : cement ratio of the grout is varied according to the resistance to the passage of the grout and can range from 0·4 to 1·0 or more. Grouting also forms part of the process for producing post-stressed concrete. Channels are arranged in the concrete as cast through which steel tendons are subsequently passed, tensioned, and fixed temporarily by end wedges. The channels are then grouted with a neat cement grout or sometimes with a mixture of cement and fine aggregate.

Chemical consolidation processes are used for solidifying loose ground in engineering constructional work and can be applied to deposits of gravel and sands, but not to clays and silts. One process depends on the reaction between calcium chloride and sodium silicate solutions precipitating a solid gelatinous mass of hydrated calcium silicate and silica gel which form a cementing agent. Perforated pipes are driven into the formation to be consolidated and the two solutions injected in separate and successive operations so that they react *in situ*. A strength of several hundred pounds per square inch is obtained in the consolidated material.

[1] See Consultating Engineer, Oct. 1969.

Another process makes use of a mixed solution of sodium silicate and sodium bicarbonate solution which precipitates silica gel after an interval of half an hour or so, permitting the mixed solution to be injected in one process. The presence of ground water salts in sufficient concentration, or seawater, may interfere with this process.

A more recent process is based on the chrome-lignin reaction. Lignin is a by-product in the sulphite process of paper making and is obtained in the form of a soluble powder or a solution. When mixed with a dichromate salt, usually sodium dichromate, a gel is formed after a time that can be controlled by adjustment of the concentrations of the reagents, by the addition of certain salts such as ferric chloride, or by alteration of the pH value of the mixed solution with sulphuric acid. The mixed solution can therefore be injected in one operation and is claimed to penetrate finer grained materials, e.g. sand of about 100-mesh size, than can be treated by other processes. The chromium salt is toxic and cannot be used where there is any risk of contaminating sources of water supplies.

Other materials which have sometimes been used include sodium silicate-aluminium sulphate and, where only waterproofing was required, bituminous emulsions containing an agent that causes a delayed 'break' in the emulsion.

Colloidal concrete

Concrete can be made by first placing the aggregate in position and then grouting. In one form of this process coarse aggregate is laid and the voids then filled with a grout composed of cement and fine sand with sufficient water to form a fluid mixture. This is mixed in a special 'colloid' mixer which so disperses the cement that the solids remain in suspension for a much longer period than would be the case for grout mixed in an ordinary mixer so that it can be pumped into position without segregation. The method has been used to a limited extent for concrete paving.

Another process is somewhat similar, but the mortar grout consists of Portland cement, pozzolana, fine sand and an intrusion aid which helps to stabilise the suspension and which contains a small amount of aluminium powder causing the mortar to expand slightly. It has been particularly used for repair work on concrete structures and is known as Prepakt concrete.

Gunite

A cement mortar applied by projecting it from a pneumatic cement gun is known as gunite, or as shotcrete in the U.S.A. The sand and cement are normally mixed dry and receive the water at the discharge nozzle, though the use of wet mixing has been introduced as an alternative in shotcreting. This process is much used for the repair of concrete structures and also in new work such as concrete shell roofs and other thin reinforced concrete sections or for the mortar covering over the cables in circular prestressed concrete tanks.

Heavy concretes for X-ray and nuclear radiation shields

Heavy concretes and plasters are used for surrounding spaces in which X-ray equipment is used in order to prevent transmission of radiation outside and give protection. Barytes (barium sulphate) aggregates are used in mixes such as 1 : 2 : 3 cement : barytes sand (minus 14 mesh down to 52 mesh) : coarse barytes ($\frac{3}{16}$–$\frac{3}{4}$ inch) by volume, or 1 : 1 : 1 : 3 cement : barytes fines (minus 100 mesh) : barytes sand : coarse barytes, and give concretes with weights of about 210 lb/ft^3 (d = 3·4). Strengths of about 2000 lb/in^2 at seven days are obtained. Denser concretes with weights up to 225 lb/ft^3 (d = 3·6) and considerably higher strengths are also obtainable.[1] For plastering, mixes of 1 : 1 : 3 by volume cement : fines : sand are used and give a product with a dry weight of about 198 lb/ft^3 (d = 3·2). Lead sulphide is sometimes present as an impurity in barytes, but amounts up to 0·25 per cent do not delay the set. Lead oxide or lead compounds soluble in lime water can affect the set in amounts as low as 0·01 per cent, but these are not usually present. Heavy concretes have also attracted much attention for the construction of biological shields for nuclear reactors. The absorption of γ rays by concrete is closely related to its density so that the thickness of the shield required can be reduced by the use of a heavy concrete. With ordinary aggregates a concrete has a unit weight of about 150 lb/ft^3 (d = 2·4). Concrete with aggregates of barytes ($BaSO_4$), magnetite (Fe_3O_4), limonite ($2Fe_2O_3.3H_2O$) and ilmenite ($FeTiO_3$) can give unit weights of about 225 lb/ft^3 (d = 3·6) while with iron punchings or steel shot a value well over 300 lb/ft^3 (d = 4·8) is obtainable.[2]

Non-calcareous cements

Cements can be obtained in which lime is replaced by the next two elements, strontium and barium, in Group II of the Periodic Classification of the elements, and similarly silica by germanium. The compounds which magnesium, the element preceding lime in Group II, forms with silica and alumina are inert and have no cementing value and the same applies to the compounds of lime and titanium, the element beyond silica in Group IV. Germanium in contrast forms a compound, calcium orthogermanate, with cementing properties[3] similar to calcium orthosilicate. A tricalcium germanate also exists. Strontium forms the silicates[4] $3SrO.SiO_2$, $2SrO.SiO_2$ and $SrO.SiO_2$. The pure compound $3SrO.SiO_2$ inverts on slow cooling to another polymorph which has only slight cementing properties in water, but when stabilised in the high-temperature form by small amounts (e.g. 2 per cent) of alumina and ferric oxide or magnesium oxide it has good cementing properties. The orthosilicate is reported by Braniski[5] to have cementing properties, but not the monosilicate. Various aluminates exist, $4SrO.Al_2O_3$, $3SrO.Al_2O_3$, $SrO.Al_2O_3$ and $SrO.6Al_2O_3$. The formation of $SrO.2Al_2O_3$ has been both reported and denied and the formula $5SrO.Al_2O_3$ instead of $4SrO.Al_2O_3$ has been suggested for the most basic compound.[6] The

[1] L. P. Witte and J. E. Backstrom, *Proc. Am. Concr. Inst.* **51,** 65 (1955); K. A. S. Halstead. V. E. Vaughan and E. L. Cameron, *Can. J. Technol.* **30,** 334 (1952).
[2] H. S. Davis, *Mat. Res. Stand.* **7,** 494 (1967).
[3] V. F. Zhurovlev, *Dokl. Akad. Nauk SSSR* **50,** 1145 (1948).
[4] *Phase Diagrams for Ceramists*, Fig. 296, American Ceramic Society (1964); R. W. Nurse, *J. appl. Chem.* **2,** 244 (1952); A. Braniski, *Zement-Kalk-Gips* **10,** 398 (1957).
[5] A. Braniski, *Zement-Kalk-Gips* **14,** 17 (1961).
[6] F. Massazza, *Chimica. Ind. Milano* **41,** 114 (1959); P. S. Dear, *Bull. Virginia Polytech. Inst.* **50,** No. 6 (1957).

tri- and mono-strontium aluminates have cementing properties but no information is available in the tetra-compound. The compound $SrO.2Al_2O_3$ was also found by Braniski to be cementitious though less so than the mono-aluminate. Barium is shown in the available phase equilibrium[1] diagram to form $2BaO.SiO_2$, $BaO.SiO_2$ and two less basic silicates but there is evidence[2] of the existence of $3BaO.SiO_2$ stable between 740° and 1800°. Both the tri- and the di-barium silicates have cementing properties but, in contrast to the calcium compounds, the dibarium silicate is the more important cementitious agent. The less basic silicates have no cementing properties. Solid solutions are formed between the orthosilicates of calcium, strontium and barium. There are three aluminates of barium,[3] $3BaO.Al_2O_3$, $BaO.Al_2O_3$ and $BaO.6Al_2O_3$ of which the first two have cementing properties though the reaction of the tri-barium compound is said to be violent. A claim[4] has also been made for the existence of a compound $2BaO.Al_2O_3$. Compounds analogous to $4CaO.Al_2O_3.Fe_2O_3$ with Sr or Ba substituted for Ca also exist.[5]

The hydration products of the strontium and barium aluminates and silicates bear some analogies to those of the calcium compounds. Hydrated strontium silicates of 3 : 1, 2 : 1, 3 : 2, 1 : 1 and 1 : 2 SrO : SiO_2 ratio have been prepared under hydrothermal conditions[6] and a hydrated monobarium silicate has been prepared.[7] Hydration of $2BaO.SiO_2$, after initial formation of a direct hydrate, produces $BaO.SiO_2.6H_2O$[8] at room temperatures and in the case of the strontium compounds it has been indicated that di- and mono-strontium silicate hydrates are formed.[9] Strontium forms an aluminate hydrate $3SrO.Al_2O_3.6H_2O$ of cubic symmetry like $3CaO.Al_2O_3.6H_2O$ and at 1° C a monohydrate $SrO.Al_2O_3.10H_2O$ which decomposes at room temperature[10] to form the cubic hydrate.

There are a number of barium aluminate hydrates[11], $BaO.Al_2O_3.aq$ existing in different degrees of hydration, $2BaO.Al_2O_3.5H_2O$, and $7BaO.6Al_2O_3.36H_2O$. The di-barium compound seems to be stable in contact with a saturated, or nearly saturated, barium hydroxide solution, but the other compounds form only as metastable phases in the system $BaO–Al_2O_3–H_2O$ at 30° and are eventually hydrolysed into $Ba(OH)_2$ and hydrated alumina. A compound $3BaO.Al_2O_3.7H_2O$ has been reported[12] but its existence needs confirmation. The hydrated barium aluminates, like barium hydroxide, have a considerably higher solubility in water than the corresponding Ca and Sr compounds. Cements composed essentially of barium aluminate without any silicate compounds harden in air

[1] *Phase Diagram for Ceramists*, Fig. 210, American Ceramic Society (1964).
[2] A. Braniski, *Zement-Kalk-Gips* **14,** 17 (1961); **21,** 91 (1968).
[3] N. A. Turopov and F. Ya. Galakov, *Dokl. Akad. Nauk SSSR* **82,** 69 (1952).
[4] E. Calvet, H. Thibon and J. Dozoul, *Bull. Soc. chim. Fr.*, 1964 (8), 1915.
[5] A. Braniski, *Revue Matér. Constr. Trav. publ.* **560,** 142 (1962).
[6] E. T. Carlson and L. S. Wells, *J. Res. natn. Bur. Stand.* **51,** 73 (1953).
[7] H. Funk, *Z. anorg. allg. Chem.* **296,** 46 (1958).
[8] H. Uchikawa and K. Tsukiyama, *Sym., Tokyo 1968.*
[9] A. Braniski, loc. cit.; *Sym., Washington 1960*, 1075.
[10] E. T. Carlson, *J. Res. natn. Bur. Stand.* **54,** 329 (1955); **59,** 107 (1957); G. Maekawa, *J. Soc. chem. Ind. Japan* **40,** 751 (1943).
[11] E. T. Carlson and L. S. Wells, *J. Res. natn. Bur. Stand.* **41,** 103 (1948); **45,** 381 (1950); E. Calvet, H. Thibon and J. Dozoul, loc. cit.; H. Uchikawa and K. Tsukiyama, loc. cit.
[12] G. Maekawa, *J. Soc. chem. Ind. Japan* **45,** 236 (1942).

but are not resistant to water. They have however excellent refractory properties. Portland and Portland blastfurnace type cements with the lime partially or completely replaced by strontium or barium can be made.[3] The barium cements of these siliceous types have a high resistence to sea water and sulphate solutions on account of the insoluble nature of the reaction product, barium sulphate.

The main interest in the barium and strontium cements lies in the highly refractory concretes[4] that can be made from the pure high-alumina type cements analogous to the white calcium aluminate refractory cement. These cements, based respectively on $BaO.Al_2O_3$, $SrO.Al_2O_3$ and $SrO.2Al_2O_3$, and $CaO.Al_2O_3$ and $CaO.2Al_2O_3$, are substantially pure materials containing only one or two per cent of silica, magnesia and ferric oxide. The melting points of the various compounds are

	Melting point (°*C*)
$CaO.Al_2O_3$	1605 (*a*)
$CaO.2Al_2O_3$	1789 (*b*)
$SrO.Al_2O_3$	1790
$SrO.2Al_2O_3$	c.1800
$BaO.Al_2O_3$	1830

(*a*) Incongruent to form CA_2 and liquid.
(*b*) Incongruent to give CA_6 and liquid.

On economic grounds the barium cements offer more attraction than the strontium cements. With a very refractory aggregate, such as deadburnt magnesite, the refractoriness of the barium aluminate concrete may be raised by 50–100° C above that of the calcium aluminate concretes.

Crazing

The crazing of concretes and cement products is a well-known defect to which such materials are subject. It consists in the development of a network of fine hair cracks over the surface. These cracks when first formed may not be visible when the concrete is dry, but they show on wetting. At a later stage the crazing becomes visible in the dry condition as dirt collects along, and outlines, the cracks. For a long period crazing was considered to be due to the shrinkage which occurs when a concrete dries; it was attributed to the use of too wet mixes, to excessive trowelling bringing the cement to the surface and forming an enriched skin having a higher shrinkage than the body of the material, to too rapid drying, and to other causes. That all such causes can lead to crazing in a concrete product is unquestionable, but nevertheless there is another important cause, carbonation. Crazing may arise in all kinds of cement products, but is most serious in the case of materials such as cast stone and similar precast concrete products in which the decorative appearance of the surface is of prime importance. Crazing in its typical form is limited to the surface of a material and does not penetrate far into the

[1] A. Braniski, *Zement-Kalk-Gips* **10,** 176 (1957); **14,** 17 (1961); **22,** 513, (1969); *Revue Matér. Constr. Trav. publ.* **560,** 142 (1962).
[2] A. Braniski, *Sym., Washington 1960,* 1075; *Zement-Kalk-Gips* **20,** 96 (1967).

mass. It can be distinguished from the major shrinkage cracks penetrating the entire mass which occur in concretes the ends of which are restrained. Such cracks are due to the drying shrinkage. In intermediate cases the line of division between crazing and major shrinkage cracking may be difficult to draw, but in general we can regard crazing as a surface defect.

The conclusion that crazing is due solely to the normal drying shrinkage has never been supported by much direct experimental evidence, and was mainly a deduction from the known overall initial shrinkage of cement products. Certain of the established facts relating to crazing have, however, been difficult to reconcile with the drying shrinkage theory. In many cases crazing does not develop during the first drying of the material, when the concrete is relatively weak, and the maximum drying movement occurs. Frequently the trouble does not start for a year or more, and not until long after the initial drying has occurred. Observation of sections cut out of concretes has shown that the depth of the crazing coincides with the depth of the carbonated surface layer,[1] thus strongly suggesting that crazing is in some way connected with the process of atmospheric carbonation. The depth of carbonation can be easily observed on a freshly cut surface by treating with phenolphthalein, when the carbonated skin remains uncoloured while the inner portion, in which free calcium hydroxide still remains, is coloured pink. A more satisfactory and less fugitive reagent for this purpose is Naphthol Green B, which does not colour the carbonated skin, but stains the inner portion green. Sections of a concrete (a cast stone) are shown in Plate XVI (ii), which depicts the two halves of a block, the lower half treated with phenolphthalein and the upper with Naphthol Green B. The skin which has been formed by atmospheric carbonation of the outer layers is very clearly shown.

If crazing, which is clearly due to a shrinkage of the surface skin with respect to the underlying mass, is to be attributed to carbonation then thus must be accompanied by shrinkage. That this was so was first shown by Brady[2] who obtained the results given in Table 119 for the shrinkage of 1 : 3 mortar prisms

TABLE 119 Per cent contraction of cement mortars stored in air and carbon dioxide
(*after Brady*)

Time	Portland cement		High-alumina cement	
	In air	In CO_2	In air	In CO_2
1 day	−0·0022	−0·037	−0·0007	−0·011
4 days	−0·0050	−0·060	−0·0018	−0·025
8 days	−0·0064	−0·065	−0·0022	−0·034

¼ in. thick. These prisms, after maturing for seven days in moist air, were partially dried in an atmosphere of 60 per cent relative humidity and then stored in air or carbon dioxide at the same humidity. It was evident from this data that

[1] *Annual Report, Building Research Board*, 1931, p. 39.
[2] F. L. Brady, *Cement* **4**, 1105 (1931): *Annual Report, Building Research Board*, 1932, 34; 1933, 33; 1934, 38.

after the initial drying shrinkage of a concrete is complete a further shrinkage of the skin of the material can take place as a result of the action of atmospheric carbon dioxide. Since all cement products exposed to air eventually form a carbonated skin, but all do not craze, the conditions under which carbonation occurs must also be a determining factor. The rate at which atmospheric carbonation occurs varies with the moisture content of the material and the humidity of the atmosphere. Carbonation at high and low relative humidities produces less shrinkage than when it occurs at humidities between about 30 and 80 per cent. Calcium carbonate also can exist in three polymorphic forms, calcite, vaterite and aragonite. Though calcite is the stable form, the initial product of the atmospheric carbonation of set cement and of calcium hydroxide can be vaterite or aragonite which later transforms to calcite.[1] The texture and physical nature of the surface carbonated layer and its proneness to crazing may possibly be influenced by such differences in the carbonation process. The very variable behaviour of concretes in respect of crazing must be attributed to the multiplicity of factors involved.

Carbonation shrinkage

The discovery as related above, of carbonation shrinkage, has an obvious bearing on drying shrinkage measurements generally. Where these are made on thin specimens the results must be influenced to a varying and uncertain degree by the extent of carbonation. In considering the behaviour of mortar renderings and other thin coatings it may not be undesirable that this should be included, since it must occur in practice. Carbonation in large dense concrete members will, however, be limited to surface layers and the shrinkage of a heavily carbonated specimen will not be a true indication of the behaviour of the larger members. More critical will be the influence of carbonation on the shrinkage of porous concrete blocks and their susceptibility to shrinkage cracking in practice.

The amount and rate of carbonation shrinkage is influenced by a number of factors, such as the dimensions and porosity of the specimen and the carbon dioxide concentration. The depth of carbonation for a given mortar and exposure conditions seems to increase in proportion to the square root of the time.[2] The factor of major importance however is the moisture content. This in turn is determined by the relative humidity of the atmosphere and the extent to which the specimen has attained equilibrium with it when carbonation occurs. The form of shrinkage curve when drying and carbonation of the specimen occur comtemporaneously differs therefore from that obtained when they occur in succession. Maximum carbonation shrinkage in specimens previously brought into moisture equilibrium occurs at about 50 per cent relative humidity. Some carbonation, but no carbonation shrinkage, occurs at 100 per cent or at 25 per cent or lower relative humidity. These effects are illustrated in Fig. 145 for 1 : 4·08 cement-sand mortars (w/c 0·54) moist cured for seven days.[3] The specimens were subsequently stored either in CO_2-free air at various relative humidities

[1] F. Schroeder, *Tonind. Ztg. keram. Ddsh.* **86,** 254 (1962); W. F. Cole and B. Kroone, *Proc. Am. Concr. Inst.* **56,** 497 (1960).
[2] R. Kondo, M. Daimon and T. Akiba, *Sym., Tokyo 1968.*
[3] G. J. Verbeck, *Spec. tech. Publ. Am. Soc. Test. Mater.*, No. 205, 17 (1958).

until drying shrinkage was complete and then in a 100 per cent carbon dioxide atmosphere at the same relative humidity, or straightway placed in a carbon dioxide atmosphere at the various relative humidities so that carbonation and drying were proceeding simultaneously. In this second condition, much of the carbonation must occur at humidities within the specimens which are higher than that of the controlled surrounding atmosphere because of the lag in establishment of moisture equilibrium.

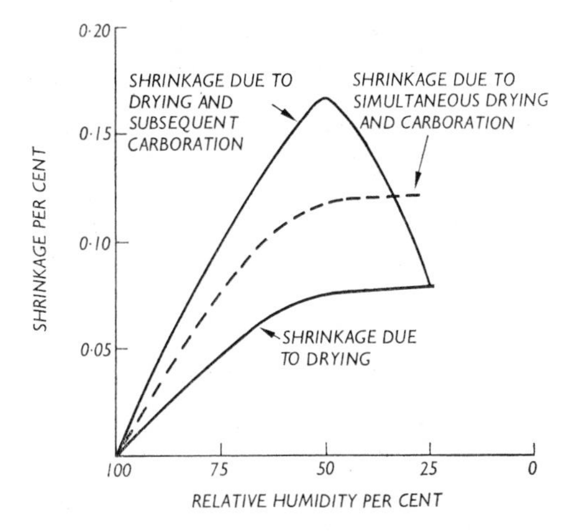

FIG. 145 Effect of sequence of drying and carbonation on shrinkage.

The relation between the amounts of carbonation shrinkage and drying shrinkage varies much with the porosity of the concrete, the size of the test specimen, the initial curing conditions and the concentration of carbon dioxide in the atmosphere. For exposure to 100 per cent carbon dioxide the maximum carbonation shrinkage of the mortars shown in Fig. 145 is rather greater than the drying shrinkage while for very porous mortars it can be more than twice as great.[1] For exposure to normal air (0·03 per cent CO_2) Shideler[2] found that the carbonation shrinkage of lean concrete blocks after one year was about one-half of the drying shrinkage, or a third of the total shrinkage, while for similar blocks cured in high pressure steam it was lower. Carbonation reduces the reversible moisture movement of concretes cured at normal temperatures or in steam at atmospheric pressure, but increases it for autoclaved concretes or sand-lime bricks. The volume stability of concrete blocks in service should be improved if the carbonation shrinkage is completed before the blocks are used. This can be achieved by pre-carbonating the blocks in waste flue gases. Such a process was tried as a remedy for surface crazing by block manufacturers in England during the 1930's, and in recent years it has been explored in the U.S.A. as a means of reducing shrinkage in service. There are many variables such as the type of aggregate, the curing conditions, and the time and temperature of carbonation which influence

[1] K. Kanimura, P. J. Sereda and E. G. Swenson, *Mag. Concr. Res.* **17** (50), 5 (1965).
[2] J. J. Shideler, *J. Res. Dev. Labs Portld Cem. Ass.* **5** (3), 36 (1963).

the results so obtained, but under the most favourable conditions the potential shrinkage may be reduced by 30 per cent or more.[1]

The mechanism of carbonation shrinkage is still not settled. The conversion of calcium hydroxide to calcium carbonate is accompanied by an increase of 11 per cent in solid volume if calcite is formed and 3 per cent if aragonite is formed. Clearly the change in solid volume in the reaction

$$Ca(OH)_2 + CO_2 \rightarrow CaCO_3 + H_2O$$

cannot account for shrinkage, but nevertheless it has been found[2] that compacts of calcium hydroxide show carbonation shrinkage curves similar to those of set Portland cement though of somewhat lesser magnitude. Carbon dioxide can also attack and decompose all the hydration products of cement with the formation of silica and alumina gels. Carbonation shrinkage occurs with hydrated cements containing no calcium hydroxide such as high-alumina cement and autoclaved Portland cements containing added silica. Though the initial reaction with set Portland cement may be represented by the above equation, the hydrated silicates and aluminates must also become involved as carbonation proceeds. The calcium hydroxide reaction liberates 1 mol H_2O per mol CO_2 combined and in the early stage of carbonation this is the ratio found. As carbonation proceeds further however with Portland cement, or from the start with autoclaved pastes containing no calcium hydroxide, the amount of water released per mol carbon dioxide combined drops much below unity. Carbonation shrinkage can probably best be explained by the fall in non-evaporable water content of the cement gel as carbonation proceeds[3] since, even though the gel changes in chemical composition, such a loss should cause shrinkage. It has been suggested by Swenson and Sereda that carbonation promotes the polymerisation and dehydration of the silicate ions. An alternative explanation advanced by Powers[4] attributed the shrinkage to the solution of calcium hydroxide crystals that are under pressure, and the formation of calcium carbonate in a stress-free condition. This hypothesis is at variance with other evidence and in any event can provide no general explanation of carbonation shrinkage. What is clear is that while carbonation of calcium hydroxide contributes to the shrinkage of set cement it is not the major cause.

INFLUENCE OF CARBONATION ON STRENGTH OF CEMENT PRODUCTS

Although the atmospheric carbonation of cement products leads to a decomposition of the hydrated cement compounds, the strength is considerably increased. It is evident that the calcium carbonate formed is in itself an excellent cementing medium, as is indeed apparent from the hardening of fat lime mortars as they carbonate.

[1] H. T. Toennies and J. J. Shideler, *Proc. Am. Concr. Inst.* **60,** 617 (1963); J. J. Shideler, loc. cit.
[2] F. G. Swenson and P. J. Sereda, *J. appl. Chem.* **18,** 111 (1968).
[3] C. M. Hunt and L. A. Tomes, *J. Res. natn. Bur. Stand.* **66A,** 473 (1962); K. M. Alexander and J. Wardlaw, *Aust. J. appl. Sci.* **10,** 470 (1959).
[4] T. C. Powers, *J. Res. Dev. Labs Portld Cem. Ass.* **4** (2), 40 (1962).

The rate of carbonation and increase in strength of 1 : 6 lime-sand and 1 : 3 Portland cement-sand mortars (proportions by weight) is illustrated by the data in Table 120[1] on the tensile strength of briquettes stored in air free from carbon dioxide, and in carbon dioxide respectively, for 25 days, following an initial 3-day period in moist air.

TABLE 120 Carbonation of mortars

	Relative humidity of atmosphere (per cent)	Lime mortar		Cement mortar	
		Air	CO_2	Air	CO_2
Tensile strength (lb/in^2)	Saturated	48	56	576	579
	80	—	154	544	679
	50	19	167	522	740
Calcium carbonate content expressed in terms of CaO (per cent)	Saturated	0·3	0·6	0·5	1·9
	80	0·7	3·8	—	2·7
	50	0·5	3·1	0·5	4·0
Free calcium hydroxide content expressed in terms of CaO (per cent)	Saturated	6·3	5·7	2·5	1·4
	80	6·4	3·0	—	0·8
	50	7·3	4·1	2·6	0·5

The compressive and flexural strengths of Portland cement concrete stored in carbon dioxide may be raised by 30 to 100 per cent above that of comparison specimens stored in carbon dioxide-free air.[2] The compressive strength of lightweight concrete blocks made with lean mixes on the other hand showed only a variable and small increase in strength by curing in carbon dioxide in tests reported by Shideler.[3] The strength of Portland-slag cements[4] is increased by carbonation for cements with slag contents of less than 60 per cent but decreased with cements of higher slag content.

It is evident from the data in Table 120 that in any long-age tests on air-stored mortar or concrete specimens of small dimensions, the results are likely to be influenced considerably by atmospheric carbonation. With specimens of larger dimensions, such as cubes or cylinders, the carbonation of dense concretes is not likely to proceed beyond a surface layer, up to about ¼ inch thick, and the main body of the material remains unaffected.

Soluble alkalis in Portland cement

Though the total content of soda and potash in Portland cement is small, and that part which is easily soluble in water still smaller, they are jointly responsible

[1] G. E. Bessey, *J. Soc. chem. Ind.* **52,** 287T (1933).
[2] I. Leber and F. A. Blakey, *Proc. Am. Concr. Inst.* **53**, 295 (1957); G. W. Washa and R. L. Fedell, ibid. **61,** 1109 (1964).
[3] J. J. Shideler, loc. cit.
[4] W. Manns and K. Wesche, *Sym., Tokyo 1968.*

for a number of troubles in building. Efflorescences on brickwork or masonry are usually composed of alkali or calcium salts; whilst in the case of brickwork these may be derived from the brick, a cement mortar also contributes part of its soluble alkali content to them. The staining of limestones set in cement mortar, and the failure of paint films on concrete, have also been traced to the content of water-soluble alkalis present in the cement. Further, as we shall discuss in the next chapter, a reaction can occur between the alkalis and some types of aggregates that can cause serious expansion of concrete.

Portland cements differ appreciably in their total alkali content, and more considerably in the proportion readily soluble in water. The alkalis are introduced into the cement from both the raw materials and the coal ash. A proportion of the alkalis is volatilised in the kiln and carried away in the dust, the amount so removed increasing with harder burning. As we have seen earlier, potassium and sodium are present in cement clinker as a solid solution of the sulphates which is readily soluble in water, but the amount of sulphur trioxide available is insufficient to combine with more than a part of the alkalis. The remaining soda and potash is combined in the calcium aluminate and silicate compounds from which it is released into solution more slowly.

On shaking a cement with water, it is found that considerable solution of alkalis takes place quite rapidly and that the amount only increases slowly on shaking for a prolonged period. In Table 121 there are shown the total alkali content and the amounts dissolved on shaking 15 g cement with 150 ml water for 48 hours at ordinary temperatures.

TABLE 121 Proportion of alkalis in Portland cement dissolved in 48 hours

Cement No.	1	2	3	4	5	6	7
Total Na_2O	0·17	0·40	0·21	0·36	0·36	0·13	0·34
Total K_2O	0·40	0·66	0·36	0·60	0·82	0·83	0·46
Soluble Na_2O	0·07	0·08	0·05	0·07	0·09	0·09	0·10
Soluble K_2O	0·08	0·28	0·15	0·29	0·35	0·48	0·43
Total Na_2O+K_2O	0·57	1·06	0·57	0·96	1·18	0·96	0·80
Soluble Na_2O+K_2O (as percentage of total alkalis)	26	34	35	38	38	60	66

It is observed that both the amount of alkalis dissolved in 48 hours and the proportion it bears to the total varies considerably with different cements. For short periods of extraction up to two hours it is generally found that a higher proportion of the potash than of the soda is dissolved. This reflects the tendency for the sulphate solid solution to contain more potash than soda. When the sulphur trioxide content of the clinker is relatively high, so that a substantial proportion of the alkalis are present in this form, up to one-half the K_2O and one-quarter of the Na_2O can pass into solution in a few minutes.[1] Conversely

[1] G. L. Kalousek, C. H. Jumper and J. J. Tregoning, *Rock Prod.* **44** (4), 52 (1941); W. J. McCoy and U. L. Eshenour, *Sym., Tokyo 1968*; *J. Materials* **3** (3), 684 (1968).

with low SO_3 contents the proportion may be only 10 per cent or less. The data in Table 121 show that after 48 hours there is no consistent relation between the proportion of the potash and soda dissolved and this may be taken to indicate that at this time the soluble alkalis are also being derived from the silicate and aluminate compounds. After 28 days most, and often all, of the alkalis in the cement have passed into solution.[1] A procedure for determining alkalis readily soluble in water is given in ASTM C114–65 in which from 25 to 150 g of cement, according to the alkali content, is treated for 30-minute periods successively with 250 ml, 150 ml and 100 ml water and the extracted alkali estimated. The soluble alkalis by this method on different cements range from a few per cent up to 70 per cent of the total content.

The total alkali content of Portland cements varies in the case of British cements from about 0·1 to 0·5 per cent Na_2O, and 0·2 to above 1 per cent K_2O, there being a general tendency for the K_2O to be low when the Na_2O is high. Contents of the same order are found in cements in other countries.

Staining of limestone and production of efflorescences

Limestone masonry often develops disfiguring brownish or yellowish stains which appear soon after the stone is set in the building and which, in sheltered positions, may persist for years. This stain is to be distinguished from the black soot deposit which slowly forms in polluted atmospheres. It has been shown[2] to be due to the action of alkali compounds contained in the mortar in which the stone is set. Most limestones contain a certain amount of organic matter which is soluble in alkali hydroxide, carbonate, and other alkali salt solutions. The alkali compounds in a cement mortar liberate alkali hydroxides which are conveyed by water into the stone, and there dissolve the organic matter and carry it to the surface; on evaporation it is deposited, forming a brown discoloration. The staining produced in this way on limestones set in cement mortar is fairly well correlated with the contents of soluble alkalis in the cement used. Hydrated high-calcium limes which contain practically no alkalis produce no staining, whilst grey Portland cements produce an amount varying with the soluble alkali content. White Portland cement and high-alumina cement, which usually have relatively low soluble alkali contents, produce less staining and sometimes none. It has similarly been found that the degree of efflorescence[3] formed on limestones, or on bricks which themselves are low in soluble salts, is related to the soluble alkali content of the mortar in which they are set.

Cements may be tested in a simple manner for their staining power on limestones, and for the production of efflorescence. A 1 : 5 cement-coarse sand mortar is made into a 2-inch cube and dried when 24 hours old, so as to obtain a permeable mortar. It is placed with its base in water and on the top of it is put a cube of limestone with a wetted filter paper or blotting-paper in between, to

[1] F. E. Jones, *Natn. Bldg Stud. Res. Pap.*, No. 15, 1952; J. L. Gilliland and T. R. Bartley, *Proc. Am Concr. Inst.* **47,** 153 (1951).
[2] F. O. Anderegg, H. C. Peffer, P. R. Judy and L. Huber, *Purdue Univ. Publs Engng Dep.* **12** (6) (1928).
[3] A discussion on efflorescences is given by R. J. Schaffer, *Building Research Special Report*, No. 18 (1949). H.M.S.O. London.

ensure capillary contact. The water passes up from the base of the mortar cube and carries any soluble products extracted from the mortar into the limestone. Evaporation from the sides of the limestone cube leaves a stain and an efflorescence by the intensity of which the cement may be judged. The efflorescence is more easily seen if a cube of red brick, itself free from soluble salts, is substituted for the limestone.

An efflorescence of magnesium sulphate on mortars can be formed by the following, somewhat unexpected, reaction:[1]

$$CaSO_4.2H_2O+MgCO_3.3H_2O+2H_2O \rightleftharpoons CaCO_3+MgSO_4.7H_2O$$

Magnesium carbonate can be formed by carbonation of magnesium oxide or hydroxide and is stable in water in contact with air containing not less than 0·04 per cent CO_2. The reaction of carbon dioxide and water on calcium sulphoaluminate also decomposes this compound, making calcium sulphate available for the reaction with magnesium carbonate. Such a reaction has been found to occur in masonry cements containing a magnesian or dolomitic lime, but is much less likely in Portland cement, or mixes of cement with high-calcium limes, in which the content of MgO is low. This reaction in a sense is a reversal, produced by carbon dioxide, of the reaction of magnesium sulphate solutions on Portland cement in which magnesium hydroxide and calcium sulphate are ultimately produced with intermediate formation of calcium sulphoaluminate (see p. 346).

Effect of cement on paint films

It is well known that concretes, mortars, and other cement products cannot be successfully painted with paints based on drying oils until they are well matured and dried. While the cause of this may be purely physical, and due to the hydrostatic pressure of water contained in the material distorting and blistering the paint film, the chemical action of the cement is probably the most fruitful source of failure. The view has long been prevalent that the free lime in set Portland cement causes the chemical destruction of the paint film, but this is only a half-truth.[2] Free calcium hydroxide, like any other caustic alkali, saponifies the linseed oil base of paints, forming a calcium soap. This calcium soap, is, however, insoluble and a layer of it appears to be formed where it comes in contact with the oil film; this largely protects the remainder of the film from further action. Drying oils such as linseed oil are readily attacked by solutions of caustic soda and potash, and the soaps formed are soluble in water. Any alkali salt in the presence of calcium hydroxide and moisture forms alkali hydroxide which attacks and quickly destroys any film containing linseed oil with which it comes in contact. This is the cause of the stickiness and softening of linseed oil paints on cements. Calcium hydroxide alone has no such destructive action, and alkali salts alone, since they cannot produce saponification, are without action. Though the amounts of water-soluble alkalis in Portland cement are low, they are sufficient to act in the above way. The alkalis are liberated from the cement as hydroxides and will remain as such as long as there is available uncarbonated lime. A specially prepared alkali-

[1] D. L. Bishop, *J. Res. natn. Bur. Stand.* **30**, 361 (1943).
[2] H. M. Llewellyn, *Building Research Bulletin*, No. 11 (1934). H.M.S.O. London.

free Portland cement has been shown to have no effect, even when wet, on a paint. Though white Portland cements have lower soluble alkali contents than ordinary cements, the content is usually sufficient to cause trouble.

Various priming solutions, or 'petrifying' liquids, have been recommended to neutralise the lime in set cement before painting. Solutions of zinc sulphate or magnesium silico-fluoride have been used for this purpose, but they afford no effective substitute for thorough drying of the concrete, though they may be useful as final safeguards on surfaces which are only faintly alkaline. Tung oil is considerably more resistant to saponification on concrete surfaces than linseed oil and a priming coat of a paint with a tung oil-resin basis affords an insulating film on which a linseed oil or alkyd gloss paint can be applied.

Chlorinated rubber paints, some emulsion paints and cement base paints are not affected in this way.[1]

Chlorinated rubber paints are bound by a non-saponifiable resin and are suitable for use on alkaline surfaces provided that no large amounts of water remain. They can be either glossy and impervious, or matt and permeable to enable moisture to escape. Various other 'exterior masonry paints', based on hydrocarbon polymers or non-saponifiable media, are also available.

Emulsion (latex) paints have in recent years become widely used on plaster, renderings and concrete. The early ones, based on polyvinyl acetate (PVA) with simple ester plasticisers such as dibutyl phthalate, were not as resistant to alkali attack on fresh cement surfaces as was claimed, and suffered loss of colour and efflorescence (calcium carbonate) or loss of adhesion. Copolymers, internally plasticised by the presence of various acrylate and other monomers are somewhat better; PVA/caprate copolymers are highly resistant to alkali, as are PVA/versatate[2] polymers and some wholly acrylic polymers (without PVA). The latter two are useful in exterior paints for renderings and concrete.

Thick-textured paints, containing sand, mica or asbestos, and based on either emulsions or solutions of various alkali-resistant polymers, are also extensively used on external renderings, and a further development is the use of emulsions of the above types as binders for fairly coarse granules of coloured minerals or glasses.

Cement paints

The term cement paint is applied to paints based on Portland cement with the addition of pigments, fillers, accelerators, and water-repellent substances. These paints, which have largely superseded the former simple cement washes, are supplied as a dry powder and mixed with water before use. For the lighter colour paints a white Portland cement is used, but for darker colours ordinary cement can be substituted as base. In general the lighter colours are to be preferred, and are practically essential for external use, because the colour is much less conspicuously affected by the formation of sporadic 'fade' films of calcium carbonate over the surface. Titanium oxide and zinc sulphide are used to improve the opacity of the white or lighter-coloured paints. Other pigments must be of the lime-fast

[1] *B.R.S. Digest* (Second Series), No. 21, revised 1968; Nos. 55, 56 and 57, 1965. Report A.M.C.I. Committee 616, *Proc. Am. Concr. Inst.* **53,** 817 (1957).
[2] 'Versatic acid' is a commercial term applied to branch-chain C9-C11 fatty acids.

type and conform to BS 1014 : 1961. Hydrated lime and silicious fillers are also incorporated. Calcium or aluminium stearate, and similar materials, are added to improve the waterproofing qualities of exterior paints and an accelerator, usually calcium chloride, to ensure that the paint will set before it dries out.

There is no British Standard for cement paints, but a U.S. Federal Specification (TT–P–21) lays down the following requirements:

Component	Per cent by weight			
	Type 1		Type 2	
	Max	Min	Max	Min
Portland cement	—	65	—	80
Hydrated lime	25	—	10	—
Water repellents (calcium or aluminium stearate)	1	0·5	1	0·5
Hygroscopic salts (calcium or sodium chloride)	5	3	5	3
Titanium oxide and/or zinc sulphide	5	3	5	3
Carbonates as CO_2	Not more than 3 per cent			
Silicious aggregate	Addition of 15–40 lb per 60 lb of paint of sand between No. 20 and No. 100 meshes			

The pigments are also required to be lime-fast and the hydrated lime must not contain more than 8 per cent of unhydrated CaO and MgO.

Additions like casein or polyvinyl acetate emulsion are sometimes made to confer a greater flexibility on the paint coating. When casein, or similar adhesives, are incorporated a bactericidal or fungicidal agent is also added and the hydroscopic salts normally used as accelerators may be omitted. The addition of organic binders like casein is not permitted in the U.S. Federal Specification, but paints containing casein have been used with success in Britain. Such an addition may have disadvantages under moist conditions, and, owing to their increased drying shrinkage, paints containing casein cannot be used on surfaces which are friable or not strong enough to restrain this movement.

Cement paints[1] are mostly used for application to concrete and concrete products, cement and cement-lime renderings, and brickwork, but not as a rule to gypsum plasters or over old paintwork of other types. The background material requires to be well wetted to stop the paint drying before it has had time to hydrate. A hard matt surface of good durability is obtained which gradually 'chalks' or dusts over a period of years. Cement paints are used externally to prevent rain penetration, as well as for decorative purposes, and can retain some waterproofing qualities for as long as ten years on vertical surfaces. They cannot be expected to stop leaks against any appreciable external water pressure. Old surfaces can be repainted without removing the old paint but must be free from dust especially if emulsion paints are used for recoating.

[1] G. W. Mack, *Building Research Congress, Div. 2*, 132 (1951); Report A.C.I. Committee 616. *Proc. Am. Concr. Inst.* **46,** 1 (1950).

Corrosion of steel in concrete

Under most conditions well-made concrete provides good protection against corrosion of reinforcing steel but, nevertheless, corrosion remains the most common cause of deterioration of reinforced concrete. Various examples and causes of this are discussed elsewhere and our main purpose here is to consider the mechanism of corrosion.

The corrosion of steel in concrete has many similarities with corrosion under other circumstances. The presence of an electrolyte (the aqueous phase in concrete) and access of oxygen are required. The final result of the corrosion process is the formation of a thick rust layer which exerts sufficient tensile forces within the concrete to crack and cause spalling of the concrete cover.

In an alkaline solution, such as the calcium hydroxide solution in set cement, a protective oxide film forms over the steel rendering it passive. For corrosion to occur this protective film must be impaired and there must be access of oxygen. The stability of the film depends on the maintenance of a certain minimum pH value and under such conditions, access of oxygen will not cause corrosion. The access of carbon dioxide reduces the pH to 10 or lower and the film is then impaired and access of oxygen will cause corrosion. The presence of chloride ions raises the pH required to stabilise the passive film to a value which may exceed that of a saturated calcium hydroxide solution so stimulating corrosion.

The corrosion of steel is an electrochemical process associated with the presence of anodic and cathodic areas arising from inhomogeneities in the surrounding liquid medium or even in the steel surface itself. The existence of such anodic and cathodic areas on the outer surface of corroded reinforced concrete has been demonstrated by Stratfull.[1]

The reactions involved in the corrosion process are:

At the anode,

$$2Fe(metal) \rightarrow 2Fe^{++} + 4e^-$$
$$2Fe^{++} + 4OH^- \rightarrow 2Fe(OH)_2$$

followed by oxidation of the ferrous ion to the ferric state.

At the cathode,

$$2H_2O + O_2 + 4e^- \rightarrow 4OH^-$$

The extent to which these reactions proceed depends upon the conductivity of the electrolyte and the difference in potential between anodic and cathodic areas. The overall reaction of conversion of iron to rust may be stopped by retarding the cathodic process and the rate at which oxygen reaches the cathode therefore controls the velocity of the anodic reaction. Under the action of the corrosion current, the electrodes are polarised and the actual compromise potential limits the corrosive effects. For steel in concrete, the strong polarisation of the anodic zones raises the potential to a value close to that of the cathode and as predicted by the Pourbaix E–pH diagram, the surface of the steel is passivated

[1] R. F. Stratfull, *Corrosion* **13,** 173 (1957).

by the formation of an oxide layer.[1] This accounts for the good durability found with reinforcing bars partially covered by millscale in sound, well-compacted concrete.[2] In permeable concrete in which the oxygen availability is increased and the pH reduced by atmospheric carbon dioxide, the millscale will stimulate corrosion of the uncovered zones as a rate determined by the oxygen access at the cathode.

The cathodic process, requiring a supply of oxygen, is affected by the different oxygen solubilities in solution of differing solid contents and by the extent of water saturation of the concrete. Under marine conditions where variable amounts of chloride are deposited on the surface of reinforced concrete members, differences in the chloride concentration of the pore liquid arise and since the oxygen solubility is a function of the solution concentration,[3] areas of high and low oxygen availability are produced and a differential oxidation cell is formed. The function of the chloride ion in promoting a differential cell of this type offers a tenable explanation of the rapid deterioration of reinforced concrete of inadequate quality under marine conditions.[4] On the other hand, Lewis and Copenhagen[5] regarded the driving force for corrosion as arising from the concentration cell set up by the differences in chloride content of the pore liquid coupled with the low pH and high oxygen availability in the porous (anodic) zones in the concrete. On this basis, there are two opposing cells, one due to differential concentration effects which tend to make the steel in the porous zones more anodic and the differential aeration cell which has the opposite effect. The work of Schaschl and Marsh[6] would appear to have established however, that where these two cells are in opposition, the differential oxidation cell exerts the greater influence. The other effects of chloride ion should not be overlooked. The strong anodic polarisation observed in chloride-free concrete does not occur when chloride is present and the corrosion is no longer controlled by anodic polarisation but depends on the velocity of the cathodic reaction which is governed by the availability of oxygen. At high chloride concentrations the resistivity of the electrolyte is greatly reduced permitting greater corrosion currents, the pH of the pore liquid is lowered and the threshold pH required to protect the steel[7] is raised.

It will be clear that the prime factor determining the protection of reinforcing steel from corrosion is the permeability of the concrete to carbon dioxide and oxygen. In various studies it has been found that the dividing line between corroded and non-corroded reinforcement is set by the depth[8] to which carbonation has occurred and the alkalinity of the concrete has thereby been reduced. The severity of the corrosion in the carbonated zone is in turn determined by factors such as the moisture content of the concrete, the nature of the exposure conditions and the presence in the concrete of chlorides or other electrolytes.

[1] D. A. Hausmann, *Mater. Protn* **6** (11), 19 (1967).
[2] C. A. Lobry de Bruyn, *Proc. Building Research Congress, Pretoria, July 1964*, Vol. 1.
[3] H. H. Uhlig, *Corrosion Handbook*, Wiley, New York, 1955.
[4] H. F. Finley, *Corrosion* **17** (3), 104t (1961).
[5] D. A. Lewis, W. J. Copenhagen, *Corrosion* **15** (7), 382t (1959).
[6] E. Schaschl and G. A. Marsh, *Corrosion* **16** (9), 461t (1960).
[7] R. Shalon and M. Raphael, *Proc. Am. Concr. Inst.* **55**, 1251 (1959).
[8] See M. Hamada, *Sym., Tokyo 1968*, for surveys of depth of carbonation of concrete in service and its relation to the environmental conditions; A. Meyer, ibid.

Corrosion of non-ferrous metals in concrete

Various non-ferrous metals and alloys can be attacked by the alkaline solutions present in damp or wet Portland cement concrete.[1]

Lead when exposed to the atmosphere rapidly develops a protective film of basic lead carbonate, but when embedded in cement mortar this film is not formed and corrosion occurs which may ultimately cause perforation of the metal. It is therefore necessary to protect lead pipes embedded in concrete or cement with a bituminous coating or by wrapping in bituminous felt. Lead damp-proof courses in walls where atmospheric carbonation cannot penetrate far also require protection with a coat of bitumen.

Zinc is resistant to attack in solution of pH 7 to 11 but at higher pH values attack occurs with evolution of hydrogen. The pH value of the solution in a wet concrete is between 12 and 13 but despite this it is only occasionally that corrosion of zinc in concrete has been found. Cases have been reported of concrete containing embedded zinc plates or rods being cracked by the increased volume of the corrosion product. Similarly galvanised reinforcement has been used normally with satisfactory results, but in some cases the bond between concrete and steel has been reduced apparently because the hydrogen evolved caused the concrete layer adjacent to the reinforcement to become porous. This variable behaviour appears to be linked with the amount of soluble chromate in the cement.[2] It has been found that a very small amount of chromate, about 70 parts CrO_3 per million of the solution is sufficient to prevent hydrogen evolution. This concentration in the solution in a concrete would be produced by about 0·003 or 0·004 per cent soluble CrO_3 in the cement. The content of chromium in Portland cements can be above or below this value and some may be present in the aggregate. The use of galvanised reinforcement, to protect the steel from corrosion, has become increasingly common and it is often chromised as a precaution. Attack on zinc is increased by chlorides and calcium chloride should not be used in the concrete. Solid zinc plates or rods embedded in concrete can be protected with a bituminous coating if necessary.

Copper is not corroded in concrete to any significant extent. Copper damp-proof courses and copper pipes have long been laid in concrete or mortar without any troubles from corrosion arising. Serious corrosion has, however, been found in district heating schemes where copper pipes were embedded in aerated concrete made with nitrogenous foaming agent which released a small amount of ammonia on mixing with Portland cement. The cause was attributed to stress corrosion which arose from residual stress in the pipes and was accelerated by traces of ammonia and the use of copper containing very small amounts of phosphorus. Ammonia-producing foaming agents should be avoided in such circumstances or suitably annealed phosphorus-free copper pipes used.

[1] *Bldg Res. Stn. Dig.*, Nos. 110, 111 (First Series), 1958; F. E. Jones, *Chemy Ind.*, 3 Aug. 1957, 1053; P. E. Halstead, ibid., 24 Aug. 1957, 1132; H. Woods, *Spec. tech. Publ. Am. Soc. Test. Mater.*, No. 169A, 230 (1966).

[2] See L. H. Everett and K. W. J. Treadaway, *Proc. 8th int. Congr. Hot Dip Galvanising, London 1967*; A. Hofsoy and I. Gukild, *Proc. Am. Concr. Inst.* **66**, 174 (1969). *Building Research Station Digest* No. 109 (1969).

Aluminium, like zinc, reacts with alkali hydroxide solutions with evolution of hydrogen. Surface attack[1] occurs initially on aluminium embedded in concrete and under unfavourable conditions the build-up of the corrosion products can cause cracking of the concrete. Anodising gives no protection and some of the alloys are less resistant to corrosion than pure aluminium. Bituminous coatings can be used for protection. Much more serious corrosion[2] can occur if the concrete contains calcium chloride and particularly in reinforced concrete if there is electrical contact between the steel and the aluminium. Various cases of serious cracking or spalling of concrete over aluminium conduits in such cases have been reported. The presence of 1 per cent calcium chloride by weight of cement is sufficient to cause galvanic corrosion. Sodium chloride would have the same effect so the use of aluminium in concrete in, or near, seawater should be avoided. Certain organic protective coatings have been recommended by McGeary when it is impracticable to avoid contamination by chlorides.

[1] F. E. Jones and R. D. Tarleton, *Natn. Bldg Stud. Res. Pap.*, No. 36, H.M.S.O. London 1963; L. H. Everett, *Builder*, 30 March, 1962.

[2] C. E. Monfore and B. Ost, *J. Res. Dev. Labs Portld Cem. Ass.* **7** (1), 10 (1965); F. L. McGeary, *Proc. Am. Concr. Inst.* **63**, 247 (1966).

18 Concrete Aggregates

The aggregates used for mortars and concretes can be conveniently divided into dense and lightweight types. The former class includes all the aggregates normally used in mass and reinforced concrete, such as sand, gravel, crushed rock and slag. The lightweight class includes pumice, clinker, foamed slag, expanded clay, shale, and slate, exfoliated vermiculite, and expanded volcanic glasses such as perlite. The density of ordinary concrete ranges from about 140 to 150 lb/ft^3 (2240 to 2400 kg/m^3), and that of most lightweight concretes from 40 to 100 lb per/ft^3 (640 to 1600 kg/m^3), though still lighter materials are obtainable. The saving in weight by the use, where circumstances permit, of lightweight aggregates is thus considerable.

Dense aggregates

Dense aggregates consist of natural sands, gravels, and crushed rocks,[1] together with artificial products such as air-cooled blastfurnace slag and broken brick.

Sands and gravels

Sands and gravels are derived from the weathering of rocks and are composed of the more resistant minerals which have been able to withstand for a long period the destructive effects of weather and transport. Most sands and gravels used in mortars and concrete are obtained from river and glacial deposits. Other sources are crushed friable sandstones and sea sands. The latter need to be well washed on account of the presence of soluble salts. The division between sand and gravel, or ballast, is an arbitrary one based on size; for concrete, material which will

[1] The nomenclature of British aggregates is described in BS 812 : 1967 and that of the mineral constituents in the same specification and also in ASTM C294–67. For methods of petrographic examination see ASTM C295–65 and L. Dolar Mantvan, *Highw. Res. Rec.*, No. 120 (1966), and for a mineralogical description of British aggregates, F. E. Jones, *Natn. Bldg Stud. Res. Pap.*, No. 15 (1952). For discussion of the properties and testing of aggregates see *Spec. tech. Publ. Am. Soc. Test. Mater.*, No. 169A (1966); *Proc. Am Concr. Inst.* **58**, 513 (1961); *Proc. Sym., Sands for Concrete*, Sand and Gravel Association of Great Britain (1964); *J. Br. Granite Whinstone Fed.* **5** (1) (1965); *Proc. Instn Civ. Engrs*, Part I **4** (3), 353 (1955); A. J. Newman, *Sym., Concrete Quality*, p. 53, Cement and Concrete Association, 1966.

pass a $\frac{3}{16}$-inch mesh is usually classed as sand and larger material as gravel. Many deposits of sands contain larger material, while many gravels contain a proportion of sand.

The particles of sands generally consist each of an individual mineral, and by far the most common mineral is quartz. Other minerals remaining from the weathering of rocks are present in small amounts, and, of these, felspars, and the clays derived from them, rank next in importance. Flakes of mica and black grains of iron ores may be present in small amounts. In sands occurring in coal-mining districts a small proportion of fine coal particles is sometimes present.

Gravels are composed of a variety of minerals, and the individual pebbles are often heterogeneous. Flint and chert (both minutely crystalline forms of silica distinct from quartz), quartz, quartzite, granite, etc., are the common constituents of gravels, but sandstone and limestone gravels are also found. Gravels are used in both the crushed and uncrushed condition. The former tend to be sharper and more angular in character than the rounded and smooth uncrushed material. It has often been considered that sharpness is a desirable characteristic in an aggregate, and that it ensures a better bond between cement and aggregate. Sharp aggregates give less workable mixes, requiring somewhat more water than rounded aggregates, and for a given workability the compressive strength tends to be reduced; the transverse strength may, however, be greater.

A certain amount of clay, varying from almost nothing to 10 per cent or more, is a common constituent of sands. An old requirement, still sometimes made, is that the clay present as lumps shall not exceed 1 per cent, but the test for this is unsatisfactory and it is not included in the British Standard. The amount of clay, silt and fine dust is limited in all standard specifications for aggregates. For natural aggregate the material is dry sieved on a 200 mesh (75 μ) sieve by the method described in BS 812 : 1967 or by agitating the aggregate in water and decanting the washwater through a 200 mesh sieve as described both in BS 812 and ASTM C117–67. A pipette sedimentation method in which a sample of the aggregate is dispersed in a solution containing 0·8 g sodium oxalate per litre is also described in BS 812 for determining the amount of material finer than 0·02 mm in size. The amount passing a 200-mesh sieve is limited in BS 882 : 1965 to 1 per cent for coarse aggregate, 3 per cent for natural sand and 15 per cent for crushed stone sands. The ASTM specification C33–67 has a maximum limit of 5 per cent for natural sands and 7 per cent for crushed stone sands. The fine material in a crushed stone sand differs from that in natural sand in that it consists largely of stone dust and not clay. A higher content of it can therefore be tolerated and may be advantageous by improving the plasticity of concrete mixes containing angular crushed rock aggregates. The ASTM limit seems in many cases to be too restrictive.

The effect of clay in a concrete aggregate depends on the manner of its distribution. It is much more injurious to the strength of the concrete if present as a film enveloping the sand grains than if distributed as fine particles throughout the mass. In the former case the adhesion of cement to sand is weakened and hardening delayed, whereas the presence of fine particles of clay has been claimed to have no ill effect, and even to increase the strength of concrete.[1] As a safeguard

[1] D. A. Parsons, *J. Res. natn. Bur. Stand.* **10,** 257 (1933).

in practice it is, however, necessary to limit the content of clay permissible in concrete aggregates.

The presence of a certain amount of clay in sands for use in mortars and renderings where high strengths are not required may be definitely advantageous. It increases the workability of the mortar and so assists the craftsman in placing the material in position. Accordingly up to 5 per cent of material passing the 200-mesh sieve is allowed in BS 1199 and 1200 : 1955 for natural sands for use in renderings and brickwork mortars.

Some sands contain a certain amount of organic matter, derived usually from associated loam, and this in some cases seriously retards the hardening of a concrete. A colour test has long been used to determine the quality of a sand in this respect and is described in ASTM C40–66. It was also formerly used in BS 812 but was replaced by another test in the 1967 revision because of its doubtful value. A failure to pass the colour test has in any event only been taken to indicate that further tests should be carried out on the strength of concrete or of mortars containing the sand in question (as in ASTM C87–68) and not to imply that it is necessarily unsuitable for use. Unfortunately certain naturally occurring organic impurities that are deleterious and some harmful industrial contaminants give negative results in the colour test. The colour test as described in ASTM C40–66 is carried out as follows. A 12 oz graduated bottle is filled to the 4½ oz level with the sand and a 3 per cent solution of sodium hydroxide added until the combined volume is 7 liquid oz. The bottle is stoppered, shaken vigorously and allowed to stand for 24 hours. The colour of the liquid is then compared with that of a standard solution. This is freshly prepared by dissolving 0·25 g potassium dichromate in concentrated sulphuric acid (sp. gr. 1·84). The former British Standard test was very similar but for the colour standard used a freshly prepared solution obtained by adding 2·5 ml of 2 per cent tannic acid solution in 10 per cent alcohol to 97·5 ml 3 per cent sodium hydroxide solution, shaking, and allowing to stand 24 hours. Alternatively, coloured glass of the correct colour could be used as standard. The new test described in BS 812 : 1967 which is suitable only for laboratory use, depends on the determination of the pH value of a mortar composed of 50 g sand, 5 g Portland cement and 12·5 cc water. The pH of this mortar is determined one hour after mixing. Not all Portland cements are suitable for the test and the suitability of that to be used has to be determined by measuring the pH, (1) in the mortar mix given above using standard sand, (2) in a similar mix made with a tannic acid solution (1·6 g in 100 ml distilled water) instead of water.

The cement is suitable for use in the test if the pH of the mortar mixed with water (1) is not less than 12·40 at 20° and that of the mortar mixed with the tannic acid solution (2) not more than 12·20. A sand may be deemed to be satisfactory if the pH of the mortar made from it is not less than 12·40 and unsatisfactory if it is 12·20 or lower. If the pH value fall between these limits the test is repeated on a sample of the sand heated to 600° for 1 hour. If this treatment does not raise the pH by more than 0·2 the sand may be deemed to be satisfactory, and unsatisfactory if the pH is raised by more than 0·2.

The presence in either gravel or crushed stone aggregates of fragments of soft shales and other soft materials is usually considered objectionable, on account of their weak and friable nature and their poor resistance to weathering. The presence of soft particles seems definitely to reduce the abrasion resistance of a concrete, but the influence on weathering is erratic, for this seems to depend more on the quality of the cement paste in the concrete than on the presence of scattered fragments of easily weathered materials. In some specifications for aggregates, as in ASTM C33–67 for example, it is the practice to limit the content of soft particles to 5 per cent. There is, however, no very satisfactory test method for defining what constitutes an objectionable soft material though a scratch hardness test has been tentatively adopted in ASTM C235–68. The more general assessment of the mechanical properties of coarse aggregates is made by various forms of abrasion, impact hardness and aggregate crushing tests (BS 812, ASTM C131 and C535).

Crushed rock aggregates

The rocks which are crushed for use as concrete aggregates usually belong, geologically, to the types granite, dolerite, basalt, sandstone and limestone and occasionally quartzite and diorite. Commercially, however, the name under which a rock is marketed often bears no relation to its geological description, and the terms granite, whinstone and basalt are loosely used to cover various types of rocks. An older term, trap, corresponded roughly with whinstone. The general characteristics of the rocks classified according to their correct geological nomenclature are as follows:

Granites are igneous rocks formed from molten magmas which have solidified well below the earth's surface under conditions of slow cooling and high pressure. They are well crystallised and the crystalline structure, which is variously described as coarse, medium or fine grain, is visible to the naked eye. Diorites belong to the same general class or rocks, but are less acid in composition.

Dolerites are igneous rocks which have solidified below the earth's crust, but under conditions such that cooling was more rapid. They are found in dykes and sills, which are wall or sheet-like masses. In appearance they are dark coloured and fine grained, and in structure microcrystalline.

Basalts are igneous rocks formed from lava flows. They are very similar to dolerites and over part of their range the two terms are synonymous. Rocks of the basalt-dolerite type are termed dolerite if they are composed of discrete crystals definitely detectable and basalt if they show a vesicular structure.

The igneous rocks can be classified according to their relative contents of acidic and basic oxides as follows:

Group	Percentage SiO_2	Examples	Minerals
Acid	> 66	Granite	Quartz, felspar, mica and hornblende
Intermediate	52–66	Diorite	Felspar, mica and hornblende
Basic	< 52	Dolerite Basalt	Felspars and more basic minerals such as augite and olivine

Sandstones are sedimentary rocks composed of quartz grains, and sometimes some additional minerals, cemented together. Various types of sandstones are distinguished according to the nature of the intergranular cement. Silicious sandstones contain an amorphous silica cement, while argillaceous sandstones are similarly cemented, but contain also dispersed clay minerals. Calcareous sandstones are characterised by a cementing material of calcium carbonate, and ferruginous sandstones by a hydrated iron oxide as cement. The more coarse-grained sandstones are called grits. Quartzite is a silicious sandstone in which the grains have been closely welded together by pressure, but the same term is also applied to metamorphic rocks produced from sandstones which have undergone recrystallisation under the influence of heat and pressure.

Limestones are sedimentary rocks composed of calcium carbonate. For the most part they are derived from the calcareous remains of marine or fresh water organisms embedded in a calcareous mud. They vary in type from the soft chalks to hard crystalline rocks. The oolitic limestones are a special class formed by the deposition of calcium carbonate around minute nuclei. Dolomitic limestones are limestones in which calcium carbonate is replaced in part or whole by dolomite $CaMg(CO_3)_2$. Marble is a metamorphic highly crystalline variety of limestone.

The use of limestone as a concrete aggregate has sometimes been suspect on account of the unsuitability of the poorer grade rocks, and also because of a widespread fallacy that limestone concrete is less resistant to the action of fire than concrete made from other aggregates. Hard limestones form satisfactory aggregates, but materials which are at all shattered or heavily veined with clay are not suitable. As mentioned elsewhere, the use of limestones may not be beneficial in concrete products which are to be cured in high-pressure steam.

Commercial nomenclature. The term granite is widely used to cover a variety of rocks with a crystalline or granular structure and includes granites, some dolerites, quartzite and even some sedimentary rocks such as dense limestones. Whinstone is a term used in some cases to cover any fine-grained dark-coloured igneous rock such as dolerite and basalt, whilst in others it includes also dark-coloured limestones with no distinct bedding; quartzites are also sometimes classed under whinstone. Basalt is a term in common use in connection with concrete aggregates and covers basalt and often dolerite.

Air-cooled blastfurnace slag

The term blastfurnace slag[1] is limited to the slags obtained in the production of pig iron and does not cover slags arising in the production of steel or of non-ferrous metals. The main requirement in blastfurnace slag to be used as aggregate is that it shall be stable, but it is also desirable that it should be dense and have a crystalline rather than a glassy texture. These properties are dependent on the composition of the slag and the rate at which it is cooled.[2] The molten slag may be cooled in large ladles, holding up to 15 tons and the 'ball' tipped, after solidifying, on to a storage ground. In a second method, the slag is run into pits or

[1] Detailed information on the properties and uses is to be found in *Hochofenschlacke*, by F. Keil, Verlag Stahleisen, Düsseldorf (1963). For a general review, see G. W. Josephson, F. Sellers and D. G. Runner, *Bull. U.S. Bur. Mines*, No. 479 (1949). Information collected from the older literature on the use of slag as an aggregate is to be found in the pamphlets prepared by the National Slag Association, Washington.

[2] T. W. Parker, *Chemy Ind.* **60** (5), 59 (1941).

canals in layers of 2–4 inches thick, further layers being added after the previous one has solidified. Alternatively, the slag is occasionally run into iron moulds. A third method is to pour the slag on a bank. The 'ball' method gives the slowest rate of cooling and the 'bank' method the fastest. The strength of slag aggregate concretes[1] compares well with that of concretes with natural dense aggregates.

Slag is composed essentially of lime, alumina and silica with variable amounts of magnesia and small amounts of sulphur, alkalis, titania, etc. The mineral constituents that can be formed by the combination of these oxides have been discussed in Chapter 15. Most of these minerals are inert materials and the only constituents that may be objectionable are calcium orthosilicate ($2CaO.SiO_2$), the sulphur compounds, or a high ferrous iron content.

Some slags of high-lime content, such as haematite slags, disintegrate during cooling or very soon afterwards and there is no risk that such materials would be used as a concrete aggregate. The great majority of the remaining slags are stable, but there is a small intermediate group which may decompose slowly and must, therefore, be avoided. Most slag producers rely primarily on their experience of their own slags to detect unstable material from its appearance during tapping, or after solidification, or from its chemical analysis. This prediction is only rarely at fault, but the experience of one works cannot be applied unchanged to another and some form of test is, in any event, desirable for the user.

The inversion of $2CaO.SiO_2$ from the β to the γ form is accompanied as we have seen earlier (p. 457), by an increase in volume that leads to the disintegration or dusting, of a slag. It can only occur in slags of fairly high-lime content and some method of limiting this, or the lime : silica ratio, has long been sought. A limit of 42–43 per cent is often taken as a lime content above which there may be risk of a slag being unstable. It is possible, however, from what is now known of the phase equilibrium in the systems concerned, and of the 'suspended equilibrium' that can arise from failure of the liquid and solid phases to maintain equilibrium during cooling, to calculate certain limiting lime contents below which calcium orthosilicate cannot occur.[2] These are used in BS 1047 : 1952, according to which a slag conforming to either of the composition limits (*a*) and (*b*) below is acceptable.

$$(a)\quad CaO + 0{\cdot}8MgO \leqslant 1{\cdot}28SiO_2 + 0{\cdot}4Al_2O_3 + 1{\cdot}75S$$
$$(b)\quad CaO \leqslant 0{\cdot}9SiO_2 + 0{\cdot}6Al_2O_3 + 1{\cdot}75S$$

The amounts of the oxides in these formulae are expressed in terms of their percentage weights in the slag.

Slags which fail to conform to either of these formulae are not, however, necessarily unsound. The conditions of cooling may be so rapid that, in a composition from which $2CaO.SiO_2$ could potentially crystallise, it fails to do so and to use these formulae alone would lead to the rejection of sound slags. A slag which falls after cooling slowly in a 7-ton ladle, for example, may be quite stable if cooled more rapidly by pouring in thin layers into a shallow pit or on a slag bank. Much attention has, therefore, been paid to direct tests for the presence of

[1] W. H. Gutt, W. Kinniburgh and A. J. Newman, *Mag. Concr. Res.* **19** (59), 71 (1967); P. C. Aitcin, *Revue Matér. Constr. Trav. publ.* **608,** 185; **609,** 249 (1966).
[2] T. W. Parker and J. F. Ryder, *J. Iron Steel Inst.* **146,** II, 21P (1942).

this compound. The γ form of $2CaO.SiO_2$ shows a characteristic yellow to brown fluorescence under ultra-violet light[1] and this test has been much used and appears, for example, in the German specification (DIN 4301 : 1962) for slag aggregates. A quartz mercury vapour lamp is used with a nickel oxide glass filter giving ultra-violet light of 300–400 $\mu\mu$ wavelength. Freshly fractured surfaces of stable slags exhibit a uniform violet or lilac fluorescence, though glassy slags may tend to appear reddish and old bank slags dark violet to brown. Disintegrating slags show yellow or brown patches against a general violet background. This, according to Guttmann, is not to be confused with a uniform attenuated distribution of minute points of yellow or brown fluorescence which is occasionally observed with stable slags. From an examination of a large number of British slags Parker and Ryder concluded that, while the test detects many unsound slags, and is of use for initial sorting, it fails to detect a substantial minority, and is not, therefore, a completely safe guide. In place of this test these authors developed a microscopic test, which is incorporated in BS 1047 : 1952. Sample pieces of the slag are impregnated in vacuo with resin (Bakelite R0014) which is hardened by heating for 4 hours at 50° followed by 24 hours at 100°. The specimen is then polished in a similar manner to that used for cement clinker and etched by placing face downwards for one minute in a 10 per cent solution of anhydrous magnesium sulphate at about 50°. This reagent selectively etches calcium orthosilicate and in slags liable to 'falling' there can be seen, at a magnification of 100–400 diameters, etched crystals having parallel or cross-hatch striations across their surface, as shown on Plate 17 (i). White highly reflecting crystals, as shown in Plate 17 (ii) may also be observed. These are calcium sulphide and without significance in regard to 'falling'. Slags which show no crystals of $2CaO.SiO_2$ are acceptable even though they do not conform to the composition limits (*a*) or (*b*) given earlier.

Another, rather rare, form of unsoundness—known as iron unsoundness—was ascribed by Guttmann to a high content of ferrous sulphide and leads to a rapid breakdown of the slag in water. Though apparently other factors can also play a part it is readily detected by immersing pieces of the slag in water for 14 days as required in the British Standard. In the German specification only a 2 days immersion is required.

The presence of calcium sulphide, which in fresh slag gives rise to a perceptible odour, is not a source of risk unless its content becomes excessive. In slag that has weathered some oxidation to sulphate may have occurred and, as with all concrete aggregates, only a limited amount can be permitted. The British Standard limits the content of acid soluble SO_3 to 0·7 per cent and that of total sulphur to 2 per cent S.

The density and structure of slags vary with their composition and rate of cooling. In general, the more acid slags are denser, but more glassy, than the more basic materials. The latter, though well crystallised, are the more prone to have a vesicular structure owing to the evolution of gases during cooling. It is normally advantageous to use the production methods giving slower cooling for the more acid slags, so as to promote crystallisation and get a less brittle product, and more rapid cooling for the slags of higher lime content, in order to obtain a

[1] A. Guttmann, *Stahl Eisen* **46,** 1423 (1926); **47,** 1047 (1927).

fine rather than coarsely crystalline material. The apparent specific gravity of slags varies widely, according to the extent of pinholing and honeycombing, but for the majority falls between 1·9 and 2·8. This may be compared with average values of about 2·4 for dense limestones and 2·7 for granites. The true specific gravity of slags varies from about 3·0 to 3·1. The unit weight of slag aggregates is often less than that of natural aggregates because of some degree of pinholing and a test of the weight per cubic foot provides an easy means of eliminating honeycombed slags that are unsuitable for use as a hard aggregate. There is no very close relation between the abrasion resistance or strength of concrete and the unit weight of coarse slag aggregate,[1] but as a practical working rule a minimum unit weight of 78 lb/ft^3 (1249 kg/m^3) is specified in the British Standard for slag above $\frac{3}{16}$ inch and up to $1\frac{1}{2}$-inch size, and 70 lb/ft^3 (1120 kg/m^3) in the ASTM specification. The British Standard also limits the absorption, after immersion for 24 hours in water at 15–25° of slag previously dried for 24 hours at 100–110°, to 10 per cent by weight. The ASTM specification (C33–67) includes a sodium or magnesium sulphate crystallisation test for soundness and an abrasion test, though unfavourable results in either may be ignored if service records or concrete strengths show the material to be satisfactory. No limitations are imposed on composition.

Slag aggregates are considerably used for the larger precast concrete products such as flags, kerbs, channels, fence posts, blocks, etc., but not usually for the smaller products such as tiles or for architectural features such as cast stone. They are also used in all classes of plain and reinforced concrete work. Some objection to the use of slag in reinforced concrete has sometimes been raised on account of its sulphur content. Experience and many tests[2] have, however, shown that where the physical properties of the slag are suitably safeguarded by the existing specifications, and the quality of the concrete in regard to density and impermeability is similar to that necessary with other aggregates for reinforced concrete, no corrosion occurs. The evidence also shows that in good slag concretes, only negligible oxidation of sulphides to sulphates occurs.[3] Slag aggregates are not very suitable for use with high-alumina cement. The setting and hardening of this cement is adversely affected by mixing waters containing alkalis and a certain amount of lime may be liberated into solution by a slag. The effect varies with different batches of high-alumina cement and with different slags but it can lead to a reduction in strength.

Broken brick

Crushed brick is sometimes used as a concrete aggregate and can form a useful material. It tends to yield concretes rather light in weight and varying from 130 to 140 lb/ft^3 (2080 to 2240 kg/m^3), according to the porosity of the aggregate. At the lower weights the strength is rather less than that of a normal gravel

[1] F. Hubbard and H. T. Williams, *Proc. Am. Soc. Test. Mater.* **43,** 1088 (1943).
[2] H. Burchartz, *Stahl Eisen* **37,** 626 (1917); **40,** 814 (1920); **41,** 193 (1921); T. W. Parker, *Trans. Instn Engrs Shipbldrs Scot.* **94,** Paper 1139 (1951); Report of Committee on Aggregates, *Proc. Am. Concr. Inst.* **27,** 183 (1931); L. H. Everett and W. H. Gutt, *Mag. Concr. Res.* **19** (59), 83 (1967).
[3] H. Burchartz and E. Diess, *Arch. EisenhuttWes.* **8** (5), 181 (1934–1935).

aggregate concrete, but at the higher weights it is often greater.[1] Owing to their absorptive properties it is usually necessary to wet brick aggregates before use. There is a risk with some crushed brick aggregates that the sulphate content may be undesirably high. This can arise from the use of bricks that have a high calcium sulphate content or of rubble from demolished buildings that is contaminated with gypsum plaster. Rubble from demolitions can be screened, as has been done in Germany, to remove all material below about 1 inch size, since this contains the higher amounts of sulphates, and the larger material crushed and graded for use as aggregate. In other cases it may be sufficient to reject only the sand size fraction.

A limit of 1 per cent has been imposed in some countries, e.g. Germany, on the SO_3 content of brick aggregates, but Gaede[2] has suggested that this value ought to be varied with the cement content of the mix. With progressive additions of gypsum to an aggregate he found a critical range of SO_3 contents over which the strength dropped rapidly and then remained more or less constant at the reduced value as further gypsum was added. Concretes made with Hochofen cement suffered less reduction in strength than those with Portland cement. He concluded that the limit of 1 per cent SO_3 ought to be reduced to below 0·5 per cent. These tests were made, however, by adding finely ground gypsum or magnesium sulphate to a gravel-sand aggregate and in consequence the effect of the sulphur trioxide was probably exaggerated. The results obtained by Newman gave no indication of any serious deleterious effects for sulphur trioxide contents of up to 1 per cent in brick aggregates in concrete mixes as lean as $1 : 2\frac{1}{2} : 7\frac{1}{2}$ by volume. The sulphate in this case was present in the bricks and not as contamination with gypsum plaster to which Gaede's results might apply more closely.

Physical properties of aggregates

Apart from the size and grading, discussed in Chapter 19, there are various other properties of aggregates which have an influence on concrete.[3] There is no close relation between the porosity of aggregates and the durability of concrete and though, in general, it is wise to restrict the permissible absorption to a maximum of 10 per cent by weight for dense aggregates, it does not necessarily follow that aggregates of higher absorption will be defective. It is now usual when there is any doubt about the durability of an aggregate to rely more on freezing tests on concrete than on measurements of the porosity or soundness of the aggregates (see p. 617). Observations of the weathering of an aggregate on exposed rock faces can also be helpful.

The thermal expansion of aggregates[4] varies considerably. Common ranges are from about 2 to 5×10^{-6} per °F for limestones, 3 to 5 for igneous rocks, and

[1] A. J. Newman, *J. Instn munic. Engrs* **73** (2), 113 (1946).
[2] K. Gaede, *Dt. Aussch. Stahlbeton* **109** (1952); **126** (1957).
[3] See report on significance of tests and properties of concrete and concrete-making materials, *ASTM Spec. Tech. Publ.*, No. 169A (1966); Symposium on mineral aggregates. *Spec. tech. Publ. Am. Soc. Test. Mater.*, No. 83 (1948); F. E. Jones, *Sym., London 1952*, 368; Review of methods of testing aggregates. *Proc. Instn civ. Engrs*, Part I **4**, 353 (1955).
[4] D. G. R. Bonnell and F. C. Harper, *Natn. Bldg Stud. tech. Pap.*, No. 7 (1951); W. H. Johnson and W. H. Parsons, *J. Res. natn. Bur. Stand.* **32**, 101 (1944); G. J. Verbeck and W. E. Hass, *Proc. Highw. Res. Bd* **30**, 187 (1950).

6 to 7 for quartz or flint sands and gravels, though individual aggregates may fall outside and extreme values range from 0·5 to 9. These values apply to pieces of rocks containing many crystals, but single crystals of minerals often show anisotropic properties. An extreme example is calcite, which has a coefficient of about 24×10^{-6} per °F parallel to the *c*-axis and a negative coefficient, −4, at right angles to it. Coarse grained limestones have been found[1] to show a non-linear behaviour, expanding with rise in temperature at a very low rate below 20° and at a much higher rate above. In contrast fine and medium grain limestones showed intermediate rates of expansion which were almost linear with temperature rise. There has been some controversy as to the part which differential thermal expansion between aggregate and set cement may play in the deterioration of concrete (see p. 621).

The specific heat and thermal conductivity of aggregates are not of much importance in most uses of concrete, but they become significant in large concrete masses such as dams where they influence the temperature rise and rate of cooling of the mass, and in concrete exposed to high temperatures.

Aggregates must obviously be sufficiently strong to permit of the strength desired in a concrete being developed, or the resistance to abrasion being sufficient for the conditions of wear to which it is to be exposed.

Shrinkable aggregates

Most dense aggregates show little change in dimensions with changes in water content, but some have been found in which the shrinkage on drying is large enough to cause distress in concrete in which they have been used. Plain concrete may show progressive cracking which is later accentuated by frost action if the concrete is exposed to the weather. Deterioration has been particularly evident in relatively thin concrete products such as facing slabs, kerbs, cills and coping units. Reinforced concrete members such as beams and floor slabs may show excessive deflection. The asymmetric distribution of the reinforcement results in the shrinkage of the concrete being more restrained on the lower than the upper side leading to bending. Cracking of the concrete along the line of the reinforcement may expose the steel to corrosion, particularly in concrete exposed to the weather. Precast concrete bricks and blocks may show drying shrinkage too high to comply with the requirements of British Standards 1180, 1217 and 2028. All these effects have been observed and may sometimes have been attributed to poor quality concrete or other causes.

Shrinkable aggregates have now been reported from various countries, for example from Scotland, northern England,[2] Denmark,[3] South Africa,[4] Japan[5]

[1] R. D. Harvey, *Mater. Res. Stand.* **7,** 502 (1967).

[2] L. C. Snowden and A. G. Edwards, *Mag. Concr. Res.* **14** (41), 109 (1962); A. G. Edwards, *J. Br. Granite Whinstone Fed.* **6** (2), (1966); *Bldg Res. Stn Dig.*, Second Series No. 35 (1963), H.M.S.O. London.

[3] P. Nepper-Christensen, *Meddr. dansk geol.* **15,** 548 (1965).

[4] N. Stutterheim, *Trans. S. Afr. Instn civ. Engrs* **4** (12), 351 (1954); *R.I.L.E.M. Sym. Concrete and Reinforced Concrete in Hot Climates*, Haifa, 1960; H. Roper, *J. Res. Dev. Labs Portld Cem. Ass.* **2** (3), 13 (1960).

[5] E. Hishioka and Yu Harada, *Proc. Japan Cem. Engng Ass.* **12,** 51 (1958).

and the U.S.A.[1] The types of aggregate involved have been dolorites, basalts, sandstones, limestones, mudstones and gravels derived from them; for example, the moisture movement of a series of Scottish dolerites was found to vary from almost zero to 0·07 per cent and higher values were recorded on a mudstone. Even from different positions within a single quarry the moisture movement of the dolerite rock varied in one case from 0·02 to 0·07 per cent. Some of the most extreme examples have come from South Africa with fine-grained sandstones showing movements of 0·1 to 0·8 per cent. In the U.S.A. movements of up to 0·06 per cent have been recorded on some dolerites, and up to 0·1 per cent on argillaceous limestones, though there is little record there of deterioration of concrete directly ascribed to this cause. It may well be, however, that some shrinkable aggregates have been rejected, without recognition as such, by freezing tests. Thus in a series of tests on limestone aggregates,[2] those showing the lowest frost resistance in concrete were characterised by contents of montmorillonite up to 4 per cent. There is no evidence to indicate whether concrete containing shrinkable dolerite aggregates show a poor resistance in laboratory freezing tests.

An example of the severe shrinkage cracking of a concrete containing a highly shrinkable aggregate is shown in Plate 18 (i). This type of cracking could be mistaken as arising from other causes so that deterioration from shrinkable aggregates may occur more generally than is recognised. It would not be expected to occur in concrete kept continuously wet and it will increase in severity with the degree of drying of the concrete. Whereas the drying shrinkage of a dense concrete with a non-shrinking aggregate is of the order of 0·03–0·04 per cent, that of concretes with shrinkable aggregates may exceed 0·1 per cent. A progressive increase in both wet and dry length of concrete specimens is also found on successive cycles of wetting and drying, amounting to as much as 0·1 per cent, as compared with little change with non-shrinking aggregates. Though both the fine and the coarse fraction of a shrinkable aggregate contribute towards the shrinkage of the concrete it is the coarse fraction that has the dominant effect. Air-entrainment appears to have some beneficial effect on the durability of concrete exposed to the weather but there is no evidence that it decreases deflection of beams or floor slabs. In some geographical regions where the rocks are predominantly of the shrinkable type it may be necessary to use some which show moderate movements. Recommendations for the use of the resultant concretes of various shrinkage levels are given in *Building Research Station Digest* (2nd Series), No. 35.

The cause of the abnormal moisture movement of shrinkable aggregates is still not clear. It seems to be associated with the presence of a very fine-grained matrix with a high surface area. In the case of some of the South African sandstones the presence of swelling clays such as montmorillonite has been detected[3] but in others the clay constituents were of the non-swelling type such as kaolinite or only present in small amounts. No general correlation has been established between shrinkage of aggregates and any single mineralogical or chemical

[1] R. Rhodes and R. C. Mielenz, *ASTM Sym. Mineral Aggregates:* Spec. tech. Publ. No. 83, 2 (1948).
[2] K. Mather, *Min. Engng* **5** (10), 1022 (1953); *Tech. Memo. U.S. WatWays Exp. Stn*, No. 6–371 (1953).
[3] N. Stutterheim, loc. cit.; H. Roper, J. E. Cox and B. Erlin, *J. Res. Dev. Labs Portld Cem. Ass.* **6** (3), 2 (1964).

characteristic. The largest group of aggregates showing shrinking characteristics in Great Britain are dolerites and they warrant therefore some further discussion.

Properties of dolerite

The term dolerite is to be interpreted here as including doleritic and basaltic types of rock, generally known in Great Britain as whinstone.

Dolerite occurs in various parts of England, notably in Staffordshire, Derby, Shropshire and Northumberland; it is found in North Wales and is widely distributed in many parts of Scotland, including the Lowlands and Argyllshire, and it is also found in Ireland. Its extensive occurrence, combined with its strength and hardness, have caused it to be used widely for concrete. While dolerites have been used over many years with satisfactory results in precast concrete products such as flags, kerbs, pipes, etc., and also in concrete cast *in situ*, some serious failures have resulted in concretes made from it. The failure of dolerite concretes in precast concrete products exposed to weather takes the form of cracking and disintegration which begins to develop from six months to a year after manufacture. In kerbs, for instance, the cracks tend to run parallel to the edge and on examination it is found that there has been a complete loss of adhesion between cement and aggregate. In concrete cast *in situ* it has sometimes been reported that the mix was slow in setting and hardening and that at a later date cracks commenced to appear.

The causes of these failures and defects was for long unexplained but the major cause must now be ascribed to the shrinkage characteristics of the aggregates used. In some British dolerites readily oxidisable ferromagnesium minerals occur as degredation products of olivine and in the case of a rock containing an unusual amount of one of these, chlorophaeite, some evidence was obtained that the oxidation was accompanied by an expansion in volume. This may therefore occasionally be an operative factor. Some dolerites tend to form very angular or flaky particles on crushing and so to give harsh concrete mixes with a corresponding tendency for high-water content to be used resulting in a low-frost resistance.

Many dolerites have a good service record in concrete and, even in the case of those that have proved unsatisfactory, it is usually only in concretes of thin section that serious troubles have arisen. Some dolerites that have proved unsound have been satisfactory when used only as a coarse aggregate, but in other cases the coarse material has led to failures.

Chemical properties of aggregates

The chemical properties of aggregates are important only in so far as an aggregate contains constituents that can undergo chemical change, or react with the cement in such a way as to cause deterioration or to interfere with its setting and hardening. The effect of sulphates in aggregates, and of materials which interfere with the hardening of cement, are discussed elsewhere and we have to consider here certain other aspects of the reactivity of aggregates. It does not necessarily follow that because some reaction occurs between cement and the surface of an aggregate, it is harmful, and it may indeed improve the bond between them. Some ferric iron compounds in aggregates, for example, appear to react with lime without deleterious effect, but there is evidence that oxidation of ferrous iron is one factor contributing to the deterioration of some dolerite concretes. Oxidation of pyrites or marcasite is also occasionally a cause of popping or spalling with some aggregates. Pyrites can occur in two forms, reactive and non-reactive and these can be distinguished by immersing the mineral in limewater. Within a few moments the reactive form produces a blue-green precipitate of ferrous hydroxide which, on standing, is oxidised to ferric hydroxide and turns brown. Such reactive pyrites is sometimes found in Thames river gravel causing the formation of brown stains on concrete or even blistering of the surface. Ferric hydroxide and calcium sulphoaluminate have been found present in the stain.[1] Another iron sulphide, pyrrhotite, found in some aggregates in Sweden, readily undergoes oxidation and has been reported[2] to cause expansion and deterioration of the concrete.

The reaction of active forms of silica with lime improves strength, but that with alkalis has been the cause of much deterioration. These are the matters we have now to consider.

Alkali-aggregate reaction

The problem of alkali-reactive aggregates first came to the fore in 1940, when Stanton[3] ascribed the deterioration of certain concretes in California to a reaction between sodium and potassium hydroxides released from the cement and a reactive form of silica in the aggregates. Since then many similar cases of expansion and cracking of concrete have been assigned to the same cause. The reaction leads to the formation of an alkali silicate gel and sets up expansive forces. The action is a relatively slow one and though in some cases the first signs of distress have appeared within a year of the placing of a concrete, in others it has been much longer delayed. Examples of deterioration from this cause have occurred in the U.S.A. in almost all forms of concrete exposed to weather, e.g. roads, bridges, dams, and buildings. In the first instance, expansion may occur without cracking and be evidenced by the closing of expansion joints or displacement of kerbs in roads, or the misalignment of different parts of a structure. This is followed by random pattern (Plate 18 (ii)) cracking and in some cases by the exudation through cracks and pores of a soft viscous gel which, on exposure,

[1] H. G. Midgley, *Mag. Concr. Res.* **10** (29), 75 (1958); R. C. Mielenz, *Highw. Res. Rec.*, No. 43, 8 (1963).

[2] T. Hagerman and H. Rosaar, *Betong* (Sweden) **40** (3), 151 (1955); H. Rosaar and E. Vessby, *Nord. Betong* **6** (3), 247 (1962).

[3] *Proc. Am. Soc. civ. Engrs* **66,** 1781 (1940). A large literature has grown up much of which can be traced through the following publications: *Proc. Am. Soc. Test. Mater.* **43,** 199 (1943); **48,** 1055 (1948); W. C. Lerch, *Proc. Am. Soc. Test. Mater.* **53,** 978 (1953); *Highw. Res. Bd Res. Rep.*, No. 18C (1958); ibid. *Spec. Rep.* **31** (1958); Per Bredsdorff *et al., Sym., Washington 1960*, 749; F. E. Jones, *Natn. Bldg Stud. Res. Pap.*, Nos. 14, 15, 17 (1952), 20, 25 (1958); H.M.S.O. London; Reports of Danish Committee on Alkali Reactions in Concrete, Danish National Institute of Building Research, Copenhagen; H. E. Vivian *et al., Bull. Commonw. scient. ind. Res. Org., Melbourne*, No. 229 (1947), No. 256 (1950); *Aust. J. appl. Sci.* **2,** 108, 114, 123, 484 (1951), **2,** 228 (1952), **6,** 78, 88, 94, 100 (1955).

hardens and turns whitish. Examination of cores shows the presence of this gel within the concrete and rims of alteration products round reactive aggregate grains with gel deposits in close proximity. In the examination of dried white surface deposits the isotropic gel can usually be distinguished from crystalline efflorescences of calcium hydroxide or carbonate by the use of a low power binocular microscope.

Both a relatively high content of alkali in the cement and the presence of particular reactive constituents in the aggregate are necessary for this reaction to occur and the trouble thus arises from a combination of two incompatible materials. It does not take place when high-alkali cements are used with other aggregates, or when reactive aggregates are used with cements of sufficiently low alkali content. The content of alkalis in cements varies from below 0·4 per cent Na_2O+K_2O to above 1 per cent and, as has been seen earlier (p. 547), a substantial proportion passes into solution in water quickly. It is usual to calculate the total alkali (R_2O) as equivalent Na_2O, i.e. as the percentage content of Na_2O plus 0·658 times the percentage content of K_2O, assuming that at equivalent concentrations KOH and NaOH are equal in their effect. This is probably not strictly true for there is evidence[1] that though the substitution of potash for soda accelerates expansion it decreases the ultimate expansion. Cements with an R_2O content below 0·6 per cent have been found as a rule to cause little expansion with reactive aggregates, but there are exceptions.[2] Such cements are commonly termed 'low-alkali' cements. Occasionally an aggregate itself may contain soluble alkali salts, or, as in the case of zeolites, produce them by base exchange.

Opal, chalcedony, tridymite, cristobalite, glassy cryptocrystalline volcanic rocks of acid to intermediate composition such as rhyolites and andesites and their tuffs, some zeolites and certain phyllites (metamorphic schists), have all provided examples of alkali reactive aggregates in the U.S.A. Opaline silica, probably the most reactive of all, is found in some cherts, shales and impure limestones. Aggregates of similar types have since been found reactive in Australia and New Zealand, but, apart from Denmark, there are relatively few authenticated cases of troubles arising from alkali-aggregate reaction in Europe. In Denmark[3] serious deterioration from alkali-aggregate reaction has occurred in bridges, marine work and other structures in Jutland where the sedimentary gravels and sands contain porous opaline silica and microcrystalline chalcedony in flint pebbles and particles derived from a hard silicified chalk. A few cases of alkali-aggregate reaction with phyllitic rocks have been reported from Norway and Sweden and it has been suspected in Holland with gravels from the Rhine and the Meuse containing flint. A survey[4] of the common British aggregates has failed to reveal any containing alkali-reactive constituents, but care is needed when using aggregates of geological types which might contain such constituents and of which no previous service experience is available. It might have been expected that flint, which is such a large constituent of the Thames valley gravels,

[1] C. E. S. Davis, *Aust. J. appl. Sci.* **9**, 52 (1958).
[2] C. E. S. Davis, *Aust. J. appl. Sci* **2** (1), 123 (1951); A. D. Conroy, *Proc. Am. Soc. Test. Mater.* **47,** 1000 (1947); **52,** 1205 (1952); D. O. Woolf, *Publ. Rds, Wash.* **27** (3), 50 (1952).
[3] G. M. Idorn, *Durability of Concrete Structures in Denmark*, Technical University, Copenhagen (1967).
[4] F. E. Jones, loc. cit.

would be reactive. Long experience with this aggregate has shown no such troubles, but it appears that it consists of a microporous mass of silica and does not contain opal.[1] Some flints, and chelcedony, in the U.S.A. have been found reactive. The difference between reactive and unreactive forms of silica is to be sought in their crystal structures.[2] An unreactive silica such as quartz has an orderly arrangement of the silicon-oxygen tetrahedra whereas the reactive forms such as opal are characterised by a random network of tetrahedra with irregular spaces between the groups of molecules. The reactive forms have in effect a high internal surface area, making them much more susceptible to surface hydration and to breakage by alkali cations of the silicon-oxygen bonds which bind together the silicon-oxygen tetrahedra. The silica is thus peptised to form a silica or an alkali silicate gel, which is capable of unlimited swelling as it adsorbs water. Lime has an inhibiting effect[3] on these processes because the calcium-silicate or calcium-alkali silicate gels are relatively insoluble and, like set cement itself, have only limited swelling properties.

The mechanism of the expansion caused by alkali-aggregate reaction is not yet entirely solved. The simplest theory[4] ascribes it to the direct enlargement of the affected pieces of aggregate producing pressures which rupture the concrete just as hydration of hard burnt lime or magnesia produces unsoundness in cement. Other theories concentrate attention more on the properties of the reaction product than on the immediate growth of aggregate particles. The gels formed are alkali silicates containing a certain amount of lime; thus samples taken from deteriorated concrete have shown percentage oxide contents such as:

(i) $82SiO_2$, $4Na_2O$, $2K_2O$, $1CaO$, $10H_2O$
(ii) $53SiO_2$, $13Na_2O$, $5K_2O$, $5CaO$, $21H_2O$

The composition of the reaction product can vary from a high calcium-alkali-silica gel of a non-expansive type to a predominantly alkali-silica gel with large expansive properties. The amount of active silica available and the relative local concentration of calcium and alkali hydroxides in the liquid phase within the concrete must be the major factors determining the composition of the gel that is formed and the extent to which it has expansive properties. Solid calcium hydroxide will always be available in the set cement, but its solubility decreases markedly as the concentration of alkali hydroxide increases so diminishing more than proportionally the ratio of lime to alkali in the solution. It is to these effects that we must ascribe the existence of a maximum limit of the alkali content of a cement below which expansion does not occur and why, though a general value such as 0·6 per cent can be established, some exceptions will nevertheless occur.

Broadly two main theories have been advanced to explain the expansion. One ascribes the expansive pressure to the adsorption of water by the gel and the other to osmotic pressure effects. The adsorption theory essentially directs attention to the behaviour of the solid gel while the osmotic pressure theory is concerned with

[1] H. G. Midgley, *Geol. Mag.* **88** (3), 179 (1951).
[2] T. C. Powers and H. H. Steinour, *Proc. Am. Concr. Inst.* **51,** 497,785 (1955); W. C. Hansen, *J. Materials* **2,** 408 (1967).
[3] R. C. Pike and D. Hubbard, *Bull. Highw. Res. Bd,* No. 171, 16 (1958); No. 175, 39 (1960); *J. Res. natn. Bur. Stand.* **59,** 127 (1957).
[4] L. S. Brown, *Bull. Am. Soc. Test. Mater.* **205,** 40 (1955).

the hydraulic pressure developed across a semipermeable membrane. On the adsorption theory[1] the swelling will be determined by the composition of the lime-alkali silicate gel and increase as the lime-alkali ratio in the gel decreases. On the osmotic pressure theory[2] it was suggested by Hansen that the cement paste round a reactive particle acts as a semipermeable membrane allowing alkali hydroxides and water to diffuse through to the particle but preventing the complex silicate ions produced by the alkali-aggregate reaction from passing out. An osmotic pressure cell is thus set up, with the alkali-silicate solution inside drawing the cement-liquid phase through the membrane. In later developments the distinction between the adsorption and osmotic pressure theories has tended somewhat to be lost. The semipermeable membrane is considered to be formed by the lime-alkali silicate gel initially formed round the particles. This permits alkalis to diffuse through it more readily than lime. The higher the alkali content of the gel membrane the slower is the diffusion of lime through it compared with that of the alkalis so that at the reaction site at the particle surface the further silicate gel that is formed is high in alkali and can swell by imbibition of water passing through the membrane. This preferential diffusion of alkali is also invoked to explain why an alkali-silicate gel can persist in a concrete in which ample quantities of calcium hydroxide are present. It has been demonstrated that concrete in contact with sodium-silicate solutions can act as a semipermeable membrane and that pressures of over 500 lb/in^2 can be developed.[3] Mixtures of cement and alkali-reactive opal aggregate can produce pressures[4] of over 2000 lb/in^2, far above the tensile strength of concrete. It is not essential however to invoke the formation of a semipermeable membrane to explain the production of pressure when a gel capable of swelling is also present. Thermodynamically the cause is a common one, a difference in partial free energy between water in two parts of the system causing water to move in such a direction as to reduce this difference. Osmotic pressure and swelling pressure are both manifestations of this. The mechanics can still remain capable of argument.

It has been found that there is a certain content, known as the 'pessimum' content, of reactive material in an aggregate that leads to maximum expansion. This 'pessimum' content may be as low as 3–5 per cent in the case of opal, whereas with less active materials it may be 10 or 20 per cent or even rise to 100 per cent. With very active materials the maximum expansion of a concrete tends to increase with the particle size, but with less active materials the reverse is true. Very finely divided material, e.g. of cement fineness, may in either case cause no expansion. Thus finely ground opalene chert added to the extent of about ten times the weight of alkali in the cement was found by Stanton actually to prevent expansion with a reactive aggregate. The probable reason why the expansion can be insignificant when the proportion of reactive material is high, or when it is very fine, is that the alkali-silica reaction products are characterised by an alkali

[1] H. E. Vivian, loc. cit., 1950, p. 60.
[2] W. C. Hansen, *Proc. Am. Concr. Inst.* **40,** 213 (1944); W. H. Parsons and H. Insley, ibid. **44,** 625 (1948); G. J. Verbeck and C. Gramlich, *Proc. Am. Soc. Test. Mater.* **55,** 1110 (1955).
[3] D. McConnell, R. C. Mielenz, W. Y. Holland and K. T. Greene, *Proc. Am. Concr. Inst.* **44,** 93, 632–1 (1948).
[4] R. G. Pike, *Bull. Highw. Res. Bd,* No. 171, 34 (1958).

to silica ratio so low as to minimise the uptake of water and swelling. Thus in the case of the Danish aggregates[1] no deleterious expansion was found to occur when the reactive flint content was less than 2 per cent nor when it exceeded about 20 to 30 per cent for $\frac{1}{8}$–$\frac{1}{4}$ mm aggregate and over 50 per cent for $\frac{1}{2}$–2 mm aggregate. There is in fact a complex relation between the quantity and fineness of the reactive material, the alkali content of the cement, and the degree of expansion. Less effect is also found with porous aggregates and concretes, in which space is available to accommodate the reaction product, than in dense concretes. It will be evident that the degree of reaction and the degree of expansion are not synonymous terms. Thus, as discussed later, pozzolanas which are reactive silicate materials are often a corrective for alkali-aggregate expansion. The expansive effect, as we have seen, is bound up with the ratio of alkali to reactive silica, which determines the amount of alkali available to each reactive particle, and the relative local concentration of alkali and lime since the latter reduces the tendency of the reaction product to form a swelling gel. Too small amounts of alkali can cause little gel formation while too large amounts produce a more fluid reaction product less capable of exerting pressures. Water is also essential to the expansion which does not occur in dry concrete.

Various methods of testing aggregates for potential alkali reactivity have been developed. A skilled petrographer can detect the presence of opal and with experience form a good impression of the probability of any troubles arising.[2] A long period test which has been used is that originated by Stanton, and later standardised in ASTM C227–67; the expansion of 1 : 2·25 mortar bars, made from the aggregate in question and stored in closed cans over, but not in contact with, water, is measured for periods up to 12 months, and longer if necessary. A storage temperature of 100°F is specified. The cement to be used may be that to be used for a particular job with the aggregate in question, or the test may have the more general purpose of classifying the aggregate in which case tests are done with both high- and low-alkali cements. Cement-aggregate combinations which show an expansion greater than 0·05 per cent at three months or 0·1 per cent at six months are considered capable of harmful reactivity.

A rapid test (ASTM C289–66) that has been found useful depends on the relation between the silica dissolved and the reduction in alkalinity when 25 g of crushed aggregate, 50–100 mesh, is immersed in 25 ml 1 N NaOH solution at 80° for 24 hours in an air-tight stainless steel vessel. The solution is filtered, analysed for dissolved silica and the reduction in alkalinity determined by titration with HCl using a pehnolphthalein indicator. The basis[3] of this test is that reactive materials seem to release more than the molar equivalent of silica into solution for a given reduction in alkalinity and that this greater proportional release of silica is associated with the formation of a swelling gel in alkali-aggregate reaction. While neither of the quantities measured is alone an index of expansive reactivity a plot of S_c, the quantity of dissolved silica, against R_c,

[1] J. Jessung, A. Kjaer, G. Larsen and E. Truds, *R.I.L.E.M. Sym. Durability of Concrete Preliminary Report*, p. 103, Prague 1961; see also Per Bredsdorff *et al.*, loc. cit.

[2] *Spec. Rep. Highw. Res. Bd*, No. 31 (1958); B. Mather, *Proc. Am. Soc. Test. Mater.* **48**, 1120 (1948); **50**, 1288 (1950); ASTM C295–65.

[3] R. C. Mielenz and L. P. Witte, *Proc. Am. Soc. Test. Mater.* **48,** 1071 (1948).

the reduction in alkalinity, both expressed in millimols per litre, enables regions to be defined in which there fall three classes of aggregates as shown in Fig. 146. Aggregates for which the test results fall in the zone marked 'potentially dangerous' require testing further by the mortar bar test. These are often highly reactive materials causing little expansion where used as the whole aggregate or in a proportion well in excess of the 'pessimum' content. There are anomalies[1] in the results given by this rapid test with aggregates containing carbonates which can give spurious results and in any event its reproducibility is not all that could be desired. Some aggregates are found falling in the innocuous zone which prove to be expansive but usually the result is on the safe side. Thus the flint aggregates of the London basin which are not expansively reactive and which, from long experience, are known to be sound aggregates are indicated as harmful by this rapid test, as are also some sound quartzites.

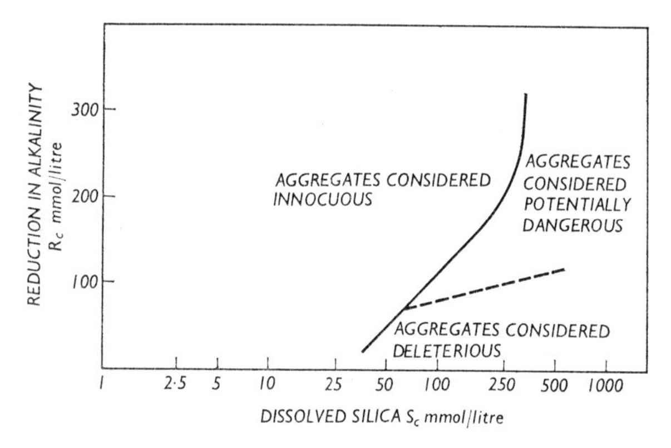

FIG. 146 Rapid chemical test for alkali reactive aggregates.

When it is necessary to use aggregates that are alkali-reactive it is the usual practice to require a cement with an alkali content below 0·6 per cent R_2O, or to make use of a pozzolana, or both. Pozzolanic cements containing 15–40 per cent or more pozzolana, and indeed of opal itself ground to cement fineness, have been found effective in eliminating expansion, or reducing it to acceptable values.[2] The percentage by solid volume of pozzolana, used to replace cement, required to meet the requirements of the test outlined below was found by Pepper and Mather to vary from 20–30 per cent with calcined shale, 40–45 per cent with pulverised fuel ash, 20 per cent with uncalcined diatomite, and 30–35 per cent with volcanic glasses. Pozzolanas are clearly not all equally effective. They vary in their activity as possolanic materials and also in the extent to which they release alkali to a lime solution. The first of these factors, and probably the second, must

[1] R. C. Mielenz and E. J. Benton, *Bull. Highw. Res. Bd*, No. 171, 1 (1958); B. Chaiken and W. J. Halstead, ibid., No. 239, 24 (1960).

[2] L. Pepper and B. Mather, *Proc. Am. Soc. Test. Mater.* **59,** 1178 (1959); *Tech. Rep. U.S. Army Exp. Stn*, No. 6, 481 (1958); W. Lerch, *ASTM Special Technical Publication*, No. 99, 153 (1950).

determine the relative value of different pozzolanas for this particular purpose. The use of Portland blastfurnace cements[1] is also effective in preventing expansion with alkali-reactive aggregates even though the cement contains more than 0·6 per cent alkalis. Seawater or other waters containing significant amounts of alkali salts should not be used as mixing water with reactive aggregates.

The effectiveness of pozzolanas in reducing expansion can be tested by using a crushed borosilicate glass (e.g. Pyrex) as the aggregate in the mortar bar expansion test.[2] The crushed glass is graded as follows:

Sieve size		
Passing	Retained on	Percentage by weight
No. 4 (4·76 mm)	No. 8 (2·38 mm)	10
No. 8 (2·38 mm)	No. 16 (1·19 mm)	25
No. 16 (1·19 mm)	No. 30 (0·595 mm)	25
No. 30 (0·595 mm)	No. 50 (0·297 mm)	25
No. 50 (0·297 mm)	No. 100 (0·149 mm)	15

Tests may be made on the premixed pozzolanic cement (ASTM C595–68) or on a pozzolana in which case it is used to replace 20 or 30 per cent by volume of a high-alkali Portland cement. The borosilicate glass is a very reactive aggregate giving a rapid expansion with a high-alkali cement so enabling a quick assessment to be made of the reduction in expansion brought about by a pozzolana. Acceptable pozzolanas reduce the expansion by at least 75 per cent at 14 days. This is equivalent to the reduction obtained in this test using cement alone when a low-alkali cement is compared with a high-alkali one. Other, long period, tests can be made if desired on the particular combination of cement, pozzolana and aggregate to be used in practice. The rapid chemical test for aggregates cannot be used for assessing the effect of pozzolanas, or of the granulated slag component of Portland blastfurnace cements.

The mechanism of the action of pozzolanas in reducing expansion is not clear, but there is a suggestion[3] that it is related to their ability to take up Na or K from solution. This may result in the reaction product being distributed throughout the concrete, rather than in localised masses around aggregate particles capable of producing osmotic or swelling pressures. This theory is open to doubt, and can only at best be a partial explanation, for natural pozzolanas, such as the Italian, which release alkali to a lime solution, are effective in reducing expansion.

Some gas entraining agents, such as aluminium powder, and air-entraining agents, proteins, etc., also reduce expansion, probably because they provide dispersed voids into which reaction products can grow without causing a development of stress.

In the interior of large mass concrete structures the alkali-aggregate reaction alone can be responsible for deterioration, but in thin members, or at the surface of concretes, its effects may be accentuated by frost action or by moisture or thermal changes. Acting together these may produce cracking that would not

[1] B. Mather, *Proc. Am. Concr. Inst.* **54,** 205 (1958).
[2] W. T. Moran, *Proc. Am. Concr. Inst.* **47,** 43 (1951).
[3] R. C. Mielenz, K. T. Greene, E. J. Benton and F. H. Geier, *Proc. Am. Soc. Test. Mater.* **52,** 1128 (1952).

arise from any single cause alone. It is not always easy to differentiate the effects of alkali-aggregate reaction from that of frost and other agencies, but the presence of the alkali silicate gel at the surface or within the concrete, and the occurrence of reaction rims round particles of aggregate, are diagnostic signs.

More than one cause of deterioration seems to be involved with some aggregates in Kansas, Nebraska and neighbouring areas, of the central plains of the U.S.A.[1] The available aggregates in these regions, derived from major rivers, contain little coarse material, only some 5–15 per cent above $\frac{3}{16}$ in. size, and are known locally as 'sand-gravel' aggregates. This is a special use of the term 'sand-gravel' which is more commonly used to designate any mixture of sand and gravel without any implication of potential unsoundness. The Kansas-Nebraska aggregates are composed essentially of coarse-grained granite and individual quartz and feldspar particles with smaller amounts of various other rocks including siliceous limestone, acid volcanic glass, chert and sometimes opal. Indications of alkali-aggregate reaction have been found in the deteriorated concrete and those aggregates that contain opal are definitely alkali-reactive. Others, however, do not respond to the usual tests for alkali-reactive aggregate and the deterioration observed in concrete is not related to the alkali content of the cement. The replacement of 15 to 35 per cent of the cement by some pozzolanas has been found beneficial in some laboratory tests but Lerch did not find them so in some field trials on road slabs. A form of expansion test which has been found useful in detecting such aggregates is described in ASTM C342–67. A 1 : 2·25 mortar bar made with aggregate graded from No. 4 to No. 100 sieves is cured for 24 hours in moist air at 23°C (73·4°F) and then in water at the same temperature for 28 days, the first measurement of length being made after 24 hours in water. The bars are then stored in water at 55°C (131°F) for 7 days followed by 24 hours in water at 23°C and measured again. The bars are then dried at 55°F for 7 days, cooled to 23°C for 24 hours and remeasured. The specimens are then stored again in water at 23° C and the length remeasured after 1, 7 and 28 days and longer periods up to a year. This test is a development of one proposed earlier by Conroy. Other tests involving temperature cycling have also been proposed. The fact that expansion does not occur in the ordinary mortar-bar expansion test at constant temperature used for testing alkali-reactivity and that temperature cycling is required to bring out the expansive properties of these aggregates suggests that physical as well as chemical caused may be involved. It has been suggested[2] that the deterioration is a combined effect of alkali-silica reaction and severe drying shrinkage arising from the climatic conditions.

Alkali-carbonate rock reaction. An expansive type of reaction[3] in concrete under moist conditions has been found to occur between alkalis from the cement and certain, not very common, types of dolomitic limestones which from their

[1] C. H. Scholer and W. E. Gibson, *Proc. Am. Concr. Inst.* **44,** 1009 (1948); C. H. Scholer and G. M. Smith, *Proc. Am. Soc. Test. Mater.* **54,** 1165 (1954); A. D. Conroy, *Proc. Am. Soc. Test. Mater.* **52,** 1205 (1952); W. Lerch, *J. Res. Dev. Labs Portld Cem. Ass.* **1** (3), 42 (1959); *Highw. Res. Bd spec. Rep. 31 (1958)*; G. W. de Puy, *Highw. Res. Rec.*, 1966 (124), 41.

[2] D. W. Hadley, *J. Res. Dev. Labs Portld Cem. Ass.* **10** (1), 17 (1968).

[3] E. G. Swenson, *Bull. Am. Soc. Test. Mater.*, No. 266, 48 (1957); *Highw. Res. Rec.*, No. 45 (1964). Symposium on alkali-carbonate rock reactions.

physical properties appear to form satisfactory aggregates. This reaction is quite distinct from the alkali-siliceous aggregate reaction, and aggregates subject to it are not detected by the methods outlined in the previous section for reactive silica. The first example of deterioration of concrete assigned to this cause was found at Kingston,[1] Ontario in 1955, but other cases have since been reported from the U.S.A.[2] No examples have so far been reported in Europe though they may of course have not been detected. The signs of distress have been map cracking or grosser cracking of concrete paving slabs, buckling of concrete floors and other indications of expansion. Moisture is essential to the reaction and it does not occur in dry concrete. The age at which the symptoms of distress have appeared in concrete has been very variable. Thus an expansion of over 0·2 per cent has been recorded in six months while in other cases it has only become evident after years. The rocks involved are limited to certain fine-grained, argillaceous, dolomitic limestones and they seem to fall within a range of composition, intermediate between the end members of the calcite-dolomite series, which are not very common. The rocks which expand excessively contain about 40 to 60 per cent dolomite and some 5 to 20 per cent of acid-insoluble material consisting predominantly of illite and similar clays, degraded mica and some fine quartz. They are invariably fine grained and generally of low porosity.

The expansion of concrete containing such rocks as coarse aggregate increases with the alkali content of the cement and to keep it down to an acceptable level it is necessary to limit the total cement alkali to a maximum of about 0·4 per cent. The partial replacement of cement by pozzolanas, which is helpful in the case of alkali-silica reactions, has little effect on the ultimate expansion. The size of the coarse aggregate has some effect, the expansion decreasing with size. The sands from these crushed rocks show the least expansion. The most reliable method for detecting aggregates of this type is to measure the expansion of concrete specimens maintained at nearly 100 per cent humidity at 73° F (23°C) using a high-alkali cement. Cracking begins to occur at linear expansions between 0·05 and 0·1 per cent. The rocks themselves can be tested by measuring the expansion of prisms cut from the rock when immersed[3] in a 1 or 2 M solution made up of equal parts NaOH and KOH. The expansion found with reactive rocks is very similar to that of concrete made from them.

In some dolomitic aggregate concretes reaction rims[4] have been found round the coarse aggregate particles and to be more siliceous than the aggregate particles. Their formation seems to have some detrimental effect on strength, but it is doubtful if it is associated with expansion. The silica is probably dissolved by alkali from the fine argillaceous and siliceous materials within the rock and redeposited in the rim.

The mechanism of the reaction is still not entirely clear. The reactive rocks have in common the invariable presence of dolomite, calcite and an acid insoluble

[1] E. G. Swenson and R. F. Legget, *Can. Consult. Engr* **2,** 38 (1960).
[2] D. W. Hadley, *Highw. Res. Rec.*, No. 45, 1, 196, 235 (1964); W. I. Luke, *Proc. Am. Soc. Test. Mater.* **64,** 1187 (1964).
[3] E. G. Swenson and J. E. Gillott, *Bull. Highw. Res. Bd*, No. 275, 18 (1960).
[4] J. Lemish, F. E. Rush and C. L. Hiltrop, *Bull. Highw. Res. Bd*, No. 196, 1 (1958); No. 239, 41 (1960); No. 275, 32 (1960); *Highw. Res. Rec.*, No. 3, 1 (1963).

fraction though the proportions may vary. In the presence of alkali a dedolomitisation reaction occurs which can be expressed by:

$$CaCO_3.MgCO_3+2NaOH = CaCO_3+Mg(OH)_2+Na_2CO_3$$

The dolomite crystals in the reactive rocks contain inclusions of clay which is considered not to be fully hydrated. This 'unwetted' clay becomes exposed to moisture as a result of the above reaction and its hydration is considered to lead to the build-up of expansive forces.[1] It has also been suggested that osmotic forces operating through a clay membrane may be involved.[2]

Lightweight aggregates

Lightweight aggregates[3] are materials used to obtain concretes with weights usually below about 100 lb/ft^3 (1600 kg/m^3), though products up to 120 lb are used for structural concrete. At the heavier end quite high strengths may be obtained, but when the weight is below about 40 lb the material is usually suitable only for non-load-bearing purposes. Some lightweight aggregates are also used in plasters and for heat-insulating screeds.

The materials can be divided into two main groups, natural and artificial. Pumice is the only representative of the natural group. The artificial aggregates can be sub-divided into two classes; the first of these consists of a waste product, furnace clinker (or 'cinders' in the U.S.A.), which is cheap but very variable in quality. Small coke, known as coke breeze, was at one time used but has now been abandoned. The second class consists of materials made by processing artificial or natural products and these are much more uniform in quality. They include foamed, or expanded, slag, expanded clays, shales and slates, together with two very light materials, exfoliated vermiculite and expanded perlite. In addition, sintering processes have been developed for producing a lightweight aggregate from pulverised fuel ash (fly-ash).

Concretes made with lightweight aggregates vary in characteristics over a wide range. The lightest concretes, made with exfoliated vermiculite and expanded perlite, can range in weight from 15–50 lb/ft^3 (240–800 kg/m^3) with a thermal conductivity[4] from 0·5 to 1·5 Btu/ft^2/h/in./°F. Though in the higher weights these materials can be used for concrete blocks, their main use is for insulation purposes, for except in the heaviest products the compressive strength is low, only one or two hundred lb/in^2 (0·7–1·4 N/mm^2). The drying shrinkage is also high from 0·2–0·4 per cent. For roof insulation concrete screeds of 20–30 lb/ft^3 (320–480 kg/m^3) are commonly used.

[1] J. E. Gillott, *Can. J. Earth Sciences*, **I**, 121 (1964); E. G. Swenson and J. E. Gillott, *Mag. Concr. Res.* **19** (59), 95 (1967).

[2] D. W. Hadley, loc. cit.

[3] A. Short and W. Kinniburgh, *Lightweight Concrete.* C.R. Books, London (1968); C; Hobbs, *Chemy Ind.*, 1964, 594; D. C. Teychenne, *Concrete* **I** (4), 111, (6), 207 (1967) Report A.C.I. Committee 213, *Proc. Am. Concr. Inst.* **64**, 433 (1967); D. W. Lewis *Spec. tech. Publ. Am. Soc. Test. Mater.*, No. 169A, 359 (1966); *R.I.L.E.M. Sym. Testing and Design Methods for Lightweight Aggregate Concretes, Budapest*, Kultura Book Exp. Dept. (1967).

[4] 1 Btu per square foot per hour per inch per °F equals 0·144 S1 units (joules per square metre per second per metre per °C).

The medium weight lightweight concretes made with sintered pulverised fuel ash, expanded clay, shale or slate, foamed slag, clinker or pumice can range in weight from 40–100 lb/ft^3 (640–1600 kg/m^3) with thermal conductivities from 1–4 Btu. For the manufacture of precast concrete blocks semi-dry mixes which are relatively lean, e.g. 1 : 7 to 1 : 10 by volume, and not fully compacted, are used. Some typical properties are shown in Table 122. These aggregates are also used for insulating roof screeds in mixes often about 1 : 8 to 1 : 10 by volume and in rather richer mixes for insulating floor screeds.

TABLE 122 Semi-compacted blocks made with lightweight aggregates

Aggregate	Density of dry concrete		Compressive strength		Drying shrinkage	Thermal conductivity	
	kg/m^3	lb/ft^3	N/mm^2	lb/in^2	per cent	SI units	B.T.U. in/ft^2 h° F
Pumice	800	50	4·2	600	0·04–0·08	0·15–0·3	1–2
Clinker	960–1520	60–95	2·1–7	300–1000	0·04–0·08	0·35–0·6	2·5–4
Expanded Clay or Shale	960–1200	60–75	5·6–8·4	800–1200	0·04–0·07	0·3 –0·45	2–3
Sintered pulverised fuel-ash	1120–1280	70–80	4·2–10·5	600–1500	0·04–0·07	0·3 –0·45	2–3
Foamed slag	960–1520	60–95	2·1–7	300–1000	0·03–0·07	0·2 –0·45	1·5–3

Conversions rounded off: 1 N/mm^2 = 10 kg/m^2.

The heaviest lightweight concretes range from 80–120 lb/ft^3 (1280–1920 kg/m^3) in weight, or even more if sand is used for part of the fine aggregate. Such concretes are made with appropriately graded aggregates from $\frac{3}{4}$ in. or $\frac{1}{2}$ in. down in mixes not leaner than about 1 : 6 by volume. They are used for structural purposes in reinforced or prestressed concrete and strengths as high as 6000 lb/in^2 (42 N/mm^2) or more at 28 days are obtainable with mixes not leaner than about 1 : 4 by volume or about 700–800 lb cement per cubic yard (420–480 kg/m^3) concrete; the drying shrinkage at this cement content ranges from about 0·05 to 0·08 per cent. Though relatively heavy, these concretes still show a useful saving in weight compared with concrete made from dense aggregates weighing 140–150 lb/ft^3 (2240–2400 kg/m^3).

The frost resistance of lightweight aggregate concretes of medium weight is relatively good compared with that of concrete of similar strength containing dense aggregates. This is to be ascribed to their pore structure and the ease with which they loose water by drainage and drying. The heaviest[1] lightweight aggregate concretes show a frost resistance comparable with that of dense aggregate concrete of similar strength and similarly influenced beneficially by air-entrainment.

Lightweight aggregates are covered by a number of British specifications viz. BS 3797 : 1964 *Lightweight Aggregates;* BS 3681 : 1963 *Methods for the Sampling and Testing of Lightweight Aggregates;* BS 877 : 1967 *Foamed Slag Aggregate;* and BS 1165 : 1966 *Clinker Aggregate.* The corresponding ASTM specifications are C330–68T *Lightweight Aggregates for Structural Concrete;* C331–64T *Lightweight Aggregates for Concrete Masonry Blocks;* and C332–66 *Lightweight Aggregates for Insulating Concrete.* BS 3797 : 1964 controls the bulk density (oven-dry) and grading of the aggregates. For aggregates other than vermiculite and perlite the bulk density must not exceed 75 lb/ft^3 per fine aggregate and 60 lb for coarse aggregate. The maximum values for vermiculite and perlite are 8 and 15 lb respectively for the whole aggregate. The standard includes limits of 1 per cent on the SO_3 content and 4 per cent for the loss on ignition. The latter requirement does not apply to vermiculite. Foamed slag and clinker aggregates are covered by separate specifications as noted above. The ASTM specifications also limit the bulk density (oven-dry) and control the grading. For aggregate for structural concrete or concrete blocks the loose weight of fine aggregate must not exceed 70 lb/ft^3, 55 lb for coarse aggregate, and 65 lb for the combined aggregate. For insulating concretes the limits for vermiculite are 6–10 lb/ft^3 and for perlite 7½–12 lb. Provision is also made in the ASTM standards C330 and C331 for testing the drying shrinkage characteristics of an aggregate by measurements on a concrete made from 1 vol cement to 6 vols combined aggregate. After 7 days moist curing the shrinkage on storing 100 days in air at 73·4°F (23°C) and 50 per cent relative humidity must not exceed 0·10 per cent. The control of shrinkage is exercised in Great Britain through limits placed on the concrete products as offered for sale whereas the ASTM specifications for concrete blocks contain no limitations on shrinkage but only on the moisture content of the block as delivered.

Concretes made from lightweight aggregates usually have a higher drying shrinkage and moisture movement than heavy concretes and are more prone to shrinkage cracking. The drying shrinkage can be reduced by drying of precast products in the course of manufacture, but this is not applicable to *in-situ* work.

BS 2028 : 1364–1968 for precast concrete blocks classifies heavy concrete blocks as those having a density not less than 1500 kg/m^3 (93·6 lb/ft^3) and lightweight concrete blocks as those with a density below this. The lightweight concrete blocks cover those made with lightweight aggregate or aerated concrete (see p. 587). The BS limits for drying shrinkage vary according to the density and strength of the blocks as shown in Table 123.

Most lightweight concretes can be easily nailed and cut, with the exception of the heavier ones, and this is a property of some importance in their use. They have good heat insulating properties, but the thermal conductivity increases

[1] P. Klieger and J. A. Hanson, *Proc. Am. Concr. Inst.* **57,** 779 (1961); R. Landgren, *Proc. Am. Soc. Test. Mater.* **64,** 830 (1964).

roughly as the density and is considerably affected by moisture content, the value increasing by 50–100 per cent from the fully dry to the water saturated state. The conductivity is about 1–2 Btu/ft^2/h/in/°F for normal air-dry materials weighing 50 lb/ft^3, 2–3 at 75 lb, and 3–5 at 100 lb.

TABLE 123 BS limits for drying shrinkage of precast concrete blocks

Type of block	Density		Compressive strength		Maximum drying shrinkage per cent
	kg/m^3	lb/ft^3	N/mm$_2$	lb/in^2	
Heavy	Not less than 1500	93·6	$<$ 10·5	$<$ 1500	0·05
			$\geqslant$ 10·5	$\geqslant$ 1500	0·06
Lightweight load-bearing	$>$ 625	$>$ 39	$<$ 7	$<$ 1000	0·07
	$\leqslant$ 625	$\leqslant$ 39			0·09
	$<$ 1500	$<$ 93·6	$\geqslant$ 7	$\geqslant$ 1000	0·08
Lightweight non load-bearing	$>$ 625	$>$ 39	Independent of strength		0·08
	$\leqslant$ 625	$\leqslant$ 39			0·09

1 Newton (N) /mm^2 = 10 kg/cm^2 = 142 lb/in^2.

There are various classes of blocks with differing strength requirements laid down in the British Standard, the minimum compressive strengths varying from 400 to 1000 lb/in^2 for load-bearing lightweight blocks and from 500 to 5000 lb/in^2 for heavy blocks normally made with dense aggregates. Practice in the U.S.A. favours the more fully compacted and heavier type of load-bearing block. For solid blocks (ASTM C145) a minimum compressive strength of 1200–1800 lb/in^2 is required, and for hollow blocks 700–1000 lb/in^2 calculated on the gross area, the precise minimum value in both cases depending on the use to be made of the block. For non-loadbearing hollow blocks the minimum is lower, 350 lb/in^2 calculated on the gross area (ASTM C129).

Pumice

Pumice is a product of volcanic origin produced from molten lavas from which considerable volumes of dissolved gases have been released during solidification. Lavas from which no release of gas occurs solidify as glassy, or partially devitrified solids. Pumice is a solid lava so swollen by escaping gases as to form a kind of solidified froth. It is a glassy mass honeycombed with cavities and sometimes contains a certain amount of crystalline matter. It varies from white to yellow or brown in colour.

Deposits of pumice are found in many different parts of the world; the chief European sources are Greece and Germany. It is quarried in the neighbourhood of Coblenz on the Rhine, where it occurs as layers several feet thick separated by thin bands of other materials. Raw pumice contains a certain amount of fine volcanic dust, shale, and other extraneous matter. The volcanic dust and any clay present can be removed by washing, and the shale by some from of flotation process; the removal of the latter heavy material reduces the volumetric weight of the pumice.

Pumice concretes usually range from 40 to 80 lb/ft^3 in weight, but materials up to 100 lb can be produced in fully compacted concrete.

Foamed slag

Foamed, or expanded, slag is manufactured by the treatment with a limited amount of water of molten blastfurnace slag. The process is so conducted as to give a fairly rapid chilling, but less than in the manufacture of granulated slag (see p. 456) in order to obtain a stronger material. Even slags which might 'fall' on normal air cooling are stabilised by the process.[1] The British Standard for foamed slag limits the bulk density of the material (oven-dry) to 50 lb/ft^3 for coarse aggregate and 70 lb for fine aggregate, and also controls the grading. The SO_3 content must not exceed 1·0 per cent. The German specification (DIN 18152) limits the total sulphur to 1·5 per cent expressed as sulphur trioxide. In the U.S.A. the material is usually known as expanded slag and sold under various trade names. It is subject to the general provision of the ASTM specifications C330–68T and C331–64T for lightweight aggregates, but no more specific requirements.

Expanded clay, shale and slate aggregates

Suitable clays when rapidly heated to a temperature such as to cause surface fusion expand considerably and become bloated. Some shales and slates behave in a similar manner. Though sintering grates are used, the process is more often carried out in a rotary kiln, the raw material being fed as a crushed solid. Little is at present known about the mechanism of bloating; the gases and fluxes necessary to cause it appear to arise not from the clay minerals but from other substances present.

Sintered pulverised fuel ash (fly ash)

In the production of lightweight aggregates from pulverised fuel ash the material is first formed into pellets by spraying with water in an inclined rotating shallow pan. The pellets are then sintered at 1000–1200°C on a travelling sintering grate. Some 7–8 per cent of fuel is needed and, if necessary, additional ground coal or coke must be added to the raw material. Vertical shaft kilns have also been tried and with their greater fuel economy will sinter material with a fuel content of 4–5 per cent. At higher fuel contents serious difficulties arise from slagging of the raw material and at lower contents they fail to burn. The use of a shaft kiln is therefore only practicable if the fuel content of the ash can be kept continuously between these narrow limits. The use of a rotary kiln for sintering is also another method that has been tried.

[1] T. W. Parker, *Iron Steel Inst. Spec. Rep.*, No. 19 (1937).

Exfoliated vermiculite

Vermiculite is a mica-like material which when heated to 600–900° exfoliates, giving a product with a loose weight of 4–15 lb per c ft according to the grading. Apart from lightweight concretes, it is also used in cement renderings and gypsum plasters, and as a loose filling for heat insulation.

Expanded perlite

A number of volcanic glasses such as perlite, obsidian and rhyolite expand when heated rapidly to the point of incipient fusion. The product obtained from perlite generally has a loose weight of 7–15 lb/ft^3 though still lighter material can be obtained. It is used for similar purposes to vermiculite.

Clinker

The term clinker is used to cover the partly fused or sintered ashes arising from the combustion of coal. Ashes from ordinary domestic grates are in a too finely divided condition to be of use, but in large furnaces, such as those used for steam raising at electricity power stations, the ashes become sintered or fused together at the higher temperatures which are attained. In the building industry the term 'breeze' was formerly used to describe some of these materials but is falling out of use. In the United States the term cinders is used. The supply of clinker aggregates is declining owing to the use of pulverised coal.

Clinker aggregates vary very widely in quality, and the poorer grades have been the cause of many building troubles and failures. Poor quality clinker aggregates containing a large proportion of combustible matter yield concretes with a high drying shrinkage and moisture movement. When used in the form of partition slabs the large shrinkage which takes place as the partition dries out after plastering frequently leads to cracking of the plaster. There is, however, another and much more serious trouble which sometimes arises with clinker aggregates which consists of a progressive expansion of the hardened concrete. This expansion has sometimes caused such severe cracking of clinker concretes, cast *in situ*, that walls become fissured and unsafe. A similar expansion in floors, partitions, or screedings on flat roofs, has led to gradual displacement of walls, parapets and chimney-stacks to such a degree as to make reconstruction essential. Clinkers producing an expansion in concrete are termed unsound.

This type of unsoundness arises from the nature of the coal burnt and though prevalent in Great Britain is apparently rare in the U.S.A. and some European countries, though it has been reported in Poland.[1] Other troubles arising in clinker aggregate concretes are popping and spalling and staining.

Unsoundness in clinker aggregates has been found to be due to the presence of certain types of coals in an unburnt or very slightly burnt condition,[2] but by no means all coals act in this way and even the dangerous coals cease to cause expansion when carbonised to a degree approaching that of coke. The total

[1] W. Roznak, *R.I.L.E.M. Sym. Lightweight Aggregate Concretes, Budapest,* Kultura Exp. Dept., 1967, 321.
[2] F. M. Lea, *Building Research Technical Paper,* No. 7 (1929).

content of combustible matter in a clinker is not, therefore, a measure of unsoundness, since this combustible matter may vary in type from unburnt coal to well-carbonised material, and the coal itself may belong to the dangerous or non-dangerous class. Apart from this specific unsoundness the general quality of clinker aggregates decreases as the content of combustible matter contained in them increases. Thus the moisture movement of the concrete rises with the combustible content of the aggregate. The amount of combustible matter in well-fused clinkers may fall as low as 1–2 per cent and is never high; in less well-burnt materials, such as many of the more lightly sintered products, it rises very much higher. Values of 15–25 per cent are very common and a figure as high as 50 per cent may be found. The combustible content of clinker can be determined with sufficient accuracy by igniting the finely ground material, previously dried at 100°, for 3–4 hours at 800° followed by 1–2 hours at 1000°.

The expansion of clinker concretes containing coals of the dangerous types has been found to be due to two factors. All coals expand on wetting and contract on drying, the movement varying from about 0·02 to 0·01 per cent on non-dangerous coals up to 1–2 per cent on dangerous coals. This movement alone would not give rise to an expansion in a concrete, since with dangerous coals it reaches its maximum value in a few hours. In the presence, however, of Portland cement, calcium hydroxide, or other substances producing hydroxyl ions in solution, the swelling of the dangerous coals continues over a period of days and attains a value considerably greater than that reached in water. This movement causes an expansion in the concrete during the early period of hardening, and it is then followed by another action which proceeds slowly and progressively for a prolonged period. This further action is due to the absorption of oxygen by the coal. All coals when exposed to the air at ordinary temperatures take up oxygen at a rate which decreases with time, but is much increased by the presence of free calcium hydroxide. The oxygen absorbed remains mostly attached to the coal substance and is not evolved as gaseous oxides of carbon. The influence of lime in increasing the absorption rate is probably connected with the removal by it of carbon dioxide films from the surface of the coal particle. This absorption of oxygen, which proceeds considerably more rapidly with dangerous than with non-dangerous coals, leads to an increase in the volume of the coal and thus produces an expansion of a concrete in which it is contained. Under dry conditions the volume increase so caused may be more than counteracted by the drying shrinkage of the concrete and no expansion occur. If such a concrete becomes wetted, however, at a later date a large expansion will occur.

There is no sharp line of demarcation between coals of dangerous and non-dangerous types and a continuous series of materials showing a gradual trend in properties is to be found. In general, those properties of coals which cause expansion in concrete tend to increase as the oxygen content and natural moisture content of the coal increases. The oxygen contents of the more dangerous coals usually exceed 10 per cent. These features, high oxygen and moisture contents, are characteristic of coals which give rise to spontaneous combustion in mines. Apart from actual tests in concretes, dangerous coals can be detected by a measurement of the moisture they absorb. A 0·5 g sample of the coal is ground to pass a 100-mesh sieve, dried *in vacuo* over sulphuric acid, and then exposed in a thin layer to saturated water vapour at 25°. The weight of moisture taken up increases with time, but, for comparison purposes, a period of 48 hours may be used. Non-dangerous coals absorb from 1 to 8 per cent moisture in this time, while the dangerous coals show values from 12 to 17 per cent. Some coals of intermediate properties fill in the gap in the series.

The detection of unsoundness in clinker aggregates require a special test, since the presence of less than 4 per cent of some coals is sufficient to cause failure of the resulting concrete. A simple form[1] of test employs a pat, similar to that used in the pat tests on cements, about 3 inches in diameter and $\frac{1}{2}$ inch deep at the centre, tapering to a thin edge at the circumference. The clinker is ground to pass a 72-mesh and 5 parts of it mixed with 1 part of a special cementing mix. This mix is made up by intimately mixing equal parts by weight of Portland cement and plaster of Paris. The clinker-cement mix is gauged with water to a plastic consistence, rolled into a ball and placed in the centre of a clean glass plate. The plate is tapped until the ball has spread out into a pat of about 3-inch diameter which is finally shaped by stroking with a knife from the outside edge towards the centre. About 50 g of clinker and cement are required for one pat. It is necessary that the gauging and making of the pat be completed rapidly as the mix sets within a few minutes. The finished pat is kept in moist air for 3–4 hours and then immersed in water. It is examined at intervals from 1 to 7 days and if it shows the presence of fine radial cracks at the edge, or the edges are lifted in a marked manner away from the plate, or the pat has become loose from the plate, the clinker is unsound. A less rigid test can be made with a mix of 1 part of the cementing agent to 3 parts of clinker instead of 5. When testing a heap of clinker aggregate, it is of course essential that the material be properly sampled and the sample reduced in amount by quartering, with intermediate grinding, until the required final sample is obtained. It may be noted that in clinker aggregates the great proportion of any unsound material present occurs in the finer fractions passing a $\frac{1}{8}$-inch mesh.

Popping and spalling[2] of clinker concretes arise from the slow hydration of particles of hard burnt lime, magnesia, or calcium sulphate, or the corrosion of iron particles present in the clinker. Though this defect is not of any structural significance it often occurs after clinker blocks have been built into a wall, displacing the plaster finish or marring a painted finish. Staining of clinker concrete which causes blemishes on painted blocks arises from the oxidation of iron compounds, such as pyrites. Storage of clinkers in the open for a period of one to two months, keeping them wet, will often do much to reduce popping and staining. If the latter is particularly troublesome treatment with a dilute suspension of hydrated lime in water, about 10–15 lb of lime per cubic yard of clinker, followed by moist storage, has been found effective.

The Le Chatelier soundness test (p. 367) used for cements has been found suitable[3] for testing the liability of clinker aggregates to cause popping of concrete. The clinker, ground to pass an 80 mesh sieve, is mixed with 10 per cent of

[1] F. M. Lea, *Bldg Res. Bull.*, No. 5 (1928); BS 3681 : 1963.
[2] S. G. Sexton, *Proc. Am. Concr. Inst.* **44,** 361 (1948); H. W. Gonell, *Bautenschutz* **8** (12), 137 (1937).
[3] B. Uranovsky, *Bull. natn. Bldg Res. Inst. Un. S. Afr.*, No. 18 (1960).

ordinary Portland cement, or 5 per cent rapid-hardening Portland cement to give the mass sufficient rigidity. The mix is gauged and filled into the Le Chatelier mould in the usual way and steamed for three hours. The expansion should not exceed 6 mm. A method for testing for the presence of staining materials in clinker or in other lightweight aggregates is given in ASTM C331–64T.

It used to be considered that the sulphur compounds in clinker aggregates were responsible for troubles experienced with them, but it has been shown that in plain concrete they are rarely a source of trouble provided the maximum content is limited. In reinforced concrete they do, however, in combination with other factors, cause accelerated corrosion of the reinforcement. Here the action of the sulphur compounds is facilitated by the usual permeable nature of the concrete, and by the tendency shown by clinker aggregates to absorb moisture from the atmosphere and maintain the concrete in a moister condition than the surroundings.[1] Clinker aggregates are banned in Great Britain for reinforced concrete, and their use in any position in contact with steelwork is also prohibited.

Various limits are placed on the combustible content of clinker aggregates in different countries. The British Standard (BS 1165 : 1966) requires that the loss on ignition shall not exceed 10, 20 or 25 per cent according to the purpose for which the aggregate is to be used. The ASTM specification (C331–64T) allows up to 35 per cent and the German (DIN 18152 : 1952) 20 per cent. A limit of 1·0 per cent is imposed on the acid soluble SO_3 content in the British Standard, while the German limits the total sulphur expressed as SO_3 to 1·5 per cent.

Clinker concretes made with aggregate of low combustible content weigh about 80–90 lb/ft^3, but with combustible contents of 20–25 per cent the weight is less, about 70 lb. Finely crushed clinker shows some pozzolanic properties and a mill-run lime-ash mortar is used in some parts of Great Britain as a bricklayer's mortar. Though slow in developing strength, such mortars ultimately become very hard.

Sawdust aggregates

Some use has been made of sawdust aggregate in flooring and in various forms of precast products, but there are two difficulties that have prevented its widespread use.[2] Sawdusts usually affect adversely the setting and hardening of Portland cement owing to the tannins and soluble carbohydrates they contain. With many softwood sawdusts, such as spruce, pine and poplar, the addition of lime to the mix in an amount equivalent to about $\frac{1}{3}$–$\frac{1}{2}$ the volume of cement used will counteract this, but the method is less effective with cedar, larch, or Douglas fir and with sawdusts of high tannin content, and most hardwoods. Many other more drastic treatments have been tried, such as boiling in water and ferrous sulphate solutions, but the cost of these limits their application. Since none of the treatments completely removes the adverse effects it is also useful to add calcium chloride, to the extent of about 5 per cent by weight of the cement, to accelerate setting.

[1] F. L. Brady, *Bldg Res. spec. Rep.*, No. 15 (1930).
[2] T. W. Parker, *Chemy Ind.*, 1947, 593; L. W. Neubauer, *Agric. Engng* **21** (9), 363 (1940); *Concrete* **45** (4), 35 (1941).

The shrinkage and moisture movement of sawdust concretes is also high, 0·2–0·5 per cent, and, though many attempts have been made to reduce this by treatment with various salts, a process known as mineralisation, it is usually necessary to allow, when fixing, for subsequent movements. The most practicable mixes are about 1 : 2 to 1 : 3½ cement : sawdust by volume, with corresponding weights of about 75–50 lb/ft^3 (800–1200 kg per m^3). Sawdust cement cannot be hardened by steam curing immediately after making, but if the concrete is first allowed to harden for a week the shrinkage can be much reduced by high-pressure autoclaving.

Amongst the uses tried have been precast products, jointless flooring made with a 1 : 1½ : 2 cement : sand : sawdust mix, and also as a roofing tile of limited life. Sawdust concrete has also been used for nailing blocks. Wood aggregates are used commercially in wood wool slabs made from wood shavings with Portland cement, or gypsum plaster, as a binding agent and in precast blocks made from chemically stabilised graded wood particles, sand and cement.

Single-sized aggregate concrete

One method of reducing the weight of concrete, and obtaining certain other benefits such as an economy in cement and the elimination of transfer of moisture by capillary suction, is to omit all fine aggregate and use only coarse aggregate graded between ¾ and ⅜ inch. Such concrete is known as 'no-fines' in Great Britain.[1] The aggregate may be gravel, crushed rock, air-cooled slag, crushed brick, or clinker, with mixes of about 1 : 8 by weight. The heavier concretes with gravel and crushed rock aggregates weigh 105–125 lb/ft^3 (1680–2000 kg per m^3) and have thermal conductances from 5 to 6 Btu per square foot per inch thickness per °F. With slag the weight falls at the lower end of this range, but the conductivity is only a little over 4 units. Clinker gives concretes of 80–90 lb weight and conductivities of 3–4 units. The strength with all aggregates is low, some 700–1400 lb/in^2, but so also is the drying shrinkage, about 0·02 per cent.

Aerated concrete

The term 'aerated concrete' covers the materials variously described as gas, cellular, or foamed concrete. Essentially it is a material in which lightness in weight is obtained by the entrainment of a large volume of air or gas in the mix instead of by the use of a lightweight aggregate.[2] It must not be confused with air-entrained concrete in which a much more limited amount of air is entrained for purposes described elsewhere (p. 598).

There are three main processes for making aerated concrete.

(*a*) A small amount of aluminium powder, about 0·2 per cent by weight of the cement, is added to a slurry of Portland cement and finely ground siliceous aggregate, such as ground sand, pulverised fuel ash, ground burnt shale, or ground

[1] R. H. Macintosh, J. D. Bolton and C. H. D. Muir, *Proc. Instn civ. Engrs* **5**, Part I, 677 (1956).

[2] For detailed information on properties, see *Bldg Res. Stn Dig.* (Second Series), Nos. 16 and 17 (1961); *R.I.L.E.M. Sym. Lightweight Concrete, Gothenburg 1960*; *Sym. Autoclaved Silicate Products*, Soc. Chem. Ind., London 1967.

slag or a mixture of these. The reaction of the aluminium with lime and alkalis released from the cement causes evolution of bubbles of hydrogen which are entrapped in the mass and cause it to swell. Wetting agents and additional alkali may be added to stimulate the reaction. In an alternative process lime is used instead of cement. The slurry is passed into moulds where it rises like a dough and hardens. After some 3–6 hours the set material is cut to the shape required by tensioned wires. It is then cured in high-pressure steam for 14–18 hours at 150–200 lb/in^2 in gauge pressure. These aluminium powder methods were originally developed in Sweden but manufacture is now widespread in Europe including many plants in Great Britain, though only to a small extent in North America. An alternative method at one time used in Germany was to add hydrogen peroxide and bleaching powder which react together to evolve hydrogen.

(*b*) A foaming agent, of a protein type, is used to entrap air in the mix. In one method the agent is added to the cement, aggregate and water and the whole vigorously mixed. In an alternative process a stiff foam is first prepared and then mixed with cement grout and finely ground aggregate.

(*c*) Finely ground silicious material, lime and asbestos fibre are mixed with excess water to form a stable cream and the mix, after pouring into moulds, hardened in high-pressure steam. The product which is very lightweight—10 to 25 lb/ft^3—is used for heat insulation.

The foam methods (*b*) are used for *in situ* thermal insulation purposes, such as roof screeds, insulation under the floors of cold stores and around hot-water and steam pipes. The strength of these air-cured products is only about half of that of the autoclaved products of the same density and the drying shrinkage is several times as great. The method in which the foam is formed in the concrete mixer is only suitable for the heavier types of aerated concrete weighing 75–100 lb/ft^3. The thermal conductivity of an air-dry 90 lb weight material is about 2·5 Btu in/f^2/h°/F. The preformed foam method when used with cement alone gives a material of about 25 lb/ft^3 weight, or even lighter, but if finely ground aggregate is added the weight rises to 45 to 60 lb.

The autoclaved aerated precast concretes made by the aluminium powder method (*a*) are used in the form of blocks and reinforced wall, roof and floor slabs. The reinforcement may be coated for protection against corrosion. The density normally lies between about 30 and 50 lb/ft^3. The compressive strength is related roughly linearly to the density rising from about 400 lb/in^2 for material weighing 30 lb/ft^3 to 600–800 lb for 50 lb weight material. The thermal conductivity is also related to the density. Typical values are:

Density lb/ft^3	20	30	40	50	60	
Thermal Conductivity	0·5	0·75	1·0	1·4	1·8	Btu in/f^{2}h°F

These values refer to aerated concretes in equilibrium with the moisture of the atmosphere of a room and containing about 2 per cent moisture by volume. The moisture content of aerated concretes used externally will be higher to a degree depending on whether or not they are protected with an external rendering. The conductivity rises with the moisture content and compared with oven-dry material, increases by a factor of the order of 1·3 for 1 per cent moisture by volume, 1·75 for 5 per cent and 2·1 for 10 per cent. The permissible drying shrinkage of aerated concrete blocks under the conditions of the BS 2028, 1364 : 1968 test varies from 0·07 to 0·09 per cent according to the density of the block as previously given in Table 123. The drying shrinkage-relative humidity curve for autoclaved aerated concrete is a roughly linear one down to 17 per cent relative humidity,[1] where the shrinkage is nearly twice that at 45 per cent relative humidity. These materials are resistant to atmosphere carbonation which in air-cured lightweight aggregate concretes tends to cause the maximum shrinkage on drying to occur at around 45–50 per cent relative humidity, where carbonation is most rapid, with little subsequent change, or even a reduction in shrinkage, at 17 per cent relative humidity. The conditions in the British Standard test correspond to a 17 per cent relative humidity.

Asbestos-cement products

An aggregate used for certain types of concrete products which merits some special consideration is asbestos. This material is a fibrous mineral with a silky lustre. The commonest type of asbestos is a fibrous variety of serpentine known as chrysotile, a hydrated magnesium silicate. The fibres are short, only a few inches in length, but of high tensile strength and flexibility. Originally the term asbestos applied to another type of fibrous mineral belonging to the amphibole group in which the fibres were of much greater length, up to several feet. Most of the asbestos used in asbestos-cement products is of the serpentine type. On account of its infusibility and low-heat conductivity asbestos is much used in a matted form as a fire-resistant material.

Asbestos is used only for cement products made in a factory,[2] since the production of asbestos-cement materials requires a special process quite different from that of ordinary concrete production. The asbestos is freed from other mineral matter, broken down into fine fibres and mixed with Portland cement and a very large excess of water. The proportion of asbestos varies from 8 to 15 per cent by weight and that of the cement from 92 to 85 per cent. The suspension of asbestos and cement in water is brought into contact with a rotating cylinder covered with a fine wire gauze through which the water passes. The cement-coated asbestos fibres adhere to the outside surface of the mesh and are carried upward by the revolving cylinder. A roller is fixed above the cylinder and between them passes an endless moving felt sheet or blanket. This is pressed down on to the cylinder and picks up the asbestos fibre from it and carries it forward to a moulding cylinder. The blanket may be caused to pass over a suction plate to remove some of the water from the solid layer, but this part of the process is sometimes considered undesirable and is omitted on some plants. The moulding cylinder which has a polished metal surface presses against the travelling blanket and itself rotates. The fibre is transferred to it as a thin layer which is built up during several rotations and compacted by another small roller pressing against it. When the required thickness is obtained the machine is stopped and the formed layer cut through and opened out as a sheet. This is cut and pressed into the

[1] W. Kinniburgh, *R.I.L.E.M. Sym. Lightweight Concrete, Gothenburg 1960*, p. 85.
[2] For details, see *Fundamental problems in asbestos cement production*, R. Hayden, Zementverlag., Berlin 1942; *Asbestos*, G. E. Howling, Imperial Institute, London, 1937; *Asbestos*, O. Bowles, U.S. Bureau of Mines Bulletin 403, Washington, D.C., 1937.

desired shape. Pipes are built up in a similar manner by deposition on a rotating mandril.

In an alternative process the cement, asbestos and a small amount of fine silica are mixed in the dry state and spread on to a conveyor belt where the mixture is sprayed with hot water and then compressed under rollers. High-pressure steam curing is used for many products made by this process. The hydration products are those to be expected from Portland cement.[1]

Asbestos-cement products are manufactured for a large variety of purposes such as flat and corrugated sheets, tiles, gutters, drain-pipes, pressure pipes, flues and hearth-casings. The presence of the asbestos fibres matted together confers on asbestos-cement products a higher tensile strength than is found with concrete and they have also a lower density. Asbestos-cement products can therefore be made thinner and much lighter than corresponding concrete materials.

[1] G. L. Kalousek, F. V. Camarda, J. E. Koponda and Z. T. Jugovic, *Mater. Res. Stand.* **6** (4), 169 (1966).

19 Resistance of Concrete to Natural Destructive Agencies

The subject of this and the following chapter is the behaviour of concrete in use, and the various conditions which may be encountered in practice that have a destructive influence on it. These potentially destructive agencies are numerous and include fire and frost, seawater and groundwaters, as well as oils, fats, and solutions of various salts to which concrete may become exposed in industry. Briefly it is the weathering of concrete, interpreting weathering in its widest sense to include both natural agencies and artificial conditions, which we have to consider.

The resistance of concrete to attack by chemical or physical agencies is not necessarily related to the mechanical strength of the material, and it does not follow that because a concrete has ample strength to fulfil the purpose for which it was designed that it will resist attack and disintegration in any environment in which it is placed. There are indeed conditions under which the strongest concrete that can be made will suffer attack while a weaker material, made from materials which are chemically more resistant to the destructive elements concerned, will prove permanent. It is nevertheless true that when using any given cement the less permeable the concrete produced the greater will be the resistance to attack. Since many factors, such as grading of the aggregate, the cement content of the mix, the water : cement ratio, etc., which influence favourably the density and watertightness of a concrete have a beneficial effect on the strength, it is also generally true that the stronger the concrete made from any one cement the more permanent will it prove. A dense Portland cement concrete, for instance, may remain almost immune from attack under conditions where one of poorer quality suffers disintegration and failure within the course of a few years. There are exceptions as we shall see later when discussing air-entrained concrete. In discussing in this chapter the extent to which concretes may be expected to remain permanent under any specified conditions, it must be emphasised that the standard of reference is a good concrete.

A detailed discussion of the general principles of concrete-making lies outside the scope of this book, but it may be well to review them very briefly.

A concrete is composed of a coarse aggregate forming the bulk of the mix, a fine aggregate[1] filling the voids between, and cement and water to bond the whole together. The sand, or fine aggregate, and the cement may together be regarded as a mortar in which the coarse aggregate is set. The properties of the concrete depend primarily on the quality and amount of this interstitial mortar and only secondarily on the coarse aggregate. The latter must be hard so as not to break under the pressure to which it is subjected when the concrete is stressed, and sufficiently impermeable not to act as a channel by which water may pass into the concrete.

The first factor of importance in obtaining a dense concrete is that the aggregate shall be so graded in size that the amount of voids left unfilled between the particles shall be as low as possible. Such a condition cannot be obtained by mixing sand grains all of one size with a coarse aggregate, but requires that the materials used shall contain particles of sizes varying over a wide range. The sand must be present in sufficient amount to fill the large cavities which will be left when the fragments of coarse aggregate, which will normally vary from about $\frac{3}{16}$ to 1 or 2 inches in size, are packed closely together. The sand itself will consist of particles from about $\frac{3}{16}$ inch in size down to about 0·006 inch (100-mesh).

Much work has been done on the grading of concrete aggregates to determine the optimum grading. It is not possible to lay down any single grading to which it would be practicable to adhere in actual work, and an almost infinite series of different gradings could be devised which would be regarded as satisfactory. Thus BS 882 and 1201 : 1965 gives four permissible grading zones for sands varying from coarse to fine. This specification allows a sand to contain from 5 to 50 per cent of material passing a No. 52 sieve, whereas ASTM C33–67 places the maximum at 30 per cent. The main bulk of the sand, which lies between a No. 52 sieve and a $\frac{3}{16}$-inch mesh, should contain particles of varying size and not consist predominantly of any one size. The object of grading is to obtain a concrete which can be satisfactorily placed in position with the minimum water content, and this requires a progressively lower sand content in the concrete mix as the sand becomes finer. The use of aggregates in which some particular limited size range is absent and what are known as gap-graded aggregates can, on occasions, be advantageous.

The proportions of cement and aggregate in concrete are expressed as 1 cement : x fine aggregate : y coarse aggregate by volume or by weight, or alternatively as the weight of cement per unit volume of the mixed concrete. The latter is calculated from the absolute volumes of the constituents, cement, aggregate and water, with an allowance for entrapped, or deliberately entrained air. In proportioning by volume allowance has to be made for the bulking of sands, that is the increase in overall volume of a damp sand compared with its volume in a dry or inundated condition. On average the bulking is about 25 per cent but it can be as high as 40 per cent. When the water : cement ratio is specified allowance has to be made for the excess or deficiency of moisture in the aggregate over and above that required to saturate it.

[1] The distinction between coarse and fine aggregate is made at a quite arbitrary size point, usually $\frac{3}{16}$ inch. All proportions in this book are by weight unless specifically stated by volume, and the aggregates are sand and gravel unless otherwise stated.

The properties of the cement-sand mortar, which binds the coarse aggregate together, depend on the proportion of cement it contains, the amount of water used, and on the fineness of the sand. The finer the sand, the greater is the surface area it possesses, and hence the larger the proportion of cement required to cover that surface. This is one reason why a sand should not be excessively fine and contain much material passing a 100-mesh. An increase in the amount of water above that necessary to give a workable mix renders the mortar weaker and more permeable and increases the volume of voids which will be left empty when the concrete dries out and the excess water is removed. If the proportion of the sand in a concrete is not sufficient, then voids between the fragments of coarse aggregate must remain; if it is present in excess, that is if the mix is 'oversanded', the fragments of coarse aggregate will be separated more than necessary by the excess mortar and the mortar itself will be leaner in cement content, assuming that the proportion of cement to the total aggregate remains unchanged. The coarse aggregate in a concrete is normally quite inert and impermeable, and it is the cement mortar which is the point of attack by most destructive agencies, and which forms the channel by which waters can permeate into the concrete. The presence of an excess of this mortar in a weaker condition tends therefore to lower the resistance of the concrete to attack. The proportion of sand required decreases as the maximum size of the coarse aggregate increases; it is also lessened somewhat when air-entraining agents are used or when the concrete is placed by vibration. As a rough working rule about 35–40 per cent by volume of the aggregate should consist of sand when the maximum size of coarse aggregate is $\frac{3}{4}$ inch, but this proportion has to be adjusted to suit the particular aggregates used.

Water to be used for mixing mortars or concretes must not contain dissolved substances which affect the setting time unduly or retard the hardening. In general, the only impurities in natural waters that are likely to be objectionable are dissolved salts and the permissible limits for these are quite wide. It can be taken as a rough rule that any water that is fit for drinking can be used in concrete, and that well waters with up to 100 parts sulphur trioxide or 50 parts chlorine per 100 000 are not objectionable. Indeed for plain concrete seawater has not infrequently been used. It tends, however, to cause dampness and surface efflorescences and must not be used where surface finish is of importance or where the mortar or concrete is to be covered with plaster or any decorative finish. The use of seawater in reinforced concrete should be avoided for it increases the risk of corrosion of the reinforcement. It must never be used for mixing high-alumina cement as it has a very adverse effect on the strength. Organic impurities in water may retard setting and hardening of cement but, except in artificially contaminated waters, they do not usually occur to an objectionable extent. In case of doubt a setting time test can be made and the results compared with those with a pure water.[1] Occasional difficulties have been reported[2] from the presence of considerable amounts of algae in water, leading to considerable entrainment of air in the concrete and substantial reduction in strength.

[1] See BS 3148 : 1959.
[2] B. C. Doell, *Proc. Am. Concr. Inst.* **51** 333 (1955).

The mixing and placing of a concrete also play an important part in determining the quality of the product. Satisfactory mixing so as to produce a uniform dispersion of the cement throughout the mix, coating all the particles of aggregate, coarse and fine, is too obviously necessary to need more than a mention. The ease with which concrete can be placed depends on the workability of the mix, that is, the ease with which the material flows while at the same time it remains coherent and resists segregation. Workability is a factor easily appreciated in practice and there are various methods used for its measurement, but none of them really assess all that is involved in this property for it is a complex one.

Fresh concrete has a certain resistance to flow, in other words a certain yield stress. At higher stresses its flow is determined by the frictional resistance between its layers, by the viscosity of the cement paste, and by the separation between the particles which depends on the amount of cement paste and the grading of the aggregate. It is the yield stress, which is related to the consistence of the mix, that determines the cohesiveness of the concrete and its resistance to segregation and bleeding. It is the ability to flow under stress that determines the ease with which the concrete can be placed in position, worked round reinforcement and compacted. The workability of concrete is measured by various tests, such as the slump test,[1] the compacting factor test,[2] the ball penetration test[3] and the Swedish Ve Be test[4] which is particularly useful for very dry mixes.

A harsh and unworkable mix can always be placed more easily by adding more water, but only at the expense of lowering the strength and increasing the permeability of the concrete and rendering it more liable to segregation. The workability of concrete is influenced by the type and grading of the aggregate as well as by other factors. Wet concrete mixes tend to segregate during placing, leaving an excess of coarse aggregate at the bottom and an excess of fine material higher up, while at the top of the lift is formed a layer of cement, fine sand, and water known as laitance. Concrete mixes which are too dry cannot be easily rammed into position and tend to leave voids and honeycombed patches, but drier mixes can be placed satisfactorily if vibration methods of compaction are used. The resistance of a concrete to attack is much diminished by the presence of honeycombed patches, segregated layers, and layers of laitance, all of which form weak points at which attack can commence. The joints in concrete work where one day's work is joined on to the next also often form planes of weakness owing to lack of proper bonding between the hardened concrete and the material placed subsequently.

The amount of water required to produce a given slump in a concrete mix increases with the temperature of the mix. Thus an increase from 60° to 100° F raises the water requirement by the order of 10 per cent. This results in a lower ultimate strength and an increase in the subsequent drying shrinkage. Even at the same water : cement ratio the ultimate strength is lower for higher mixing temperature,[5] possibly because of some adverse effect of the more rapid setting.

[1] BS 1881 : 1970; ASTM C143–66.
[2] W. H. Glanville, A. R. Collins and D. D. Matthews, *Rd Res. Tech. Pap.*, No. 5, H.M.S.O. 1947; BS 1881 : 1970.
[3] ASTM C360–63.
[4] V. Bährner, Report No. 1, Svenska Cement Förenigen, 1960; BS 1881 : 1970.
[5] W. H. Price, *Proc. Am. Concr. Inst.* **47**, 417 (1951); P. Klieger, ibid. **54**, 1063 (1958).

This influence of temperature can be appreciable in hot climates, and for concrete that is subsequently to be exposed to severe weather conditions or erosion some authorities have stipulated a maximum temperature of 85° or 90° F for the mixed concrete.[1]

Concrete can be damaged by lack of proper curing. It is essential for the development of high strengths that the concrete should be kept moist for a period and not allowed to dry out rapidly. So-called 'curing' compounds[2] are sprayed on to exposed concrete surfaces for this purpose or the surface is protected from sun and wind by some form of covering. The exposure of concrete to frost while it is still saturated with water, and before it has had time to develop sufficient strength to resist the disruptive effects of freezing, may be entirely destructive and render replacement essential. In other cases, perhaps more insidious, it may render the material weaker and more porous and much more susceptible to any subsequent attack. Once a concrete has been cured under moist conditions for a sufficient period, its resistance to attack by chemical action is increased by allowing it to remain in air and dry out. A film of calcium carbonate forms over the surface of the material, blocking the pores and producing a hard and dense surface skin.

An additional factor is involved in reinforced concrete where steel bars are embedded in the material. The function of the concrete here is not only to provide a medium to withstand the compressive stresses to which the reinforced concrete member may be subject, but also to protect the steel reinforcement against corrosion. Any corrosion of the reinforcement results in the formation of a film of iron oxide over the metal occupying a volume about 2·2 times that of the iron from which it is formed. The expansion which thus occurs results eventually in the flaking off or cracking of the concrete overlying it. The corrosion hence damages not only the steel but also the concrete. The degree of protection afforded to the reinforcement depends on the impermeability and thickness of the concrete covering it. A concrete of high quality is essential for reinforced concrete work in order to prevent access of moisture and air to the steel. When this is obtained the reinforcement is, under ordinary conditions, permanently protected. Reinforced concrete which is exposed to seawater, or to attack by other aggressive waters, requires, however, a degree of protection far higher than that necessary under less severe conditions of use. Reinforced concrete members may show fine cracks arising from deflection under load or from shrinkage of the concrete. If the width of the cracks is sufficiently small they tend to become sealed by autogenous healing. Provided the width at the surface is not more than about 0·2 mm the presence of such cracks does not usually lead to any progressive corrosion of the steel, though the critical width depends on the thickness of the concrete cover and the exposure conditions. For internal members a crack width of 0·3 mm may be considered tolerable. Reinforced concrete may also suffer attack from one source which is without influence on plain concrete, namely, electrolysis caused by stray electric currents. In buildings or structures of any type where, through electrical leakages, the reinforcement serves as an earth, conveying

[1] Recommended Practice for Hot Weather Concreting, *Proc. Am. Concr. Inst.* **55**, 525 (1959); U.S.A. Bureau of Reclamation, *Concrete Manual 1966*, 279.

[2] ASTM C156–65; Discussion, *Proc. Am. Concr. Inst.* **48**, 701 (1952).

current, corrosion of the reinforcement occurs in damp positions owing to direct electrolytic action.

Admixtures

Admixtures[1] in the most general sense are materials that are added when mixing to mortar or concrete in order to modify the properties of the product either in its fresh or hardened state. Some types are used to improve workability, to reduce the amount of water needed in mixing, to retard or accelerate setting and to improve the frost resistance. Other types are used to reduce permeability or to improve specific mechanical properties of the concrete such as abrasion resistance. Yet other agents are added for such purposes as to produce expansion or inhibit shrinking, as in expanding or non-shrinking cements, to produce lightweight products as in aerated concrete, to improve chemical resistance as in pozzolanic cements, to reduce alkali-aggregate expansion, or to impart fungicidal, germicidal and insecticidal properties to a concrete. Some of these types of additive are discussed elsewhere and in the present chapter we shall be concerned with those that are added mainly in small amounts to influence the properties of the plastic mix or to improve certain properties of the hardened concrete such as frost resistance and water tightness.

The classification of admixtures according to their technical properties is an arbitrary one since there is overlap between them, but we can make a more fundamental distinction between certain broad classes according to the mechanism of their action. This is particularly helpful in dealing with the surface-active materials used as air-entraining, set-retarding and water-reducing agents.

Air-entraining agents[2] are essentially materials that entrap air in a mortar or concrete and alter the capillary structure of the set cement paste by blocking capillaries. The agents consist of a non-polar hydrocarbon chain or some other hydrophobic group joined to a polar or hydrophilic group such as $—COO^-$, $—SO_4^{--}$ or $—NH_3^+$. At the air-water interface the polar groups are orientated towards the water phase lowering the surface tension, promoting bubble formation and counteracting the tendency for the dispersed bubbles to coalesce. At the solid-water interface, where directive forces exist in the cement surface, the polar groups become bound to the solid with the non-polar groups orientated towards the water making the cement surface hydrophobic so that air can displace water and remain attached to the solid particles as bubbles. Air-entraining agents do not affect the setting time of cement but they do reduce the water content required in a mix.

[1] For a general discussion, see *R.I.L.E.M. International Symposium on Admixtures for Mortar and Concrete, Brussels 1967*; Report of A.C.I. Committee 212, *Proc. Am. Concr. Inst.*, **60,** 1481 (1963); W. E. Grieb, G. Werner and D. O. Woolf, *Bull. Highw. Res. Bd*, No. 310 (1962); ASTM C494–68; *Spec. tech. Rep. Am. Soc. Test. Mater.*, No. 196A, 556 (1966); W. Albrecht, *Betonsteinzeitung* **31,** 59 (1965); H. E. Vivian, *Sym., Washington 1960*, 909; W. G. Hime, W. F. Mivelaz and J. D. Connolly, *Spec. tech. Publ. Am. Soc. Test. Mater.*, No. 395, 18 (1966); Report of Concrete Society Sub-Committee on Admixtures, *Concrete* **2** (1), 39 (1968); R. C. Mielenz, *Sym., Tokyo 1968*.

[2] R. C. Mielenz, V. E. Woldokoff, J. E. Backstrom and H. F. Flack, *Proc. Am. Concr. Inst.* **55,** 95, 261, 359, 507 (1958); also see T. C. Powers, *J. Res. Dev. Labs Portld Cem. Ass.* **6** (3), 19 (1964) and **7** (1), 23 (1965), for many references to the literature.

A list of air-entraining agents used in commercial products has been given earlier (p. 14), but it should be noted that some of the materials listed there, e.g. calcium lignosulphonate, should be classified as set-retarders and water-reducers as discussed below. Amongst the common air-entraining agents are the alkali salts of pinewood resin extracts (Vinsol resin), and sodium abietate also obtained from pinewood resins, the synthetic detergent alkyl aryl sulphonates, the triethanolamine salts of sulphonated aromatic hydrocarbons, and the calcium salts of proteins.

The water-reducing agents, in contrast to the air-entraining agents are materials which when absorbed on the cement surface make the particles hydrophilic. The active groups in the molecules which promote this adsorption are HO—C—H; OH; COOH; and HO—C—C=O. They are mainly anionic agents giving a negative potential to the particles so that water dipoles become orientated in a sheath which prevents close approach of the particles thus facilitating mobility. They also much influence the form of crystallisation of the calcium aluminate hydration products, the rate of hydration, and the establishment of a structure in the cement paste, as discussed earlier (p. 257). As a result of these influences they commonly act also as retarders of set. The anionic agents can be classified in four main groups:

1. Lignosulphonic acids and their salts (Ca, Na, Mg and NH_4).
2. Modifications and derivatives of (1).
3. Hydroxylated carboxylic acids such as gluconic acid, and their Ca, Na or triethanolamine salts.
4. Modifications and derivatives of (3).

Various non-ionic agents are also used, e.g. carbohydrates, such as sugars, polysaccharides and dextrins, and water-soluble derivatives of cellulose and of silicones.

Class (1) usually entrain small quantities of air (2–3 per cent) and act as retarders, whereas class (3) do not entrain air but do retard setting. Classes (2) and (4) are materials that are modified as by the addition of materials, such as an air-entraining agent to increase the air content of the concrete, or accelerating agents such as calcium chloride or triethanolamine to offset the retarding effect of the water-reducing agent. This retarding effect in the case of the lignosulphonates, which chemically are complex impure polymers,[1] arises at least in part from the presence of sugars. The materials are refined to reduce this content to a desirable proportion.

It is clear that there is an overlap in properties between the hydrophobic air-entraining agents and the hydrophilic water-reducing agents. Both reduce the water content required in mixing for a given workability, the latter more than the former, and some of the water-reducing agents entrain a limited amount of air. The water reduction arises in the case of the air-entraining agents from the increase in paste volume produced by the entrained air and its lubricating action on the mix, and perhaps also by the lubricating action of the adsorbed hydrophobic films or the cement particles. The air-entraining agents actually increase the viscosity of the cement paste, but the influence of this on workability is more than offset by these other effects. In the case of the water-reducing agents the

[1] See W. J. Halstead and B. Chaiken, *Bull. Highw. Res. Bd*, No. 310 (1962) for analyses.

water reduction arises from the properties of the adsorbed layers at the particle surface. Air-entraining agents affect the capillary structure of the set cement paste and because of their hydrophobic nature reduce capillary suction. Water-reducing agents do not affect the capillary structure of the paste otherwise than by the reduced w/c ratio. The action of air-entraining agents is little influenced by the particular cement used, but it is by the aggregate grading. The action of water-reducing agents on the other hand is dependent on the particular cement used, particularly in their retardation effects. An overdose of an air-entraining agent leads to an excessive air content, that of a water-reducing agent to excessive retardation of set.

Air-entrained concrete

There has been a large development in the use of air-entrained concrete. It arose first in the U.S.A. from the serious surface scaling of concrete roads treated in winter with common salt, or calcium chloride, to melt ice on the surface. It was found, accidentally, that the difference in performance of different cements was caused by small amounts of grease in the cement which led to entrainment of air in the concrete during mixing. This was followed by the development of air-entraining agents for intergrinding with the cement or as an addition to the concrete mix. These materials reduce the surface tension of water and stabilise very small air bubbles in the mix in a manner analogous to the formation of a foam on soapy water. The effect of these small air bubbles is quite different to that of larger air voids left in a concrete owing to inadequate compaction.[1] The entrainment of air increases frost resistance, and to some limited extent, resistance to other aggressive agents such as sulphate waters; it improves the workability and reduces segregation and bleeding of the concrete mix. This latter property has led to the widespread use of air-entrained concrete in mass concrete, such as the interior of dams,[2] as well as for roads and other constructions.

The nature of air-entraining agents has been described earlier (p. 14); only a very small amount, 0·02–0·06 per cent by weight of the cement, is needed. The amount of air entrained varies to some extent with the richness of the concrete mix, the type and grading of the aggregate, and other factors, and for this reason some users prefer to use the air-entraining agent as an additive to the concrete mix, so that the amount can be adjusted to suit the particular conditions. Though this flexibility is lost with an air-entraining cement containing the agent already interground, the latter avoids the difficulties in control that can arise when small additions have to be made at the concrete mixer. Both methods are used. The aim in either case is to entrain some 5–6 per cent by volume of air in a concrete with aggregate up to 1 inch size, as compared with 1 per cent or less in ordinary dense concrete. This causes a decrease in strength, which is related to the volumetric ratio of (water + air) to cement in a roughly equivalent way to the water : cement ratio in ordinary concrete, but this can be partially offset by taking advantage of the improved workability of the concrete and reducing slightly the

[1] Many references to the literature are given by T. C. Powers, *J. Res. Dev. Labs Portld Cem. Ass.* **6** (3), 19 (1964); **7** (1), 23 (1965).
[2] W. Lerch, *6th int. Congr. large Dams, New York 1958*, Paper R98.

sand and water-content of the mix. In practice a reduction of some 10–15 per cent in compressive strength may be expected. There is no advantage in respect of durability in increasing the air content above about 6 per cent and it causes a further decrease in strength. This optimum content increases, however, if the maximum aggregate size is less than $\frac{3}{4}$ inch and rises to about 7–8 per cent for $\frac{3}{8}$-inch aggregate and about 9–10 per cent in mortar of $\frac{3}{16}$ inch down.[1] It falls to 3–4 per cent with 3-inch aggregate. Mortars entrain much more air, over twice as much by volume, as concrete for the same amount of air-entraining agent and the increased plasticity this gives is taken advantage of in masonry cements and in rendering mixes.

Some sands have been found which cause the entrainment of large and variable amounts of air in concrete. This is due to the presence of organic matter of a type that is not shown up in the ordinary caustic soda colour test, but which has strong foaming properties. The air bubbles are less stable than those produced by air-entraining agents. The remedy used in some Canadian constructional work[2] was to add an excess of an air-entraining agent, to produce excess 'stable' air in the concrete, and then to add sufficient of a foam control agent to eliminate the excess air and leave the air content required. Materials such as tributyl phosphate and some higher alcohols have anti-foaming properties. Air-detraining agents or anti-foam admixtures are also sometimes used to control the air content of concrete when using other admixtures.

The effect of air-entraining agents on frost resistance is discussed later, but it is entirely dependent on the entrained air, for if this is all subsequently sucked out *in vacuo* no improvement is found. Concrete resistant to moderate frost exposure can be made without such aid, but air-entrainment is obviously beneficial in offsetting adverse conditions and in affording a wider margin of safety. The permeability of concrete is not materially influenced by entrained air,[3] but the improved workability of the mix results in greater uniformity of the concrete as placed thus tending to reduce the permeability. The concrete, once it has dried, absorbs water less readily and this probably accounts for such improvement as is found in its resistance to sulphate salts and seawater.[4]

The thermal expansion and drying shrinkage of concrete are not significantly affected by air-entrainment up to 6 per cent but at higher air contents the shrinkage increases by up to one-third or thereabouts.[5]

Air-entraining Portland cement is covered by ASTM C175–68 and air-entraining agents[6] by ASTM C260–66T with methods of test given in C233–66T. In the cement specification the amount of the air-entraining agent present is controlled by a test on a 1 : 4 standard sand mortar (C185–59) in which the weight per 400 ml of mortar is determined and the air content calculated from the theoretical weight of the air-free material.

[1] P. Klieger, *Proc. Highw. Res. Bd* **31,** 177 (1952); *Bull. Highw. Res. Bd,* No. 128 (1956).
[2] M. F. Macnaughton and J. B. Herbick, *Proc. Am. Concr. Inst.* **51,** 273 (1955).
[3] R. F. Blanks and W. A. Cordon, *Proc. Am. Concr. Inst.* **45,** 469 (1949).
[4] W. C. Hansen, *Bull. Res. Dev. Labs Portld Cem. Ass.,* No. 175 (1965); V. V. Stolnikov, *Investigations into Concrete for Hydraulic Structures,* Moscow 1962.
[5] P. Klieger, *Proc. Am. Concr. Inst.* **45,** 149 (1949).
[6] A British Standard is in course of preparation. Test methods are discussed in *R.I.L.E.M. Sym. Admixtures for Mortars and Concretes, Brussels 1967,* Topic 5.

Three main methods are available for determining the air content of freshly-mixed concrete. The gravimetric method is similar in principle to that mentioned above, and involves a knowledge of the specific gravity of the aggregates and cement and is not as accurate as desirable, or very practicable for field determinations. Either a direct volumetric method or a pressure method is, therefore, normally used, and the latter is on the whole the more widely preferred. In the volumetric method the concrete is inundated in water, the entrained air allowed to escape by some process of stirring, or of rolling the container, and the volume of liquid that has to be added to restore the original volume of the concrete plus water measured. In the pressure method, the concrete is placed in a closed vessel and subjected to pressure to compress the entrained air and the volume of air determined by the application of Boyle's law.

For the gravimetric method (ASTM C138–63), the weight per cubic foot of the concrete (w) is determined and the theoretical weight (T), with no air voids, calculated from the total weights and absolute volumes of the constituents.

The percentage air content by volume (A) is then:

$$A = \frac{T - w}{T} \times 100$$

The volumetric method (ASTM C173–68) depends on the complete removal of all entrained air and the measurement of the volume of water required to replace it. A sample of concrete is placed in a flanged cylindrical steel bowl on to which an upper vessel, with a neck carrying a reference mark, and a cap, can be clamped. The concrete is placed in the lower container, and its volume gauged, and the upper vessel clamped on and filled with water. The entrained air is caused to escape by a process of tumbling and rolling. Water and *iso*propyl alcohol are then added until the liquid level is restored. This alcohol is added because it has the beneficial effect of destroying the foam which forms on the surface of the water.

In the pressure method (BS 1881 : 1970 and ASTM C231–68) a sample of the concrete is placed in a flanged cylindrical bowl on to which can be clamped a conical cover carrying a graduated glass tube with a pressure gauge at the top and connected through a side arm to a source of air pressure. The bowl, of known volume, is filled with concrete, the cover and attachments clamped on, and water added through the glass tube until it is about half full. The pressure gauge is attached and the change in volume noted on applying a pressure which is pre-determined from the calibration. The apparatus is calibrated, so as to read percentage of air directly, by tests in which only water and a known volume of air are present. Minor correction factors for the compression of air held in pores in the aggregates and for certain characteristics of the apparatus are introduced. The former varies from about 0·15 to 1 per cent of air for different natural aggregates, depending on their porosity and degree of saturation. It is determined by a separate test on a sample of inundated fine and coarse aggregate in approximately the same proportions as in the concrete.

In a variant of this pressure method, instead of filling the apparatus with water, a known volume of air at a known pressure is released from a chamber into the concrete container and the reduction in pressure determined.

In all these methods the concrete may be rodded or vibrated in the container to simulate the method of compaction to be used in practice.

The air content of hardened concrete can be determined by microscopic measurement on plane ground random sections. Linear traverses are made and a binocular microscope with a $\times 30$ or $\times 50$ magnification is advantageous (see p. 678).

Water-reducing and set-retarding agents

Water-reducing agents enable an economy to be made in cement because they reduce the amount of water needed to obtain the required slump and workability of the concrete mix. A given water : cement ratio can thus be obtained with a reduced cement content. Alternatively they produce an increased slump and workability if the water : cement ratio is not changed. When they are also retarders of set, as are Classes I and III, they lengthen the time by some 2–6 hours during which concrete can be vibrated and become monolithic with the next layer placed on it. This is often particularly important in hot weather or where the nature of the construction prolongs the time between the placing of successive layers of concrete. A method for measuring the time of setting of concrete is given in ASTM C403–68. The amount of water reduction varies over a range of about 5–15 per cent with different agents, cements and concrete mixes. The retardation of set beyond the normal setting time varies with the particular admixture used and the cement. This variation with the cement, the reasons for which have been mentioned in Chapter 10 (p. 258), is the main difficulty in the use of set-retarding agents. Too low an SO_3 content in a cement, for example, can lead to excessive delay in hardening of concrete with a lignosulphonate admixture.[1] Preliminary trials are required with the particular cement and concrete mix to be used on a job. Over-dosing with a set-retarder can lead to excessive retardation of setting though not necessarily reducing the ultimate strength. The optimum rate of addition of these additives is recommended by their producers but requires adjustment to suit the particular Portland cement used. When in the form of powders the normal rate is from a few ounces up to 1 lb per cwt of cement and in the form of liquids up to about $\frac{1}{2}$ lb.

Water-reducing agents that do not entrain air tend to increase bleeding because of the greater fluidity of the concrete mix. When they also entrain air they reduce bleeding. The strength of concrete at 1 day is about normal and at 28 days it may be increased by up to 20 per cent, depending on the cement and the additive. This increase is rather more than can be attributed to the reduction of the water-cement ratio alone and may perhaps be due to the effect of the agent on structure formation in the set cement. These agents have no consistent effect on drying shrinkage but reduce permeability. The resistance to frost action is increased in so far as the water : cement ratio is lowered and air entrained. In the modified types containing also an air-entraining agent the resistance becomes similar to that of an air-entrained concrete. Much data on the effect of water-reducing and set-retarding agents on the properties of concrete is given in ASTM Special

[1] L. H. Tuthill, R. F. Adams, S. N. Bailey and R. W. Smith, *Proc. Am. Concr. Inst.* **57**, 1091 (1961); R. Bauset, *Sym., Tokyo 1968*.

Technical Publication No. 366 (1960) and in the papers to the RILEM Symposium on Admixtures (1967), where methods of test are also discussed. A specification[1] for water-reducing and set-retarding admixtures has been issued by the ASTM (C494–68). These water-reducing workability aids are used in a wide variety of concrete construction, in pumping concrete, cement grouting and in precast concrete manufacture.

Water-retaining and bonding admixtures

Water-retaining admixtures, such as methyl cellulose, increase the water retentivity of a mortar against suction and are useful in mortar to be applied to backgrounds of high suction. Various polymer resin emulsions, such as polyvinyl acetate are used to improve bonding to a concrete background. The improved bond obtained with the PVA emulsions is not however resistant to wet conditions. Natural rubber latex, styrene-butadiene latex and acrylic emulsions are also used to improve bond and for other purposes such as improving chemical resistance.

Workability aids

There are various other materials which can be added to a concrete mix to improve its workability.

The addition of a fine powder, inert or active, helps in concrete mixes that are harsh and deficient in fines by increasing the amount of paste and so facilitating movement between the coarser particles of aggregates. This benefit can, however, often be offset by an increase in the amount of water required. Some 2–3 per cent of diatomite,[2] by weight of cement, has sometimes been added to improve plasticity and reduce segregation, while pozzolanic cements often give more workable concretes than Portland cement. The partial replacement of cement by lime in mortars is also a well-known means of increasing their workability. Fine mineral additions are also used in cement grouting and in the mortar for Prepakt concrete.

Integral waterproofers

Many materials are sold, usually under proprietary names, for incorporating in mortars and concretes with the object of making the concrete less permeable to water. They are somewhat of a jungle because the makers rarely disclose their composition. Apart from the chlorides, which essentially are only accelerators, the commercial waterproofers can be divided into water-repellents, hydrophobic liquids and inert fillers. They take the form of clear or cloudy liquids, thick liquids, pastes or dry powders and as a class are known as integral waterproofers. They are added either to the gauging water or direct to the concrete mix, or

[1] A British Standard is also being prepared. Test methods are discussed in *R.I.L.E.M. Sym. Admixtures for Mortars and Concrete, Brussels 1967*, Topic 5, and in *Bull. int. Commn large Dams*, No. 20, Paris 1968.

[2] R. E. Davis and A. Klein, *Spec. tech. Rep. Am. Soc. Test. Mater.*, No. 99, 93 (1949); V. L. Eardley-Wilmot, *Diatomite*, Dep. Mines Can., Ottowa 1928.

sometimes premixed with the cement. They can be grouped roughly into the following classes:

1. Inert materials acting as workability aids and pore fillers such as hydrated lime, clays, ground silica, silicates, talc (soap-stone), chalk, barium sulphate, etc.
2. Active materials acting as pore fillers such as sodium and potassium silicates, silicofluorides, iron filings with ammonium chloride, and diatomaceous silica.
3. Soluble chlorides and sulphates such as calcium and aluminium chlorides, alum, alkali sulphates, and gypsum. These materials may all be regarded as active additions which in some cases act also as pore fillers.
4. Inert water-repellent materials such as calcium soaps, waxes, mineral oils, bituminous emulsions, coal-tar residues, glue, and cellulose with wax in ammoniacal copper solution.
5. Active water-repellent materials such as sodium, potassium, and ammonium soaps, free fatty acids, butyl stearate, vegetable oils, and resins.
6. Protein decomposition products.

The materials classed as active are those which react in some way with the cement. Thus the soluble alkali soaps react to form the insoluble calcium soaps, the sodium silicate to form calcium silicate, the alkali sulphates to form gypsum and calcium sulphoaluminate, the chlorides to form calcium chloraluminate. The iron filings, if used alone, should be regarded as inert, but if, as usual, they are mixed with ammonium chloride, rusting occurs with formation of the more voluminous iron oxides which fill the pores. This particular material is used as a surface hardener as well as a waterproofer.

When calcium chloride is present it has usually been added to offset any adverse effect of other constituents on setting and hardening of the cement.

A large number of proprietary integral waterproofers are mixtures of materials from the different groups given above. The compositions of some of them are as follows:

1. Solutions of calcium chloride in water, containing from 20 to 40 per cent $CaCl_2$.
2. Turbid liquids or pastes containing calcium chloride, aluminium chloride, together with sodium, potassium, ammonium, or calcium stearate or stearic acid. The amount of combined fatty acids present ranges from 4 to 13 per cent in various materials of this class for which analyses have been published.
3. Pastes or thick liquids containing alkali, ammonium, or calcium stearates and mixtures thereof. The fatty acid content of some samples of these materials varied from 5 to 23 per cent.
4. Pastes or thick liquids containing calcium or ammonium oleate and calcium chloride or unsaponifiable oils.
5. Pastes or solids containing calcium or ammonium soaps, hydrated lime and gypsum.
6. Talc, barium sulphate and free fatty acids in the form of a powder.
7. Liquids containing alkali and calcium silicates and chlorides and drying oils such as linseed oil and tung oil or gelatin or casein.
8. Pastes of hydrated lime and petroleum jelly.

9. Sodium silicate and glue.
10. Bituminous emulsions.
11. Cellulose solutions.

The stearates are probably the materials most used in proprietary compositions, but the silicate-vegetable oil or fatty acid, and silicate-protein types are also prominent. Aluminium, ammonium and sodium stearates can be obtained commercially as chemical compounds, but they should not be used in amounts greater than about 0·2 per cent by weight of the cement or excessive frothing may occur. Butyl stearate can be added as an emulsion. This does not cause frothing and up to 1 per cent can be used. The stearate soaps reduce the rate of hydration and of strength development of concrete, unless an accelerator is also added, but butyl stearate is free from this adverse effect.

There has been much controversy as to the value of integral waterproofers. Methods of test have been described by Kocataskin and Swenson.[1] There is ample evidence that in laboratory tests they do not usually reduce, and may increase, the permeability of concrete to water under pressure,[2] except in the case of the addition of inert fillers to lean, though not to rich, concretes. Many of these materials, however, also act as workability aids and make a concrete less liable to segregation in placing. This in itself is important, since in practice leakage of water is often due more to weak patches in a construction than to the general permeability of concrete as a whole. Further it may aid in controlling the uniformity of the mixed concrete by making it less susceptible to fluctuations in other variables. Concrete can be made very impermeable without such additions, but many users seem to find them of benefit in retaining walls, cellars, tunnels, tanks, etc., where water pressures are low. This does not mean that a really porous concrete can be waterproofed, or that waterproofing agents are any substitute for good mix design and workmanship, but they may be useful as a safeguard against casual fluctuations in these standards.

Many waterproofers, like the stearates, are water repellent and reduce the capillary forces by which water is drawn through a mortar or concrete and thus the rate of absorption under zero head. This action, sometimes called damp-proofing as distinct from resistance to water under pressure, is of value and the incorporation of waterproofers in the undercoat of a rendering, or sometimes in mortars, can be advantageous. It will also reduce capillary transmission of water, but not diffusion of water vapour, through concrete in contact with the ground. This water-repellent action is not effective against water pressures in excess of a few centimetres.

The claim is sometimes made for integral waterproofers that they increase the resistance of concrete to chemical attack. Some limited benefit may result under some conditions from the decreased rate of absorption under zero head, but the results of many tests have been disappointing and they cannot be relied on as a protective measure.

Waterproofers should not be added to high-alumina cement as they may seriously affect the strength.

[1] F. Kocataskin and E. G. Swenson, *Bull. Am. Soc. Test. Mater.* 67 (1958).
[2] C. H. Jumper, *Proc. Am. Concr. Inst.* **28,** 209 (1932); report by A.C.I. Committee, ibid. **60,** 481 (1963).

Accelerators and retarders

Calcium chloride, or proprietary agents based on it, is sometimes added to concrete to increase the rate of strength development. This may be done to counteract the effect of low temperatures in winter, to permit of earlier removal of formwork, to speed up repair work, or to reduce the time required for moist curing. It can be added at the mixer in the form of solid calcium chloride flake (70–72 per cent $CaCl_2$), but an aqueous solution is preferable; the granular form dissolves too slowly to be used as solid. About 2 per cent flake by weight of the cement is the maximum desirable, though up to 3 per cent has sometimes been used. In the case of reinforced concrete the addition of calcium chloride causes initially some superficial corrosion of the reinforcement but this does not progress with age provided the concrete is dense.[1] The effect, therefore, is not serious with good concrete work, but the margin of safety against inferior workmanship is reduced and it should not be used with steam-cured concrete. The risk of corrosion is much more serious with prestressed concrete[2] and the use of calcium chloride is not permitted.

The effect of calcium chloride[3] on the rate of strength development varies for different Portland cements, and water : cement ratio and temperature. An addition of 2 per cent calcium chloride to an average Portland cement concrete at 60–70° F increases the strength by some 400–800 lb per sq in. at 1 day, 700–1000 at 3 days and 7 days and 500–1000 at 28 days. With rapid-hardening Portland cement the increase may be 1000 lb or more at 1 day, falling to 400–700 at 7 days and only 100–200 at 28 days. In general the slower the rate of hardening of the cement the later is the age at which the maximum effect appears. At lower temperatures, e.g. 32–40° F, the increase in strength at 1 day is about halved, at 3 days little changed, and at 7 and 28 days increased. The initial setting time is reduced with a 2 per cent addition and with larger amounts can become very short. Data on the effect of calcium chloride on drying shrinkage are conflicting but in general it seems to be increased. Resistance to sulphate attack is reduced, and alkali-aggregate expansion with high-alkali cements is increased, though when this is controlled by the use of low-alkali cement or pozzolanas this effect becomes unimportant.

Various mixes are sold under trade names for accelerating the set of cements for use under circumstances where a very rapid set is required, such as for stopping leaks against water under pressure or rendering of surfaces through which water is percolating. These products are composed of aqueous solutions of various salts, such as alkali carbonates, aluminates and silicates, or mixture of various chlorides such as those of calcium, aluminium and sodium.

Other proprietary compositions are sold for retarding the hardening of cement. They are usually applied as a liquid to the interior surface of the formwork in which the concrete is to be placed and they retard the hardening of the surface

[1] J. C. Blenkinsop, *Mag. Concr. Res.* **15** (43), 33 (1963).
[2] R. F. Legget, *Proc. Instn civ. Engng* **22,** 11 (1962); G. E. Monfore and G. J. Verbeck, *Proc. Am. Concr. Inst.* **57,** 491 (1960); A. Baumel and H. J. Engel, *Arch. Eisenhütt Wes.* **30,** 417 (1959).
[3] J. J. Shideler, *Proc. Am. Concr. Inst.* **48,** 537 (1952); T. E. Stanton, *Proc. Highw. Res. Bd* **30,** 232 (1950). Much earlier literature can be traced through these papers.

layer of cement. On striking the forms the surface cement can be removed by brushing, thus exposing the aggregate at the surface of the concrete and producing an agreeable texture. Sugar, and carbohydrate derivatives, are often the basis of such liquid compositions.

SURFACE TREATMENTS FOR WATERPROOFING AND PROTECTING CONCRETE

There are a large variety of materials which are applied to the surface of concrete either to waterproof the surface or to render it resistant to attack by chemical agencies. The conditions which the surface dressing has to withstand are much more severe in the latter event and materials require to be selected specially suited to each individual case. It is not sufficient that the material used should be resistant to attack; it must also provide a coating that remains free from cracks and pinholes through which liquids or gases can gain access to the underlying concrete. With the developments in plastics over the last decade the variety of protective materials available has much increased and it is only possible here to give a general outline. A detailed account is to be found in the *Guide for the Protection of Concrete against Chemical Attack by Means of Coatings and other Corrosion-Resistant Materials* issued by the Committee of the American Concrete Institute and other publications.[1] Some of the surface treatments which may be used to protect concrete against chemical attack in lesser or greater degree are given below but the list is far from exhaustive. No single one of them is suitable for use in all circumstances.

A Surface treatments

Aqueous solutions of sodium silicate, or magnesium or zinc silico fluoride.
Drying oils such as linseed or tung oil often with a volatile thinner.
Mineral oils with or without volatile thinners.

B Thin surface coatings

Bituminous paints and solutions.
Coal-tar paints and solutions.
Chlorinated rubber paints.
Neoprene paints.
Epoxy-ester paints.
Synthetic resin lacquers such as the vinyls and urethanes.

C Thicker surface coatings

Bituminous paints or pastes containing fillers and volatile organic solvents for cold application.
Bituminous emulsions, pastes with fillers and sometimes asbestos fibre.
Bituminous-rubber emulsions with fillers.
Coal-tar pitch in coal tar for hot application.

[1] Report of Committee 515, *Proc. Am. Concr. Inst.* **63,** 1305 (1966); T. Hop and Z. Mrodynski, *Bldg Sci.* **2** (2), 147 (1967); W. L. Dolch, *Tech. Rep. U.S. Corps Engrs, Ohio River Div. Labs,* No. 2–29 (1963).

Bituminous enamels with fillers for hot application.
Coal tar–epoxy two-part systems.
Two-part synthetic resin systems with or without glass fibre reinforcement.

D Linings and sheets
Chemical-resistant asphalt mastic.
Rubber mastics.
Rubber latex–high alumina cement–sand mortar.
Synthetic resin emulsions–high alumina cement–sand mortar.
Synthetic resin emulsions–Portland cement–sand mortar.
Synthetic resin-filler mortars with glass fibre reinforcing.
Sheet linings of natural or synthetic rubber, polyvinyl chloride, polyester–glass fibre, tetrafluoroethylene and chloro-trifluoroethylene.
Clay tile, glass and other inert materials.

E Gas treatment
Precast concrete products treated with SiF_4 producing a coating of hydrated silica and alumina, along with calcium fluoride, that is resistant to many aggressive agents.

Some of the materials mentioned, particularly under (*D*), are used only for protection against aggressive agents arising in manufacturing industry, such as are discussed in the next chapter; the types used under conditions of natural exposure are limited to the less costly products.

The silicon fluoride treatment (*E*), known as the Ocrate process,[1] has been used for such purposes as the treatment of sewer pipes and of concretes exposed to aggressive conditions in industry. The process is operated in various countries, e.g. Holland, Germany, South Africa, Australia and the U.S.A.

It is not possible to discuss in detail all the differences in properties that are to be found between different trade products, but the following notes outline the broad properties of the various types of protective materials.

Many of the materials will not adhere to a wet concrete surface, but coal-tar products are less sensitive than bituminous ones, while emulsions as a class have the important advantage that they can be so applied. Though in paint form they are not sufficiently impermeable, the thicker emulsion coatings obtained from pastes are more adequate in this respect. Some of the synthetic resin solutions give coatings that adhere well to concrete, e.g. styrene-butadiene copolymers, while others, such as polyvinyl chloride, give poor adhesion and require special primers. Amongst the two-part synthetic resin systems the epoxy and coal tar–epoxy resins have particularly good adhesion. Mortars with epoxy and polyester resin binders are used for patching and repair of concrete.[2]

[1] W. Wittekindt, *Zement-Kalk-Gips* **5,** 203 (1953); **7,** 337 (1954); B. P. Hofstede, *Fourth Int. Congr. Precast Concrete Industry, Paris 1963*, Paper No. 15; R. Miller, *Mater. Protn* **6,** 29 (1967).

[2] Report of A.C.I. Committee 403, *Proc. Am. Concr. Inst.* **59,** 1121 (1962); *ASTM spec. tech. Rep.*, 169A, 565 (1966); *R.I.L.E.M. Bull.* **28,** 4 (1965); *R.I.L.E.M. Sym. Synthetic Resins in Building Construction, Paris 1967*; *R.I.L.E.M. Bull.*, No. 37 (1967); *Materials and Structures*, Nos. 1 and 3 (1968); *Epoxies with Concrete*, A.C.I. Special Publication No. 21 (1968).

Treatment with sodium silicate and silicofluoride only affords protection against mild conditions of attack either by aqueous solutions or organic liquids. Drying oils can exert a protective influence over a period of some years against dilute aqueous solutions of aggressive salts.[1] Bituminous materials are not resistant to oils, fats and many organic acids though materials based on coal-tar are better in this respect. In thin surface coatings these materials will not prevent the penetration of water into concrete indefinitely, but they can postpone the commencement of attack. Thus in a series of German tests[2] various proprietary bituminous preparations were applied as surface coatings to cement mortars which were immersed in 5 per cent sodium and magnesium sulphate and 0·5 per cent sulphuric acid solutions. In the salt solutions most of the coatings suffered little up to three years though slight damage was then becoming apparent. In the acid solution most were destroyed in a few months and all within three years except for a thicker coating of a hot applied asphalt. In tests on concrete cylinders in the highly sulphate waters of Medicine Lake[3] various treatments with linseed oil, bituminous and tar solutions afforded initial protection to the concrete, but little effect remained after five years. In acidic peat waters where deterioration of concrete was much slower a protective effect was apparent up to 17 years. In 5 per cent ammonium sulphate solution the protective effect of thin bituminous coatings disappears within two years. As a rough generalisation thin surface coatings with bitumen or tar products will retard the action of aggressive salt solutions, but deterioration will occur sooner or later according to the strength of the solutions. Nevertheless the initial period of protection may be useful. Chlorinated rubber treatments are often effective in protecting a concrete against aqueous solutions of salts and dilute acids, but not oils. Those with the highest chemical resistance will not stand outside exposure to weather. Neoprene and epoxy-ester paints have, for their thickness, a fair chemical resistance. Synthetic resin lacquers specially prepared for the treatment of concrete surfaces often afford good protection against the action of vegetable oils, fats, acids and aqueous solutions, but not always to alcohol, organic solvents and strong alkali solutions. Wax coatings applied hot give protection against aqueous solutions, but they are not resistant to oils and fats. These coatings are, however, difficult to apply satisfactorily.

Thin surface coatings have only a limited resistance to abrasion and a limited life when exposed to weather and indeed under most conditions of use periodic renewal is necessary if they are to be retained intact. An effective life of up to 15 years has been obtained with thin bituminous films (e.g. 0·002 inch) when applied to surfaces to which a good bond can be obtained. When exposed to aggressive running water in the interior of pipes there is evidence that bituminous paints can give protection up to 10 years, while with non-aggressive waters the effective life can exceed 25 years. For external protection of concrete, or the interior lining of pipes carrying aggressive waters, the thicker coatings listed

[1] R. D. Hughes, *Concrete Bridge Decks, Deterioration and Repair*, Kentucky Department of Highways 1966.

[2] H. Burchartz and H. W. Gonell, *Deut. Ausschuss für Eisenbeton*. Vol. 72 (1933).

[3] D. G. Miller and P. W. Manson, *Tech. Bull. U.S. Dep. Agric.*, No. 358 (1933); *Tech. Bull. Minn. agric. Exp. Stn*, No. 180 (1948); E. C. E. Lord and D. G. Runner, *Publ. Rds, Wash.* **23,** 282 (1944).

under (*C*) are usually required if a long life is needed. These can be applied in individual layers varying from $\frac{1}{64}$ to $\frac{1}{8}$ inch thick, according to the type of material, and, in the thicker ones, a layer of hessian can be incorporated between two coats to reinforce them. The effectiveness of these coatings depends much on the bond obtained to the concrete and if this is inadequate they tend to crack and become displaced and they may then be worse than the thin paint treatments.

Thick bituminous coatings can be used for protection against mineral acids up to a few per cent concentration if the fillers are of an acid-resistant type while synthetic resin compositions and the materials listed under (*D*) have a wide use for a variety of industrial conditions of greater severity and for jointless floor coverings. Thus both epoxy and polyester mortars have been used for the repair of sewers damaged by bacterial sulphuric acid attack and the resins for linings of concrete pipes carrying corrosive industrial effluents.[1] Linings of P.V.C. sheet, polyester–glass fibre sheet and pitch–epoxy coatings have also been used in sewers to prevent sulphuric acid attack. The silicon tetrafluoride treatment (Ocrate process) is also applied for the treatment of concrete sewer pipes[2] to safeguard against acid attack. Phenol-formaldehyde resins have also been used for lining concrete pipes carrying chemical wastes.[1]

Surface waterproofers for application to concretes to decrease the permeability to water include many other materials besides those mentioned above. They may be divided roughly into materials which are transparent and those which form a visible solid coating over the surface. The transparent materials are solutions or emulsions of hydrophobic substances which act by lining the pores on the surface and rendering them water-repellent rather than by forming a complete surface film. While many surface waterproofers may be effective, at least for a period, when applied to concrete, the transparent materials cannot be expected to prove satisfactory on renderings. In this case fine cracks in the rendering are almost inevitable and the transparent films fail to bridge them. The materials which form visible solid coatings give temporary protection even to cracked renderings, but are liable to crack and lose their efficacy if further shrinkage of the rendering occurs. In this class of materials there may be grouped cement paints, certain pigmented bituminous emulsions and special paints composed of aqueous emulsions of oils, resins and pigments. Ordinary linseed oil paints are not permanent.

Among the surface treatments producing transparent surface films are the following:

Silicones.
Aqueous solutions of alkali soaps.
Solutions of aluminium, calcium and other soaps in paraffin oils.
Solutions of wax in volatile solvents.
Successive treatment with alkali soap solutions and a solution of alum (Sylvester process).
Successive treatment with solutions of barium chloride and sodium sulphate.
Aqueous solutions of sodium silicate and silicofluorides.

[1] G. Lorentz, *Betonsteinzeitung* **27** (5), 214 (1961); R. W. Childers, *Mater. Protn* **2** (3), 18 (1963).
[2] H. Schremmer, *Zement-Kalk-Gips* **17**, 417 (1964).

Mixtures of vegetable and mineral oils.
Butyl stearate.
Butyl oleate.
China wood oil in a volatile solvent.
Linseed oil.

Jumper found that the coatings produced on concrete by the solutions of wax and alkali soaps broke down rapidly in water, and that of a variety of transparent coatings examined the linseed oil and China wood oil were the most effective. Experience[1] in North America has also indicated that linseed oil impregnations can have an effective life of a few years. No coating of this type will withstand any appreciable water pressure.

Metallic coatings of zinc and lead applied by the spray process have also been found effective for the protection of concretes against aqueous solutions of sulphate salts. Such coatings have been found to remain watertight up to a pressure of 15 atmospheres.[2]

Surface hardeners for concrete

A concrete surface can be hardened and rendered more resistant to abrasion, or less liable to dust, by suitable treatment. The surface of the hardened and air-dry concrete may be treated with a solution of sodium silicate, aluminium, or zinc, sulphate, or silicofluorides, or with a drying oil like linseed oil or tung oil; alternatively a proportion of carborundum or fused alumina, or one of the finely divided iron-ammonium chloride preparations, may be incorporated in the surface layer of the concrete while placing.

Carborundum, fused alumina, and finely divided iron are effective in rendering a concrete surface less slippery and more resistant to abrasion, but they are not very effective in rendering the surface less dusty. Treatment with a solution of sodium silicate hardens a concrete surface and also renders it less dusty; treatment with the sulphate solutions and silicofluorides is also effective. Sodium silicate for use in the surface treatment of concrete should have a molecular ratio of silica to soda of three or four to one and should not be of the more alkaline types with lower $SiO_2 : Na_2O$ ratios. The sodium silicate is obtained commercially as a solution containing about 30 per cent SiO_2 and 9 per cent Na_2O, or in the form of a powder containing about 60 per cent SiO_2 and 20 per cent Na_2O soluble in hot water. The solution is usually diluted with 3–4 times its own volume of water and applied in several coats.

Tung oil and linseed oil are applied to concrete surfaces either neat, hot, or thinned with turpentine or white spirits. The treatment gives a hard surface and freedom from dust. Floor paints also have a reasonable durability if the conditions of wear are not heavy. Oil paints with a tung-oil medium, or bituminous paints, can be used, but paints containing synthetic resins, particularly polyurethanes or epoxy-esters, or chlorinated rubber have a greater resistance to wear. None of these surface treatments are effective on a weak or friable concrete surface.

[1] R. B. Young, *Proc. Am. Concr. Inst.* **33,** 367 (1937); R. D. Hughes, loc. cit.
[2] Deutscher Beton Verein, *Beton Eisen* **27** (13), 261 (1928).

Resistance of concrete to frost

The resistance of concrete to frost is of much importance in northern Europe, Canada, and the northern parts of the United States where much damage has resulted from its action. Though less serious in Britain it has also led to severe failures of kerbstones and other exposed features, and is often the final cause of destruction of concrete already weakened from other causes.

Water expands about 9 per cent in volume during freezing and when it is trapped in cavities the pressures that may be exerted are very large. In a wet concrete the water enclosed in the pores of the material tends on freezing to force the particles of mortar apart or to set up severe internal stresses. The effect is intensified by repetitions of the processes of freezing and thawing. Minute cracks first develop and become filled with water which on subsequent freezing produces further cracking and fissuring. The first appearance of frost action is usually a flaking of the surface which works progressively further into the mass, though in cases of very poor concrete deep fissures may also appear.[1] Dry concrete is not affected by frost.

The requirements for normal concrete which is to be frost-resistant are low permeability and absorption, and adequate cover to any reinforcement. Beyond this it is necessary that local weaknesses, as discussed on page 594, should be avoided for defects usually start at points where the concrete is too porous or at segregated layers and laitance bands. The severest conditions for frost action arise when concrete has more than one face exposed to the weather and is in such a position that it remains wet for long periods. Examples are to be found in road kerbs, parapets, and concrete members in hydraulic or seawater structures just above water level. The special problem of road slabs is dealt with later.

There are various explanations of frost damage. In the oldest theory the damage it attributed directly to the empty space available being insufficient to accommodate the additional solid volume produced when the water in the concrete freezes. Damage is related to the degree of saturation as discussed later. In the later theories a less direct mechanism for the production of pressure is postulated. The frost heave theory[2] ascribes the production of pressure to the growth of ice lenses parallel to the surface of the concrete owing to the migration of water from capillaries where the freezing point is depressed. This is similar to the theory of frost heaving in soils.[3] A thermodynamic treatment of this has been given by Everett.[4] In an alternative theory,[5] the action is ascribed to the generation of a water pressure within the capillary cavities as the ice crystals grow. This hydraulic

[1] For discussion of typical frost action and its causes, see R. B. Young, *Bull. Ont. hydro-elect. Pr Comm.*, No. 27, 230, 276 (1939); F. H. Jackson, *Proc. Am. Concr. Inst.* **43,** 165 (1947); **52,** 159 (1955); W. A. Cordon, *Am. Concr. Inst. Monogr.*, No. 3 (1966); F. M. Lea and N. Davey, *J. Inst. civ. Engrs* **32,** 248 (1949); A. R. Collins, ibid. **23,** 29 (1944); various authors, *R.I.L.E.M. Bull.*, Nos. 40 and 41 (1958); *R.I.L.E.M. Sym. Winter Concreting, Copenhagen 1956*; P. Nerenst, *Sym., Washington 1960*, 807; M. Venuat, *Revue Matér. Constr. Trav. publ.* **573,** 181; **574/5,** 229 (1963).

[2] A. R. Collins, loc. cit.

[3] E. Penner, *Spec. Rep. Highw. Res. Bd*, No. 40, 191 (1959); *Bull. Highw. Res. Bd.* No. 225, 1 (1959); *Proc. int. Conf. Low Temperature Science, Japan 1966*, **1** (Part 2), 1401 (1967).

[4] D. H. Everett, *Trans. Faraday Soc.* **57,** 1541 (1961); D. H. Everett and J. M. Haynes, *R.I.L.E.M. Bull.*, No. 27, 31 (1965).

[5] T. C. Powers, *Proc. Am. Concr. Inst.* **41,** 245, 272–1 (1945); **43,** 933 (1947).

pressure can only be relieved by flow of water into other spaces, since the ice seals the exterior, and the pressure required to force water through the fine capillaries rises as the length of the flow path increases. Local pressures eventually exceed the tensile strength of the concrete and cause breakdown. On all these theories the permeability, rate of absorption and degree of saturation of the concrete are important factors. A high permeability would allow easier flow of water, and, since it will arise from larger capillaries, more room for ice growth, but such a concrete would readily become more highly saturated. In practice, frost resistance is high for dense impermeable concretes and for very porous concretes, e.g. some no-fines and lightweight concretes, and lower for the more permeable 'dense' concretes. This, it is clear, is readily explainable on either theory.

Powers[1] has extended his theory to include also the forces that arise when all the water in a capillary cavity is frozen, the excess having been pumped out by the hydraulic pressure generated, and the ice crystals then grow further as a result of diffusion of water from the surrounding cement gel. This leads to 'lens' effects but on a microscopic scale rather than the gross scale visualised by Collins. In air-entrained concrete the function of the air voids, according to Powers, is to limit the hydraulic pressure by providing release spaces, and to limit the time during which capillary ice can grow by diffusion of gel water, for this in turn is dependent on the existence of differential pressures. In a more recent modification[2] of the theory, stress is laid on the production of localised osmotic pressure at the sites of ice formation rather than on the development of water pressure in all the pores.

Valenta[3] has found that not only the compressive, but also the flexural and tensile, strength of the concrete is greater in the frozen than the unfrozen condition, a result not easy to reconcile with the hydraulic pressure theory.

The resistance of concrete to frost depends on the quality of the cement mortar and of the aggregate.[4] Fairly wide variations in the quality of the coarse aggregate can be tolerated, for there is no close correlation between the durability of aggregates and the concrete made from them, but argillaceous or soft porous limestones, gravels containing too much weak particles such as porous chert, shales and friable sandstones, are susceptible to breakdown by frost action. There is no close relation between the total porosity or absorption of aggregates and their frost resistance, for it is the character of the pores which is of predominant importance. The larger pores are not necessarily detrimental, and it is the finer pores, e.g. below about 0·005 mm in diameter, which, if present in abundance, have most influence in reducing the frost resistance of concrete, for they tend to remain full of water. This leads to an expectation that aggregates with a large internal, but accessible, surface area should tend to be less frost-resistant and there is some evidence to support this.[5] The incorporation of fine limestone has

[1] T. C. Powers and R. A. Helmuth, *Proc. Highw. Res. Bd* **32,** 285 (1953); R. A. Helmuth, *Sym., Washington 1960*, 855.

[2] R. A. Helmuth, *Sym., Washington 1960*, 829.

[3] O. Valenta, *Proc. R.I.L.E.M.-C.I.B. Symposium on Moisture Problems in Buildings, Helsinki 1965*, Vol. 1 (ii), Paper 14.

[4] G. Verbeck and R. Landgren, *Proc. Am. Soc. Test. Mater.* **60,** 1063 (1960); R. D. Walker, *National Cooperative Highway Research Programme, Report No. 12* (1965); H. T. Arni, *ASTM spec. tech. Publ.*, No. 169A, 261 (1966).

[5] R. L. Blaine, C. M. Hunt and L. A. Tomes, *Proc. Highw. Res. Bd* **32,** 298 (1953).

been claimed to improve frost resistance because of its strong bond to the cement.[1]

The frost resistance of aggregates is now generally assessed from freezing and thawing tests on concrete made from them, but a sodium, or magnesium, sulphate crystallisation test[2] has also been used as a simulated frost test on aggregates. It has not proved a reliable index of behaviour of aggregates in concrete. It has been suggested that the test might be used to accept aggregates and that failure under it should be taken as no more than an indication that further investigation by freezing tests on concrete, or study of performance in actual use is needed. Unfortunately the test does not appear to be reliable even for this limited purpose since some aggregates that pass it can fail under frost.[3]

The effect of the type of cement on frost resistance is still a matter of controversy. Some workers have claimed that cements with the higher tricalcium aluminate or alkali contents have lower frost resistance; various French workers[4] have claimed that Portland blastfurnace cement has a higher frost resistance than Portland cement though others have found no significant difference. It can only be concluded that, apart from air-entrainment, the cement itself is a minor factor in determining frost resistance.[5]

With sound aggregates, failure of concrete under frost occurs by the weakening and breakdown of the cement mortar which binds the coarser aggregate together. It is the quality of this mortar, therefore, as affected by the cement content, water content, air content and grading of the mix, which determines the resistance of the concrete.

It was long ago suggested by Hirschwald, Kreuger and others that since water expands about one-tenth of its volume in freezing, damage by frost should not occur until about 90 per cent of the pore space in a material is filled by water. The ratio of the water absorbed by a material to its total pore space, as determined from measurements of true and apparent density, has often therefore been regarded as an index of resistance to frost action and called the saturation coefficient. Schurecht used a somewhat different ratio, that of the absorption occurring after 48 hours' total immersion in water to that obtained on 48 hours' immersion followed by 5 hours' boiling.

Tests on different concretes have failed to show any definite correlation between frost resistance and either of the above ratios,[6] though Venuat[7] found some relation with the capillary absorption of air-dry concretes placed in wet sand.

[1] J. Farran and J. C. Maso, *Revue Matér. Constr. Trav. publ.* **586/587,** 195 (1964).
[2] ASTM C88 : 1963. A very similar test is to be found in BS 1438 : 1948, *Media for Percolating Filters*, Appendix B.
[3] D. L. Bloem, *ASTM spec. tech. Publ.*, No. 169A, 497 (1966); see also *Highw. Res. Bd, spec. Rep.*, No. 80 (1964) for review of literature on methods of identifying aggregates that are not resistant to freezing.
[4] J. Chapelle, *R.I.L.E.M. Bull.* **40,** 30 (1958); *Bull. int. Commn large Dams*, No. 15, Paris 1960.
[5] F. H. Jackson, *Proc. Am. Concr. Inst.* **52,** 159 (1955); C. C. Oleson and G. Verbeck, *Bull. Res. Dev. Labs Portld Cem. Ass.*, No. 217 (1967).
[6] J. Tucker, G. W. Walker and J. A. Swenson, *J. Res. natn. Bur. Stand.* **7,** 1067 (1931); F. B. Hornibrook, H. Freiberger and A. Litvin, *Proc. Am. Soc. Test. Mater.* **46,** 1320 (1946); see also B. Warris, *Swed. Cem. Concr. Res. Inst., Proc. No. 36* (1964).
[7] M. Venuat, *Rev. Mater. Constr.* **573,** 181; **574/5,** 1229 (1963).

One of the main difficulties in attempting to relate frost resistance to the absorption, or degree of saturation, lies in the definition of these terms. Porosity arises from:

(*a*) Voids in the aggregate particles themselves;
(*b*) Spaces between the aggregate particles and the cement paste;
(*c*) Capillaries in the set cement paste;
(*d*) Ultra-fine pores in the set cement gel;
(*e*) Air bubbles entrapped in the concrete in making.

These are not all of equal significance. Thus spaces under aggregate particles may not, once the water originally filling them has dried out, be easily filled again by simple contact with water, though if subjected to one-sided water pressure they may fill readily and are regarded as one cause of low frost resistance in hydraulic structures. Very small air bubbles, accidentally or deliberately entrained in the concrete, are very difficult to fill with water.

The porosity of the cement paste itself arises from the excess water present in the mix, and the contraction of the system (cement and water) during hydration, the capillaries becoming partially emptied during hydration unless extra water can flow in to fill them. This 'self-dessication' of the paste is probably an important factor in contributing to frost resistance. The water in the capillaries can, however, come and go in response to changes in external conditions and the amount that can do so is an index of the porosity of the hardened paste.

As we have seen earlier (p. 271), Powers[1] distinguishes between the non-evaporable and the evaporable water in set cement. The former is regarded as proportional to the extent of hydration of the cement while the latter includes not only water in pores which freezes at 0°, and water in capillaries which will freeze at lower temperatures, depending on the capillary diameter, but also water under the action of surface forces (as in (*d*) above) which for practical purposes can be regarded as non-freezable and the amount of which is related by a roughly constant factor to the extent of hydration of the cement and thus to the non-evaporable water.

For practical purposes the freezable water in the set cement paste can be identified with the capillary water and the great bulk of this is frozen by −12°, though supercooling to at least this temperature can occur if there is no seeding from external ice crystals. Of the water in a saturated set cement freezable at −30°, Powers[2] found some 60 per cent was frozen by −4° and over 80 per cent by −12°. He also obtained an equation:

$$w_f = w_t - a w_n$$

relating the weight of water (w_f) in a saturated cement paste freezable at a given temperature, to the total water content (w_t) and the non-evaporable water (w_n). The constant a varies with the temperature of freezing. It is inherent in this equation that below certain initial water : cement ratios there will be no water freezable at any selected temperature in a set cement paste unless it subsequently gains additional water from external sources. For an average set cement at an age of 90 days and a freezing temperature of −15°, Powers found this water : cement

[1] T. C. Powers, *Bull. Am. Soc. Test. Mater.*, No. 158 (1949).
[2] T. C. Powers and T. L. Brownyard, *Proc. Am. Concr. Inst.* **43,** 933, 971 (1947).

ratio to be about 0·30–0·35. Water in excess of this ratio, introduced initially or gaining access later, will be freezable at $-15°$. In practice it is found that the frost resistance of non-air-entrained concrete with a water : cement ratio of 0·50–0·55 is usually high, but it decreases progressively with further increase in this ratio.[1] Higher values are, of course, acceptable for many conditions of exposure, while for stringent conditions lower ones may be needed.

The conception of freezable water has been used by Whiteside and Sweet[2] in defining the degree of saturation of a concrete as:

$$S = V_f/(V_f+V_A)$$

where V_f and V_A are the volumes of freezable water and air per unit volume of concrete. In freezing tests they found rapid deterioration of concrete occurred when $S > 0{\cdot}91$ and a high durability when $S < 0{\cdot}88$. Too much significance should not be attached to these precise figures, since there are various assumptions inherent in the data of Powers that are used. In some tests in Sweden, Bergström[3] found that the frost resistance varied with the product of $(w/c-0{\cdot}2)$ multiplied by the per cent volume content of the cement paste in the mortars tested.

A more empirical approach to the prediction of frost resistance is to be found in many studies of its relation to absorption. It has long been known that the absorption as measured by methods which involve drying at about 100° and immersion in water for 24 hours, or boiling, does not correlate satisfactorily with frost resistance. The reason why such a gross value, which makes no distinction between the different classes of voids, does not afford an adequate index, will be evident from the foregoing discussion. Various attempts have been made to develop absorption test procedures in which only the water most readily lost or gained is measured and these have given improved correlations. Thus Hansen[4] found a relation between frost resistance and the absorption on immersion in water for 48 hours of specimens previously cured for 14 days in moist air followed by 14 days in air at 50 per cent relative humidity and 75° F. Since the specimens are only partially dried in this procedure, the absorption determined represents only the more readily evaporable water. A limit of absorption of about 0·75 per cent in this test was found to give a useful criterion for frost resistance.

In tests on concrete kerbs[5] the amount of water taken up in 10 minutes' immersion in water, after previous drying at 100°, has been found to be a more reliable guide than the 24-hour absorption. The absorption values ranged from 1 to 4 per cent for the 10-minute immersion and kerbs showing good performance in practice gave values below about 2 per cent, whereas in the 24-hour procedure the values ranged from 5 to 12 per cent for good and bad kerbs alike. Though much quicker to carry out, this method is less well founded theoretically, since

[1] J. A. Loe and F. N. Sparkes, *3rd Congr. int. Ass. Bridge Struct. Engrs* (1948), 201; E. Gruenwald, *Proc. Am. Soc. Test. Mater.* **39,** 810 (1939); Combe. *4th int. Congr. large Dams* (1951).

[2] *Proc. Highw. Res. Bd* **30,** 204 (1950).

[3] S. G. Bergstrom, *Bull. Swed. Cem. Concr. Res. Inst.*, No. 22 (1955).

[4] *Proc. Am. Concr. Inst.* **39,** 105 (1943); *Proc. Highw. Res. Bd* **20,** 568 (1940).

[5] J. A. Loe, *Rd Res. tech. Pap.*, No. 18 (1950), H.M.S.O. London; P. J. F. Wright, *Engineer* **211,** 855 (1961).

it is the less, and not the more, readily evaporable water that is measured. This test, with a maximum permissible absorption of 2·5 per cent by weight, is used in BS 340 : 1963 for precast concrete kerbs.

Air-entrainment is now standard practice in many countries to improve the frost resistance of concrete. Numerous laboratory test and exposure trials have shown its value under severe frost conditions. The fine air bubbles entrained in the concrete are very difficult to fill with water, so reducing the degree of saturation, and their intimate distribution ensures that the distance between neighbouring air spaces is so small that pressures created by the growth of ice crystals are reduced. It is usual to aim at an air content of 5–6 per cent by volume in the concrete. For good frost resistance[1] the distance between individual air bubbles should not exceed about 0·01 inch though this may not always be a sufficient condition.[2] This indicates that occasional gross air voids cannot have the same beneficial influence as small air bubbles, closely spaced, even when the total void space is the same.

Despite the great value attached to air entrainment, it must not be forgotten that much frost-resistant concrete has been made without the deliberate addition of air-entraining agents. The benefits of air-entrainment seem to be most pronounced with concretes of high w/c ratio and in giving protection against damage by de-icing salts.

Frost scaling of concrete road surfaces

Deterioration of concrete road surfaces arising from the use of calcium chloride, or common salt, to remove ice in winter proved a serious problem in many countries. It has been largely overcome by the use of air-entrained concrete, though trouble is still experienced under very severe climatic conditions and on concrete bridge decks[3] where it is also accentuated by corrosion of the reinforcement owing to the penetration of chlorides. Though serious deterioration has occurred the problem has been less widespread in Great Britain than in the U.S.A. and Canada, partly perhaps owing to the less severe climatic conditions and partly to the tendency for drier concrete mixes to be used in road construction. As Collins[4] has shown the damage increases as the water : cement ratio of the concrete increases over the range 0·4 to 0·6. The use of silicone surface treatments[5] has been shown to be ineffective, or even to increase damage, but a boiled linseed oil–petroleum spirit treatment[6] has been found useful on bridge decks where deterioration is the more severe.

It is common experience that concrete frozen in a common salt or a calcium chloride solution suffers more rapid deterioration than when frozen in water, but

[1] P. Klieger, *Spec. tech. Publ. Am. Soc. Test. Mater.*, No. 169A, 530 (1966).
[2] V. Danielsson and A. Wastesson, *Proc. Swed. Cem. Concr. Res. Inst.* **30,** 1958.
[3] B. Ost and G. E. Monfore, *J. Res. Dev. Labs Portld Cem. Ass.* **8** (1), 46 (1966); Symposium, *Bull. Highw. Res. Bd*, No. 323 (1962).
[4] A. R. Collins, *Rds Rd Constr.* **18** (No. 209), 98 (1940).
[5] P. Klieger and W. Perenchio, *Highw. Res. Rec.* **18,** 33 (1963); D. J. T. Hussell, ibid., p. 13.
[6] R. D. Hughes, *Concrete Bridge Decks, Deterioration and Repair*, Kentucky Department of Highways 1966; see also R. S. Yamasaki, *J. Paint Technol.* **39,** 394 (1967) for a review of work on protective coatings.

the cause of this is still uncertain. Arnfeldt[1] found that a 3·5 per cent calcium chloride solution was more destructive than weaker or stronger solutions, and that other solutions, e.g. common salt, ethyl alcohol, urea, causing a similar depression of freezing point were equally active. This indicates that the action is primarily a physical and not a chemical one and it has been suggested[2] that the damage is caused by the absorption of heat during the melting process with consequent rapid cooling of the upper layers of the concrete. The resulting sharp change in temperature is thought to produce stresses in the frozen surface layers of fine mortar which cause deterioration. Unfrozen water present in a deeper zone parallel to the surface of the concrete will also, as a result of the heat abstraction, be caused to freeze quickly increasing the risk of spalling of the overlaying concrete. In an alternative theory advanced by Powers[3] the damage is attributed to osmotic pressure effects caused by the build-up of the de-icing agent in the capillaries of the cement paste to a greater concentration than in the gel pores, resulting in water being drawn from the gel pores into the capillary spaces. The de-icing agent is considered to have two opposite effects. By reducing the amount of ice that can form in the capillaries of the cement paste at a given temperature, it tends to reduce the pressure, either hydraulic or osmotic, generated on freezing. On the other hand it increases the osmotic pressure arising from the difference in the salt concentration between capillaries and gel pores. These opposite effects are invoked to explain the observation that the maximum destructive effect occurs at an intermediate salt concentration.

Freezing and thawing tests

Laboratory freezing and thawing tests on concretes are carried out to evaluate the relative resistance of different aggregates or cements, to test the influence of admixtures and particularly air-entraining agents, and to assess the durability of particular mixes to be used on constructional work.

In principle freezing and thawing tests are simple; specimens are prepared, cured under some specified conditions, then subjected to cycles of freezing and thawing and the effect produced evaluated by visual inspection or some physical measurement. While there are broad similarities between the methods used in different laboratories, the details vary so much that comparison of results is often impossible.[4]

The specimens are commonly prisms, about 3-in. by 3- or 4-in. cross-section and 12–16 in. long, but 4-in. cubes, or 6-in. by 3-in. cylinders, are sometimes used. For aggregates above 1 in. size these dimensions are increased. Specimens are usually cured in water or saturated moist air for periods varying from about 9 to 90 days before being subjected to the freezing and thawing cycles, but some laboratories store them for an intermediate period in dry air followed by a final 2 days or so in water. Freezing is usually carried out in air

[1] H. Arnfeldt, *Meddn St. Väginst.*, 66 (1943); G. Verbeck and P. Klieger, *Bull. Highw. Res. Bd*, No. 150 (1957).

[2] E. Hartman, *Zement-Kalk-Gips* **10**, 265, 314 (1957).

[3] T. C. Powers, *Proc. Am. Soc. Test. Mater.* **55**, 1132 (1955); *Cem. Lime Gravel* **41** (6), 181 (1966).

[4] See Report on Frost Resistance of Concrete, R70, *5th int. Congr. large Dams*, Paris 1955, for details of the various test procedures and various papers on factors influencing frost resistance; also *Bull. int. Commn large Dams*, 1–15, Paris 1960.

and thawing in water, but some laboratories freeze in water by enclosing the specimens in metal containers or rubber bags filled with water, and immersing the containers in the freezing bath. Thawing is also occasionally carried out in air. Freezing in water causes more rapid deterioration than freezing in air.

The faster the rate of freezing, that is the rate of temperature drop of the specimen below 0°, the more rapid is the rate of deterioration, but the degree of discrimination between concretes of differing frost resistance also decreases. In actual practice the temperature of exposed materials when frozen does not normally drop at a rate greater than about 1·5° C per hour in England at low altitudes and 5° per hour at high altitudes in mountains in other countries. The rates occurring in the test procedures of different laboratories vary greatly, from about 1° per hour to 10° in 10 minutes or an equivalent rate of 60° per hour. This rate depends on the temperature of the freezing chamber, the freezing medium (fluid or air), and the relation between the load placed in the chamber at one time and its capacity. A common temperature for the freezing bath is −15° to −20°. Thawing is carried out in water at temperatures varying from 5° to 25°; there is evidence that increase in the temperature, or time, of thawing tends to reduce the rate of deterioration and this has been ascribed to the effect of autogeneous healing.[1] Rather surprisingly it has been found[2] that specimens which are dried after thawing and then rewetted again before freezing suffer more damage than when the freezing and thawing cycle is uninterrupted. The duration of the cycle is also very variable, the periods of freezing and of thawing each varying from less than one hour up to 24 hours. Some laboratories have installed automatic apparatus[3] which permits of a complete cycle every few hours, whereas laboratories using manual handling often work on one cycle per day.

The degree of saturation of the concrete at the commencement of the freezing and thawing cycles also has much influence on the rate of failure.[4] Thus even the best concretes, if saturated by immersion in water *in vacuo*, fail rapidly. Since, however, the permeability and rate of absorption of concretes determine the degree of saturation that will occur in practice, a laboratory procedure which aimed at obtaining standardisation by complete saturation would be unlikely to bring about the differences found in practice between concretes of varying quality. Recourse must, therefore, be made to some defined procedure in curing specimens before commencing the tests. In tests on matured and dried concrete soaked for various periods in water, Walker[5] found that the number of cycles of freezing and thawing needed to reduce the dynamic modulus of elasticity by 30 per cent in the particular form of test used was 80 cycles for 24 hours or 7 days soaking, 70 for 28 days, and only 35 for 5 months. Relatively small differences in the air content of the concrete have a marked influence on the frost resistance.

[1] See F. B. Hornibrook, H. Freiberger and A. Litvin, *Proc. Am. Soc. Test. Mater.* **46,** 1320 (1946); M. O. Withey, *Proc. Highw. Res. Bd* **24,** 174 (1944); F. B. Hornibrook, discussion on paper by F. V. Reagel, ibid. **20,** 587 (1940); B. Myers, *Proc. Am. Soc. Test. Mater.* **39,** 930 (1939) found a reverse effect.

[2] D. A. Parsons, *J. Res. natn. Bur. Stand.* **10,** 257 (1933); L. Schuman and E. A. Pisapia, ibid. **14,** 724 (1935).

[3] *Bull. Highw. Res. Bd*, No. 259 (1960).

[4] See R. C. Valore, *J. Res. natn. Bur. Stand.* **43,** 1 (1949); *Proc. Am. Concr. Inst.* **46,** 417 (1950).

[5] S. Walker, *Circ. natn. Sand Gravel Ass.*, No. 26 (1944).

For normal, not air-entrained, concretes, the volume of air entrained in the concrete varies with the type of concrete mixer used, and it was reported by Withey,[1] for example, that specimens containing about 1·5 per cent air by volume had a markedly superior resistance in frost tests to those with lesser contents. Further, prolonged moist curing for some two years raised the water content of specimens by about 1·5 per cent by volume, compared with others moist cured for 28 days, and caused very rapid failure of good concrete in frost tests. Many laboratories make a practice of measuring the air content of the concrete before test.

The age also has an influence. Thus, if the frost tests are begun at short ages, e.g. 14 days, the more rapid-hardening cements have an advantage over the slower, but in practice this would only be of importance for concrete exposed to frost within a few weeks of placing.

With an increasing period of preliminary curing in water the resistance in frost tests usually rises to a maximum and then decreases again. The rise is to be attributed to the development of strength, and the subsequent fall to the increasing extent of saturation of the specimen. This maximum may occur at from one to many months water curing. Many laboratories standardise on an age of 28 days' curing in water (or saturated lime water) or moist air ('fog'), but others vary from 9 to 90 days.

Concretes which are allowed to dry before commencing the frost resistance tests show an enhanced resistance, because of the slowness with which the pores refill with water, but the onset of damage is only postponed, not prevented.

The degree of saturation of the aggregate used in making the concrete also influences tests results because of its effect on the degree of saturation of the concrete. Thus concrete made with a saturated gravel coarse aggregate (1·1 per cent moisture) was found in an American series of tests[2] to require only about half the number of cycles of freezing and thawing to produce a decrease of 30 per cent in the dynamic modulus of elasticity as a similar concrete in which the same gravel containing 0·8 per cent moisture was used. Walker[3] found a similar ratio between the frost resistance of concretes made with fully and half-saturated gravel aggregate containing chert. Some laboratories use dry aggregates, allowing for the water absorbed by the aggregate (e.g. the 30-minute absorption value) in computing the net water : cement ratio; a few use the drastic condition of previous vacuum soaking of the aggregate; whilst many wet the aggregate the day before use, in accordance with the normal definition of 'saturated' aggregate.

Much work[4] has been done in the U.S.A. in an attempt to standardise test procedures, and this has led to the development of the two ASTM Methods of Test outlined in Table 124 (p. 620). The C290 method involves freezing and thawing in water and the C291 method freezing in air and thawing in water. Both methods use rapid freezing and thawing.

The temperatures are measured at the centre of control specimens. The specimens are prisms made in accordance with ASTM C192–68 which specifies that

[1] M. O. Withey, *Proc. Highw. Res. Bd* **24,** 174 (1944).
[2] M. O. Withey, loc. cit.
[3] Loc. cit.
[4] H. T. Arni, *Spec. tech. Rep. Am. Soc. Test. Mater.* 169A, 261 (1966).

aggregates shall be brought to a 'saturated' condition (24 hours' immersion) before use and that the concrete shall be cured under moist conditions for 14 days or such period as required.

TABLE 124 ASTM methods of test of resistance of concrete to freezing and thawing

	C290–67	C291–67
Freezing medium	Water	Air
Minimum temperature	0° F	0° F
Thawing medium	Water	Water
Maximum temperature	40° F	40° F
Time of freezing	2–4 hr†	≯ 3 hr
Time of thawing		≯ 1 hr

† Total cycle time.

There is not sufficient evidence to indicate that any particular procedure is to be preferred as giving results more closely in accord with practice, nor is there any certainty that different concretes are placed by different procedures in the same relative order of frost resistance. However, no great degree of refinement of results is to be looked for in frost testing and many laboratories are reasonably satisfied that the frost test they use gives a broad indication of relative frost resistance. It will be evident from the earlier discussion that even with a standardised procedure it is difficult to establish a standard criterion for judging test results since small differences in methods have an appreciable effect. Individual laboratories therefore tend to set their own standards based on the behaviour in the test of types of concrete or aggregates of known performance in practice.

No direct proportionality between the results of frost resistance tests and durability in service can be expected since the severity of exposure to frost can vary widely. It must suffice if the results of laboratory tests place a series of concretes in a relative order of frost resistance.

The kind of deterioration observed in frost tests varies somewhat with the type of concrete and the method of freezing. Specimens frozen in water tend to show much more surface crumbling and scaling than those frozen in air, whereas the latter exhibit more general cracking and disintegration. Pitting or spalling may occur in either case from unsound particles of aggregate near the surface. The estimation of the progress of deterioration is most commonly made by non-destructive tests on three to six replicate specimens, e.g. loss in weight, expansion, change in modulus of elasticity (static or more usually dynamic) and ultrasonic wave velocity. Loss in weight is not a good criterion for specimens frozen in air, but is used in conjunction with water freezing, though in the earlier stages of deterioration it may show little relation to expansion or changes in E value. The dynamic modulus of elasticity, measured from the resonant vibration frequency, is the commonest method. The reduction in E is fairly well related to the fall in transverse strength and the latter is much more sensitive to the deterioration arising in frost tests than is the compressive strength. Since no single method of non-destructive testing is an entirely reliable index to the progress of deterioration, it is a fairly common practice to use more than one and at the end of the freezing test to carry out a strength test.

When making comparative tests on aggregates it is the practice of American laboratories to use air-entrained concretes with 4–5 per cent air content with the object of ensuring that a frost-resistant cement paste is obtained, so that the frost resistance of the concrete will reflect the durability of the aggregate.

Thermal expansion and the durability of concrete

We have seen earlier that the thermal expansion of different aggregates varies considerably. It has been suggested that differential thermal movements, either between the aggregate and the cement paste, or between different aggregates in the same concrete, is one factor leading to deterioration.[1] The thermal expansion of neat cement pastes is about $8–10 \times 10^{-6}$ per °F and aggregates rarely have expansions larger than this. Some aggregates, however, have expansions as low as $1–2 \times 10^{-6}$ and these may have more influence on the durability of a concrete. Thus failures with marble and calcite aggregates in cast stone have been reported; in the case of coarse calcite crystals, e.g. $\frac{1}{4}$–$\frac{3}{8}$ inch size, the differential expansion in two directions (see p. 566) must be a source of internal stress.[2] Crystals of orthoclase and microline felspars also have a very low thermal expansion in one direction. Various workers[3] have examined the effect of combining in a concrete coarse and fine aggregates differing in thermal expansion, so setting up a differential thermal movement between mortar and coarse aggregate. While there is evidence that this contributes to deterioration of concrete, most failures of concrete arise from a combination of causes and it is questionable if this individual factor alone is normally a cause of serious distress.

Concrete durability: General

We have seen in a previous section the effect of frost, and of certain aggregates, on the durability of concrete and in sections to follow the more specific problems that arise in seawater and sulphate-bearing and acid waters are discussed.[4] Under many conditions, however, the deterioration of concrete is not due to any single cause, but arises from the combined action of a number of potentially destructive agencies. Particular interest attaches, therefore, to a general field study of the durability of concrete that was commenced in the U.S.A. in 1940. This investigation, known as the Long-Time Study of Cement Performance in Concrete, and carried out under the auspices of the Portland Cement Association, is primarily

[1] See F. E. Jones, *Sym., London 1952*, 368, for a general review.

[2] J. C. Pearson, *Proc. Am. Concr. Inst.*, **38**, 29, 36—1 (1942); **40**, 33 (1944); D. W. Kessler, A. Hockman and R. E. Anderson, *Bur. Stand Rep.* BMS 98 (1943).

[3] E. J. Callan, *Proc. Am. Concr. Inst.* **48**, 485 (1952); S. Walker, D. L. Bloem and W. G. Mullen, ibid., 661; R. F. Blanks, *Proc. Am. Soc. civ. Engrs* **75**, 441 (1949); R. Rhoades and R. C. Mielenz, *ASTM spec. Publ.*, No. 83, 36 (1948).

[4] A general review has been given by H. Woods, *Durability of Concrete Construction*, Am. Concr. Inst. Monogr. No. 4, 1968. See also RILEM *Internat. Sym. on Durability of Concrete*, Prague 1961 and 1969.

concerned, as its title implies, with the influence of the cement on concrete durability. But the scope of the investigation is such that the influence of other factors such as the mix proportions, the type of aggregate, and constructional practices are also involved.

The cements used, numbering 27, were all of the Portland cement type, covering the five types of the ASTM specifications: I, normal, II, modified, III, high early strength, IV, low heat, V, sulphate resisting; in the case of the first three types examples of air-entrained cement were also included. These cements were used in a variety of concrete structures, including roads, reinforced concrete piles, and thin section concrete slabs for parapets and walkways. The concrete roads have been built in three locations, providing a wide range of climatic conditions. The concrete piles are installed in sea- and freshwater with severe winter conditions, and also in seawater in the mild climate of northern Florida and southern California. The thin section concrete slabs have been used in parapets and walkways on dams at high altitudes where natural weathering is severe. Test slabs, concrete boxes and columns have also been exposed to natural weathering at two experimental 'farms', one in Illinois with severe winter conditions and one in Georgia with mild conditions. Concrete prisms have in addition been exposed in a mild climate to sulphate soils. The aggregates used in the field projects were those normal to the locality, but in the two experimental farms up to three combinations of fine and coarse aggregate and up to six concrete mixes were used in the test specimens. Very detailed examination has been made of the physical, chemical and mineralogical properties of the cements and records kept of all phases of their production in order that these various factors might eventually be related to service performance. The original exposure trials were commenced in 1941–1943 and various extensions have since been made. Some of the results will be discussed in appropriate later sections, but it may be useful to review the conclusions[1] as a whole here.

In road slabs exposed to severe frosts and in which calcium, or sodium, chloride was used to remove ice in winter, no relationship has appeared between general durability and the type of cement; the effect of air-entrainment in reducing surface scaling has overshadowed all other variables. In concrete specimens exposed in sulphate soils the cement content of the mix has proved of primary importance and the C_3A content of the cement second with air-entrainment having some beneficial influence. In the seawater tests corrosion of the reinforcement resulting from the inadequacy of the $1\frac{1}{2}$-inch cover used has been the major cause of deterioration and more pronounced in a mild than in a severe climate. Softening and disintegration of the concrete has occurred in addition in piles exposed to seawater in a severe winter climate, and air-entrainment has had some beneficial effect on this. The thin concrete slabs exposed to atmospheric conditions at high altitudes have shown little deterioration. On the test 'farms' only the thin-walled hollow concrete boxes exposed to a severe winter climate have suffered damage. The deterioration here has occurred primarily in concretes of high water : cement ratio, and no very consistent effect of the type of cement has appeared. The beneficial effect of air-entrainment on these poorer concretes has

[1] The twenty year report gives references to the earlier reports. W. C. Hansen, *Bull. Res. Dev. Labs Portld Cem. Ass.*, No. 175 (1965).

been marked, but many of the better concretes without air-entrainment were sound after twenty years.

These trials, originally concerned only with Portland cements, were extended in 1958 to include Portland blastfurnace cements and pozzolanas, but no significant differences in performance have yet appeared. In comparing the resistance of concretes to various destructive agencies we have also to consider the other types of cement that are available. Their properties have been discussed in previous chapters, and we shall need to refer to them again in the discussion that is to follow. It may, however, assist the reader if, at this stage, the main features of the different types of cements are summarised in tabular form. Such a summary, as given in Table 125 (p. 624), can only indicate some broad distinctions in properties. It must inevitably hide the variations that may occur between different cements of the same type, which can be sufficient to cause some overlapping in the classification. It must also be noted that the comments under, for example, resistance to chemical attack, apply to the inherent resistance of the cement itself. The actual resistance of a concrete depends much also on its general quality. Thus, even when the resistance to chemical attack is given as 'low', a well-made dense concrete may under many conditions only suffer deterioration very slowly. Conversely, a poor quality, porous concrete, even though made with a cement of high chemical resistance, may deteriorate rapidly. This will become apparent in some of the following sections.

Concrete in seawater

Concrete is very widely used in the construction of harbours, docks, breakwaters and other structures exposed to the action of seawater, and its permanence in maritime works is therefore a matter of much importance. The destructive action of seawater on concretes has attracted attention in most countries and a very extensive literature relating to it exists.[1] Concretes in seawater may suffer attack owing to the chemical action of the dissolved salts, to crystallisation of salts within the concrete under conditions of alternate wetting and drying, to frost action, to mechanical attrition and impact by waves, and to corrosion of reinforcement embedded in it. Attack in any one of these ways renders the material more susceptible to the action of the remaining potential agents of destruction.

The approximate amounts of the more important constituents in seawater in grams per litre are shown in Table 126.

The chemical action of seawater on concrete is mainly due to the presence of magnesium sulphate, as was first realised by Vicat, the founder of much of our knowledge of concretes. Vicat began his experiments in 1812 when, as he says, there was a 'chaos of opinion', and the first results of his investigations were published in 1818, to be brought to completion in his memoir of 1857 entitled *Recherches sur les Causes Physiques de la Destruction des Composés Hydrauliques par l'Eau de Mer*. For this memoir Vicat was awarded the two prizes of 2000 francs each which had been offered in 1853 by the Société d'Encouragement pour l'Industrie Nationale for the best studies on mortars for marine work and means

[1] A bibliography of the older work is to be found in a paper by W G. Atwood and A. A. Johnson, *Proc. Am. Soc. civ. Engrs* **49**, 1038 (1923).

TABLE 125 Comparative properties of cements

	Rate of development of strength	Rate of heat evolution	Resistance to shrinkage cracking	Inherent resistance to chemical attack		
				Sulphates	Weak acid	Pure waters
Portland cements:						
Rapid hardening	High	High	Low	Low	Low	Low
Ordinary	Medium	Medium	Medium	Low	Low	Low
Low-heat	Low	Low	High	Medium	Low	Low
Sulphate-resisting	Medium	Low to medium	Medium	High	Low	Low
Cements containing blast-furnace slag:						
Portland blastfurnace	Medium	Low to medium	Medium	Medium to high	Medium to high	Medium
Supersulphated	Medium	Very low	Inadequate information	High	Very high	Low
High-alumina cement	Very high	Very high	Low	Very high	High	High
Pozzolanic cements	Low	Low to medium	High	High	Medium	Medium

TABLE 126 Composition of seawater

	Mediterranean[1]	Atlantic[1]	Mean seawater[2]
Na	11·56	9·95	11·00
K	0·42	0·33	0·40
Mg	1·78	1·50	1·33
Ca	0·47	0·41	0·43
Cl	21·38	17·83	19·80
Br.	0·07	0·06	—
SO_4	3·06	2·54	2·76

by which materials resistant to attack could be recognised. Michaëlis[3] also carried out much pioneer work on the action of seawater on cements.

As we have seen in previous chapters, magnesium sulphate reacts with the free calcium hydroxide in set Portland cement to form calcium sulphate, at the same time precipitating magnesium hydroxide; it also reacts with hydrated calcium aluminate to form calcium sulphoaluminate. These have often been assumed to be the actions primarily responsible for the chemical attack of concretes by seawater. These reactions do occur but the picture they give is too simple to account for all the facts. It is, for instance, a common observation that deterioration of concrete in seawater is often not characterised by the expansion found in concretes exposed to sulphate solutions, but takes more the form of an erosion or loss of constituents from the mass. Though laboratory data are conflicting it seems that the presence of chlorides must retard the swelling of concrete in sulphate solutions. It is also found that concretes that have suffered deterioration have lost part of their lime content. Both calcium hydroxide and calcium sulphate are considerably more soluble in seawater than in plain water and this, when combined with the conditions produced by wave motion, must lead to an increased leaching action. Again calcium sulphoaluminate, though one of the initial products of reaction of set cement with magnesium sulphate, is unstable in the resulting solution and eventually decomposes again to form hydrated alumina, gypsum and magnesium hydroxide. The hydrated calcium silicates in set cement are also decomposed by magnesium sulphate to give hydrated silica, gypsum and magnesium hydroxide, though some of the latter can combine again to give a hydrated magnesium silicate. It is indeed one of the theories of the beneficial action of pozzolanas that they increase the amount of the silica gel that remains and which itself can form a bonding agent. This is illustrated by the firm condition of some of the old Roman lime-pozzolana concretes which now, after nearly 2000 years' exposure to seawater, have a very low lime content. Analyses of concrete progressively attacked by seawater show a progressive rise in the magnesium hydroxide content and a fall in that of lime, but the sulphate content tends first to increase and then to decrease again.

[1] T. Schloesing, *C. r. hebd. Séanc. Acad. Sci., Paris* **142,** 320 (1906).
[2] Dittmar.
[3] *The Behaviour of Hydraulic Cements in Sea-water*, Berlin, 1895. Abridged account in *Proc. Instn civ. Engrs* **129,** 325 (1896–1897, Part III).

We must, therefore, picture the chemical action of seawater as one of several reactions proceeding concurrently. Leaching actions remove lime and calcium sulphate while reaction with magnesium sulphate leads to the formation of calcium sulphoaluminate which may cause expansion, rendering the concrete more open for further attack and leaching. The deposition of magnesium hydroxide blocking the pores of the concrete probably tends to slow up the action on dense concretes though on more permeable materials it may be without much effect. The relative contributions to deterioration of expansion and leaching will depend on the conditions. Concrete of not too massive dimensions exposed to the open sea is more likely to show the effects of leaching than of expansion. Structures such as dock walls, on the other hand, into which the seawater may percolate, but in insufficient quantity to cause any considerable degree of leaching, may show the effects of expansion. Examples of both are to be found. The rate of chemical attack is increased by temperature and both the rate and its effects are influenced by the type of cement. With high-quality concrete and, particularly for warm or tropical waters, an appropriate choice of cement, the deterioration from chemical action can be kept very slow, but conversely with inferior concrete it has often caused much destruction.

Another factor that has attracted some attention is the effect of carbon dioxide. In normal seawater only small amounts of carbonate and bicarbonate are present, about 10 and 80 parts per million respectively, and a small amount of free dissolved carbon dioxide. The pH varies[1] between about 7·5 and 8·4, an average value for seawater in equilibrium with the carbon dioxide in the atmosphere being 8·2. In normal seawater some gradual carbonation of set cement occurs, and may indeed help by the formation of a protective surface skin, but it is doubtful if the free carbon dioxide content plays more than a minor part in the leaching of lime from a concrete. Under exceptional conditions, seawater can contain abnormal amounts of dissolved carbon dioxide and then become much more aggressive, for carbonic acid behaves as a much stronger acid[2] in seawater than in plain water. These conditions can arise in sheltered bays and estuaries if the sea-bed is covered by organic matter which in its decay produces carbon dioxide. Unless this is accompanied by a corresponding rise in the calcium bicarbonate concentration, aggressive carbon dioxide will be available to attack concrete. Such a case[3] was found in a concrete dock at Newport News, U.S.A., where rapid deterioration occurred as a result of expansion of the concrete by sulphate attack and weakening of the concrete by leaching. The conclusion was reached that when the pH of the seawater is above 7·5 there is little likelihood of leaching by carbon dioxide, that at a pH of 7 the content of aggressive carbon dioxide may be near the tolerable limit, and that below this pH the content is almost certain to be excessive and cause damage even to well-made Portland cement concrete.

It has been suggested[4] that bacteria play a part in the deterioration of concrete in the Black Sea and that those which oxidise sulphur to sulphuric acid have the most effect. The part played by bacterial oxidation of hydrogen sulphide in the

[1] H. V. Sverdrop, M. W. Johnson and R. H. Fleming, *The Oceans*. Prentice-Hall, New York (1942).
[2] D. M. Greenberg and E. G. Moberg, *Bull. natn. Res. Coun., Wash.*, No. 89, 73 (1932).
[3] R. D. Terzaghi, *Proc. Am. Concr. Inst.* **44**, 977 (1948).
[4] L. I. Rubenchik, *Chem. Abstr.*, 1942, 6770.

corrosion of concrete sewers is well recognised, but it is difficult to visualise how the very special conditions that lead to this could be set up in seawater structures. The conditions under which hydrogen sulphide can occur in seawater are similar to those for abnormal amounts of carbon dioxide, i.e. sheltered waters and decaying vegetation, and it is not clear whether any significant effect of bacteria, in comparison with direct chemical action, has been proved. It has also been suggested[1] that molluscs can damage concrete by the liberation of some form of ammonium carbonate.

Apart from these chemical actions, deterioration of concrete in seawater arises also from frost action and in the case of reinforced concrete from corrosion of the reinforcement. Indeed in severe climatic conditions such as on the Norwegian coast or northern Maine in the U.S.A., and northern regions generally, frost action is the main factor causing deterioration. The concrete in the tidal zone undergoes many cycles of freezing and thawing, up to two or three hundred in a year, as it is alternately immersed in the sea or exposed to air at low temperatures. Freezing in seawater is also much more detrimental than in freshwater. Under Norwegian conditions Lyse[2] found that up to 10–12 per cent entrained air was necessary for maximum durability or about twice as much as for freshwater. Conversely, corrosion of reinforcement which is a major source of deterioration becomes progressively a more serious risk as the temperature rises, as does also chemical attack on the cement. Corrosion has been found with particular frequency on the underside of deck slabs which are subject to deflection from live loads and cracking which facilitates access of the seawater to the reinforcement.

Long experience has shown that the most severe attack of seawater on concretes occurs just above the level of high water, that the portion between low- and high-water marks is less affected, and that the parts below the low-water level which are continuously immersed are rarely damaged. Concrete is not attacked by seawater unless it can penetrate into the mass. Although no concrete is strictly impermeable, a good concrete is so dense that the rate of penetration of seawater into it is negligible when it is completely immersed and not subjected to an excess hydrostatic water pressure on one side. In concrete which is just above the water level, however, the seawater tends to rise in the material by capillary action and, by evaporation of its upper surface, to draw more water continuously through the mass. Under these circumstances the seawater may very slowly attack the cement chemically while the crystallisation of the salts in the concrete probably tends also to have a disruptive action. The alternate wetting and drying of the surface accentuates the disintegration. It is also the concrete above water level which is subject to the action of frost and is most open to wave action and attrition[3] Reinforced concrete is much more subject to attack than plain concrete, for if any corrosion of the reinforcing bars occurs, owing to penetration of the seawater to it, the consequent expansion cracks the overlying concrete and opens the way to further attack.

The Committee of the Institution of Civil Engineers appointed to investigate the deterioration of structures exposed to the action of seawater reported in 1920[4]

[1] F. W. Freise, *Concr. constr. Engng* **33,** 535 (1938).
[2] I. Lyse, *Proc. Am. Concr. Inst.* **57,** 1575 (1961).
[3] R. T. L. Allen and F. L. Terrett, Proc. 11th Conference on Coastal Engineering, London 1968, Amer. Soc. Civ. Eng., 1969 (II) 1200.
[4] *First Report.*

W

that 'it is taken by most engineers as definitely determined that properly constituted Portland cement concrete, employed as it should be, may be relied upon to produce sound and permanent work'. The long list of structures, built in most cases with what would be called good materials and careful workmanship, which have suffered damage indicates, however, that the implications of the term 'properly constituted' need most critical consideration. Wig and Ferguson,[1] after examining a large number of structures in seawater in the United States, reported in 1917 that most of the reinforced concrete structures showed evidence of deterioration and failure, but that the deterioration of plain concrete only occurred very slowly when only chemical action without surface abrasion was involved. In a later report, Hadley[2] concluded from an inspection of seawater structures on the Pacific coast of the U.S.A. that the causes of deterioration were erosion of bad concrete, unsoundness caused by faulty aggregates, and corrosion of reinforcement due to insufficient cover, but not chemical attack. The latter was dismissed because of the absence of sulphate expansion, but, as we have seen, this is far from being a typical sign of seawater action. An illustration[3] of this is to be found in a bridge in San Francisco Bay which was built in 1929 with a Portland cement containing about 12 per cent C_3A and which developed serious corrosion of reinforcement and cracking of the concrete with little visual evidence of sulphate attack. Analysis of the concrete showed up to 7 per cent MgO, calculated on the cement which had contained 1·7 per cent. Clearly considerable chemical attack had occurred. The condition, varying from good to deteriorated, of many other seawater structures round the coast of the U.S.A. was reported at a discussion[4] in 1958.

The results of a survey[5] of the condition of seawater structures in Norway was published in 1936. Concrete below water level protected from evaporation and freezing was intact and damage had occurred mainly between low- and high-water level. Even in this latter zone, concrete that had been protected by leaving the timber formwork permanently fixed in position had stood up well. Failures were common in exposed reinforced concrete work from rusting of the reinforcement. A further survey[6] has been made in recent years. Corrosion of reinforcement was found to be common in the upper parts of piles despite a cover of 7 cm. Below water level the condition was usually good. Damage from corrosion was also found in girders and beams supporting the deck slabs, due in some cases to insufficient cover to the reinforcement. The deck slabs themselves were usually in good condition. Similar surveys which have been made in Sweden and Denmark[7] have led to a classification of the environmental conditions in the Scandinavian countries in terms of the relative severity of the various factors

[1] *Engng News Rec.* **79,** 532 et seq. (1917). See also Atwood and Johnson, *Marine Piling Investigations*, National Research Council, U.S.A. (1924).
[2] *Proc. Am. Soc. civ. Engrs* **67,** 33, 512, 687, 918, 1150, 1456, 1757, 1893 (1941).
[3] *Bull. Highw. Res. Bd*, No. 182 (1957).
[4] *Proc. Am. Concr. Inst.* **54,** 1309 (1958).
[5] Concrete in seawater, *Meddr. Betongkom. norsk Ing.-Foren.*, No. 3 (1936).
[6] O. E. Gjorv, *R.I.L.E.M. Sym. Behaviour of Concrete in Seawater, Palermo 1965; Materials and Structures*, **2,** 467 (1969); *Durability of Reinforced Concrete Wharves in Norwegian Harbours*, Ingenior Fortaget A/S, Oslo (1968).
[7] G. M. Idorn, *R.I.L.E.M. Bull.*, No. 16, 59 (1962); K. E. C. Nielsen, *R.I.L.E.M. Sym. Behaviour of Concrete in Seawater, Palermo 1965.*

involved. Assessing aggressiveness on the scale: No influence 0, Mild influence 1, Perceptible influence 2, Severe influence 3, the portential influence of the major aggressive factors was assessed as shown in Table 127.

TABLE 127 Relative importance of weathering factors in the Scandinavian countries

Country	Salinity of seawater	Cycles of freezing and thawing	Tidal variation	Alkali aggregate reaction
Sweden	0–1	0–2	0	0
Norway	0–2	2–3	2–3	0
Denmark	2–3	1–3	0–3	3

The salinity is low in the Baltic and below normal in many Norwegian harbours situated in estuaries or long fiords. The chemical activity of the seawater along much of the Norwegian coast is also reduced by the low average temperature of the seawater. Along much of the coast of Sweden there is very little tide so that while concrete above water may be exposed to prolonged freezing temperatures there are only few cycles of freezing and thawing.

Surveys of seawater structures on the German coasts have shown less differentiation between cements than have exposure trials on cubes and slabs. Thus Seidel[1] concludes that experience of seawater structures has shown that well-made concrete with between 670 and 840 lb cement/yd^3 (400 and 500 kg/m^3) of Portland, Eisenportland or Hochofen cement has remained undamaged after as much as 40 years' exposure. With the advances that have been made over that period in concrete technology it is to be expected, states Seidel, that concrete with a lower cement content may be expected to have the same durability.

Reference to the many reports presented to the various international congresses[2] on navigation and testing materials, affords many examples of serious damage to reinforced concrete structures in seawater within a relatively short period of years after construction. Deterioration of plain concrete has also been far from uncommon, particularly in positions where frost action has been severe, or when careful attention has not been paid to the composition and placing of the concrete.

A long series of exposure trials on the resistance of concrete to seawater have been carried out by various national associations and authorities. The earliest of these in the Scandinavian countries, Germany and France were started in the last decade of the 19th century, to be followed by many more in the present

[1] K. Seidel, *Dt. Aussch. Stahlbeton*, 134 (1959).
[2] See the various reports on concrete in seawater presented to the International Congresses of Navigation held in London in 1923, Venice in 1931 and Lisbon in 1949; also the meetings of the International Association for Testing Materials at Amsterdam in 1927, New York in 1912, and Copenhagen in 1909.

century. Because of advances in concrete technology we may pass lightly over the earlier trials. They showed that in northern climates the main destructive agent was frost, and that concrete in the tidal range was much more heavily damaged than concrete completely immersed. They also showed that in the Baltic, with small tidal movements and a dilute seawater containing only 0·5 per cent salts, the damage was much less marked than on the west coast of Denmark and in Norway. Use of Moler cement (containing a diatomaceous earth pozzolana) was promoted in Denmark by Poulsen[1] to give added chemical resistance, and proved beneficial. It is now known that the flint aggregates used in Denmark are alkali-reactive and the beneficial effect of the Moler in the Danish climate is now to be ascribed to its action in reducing the alkali-aggregate reaction expansion. The results of the German trials started in 1894 and 1902 stimulated the use of trass in seawater concrete.[2]

Though the Scandinavian trials showed that the chemical action of seawater was of secondary importance in northern regions this does not hold for warmer waters. During the destruction of the harbour works of Heligoland in accordance with the Treaty of Versailles the condition of the concrete, then some 13–19 years old, was examined by Grun.[3] On the whole, little sign of mechanical damage was apparent but there was considerable damage from chemical attack.

Work in France at the Laboratoire des Ponts et Chaussees at Boulogne commenced in 1886 and continued over a long period of years. In reports published in 1924–25 Feret[4] stated that in general the behaviour of the Portland cements was relatively bad but that favourable results should be obtained from the addition of suitable pozzolana or of cements containing granulated slag. There has also been a long and successful experience in France[5] with a special sulphate-resisting type of Portland cement made from a silicious limestone with a low content of alumina (2·5 per cent) and iron oxide (1 per cent).

In Italy where both tradition and practice are averse to the use of Portland cements alone in seawater, the use of pozzolanic cements for concrete, or in mortars of lime-pozzolana mixes with a small proportion of cement added to increase the rate of hardening, has been general. In recent years pozzolanic cements with a very high resistance to sulphate attack, based on a sulphate-resisting type Portland cement and a very active high-silica (90 per cent) pozzolana have been introduced.[6]

Turning to the more modern trials[7], a series of tests on Portland, Eisenportland (30 per cent slag) and Hochofen (60–70 per cent slag) Erz and high-alumina cements, and of trass substitution were commenced in Germany in 1929. Concrete cubes of 20 cm side made with 450, 540 and 675 lb cement/yd^3 (265, 320 and 400 kg/m^3) with w/c ratios of 0·66, 0·56 and 0·44 were exposed between

[1] A. Poulsen, *5th int. Congr. Test. Mater., Copenhagen 1909*, xi, 4; *Proc. Instn civ. Engrs* **200,** I, 409 (1914–15); *13th Congr. perm. int. Ass. Navig. Congr., London 1923,* Communication No. 53.

[2] M. Gary, *Mitt. MaterPrüfAmt Grosse-Lichterfelde* **37,** 132 (1919).

[3] R. Grun, *Tonind.-Ztg. keram. Rdsh.* **52,** 500 (1928).

[4] R. Feret, *Annls Ponts Chauss.* **92** (iv), 5 (1922).

[5] R. Dubrisay and H. Lafuma, *17th Congr. perm. int. Ass. Navig. Congr., Lisbon 1949,* Section 2, Communication 2, p. 55.

[6] A. Rio and A. Celani, *R.I.L.E.M. Sym. Behaviour of Concrete in Seawater, Palermo 1965.*

[7] For survey of test methods see Materials and Structures **3,** (14) 107 (1970).

tide levels at Wilhelmshaven. An examination[1] after 19 years showed that the two leaner mixes had disintegrated, and the richest mix suffered attack, with the Portland and Eisenportland cements and Erz cements. The two leaner mixes with Hochofen cement had deteriorated, but the rich mix showed only a relatively small degree of attack. The substitution of 30 per cent trass for cement in the Portland and Eisenportland cement mixes had some, but insufficient, beneficial effect. With high-alumina cement some deterioration occurred in the leanest mix, but the medium and rich mixes had remained sound and showed little sign of deterioration. In their report Eckhardt and Kronsbein summarised the conclusion to be drawn from these and earlier trials and German experience with concrete in seawater. They concluded that mixes should not be leaner than 675 lb/yd^3 (400 kg/m^3), that the cements of lower lime content than Portland cement are preferable, but that trass must be used as an addition to the mix and not as a substitution for cement, and that high-alumina cement had the properties most desired for seawater construction, i.e. rapid-hardening when placed *in situ* and high resistance to attack.

These results were re-examined[2] later in relation to the porosity of the various concretes and of the cement pastes. The amended conclusion reached was that a satisfactory resistance of concrete to seawater should be obtained with any of the cements if the w/c ratio was 0·40–0·45 and probably up to 0·55. With the more resistant cements higher water : cement ratios should be permissible. Plotting the durability against the pore space in the cement paste Wesche concluded that the highest intrinsic resistance was shown by:

1. High-alumina cement, followed in order by
2. Blastfurnace cements with 30 per cent trass substitution,
3. Blastfurnace cements without trass, ordinary Portland cement with or without trass.

These conclusions refer specifically to plain concrete.

Systematic tests on mortars and concretes made with a wide variety of cements and immersed in the sea between tide levels in open crates in the substructure of a jetty at Ostend were commenced in 1934. After 30 years the results[3] showed that the poorest performance was given by ordinary and rapid-hardening Portland cements and by cements in which trass had been substituted for 33 or 50 per cent of the Portland cement. The best performance in mortars was given by high-alumina cement and cements with a high slag content, e.g. ciment permetallurgique containing over 70 per cent granulated slag, ciment de haut-fourneau normal with 30–70 per cent granulated slag, and supersulphated cement. The concrete specimens of a 1 : 1·8 : 3·6 mix with a cement content of about 590 lb cement/yd^3 (350 kg/m^3) and a water : cement ratio of 0·45 showed varying effects. The cubes with the Portland cements and the trass mixes had been heavily attacked, becoming spherical in shape. The cubes from the slag cements noted

[1] A. Eckhardt and W. Kronsbein, *Dt. Aussch. Stahlbeton*, vol. 102 (1950).
[2] A. Hummel and K. Wesche, *Dt. Aussch. Stahlbeton*, 124 (1956); K. Wesche, ibid. **168,** 33 (1965); *R.I.L.E.M. Bull.*, No. 32, 291 (1966); H. Kramer, *Zement-Kalk-Gips* **21,** 134 (1968); F. W. Locher, *Beton* **18,** 47 (1968).
[3] F. Campus, R. Dantinne and M. Dzulynski, *R.I.L.E.M. Sym. Behaviour of Concrete in Seawater, Palermo 1965*; F. Campus, *Memoires C.E.R.E.S. Univ. of Liege* No. 24 (1968).

above showed surface corrosion and those from the high-alumina cement were intact though there was some retrogression in strength, doubtless from conversion.

It was concluded from the experience over the whole 30 years of the trials that attack by seawater proceeds progressively inwards at a rate determined by the initial porosity of the mortar or concrete. The use of cements resistant to chemical attack helps to offset the effects of porosity but nevertheless a prime requirement for seawater work is low porosity, permeability and absorption. A minimum cement content for mortars of 760 lb/yd^3 (450 kg/m^3) and preferably 1000 lb was recommended, and for concrete 590 lb/yd^3 (350 kg/m^3). Cements rich in blastfurnace slag of known quality were preferred.

Some French tests[1] on plain concrete were started in 1955 to compare the behaviour of various cements. The specimens, $7 \times 7 \times 28$ cm concrete prisms, were completely or half-immersed in tanks, through which the seawater flowed, at Marseilles on the Mediterranean. After 8 years the completely immersed specimens made with Portland cements of not above 8 per cent C_3A content, with ciment de latier au clinker (more than 80 per cent slag and less than 20 per cent Portland cement) and with supersulphated cement showed a good performance with cement contents as low as 335 lb cement/yd^3 (200 kg/m^3). A Portland cement of 12 per cent C_3A content was much more affected in concrete of this cement content and also in one of 500 lb cement/yd^3. In the half-immersed specimens a content of at least 505 lb cement (300 kg/m^3) was required with Portland cements of not above 8 per cent C_3A content for good performance. Both the ciment de latier au clinker and the supersulphated cement showed a poor behaviour, but high-alumina cement behaved well even in a concrete of 335 lb cement/yd^3. It was concluded that in a climate where evaporation and salt crystallisation is an important factor, as in the Mediterranean, a Portland cement of moderate C_3A content was preferable to slag cements of high slag content.

Tests on 1 : 2 : 3 concrete cylinders exposed in the tidal zone in northern Norway were started in 1929. They showed that after six years' exposure in the tidal zone those made with Portland cement and Eisenportland cement had suffered damage, but those with high-alumina cement, Moler cement (1·4 : 2 : 3 mix, the diatomaceous earth pozzolana being treated as an addition), and with a 30 per cent addition of trass to Portland cement, had remained intact. After 21 years only the specimens made with high-alumina cement and with Moler cement still remained free from deterioration.

No damage had occurred after six years to concrete permanently below water level with any of the cements. These results agreed with experience of Moler cement in seawater structures in Denmark, trass cements in Germany, and high-alumina cement in France. Erz cement had also shown good performance and specimens exposed in 1908 were still in good condition after 28 years. It was concluded that the use of high-alumina or Erz cements, or of pozzolanic additions, was to be preferred but that, where this was not practicable, good performance could be secured with Portland cement provided it was used in mixes not leaner than 600–700 lb cement/yd^3 (356–415 kg/m^3) for exposure above low-tide level and the concrete was protected by permanent shuttering; for concrete per-

[1] J. Brocard and R. Cirodde, *R.I.L.E.M. Bull.*, No. 323 (1966).

manently immersed 500 lb/yd^3 (297 kg/m^3) was adequate. For reinforced concrete the minimum cover should be 5 cm and some surface coating was desirable.

Another series of tests[1] on reinforced concrete piles was started in Norway in 1936, the piles with a cover of 5 cm to the steel being exposed in the tidal zone on the coast at Trondheim, and a final report was published in 1964. The worst corrosion of steel was found at the top of the piles above high-water level but the primary cause of damage to the piles as a whole was frost action. The least damage to piles containing about 530 lb cement/yd^3 (313 kg/m^3) (w/c 0·56) was found with a sulphate-resisting Portland cement (3 per cent C_3A) and with one Portland blastfurnace cement. With another Portland blastfurnace cement of higher lime, and presumably lower slag content, and with three other Portland cements with C_3A contents from 6 to 11 per cent there was much more damage to the piles. Raising the cement content of the concrete with a 6 per cent C_3A Portland cement to about 700 lb/yd^3 (417 kg/m^3) (w/c 0·47) gave piles which were little damaged. A 60 per cent addition of trass to the 530 lb Portland cement mix had a similar favourable effect. High-alumina cement piles with 530 lb cement/yd^3 and a w/c ratio of 0·62 suffered severe damage.

Tests were commenced in 1924 at Portsmouth, New Hampshire, on reinforced concrete columns 14 inches square and 13 feet long immersed in the sea in the tidal zone. After 12 years the columns made with a $1:1\frac{1}{2}:3$ Portland cement concrete and with a 2 in. cover over the reinforcement were generally in good condition, but about half the columns with a $1:2:4$ mix had deteriorated or failed.[2] Further tests have been undertaken by the Waterways Experiment Station in the U.S.A. since 1935 and they also form a part of the Long-Time Study of Cement Performance in Concrete that was started in 1940. In both series the test specimens have been exposed between tide levels to a severe winter climate in northern Maine and to warm seawater in Florida. The former tests, which have been made on plain concrete, have been designed as much to study the influence of different aggregates as of cements.[3] In warm seawater well-made concrete containing 500–600 lb cement/yd^3 (300–350 kg/m^3) had not been seriously affected up to 10 years except in the case of Portland cements containing above 12 per cent tricalcium aluminate. At the Maine exposure station, where frost action was predominant, it was found that the Portland cement concretes would not withstand the exposure for more than a year unless air-entrainment was used, but that many of the air-entrained concretes remained in good condition after 11 years. High-alumina cement concretes were unaffected.

The tests carried out as part of the Long-Time study have been made on reinforced concrete piles 12 inches square with $1\frac{1}{2}$-in. cover to the reinforcement. Two cement contents, 470 and 660 lb/yd^3 (280 and 390 kg/m^3) were used with a 2 in. slump (w/c ratios 0·62 and 0·44), and also in the case of the 660 lb mix with an 8 in. slump. The major source of deterioration in Maine, Florida and southern California was rusting of the reinforcement causing cracking of the concrete. Rusting was the more severe in the two mild climates than in Maine,

[1] O. E. Gjorv, I. Gukild and H. P. Sundh, *R.I.L.E.M. Bull.*, No. 32, 305 (1966).
[2] F. H. Fay, *17th Congr. perm. int. Ass. Navig. Congr., Lisbon 1949*, Section II, Communication 2, p. 37.
[3] H. K. Cook, *Proc. Am. Soc. Test. Mater.* **52,** 1169 (1952); T. B. Kennedy and K. Mather, *Proc. Am. Concr. Inst.* **50,** 141 (1954).

though it was difficult there to distinguish between the effects of corrosion cracking and frost deterioration. Corrosion was much less in the 660 lb than the 470 lb cement mix in Florida and California and it was also less in the 2 in. than the 8 in. slump concrete. Serious deterioration in the form of softening and disintegration of the concrete from frost action occurred in Maine and was much more severe in the leaner concrete. Air-entrainment was moderately beneficial in the severe climate of Maine and it also reduced corrosion cracking in the warm climate of Florida. The composition of the Portland cement, as classified by the five ASTM types, was without any significant influence except for the cement of the highest C_3A content (12 per cent) which showed strong evidence of seawater attack in Florida. The general[1] conclusions from these tests after 20 years' exposure was that for reinforced concrete in seawater a $1\frac{1}{2}$ in. cover was inadequate for any concrete, that the minimum cement content should be 660 lb/yd^3 (390 kg/m^3) and the maximum water : cement ratio 0·45. Of the three forms of deterioration, rusting, chemical attack and frost action, air-entrainment increased the resistance in the case of the first and the last. Another series of tests[2] in Maine on stressed reinforced concrete beams confirmed that air-entrainment had a significant beneficial influence, but showed an unexpectedly good behaviour of beams with only a $\frac{3}{4}$-in. cover.

In 1929 a large series of tests were commenced in Great Britain by the Sea Action Committee of the Institution of Civil Engineers in collaboration with the Building Research Station on the resistance of reinforced concrete to attack by seawater. Short reinforced concrete piles 5 inches square and 5 feet long containing reinforcing bars with 1 or 2 inches of cover were exposed in the tidal zone at Sheerness and another set were partially immersed in outdoor tanks of an artificial solution of three times normal seawater concentration with arrangements for simulating tidal action. Concrete cylinders (6 × 3 inches) were also immersed in a similar solution. The concrete mixes tested were 1 : 2·6, 1 : 5 and 1 : 9 by weight (1000, 600 and 360 lb/yd^3 or 213, 356 and 593 kg/m^3) and the cements included Portland, Portland blastfurnace, pozzolanic (40 per cent pozzolana) and high-alumina cements. The effect on the strength of the concrete cylinders of complete immersion in the solution of three times normal seawater concentration is illustrated in Fig. 147 for some of the cements. The Portland cements showed a variable behaviour which was not related to their calculated content of $3CaO.Al_2O_3$; this was 8, 8, 10 and 9 per cent respectively for cements L, M, N and D. The pozzolanic, Portland blastfurnace and high-alumina cements all showed a high resistance to direct attack in the 1 : 2·6 and 1 : 5 mixes.

The reinforced concrete piles suffered very much less from direct attack of the concrete than the small cylinders and deterioration arose primarily from corrosion of the reinforcement with consequent cracking, or spalling, of the concrete. With a 2-in. cover to the reinforcement and a 1 : 5 mix no cracking of piles occurred up to 10 years with high-alumina and Portland blastfurnace cement and with the pozzolanic cements composed of ordinary, or rapid-hardening, Portland cement

[1] I. L. Tyler, *Proc. Am. Concr. Inst.* **56,** 825 (1960); W. C. Hansen, *Bull. Res. Dev. Labs Portld Cem. Ass.*, No. 175 (1965).

[2] E. C. Roshore, *Proc. Am. Concr. Inst.* **64,** 253 (1967).

and trass but some cracking occurred with the ordinary and rapid-hardening Portland cements alone.

From 1939 onwards, after the first 10 years, the storage conditions for the piles could not be maintained and those at Sheerness remained standing in empty tanks, either dry or in a few inches of rainwater, and the others in about a foot of stagnant artificial solution of one and a half times normal seawater concentration. At the end of about 20 years some cracking had occurred in the 1:5 mix with all cements except for one high-alumina cement and one trass cement. In the rich (1:2·6) mix none of the piles with a 2-in. cover had cracked after 10 years and there was little change after 20 years.

The general conclusion[1] reached from these tests was that in the temperate climate of southern England high-alumina cement concrete had given an excellent performance and that Portland blastfurnace and trass cements had proved rather better than the Portland cements. It must, however, be noted that no sulphate-resisting Portland cement, which might be expected to give a performance at least comparable to the Portland blastfurnace and trass cements, was included in these trials. The primary cause of deterioration of reinforced concrete piles was corrosion of the reinforcement and not disintegration of the concrete from the chemical or physical effects of the salts in seawater. The latter nevertheless had an influence on the incidence of corrosion since this became progressively greater as the resistance of the cement to direct attack decreased. The cement content of the concrete and the thickness of cover were of major importance in determining the durability of the piles. Thus with Portland and trass-Portland cements the order of decreasing resistance to corrosion cracking was:

	W/c ratio about
1. 1000 lb cement/yd^3 with 2 in. cover	0·35
2. 600 lb cement/yd^3 with 2 in. cover	0·53
3. 1000 lb cement/yd^3 with 1 in. cover	0·35
4. 600 lb cement/yd^3 with 1 in. cover	0·53
5. 360 lb cement/yd^3 with 2 in. cover	0·90

With high-alumina and Portland blastfurnace cements there was little difference between (2) and (3).

The general conclusions that have been drawn from tests and experience with concrete in seawater and present general practice may be briefly summarised as follows:

The prime essential for all seawater concrete is that a dense product of low porosity should be obtained. The use of too wet mixes is dangerous, but mixes that are so dry that they cannot be adequately compacted must also be avoided. Vibration compaction is to be recommended. For precast Portland cement concrete products the resistance to deterioration is increased by a long period of hardening in air subsequent to the usual periods of wet curing. For concrete permanently under water a minimum cement content of 500 lb/yd^3 (300 kg/m^3) is desirable. Portland cement should give a satisfactory performance, but sulphate-resisting Portland cements, pozzolanic, or slag-containing cements, are to be

[1] F. M. Lea and C. M. Watkins, *Natn. Bldg Stud. Res. Pap.*. No. 30, H.M.S.O. London (1960)

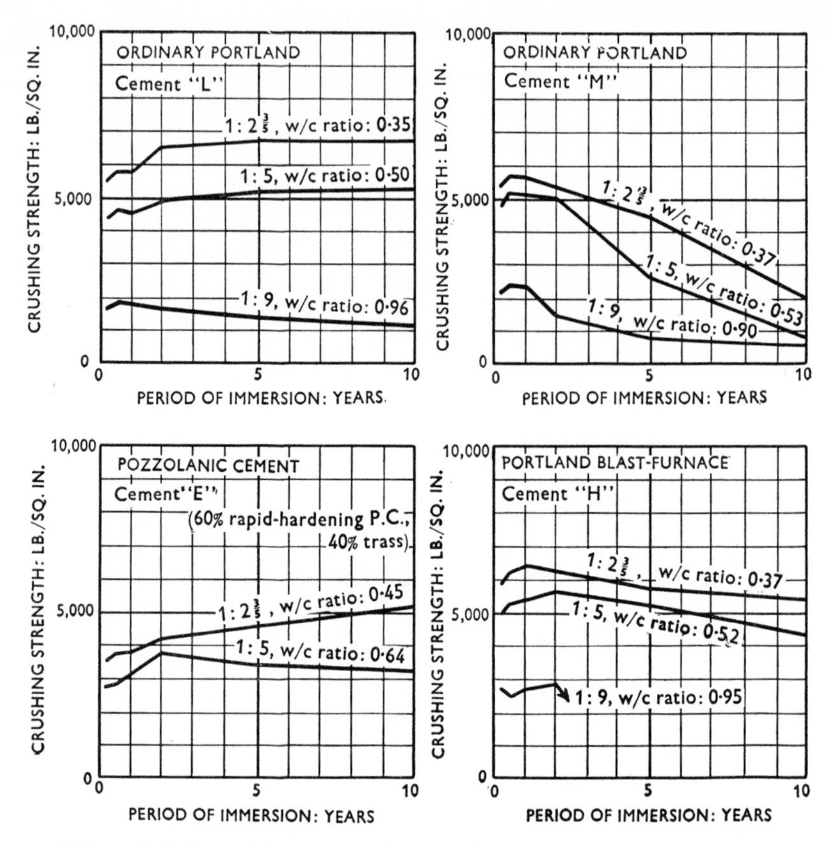

FIG. 147 Effect of immersion in artificial seawater of three times normal

preferred, particularly for work in tropical or semi-tropical regions. Concrete between tide levels, or immediately above high-water level, is subject to rapid freezing as the tide recedes in cold water climates, and to rapid evaporation in hot climates. A minimum cement content of 600 lb/yd^3 (350 kg/m^3) is necessary. There is not universal agreement as to the relative merits of different types of cements. For cold or temperate, but not semi-tropical or tropical, climates, well-made high-alumina cement concrete, not leaner than a 1:5 weight mix, is probably the most immune to attack, but it is more costly and is more susceptible to the effects of lack of care in making. The w/c ratio must not exceed 0·5 in order to minimise the effect of conversion on strength (see p. 516). In severe winter climates Portland cement concrete is likely to suffer damage unless protected from too rapid freezing by permenent shuttering. Air-entrained concrete should be used. Cements based on blastfurnace slag with a preference for those containing 50 per cent or more of slag, such as the Hochofen or supersulphated type, have a good record both in tests and practice. The use of about 30 per cent pozzolana by weight of the Portland cement is advantageous, but for severe

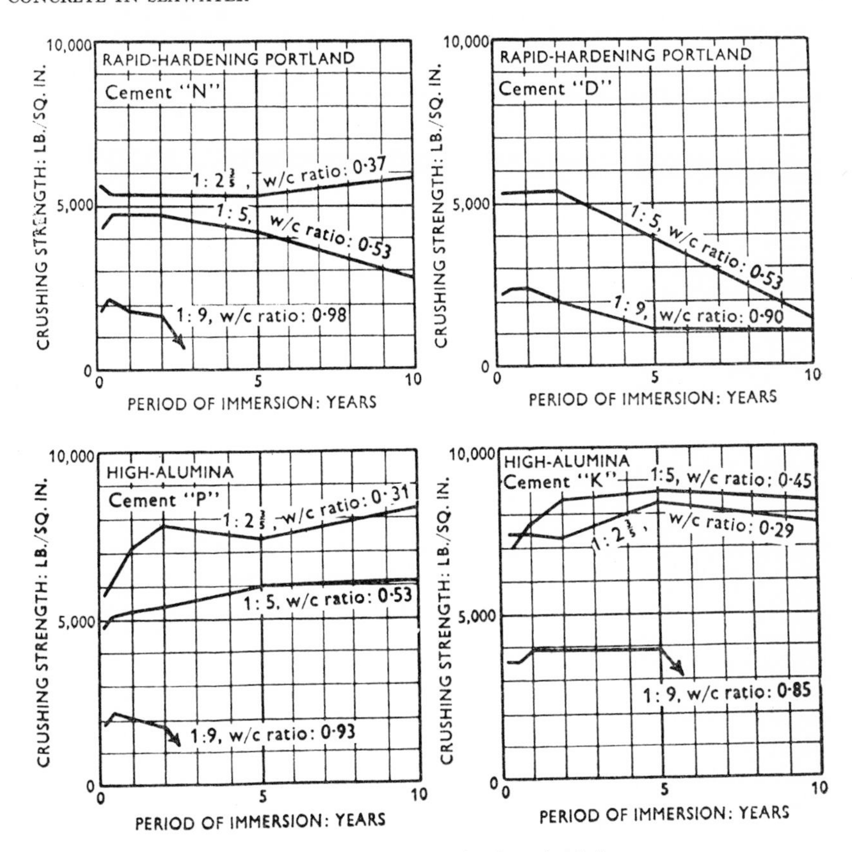

concentration on the strength of concrete cylinders 6 × 3 in.

winter climates this should be as an addition to, and not a substitution for, the Portland cement. In warm climates, such as the Mediterranean, or tropical climates, where chemical attack is a dominant factor, the slag-containing and pozzolanic cements, or sulphate-resisting Portland cement, are to be preferred; in this case experience indicates that the pozzolana can be treated as a substitution for up to 30 per cent of the Portland cement. The mildest conditions of exposure are those found in temperate waters, such as around the British Isles, where much well-made Portland cement concrete has shown a good record, but advantage is still to be gained from the use of more resistant cements.

Reinforced concrete above low-tide level is much more vulnerable than plain concrete owing to the additional risks of corrosion of the reinforcement. The minimum cover of concrete over the reinforcement should be 2 inches, and for tropical waters an even thicker cover is desirable. Mixes should not be leaner than 600 lb cement/yd^3 (350 kg/m^3) and the w/c ratio should not exceed 0·45. The comments made above on the various types of cement apply equally to reinforced concrete. Coatings of tar and pitch have often been applied to protect reinforced concrete from deterioration in seawater but the general consensus of opinion is that their effective life is too short and that in some cases they only hide the damage that is taking place. Coatings applied to the underside of concrete decks have, however, been found useful. The impregnation of concrete piles with hot bitumen under reduced pressure has been extensively used by the Los Angeles Harbour Board[1] with very effective results in stopping corrosion of reinforcement.

Seawater may be used for mixing plain concrete for seawater structures except for high-alumina cement. It should not, unless quite unavoidable, be used in reinforced concrete, for it much increases the risk of corrosion of the reinforcement.

Concrete in coastal environments

Deterioration of reinforced concrete from the effect of seawater salts can occur in structures up to a few miles inland from the coast. It arises from corrosion of the reinforcement, not attack on the concrete itself, and is caused by sea spray carried inland by the prevailing winds. The salts gradually accumulate in the concrete and by wetting and drying gradually migrate inward to the steel. This type of damage has been reported most frequently in tropical or semi-tropical countries. In South Africa,[2] for example, serious corrosion has occurred in balustrading of bridges with $1\frac{1}{4}$-in. cover to the reinforcement and also in the main beams. In such conditions a minimum cover of 2 in. is needed as for concrete structures in seawater. Similar troubles have been reported from Italy, California, and the south Pacific.

Many cases of corrosion of reinforcement in inland areas from chlorides present in the soil and groundwaters have also been reported. Thus an examination of 239 reinforced concrete bridges in California[3] showed that corrosion had occurred in 28 per cent of them and that the amount was related to the chloride content of the soils and groundwaters in the environment. We have already noted the effect of chloride de-icing salts in causing corrosion in reinforced concrete bridge decks.

Groundwaters and soils containing sulphate salts[4]

In parts of Great Britain and in many extensive regions abroad clays and soils are found containing considerable quantities of mineral sulphates, notably those of magnesium, calcium, sodium and potassium. Clays containing 5 per cent or more of gypsum and sometimes appreciable quantities of other sulphates are, for instance, found. Thus calcium, magnesium and alkali sulphates occur in parts of

[1] C. M. Wakeman, E. V. Dockweiler, H. E. Stover and L. L. Whiteneck, *Proc. Am. Concr. Inst.* **54,** 841 (1958).
[2] S. Halstead and L. A. Woodworth, *Trans. S. Afric. Instn civ. Engrs* **5** (4), 1955; D. A. Lewis and W. J. Copenhagen, *S. Afr. ind. Chem.* **11** (10) (1957); D. F. Griffin, *Mater. Protn* **4** (11), 8 (1965).
[3] J. L. Beaton and R. F. Stratfull, *Highw. Res. Rec.* **14,** 60 (1963).
[4] The performance of concrete in sulphate soils was reviewed at a Symposium held in Toronto in honour of Prof. T. Thorvaldson in 1967 and reported in *Performance of Concrete*, Canadian Building Series No. 2, edited by E. G. Swenson, Univ. of Toronto Press 1968.

the London, Gault, Weald, Lias, Oxford and Kimmeridge formations.[1] In parts of the Keuper Marl gypsum is the predominant sulphate. Across large areas in the United States and Canada there occur the so-called 'alkali' soils. These soils often contain several per cent of sodium, potassium, magnesium and calcium sulphates. Gypsiferous territories and sulphate-bearing soils occur also in France and other parts of Europe and elsewhere. The problem of the resistance of concretes to sulphate salts is therefore one of widespread interest.

In Great Britain the distribution of sulphate salts in a clay is often very irregular and may vary much between points only 50 feet apart. There is also considerable variation with depth, for, as a consequence of gradual leaching by rainwater, the top 2 or 3 feet are often relatively free from sulphates. In dry regions on the other hand where rates of evaporation are high, as in the Canadian prairies, there may be a larger concentration near the surface. The groundwaters which are present in any clays containing soluble sulphates become more or less heavily charged with them.[2] Calcium sulphate has only a limited solubility, about 0·2 per cent, and this sets a limit of about 120 parts SO_3 per 100 000 to the concentration to which it can give rise, but the other sulphates are much more soluble. In Britain, concentrations of sulphates in groundwaters above 500 parts SO_3 per 100 000 are rare, but in the alkali soils of North America values up to 1 per cent have been found. The action of soluble sulphates on cements has been discussed in an earlier chapter, and we have seen that it results in the formation of calcium sulphoaluminate and gypsum accompanied by a marked expansion. In general, concrete attacked by calcium or sodium sulphate becomes eventually reduced to a soft mush, but when magnesium sulphate is the main destructive agent the concrete remains hard though it becomes much expanded.

The disintegration of concretes in sulphate soils first attracted attention in France during the latter years of the nineteenth century, and in the United States and Canada early in the present century. The investigations to which it led in France resulted finally, as we have seen elsewhere, in the production of high-alumina cement. In the alkali soil regions of the North American continent troubles were experienced with concrete sewers, drain-pipes used for land drainage, foundations, culverts and other concrete structures. Many extensive investigations on the resistance of concrete in sulphate soils have been carried out in North America,[3] notably those of the Bureau of Reclamation, the Bureau of Standards, the U.S. Department of Agriculture and the American Portland Cement Association, and the Engineering Institute[4] and the National Research Council of Canada.

The work conducted by the Bureau of Standards[5] included trials on reinforced concrete blocks 10 × 10 × 30 inches made from 1 : 1½ : 3 and 1 : 2½ : 5 mixes, and also on drain-pipes, buried in the ground at eight different sites in the United States. At some sites the test pieces were permanently below water level, whilst

[1] G. E. Bessey and F. M. Lea, *Proc. Inst. civ. Engng*, 1953, Part I, 159.

[2] Rapid methods of analysis are given by S. R. Bowden, *Bldg Res. Stn curr. Pap.*, No. 3–68.

[3] See W. C. Hansen, *Highw. Res. Rec.*, No. 113, 1 (1966) for a review of the North American field tests.

[4] A. S. Dawson, *J. Engng Inst. Can.*, Oct. 1920; G. M. Williams, *Can. Engr*, Feb. 7th and Nov. 14th, 1922.

[5] *Technological Papers*, Nos. 44 (1915), 95 (1917), 214 (1922), 307 (1926).

at others they were in regions of varying water level. The content of sodium, calcium and magnesium sulphates varied from about 0·1 to 5·0 per cent in the soils and from 0·1 to 1·0 per cent in the groundwaters. From the experience gained from eight years' exposure it was concluded that seepage waters and alkali soil conditions may be encountered which will disintegrate Portland cement concrete of the best quality. In reinforced concretes the destruction may be aided and accelerated by corrosion of the steel and consequent cracking of the concrete. The resistance to attack increased with the richness of the mix and with the period of air curing prior to exposure to the sulphate waters. Even a 1 : 1½ : 3 concrete was completely disintegrated within eight years under the most severe conditions in groundwaters containing more than 1 per cent of sulphate salts, and suffered considerable surface attack in some cases when the average concentration was as low as 0·25 per cent.

The tests carried out by the American Portland Cement Association[1] were begun in 1921, when some 2000 concrete cylinders 10 inches diameter and 24 inches long were exposed at an age of 2–3 months in some alkali soils and water of high-sulphate content. Concretes with a slump of 3–4 inches, for any particular mixture, gave the best results in these tests. Places at which honeycombing had occurred in the drier mixes, and segregation in the wetter mixes, offered points for preferential attack which were not present in the mixes of intermediate consistence. The effect of differences in consistence was more pronounced with lean than rich mixes. Integral waterproofing compounds in general lowered the resistance of the concrete. Boiled linseed oil applied in three coats to the dry surface of the concretes almost entirely prevented sulphate attack up to seven years' exposure. Asphalt, tar, and pitch coatings gave favourable results, but in many cases did not adhere well enough to the concrete to exclude the sulphate waters.

Very comprehensive tests on the influence of the nature of the cement and other factors on the resistance of concretes to attack by sulphate waters have been carried out by D. G. Miller for the U.S. Department of Agriculture.[2] In these tests small concrete cylinders 2 inches diameter and 4 inches long have been immersed in sodium and magnesium sulphate solutions in the laboratory and in the waters of Medicine Lake, S. Dakota, for periods up to 25 years.

This lake contains salts to an extent which has varied from about 2·3 to 7·4 per cent over the years, averaging about 5 per cent, of which some two-thirds is $MgSO_4$, one-quarter Na_2SO_4, about one-twentieth $CaSO_4$ together with 0·1 per cent NaCl and minor amounts of carbonates, bicarbonates, etc. This investigation, like others, has shown the great importance for maximum durability of a high-quality concrete with low permeability and freedom from honeycombing. Perhaps the most notable feature of the results obtained by Miller is the difference in behaviour of different cements and the effect of various treatments. Concrete made from the most highly resistant group of Portland cements retained over half its strength after 10 years' exposure in Medicine Lake, whereas that from the least resistant group dropped to a similar proportion in one year, and then com-

[1] *Report of Director of Research*, Nov. 1928, p. 70.
[2] D. G. Miller and P. W. Manson, *Tech. Bull. U.S. Dep. Agric.*, No. 358 (1933); *Tech. Bull. Minn. agric. Exp. Stn*, No. 194 (1951).

pletely disintegrated. There was a broad correlation between durability and the calculated content of $3CaO.Al_2O_3$ in the cement, and Miller concluded that, for use in concrete exposed to sulphate waters, Portland cements should not contain more than 5·5 per cent $3CaO.Al_2O_3$. Neither air-entraining agents nor other additives produced any really significant benefits, and some were deleterious. Steam curing of concrete at temperatures up to 90° had little effect on the sulphate resistance, but a period of 48 hours in steam at 100° was very beneficial and curing in steam under pressure at higher temperatures increased the resistance to attack to such a degree that under the best conditions almost complete immunity to sulphate action up to 17 years was obtained. A period of six hours at 175° (120 lb steam pressure) was the best. Surface treatment of concrete cylinders with boiled linseed oil and thin bituminous and tar coatings afforded appreciable protection for a number of years, but after five years little protective effect remained. High-alumina cement concrete immersed in the moderately cool waters of Medicine Lake proved immune from attack over 20 years.

Thorvaldson, working in association with the National Research Council of Canada, also carried out very extensive investigations on the resistance of cements to sulphate waters. This work, which was for the most part limited to studies in the laboratory of the resistance of cement mortars, has been discussed in earlier chapters. Thorvaldson, like Miller, found a very marked difference in the resistance of different Portland cements to attack and that steam curing under sufficiently high steam pressure confers a high degree of immunity.

Tests on large concrete prisms partially buried in alkali soils at Sacramento, California, are also being carried out as part of the American Long-Time Study of Cement Performance.[1] The specimens are exposed in two soils, each containing about 10 per cent soluble salts, but in one these are composed of about 85 per cent sodium sulphate and 10 per cent sodium carbonate with little magnesium sulphate, and in the other about one-third of the sodium sulphate is replaced by magnesium sulphate. The exposure beds are flooded periodically in the summer, resulting in crystallisation of salts by evaporation over the exposed surfaces. After some five years under these severe conditions the factor of major importance was the content of cement in the concrete and the type of Portland cement was secondary to this. Lean concretes of about 375 lb cement/yd^3 deteriorated badly and many disintegrated completely in a few years. With medium concretes of about 520 lb cement content the deterioration after five years varied from slight to severe, whilst most of the rich mixes with about 660 lb cement showed only slight deterioration. In the concretes of low and medium cement content there was a general relation between degree of deterioration and the calculated $3CaO.Al_2O_3$ content of the cement. From the results in the 660 lb concrete mix after 20 years it was concluded that a limit of 5·5 per cent should be placed on the C_3A content, corrected for minor components, and about 3·5 per cent as determined by X-ray diffraction. These values correspond roughly to about 6·5 per cent calculated by the Bogue method. This may be compared with limits of 5·5 per

[1] F. R. McMillan, T. E. Stanton, I. L. Tyler and W. C. Hansen, *Concrete Exposed to Soils*, Am. Concr. Inst. Spec. Publ. (1950); W. C. Hansen, *Bull. Res. Dev. Labs Portld Cem. Ass.*, No. 175 (1965).

cent suggested by Miller, the 5·0 per cent of the ASTM specification (C150–68) for sulphate-resisting Portland cement and the 3·5 per cent of BS 4027 : 1966. These limits serve as the best guide available for selecting Portland cements of relatively good performance, but they are not free from anomalies (see p. 354).

The U.S. Bureau of Reclamation have concluded[1] from their trials and field experience that the C_3A content is the greatest single factor influencing the resistance of Portland cement concretes to sulphate attack, with cement content second, thus reversing the order placed on these variables from the Sacramento tests mentioned above. Increasing the cement content improves resistance, but some 700 lb cement/yd^3 (415 kg/m^3) was reported to be required with a Type II cement to give as good a resistance as 420 lb (250 kg/m^3) with a Type V cement. Air entrainment only slightly improved the resistance, but air drying following wet curing increased the resistance of precast concrete products.

There is evidence[2] that concretes with calcareous or blastfurnace slag aggregate have a somewhat higher resistance to sulphate attack than those with silicious aggregates. Concrete under bending stresses has a lower resistance than unstressed concrete and this may perhaps be due to incipient cracking.[3]

Various methods for testing the relative resistance of cements to sulphate waters have been developed (see Chapter 12) and though no short-period test has been found free from anomalies, or difficulties in reproducibility, a reasonable assessment can be made. In general, tests on lean mortars or concretes, or on specimens of small dimensions, tend to emphasise the effect of the type of cement and to overestimate the difference in performance in concrete of high quality.

The work carried out in various European countries on the resistance of pozzolanic cements to sulphate waters, which has been discussed in previous chapters, has shown that the resistance to attack of ordinary Portland cements is increased by the use of pozzolanas and that under the less severe conditions their use provides a resistant concrete. A higher resistance is obtained from sulphate-resisting Portland cement when the main sulphate present is magnesium sulphate and this is increased by a pozzolana containing a high content (90 per cent) of reactive silica, but not by one containing a high content of alumina. Experience with supersulphated cement shows that it is comparable to sulphate-resisting Portland cement in resistance to sulphate attack and that properly designed high-alumina cement concrete is the most resistant.

It will be evident that there are now available the results of many large-scale tests and laboratory experiments from which certain definite conclusions can be drawn. These conclusions can, however, only be used as a general indication of the behaviour to be expected, for local conditions will modify them in varying degrees. Thus a soil which is relatively high in sulphates, but which is dry for long periods of the year, or which is so drained that the solutions formed by the leaching of the soil are removed from the vicinity of the concrete, will be much less destructive

[1] B. P. Bellport, *Performance of Concrete*, Can. Bldg Ser. No. 2, 77 (1968); L. H. Tuthill, *Spec. tech. Rep. Am. Soc. Test. Mater.*, No. 169A, 275 (1966).

[2] W. Gutt, W. Kinniburgh and A. J. Newman, *Mag. Concr. Res.* **19** (59), 71 (1967); J. Y. Karpinski, *Revue Matér. Constr. Trav. publ.* **568,** 19 (1963).

[3] V. V. Stolnikov, *Trans. 8th int. Congr. large Dams* **3,** 403 (1964); C. Goria and M. Applano, *Industria ital. Cem.*, 1949 (5/6), 77.

TABLE 128 Concrete for sulphate soils

	Total SO_3 in soil per cent	SO_3 in ground-water pts per 100 000	Cement	Concrete needed: Minimum cement eontent lb/yd³	Concrete needed: Minimum cement eontent kg/m³	Concrete needed: Maximum free w/c ratio†
II	0·2	30–120	Ordinary Portland or Portland blastfurnace	550	330	0·50
			Sulphate-resisting Portland	475	280	0·55
			Supersulphated	515	310	0·50
III	0·5–1·0	120–250	Sulphate-resisting Portland, supersulphated or high alumina	550	330	0·50
IV	1–2	250–500	Sulphate-resisting Portland or supersulphated	625	370	0·45
			High alumina	575	340	0·45
V	Over 2	Over 500	(i) High alumina (ii) Sulphate-resisting Portland or supersulphated as in IV with surface protection	625	370	0·40

† Total water in concrete less that absorbed by the aggregate.

than one in which the sulphate content is lower, but the physical conditions are more unfavourable. Regions in which alternate wetting and drying occur fairly frequently are likely to be particularly destructive. It is also often found that the salt content of waters and soils varies widely between points quite close together; this increases the difficulty of assessing in advance the conditions to which a concrete will be subjected. The salt content of the soil is only important in so far as it represents the reserve supplies available for maintaining, or increasing, the salt content of the waters contained in it, or draining from it. Despite these difficulties, an examination in suspected cases of the general ground conditions affords a guide to the precautions which are required to ensure the permanency of a concrete.

The least severe condition of exposure to sulphate action is that of concrete completely buried where the excavation for the structure does not form a channel along which flow of groundwater is likely to occur, e.g. foundations and concrete piles. The most severe condition is when concrete is exposed on one side to water pressure and on the other side to the air, promoting evaporation and so tending to increase the volume of the sulphate-bearing waters drawn through the concrete. The use of precast concrete products is advantageous, where practicable, since the quality of the concrete can be controlled more easily than with *in situ* work and it can be well matured before exposure. Lean concretes made from 'all-in' aggregate of the types often used for haunching or bedding pipes are very vulnerable to attack. The use of sulphate-resisting Portland cement, pozzolanic cements or supersulphated cement for conditions of moderate severity, and of high alumina (in temperate climates) for severe conditions, will help to give durable structures, but none of these cements is immune from attack in low-quality concrete. Under conditions, such as in Great Britain, where local and seasonal variations of the sulphate concentration in groundwaters is high, it is necessary to be conservative in defining the meaning of 'moderate severity' and 'severe'. Concentrations of SO_3 below about 30–50 parts per 100 000 can usually be ignored. For cast *in situ* concrete an average concentration of above about 120 parts SO_3 per 100 000 may be taken as severe and for dense precast products 200 parts. Asbestos cement pipes, and the densest precast concrete pipes with an absorption on 24 hours' boiling of only 3–5 per cent by weight, are not likely to suffer seriously over periods of 10–20 years even under somewhat more severe conditions, but concrete pipes with absorption values of 8–10 per cent are vulnerable. Where the actual concentrations are known not to vary much, rather higher limits could be set. In suitable cases surface protection with one of the thicker pitch-tar or bituminous coatings can be given to Portland cement concrete. Concrete pipes have been made with a polyester-glass fibre wrapping to protect them. Reliance should not be placed on integral or surface waterproofers.

The general recommendations now current in Great Britain[1] classify the soil condition and the type of concrete needed as set out in Table 128.

The U.S. Bureau of Reclamation classify sulphate soils in relation to the degree of sulphate attack as follows:

[1] Building Research Station Digest No. 90 (Second Series) 1968.

Severity	Total SO_3 in soil per cent	SO_3 in groundwater parts/100 000
1. Negligible	0–0·083	0–12·5
2. Positive	0·083–0·17	12·5–83
3. Considerable	0·17–0·42	83–166
4. Severe	Over 0·42	Over 166

Type II Portland cement is recommended for class (3) and Type V for class (4).

The draft revision of the German regulations[1] set out in DIN 4030 (1968) and DIN 1045E require the use of a sulphate-resisting cement if the content of SO_3 in the groundwater exceeds 33 parts per 100 000. The cement may be a sulphate-resisting Portland cement ($C_3A \not> 3$ per cent), a Hochofen cement containing at least 70 per cent granulated slag, a Hochofen cement made with a sulphate-resisting Portland cement clinker and a granulated slag containing not more than 13 per cent Al_2O_3, or supersulphated cement.

A special case of sulphate attack has been found in concrete slabs on the ground where these have been laid on a bed of hardcore containing sulphate salts. When the sites have been wet and the groundwater level high, migration of sulphates has occurred into the concrete slab causing expansion. This has caused serious damage to houses in England where colliery wastes have been used as the hardcore.[2] A limit of 0·5 per cent SO_3 content should be imposed on hardcore used in this way.

Marsh waters

The waters found in areas in which considerable development of peat occurs may have an acid reaction. Peat soils are often divided into what are known as the high moorland, or mountain, and low moorland, or marsh type. The latter, which are considered in this section, are characterised by a much higher content of bases and soluble salts than the mountain type. The most notable development of marsh peat to be found in Great Britain is in the Fen district.

Marsh waters are rarely acid from organic acids arising from the decay of peat as the content of lime in the soil is adequate to neutralise them. The peats occurring in such areas may, however, contain pyrites and marcasite, and the oxidation under suitable conditions, of these iron sulphides leads to the production of free sulphuric acid. This is neutralised by the bases present in the soil giving sulphate salts, but in some cases the production of acid may be so marked that free sulphuric acid occurs in the ground waters. Thus in the well-known moor around Osnabruck in Germany, where the peat soil contains 17 per cent pyrites, free sulphuric acid has been found in the groundwaters to the extent of 8–78 milligrams per litre. The waters in districts of this type may contain therefore both sulphate salts and free sulphuric acid while they may be rich also in carbon dioxide and hydrogen sulphide. The position of the water table is of predominant importance in determining the degree to which pyritic oxidation can occur. If the pyrites remain below the water level no oxidation takes place, while in regions of fluctuating water level the oxidation proceeds most rapidly.

[1] Summarised by J. Bonzel and F. W. Locher, *Beton* **18,** 401, 443 (1968).
[2] H. J. Eldridge, *Munic. J.* **72,** 2893 (1964); F. M. Lea, *Performance of Concrete,* Can. Bldg Ser. No. 2, 56 (1968).

Large-scale exposure tests and investigations were commenced in Germany in 1909 following on some failures of concretes in marsh waters of this type in which the principal destructive agent is free mineral acid. The results of this series of tests were reported[1] in 1922, but as these did not, owing to the interference caused by the 1914–1918 war, give all the information desired, another series of trials was begun in that year.[2] The original trials included the exposure of concrete piles, pillars and pipes on a number of sites in low moor, or marsh, areas, while in the second series concrete slabs 50 × 25 × 8 cm were used as test pieces. The groundwaters in the areas in which the slabs were placed showed pH values which varied over the range 4–6 during the period, the acidity being due to free sulphuric acid and carbon dioxide. After a period of seven years' exposure $1:1\frac{1}{2}:2\frac{1}{2}$ concretes had proved comparatively resistant, though the surface was more or less strongly corroded and the aggregate exposed. Portland and Portland blastfurnace cements were found to be about equal in their resistance to the exposure conditions. Concretes made from a 1:3:5 mix were not in such good condition and had suffered more attack, showing small holes and furrows on the surface; the cementing material was more or less dissolved out of all parts which came into immediate contact with the waters. The concretes tested, it should be noted, were very dense products, the absorption of the slabs with the richer mix being only about 2 per cent and those of the leaner mix only 4–5 per cent. Various surface coatings were found effective for some years, but within seven years all had partially or completely disappeared. After four years a hot-dip coat of tar-asphalt was mostly gone, one of the two coats of a hot-applied highly acid asphalt still remained intact, a coat of a cold-applied bituminous solution was flaking, and two coats of a phenolformaldehyde resin applied in alcoholic solution had disappeared between high- and low-water levels. After the coatings had been destroyed attack on the treated slabs commenced.

In the report of 1931 Burchartz concludes that in certain marsh waters attack on any unprotected Portland cement concrete must be expected and that remedial measures must take the form of insulating the concrete as far as possible from the aggressive waters. In any case the surface layers of the concrete should be as dense as possible, and it was suggested that if trass is used it should be treated as an addition to the mix and not as a substitution for cement.

A third series of trials[3] was started in 1932 but many of the records were lost during the war. Cubes (20 cm) and cylinders (7 × 15 cm) made with concretes of 350 and 270 kg cement per m^3 (w/c 0·50–0·55 and 0·60–0·65 respectively) were exposed at Pressler Moor where the pH of the groundwaters varied from 2·8 to 3·6 over the first five years but in the later years rose into the range of 4 to 6 with an average of about 5 over the whole 20 years' exposure. The acidity arose from the presence of aggressive carbon dioxide and free sulphuric acid. The cements examined were four Portland cements, a trass-Portland cement, two Eisenportland cements, two Hochofen cements and one high-alumina cement. After five years, specimens exposed at one year old had behaved better than those exposed at an age of two months, and the larger cubes better than the cylinders.

[1] M. Gary, *Dt. Aussch. Eisenbeton* **49** (1922).
[2] H. Burchartz, ibid. **64** (1931).
[3] K. Siedel, *Dt. Aussch. Stahlbeton* **134** (1959).

Though deterioration had progressed, the observation made after 20 years presented the same general picture as those made after 5 years, coarse aggregate being deeply exposed on the surface of the specimen. Nevertheless the compressive strength of the specimen had increased indicating that the attack had not progressed to any great depth. Broadly the concretes with the highest initial strengths had suffered the least, and those with the lowest the most, visible attack. No other distinction under this condition of acid attack could be drawn between the cements except that the high-alumina cement specimen showed a less deep surface attack.

Acidic conditions can also arise when iron sulphides present in shales or sands are exposed to the air and oxidise. In the region of Oslo in Norway for example[1] there are some 'alum shales' which contain pyrrhotite ($FeS_{1\cdot14}$) which is readily oxidised producing iron sulphates and sulphuric acid when the shale is crushed and exposed to weathering. Exceptionally, pH values in the groundwater as low as 2·5 have been recorded though values of 5–6 are more normal. Deterioration of concrete in these shales has long been a problem.

An unusual case of oxidation of pyrite (FeS_2) in sands occurred in south-eastern England.[2] A tunnel was driven through these sands in compressed air and oxidation of the pyrites occurred leading to the production of sulphuric acid. The groundwater, originally neutral, which returned into contact with the concrete lining of the tunnel showed pH values down to 1·8.

Mountain waters

The waters draining from mountain or high moorland regions are often almost free from dissolved salts, but are acidic owing to the presence of carbon dioxide and humic acid arising from the decay of peat. In these regions there is an almost complete absence of bases and soluble salts in the surface soils, and the rock formations are often igneous or silicious, or dense limestones into which the waters cannot penetrate. When the rocks are composed of the less dense limestones neutralisation of the acidic constituents in the surface waters can occur.

A saturated solution of humic acid in water has a pH value of 3·6–4·1. Owing to its low solubility, about 0·19 g/litre, the quantity of free acid which can be carried in a water is small and, though it has some action on concrete, this is much less serious than that of very pure, or carbon-dioxide-containing, waters. Calcium humate which is formed by its reaction with lime is practically insoluble in water.

Pure water dissolves lime to the extent of about 1·2 g/litre and, as we have seen in earlier chapters, is also capable of leaching out lime from set cement compounds. When carbon dioxide is present in the water in a free state it increases the aggressive action of the water. Initially it reacts with the lime to form insoluble calcium carbonate, but on further action the much more soluble calcium bicarbonate is formed. A solution of calcium bicarbonate alone in water will not dissolve further quantities of calcium carbonate, and if it reacts with lime the insoluble calcium carbonate is formed. Further, a certain amount of free carbon dioxide is required to stabilise calcium bicarbonate in solution so, as we have

[1] J. Mourn and I. Th Rosengvist, *Proc. Am. Concr. Inst.* **56,** 257 (1959).
[2] F. M. Lea, *Performance of Concrete*, Canadian Building Series No. 2, 56 (1968).

seen earlier (Chapter 12), not all of it is available to attack concrete. Waters which contain only combined carbon dioxide present as carbonates or bicarbonates have, therefore, no appreciable solvent action and it is only when the water is either very pure or contains free carbon dioxide that it attacks set cements.

Carbon dioxide is present in ordinary air to the extent of about 0·03 per cent by volume and under this pressure is soluble to the extent of 0·00054 g/litre, giving a solution which has a pH value of 5·7. Some spring waters in which carbon dioxide has dissolved under pressure from underground sources may have a higher content and be more acidic. Values down to 5·0, caused solely by carbon dioxide, have also been recorded in waters in Malaya.

Moorland waters which are practically free from salts show pH values between 4 and 7. Occasionally a lower pH value is found, but this is due to the presence of small amounts of free mineral acids, mainly sulphuric. Indeed, in rivers water, dissolved carbon dioxide rarely gives rise to pH values below about 5·5 and lower values are often due to the presence of humic acid. The acidity of moorland waters fluctuates with the seasons and weather conditions and is usually greatest after heavy rain following a warm, dry period. This may be due to the increased rate of production of acids in the peat under warm conditions.

Waters which are acid owing to the presence of organic acids and carbon dioxide do not normally produce more than a surface attack on mass Portland cement concretes, though with the more acid waters over a long period of years the action may penetrate to a depth of inches. If, however, there exists a considerable hydrostatic pressure tending to force the water through a concrete mass the action may become much more serious. The destructive effect on thin-walled concrete products like pipes may also be severe.

The definition of the types of soft waters that have a solvent action on concrete is rather complicated, since it is dependent on a number of inter-related factors. However, as a broad guide, waters with a temporary hardness above 4–5 parts $CaCO_3$ per 100 000 are not likely to be very aggressive unless the free carbon dioxide content is very high, about 5 parts per 100 000 or above. As the temporary hardness decreases so does the concentration of carbon dioxide required for the water to be aggressive. For a temporary hardness of 1–2 parts/100 000 a water will not have a marked solvent action unless the free CO_2 is above about 1 part/100 000; for a temporary hardness of 0·5 part the corresponding CO_2 value is about 0·5 part, while waters of very low temporary hardness, below say 0·25 part/100 000, can be aggressive even if the free CO_2 content is negligible. The pH value, as we have seen, is not a safe guide but, in general, waters may be aggressive at pH values up to 7, or even 7·5, if the temporary hardness is below about 0·25 part/100 000 and at pH values up to 6·0–6·5 for higher values of temporary hardness. Cases have been reported of renderings completely disappearing in the course of years in pure waters with no appreciable acidity. Walsh[1] has also reported that masses of permeable concrete were completely destroyed in five years by peaty waters of pH 6·5 combined with the action of waves. A dense concrete suffered only surface attack to the depth of $\frac{1}{8}$ inch over the same period. As another example, dense concrete, composed of 90 lb cement

[1] *J. struct. Engng* **11,** 335 (1933).

to 2 ft^3 sand and 4 ft^3 crushed sandstone with a slump of 1–2 inches, forming the base of a water conduit carrying a peaty moorland water showed surface erosion after 15 years. The coarse aggregate stood proud at the surface, about $\frac{1}{4}$ inch above the matrix, but the underlying concrete remained hard and firm.

In water conduits built in Scotland in connection with a hydro-electric power scheme[1] it was found that the surface of the concrete became attacked and the aggregate exposed within less than one year's exposure to waters of an average pH of about 6·8. After a period of years the coarse aggregate near the surface became loose and commenced to be washed out. A 1 : 2 mortar was satisfactory up to about eight years, but eventually the surface became rough. A high-alumina cement concrete surface was unaffected after 14 years. In other constructional work on similar schemes in Scotland a lining of high-alumina cement concrete has been applied to aqueducts and pressure tunnels.

Some other instances of the behaviour of concrete have been cited by Terzaghi.[2] In Ontario, waters containing 0·3–0·5 parts aggressive carbon dioxide per 100 000 with a pH of 6·5–7 had only a slight effect on well-made concrete. In Germany deterioration of culverts has occurred over 20 years with water containing more than 1·7 parts aggressive carbon dioxide per 100 000. As we have seen however the concentration of aggressive carbon dioxide likely to cause damage depends on the temporary hardness of the water.

Though much other evidence confirms the high resistance in temperate climates of well-made, high-alumina cement concretes to attack by soft natural waters, experience with concrete placed *in situ* has not always been satisfactory. This cement is more sensitive to the type of aggregate and conditions of placing than Portland cement and the w/c ratio must not exceed 0·5 if the disadvantageous effects of conversion are to be avoided. Pozzolanic cements, and some Portland blastfurnace cements, are rather more resistant to attack than Portland cement,[3] but concretes made with them still suffer gradual surface erosion when exposed to running water, or leaching if the concrete is at all permeable. Trials by Rengade[4] on concrete and mortar slabs exposed to a natural running water of pH 6·6 show that supersulphated cement is susceptible to surface attack by soft waters. A large number of surface treatments were tested in the Scottish trials, but the effective life was less than 10 years, except for some bituminous enamels which were gradually wearing away after 15 years.

Considerable trouble has been experienced in the Scandinavian countries with large concrete dams exposed to pure or acid mountain waters.[5] A dam is subjected to a one-sided pressure by water and, if the concrete of which it is composed is in any degree permeable, water will continuously permeate through the mass. With waters which are not very pure and have little solvent action the pores in a concrete mass often slowly seal up, but when the water has a marked

[1] Halcrow, Brook and Preston, *J. Instn Wat. Engrs* **33** (1928).
[2] R. Terzaghi, *J. Boston Soc. civ. Engrs* **36,** 136 (1949).
[3] F. M. Lea, *Bldg Res. tech. Pap.*, No. 26 (1939); R. Grün, *Zentralblatt der Bauverwaltung* **56,** 1373 (1936); W. C. Hansen, *ASTM Bull.*, No. 232, 51 (1958); G. Goggi, *Ind. Ital. Cemento* **30** (8), 1960.
[4] E. Rengade, *Proc. 15th Congr. Industrial Chemistry, Brussels 1935*; see also R. Bidron, *Revue Matér. Constr. Trav. publ.* **443–444,** 224 (1952).
[5] Cf.B. Hellstrom, *J. struct. Engng* **11,** 210 (1933).

solvent action lime is gradually removed from the concrete which becomes progressively more porous. A concrete dam may be amply strong to resist the mechanical forces it is designed to withstand and still be incapable of resisting the slow insidious solvent action of the waters it retains. The increasing leakage of many dams in Norway and Sweden attracted much attention and formed a prominent topic at the meetings of the International Commission on Large Dams held in Stockholm in 1933 and Washington[1] in 1936. Many of the Scandinavian dams had been built with rather lean concrete mixes which proved insufficient to resist the action of the impounded waters having pH values from 5·7 upwards. Little or no trouble of this type has so far been observed in dams in Great Britain, owing probably in part to the different character of the waters in many cases, but mainly to the general use of richer mixes. Halcrow[2] has, however, cited the case of a dam in Scotland which after 24 years' exposure to waters of pH about 6·4 showed signs of deterioration which it was evident would continue. This dam was built with a facework 2 feet thick of 1:4 concrete with a hearting of 1:5 concrete. Deterioration of dams from this cause has not been observed in the U.S.A.

Concrete pipes in acidic waters

Concrete pipes and similar thin-walled concrete units require rather special consideration in regard to their resistance to chemically destructive agencies. A degree of attack which would be relatively unimportant in a mass concrete may be sufficient to destroy a pipe the wall of which is only one or two inches thick. Such precast concrete units can on the other hand be well matured before use.

Special investigations on concrete pipes have been carried out in Switzerland, Sweden, Germany and the United States, in each case following on the discovery of cases of failure caused by the action of chemical agencies.

Particular attention has been paid in Sweden to the attack of concrete pipes by acid waters containing free carbon dioxide. A careful examination of the cause of various failures experienced in Sweden had shown that they were due in all cases to natural acid waters and no case of sulphate action was found.[3] Damage has occurred to pipes laid in sand or sandy clay through which water could percolate fairly freely, but not to those laid in clay only. Attack has been found to be most severe on the tops of pipes and at the sockets. The average content of free carbon dioxide found in the waters in cases where damage had occurred was 0·05 g/litre.[4] A series of tests was carried out at the Government Testing Institute, Stockholm, on concrete pipes exposed to running water containing free carbon dioxide. Rather porous pipes made from a 1:6 mix were found to suffer rapid attack while dense pipes made from a 1:3 mix suffered only insignificant solution of the surface. The addition of a pozzolana to the 1:6 mixes was not found to increase the resistance of the pipes. It was recommended that mixes should not

[1] K. Baalsrud and Kr. Friis, *2nd int. Congr. large Dams, Washington 1936*, vol. II, 149; Survey of Scandinavian literature, ibid., 383.

[2] *Engineering* **137,** 609 (1934).

[3] R. V. Frost and E. I. Virgin, *Int. Congr. Concr. reinf. Concr., Liège 1930*, Sect. II, VII—6. *Meddelande 48 Statens Provningsanstalt,* Stockholm (1929).

[4] Failures of concrete pipes due to attack by carbon dioxide have also been reported elsewhere.

be leaner than 1:3 for pipes of cement and sand and 1:2:3 for pipes of cement, sand and gravel. Concrete pipes of sound quality could be used, it was concluded, unless the subsoil water had a very low content of bicarbonates of lime and magnesia (a very low temporary hardness), or had a rather high content of aggressive carbonic acid, and the circulation of the water at the same time was considerable.

Further work carried out later in Sweden by Granholm, Werner and Giertz-Hedstrom[1] included laboratory investigations and field trials. From tests on concrete pipes manufactured by various methods the relative aggressiveness of natural waters containing carbon dioxide was classified as shown in Table 129.

TABLE 129

Type of water	Composition of waters		Aggressiveness of water towards concrete
	Temporary hardness parts $CaCO_3$ per 100 000	Aggressive carbon dioxide parts per 100 000	
I	> 3·5	< 1·5	Practically none
II	> 3·5	1·5–4·0	Slight
	0·35–3·5	< 1·5	Slight
III	> 3·5	4·0–9·0	Appreciable
	0·35–3·5	1·5–4·0	Appreciable
	< 0·35	< 1·5	Appreciable
IV	> 3·5	> 9·0	Great
	0·35–3·5	> 4·0	Great
	< 0·35	> 1·5	Great

The final results, after 10 years, on a test pipe-line carrying waters with a very small temporary hardness, an aggressive carbon dioxide content of 2–7 parts per 100 000 and a pH of 4·5–6·5, showed that spun and dense machine pressed concrete pipes and high-alumina cement pipes had suffered only slightly, whereas more porous and permeable pipes had suffered severely. These results indicate that the densest precast concrete pipes will resist for a considerable period waters falling in the most aggressive class of Table 129. The earlier results from laboratory tests indicated that coating of the surface with a cold bituminous primer followed by two coats of hot bitumen (1·7 kg/m^2 of surface), or by a coal-tar primer followed by two coats of hot pitch (1·2 kg/m^2 of surface), gave the greatest degree of protection of a variety of surface treatments tried, but that neither entirely prevented absorption of water. In the experimental pipe-line a hot bitumen coating and a surface impregnation with a bituminous solution were the only surface treatments tested, these being applied to rather porous hand-tamped pipes. From the inspection after 10 years Granholm (1944) found that the thicker bitumen coating had cracked and become displaced in places and that the surface impregnation had been more effective though the concrete was

[1] *Engineer* **157,** 235 (1934); *Betong,* 1934 (1), 1; ibid., 1936 (1), 20; *Chalmers tek. Hogsk. Handl.,* No. 27 (1944).

exposed in places and the porous pipe laid open to attack. In laboratory tests in Denmark[1] on pipes immersed in carbon dioxide waters, coatings of hot thick coal-tar pitch gave better results than hot-bitumen or cold-tar paints but blistering had commenced after 3½ years.

A survey[2] made of asbestos-cement pipes, conforming to British Standard 486, after 10–12 years' service in England in a variety of conditions, both aggressive and non-aggressive, showed that a thin bitumen coating (0·002–0·003 inch) applied by dipping the pipes in a naphtha solution of bitumen lasted up to 10–15 years before any appreciable proportion of either the inner or outer surface was bare. No deterioration of the pipes themselves was observed in sulphate-bearing clays over 10–15 years and, rather unexpectedly, pipes carrying ferruginous acid mine waters containing some sulphuric acid, arising from oxidation of iron pyrites, and large amounts of sulphates, showed no appreciable deterioration in some 10 years. Thick coatings of hydrated ferric oxide had deposited inside the pipes, forming a protected layer. Under other conditions, however, acidic water can cause gradual softening.

A similar survey[3] made in the U.S.A. on uncoated asbestos cement pressure pipes exposed to different soil conditions for periods up to 11 years has shown their high resistance to attack by sulphate soils containing up to 2–3 per cent soluble sulphur trioxide. In a very acid peat soil with a pH as low as 2·6 some deterioration set in after a few years and softening of the surface extended to a depth of 0·20 inch after 10 years. Less acid soils with pH values from about 4 to 7 often produced slight surface softening not extending beyond a depth of 0·06 inch after 10 years. The condition of asbestos-cement pipes after periods up to 10 years in acidic groundwaters in New Zealand was examined by Penhale.[4] In the most acidic conditions where the groundwater contained 133 mg aggressive CO_2/litre and had a pH of 5·1 attack had proceeded to a depth of 0·12 in. after 6 years and 0·18 after 10 years. In another case where the aggressive CO_2 content was 61 mg/litre and pH 6·3 attack had proceeded to a depth of 0·05–0·07 in. after 10 years. It was concluded from this survey that the classification of aggressiveness of groundwater given in Table 129 was applicable where freely moving groundwaters were in continuous contact with the pipes over the ten-year period but that where the contact of groundwaters with the pipes was seasonal and movement of water slow the attack was less severe. Spun concrete pipes showed a comparable behaviour to that of the asbestos-cement pipes.

Various other surveys have been reported on the condition of concrete pipes after many years of service. Thus concrete pipes dug up in Norway after 20–30 years' service in the ground showed[5] that only in an alum shale (see p. 647) where the groundwater contained both sulphates and sulphuric acid was there severe damage. On other sites with groundwaters at the time of the inspection containing up to 8 parts free CO_2/100 000 and with pH values down to 6·4 the pipes, though showing surface attack in varying degrees, were still in serviceable con-

[1] E. Suenson, *Ingvidensk. Skr.*, 1935, B. No. 15.
[2] F. E. Jones and J. P. Latham, *Natn. Bldg Stud. Res. Pap.*, No. 15 (1952).
[3] I. A. Denison and M. Romanoff, *J. Res. natn. Bur. Stand.* **47**, 367 (1951); M. Romanoff, *Circ. U.S. natn. Bur. Stand.*, No. 579 (1957).
[4] H. R. Penhale, *N.Z. Jl Sci. Technol.*, Section B38 (3), 257 (1956).
[5] E. Harildstad, *Betongen Idag.* **23** (2), 37 (1958).

dition. This was so also for an 18-year-old pipeline where the groundwaters contained 13 parts free CO_2 and 40 parts $CaCO_3$ per 100 000 and had a pH value of 5·3. In the U.S.A.[1] pipes carrying a soft water with a temporary hardness of 0·6 parts $CaCO_3$/100 000 and a pH value of 6·7 were found to have suffered surface attack to a depth of about 0·03 in. over a period of three years.

Concrete pipes were first used in Switzerland in 1913 and by 1922 it was found that over certain areas about one-third of the pipes had suffered attack. Whilst this was probably attributable in part to the poorer quality of the pipes formerly made, it indicated the necessity of determining what conditions were to be regarded as dangerous to concrete pipes. Where damage had occurred it was found to be most serious in the regions between the high and low positions of the water table. In 1922 a most comprehensive study of the behaviour of concrete pipes was commenced by a Swiss Commission.[2] Pipes of six different makes and three different mixes, about 1 : 6·5, 1 : 4·9 and 1 : 3·9 by weight, were buried in a number of different soils of dangerous type and examined at intervals. A very careful examination of the characteristics of the soils was also undertaken. It was concluded that there are several different soil factors which may contribute to the decay of concrete pipes. In addition to acid waters and sulphate salts two other factors were also defined. One of these was the presence of considerable amounts of magnesium carbonate or bicarbonate in the soil. Magnesium bicarbonate was found to displace lime from set cement, and cases of damage due to its presence in soils were stated to have been found. The other factor was the presence of what were termed exchangeable soil acids. Certain soils which are not acidic in reaction themselves are capable of exchanging a hydrogen ion for a base ion by reaction with a salt solution and so forming an acid. Wiegner writes this reaction as

$$\text{Clay}\langle{}^{H}_{Ca} + KX = \text{Clay}\langle{}^{K}_{Ca} + HX.$$

The exchange acidity is determined by shaking 100 g soil with 200 cc normal sodium acetate solution for one hour and then titrating the solution with N/10 NaOH with phenolphthalein as indicator. The exchange acidity is defined as the number of cc N/10 NaOH required.

Five exposure sites of varying severity and with differing destructive agencies were used in the Swiss tests. Their main characteristics were as follows:

(*a*) *Aigle*. Soil strongly acid (pH 5·6–6·4, exchange acidity 60–120), 1 per cent SO_3, 2 per cent MgO. Groundwater 37 mg free CO_2, 34 mg aggressive CO_2, 428 mg sulphate SO_3 per litre; pH 5·9.

(*b*) *Stammheim*. Soil 0·5–3·75 per cent SO_3, weakly acid (pH 6–7, exchange acidity up to 200).

(*c*) *Mels*. Soil weakly acid (pH 6·3–6·9, exchange acidity 40–50). Sulphate content very variable 0–0·8 per cent SO_3.

[1] M. J. McCoy, R. J. Sweitzer and M. E. Flentje, *Proc. Am. Concr. Inst.* **54**, 647 (1958); M. E. Flentje and R. J. Sweitzer, *J. Am. Wat. Wks Ass.* **47**, 1172 (1955); **49**, 1441 (1957).

[2] *Schweiz. Verband für die Materialprüfungen der Tecknik; Bericht* No. 10 (1928); No. 35 (1937). Zurich; see also G. Wiegner, *Internat. Assoc. Test. Mat., Zurich*, vol. 1, 620 (1931).

(*d*) *Munchenbuchsee*. Soil neutral (pH 7–7·5, exchange acidity 150).

(*e*) *Schwerzenbach*. Soil 4 per cent MgO, exchange acidity in places above 20. Groundwater 14 mg free CO_2, 3·7 mg aggressive CO_2, 28 mg $MgCO_3$, 158 mg $CaCO_3$ per litre; pH 7·5.

After 10 years the ratio of the number of pipes in good condition to those in bad condition was (*a*) 0·17, (*b*) 0·45, (*c*) 0·25, (*d*) 5, (*e*) 4, indicating that the highly sulphate or acid conditions were the most severe. As was expected, the cement content of the concrete exerted a strong influence on the rate of deterioration. This is illustrated by the number of pipes on all sites in the two extreme conditions of good or bad after 10 years.

Mix (weight)	1 : 6·5	1 : 4·9	1 : 3·9
Condition:			
Good	11	14	23
Bad	17	17	9

The method of manufacture also proved important, the permeability of the pipe being a much better index to durability under adverse conditions than strength.

Some later studies made at the most severe exposure site at Aigle showed that after five years, protective bituminous coatings remained in good condition except for bituminous emulsions. Pipes made with high-alumina cement were unattacked when dense, but porous pipes deteriorated from attack by the soil acids. Asbestos-cement pipes, uncoated, suffered damage, the walls being softened, after 2½ years, but this had not progressed at five years, suggesting that the gelatinous silica produced by the action of the soil acids on the cement had formed with the asbestos a protective coating. Pipes coated with bitumen remained in good condition.

It was concluded from the Swiss tests, and from the examination of soils in which failures of concrete pipes had occurred, that the following soil conditions should be regarded as potentially dangerous and that any concrete pipes used in such soils should have protective coatings.

1. Acidity of groundwaters below pH = 6.
2. Acid-soluble sulphate contents in the soil greater than 0·2 per cent sulphur trioxide.
3. Acid-soluble magnesia contents greater than 2 per cent MgO.
4. Exchangeable soil acids equivalent to more than 20 cc N/10 NaOH per 100 g air-dry soil.

Deterioration of concrete drainpipes has also occurred in acidic peat soils in Minnesota and Wisconsin in the U.S.A. The peats in these regions have pH values ranging from 4 to 8 and sometimes smell strongly of hydrogen sulphide. In general the rate of corrosion has been found to increase with the acidity of the peat. The quality of the concrete also plays a large part and poor quality material has been unsatisfactory in peats of pH 6 or even higher, and disintegrated completely in under 20 years under more acidic conditions. Miller and Manson[1] have

[1] *Tech. Bull. Minn. agric. Exp. Stn*, No. 180 (1948).

concluded from the results of long-term field trials that under these conditions there is little difference in the performance of different Portland cements and that neither admixtures nor steam curing give any improvement. The prime requirement is a high quality concrete pipe but some additional benefit was obtained from surface treatments with linseed oil or bitumen cutbacks.

In temperate conditions the decay of vegetation gives rise to acids of the humic acid type that have a low solubility, but in tropical swamp-lands much more aggressive organic acids can be produced under partially anaerobic conditions. Thus contents of up to 1 per cent or more of lactic acid have been recorded[1] in swamp waters and been the cause of severe deterioration of concrete and steel.

In brackish waters, containing appreciable amounts of alkali chlorides, etc., carbonic acid behaves as a stronger acid[2] than in pure water and less free carbon dioxide is required to stabilise dissolved calcium bicarbonate (see p. 340). A higher proportion of the free carbon dioxide is therefore available to attack concrete and dissolve lime from it, and experience indicates that the severity of attack can be considerably increased. The presence of calcium salts, such as gypsum, in the water, has the reverse effect.

It is not to be inferred from the troubles that have been found to occur under certain conditions with concrete pipes that they are not suitable for use under most circumstances. Under normal conditions a well-made Portland cement concrete pipe may be regarded as possessing a satisfactory degree of permanence, but it is evident that under some conditions the use of any but the densest products is fraught with risk and that even the latter may not always be durable. In cases of doubt an examination of the subsoil conditions should be made. In England cases have occurred of rapid deterioration of Portland cement concrete pipes in soils containing a high content of gypsum, but no case of trouble due to attack by acid groundwaters seems to have been experienced.

The resistance of different types of Portland cement concrete pipes to attack seems to vary very considerably. The permeability of the concrete is clearly an important factor and a high absorption value must at least be regarded as suspicious. The absorption on immersing for 14 days in water was found in the Swiss tests to be of some value as an indication of resistance to attack, but the permeability was a much more satisfactory criterion. The strength was not a good criterion, though a very low value was indicative of a low resistance to attack. The Swedish tests also showed that it was the pipes of lower absorption that resisted deterioration. It was concluded from the Swiss work that mixes leaner than 675 lb cement per yd^3 should not be used for concrete drainpipes, and that 840 lb was often preferable. The evidence from various countries also indicates that pipes produced by the centrifugal and machine-pressed methods are the more durable under severe aggressive conditions.

Surface treatments of concrete pipes with bituminous or tar solutions offer resistance for a time to the attack of aggressive waters and under favourable conditions of application, which are more readily attained on precast products than on concrete cast *in situ*, may have a life of up to 15 years. Such treatments

[1] F. W Freise, *Concr. constr. Engng* **30,** 163 (1935).
[2] D. M. Greenberg and E. G. Moberg, *Bull. natn. Res. Coun., Wash.*, No. 83, p. 73 (1932); R. D. Terzaghi, *J. Boston Soc. civ. Engrs* **36,** 136 (1949).

have the additional value that the concrete is protected until it is well aged. Hot applied, thicker, coatings should have a longer life but, as we have seen, this is dependent on obtaining a good and uniform bond to prevent blistering.

In pipelines carrying non-aggressive waters of pH 7·8, two coats of bituminous paint applied over a smooth interior concrete surface have remained intact after 25 years.

Resistance of concrete to fire

Concrete, though not a refractory material, is incombustible and has good fire-resistant properties. The British Standard Definition for Fire-Resistance[1] states that the term fire-resisting should not be applied to a material, but only to structural elements of which it forms a part. In accordance with this definition we may not, then, strictly apply the term to concrete as a material, but must refer it specifically to, for instance, a concrete wall, a reinforced concrete column, a steel joist or staunchion encased in concrete, etc. It is nevertheless convenient to use the term in describing the relative value of different types of concretes in producing structural units of good fire-resistant properties, and, with this word of explanation, this may perhaps be done without ambiguity.

The fire-resistance of structural elements is measured by the length of time for which they continue to perform their normal functions as partition walls, load-bearing units, etc., when exposed to fire. The testing of such structural units requires very elaborate equipment, and special laboratories exist to carry out such tests.

The fire-resistance of a concrete structure is determined by three main factors, the capacity of the concrete itself to withstand heat and the subsequent action of water without losing strength unduly, and without cracking or spalling; the conductivity of the concrete to heat; and the heat capacity of the concrete. These factors together determine the extent to which the effects of high temperatures, persisting only for a limited time, will be restricted to the surface material, or transmitted to the interior of the mass and to embedded reinforcement or encased steelwork. When reinforcing bars or steelwork become exposed to high temperatures excessive expansion and loss of strength occur, and the steel tends to become warped and twisted out of shape. The fire resistance of a reinforced concrete structure is thus dependent not only on the type of concrete, but also on the depth of cover to the reinforcement and on the presence of surface coatings such as plaster.

A neat Portland cement on heating first expands owing to the normal thermal expansion; this expansion is opposed, however, by a contraction due to the shrinkage of the material as water is driven off from it. The contraction due to drying eventually becomes much larger than the normal thermal expansion and the material then commences to shrink. The actual temperature at which the maximum expansion is reached varies with the size of the specimen and the conditions of heating. It may be as high as 300° for air-dry specimens under conditions of fairly rapid heating. At higher temperatures the neat cement steadily shrinks, the contraction from the original dimensions amounting ultimately to 0·5 per cent or more. During this process severe cracking occurs.

[1] BS 476:1953; ASTM E119–61.

Hydrated Portland cement contains a considerable proportion of free calcium hydroxide which loses its water above 400–450°, leaving calcium oxide. If this calcium oxide becomes wetted after cooling, or is exposed to moist air, it rehydrates to calcium hydroxide accompanied by an expansion in volume which may disrupt a concrete which has withstood a fire without actual disintegration. Portland blastfurnace cement has been found more resistant to an action of this type. This may probably be attributed to the lower proportion of free calcium hydroxide present in such cements after setting, and also to combination of free lime with granulated slag during exposure to high temperatures.

In mortars and concretes the aggregates present undergo a progressive expansion on heating while the set cement, beyond the point of maximum expansion, shrinks. These two opposing actions progressively weaken and crack the concrete. The various aggregates used in concrete differ considerably in their behaviour on heating. Quartz, the principal mineral constituent of sands and most gravels and a major constituent of the acid igneous rocks, expands steadily up to 573°. At this temperature it undergoes a sudden expansion of 0·85 per cent, caused by the transformation of 'low' quartz to 'high' quartz. This expansion has a disruptive action on any concrete in which quartz forms an aggregate. Concretes with aggregates of silicious gravels, flint, and granite spall when exposed to fire and are amongst the least resistant concrete materials. Sandstones, though containing quartz, do not cause a concrete to spall so badly. The intergranular natural cementing material in sandstones shrinks on heating and thus, to some extent, tends to counteract the expansion of the quartz grains. The loss of strength in sandstone concretes on exposure to fire is, however, often high and sandstones do not form good fire-resistant aggregates.

The best fire-resistant aggregates amongst the igneous rocks are the very finely crystalline, or non-crystalline, basic rocks such as dolerites and basalts. Limestones expand steadily until a temperature of about 900° is reached and then begin to contract owing to decomposition with liberation of carbon dioxide. It has often been considered, on account of this decomposition, that limestones are of little value for making concretes resistant to fire. Actually, however, this decomposition does not occur until a temperature is reached considerably above that at which quartz aggregates spall. In actual fire tests it has been found that dense limestones form good fire-resistant aggregates, and that, except under prolonged exposure to high temperatures, only the surface material becomes decomposed. The most resistant of all concretes to fire are probably those made with a blastfurnace slag aggregate. This has been shown both by small-scale laboratory experiments and large-scale tests on structures. Broken brick also forms a good aggregate in respect of fire-resistance provided it is free from quartz.

Lightweight aggregates such as pumice, clinker, expanded clay products, etc., have in themselves a high resistance to fire, and concretes made from them have a low-heat conductivity. Their heat capacity, on the other hand, is less than that of heavier concretes.

Long series of tests on the fire-resistance of structures have been carried out by the Fire Testing Station in England, the Bureau of Standards, the Portland Cement Association, and the Underwriters' laboratories in the U.S.A., and other testing institutions. Even the most fire-resistant of concretes have been found to fail if exposed for a considerable period to a temperature exceeding 900°, while serious reductions in strength occur when a temperature above 600° is attained. If held permanently exposed to heat, any cement product must suffer considerable loss in strength and undergo gradual breakdown at temperatures which are considerably lower than 600°. Up to 300° the loss in strength is small but at 500° it can be 50 per cent or more.

The temperatures attained in the concrete of a building structure during a fire can sometimes be estimated from subsequent examination of the concrete. Bessey[1] has found that certain colour, and other, changes that occur in the aggregates in mortars and concretes provide an index to these temperatures and enable an approximate temperature scale to be established.

Electrolysis of concrete

Deterioration arising from the electrolysis of concrete[2] caused by stray leakage currents is comparatively rare. It is only in reinforced concrete that damage may occur and then only with direct, and not alternating, current. When the reinforcement acts as the anode it can be oxidised and the corrosion products cause expansion and cracking. As a cathode there is no action on the iron, but the alkalis from the cement tend to concentrate there and under severe conditions the concentration of alkali hydroxides can become so high as to attack the set cement and soften it to such a degree that the bond between concrete and steel is almost destroyed. Little or no action occurs in dry concrete and in wet concrete gradients as high as 60 volts per foot are needed to cause appreciable anode corrosion, but softening of the concrete at the cathode can occur at somewhat lower potential gradients. The resistivity[3] of moist concrete is of the order of 10^4 ohm-cm and that of oven-dry concrete 10^{11} ohm-cm. Much more serious effects occur if the concrete contains soluble salts and particularly chlorides as in concrete mixed with, or exposed to, seawater. The resistance of the concrete is much reduced and the efficiency of the corrosion process at the anode much increased. This is the reason why reinforced concrete jetties in seawater have been the most susceptible to damage by electrolysis.

[1] *Natn. Bldg Stud. tech. Pap.*, No. 4 (1950).
[2] G. Mole, *Engineering* **166**, 453 (1948); E. B. Rosa, B. McCollum and O. S. Peters, *Bur. Stand. tech. Pap.*, No. 18 (1913).
[3] G. E. Monfore, *J. Portld Cem. Ass.* **10** (2), 35 (1968).

20 Resistance of Concrete to Various Organic and Inorganic Agents

In this chapter the actions of a wide variety of substances on concrete are reviewed. Many of these substances are encountered in industrial processes in which they may come into contact with concrete, for example, by storage in concrete tanks or by spilling on concrete floors. Concrete is a material of such widespread utility, and so adaptable to many different purposes, that the problem of its chemical resistance to the conditions to which it will be exposed is sometimes overlooked until extensive repair work is found to be necessary.

The protection of concrete against chemical attack can often be effected by suitable treatment of the surface. The general types of protective agents have already been discussed in the previous chapter. There are, however, in addition many proprietary compositions marketed by firms who deal specially with the production of chemically resistant materials. These compositions are for the most part of the same general types as those mentioned earlier.

A protective agent for application to concrete must often be mechanically as well as chemically resistant. Thus while a material for use on concrete tanks may have only to withstand the action of the liquid contained in the tank, a surface treatment to be applied to a floor must be resistant also to mechanical wear and abrasion.

The present account does not set out to cover all the many possible aggressive agents. Further information is to be found in a paper by Kuenning,[1] in an American Concrete Institute report[2] on protective coatings for concrete, and in books by Kleinlogel[3] and Biczok.[4] With the developments in the plastics industry a large number of protective treatments are now available to supplement or replace older materials. When exposure to the aggressive agent is intermittent a thin surface coating of 0·005 to 0·05 in. thickness may suffice but where exposure is continuous, as in tanks holding liquids, thicker coating and sheet materials, or bricks and tile linings jointed with chemically resistant materials are often needed. Dense brick and tiles set in a mortar resistant to the aggressive agent may be

[1] *Highw. Res. Rec.*, No. 113, 43 (1966).
[2] Report of A.C.I. Committee 515, *Proc. Am. Concr. Inst.* **63,** 1305 (1966).
[3] A. Kleinlogel, *Influences on Concrete*, Frederick Ungar, New York (1950).
[4] I. Biczok, *Concrete Corrosion and Concrete Prntection*, Publishing House of the Hungarian Acaedmy of Sciences, Budapest (1964).

required for floors, or alternatively toppings of rubber latex-cement mortars, of polyvinyl acetate (or chloride) emulsion-cement mortars, or of synthetic resin and filler composition. The protective materials to be used against any particular aggressive agent must obviously be chosen from those that are resistant to the agent in question. Thin coatings are obtained from solutions, such as that of a resin in a solvent, from resin and latex emulsions, or from thermoplastic or thermosetting resins without solvents. The thicker coatings are obtained from mastics or mortars applied with a trowel or by using materials in the form of sheets.

Mineral oils

Mineral oils such as petrol, fuel oils and petroleum distillates in general, do not attack matured concrete, though they seriously affect the hardening of a green concrete. These oils contain no constituents which react chemically with set cements. Creosotes which contain phenols, cresols and similar acidic compounds may, however, have some effect on concretes. Lubricating oils which are entirely of mineral origin do not attack concrete, but if vegetable oils are present they have a definite action and cases of damage to garage floors have been recorded. Thus the case[1] has been described of a reinforced concrete floor which in the course of 20 years was severely attacked by lubricating oils with which it had become saturated. Both the reinforcement and steel joists were heavily corroded and on heating the concrete it burnt fiercely.

Though having no chemical action on concrete, it is difficult, or impossible, to make a concrete impermeable to the lighter mineral oils. Investigations, and experience, of the storage of mineral oils in concrete tanks have shown that concrete can be made sufficiently impermeable to stop loss by seepage for fuel oils of specific gravity 0·875 and above at 15°. For such tanks a fairly rich concrete (1 : $1\frac{1}{2}$: 3) is needed, while for lighter fuel oils, down to 0·85 specific gravity, a surface treatment, such as four coats of sodium silicate, has been recommended in addition.[2] The permeability of concrete to lighter oils with viscosities below 0·06 poises at 20°, such as diesel oil, kerosene, and petrol, is too high for it to be used without more substantial surface protection. In the case of petrol the quality of the fuel may also be affected by reaction with alkalis from the set cement. Various plastic thin sheet linings, such as polysulphide rubber and polyvinylidene, or synthetic resin surface treatments may be used as lining materials. Surface treatments with sodium silicate or magnesium silicofluoride are not adequate.[3]

Concrete floors or mortar joints in brick and tile floors may also be permeable to the lighter oils. Surface treatments with sodium silicate or magnesium silicofluoride solutions are recommended, though as mentioned above, their value is limited.

Organic acids

There are a number of organic acids which sometimes come into contact with concrete and exert a deleterious effect on it. They may be divided roughly into

[1] O. Colberg, *Beton Eisen* **27** (8), 160 (1928).
[2] Recommended practice for concrete and reinforced concrete. *Proc. Am. Soc. civ. Engrs* **66,** 109 (1940); M. A. Spamer, *Proc. Am. Concr. Inst.* **40,** 417 (1944).
[3] F. B. Hornibrook, *Proc. Am. Concr. Inst.* **41,** 13 (1945).

two groups. The first group includes acids of relatively low molecular weight such as lactic acid and butyric acid arising from the souring of milk and butter, acetic acid which is encountered in vinegar and in various food pickling and other industries, and other acids such as oxalic and tartaric. These acids are all soluble in water. The second group includes acids of high molecular weight such as oleic, stearic and palmitic acids, encountered as constituents of various oils and fats.

Lactic acid has a most destructive action on set cements, and trouble has often been experienced with concrete floors in dairies and cheese factories. These troubles are much enhanced when the floor is also subject to heavy abrasion by milk churns. Troubles are also experienced with tile floors laid in cement mortars. When concrete or mortar specimens are immersed in lactic acid solutions the cement is gradually dissolved away, leaving the aggregate exposed until eventually almost all the cementing material may be removed. No expansion occurs during this action. In solutions of lactic acid above about 1 per cent concentration high-alumina cement is attacked more rapidly than Portland cement, but it becomes progressively the more resistant as the solution becomes more dilute. Thus the effluent from dairies contains lactic and other acids with a pH down to 4; in plant for its treatment high-alumina cement concrete is resistant whereas Portland cement is attacked.[1] High-alumina cement has also been used for concrete floors, or for mortar jointing for clay tile flooring, in dairies on account of its higher resistance to quite dilute organic acids, but it will not withstand strongly alkaline detergents. Acid-resisting cements are also used for the jointing to tile floors in milk processing and bottling rooms.[2] Neither pozzolanic nor Portland blastfurnace cement offer any appreciable advantage over Portland cement against attack by lactic acid.[3] Portland blastfurnace cement is also liable, owing to the sulphides present in it, to give rise to offensive odours when brought in contact with the acid. The same objection may apply to supersulphated cement, even though its resistance to attack may be expected to be greater. Thin surface treatments to concrete floors are not of significant value in increasing their resistance to attack. For tanks to hold dilute solutions of lactic acid, as in whey, non-tainting bituminous paint or synthetic resin coatings are sometimes used, but in cheese making more durable linings such as asphalt mastic or white glazed clay products, preferably with acid-resistant cement jointing, may be selected.

Butyric acid seems to have a similar action on concrete to lactic acid. It occurs in sour silage.

Acetic acid attacks set cement and concretes. In a 5 per cent solution, about the maximum concentration in vinegar, it has a marked action within a few months. At this concentration high-alumina cement is even more severely attacked, but it is much the more resistant to solutions below 0·5 per cent. Pozzolanic and Portland blastfurnace cements are rather more resistant than Portland cement. For protection against the weaker solutions, acid-resistant paints have been used, but for the stronger solutions more adequate linings are required as with lactic acid.

[1] F. M. Lea and G. E. Bessey, *Concr. constr. Engng* **34,** 610 (1939).
[2] F. C. Harper, *J. Soc. Dairy Technol.* **4,** 184 (1951).
[3] Cf. E. Suenson, *Bautenschutz* **1,** 3 (1930) et seq.

Citric and malic acids have a similar action to acetic acid and in a 1 per cent solution seriously attack Portland cement concretes within a year or so. The Ocrate Process (p. 607) of treatment with silicon tetrafluoride gas substantially increases the resistance to attack.

Tartaric acid is similar in its action on concrete to lactic and acetic acids. Quite dilute solutions such as occur in fruit juices do not as a rule have very much effect, but the juices may be affected adversely. For this reason, and because sugars are also present in the juices, some surface treatment is advisable, for example with silicofluorides. Apple juice, in which malic acid occurs, can cause serious deterioration of Portland and high-alumina cement concrete.

Oxalic acid has some little action on Portland cement concretes, but its effects are not serious; indeed, it has been used for treating concrete surfaces to render them more resistant to other weak organic acids. An insoluble surface film of calcium oxalate is formed. This acid does not attack high-alumina cement. It is of course poisonous.

Organic acids of high molecular weight such as oleic, stearic and palmitic acids together with the general series of unsaturated and saturated aliphatic acids all have a very definite action on concrete. The destructive action is stated to increase with increasing molecular weight in both the stearic acid ($C_nH_{2n}O_2$) and oleic acid ($C_nH_{2n-2}O_2$) series.[1] The lowest members of the series are, however, an exception to this rule. These fatty acids of high molecular weight are encountered in industry as constituents of oils and fats. They are all insoluble in water and, at ordinary temperatures, the principal members are solids of low melting-point in the stearic acid series and liquids of high boiling-point in the oleic acid series. In general, these acids as constituents of oils attack any unprotected concrete. The disintegration produced is usually more pronounced when the concrete is exposed to air, as in a floor, than when it is continuously immersed beneath a liquid. Though Portland blastfurnace cement, high-alumina cement, and pozzolanic cements are somewhat less vulnerable than Portland cement to their action it is questionable whether the differences are sufficiently considerable to be of practical significance.

Vegetable and animal oils and fats

The vegetable and animal oils and fats are natural products composed mostly of the glycerides, or other esters, of the higher members of the several series of fatty acids, but they also contain in some cases notable amounts of the corresponding free fatty acids and alcohols. Glycerol (glycerine in commercial nomenclature) is the chief alcoholic constituent of oils and fats and occurs for example as glyceryl stearate (stearin) in tallow and lard, as glyceryl palmitate (palmitin) in palm oil and as glyceryl oleate (olein) in olive oil.

Oils of vegetable origin, even when quite fresh, usually contain appreciable quantities of free fatty acids. Animal fats, when freshly rendered, contain only small quantities of free acid, but the amount is increased on exposure to the atmosphere. Rancidity consists in the development of certain oxidised free fatty acids in oils or fats.

[1] F. W. Freise, *Concr. constr. Engng* **27**, 347 (1932).

The glycerides and the other esters are broken up by hydrolysis into their constituent alcohol and acid components. This process, termed saponification, can be produced by the action of acid or alkali solutions. When oils are brought into contact with concrete the free lime present in the set cement saponifies the material, forming a calcium salt of the fatty acid and liberating the polyhydric alcohol. This alcohol can itself often react with lime. Thus with olein there is formed calcium oleate and glycerol, and the latter then combines further with more lime to form calcium glycerolate. This is a typical example of the mechanism of the destructive action of the saponifiable oils and fats on concrete. If free acids are also present, as is often the case, they also attack the concrete and form their calcium salts.

The extent to which an oil can attack a concrete is dependent on the ease with which it can penetrate the material. The viscosity of the oil is thus an important factor and, other things being equal, any action is likely to be the less serious, the more viscous the oil. Exposed oils which have accumulated moisture and undergone oxidation become more active in attack. Thus oils which can be stored successfully in closed concrete tanks may often be very destructive to concrete floors where any protective films can rapidly be worn away and there is free access of air and moisture.

Glycerine is produced in large quantities in soap manufacture. Though free glycerine is not a constituent of hard soaps it is usually present in soft soaps. It is completely miscible with water in all proportions and is a solvent for lime. It attacks concrete by slowly combining with, and dissolving, any free calcium hydroxide present in the set cement. A solution of glycerine in water as weak as 2 per cent has a destructive action on fresh Portland cement concrete, but has little effect on a well-carbonated surface of a matured concrete. Steel tanks lined with cement mortar have been used for the storage of dilute glycerine solutions, but very particular care is required in their construction. Surface treatments with synthetic resin or other suitable paints have been used as protection against the weak solutions, but stronger ones, e.g. 10 per cent and upwards, have a very destructive effect on concrete and thicker protective linings may be needed. Concentrated glycerine containing only a few per cent of water is stated to be less destructive.

Among the commoner non-mineral oils are cotton-seed oil, palm oil, rape-seed oil, olive oil, coconut oil, linseed oil, tung oil, lard oil and fish oils. It should be assumed that in general all such oils are likely to be dangerous to unprotected Portland cement concrete, though in some cases it is possible to render their attack negligible. Some drying oils such as linseed oil and tung oil, in fact, are often used for the surface treatment of concretes, a usage which may seem incompatible with the preceding general statement. It must be remembered, however, that in such cases a dressing of oil is applied and allowed to harden by drying and that exposure to a continuous, or intermittent, supply of the fresh oils does not occur. Further, these particular oils, known as drying oils, undergo oxidation, and harden when exposed to the air, whereas the majority of other oils become rancid and develop acid products.

The problem of providing suitable floors to withstand the action of oils and fatty acids arises in many industries concerned with the manufacture of soap, margarine and fats, preserved foods, candles, lubricating oils and greases with a vegetable basis, the crushing of oil seeds, and other similar processes. The difficulties which have to be met are often much increased by the requirement that the floor shall be resistant to heavy wear by trucks; the substitution in these cases of rubber for steel tyres, where possible, would often reduce markedly the wear on the concrete.

It has been found in the soap and oil seed industries that Portland cement concrete floors are not generally suitable for use where they will be subject to the action of vegetable seed oils and acids. Even the best granolithic floors, or concrete floors floated with neat cement, soften and finally disintegrate. The higher the content of free fatty acids in the oils concerned, the more rapid is the action. Surface hardeners add to the wearing qualities of the floor when traffic is light, but do not otherwise effect sufficient improvement to be of value. It has been similarly found that concrete floors are often unsatisfactory in factories manufacturing cooked meats, animal fats, margarine, etc. These general statements apply to conditions which are severe, but in many cases where the exposure is less, and where it may be possible to remove the oils and fats from a floor soon after spilling, satisfactory service with concrete may be obtained. Composite floors in which steel plates are set into the surface concrete are also often used.

Laboratory tests show that Portland cement concretes are rapidly attacked by cotton-seed oil, but that Portland blastfurnace cement is more resistant, and high-alumina cement very much more resistant to its action. When a considerable proportion of free organic acids is present in the oil the resistance of high-alumina cement is decreased and may not be superior to Portland cement. Rape-seed oil attacks both Portland cement and high-alumina cement rapidly. Lard oil and raw linseed oil usually attack Portland cement concretes rather less rapidly than does cotton-seed oil. High-alumina cement does not seem to be more resistant to linseed oil than Portland cement, but supersulphated cement concrete specimens have been found to be unattacked after five years. Coconut oil attacks Portland cement concrete rapidly. Relatively few test data are available to show the comparative resistance of different cements to attack by oils, but it is clear that while in some cases high-alumina cement is more resistant than Portland cement this does not always hold. The addition of a pozzolana to Portland cement, or the use of Portland blastfurnace cements, has been recommended in German practice, but these cements only give somewhat increased resistance, and not immunity, to attack.

Concrete tanks have been considerably used for the storage of many oils and have been found not to suffer attack to the extent which might have been anticipated from tests on small concrete specimens, or from experience with concrete surfaces exposed also to air and moisture. As surface treatments sodium silicate and magnesium silicofluoride solutions are often used, but in the more severe cases thin resin-base coatings have been found more satisfactory. Bituminous materials which are softened by oils are unsuitable.

A series of tests on 1 : $1\frac{1}{2}$: 3 concrete tanks were made at the American Bureau of Standards[1] with cotton-seed oil, raw and boiled linseed oil, neat's-foot oil, lard

[1] J. C. Pearson and G. A. Smith, *Proc. Am. Concr. Inst.* **15**, 186 (1919); G. A. Smith, ibid. **17**, 22 (1921).

oil and coconut oil. While some of these oils caused some initial surface attack on the tanks, no progressive action was found up to three years. Concrete cylinders stored in the oils were, however, more seriously affected.

Recommendations for the treatment of concrete have been published by the American Society of Civil Engineers[1] and are shown in Table 130.

TABLE 130 Effect of oils on concrete
(*American Society of Civil Engineers*)

Oil	Effect of untreated concrete	Surface treatments
Lard and lard oil	Very slight attack	Fluosilicate, sodium silicate, linseed oil
Foot oil	,, ,, ,,	,, ,, ,, ,, ,,
Fish oils	,, ,, ,,	,, ,, ,, ,, ,,
Coconut oil	Slight attack	,, ,, ,, ,, ,,
Olive oil	,, ,,	As above and resin varnishes
Rape-seed oil	,, ,,	,, ,, ,, ,, ,,
Almond oil	,, ,,	,, ,, ,, ,, ,,
Poppy seed oil	Very slight attack	,, ,, ,, ,, ,,
Walnut oil	,, ,, ,,	,, ,, ,, ,, ,,
Soya bean oil	,, ,, ,,	,, ,, ,, ,, ,,
Peanut oil	,, ,, ,,	,, ,, ,, ,, ,,
Cotton-seed oil	Slight attack	None
Linseed oil	,, ,,	,,
Rosin oil	,, ,,	,,

The degree of attack indicated and the protective measures must only be taken as applicable to concrete tanks and not to conditions where the oil is exposed to air and moisture.

For coconut oil and glycerol the use of a synthetic resin varnish has been recommended. For cotton-seed oil it has been common practice in the United States to use concrete tanks without any special finish, while for soya bean, peanut and fish oils a surface treatment with sodium silicate has been used.[2] Though it has been found that there is some tendency for surface attack on the concrete, the resultant film of an insoluble lime soap builds up a protective coating which prevents further attack.

Turpentine has little effect on concrete, but considerable penetration occurs. Lubricating oils entirely of petroleum origin do not attack concrete, but many contain vegetable oils as also do many cutting oils used in machine shops. Such oils, when continually spilt on concrete floors, often cause gradual deterioration as well as penetrating deeply into, or even through, the concrete. High-alumina cement is more resistant.

Silage

The fermentation of greenstuffs with close packing and little access of air produces 'sour' silage. With loose packing, more rapid oxidation occurs giving

[1] *Proc.* **66** (6), Part 2, 109, (1940) Concrete Specifications Number.
[2] *Chem. metall. Engng* **33,** 631 (1926).

'sweet' silage. In the production of silage organic acids are formed and the temperature rises. Acetic, lactic and butyric acids are present in sour silage to the extent of 0·5–1 per cent, while in sweet silage the acidity is less. Unprotected Portland cement concrete is slowly attacked by silage, the acid silage being the more dangerous. The action is reported to be negligible when the silage juices have a pH of 5 or higher.[1] In some field trials[2] on precast concrete silos it was found that after eleven years use for storing silage the extent of attack varied from almost nil to extensive and deep etching up to $\frac{1}{2}$ in. deep according to the quality of the concrete used. For ordinary wet-cast units the cement content was of prime importance, the attack progressively increasing as the cement content was reduced from 840 to 450 lb/yd^3. For dry-pressed units the cement content was of less importance than the quality of the aggregate. High-alumina cement concrete, or rendering, has proved resistant to sweet silage and is likely to be fairly resistant to sour silage. Surface treatments with bituminous and coal-tar products, synthetic resin varnishes containing chlorinated rubber, and wax, etc., have been found[3] in the U.S.A. to give good protection to Portland cement, but the bituminous coatings are stated to tend to make the silage stick to the walls. Surface treatment of Portland cement renderings with silicofluoride or sodium silicate may be adequate for sweet silage.

Action of sugar solutions on concrete

The problem of obtaining concrete floors resistant to the action of sugar solutions, often hot, arises in various food and sweet manufacturing industries. Portland cement concrete floors of granolithic or other types are attacked by sugar solutions, and, under conditions where severe exposure is combined with considerable wear, have often proved troublesome in practice. The use of a pozzolana is recommended in Germany and the treatment of the concrete surface with sodium silicate or, probably better, magnesium silicofluoride solutions. Under bad conditions, however, it is questionable whether these methods are more than temporary palliatives. High-alumina cement, used either as concrete or as a screeding to a Portland cement concrete, is more resistant and has been used with satisfactory results. Though more durable than Portland cement concrete, it does not however, always afford a permanent remedy.

Concrete tanks have been used for the storage of molasses with satisfactory results, but in some cases the concrete surface has become softened and cracked. Light refined molasses are stated to have a more aggressive action than dark molasses. It has been recommended[4] that concrete should be allowed to age for at least 28 days in air before exposure and that surface treatments of magnesium silicofluoride or sodium silicate should be applied. Coatings of tar asphalt are also used.

[1] C. A. Hughes, *Proc. Am. Concr. Inst.* **36,** 553 (1940).
[2] A.C.I. Committee Report, *Proc. Am. Concr. Inst.* **57,** 797 (1961).
[3] J. R. Spraul, *Agric. Engng* **22,** 209 (1941).
[4] M. N. Clair and M. A. Morrissey, *Engng News Rec.* **111,** 775 (1933).

Action of sewage on concrete

Concrete and concrete pipes have long been used in the construction of sewers and have, in general, given satisfactory service. Certain sewage conditions occasionally arise, however, which may lead to severe attack.

Normal sewage has an alkaline reaction, but the effluents arising from industrial processes may sometimes be acid and will then attack concrete unless neutralised immediately by discharge into an excess of normal sewage. The discharge of such effluents into concrete pipes, or into any cement-jointed pipe, is clearly hazardous and the effluent should be neutralised before discharge. In one case[1] a concrete sewer had been eroded to a depth of 2 inches over a period of seven years owing to the discharge of a factory effluent which raised the acidity of the sewage to values varying from 50 to 300 parts H_2SO_4 per million under conditions of normal flow. Two miles lower down the sewer where the acidity had fallen to 0–80 parts per million the damage was only about one-quarter as great. Contents of mineral acid averaging even a few parts per million must be expected to damage concrete sewers over a long period of years. Many authorities place a minimum limit of pH 5·5 or 6 on effluents discharged into sewers. Where mineral acid effluents must be carried, materials such as glazed stoneware pipes with a proprietary acid-resistant cement jointing are required,[2] or pitch-fibre pipes provided organic solvents, etc., are absent. Deterioration of concrete has also been reported where sulphuric acid had been used as a coagulant at a sewage works and the acidity of the sewage entering the sedimentation tanks was too high.

The concentration in sewage of neutral sulphate salts is rarely sufficient to cause direct damage to concrete sewers, but deterioration has occurred where the content during dry weather flow has risen as high as 0·25 per cent SO_3. It is advisable to regard with suspicion amounts of sulphates, and salts of other sulphur acids, approaching 0·1 per cent SO_3 if this concentration is likely to be maintained over considerable periods.

Though normal sewage is usually without any action, it is well known that under some conditions considerable evolution of hydrogen sulphide may occur and give rise indirectly to severe attack of concrete and mortar.[3] Serious troubles from this cause have been experienced, for example, at Cairo, Cape Town, Melbourne, Los Angeles and other cities in the U.S.A., and in India, Germany and elsewhere. An early case in England was at Nuneaton and troubles have subsequently arisen in other cities. Prolonged field investigations have been carried out in South Africa, Australia, the U.S.A. and elsewhere.

Hydrogen sulphide itself has no marked action on concrete and the disintegration arising from it is due to oxidation with production of sulphuric acid. The attack on sewers takes place only in the roof and upper part of the sewer and the

[1] C. J. Mackenzie and T. Thorvaldson, *Engng J. Can.* February 1926.
[2] Cf. A. J. Wigley, *J. Inst. civ. Engrs*, 1947–1948 (6), 196.
[3] Numerous papers are to be found in the literature. Some of the more recent ones on the occurrence, mechanism and prevention of deterioration are: N. Stutterheim, *Corrosion of Concrete Sewers*, South African Council for Scientific and Industrial Research, Series DR12 (1959); C. D. Parker, *Sewage ind. Wastes* **23,** 447 (1951); *J. gen. Microbiol.* **8,** 344 (1953); F. M. C. Gilchrist, *S. Afr. ind. Chem.*, Nov. 1953; P. Pomeroy and F. D. Bowlus, *Sewage Wks J.* **18,** 597 (1946); S. S. Morris, *J. Instn civ. Engrs* **13,** 337 (1940) **14,** 531 (1940); A. M. Douglas, *Munic. Engr* **91** (4), 130 (1964).

part below the liquid level is not affected. Sewers running full are not attacked. The action requires (i) the generation of hydrogen sulphide in the sewage, (ii) its release into the air space of the sewer, (iii) its oxidation at the exposed surface of the concrete to sulphuric acid.

The production of hydrogen sulphide in sewage arises from the action of anaerobic bacteria on organic sulphur compounds and on sulphates and other inorganic sulphur compounds. This action appears to have its seat primarily in the slime deposits on the sewer walls and perhaps also in silt deposits. Thus an effluent which on its own may generate little hydrogen sulphide in 24 hours will commence to do so within a few hours if mixed with slime. Temperature also has a very important influence, the anaerobic bacteria becoming almost inactive below about 50° F. The rate of production of hydrogen sulphide increases progressively above this temperature up to about 100° F. This accounts for the much greater prevalence of the trouble in countries with hot climates. The sulphide generation is also affected by the pH, being a maximum at slightly alkaline value and falling to almost zero below pH 5 and above pH 10. No definite relation is to be found between the content of sulphur compounds in the sewage and the incidence of trouble, but there is some indication that thiosulphates and partially reduced sulphur compounds may be reduced the more readily.

Evolution of hydrogen sulphide from the sewage into the air space tends to increase with the age of the sewage, and thus with the length of the outfall sewer, but it is particularly influenced by turbulence.

The hydrogen sulphide evolved into the air space of a sewer dissolves in moisture films on the exposed concrete surfaces where it undergoes oxidation by aerobic bacteria to sulphuric acid. This can apparently occur either directly or by intermediate formation of thiosulphates and polythionates. Some of these sulphur-oxidising bacteria remain active up to high acid concentrations and solutions as acid as pH 1 have been observed on the surface of sewers. In the early stages of corrosion of concrete in sewers a whitish surface deposit appears, followed later by gradual softening of the cement paste. Fine aggregate then drops away leaving a very rough surface with pieces of coarse aggregate projecting, and later these fall away as the action eats deeper into the concrete. Rates of attack up to $\frac{1}{4}$–$\frac{1}{2}$ inch depth per year have been recorded. In the case of mortar joints to brick linings or manholes the mortar swells and becomes pushed out and may cause spalling and disruption of the brickwork.

Remedial measures can take any of three forms:

(i) Prevention of hydrogen sulphide evolution;
(ii) Prevention of hydrogen sulphide condensation;
(iii) Prevention of attack.

Normal sewage will commonly contain enough sulphur compounds to give rise to sulphide production if the other necessary conditions are also present, but in special cases a reduction in sulphur content by elimination of particular industrial effluents may be beneficial. Chlorination of sewage, which at practicable dosages appears to act primarily by oxidation of sulphides rather than killing of the bacteria, has been used successfully to prevent sulphide production, and the injection of compressed air into rising mains has also been found promising. The addition of lime to raise the pH of the sewage to above 10 is another possible remedy, while an addition has occasionally been made of salts of metals forming insoluble sulphides (e.g. copper, zinc, ferrous iron). Various compounds toxic to bacteria have also been considered but the cost, apart from other considerations, has usually been prohibitive. Removal of slimes and silt, reduction in detention time by increasing velocities, and avoidance of points of turbulence are other measures found useful.

Forced ventilation of sewers has been found only to be effective in so far as it dries the exposed surface of the concrete preventing absorption of the hydrogen sulphide and proliferation of the sulphur-oxidising bacteria.

Considerable attention has been given to the trial of constructions resistant to sulphuric acid. Neither sulphate-resisting Portland, Portland blastfurnace, nor pozzolanic cements offer any sufficiently improved resistance to be of value, but the use of a limestone aggregate has been found to reduce the rate of attack of sulphuric acid on Portland cement concretes because the aggregate itself is dissolved and thus assists in the neutralisation of the acid and stops its entire action being concentrated on the cement binder.

It was concluded from the South African studies that the life of a sewer under acid attack could be expected with a limestone aggregate to be three to five times that with silicious aggregates. High alumina cement has been found to give a better resistance to sewer corrosion and has been used, for instance at Cairo and Durban, but attack still occurs. There is less evidence of the resistance of supersulphated cement concrete to acid attack in sewers, but, such as there is, suggests that its performance may be at least comparable to that of high alumina cement. The most successful treatment[1] of Portland cement concrete pipes has been the Ocrate process which uses SiF_4 gas. The evidence available to date indicates that such pipes have a high resistance to attack. For brickwork in manholes acid-resistant cements of the silicofluoride-silicate-filler type have long been successfully used and more recently a good performance has been obtained with furan, epoxy, polyurethane, and polyester resin jointing compounds.

Many trials have been made on protective treatments to concrete. On old surfaces that have suffered deterioration, it is essential to clean the surface thoroughly and to hack, wire-brush and clean all attacked material. Even so, it is often difficult to get satisfactory adhesion because of the dampness of the concrete. Bitumens are emulsified and softened by fats, but coal-tar products are more resistant. Protection for some years at least has been obtained by treatment with a primer followed by two coats of a tar-base paint; by a thick (e.g. $\frac{1}{8}$ inch) coat of rubber-bitumen emulsion; and by a high-alumina cement rendering. In the U.S.A., coatings of resin lacquers containing inert fillers to a thickness of $\frac{1}{16}$ inch are similarly stated to have afforded protection over some years. A more drastic and costly method of protecting concrete pipes is to line them with, for example, a thin polyvinyl chloride sheet[2] or with an epoxy resin mortar $\frac{1}{8}$ to $\frac{1}{4}$ in. thick.[3] Polyester-glass fibre sheets have also been used for the repair of sewer damaged by acid attack.

[1] H. Schremmer, *Zement-Kalk-Gips* **17** (9), 417 (1964).
[2] J. H. Rigden and C. W. Beardsley, *Corrosion* **14** (4), 206t (1958).
[3] R. W. Childers, *Mater. Protn* **2** (3), 18 (1963).

Certain other cases of corrosion of concrete, such as of concrete cooling towers, by sulphuric acid arising from the action of sulphur oxidising bacteria have also been reported. In cases where water is recirculated the addition of small amounts of toxic agents may be practicable. Certain dyestuffs derived from 3 : 6-diamino-acridine,[1] nitro compounds such as dinitrobenzene,[2] and chlorinated hydrocarbons like *ortho*-dichlorobenzene,[3] have potent bactericidal properties in concentrations of one, or a few, parts per million.

Action of gases on concrete

The action of gases on concrete is usually negligible, but certain conditions, beside those already mentioned, are found where attack may occur.[4] Such cases sometimes arise in railway tunnels, power stations, chemical factories, domestic chimneys, etc. Sulphur dioxide and carbon dioxide are the gases usually responsible for any attack on concrete. In railway tunnels both of these gases may be present in relatively large quantities in the atmosphere, but when the tunnel is fairly dry they produce no ill effects. Under moist conditions, however, fairly rapid attack on concrete linings to tunnels, or on mortar pointing to brickwork, is occasionally experienced. The use of pozzolanic cements or Portland blast-furnace cement in preference to Portland cement has been recommended,[5] but there is evidence that high-alumina cement is more resistant than either.

Troubles may also be experienced in concrete chimneys at power stations, etc., since the temperature of the waste gases may be so low that condensation can occur. When the gases are not treated to remove sulphur compounds it is common to use brick linings, often set in acid-resistant cement. If the flue gases are washed to remove sulphur dioxide the content of this in the exit gases will be low, e.g. one-hundredth of a grain of sulphur per c. ft., but some protection still seems necessary if deterioration is to be avoided. At Fulham power station, for example, the interior of an unlined concrete chimney was treated with an acid-resisting bituminous paint applied over a sealing medium.[6] High-alumina cement renderings with a mesh reinforcement have also been widely used for linings to industrial chimneys.[7] For steel or concrete chimneys they are commonly applied by the cement gun method, but hand rendering is often used on brick chimneys. Where temperatures are below 400–500° F sand is used as aggregate. For higher temperature in ducts or flues, or where some additional thermal insulation is desired, aggregates such as calcined diatomite and expanded clay are used in place of sand.

The modern domestic boiler with high thermal efficiency also gives flue gases at a low temperature and many cases of deterioration of brickwork chimneys have occurred. The prime destructive agent here appears to be ammonium salts.

[1] T. H. Rogers, *J. Soc. chem. Ind., Lond.* **59,** 34 (1940).
[2] L. A. Allen, *Proc. Soc. appl. Bac.*, 1949, 26.
[3] R. Eliasser, A. N. Heller and G. Kirch, *Sewage Wks J.* **21,** 457 (1949).
[4] Cf. *Proc. Am. Soc. Test. Mater.* **31** (II), 818 (1931) for bibliography.
[5] K. T. Herrmann, *Revue Mater. Constr. Trav. publ.* **250,** 274 (1930).
[6] W. C. Parker and H. Clarke, *J. Inst. civ. Engrs* **9,** 17 (1937–1938).
[7] T. D. Robson, *Mod. Pwr Engng*, March 1954, 134; M. Zar, *Proc. Am. Concr. Inst.* **59,** NL15 (1962).

remedies include such measures[1] as lining new flues with stoneware pipes, socket upwards, or, in existing flues, the insertion of asbestos-cement pipes or flexible stainless steel or aluminium linings in the top part of the chimney. The admission of additional air into the flue, or additional thermal insulation, can also help.

A relatively high concentration of sulphur dioxide in the atmosphere may also occur in the neighbourhood of large power stations when the flue gases are emitted without treatment. Under these conditions Portland cement concrete may suffer surface attack. Thus, for example, Portland cement concrete posts and slabs exposed to the weather to the leeward of one large power station suffered surface attack to a depth of about $\frac{1}{2}$ inch over a period of some 10 years, but adjacent high-alumina cement concrete was quite unaffected. Because of its much higher resistance to deterioration, high alumina cement has been used in some power stations for the mortar for brickwork casings to steel stanchions, for concrete encasement of steelwork, and in gas-washing chambers. Surface protection of Portland cement concrete may be obtained with suitable paints, e.g. bituminous or chlorinated rubber, but periodic renewal is necessary.

The disintegration of concrete linings to gas-washing towers in power stations has also been reported. In general some attack on Portland cement concrete is to be anticipated when it is exposed continuously to atmospheres containing appreciable amounts of sulphur dioxide under moist conditions, or under conditions of alternate wetting and drying. Reliance has to be placed on a thick cover to reinforcement, dense cement renderings, and surface treatments as, for example, with acid-resisting bituminous paints suited to the temperature involved.

Action of some inorganic compounds and other materials on concrete

There are briefly summarised in the following section the effects on concrete of a number of inorganic compounds and certain miscellaneous materials not previously considered.[2] It should be noted that the effects mentioned refer to plain concrete and that a salt, for instance, which has no destructive action on the concrete itself may cause accelerated corrosion of steel in reinforced concrete if the concrete is at all permeable or the cover to the reinforcement is insufficient. The attack on concrete by aggressive solutions also increases much with temperature and a salt which is inocuous at ordinary temperatures may be deleterious in hot solutions. Further, strong solutions of salts which have no chemical action on concrete may cause damage by crystallisation under the surface if it is periodically exposed to free evaporation and drying. It is only possible to discuss the action of a limited number of salts, and indeed for many salts no data are available, but it may be noted that salts of strong acids, e.g. nitric and hydrochloric, and weak bases, e.g. aluminium and iron, often have some destructive action when the corresponding salt of a strong base, e.g. alkalis and lime, has little or no effect. The solutions of the salts of strong acids and weak bases are acidic in reaction. The ammonium salts are in general more destructive than analogous salts of other bases, probably because the base, ammonia, released by

[1] *Bldg Res. Stn Dig.*, No. 60, 1965.
[2] For a summary of the action of many agents see *Proc. Am. Concr. Inst.* **63,** 1305 (1966); W. H. Kuenning, *Highw. Res. Rec.*, No. 113, 43 (1966).

the reaction of ammonium salts with lime, may be lost from solution. Thus ammonium chloride, or nitrate, solutions, gradually dissolve lime from concrete, progressively weakening it, without causing any appreciable outward signs of attack when cubes are immersed in the solution and continuous leaching can occur. Magnesium salts are generally more destructive than the alkali or calcium salts because of the low solubility of magnesium hydroxide. Thus magnesium hydroxide is precipitated and lime is leached from the concrete by formation of the soluble calcium salt, as in the case of the chloride or nitrate.

Acetates. Ammonium acetate was found by Grün[1] to be aggressive to Portland cement mortars and to a limited extent to high alumina cement, with Portland blastfurnace cement coming intermediate. Sodium acetate had no action, calcium acetate a slight one, and aluminium acetate a more serious one on Portland cement, but not on the other cements.

*Alkali hydroxides.*A 10 per cent solution of sodium hydroxide was found by Dorsch[2] to have no effect on 1 : 3 mortars, of Portland, Portland blastfurnace and pozzolanic cements, immersed in the alkali solution when 7 days old. After 700 days the tensile strength was unaffected and no sign of attack was visible. With high-alumina cement there was no visible attack, but the strength progressively diminished and after 700 days had fallen to 40 per cent in 1 : 3 standard sand mortars and to 68 per cent in 1 : 3 graded sand mortars. Other evidence indicates that it is attacked by 1–3 per cent cold caustic soda. Alkali hydroxide solutions probably act on high-alumina cement by progressively dissolving the alumina gel and perhaps attacking the hydrated calcium aluminates. A 10 per cent solution of ammonia had no effect on high-alumina or any other cement up to 700 days. Despite the high resistance of Portland cement to alkali hydroxide solutions, some slow deterioration of concrete has occasionally been found in practice under long exposure to solutions of high concentration, e.g. 10 per cent. A very dense mortar should normally give sufficient protection. More rapid deterioration occurs with 20 per cent solutions.

Carbonates. Sodium carbonate solutions have little or no chemical action on dense matured Portland cement concrete, but high alumina cement may be expected to be rather less resistant. Eisenbeck[3] found a slow increase in strength of Portland and high-alumina cement mortars immersed in a 10 per cent ammonium carbonate solution. Ammonium bicarbonate solutions seem to have some action on Portland cements.

Chlorides. Solutions of sodium and potassium chloride have no effect on matured Portland cement concrete when the latter is immersed in them. Strong solutions of calcium chloride have a destructive influence;[4] cases of gradual disintegration of concrete floors owing to leakages from brine-freezing plants have occurred. The formation of calcium chloro-aluminate is probably an important factor in causing this disintegration, but the increased solubility of calcium hydroxide in calcium chloride solutions has also been suggested as a contributory factor. The use of unlined reinforced concrete tanks to hold calcium

[1] R. Grün, *Z. angew. Chem.* **51,** 879 (1939).
[2] *Cement* **6,** 381 (1933).
[3] H. Eisenbeck, *Chem. Z.* **50,** 165 et seq. (1926).
[4] M. Lawrence and H. E. Vivian, *Aust. J. appl. Sci.* **11,** 490 (1960).

chloride brine is not to be recommended, since it is difficult, even with a thick cover, to prevent corrosion of the reinforcement. Asphalt mastic, or similar, linings, can be used. All chlorides accelerate the corrosion of reinforcement if they can gain access to it and in some factories handling solutions of common salt it is very desirable to protect a reinforced concrete sub-floor with a waterproof layer under the floor surfacing. Solutions of magnesium chloride of 2 per cent concentration and upwards produce a gradual diminution in the strength of Portland cement mortars, Portland blastfurnace and pozzolanic cements are also attacked but high-alumina cement is immune. In dilute solutions (e.g. 1 per cent) magnesium chloride has no appreciable effect on Portland cement. Very strong solutions (3M) destroy Portland cement concrete.[1] Barium chloride solutions do not attack cements; ammonium chloride solutions of 0·5 per cent concentration have a destructive action on Portland and Portland blastfurnace cement concretes; high-alumina cement is affected at a 5 per cent concentration. Iron and aluminium chlorides are also very damaging to Portland cement. Grün found that high-alumina cement was less resistant than Portland cement to aluminium chloride.

Coal and Clinker. Concrete storage hoppers are sometimes found to suffer severe disintegration from the action of wet coals, and particularly of those with a high content of pyrites and marcasite. Oxidation of these sulphides in the presence of moisture results in the formation of free sulphuric acid and iron sulphates. In severe cases Portland cement concrete has been known to disintegrate to a depth of more than 2 inches in the course of a few years, while cement mortar in brickwork may similarly fail rapidly. In other cases concrete coal bunkers have lasted thirty years without trouble. A lining of high-alumina cement mortar has been found useful in reducing the destructive action. Surface treatments of the ordinary types are in general of little or no value.

Corrosion and abrasion of ash bunkers is also sometimes a serious problem and cases of rapid deterioration of reinforced concrete have been reported.[2] The water in the ash-sluicing system may become very acid from oxidation of sulphides and its effect is aggravated by the abrasive action of the clinker. Clay tile linings set in acid-resistant cement have been found effective and, under not too severe conditions, high-alumina cement concrete[3] has been used.

Inorganic acids are destructive to concrete. A 1 per cent solution of sulphuric, hydrochloric or nitric acids will corrode concrete deeply within a few months[4] and much weaker solutions affect it more slowly. Phosphoric acid, because of the insolubility of calcium phosphate, is somewhat less destructive and Grün[5] found that specimens immersed in 5 per cent solution increased slightly in strength up to six months, but this was followed by a progressive decrease and ultimate failure. Fully acid-proof linings are required for any containers for mineral acids. Concrete is sometimes used for floors in factories handling mineral acids where

[1] H. G. Smolczyk, *Sym., Tokyo 1968.*
[2] T. H. Carr, *J. Instn civ. Engrs*, 1945–1946 (4), 375.
[3] A. V. Hussey and T. D. Robson, *Conference on Materials of Construction in the Chemical Industry*, Soc. Chem. Ind. London (1950), p. 23.
[4] H. Eisenbeck, loc. cit.; J. H. P. van Aardt, *Bull. natn. Bldg Res. Inst. Un. S. Afr.*, No. 13, 44 (1955); No. 17, 1 (1959); *Zement-Kalk-Gips* **14,** 440 (1961).
[5] *Z. angew. Chem.* **43,** 496 (1930).

spillage is not heavy and the floors can be frequently washed down. Deterioration is accepted as inevitable and frequent repairs done. High-alumina cement will withstand acid conditions down to a pH of 4–5, and is therefore more resistant than Portland cement if the acid in contact with the floor can be kept very dilute, but supersulphated cement has a better resistance and will withstand acid solutions down to a pH of about 3·5.

Metal oxides. The presence of small amounts of zinc or lead oxides very seriously retards the hardening of Portland cement concretes. The use of mine-tailings in which small amounts of these oxides (or the corresponding sulphides) were present as concrete aggregates has occasionally resulted in the failure of the concrete to harden. The addition of 0·001 per cent lead oxide soluble in lime-water has been found to delay the set of Portland cement, but lead salts insoluble in lime-water do not appear to have an adverse effect. With large additions of lead oxide, such as 1 per cent, little sign of setting occurs.[1] The addition of less than 0·1 per cent zinc oxide similarly retards the set and diminishes the strength at early ages. The addition of 0·1 per cent to high-alumina cement retards the set somewhat.[2]

Nitrates. Alkali and calcium nitrate solutions in concentrations up to 10 per cent do not affect matured concretes, but aluminium nitrate has some action. Ammonium nitrate even in an 0·5 per cent solution attacks Portland cement and high-alumina cement is not immune at higher concentrations. Thus specimens immersed in a 5 per cent solution show a progressive fall in strength as rapid as with Portland cement.[3]

Phosphates. The basic phosphate salts, including even the ammonium salt, do not appear to have any serious action on concrete, but definite information is not available on the acid salts.

Sulphates. The attack of solutions of sodium, potassium, magnesium, and calcium sulphates on cements has been discussed fully in earlier sections. Various other sulphates are encountered in industry, and it may be taken as a general rule that sulphate-resisting Portland, pozzolanic and Portland blastfurnace cements will be more resistant than ordinary Portland cement, but not immune to attack, particularly by magnesium sulphate, and that high alumina will, in most cases, suffer little or no damage. Supersulphated cement is resistant up to 0·5 per cent, and with some cements up to 2 per cent, solutions of magnesium and aluminium sulphates and 0·5 per cent ammonium sulphate, but suffers attack in stronger solutions.

Ammonium sulphate is probably the most destructive of all the sulphate salts to Portland cement concrete and many troubles due to its action have been experienced in the synthetic nitrogen industry. An interesting account of the experience of the German industry in this respect has been given by Mohr.[4] The very aggressive action is probably connected with the increased solubility of calcium sulphate in ammonium sulphate solutions. A double salt $CaSO_4 . (NH_4)_2SO_4 . H_2O$ is formed. High-alumina cement can be regarded in practice

[1] B. Garre, *Zement* **16,** 469 (1927); W. Lieber, *Sym., Tokyo 1968*; H. G. Midgley, (Mag. Concr. Res., **22** (70), 42 (1970).
[2] E. Rengade, *Revue Mater. Constr. Trav. publ.* **239**, 290 (1929).
[3] F. M. Lea, *Mag. Concr. Res.* **17** (52), 115 (1965).
[4] *Bauing* **6,** 284 (1925).

as immune to attack, at any rate to solutions up to 5 per cent concentration. Aluminium sulphate solutions attack Portland cement and in concentrations of 1 per cent or more the action is very marked. Portland blastfurnace and pozzolanic cements are less attacked, but even high-alumina cement concrete is not entirely immune to a 5 per cent solution. This attack of aluminium sulphate solutions on Portland cement concretes continuously immersed may be contrasted with the surface-hardening effect that is produced by a surface treatment. The iron and copper sulphates are also destructive. Potassium hydrogen sulphate solution (5 per cent) was found by Eisenbeck to attack Portland cement, but not high-alumina cement.

Action of some miscellaneous materials on concrete

Material	Action on Concrete
Alkyl esters	Esters of low molecular weight, e.g. methyl and ethyl, and low viscosity can penetrate concrete and by their saponification cause damage.
Beer	Fresh beer has no action, but when stale it may cause slow attack. The acids produced during the fermentation process slowly attack concrete; linings are also needed in fermentation tanks to avoid any effects on the beer.
Bleaching powder	Solutions have no destructive action on good concrete. Free chlorine, and acid solutions of bleaching powder, attack concrete.
Calcium bisulphite	Solution attacks concrete. High alumina cement has been used successfully for setting tile linings.
Chlorine	Surface etching of concrete continuously exposed to water containing 5 to 10 parts chlorine per million has been reported.
Coffee and cocoa beans	During fermentation sugars and organic acids are produced which attack Portland cement concrete. High alumina cement has proved more resistant.
Detergents	Acid detergents, e.g. containing phosphoric acid, are likely slowly to affect Portland cement concrete, and to a lesser extent high alumina cement. The latter is the less resistant to detergents containing free alkali hydroxides.
Ethylene glycol	Slow attack on concrete can cause surface scaling of frozen concrete.
Formaldehyde	Aqueous solutions attack concrete strongly. Formic acid, which is readily formed by oxidation, appears to be more destructive than acetic acid. Synthetic resin enamels, or linings of glass, rubber, or asphalt mastic and tiles, are used for concrete tanks.
Ink	The acid types of ink, containing free sulphuric and organic acids, attack concrete.

Phthallates — Alkyl phthallates attack concrete and have caused deterioration of concrete floors.

Sodium borate (borax) — Slight effect.

Sodium sulphide — Solutions of several per cent concentration attack Portland cement, forming sulphides. The strength decreases markedly. High alumina cement concrete has given good service in tanks holding a 14 per cent solution.

Sodium sulphite and bisulphite — Solutions attack Portland cement. High alumina cement is stated to be more resistant.

Sodium thiosulphate ("hyposulphite") — Solutions attack concrete. Leakage of photographic solutions has been known to cause disintegration of concretes, brickwork mortars and renderings. Renderings of Portland cement mortar on "hypo" wash-water tanks have also been affected.

Tan liquors — Liquors from hide tanning contain gallic, lactic and formic acids as well as sulphuric and sulphurous acids and various salts. The acid concentration may vary from a trace to 1 per cent with pH values from 5 down to 2. These liquors will be destructive to concrete.

Trisodium phosphate — 5 per cent solution has no appreciable effect.

Urine — No action when fresh, but acidic substances are developed on ageing. Deterioration of concrete floors in animal houses and of mortar joints in liquid manure tanks has occurred. Sodium silicate or silicofluoride treatments may afford some protection to floors. High alumina cement should give more resistant mortar.

Wine — Fresh concrete affects taste. Surface treatments with a solution of 1 part tartaric acid in 3–4 parts water by weight is sometimes used to stop this initial effect in concrete tanks used for storage of wine.

21 The Examination of Concrete Failures

The deterioration or failure of concrete, apart from structural failures, may be due to a great variety of causes, and the determination of the responsible factor, or factors, requires a careful examination of the concrete, supplemented in many cases by a study of the site. A knowledge of the history of the work is also valuable, but is often difficult or impossible to obtain. Whilst definite conclusions as to the reasons for the deterioration can often be reached, it is not always possible to arrive at a solution.

Failures in concrete may be assigned to one or more of three general causes.

1. Unsuitable materials.
2. Errors in preparation, placing and curing.
3. Exposure to natural or artificial destructive agents.

Under the first of these causes, unsuitable materials, we can group defective cements, defective aggregates, incorrect proportions of cement and insufficient entrained air to give the required frost resistance, and excessive additions of admixtures. The second, errors in preparation, placing and curing, includes poor mixing and the use of too wet or dry mixes with the accompanying troubles of segregation, laitance and honeycombing, the last being aggravated by insufficient ramming. Bad jointing between two days' work, inadequate cover to reinforcement, excessive trowelling of surfaces and inadequate curing may also be grouped under this head. The third head covers all the various destructive agents which have been considered in the two previous chapters.

Visual examination of a concrete structure will usually show if segregated layers, laitance bands, honeycombed patches or bad construction joints are present and have suffered more severe attack than the surrounding concrete. The shape of the voids present in a fractured surface may give some indication of the consistence of the mix. Small bubble-holes with smooth surfaces and spherical shape are characteristic of rather wet mixes, while the presence of numerous voids of irregular shape and an uneven distribution of the fine aggregate indicates the use of a mix which has been too dry for the degree of ramming employed. The size of the larger voids arising from accidentally entrapped air, which may be up to 2 mm or more, is of a higher order of magnitude than that of bubbles of deliberately entrained air which are generally less than about 0·25 mm.

Bad grading of the aggregate is also usually indicated by the appearance of the fractured surface; the actual proportion of the different sizes of aggregate present can be estimated in the manner described later. Excessive trowelling of a surface leading to surface cracking or flaking will often be quite evident. The appearance of cracks and rust stains along the lines of the reinforcement may indicate an insufficient cover to the steel, a poor quality of concrete, very severe conditions of attack, or a combination of these factors.

The general quality of the concrete may be ascertained by carrying out crushing tests, or measuring the permeability or absorption, on sections cut from it. Though the results of these tests may show that the quality of the concrete is low, it is not always possible to assign the reason for this, for not all the factors concerned in the preparation and curing of a concrete which affect its properties adversely leave their own characteristic trace in the finished product. Further, the destructive attack that the concrete has undergone may have been so severe that it is not possible to obtain any clear indication of the original quality of the material.

The physical properties of a concrete determine the degree of resistance it offers to attack by physical or chemical agencies, but do not define the nature of the particular agencies under whose action it has deteriorated. These can usually only be determined from a knowledge of the conditions to which the concrete has been exposed and from the results of further examination of the concrete.

Petrographic examination of concrete[1]

Visual examination with the unaided eye of concrete will reveal the grosser defects noted above but an examination of existing or freshly broken surfaces with a hand lens (X10) will reveal finer detail such as of the nature of the aggregate, evidence of aggregate reaction or the distribution of entrained air voids. Sawn and finely-ground surfaces of the concrete can be examined under a binocular microscope at higher magnification (e.g. X50) or under a petrographic microscope. Reaction rims round aggregate particles, gel exudations produced by alkali-aggregate reaction cracks, signs of poor bond between cement paste and aggregate, reaction products of attack on the cement paste and other diagnostic features can be observed. By linear traversing or point count procedures the content of air voids[2] can be measured and also the relative proportions of coarse or fine aggregate and cement paste. Further examination can be made on thin sections prepared from the concrete, or on selected powder preparations, allowing more positive identification of any reaction products seen in the concrete.

Analysis of fresh concrete

The analysis of fresh concrete before it has set can be carried out by relatively simple methods which depend on physical separations. That adopted in the British Standard for methods of testing concrete (BS 1881 : 1970) depends

[1] For details see R. C. Mielenz, *Cem. Lime Gravel* **40,** 135, 179 (1965); K. Mather, *Spec. tech. Publ. Am. Soc. Test. Mater.*, No. 169A, 125 (1966); G. M. Idorn, *Durability of Concrete Structures in Denmark*, Copenhagen 1967.

[2] ASTM C457–67T.

essentially on weighing the sample of concrete successively in air and water and then thoroughly washing it through $\frac{3}{16}$-inch (4·76 mm) and 100-mesh (150 μ) sieves to separate the coarse aggregate, fine aggregate and cement. The clean aggregates recovered from these operations are weighed in water and the cement content derived from the difference between the original weight of the sample in water and that of the aggregates. The specific gravities of the aggregates and of the cement are required in these calculations. The water content is derived from the difference between the weight in air of the original sample and that of the solid constituents. Corrections are applied to make allowance for the amount of the original coarse aggregate passing the $\frac{3}{16}$-inch mesh and that of the coarse and fine aggregate passing the 100-mesh, so as to give a direct comparison with the specified concrete mix. These corrections are determined by tests on samples of the original aggregates. As an alternative to this method of physical separation an extremely rapid chemical method has been described.[1] A suspension is prepared from the concrete by washing it through a nest of sieves with recirculation of the wash waters with a pump. An aliquot is taken and the CaO content determined by flame photometry. It is claimed that the entire procedure can be completed in about six minutes. Fine aggregates containing lime will interfere with the method.

Estimation of proportion of water used in mixing

The determination on an aged concrete of the proportion of water used in its mixing is subject to many possible sources of error. In one method, due to Brown,[2] a sample of the concrete is dried at 105° and then saturated *in vacuo* with carbon tetrachloride (sp. gr. 1·593) and the amount absorbed determined and the equivalent amount of water calculated. The sample is then dried again to remove the carbon tetrachloride and the combined water determined by ignition with a correction for carbon dioxide. The original water content is taken as the sum of these two. The cement content is determined in the usual way. For dense concretes, not air-entrained, with aggregates of low porosity Brown found that the w/c value determined was correct to about $\pm 0{\cdot}02$. However much larger errors occur with air-entrained concretes, with concretes of low w/c ratio not adequately compacted, or with porous aggregates. In another method, an entirely physical one,[3] the volumetric proportions of aggregate and air voids are determined microscopically by linear traverse on polished surfaces of the hardened concrete. The volume of cement paste is taken as the difference between the sum of these volumes and the total volume of the concrete. The total evaporable water content is determined by vacuum saturation of the concrete after drying at 105°. This absorption value is taken as a measure of the water in the air voids—already determined by the linear traverse—and in the capillary pores. The total original water content of the concrete (less any lost by bleeding of the fresh concrete) is assessed by adding to the volume of the capillary pores a value for combined water in the set cement estimated according to the age of the concrete.

[1] C. A. Chaplin and R. T. Kelly, *Chemy Ind.*, 1967 (35), 1467.
[2] A. W. Brown, *J. appl. Chem.* **7**, 565 (1957).
[3] F. O. Axon, *Proc. Am. Soc. Test. Mater.* **62**, 1068 (1962); T. D. Larson and P. D. Cady, *Mater. Res. Stand.* **8**, 8 (1968)

The volume of cement is measured by subtracting the volumes of combined and capillary water from that of the cement paste. Corrections are made for the porosity of the aggregate in carrying out the calculations. The maximum error in the w/c ratio by weight was claimed by Axon not to exceed about 0·05, but Mielenz found that the estimated value was consistently low and gave only 86–93 per cent of the actual value.

Estimation of cement content of hardened concretes[1]

The estimation of the cement content of a concrete is usually based on the determination of the amount of lime or soluble silica present and, unless the composition of the cement is known, the assumption of an average value for the percentage of these oxides present in Portland cement. Though so apparently simple in principle the difficulties in practice of obtaining even approximately accurate values are often considerable. If the aggregate is free from any appreciable quantity of calcareous compounds the lime content of the concrete can be used as the basis of estimation, but when this is not so the soluble silica content has to be used. Acid extraction is required in either case to dissolve the cement and some solution of silica may occur from the aggregate. The various methods employed differ mainly in the measures taken to ensure that all the cement silica passes into solution and to reduce the extraction of silica from the aggregate.

The sampling of concrete and the preparation of a representative portion for analysis requires much care. With aggregates of normal size a sample not less in volume than a 6-inch (3·5 litres) cube should be taken. The specimen is reduced in size by preliminary crushing and this can be assisted by heating the concrete several times to 600° with intervening quenching in water. A microwave oven is very suitable. It is usually possible, after coarse crushing, to remove a portion of the coarse aggregate (and weigh it); this is advantageous, since it improves the accuracy of the final analysis, and, in the case of aggregates that are not completely insoluble, reduces the error from this source.

The acid extraction can be carried out on a sample of the coarsely crushed material or on a pulverised sample. Use of the latter speeds up the analysis and makes the determinations, especially of soluble silica, easier but it much increases the risk of solution from the aggregates.

The extraction is carried out with hydrochloric acid with, or without, subsequent treatment with sodium carbonate or hydroxide solution. The Florentin method depends on extraction with cold (below 5°) 24 per cent HCl ($d = 1{\cdot}12$) alone, but there is evidence that not all the silica is obtained in solution.[2] There is in fact the inherent difficulty with any extraction depending on acid treatment alone that whereas unhydrated cement can be dissolved with little or no separation of gelatinous silica this is not true of set cement. Steopoe had recognised that any hydrated silica produced by carbonation remained insoluble in acid treatment, but the difficulty appears to be wider than this. To ensure solution of any

[1] J. W. Figg and S. R. Bowden, *The Analysis of Concrete*, H.M.S.O. (1970); L. J. Minnick, *Spec. tech. Publ. Am. Soc. Test. Mater.*, No. 169A, 326 (1966); *Analytical Techniques for Hydraulic Cement and Concrete, Spec. tech. Publ. Am. Soc. Test. Mater.*, No. 395 (1966).
[2] P. Esenwein, *Schweizer. Arch. angew. Wiss. Tech.* **19,** 279 (1953).

gelatinous silica produced, it is necessary to introduce a subsequent treatment with alkali hydroxide or carbonate. This increases the risk of some attack on siliceous aggregates but it is adopted in most recognised methods. Approximate corrections can be made by tests on the aggregate separately.

Some of the various methods of extraction in use are as follows:

1. Extraction[1] of coarse crushed material with successive quantities of cold dilute (1 : 10) HCl, rubbing the material as needed with a rubber-covered pestle to remove any coatings of gelatinous silica. Fine solids are removed with the solution by decantation and the sediment boiled with 4 per cent sodium hydroxide solution.
2. Extraction[2] of pulverised material with 1 : 3 HCl with gentle heating followed by treatment of the residue with 1N NaOH at about 75°.
3. Extraction[3] with cold 1 : 1 HCl followed by washing with hot 0·5 NaCl and then with hot 1N Na_2CO_3.

With basic igneous rock aggregates, such as dolerite, a weaker acid is needed to reduce extraction of silica or lime from it. Cold dilute acetic acid, or 2N acetic acid buffered with 0·1N sodium acetate, may be used, but even so it is necessary to restrict contact with the acid to the minimum time required for solution of the cement. A dimethylamine citrate solution[4] (192·6 g crystalline citric acid, and 891 ml 33 per cent aqueous dimethylamine made up to 3 litres with distilled water, pH about 12·6) has been found to have much less action on a rock such as diabase than acid treatments. Though unhydrated cement dissolves almost completely in this solution, the hydrated cement does not and a correction is required. The method may nevertheless be useful with soluble rocks where other methods are likely to give very erroneous results. With aggregates such as slag, and clinker, none of these methods can be used as the aggregate is too readily attacked and reliance has to be placed on the determination of minor constituents, such as Mn_2O_3 in a slag aggregate and in the concrete. Such determinations if based on one minor constituent only can be unreliable. The risks are reduced if several constituents can be estimated giving more information than is formally necessary to solve the equations involved. The extended calculations that arise can be done on a small computer. Some cases, such as a concrete composed of Portland blast-furnace cement and a slag aggregate, can present an insoluble problem.

With high-alumina cement concrete either the lime or the alumina content may be used as the index for calculating the cement content. If the analysis of the cement is unknown a value of 40 per cent for CaO and 40 per cent for Al_2O_3 may be assumed.

In favourable cases when the composition of the cement is known and the content of soluble material in the aggregate is negligible, the error in the calculated cement content by weight of a mortar or concrete may not exceed 5 per cent. If the composition of the cement has to be assumed the possible errors rise and may reach 10 per cent by the silica method. For a large number of British Portland cements the mean CaO and SiO_2 contents are 64·5 and 21·4 per cent, the

[1] J. W. Figg and S. R. Bowden, loc. cit.
[2] ASTM C85—66; C. L. Ford, *Bull. Am. Soc. Test. Mater.* **181,** 47 (1952).
[3] P. Esenwein, loc. cit.
[4] F. Gille, *Zement-Kalk-Gips* **5,** 286 (1952).

respective standard deviations being 0·69 and 1·19 per cent. Another source of error arises when it is desired to convert the weight proportions of cement and aggregate to proportions based on weight of cement per unit volume of aggregates since a value may have to be assumed for the unit weight of the latter. The results cannot then be relied on to more than about 10 per cent for the CaO method, and 20 per cent by the SiO_2 method. In general the CaO method, where applicable, involves less error than the silica method, but the words 'where applicable' require emphasis. Calcium may be present in clay impurities, or in occasional shells or limestone pebbles in siliceous aggregates, and as a normal constituent of basic igneous rocks. The error in the silica method increases as the concrete mix becomes leaner (e.g. 1 : 10 mixes). When appreciable amounts of soluble constituents are present in the aggregate, the extent to which the errors are increased will depend on the success with which appropriate corrections can be made.

Concretes that have been exposed to the action of seawater or of very soft natural waters may have lost some of the lime from the cement by leaching, and this may also happen in ordinary brickwork mortars. The lime content is then not a measure of the original cement content.

As an alternative to the chemical methods the micrometric procedures mentioned above of linear-traverse or point-counting on a ground and polished surface of the concrete can be used particularly when the chemical estimation is likely to be unreliable. A binocular microscope is preferable, though a monocular instrument can be used satisfactorily. A magnification of about 100 permits distinction of cement paste from aggregate particles. The results obtained are of course on a volume basis and have to be converted to obtain proportions by weight. Under the best circumstances the probable error of a micrometric determination is unlikely to be less than ± 5 per cent and an error of ± 10 per cent must be regarded as more likely.

Examination of nature of aggregate

The examination of the nature of the aggregate includes both its chemical and physical characteristics. It is first necessary to ascertain the type of aggregate present, as the further examination will depend on this. In many cases the aggregate can be identified by inspection or petrographic examination, or with the aid of very simple tests, as a ballast, sandstone, limestone, igneous rock, clinker, etc. A quantitative chemical analysis may be carried out when the nature of the aggregate cannot otherwise be definitely ascertained, or when it is desired to estimate the amount of some deleterious impurity suspected to be present. Thus microscopic examination of the crushed aggregate may show the presence of some adventitious constituent. The amount of any such material present, and its nature, if still in doubt, may often be determined by chemical analysis. The particular class to which an igneous rock, limestone or sandstone aggregate belongs can often be settled most easily by cutting a thin section for examination under the microscope.

The next step will usually consist of specific tests to determine whether the aggregate is a sound material. Thus, if alkali-aggregate reaction, or high-moisture movement of the aggregate, is suspected, the appropriate tests can be made. In the case of sand and ballast aggregates the amount of clay present can only be determined after dissolving out the cement as described later. The presence of soft shaly or other similar constituents can be detected by inspection of a broken concrete, but the amount present is most easily determined by hand-picking the aggregate after dissolving the cement. The porosity and general physical characteristics of the aggregate may be determined on pieces obtained in the same way, or taken direct from the concrete as most convenient.

Determination of grading of aggregate

The concrete may be crushed in a testing machine so as to fracture the material, or it can be carefully broken up with a hammer. It is inevitable that the grading should be slightly altered during the process of breaking the concrete into lumps, but, provided it is carefully carried out, the ratio of fine to coarse aggregate, and the grading of the fine aggregate, should not be seriously affected. The broken-up sample is treated with dilute hydrochloric acid until the cement is all dissolved.[1] The extracts should be passed through a filter to avoid loss of fine sand and clay. The content of clay and silt is best determined before drying, and the sieve analysis carried out after drying the aggregate.

As noted earlier it is sometimes useful to break the concrete into pieces of about 2-inch size and heat them to 600° and quench them. Unless limestone is present a temperature of 700–900° can be used with advantage. The cement becomes dehydrated and weakened and the concrete can then be broken down by rubbing in a mortar with a rubber-covered pestle. It is possible in this way to obtain a fairly representative sample of the coarse aggregate and of the sand with some adhering cement. The cement can if necessary be dissolved as before by treatment with dilute acid.

Concretes containing limestone aggregate present a difficult problem. An acid extraction cannot be used while heat treatment at 600° may fail to weaken the concrete sufficiently to permit of it being broken down.

As an example of a badly graded aggregate containing an excessive proportion of clay the following results may be quoted. These were obtained on samples taken from concrete foundations which after many months were found to be still very weak and soft.

	Per cent
Coarse aggregate:	
On ½-inch mesh	13·6
Through ½-inch on ⅛-inch mesh	28·8
	42·4

[1] Some gelatinous silica is formed in this process and can be removed by subsequent treatment with hot dilute sodium carbonate. Since, however, this treatment disperses any clay present, it is preferable to omit it. Serious errors are not introduced thereby.

Sand:

Through $\frac{1}{8}$-inch on 16 mesh	11·4
16 ,, 30 ,,	13·6
30 ,, 50 ,,	9·9
50 ,, 100 ,,	7·3
100 ,, 180 ,,	1·7
180 ,, — ,,	6·3 Fine sand
	7·4 Clay
	57·6

This aggregate contains clay amounting to some 13 per cent by weight of the sand, and a total amount of material finer than a 180-mesh amounting to some 24 per cent of the sand. The proportion of sand to coarse aggregate is also excessive.

The aggregate grading can also be judged in the microscopic linear traverse examination of a polished surface of the concrete carried out for other purposes. Obvious complexities in deriving the equivalent of sieve sizes arise from the shape of the aggregate particles, and also because the plane of the section through the concrete examined will intercept the particles in a random manner, as will also the microscopic traverse line.

Chemical examination of concrete

The chemical examination of concrete is usually undertaken to ascertain whether there is any evidence of chemical attack. It may be known that the concrete has been exposed to aggressive waters, but it is unsafe to conclude without examination that the action of these has been the cause of failure, and that physical influences, such as frost, have not been mainly responsible. The correct identification of the cause of deterioration of a concrete is often most important when considering questions of repair or reconstruction.

It is often preferable before beginning the chemical examination of a concrete to crush the material roughly and sieve off as much as possible of the aggregate. The cementing material is thus concentrated in the fine fraction, which is finely ground before analysis.

Concrete which has suffered attack by seawater contains within it, in part or whole, the products of the action. These include calcium sulphate, calcium sulphoaluminate, and magnesium hydroxide. On prolonged action the calcium sulphate tends to be leached out and this occurs the more readily since its solubility in water is increased somewhat by the presence of sodium chloride. The calcium sulphoaluminate first formed by the action of magnesium sulphate is eventually decomposed again by the continued action of the same salt. While the content of magnesium hydroxide tends to increase steadily in a concrete attacked by seawater, that of sulphur trioxide may therefore first increase and then decrease again. This is illustrated by the following analyses of the acid-soluble matter in some concretes attacked by seawater.

Concrete	Percentage content in concrete after separating part of aggregate			Content recalculated as a percentage of cement		
	CaO	MgO	SO_3	CaO	MgO	SO_3
A	16·4	5·90	2·17	65	23·00	8·50
B	12·6	5·50	0·93	65	28·30	4·80
C	20·3	0·36	0·60	65	1·15	1·92

The sample A contains a high content of magnesium hydroxide and sulphur trioxide. This is more clearly seen when the analytical values are calculated back to a cement basis, assuming the cement to have contained 65 per cent CaO. Sample B, however, contains only a much lower content of sulphur trioxide in proportion to magnesia. Sample C represents a portion of quite unattacked concrete, the value of MgO and SO_3 being normal for a Portland cement.

The attack of sulphate groundwaters containing calcium or sodium sulphates on concretes also results in an accumulation of sulphur trioxide in the concrete. The following analyses afford an interesting comparison of the compositions of two concretes which had deteriorated under the action of sulphate groundwaters with that of two other concretes which had been disintegrated by frost action.[1]

	Percentage content in concrete after separating part of aggregate		Content recalculated as a percentage of cement	
	CaO	SO_3	CaO	SO_3
1. Concrete: Sulphate action	25·10	9·65	65	25·0
2. ,, ,, ,,	31·80	11·90	65	24·3
3. ,, Frost action	27·05	0·59	65	2·5
4. ,, ,, ,,	22·40	0·86	65	2·9

While the amount of sulphur trioxide present in samples 3 and 4 is seen, when recalculated to a cement basis, to be quite normal, that of samples 1 and 2 is extremely high.

It is sometimes desirable to be able to distinguish between that part of the sulphur trioxide in a concrete which is present as calcium sulphate, or as other soluble sulphates, and that present as calcium sulphoaluminate. Thus waters containing gypsum may have evaporated through a concrete and it may be desired to know whether the sulphur trioxide present simply represents deposited gypsum, or whether attack on the set cement to form calcium sulphoaluminate has also occurred. Alternatively, it may be suspected that the aggregate in a

[1] C. J. Mackenzie and T. Thorvaldson, *Engng J. Can.*, February 1926.

concrete contained some gypsum, or that a gypsum plaster, or a pigment having a very high gypsum content, had been mixed with the cement. A determination of the extent to which the cement has suffered attack may then be important. Calcium sulphoaluminate is soluble in water to the extent of 0·127 g SO_3 per litre, but in a nearly-saturated lime solution the solubility is less than 0·01 g SO_3 per litre. On extracting a finely ground cement or concrete with a lime solution the gypsum or other soluble sulphates are removed, but not the sulphoaluminate. A sample of 0·5 g of the finely powdered material is extracted in a stoppered flask with two successive portions of 50 ml each of a suspension containing 0·25 g CaO per 100 ml. The first extraction is allowed to stand about 4 hours with intermittent shaking and the second to stand overnight. If the soluble SO_3 approaches 19 per cent by weight of the sample a further double extraction must be carried out. The SO_3 content of the extract solutions, after filtration, is determined in the usual way. The total sulphur trioxide content of the material can be obtained by solution in dilute hydrochloric acid, and hence the content of sulphur trioxide present as calcium sulphoaluminate derived from the difference between the acid-soluble and lime-soluble values. The following analyses on some rich (1 : 1) cement mortars will illustrate this.

Mortar	Acid-soluble SO_3	Lime water-soluble. SO_3	SO_3 present as calcium Sulphoaluminate
A	1·3	0·1	1·2
B	10·1	8·5	1·6
C	6·0	1·0	5·0

Sample A is a mortar which has not been exposed to the action of sulphates. It contains a normal proportion of sulphur trioxide, most of which has been combined as calcium sulphoaluminate during the setting of the cement. Sample B is a mortar in which the cement had been mixed with gypsum, but owing to rapid drying out little reaction between cement and gypsum had occurred. Sample C is a mortar which had been disintegrated by the action of gypsum. It has a high content of sulphur trioxide combined as calcium sulphoaluminate.

Acid waters in general tend to leach out the lime from a concrete, but they often leave appreciable amounts of the corresponding calcium salt in the material. An analysis of the soluble silica and lime content, or of the total content of a concrete soluble in dilute acid and alkali, will show evidence of this action. The following analyses,[1] one on a fresh concrete and the other on a concrete that had been heavily attacked and disintegrated by waters containing aggressive carbon dioxide, afford an example. The material soluble in dilute acid and alkali has in each case been calculated to a total of 100 per cent.

[1] N. Sundius and G. Assarsson, *Tek. Meddn K. Vattenfallsstyrelsen*, Ser. B, No. 16, p. 108, July 1929.

	Fresh concrete	Concrete attacked by carbon dioxide waters
SiO_2	22·09	28·02
Al_2O_3	7·55	9·46
Fe_2O_3	5·56	9·04
MgO	1·52	0·72
CaO	60·18	40·26
CO_2	2·60	11·06
SO_3	0·50	1·44
	100·00	100·00

The leaching which has occurred in the disintegrated concrete is shown by the low content of lime and by the reduction in the ratio of lime to silica when compared with the normal ratio for cement as given by the fresh concrete. The high content of carbon dioxide indicates that the aggressive acid has been carbonic acid.

In the case of attack by sulphuric acid the lime is again partly leached from the concrete and partly converted to calcium sulphate and sulphoaluminate; some of the sulphate radical remains in the concrete and serves to identify the acid responsible for the attack. Even with the acids which form very soluble calcium salts, such as hydrochloric acid, appreciable amounts of the acid radical may be found in the concrete owing to the formation of insoluble complex salts like calcium chloroaluminate.

The examination of concretes which have deteriorated under the conditions to which they have been exposed in factories and industrial processes offers a wide variety of problems. In some cases the factor responsible for the deterioration will be evident from the nature of the manufacturing processes, but in other cases it may be necessary to decide which of a number of possible agents of attack have been responsible for the damage produced. The analysis and microscopic or X-ray examination of the deteriorated concrete will often give a clear indication of the cause of failure. In view, however, of the large variety of possible causes it is not practicable to give any detailed methods of examination, and these must be left to the ingenuity of the chemist making the investigation.

APPENDIX 1

ATOMIC WEIGHTS

	Symbol	Atomic weight
Aluminium	Al	26·98
Calcium	Ca	40·08
Carbon	C	12·01
Chlorine	Cl	35·457
Fluorine	F	19·00
Hydrogen	H	1·0080
Iron	Fe	55·85
Magnesium	Mg	24·32
Manganese	Mn	54·94
Nitrogen	N	14·008
Oxygen	O	16·00
Phosphorus	P	30·975
Potassium	K	39·10
Silicon	Si	28·09
Sodium	Na	22·99
Sulphur	S	32·066
Titanium	Ti	47·90

MOLECULAR WEIGHTS

Al_2O_3	101·96
CaO	56·08
CO_2	44·01
Fe_2O_3	159·70
SiO_2	60·09
SO_3	80·07
$CaCO_3$	100·09
$CaSO_4$	136·15
$CaSO_4 . 2H_2O$	172·18
$3CaO . Al_2O_3$	270·20
$5CaO . 3Al_2O_3$	586·28
$12CaO . 7Al_2O_3$	1386·68
$CaO . Al_2O_3$	158·04
$CaO . 2Al_2O_3$	260·00
$CaO . Fe_2O_3$	215·78
$2CaO . Fe_2O_3$	271·86
$4CaO . Al_2O_3 . Fe_2O_3$	485·98
$3CaO . SiO_2$	228·33
$2CaO . SiO_2$	172·25
MgO	40·32

APPENDIX 2

METRIC EQUIVALENTS AND OTHER DATA

1 Centimetre = 0·3937 in
1 Sq Centimetre = 0·1550 sq in
1 Sq Metre = 10·76 sq ft
1 C Centimetre = 0·0610 c in
I Litre = 1·7598 Imperial pints
1 Litre = 0·03531 c ft
1 C Metre = 1·308 c yd
1 C Metre = 35·31 c ft

1 Inch = 2·5400 cm
1 Sq Inch = 6·4516 sq cm
1 Sq Foot = 0·0929 sq metre
1 C Inch = 16·387 cc
1 Imperial Pint = 0·5682 litre
1 C Foot = 28·317 litres
1 C Yard = 0·7646 c metre
1 C Foot = 0·02832 c metre

1 British Gallon = 0·1605 c ft

1 Litre = 0·2200 British gallon
1 British Gallon = 4·546 litres

1 British Gallon = 1·2 American gallons

1 Litre = 0·2642 American gallon
1 Gram = 0·03527 ounce
1 Gram = 15·432 grains
1 Kilogram = 2·2046 pounds
1 Tonne (1000 kilograms) = 0·9842 British ton

1 American Gallon = 3·785 litres
1 Ounce = 28·350 grams
1 Grain = 0·0648 gram
1 Pound = 0·4536 kilogram
1 British Ton = 1·016 tonnes

1 British Ton = 2240 lb
1 American Ton = 2000 lb

1 Tonne (1000 kilograms) = 1·1023 American tons
1 American Ton = 0·9072 tonne

1 Kilogram = 0·0197 cwt (112 lb)
1 Cwt = 50·80 kilograms

1 Gallon = 10 lb of water at 62° F
1 C Foot = 62·3 lb of water at 62° F

1 Kilogram per sq cm = 0·098 Newton per sq mm = 14·22 lb per sq in
1 lb per sq in = 0·0703 kg per sq cm = 0·00689 N per sq mm

For practical purposes 1 kg per sq cm may be taken as equivalent to 0·1 N per sq mm

1 Kilogram per c metre = 0·06243 lb. per c ft
1 lb per c ft = 16·02 kg per c metre

1 lb per c yd = 0·593 kg per c metre
1 U.S.A. sack cement = 94 lb
1 U.S.A. barrel cement = 376 lb
1 U.S.A. gallon water per 94 lb cement = water/cement ratio of 0·0885 by weight
1 c ft water per 94 lb cement = water/cement ratio of 0·664 by weight

1 grain per British gallon = 0·830 grain per American gallon
= 1·43 parts per 100,000
1 grain per American gallon = 1·2 grains per British gallon
= 1·71 parts per 100,000
1 part per 100,000 = 0·70 grain per British gallon
= 0·58 grain per American gallon

Hardness ($CaCO_3$) Conversion Data

1 part $CaCO_3$ per million = 0·07 grains per gallon (Degrees Clark)
1 Degree Clark = 1 grain $CaCO_3$ per gallon = 14·3 parts $CaCO_3$ per million
1 French Degree = 1 part $CaCO_3$ per 100,000
1 German Degree = 1 part CaO per 100,000
= 17·8 parts $CaCO_3$ per million
1 Calorie = 4·187 Joules = 0·00397 B.T.U.
1 B.T.U. = 252·0 cal = 1055 J
1 cal per gm = 1·8 B.Th.U. per lb
1 B.T.U. per sq ft per hour per inch thickness per °F = 2903 cal per sq cm per sec per cm per °C = 0·144 joules per sq m per sec per m per °C (or watts per metre per °C)

APPENDIX 3

POWDER X-RAY DATA ON CEMENT MINERALS

The values of I represent relative intensity.

The d values for the spacings are given in Å, Angstrom units, the intensities are on an arbitrary numerical scale, 10 being the strongest and 1 the weakest. Where it is known that doublets would be shown by using a camera of very high resolving power (such as the Guinier camera), the lines are marked with an asterisk. The three strongest lines are underlined. The symbol B indicates a broad line.

CaO [1]		MgO [1]		$Ca(OH)_2$ [1]		$Mg(OH)_2$ (Brucite) [1]	
d	I	d	I	d	I	d	I
2·778	8	2·431	1	4·900	7	4·77	8
2·405	10	2·106	10	3·112	2	2·725	2
1·701	9	1·489	9	2·628	10	2·365	10
1·451	6	1·270	1	2·447	1	1·794	7
1·390	3	1·216	2	1·927	4	1·573	6
1·203	1	1·0533	1	1·796	4	1·494	5
1·1036	1	0·9665	1	1·687	2	1·373	5
1·0755	4	0·9419	2	1·634	1	1·363	1
0·9819	4	0·8600	2	1·557	1	1·310	3
0·9258	1	0·8109	1	1·434	2	1·192	1
0·8504	1			1·449	2	1·183	2
0·8131	3			1·314	1	1·118	1
0·8018	3			1·228	1	1·092	1
				1·211	1	1·034	2
				1·762	1	1·030	1
				1·1432	2	1·0067	2
				1·1275	1		
				1·0599	2		
				1·0366	1		
				1·0143	2		

$3CaO.SiO_2$ [2]		Alite [2]		α $2CaO.SiO_2$ [3]		α' $2CaO.SiO_2$ [3]	
d	I	d	I	d	I	d	I
5·901	4	3·861	3	4·69	2	4·74	2
3·862	3	3·517	1	3·87	2	4·63	1
3·510	2	3·334	2	3·41	2	3·83	2
3·346	2	3·144	2	3·15	2	3·40	2
3·227	1	3·022	8	3·02	1	3·14	2
3·022	8	2·959	6	2·88	7	2·89	2
2·957	6	2·880	2	2·83	2	2·81	2
2·891	3	2·804	1	2·76	9	2·76	10
2·818	1	2·764	10	2·71	10	2·75	10
2·776	10	2·739	9	2·54	2	2·69	8
2·730	8	2·682	3	2·46	1	2·63	2
2·670	1	2·592	9	2·35	2	2·36	2
2·602	10	2·436	3	2·305	2	2·28	2
2·549	1	2·313	6	2·208	7	2·23	2
2·449	3	2·178	9	2·129	1	2·200	6
2·326	6	2·172	6	2·072	2	2·137	1
2·304	5	2·089	1	2·052	2	2·082	1
2·277	2	2·060	1	2·008	1	2·035	6
2·234	1	2·028	1	1·971	2	1·965	2
2·185	10	1·973	5	1·935	7	1·939	6
2·159	1	1·928	6	1·870	1	1·914	1
2·125	3	1·831	5	1·837	1	1·786	1
2·083	4	1·819	5	1·806	2	1·755	2
2·045	2	1·799	2				
2·011	1	1·761	9				
1·979	6/B	1·689	2				
1·940	7	1·640	2				
1·926	6	1·623	8				
1·900	2/B	1·537	6				
1·863	1	1·522	2				
1·825	6/B	1·485	9				
1·797	3						
1·771	9 (1)						
1·752	8 (1)						
1·642	2						
1·632	8						
1·623	6						
1·543	6						
1·526	2						
1·513	2						
1·497	6						
1·481	6						

Pure $3CaO.SiO_2$ can be distinguished from Alite since at (1) it gives doublets instead of single lines.

β $2CaO.SiO_2$ [5]		γ $2CaO.SiO_2$ [4]		$3CaO.Al_2O_3$ [5]	
d	I	d	I	d	I
4·920	1	5·625	4	4·08	2
4·645	1	4·320	4	3·34	1
3·790	3	4·047	2	2·70	10
3·380	1	3·794	4	2·39	2
3·335	1	3·354	2	2·258	1
3·090	1	3·002	10	2·200	5
3·040	2	2·881	2	1·984	1
2·874	2*	2·728	10	1·951	1
2·778	10*	2·525	2	1·907	9
2·740	10*	2·508	2	1·826	1
2·714	1	2·460	2	1·556	8
2·607	10	2·320	2	1·346	5
2·544	3	2·243	2	1·206	5
2·448	4*	2·186	2	1·106	2
2·403	4	2·024	2	1·023	5
2·279	3	1·963	2		
2·189	6*	1·928	8		
1·163	4	1·878	2		
2·128	1	1·800	6		
2·088	1*	1·751	6		
2·044	2	1·685	8		
2·019	1	1·669	2		
1·982	7	1·632	6		
1·911	1	1·539	2		
1·892	4	1·524	2		
1·844	1	1·498	2		
1·806	2	1·469	2		
1·787	2	1·457	2		
1·763	1	1·443	2		
1·706	3	1·414	2		
1·632	7	1·401	2		
1·606	4	1·374	2		
1·587	2	1·352	2		
1·573	2	1·268	2		
1·550	1 B				
1·523	4				
1·483	3				
1·448	1				
1·427	1				
1·416	1				
1·406	1				
1·393	1				

$4CaO.Al_2O_3.Fe_2O_3$ [5]		$6CaO.2Al_2O_3.Fe_2O_3$ [5]		$2CaO.Fe_2O_3$ [5]	
d	I	d	I	d	I
7·24	5	7·17	5	7·36	4
3·63	3	4·87	2	5·23	1
3·39	1	3·62	4	3·88	2
2·77	8	3·37	1	3·68	4
2·67	7	2·99	1	3·05	2
2·63	10	2·76	8	2·78	7
2·57	2	2·69	1	2·71	6
2·43	1	2·65	8	2·67	10
2·20	3	2·62	10	2·60	2
2·15	3	2·56	4	2·35	1
2·04	6	2·44	1	2·22	1
1·92	8	2·41	1	2·18	3
1·86	2	2·19	6	2·07	5
1·81	4	2·14	6	1·94	7
1·73	2	2·04	7	1·90	1
1·57	4	1·91	8	1·88	2
1·53	4	1·85	3	1·84	4
1·51	1	1·81	6	1·74	3
1·50	2	1·79	1	1·66	1
1·45	1	1·72	4	1·62	2
1·42	1	1·66	1	1·59	4
1·39	2	1·60	2	1·55	4
1·33	1	1·57	6	1·54	1
1·32	2	1·56	1	1·52	2
		1·53	7	1·48	1
		1·51	2	1·46	1
		1·50	3	1·43	1
		1·45	2	1·38	1
		1·41	2	1·36	1
		1·38	3	1·34	4
		1·35	3		
		1·34	1		
		1·33	2		
		1·32	4		

$CaO.Al_2O_3$ [7]		'Pleochroite' [8]		$12CaO.7Al_2O_3$ [5]		$CaO.2Al_2O_3$ [5]	
d	I	d	I	d	I	d	I
5·54	6	5·40	1	4·89	10	6·20	3
4·66	9	4·90	3	4·24	2	4·44	7
4·04	7	4·60	2	3·795	5	3·59	4
3·71	8	4·11	5	3·200	5	3·49	10
3·41	3	3·70	8	2·999	7	3·21	3
3·29	7	3·01	7	2·680	10	3·08	5
3·19	7	2·97	1	2·553	4	2·87	4
3·06	1 B	2·87	10	2·445	8	2·74	4
2·98, 2·95	10	2·76	10	2·347	3	2·70	4
2·90	5	2·69	3	2·186	7	2·59	9
2·85	8	2·60	1	2·056	2	2·53	3
2·75	5	2·53	1	1·944	7	2·43	4
2·53, 2·50	10	2·44	5	1·898	1	2·40	1
2·43	4	2·37	6	1·894	2	2·32	3
2·42, 2·39	10	2·34	6	1·762	2	2·21	1
2·33	8	2·21	2	1·729	4	2·17	3
2·29	7	2·18	2	1·694	2	2·06	1
2·26	7	2·11	4	1·662	7	2·05	5
2·20, 2·19	9	2·03	6	1·646	1	2·00	3
2·16	4	1·92	6	1·630	4	1·96	2
2·13	8	1·89	1	1·601	7	1·93	1
2·10	7	1·84	2	1·559	1	1·90	3
2·08	1 B	1·79	7	1·522	3	1·87	3
2·01	8	1·76	8	1·497	2	1·80	5
2·00	4	1·72	4	1·475	3	1·76	5
1·956	7	1·66	4	1·393	6	1·68	4
1·921	10	1·63	5	1·357	1	1·62	3
1·909	8	1·55	4	1·339	3	1·55	1
1·852	4	1·52	5	1·322	1	1·53	5
1·830	8	1·48	6	1·307	4	1·51	2
1·802	2	1·45	1	1·292	2	1·48	2
1·780	2	etc.		1·277	2	1·45	1
1·740	7			1·263	3	1·42	1
1·721	6			1·235	2	1·40	1
1·696	6			1·210	2	1·37	5
1·677	7			1·198	1	1·35	2
1·651	8			1·186	1	1·33	4
1·617	3			1·174	2	1·31	1
1·602	2					1·29	2
						1·28	2
						1·26	2
						1·25	2
						1·22	2

Calcium silicate hydrate (I) [9]		Calcium silicate hydrate (II) [9]		Afwillite [9]		$2CaO . SiO_2$ α hydrate (C_2SH(A)) [9]	
d	I	d	I	d	I	d	I
9–14	10	9·80	9	6·45	8	5·35	3
3·06	10	4·90	2	5·74	8	4·63	1
2·81	8	3·07	10	5·08	5	4·22	9
1·83	8	2·85	5	4·73	8	3·90	8
1·67	4	2·80	9	4·15	5	3·54	8
1·53	2	2·40	4	3·91	5	3·27	10
1·40	4	2·20	1	3·75	5	3·04	3
1·17	1	2·10	1	3·28	5	2·87	8
1·11	2	2·00	6	3·19	10	2·80	8
1·07	1	1·83	9	3·05	5	2·77	3
This is the data for the poorly crystalline material. The long spacing can vary considerably and may also be undetected.		1·72	1	2·84	10	2·71	3
		1·62	1	2·74	10	2·69	2
		1·56	5	2·67	5	2·65	6
		1·40	4	2·59	5	2·60	8
		1·225	3	2·44	4	2·56	3
		1·165	3	2·35	6	2·52	6
		1·100	1	2·31	5	2·47	1
		1·045	2	2·21	5	2·41	9
		1·025	1	1·145	8	2·31	2
		1·000	1	2·064	4	2·27	2
				2·017	4	2·24	3
				1·989	6	2·18	5
				1·949	8	2·16	3
				1·924	5	2·10	2
				1·862	6	2·08	3
				1·805	8	2·06	4
				1·776	8	2·03	3
				1·724	4	2·02	3
				1·704	6	1·982	5
				1·683	6	1·956	3
				1·630	6	1·926	5
				1·604	8	1·890	3
				1·589	6	1·872	4
				1·563	4	1·842	2
				1·507	5	1·820	5
				1·413	4	1·788	8
				1·382	4	1·737	4
				1·380	4	1·712	4
				1·309	4	1·687	1
				1·345	4	1·662	4
				1·265	4	1·654	5

Hillebrandite [9] (Synthetic) (C_2SH(B))		$2CaO.Al_2O_3.8H_2O$ [5]		$3CaO.Al_2O_3.6H_2O$ [5]	
d	I	d	I	d	I
12·0 ?	1	10·70	10	5·140	9
8·10	3	5·36	8	4·453	4
5·70	3	4·25	1	3·366	6
4·74	9	4·10	1	3·149	5
4·03	4	3·94	1	2·816	8
3·51	6	3·80	2	2·571	2
3·32	6	3·58	6	2·469	3
3·00	7	2·86	7	2·300	10
2·90	10	2·78	2	2·226	1
2·80	2	2·68	6	2·043	9
2·75	8	2·54	7	1·991	1
2·67	4	2·49	2	1·817	1
2·62	4	2·45	1	1·746	4
2·44	3	2·39	6	1·714	3
2·36	7	2·24	3	1·683	5
2·23	9	2·15	1	1·599	2
2·05	6	2·10	5	1·574	1
1·95	6	2·03	1	1·484	1
1·93	6	1·97	3	1·408	8
1·85	6	1·94	1	1·366	1
1·80	9	1·84	3	1·342	2
1·75	5	1·79	1		
1·71	2	1·73	1		
1·68	1	1·67	6		
1·66	1	1·65	4		
1·62	2	1·59	4		
1·56	2	1·51	2		
1·53	3	1·44	3		
1·52	2	1·42	1		
1·46	4	1·37	2		
1·44	2	1·09	3		
1·41	2	1·06	1		
1·35	1				
1·33	1				
1·32	2				
1·17	4				
1·11	2				
1·09	2				

$CaO.Al_2O_3.10H_2O$ [5]		$CaO.Al_2O_3.10H_2O$ [5]	
d	I	d	I
14·30	10	2·06	4
7·16	10	1·94	5
5·39	4	1·87	1
4·75	4	1·83	3
4·52	3	1·79	4
4·16	3	1·75	1
3·93	1	1·71	3
3·72	5	1·64	5
3·56	7	1·60	6
3·26	6	1·56	1
3·10	5	1·52	2
2·88	6	1·47	3
2·69	5	1·40	2
2·55	7	1·38	4
2·47	5	1·27	1
2·36	6	1·24	2
2·26	6	1·18	2
2·18	6	1·07	2
2·11	4		

$4CaO.Al_2O_3.13H_2O$ [5]		Ettringite $3CaO.Al_2O_3.3CaSO_4.32H_2O$ [5]		$3CaO.Al_2O_3.CaSO_4.13H_2O$ [5]	
d	I	d	I	d	I
8·05	10	9·80	10	8·92	10
4·50	1	5·70	8	4·88	1
4·05	2	4·90	6	4·72	1
3·90	5	4·67	7	4·46	6
3·63	1	4·34	2	3·99	6
2·86	9	3·87	8	3·65	1
2·69	6	3·60	3	2·87	7
2·54	3	3·45	6	2·73	4
2·45	6	3·26	4	2·60	1
2·36	3	3·02	3	2·45	6
2·23	4	2·79	9	2·41	5
2·17	1	2·67	3	2·35	1
2·04	2	2·57	8	2·33	2
1·97	3	2·43	3	2·25	2
1·93	2	2·36	1	2·19	2
1·86	3	2·20	8	2·06	4
1·74	2	2·14	6	1·99	2
1·66	8	2·06	3	1·90	1
		1·94	3	1·87	1
		1·89	2	1·82	4
		1·84	4	1·66	5
		1·80	1	1·63	4
		1·75	4	1·58	1
		1·70	4	1·55	1
		1·66	6	1·54	1
		1·62	2	1·44	2
		1·57	4	1·42	1
		1·54	2	1·39	2
		1·50	4	1·37	1
		1·45	3	1·35	1
		1·34	3		
		1·30	3		

LIST OF REFERENCES

1 H. E. Swanson and E. Tatge, *Circ. U.S. natn. Bur. Stand.*, No. 539, Vol. 1, 1953.
2 H. Simons and J. W. Jeffery, *Sym., London 1952*, 30.
3 G. Yamaguchi, H. Miyabe, K. Amano and S. Komatsu, *J. Ceram. Ass. Japan* **65**, 99 (1957).
4 R. W. Nurse, *Sym., London 1952*, 260.
5 H. G. Midgley, private communication.
6 N. Yanaquis, *Sym., London 1952*, 117.
7 L. Heller, Ph.D. Thesis, London Univ., 1952.
8 C. M. Midgley, private communication.
9 H. F. W. Taylor and L. Heller, *Crystallographic data for the calcium silicates*. H.M.S.O. London (1956).

Name Index

AA

Subject Index